Green Energy and Technology

Climate change, environmental impact and the limited natural resources urge scientific research and novel technical solutions. The monograph series Green Energy and Technology serves as a publishing platform for scientific and technological approaches to "green"—i.e. environmentally friendly and sustainable—technologies. While a focus lies on energy and power supply, it also covers "green" solutions in industrial engineering and engineering design. Green Energy and Technology addresses researchers, advanced students, technical consultants as well as decision makers in industries and politics. Hence, the level of presentation spans from instructional to highly technical.

Indexed in Scopus.

Indexed in Ei Compendex.

Suryanarayana Doolla · Zakir Hussain Rather ·
Venkatasailanathan Ramadesigan
Editors

Advances in Clean Energy and Sustainability

Proceedings of ICAER 2022

Springer

Editors
Suryanarayana Doolla
Department of Energy Science
and Engineering
Indian Institute of Technology Bombay
Mumbai, Maharashtra, India

Zakir Hussain Rather
Department of Energy Science
and Engineering
Indian Institute of Technology Bombay
Mumbai, Maharashtra, India

Venkatasailanathan Ramadesigan
Department of Energy Science
and Engineering
Indian Institute of Technology Bombay
Mumbai, Maharashtra, India

ISSN 1865-3529 ISSN 1865-3537 (electronic)
Green Energy and Technology
ISBN 978-981-99-2278-9 ISBN 978-981-99-2279-6 (eBook)
https://doi.org/10.1007/978-981-99-2279-6

This Springer imprint is published by the registered company Springer Nature Singapore Pte Ltd.
The registered company address is: 152 Beach Road, #21-01/04 Gateway East, Singapore 189721, Singapore

Foreword

The Department of Energy Science and Engineering (DESE) at IIT Bombay is a leading interdisciplinary energy education and research center. DESE has developed several novel education programs focusing on the application of science and engineering to problems in energy. Keeping the vision of the department "To develop sustainable energy systems and solution for the future" in mind, there is a strong need of providing a common platform to the researchers in the field of energy and allied domains. DESE has been organizing the biennial conference International Conference on Advances in Energy Research since 2007 which serves as an excellent forum to present new findings, exchange novel ideas, discuss new developments, and reflect on the challenges that lie ahead.

This book is a compendium of all the papers of the submissions accepted at the 8th International Conference on Advances in Energy Research (ICAER 2022), organized during July 7–9, 2022. After a rigorous peer-review process, about 97 papers have been accepted for the proceedings of the conference.

Various aspects of energy research including, but not limited to, conventional energy, renewable energy, grid integration of renewables, electric mobility, energy storage, energy policy and economics, and energy education were covered in the conference. This conference threw light on various recent accomplishments by researchers worldwide in the areas of solar thermal, thermal storage, solar PV with new materials, novel batteries, biofuel-based transportation, and rural energy needs, to name a few. The conference also included a special session on "Industry innovations in energy" where leading experts from industry were invited to present innovative case studies from their respective industries.

Preface

The 8th International Conference on Advances in Energy Research (ICAER 2022) was organized by the Department of Energy Science and Engineering, Indian Institute of Technology Bombay during July 7–9, 2022, virtually in Mumbai, India. The conference received around 250 submissions. Of these, around 120 submissions were accepted for oral presentation after a rigorous peer review. The conference was attended by over 300 participants. This book is a compendium of selected papers presented at the conference. The Springer Nature Publications sponsored three best paper awards and seven consolation prizes. The conference hosted 10 invited lectures and presentations by academics and industry personnel from all over the world. Two special sessions on "Industry innovations in energy" and "Energy education" were also organized.

Mumbai, India

Prof. Zakir Hussain Rather
Prof. Suryanarayana Doolla
Prof. Venkatasailanathan Ramadesigan

Acknowledgments We would like to sincerely thank Indian Institute of Technology Bombay for organizing the 8th edition of International Conference on Advances in Energy Research (ICAER 2022) in the Department of Energy Science and Engineering. We would also like to thank Prof. Suneet Singh (Head of the Department) for his constant support and valuable suggestions in organizing this conference. We would also like to thank all the invited speakers, delegates, sponsors, the members of the organizing and advisory committee and most importantly the Student Organizing Committee, student volunteers, and the staff of Department of Energy Science and Engineering for their dedicated efforts in organizing this conference.

Contents

About the Editors

Suryanarayana Doolla is currently a professor at the Indian Institute of Technology Bombay, where he teaches and directs research in power electronics and power systems as a faculty member of the Energy Science and Engineering Department. He received the B.Tech. degree in electrical engineering from the Nagarjuna University in 2000, M.Tech. degree in Energy Systems Engineering from IIT Bombay, and Ph.D. degrees in Power Systems from IIT Delhi, India, in 2002 and 2007, respectively. He was with Power Research and Development Consultants (2009), Bangalore, and Machine 2 Machine (2006–2008), Hyderabad, before joining IIT Bombay.

Dr. Zakir Hussain Rather is an electrical engineer by training, with Ph.D. from Aalborg University, Denmark. He has been working in renewable energy (RE) and electric mobility sector for past 12 years, with around 4 years of RE industry experience in Europe. He, in close collaboration with the Danish National transmission system operator, Energinet.dk, has extensively worked on the Danish grid. He is currently working as an associate professor with the Department of Energy Science and Engineering, Indian Institute of Technology (IIT) Bombay, where the focus of his work continues to be on grid integration of renewables, grid integration of Electric Vehicles (EVs), power system operation under high penetration of renewable generation and EVs. He is an editor of IEEE Transactions on Sustainable Energy, editor of IETE Technical Review, guest editor of International Journal of Power and Energy Systems, and a senior member of IEEE, IEEE Power & Energy Society, IEEE Smart Grid Community, IEEE Industrial Electronics Society, and Danish Smart Grid Research Forum. He is also involved in various BIS committees on EV charging infrastructure. His areas of research interest include grid and system integration of wind and solar power, power system dynamics, EV charging infrastructure and EV grid integration, smart and micro-grids, and WAMPACS.

He is the recipient of 'IREDA-NIWE Award for Best Research Work in Wind Energy'. He is a registered expert in Joint Research Centre (JRC) European Commission. He is also the recipient of Young Faculty Award, IIT Bombay, and several best paper and industry awards.

Venkatasailanathan Ramadesigan is an associate professor in the Department of Energy Science and Engineering at the Indian Institute of Technology Bombay. He received his B.Tech. in Chemical and Electrochemical engineering from the Central Electrochemical Research Institute (CECRI), Tamil Nadu, and obtained his MS in Chemical Engineering at the University of South Carolina, Columbia, USA. He received his Ph.D. in Energy, Environmental and Chemical Engineering at Washington University in St. Louis, Missouri, USA. His research interests broadly include modelling and simulation of electrochemical energy storage and conversion systems, nonlinear parameter estimation, model-based optimization and design, advanced battery management systems, system integration, and large-scale energy storage. His current research involves modelling, simulation and optimization of lithium-ion and other metal-ion batteries, redox flow batteries, fuel cells, modelling and control of hybrid renewable energy-battery systems, advanced battery management systems, and battery recycling.

Experimental Investigation on Corrosive Nature of Acid Oil Biodiesel on Selected Automotive Materials

S. Vaishak, Purnanand V. Bhale, Mehulkumar L. Savaliya, and Bharatkumar Z. Dholakiya

1 Introduction

Increased energy demand and environmental pollution have necessitated the use of renewable energy. Biodiesel is one such alternative to conventional energy sources because of its reduced emission and direct employment in existing technology. Biodiesel is an assortment of fatty acid methyl esters (FAME) derived from a diversity of feedstocks such as animal fats, waste cooking oil, algae oil, vegetable oils etc. It is generally made by trans-esterification of these feedstocks with methanol using potassium hydroxide or sodium hydroxide as catalyst. At present, Pongamia, Mahua, Neem and Jatropha feedstock price per litre is, 1.49, 1.3, 1.67 and 1.39 times the diesel price in India. For the above reason, biodiesel production from waste materials like animal fats, fish oil, used cooking oil etc. will be more economical and environment-friendly. By-products of vegetable oil refineries, such as deodorized distillate and acid oil, have been identified as cheap and readily available feedstock for biodiesel production [1, 2]. As a second-generation source of biodiesel in the current investigation, acid oil from a vegetable oil refinery with 88 wt% free fatty acid (FFA) and 12 wt% triglycerides was used.

Exposure to biodiesel, materials in the fuel supply line undergo corrosion and degradation. These materials are mainly classified into non-ferrous alloys, ferrous alloys and elastomers. The extent of corrosion and degradation of these materials in biodiesel mainly depends on the concentration of water and FFA (unsaturated fatty acid) in it. Another important factor that determines the corrosive nature of biodiesel

S. Vaishak (✉) · P. V. Bhale · M. L. Savaliya · B. Z. Dholakiya
Sardar Vallabhbhai National Institute of Technology, Surat, Gujarat 395007, India
e-mail: vaishak1990@gmail.com

S. Vaishak
Birla Institute of Technology and Science, Pilani, Rajasthan 333031, India

M. L. Savaliya
Atmiya University, Rajkot, Gujarat 360005, India

S. Doolla et al. (eds.), *Advances in Clean Energy and Sustainability*,
Green Energy and Technology, https://doi.org/10.1007/978-981-99-2279-6_1

is its oxidation stability because the biodiesel oxidation converts esters into different monocarboxylic acids, which enhance corrosion [3]. Compared to diesel, biodiesel is more prone to oxidation and the extent of that mainly depends on the fatty acid profile of the parent feedstock, storage conditions and the presence of naturally occurring antioxidants. Due to the presence of reactive sites that are vulnerable to free radical attack, biodiesels with a high content of unsaturated fatty acids are more likely to oxidise. Peroxides and hydroperoxides are the main products of the oxidation of biodiesel and these compounds, on further degradation, form shorter-chain compounds such as alcohols, ketones, aldehydes and low molecular weight acids [4, 5]. In order to increase the oxidation stability of biodiesel, various antioxidants are added to the biodiesel, which retards the rate of oxidation by forming a stable compound with free radicals. Increased water content in the biodiesel due to oxidation promotes microbial growth, which again enhances its corrosion characteristics. Feedstock from which the biodiesel is processed is also an important factor that determines its corrosive characteristics. According to Kaul et al. [6] biodiesel from salvadora and *Jatropha curcas* are more corrosive than biodiesel from kanarja and mahua.

It is generally observed that copper-based alloys are more corrosive in biodiesel when compared to ferrous and aluminium alloys [7–10]. Common elastomers like styrene-butadiene rubber (SBR), neoprene, nitrile butadiene rubber (NBR) etc., were found to be reactive with biodiesel and degradation characteristics of these are yet to understand fully [11, 12]. The combined effect of changes in total acid number (TAN), oxidation product, increased water content, unsaturated fatty acids, presence of metal species, etc. makes the corrosion in biodiesel more complex [9]. Because of this, there are concerns that the traditional techniques, such as TAN and copper strip corrosion, are insufficient to assess the biodiesel's corrosiveness. Here the corrosive behavior of copper, aluminium and stainless steel in acid oil biodiesel was tested by a 2000 h. static immersion test at 45 ± 2 °C as per ASTM G31.

2 Experimental Procedure

2.1 *Esterification of Acid Oil to Biodiesel*

For the present study, acid oil from a vegetable oil refinery consisting of about 88 wt% FFA and 12 wt% triglycerides was used as a second-generation feedstock for biodiesel production. The esterification/trans-esterification reactions were carried out in a properly instrumentalized laboratory autoclave. The reaction setup is equipped with a temperature controller, pressure gauge, and automatic cooling systems to control reaction parameters properly. The oil sample was preheated to 60 °C before the addition of the methanol-catalyst mixture. To limit the mass transfer, the mixture was continuously stirred at 600 rpm. The biodiesel reaction was carried out at 1:10 oil to methanol molar ratio with 5% catalyst (BTSA-boron tri sulfonic acid). The

Fig. 1 Reaction scheme for esterification

Table 1 Properties of acid oil biodiesel

Properties	ASTM specification	Acid oil biodiesel
FAME content %	Min. 96.5	91.78
Density, kg/m^3	860–900	876
Kinematic viscosity, mm^2/s	1.9–6.0	4.7
Acid value, KOH/g	Max. 0.5	1.2
Pour point, °C	Report	6
Flash point, °C	> 130	176
Cloud point, °C	Report	12
Calorific value, kJ/kg	Report	39,046
Peroxide value	Report	10

reaction was carried out for 10 h at 110 °C and autogenous pressure, which was discovered to be the most ideal setting for the maximum yield.

The general reaction scheme for the synthesis of biodiesel from acid oil is depicted in Fig. 1. Table 1 shows the various characteristics of the acid oil biodiesel which are determined by standard analytical methods.

2.2 *Corrosion Testing Procedure*

Corrosion behaviour of copper, aluminium and stainless steel was tested by static immersion for 2000 h at 45 ± 2 °C. In the standard procedure, it was recommended to change the fuel weekly during the immersion test in order to replenish ionic contaminants, and oxygen depletion and to minimize bulk solution composition changes. The test coupons (25 mm × 2 mm) were made from round bars and were immersed in a 100 ml borosilicate glass bottle. A 2 mm hole is drilled near the edge of the specimen in order to hang the coupons into the fuel sample by using stainless steel wires. Before the immersion test, coupons were washed with deionized water after dipping in 10% sulfuric acid. Then these coupons were degreased with acetone and were weighed by a proper weighing balance before placing it in the oven. Photographs of the specimens were taken at the end of the test for visual analysis and the rate of corrosion was determined by weighing the sample again.

The weight loss was converted into corrosion rate using the equation:

$$\text{Corrosion rate (mpy)} = \frac{W \times 345}{D \times T \times A} \quad (1)$$

where 'T' is the exposure time (h), 'D' is the density (g/cm^3), 'W' is the weight loss (mg), and 'A' is the exposed surface area (cm^2). SEM (Model: Hitachi Series 3400N) images of the specimens were also taken to study the changes in surface morphology.

3 Results and Discussion

From the static immersion test conducted per the ASTM G31 standard procedure, the corrosion rate of stainless steel, aluminium and copper was found to be 0.0018, 0.0104 and 0.1407 (mpy), respectively (Fig. 2). The copper and aluminium were found to be more corrosive in biodiesel, whereas the corrosive rate of stainless steel in biodiesel is comparable with that in diesel. The higher content of unsaturated fatty acids, such as oleic acid and linoleic acid, which are very reactive with metals, is what causes the enhanced corrosion rate of biodiesel. Higher temperature also enhances corrosion because of the increased reaction rate and transport of relevant species [9].

Table 2 compares the corrosion rate of various metals in acid oil biodiesel with data presented in the literature. Studies by Haseeb et al. [11] shows that in palm oil biodiesel the corrosion rate of copper was 0.053 mpy, when the immersion test was carried out at 60 °C for 840 h. The reduced corrosion rate in palm biodiesel is because of the less concentration of unsaturated fatty acid in it when compared to acid oil biodiesel. Haseeb et al. [10] also studied the corrosion of stainless steel, aluminium and copper in palm oil biodiesel at a temperature of 80 °C for 1200 h. Here, a magnetic stirrer was used to continually agitate the fuels at a speed of 250 rpm in place of static immersion. In this case, a high corrosion rate was observed in palm

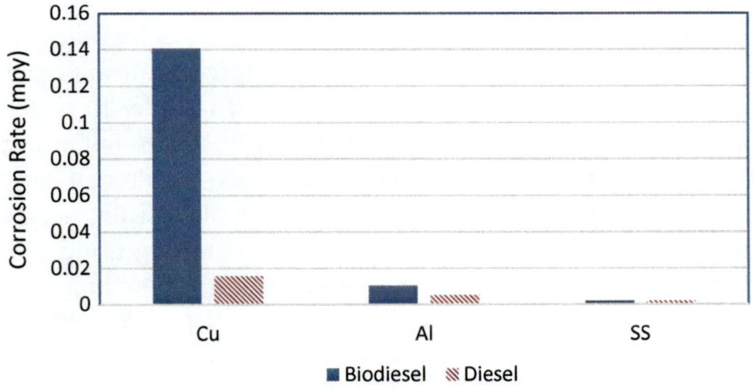

Fig. 2 Corrosion rate of Cu, Al and SS when the fuel sample is changed periodically

Table 2 Comparison of corrosion rate with data in the literature

Biodiesel	Metals	Conditions	Corrosion (mpy)	References
Palm	Copper	Room temperature, 2640 h	0.042	Haseeb et al. [8]
		60 °C, 840 h	0.053	
Jatropha curcas	Aluminium alloy	Room temperature (15–40 °C), 300 days	0.0117	Kaul et al. [6]
Kanarja			0.0058	
Mahua			0.0058	
Salvadora			0.01236	
Palm	Copper	80 °C, 1200 h, agitation rate of 250 rpm	0.586	Fazal et al. [9]
	Aluminium		0.202	
	Stainless steel		0.015	
Acid oil biodiesel	Copper	45 °C, 2000 h	0.1407	Present study
	Aluminium		0.0104	
	Stainless steel		0.0018	

oil biodiesel because of the relative motion between the fuel and metal and the high temperature of the test fuel. Kaul et al. [6] found that because of the higher concentration of unsaturated fatty acid (C18:2) *Jatropha curcas* biodiesel is more corrosive than kanarja and mahua biodiesel where the unsaturated fatty acid level is only 8%. However, increased corrosion rates of salvadora biodiesel, which has very low levels of unsaturation, are mainly because of its high sulfur content in it when compared to other non-edible oils.

The photographs of the metal surface before and after the immersion test is shown in Figs. 3 and 4, respectively. From Fig. 4, it can be observed that oxide layers were formed on the copper surface, which is exposed to both biodiesel and diesel. Oxygen content and temperature of the medium are more important parameters that affect the oxidation of copper into different compounds. Usually, copper forms a $CuCO_3$ (green oxide) at room temperature, but it turns to CuO at a higher temperature (black oxide). Because of the high temperature and oxygen content of biodiesel, Cupric oxide was the oxide layer that developed on the metal's surface. However, aluminium and stainless steel do not show any significant changes in the metal surface after immersion.

For further investigation, SEM images of the coupons exposed to biodiesel were taken and is as shown in Fig. 5. On the surface of copper, there are considerably more pits than on aluminium or stainless steel, according to the SEM photographs. It is said that when copper is exposed to biodiesel, pits are more prone to form because the CuO layer on the copper surface is destroyed, replacing oxygen ions from Cu_2O [9]. Some microbial growth were also observed on the copper surface after immersion test. This is mainly because of the hygroscopic nature of the biodiesel. In addition to this copper act as a catalyst for oxidation of biodiesel, which will again increase the water content

Fig. 3 Metal surface before immersion

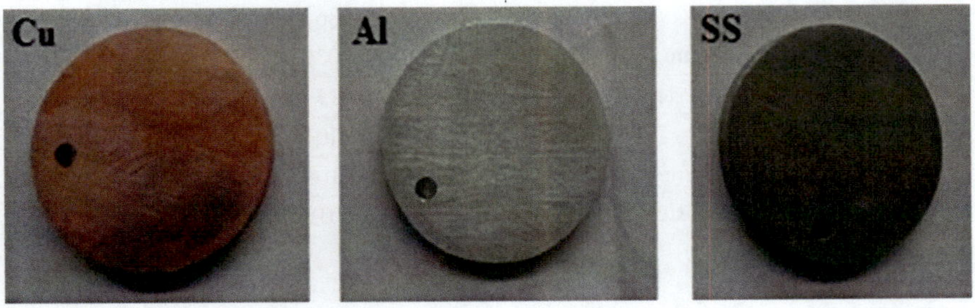

Fig. 4 Metal surface after immersion test

in the biodiesel and hence favor the microbial growth. By a phenomena called as microbiologically influenced corrosion, this increased microbial proliferation will accelerate corrosion.

It was also observed that biodiesel exposed to copper shows a significant change in color when compared to other two samples (Fig. 6). This further shows that copper is more reactive than aluminium and stainless steel with biodiesel and copper carbonate ($CuCO_3$) was found to be more dominant species in the biodiesel sample after immersion test.

GC-MS analysis of the biodiesel sample shows that the total FAME content of the prepared acid oil biodiesel was 91.78%. In that approximately 80% are unsaturated fatty acid methyl ester (Linoleic acid methyl ester-40.74%, Elaidic and Oleic acid methyl ester-37.63%). According to Knothe [13] the relative oxidation rates for

Fig. 5 SEM images of the metal surface exposed to biodiesel

Fig. 6 Colour change in biodiesel sample after immersion test **a** SS, **b** Al and **c** Cu

the unsaturated esters are linolenic > linoleic >> oleic. These unsaturated fatty acids contain highly reactive double bonds, which make the biodiesel more prone to oxidation. Hence, it is expected that TAN of biodiesel increases during storage because of oxidation. Studies show that pre-treatment of feedstock and better purification techniques can increase the oxidation stability of biodiesel. A up-streaming process such as degumming, neutralization, hydrogenation, dehydration can drastically increase the oxidation stability by removing the non-glyceride material.

4 Conclusion

The by-products of oil refineries such as deodorized distillates, soap stock and acid oils have been suggested as cheaper feedstocks for biodiesel. In the present study, acid oil from a vegetable oil refinery consisting of about 88 wt% FFA and 12 wt% triglycerides was used as a second-generation feedstock for biodiesel production. The corrosive behaviour of stainless steel, aluminium and copper in acid oil biodiesel was tested by a 2000 h. static immersion test at 45 ± 2 °C as per ASTM G31.

From the static immersion test conducted, the corrosion rate of stainless steel, aluminium and copper was found to be 0.0018, 0.0104 and 0.1407 (mpy), respectively. The number of pits on the surface of copper is significantly higher than that of aluminium and stainless steel, according to SEM pictures of the exposed metal surface. This study suggests that biodiesel from acid oil is more corrosive to copper and aluminium when compared to diesel.

References

1. Kulkarni BM, Pujar BG, Shanmukhappa S (2008) Investigation of acid oil as a source of biodiesel. Indian J Chem Technol 15:467–471
2. Kitakawa NS, Hiromori K, Ihara T, Nakashima K, Yonemoto T (2015) Production of high quality biodiesel from waste acid oil obtained during edible oil refining using ion-exchange resin catalysts. Fuel 139:11–17
3. Tsuchiya T, Shiotani H, Goto S, Sugiyama G, Maeda A. Japanese standards for diesel fuel containing 5% fame blended diesel fuels and its impact on corrosion. SAE technical paper no. 01-3303
4. Mankowski G, Duthil JP, Giustit A (1997) The pit morphology on copper in chloride and sulphate-containing solutions. Corros Sci 39:27–42
5. Schober S, Mittelbach M (2004) The impact of antioxidants on biodiesel oxidation stability. Lipid Sci Technol 106:382–389
6. Kaul S, Saxena RC, Kumar A, Negi MS, Bhatnagar AK (2007) Corrosion behavior of biodiesel from seed oils of Indian origin on diesel engine parts. Fuel Process Technol 88:303–307
7. Bhardwaj M, Gupta P, Kumar N (2014) Compatibility of metals and elastomers in biodiesel: a review. Int J Res 1(7):376–391
8. Haseeb ASMA, Masjuki HH, Ann LJ, Fazal MA (2010) Corrosion characteristics of copper and leaded bronze in palm biodiesel. Fuel Process Technol 91:329–334
9. Fazal MA, Haseeb ASMA, Masjuki HH (2010) Comparative corrosive characteristics of petroleum diesel and palm biodiesel for automotive materials. Fuel Process Technol 91:1308–1315
10. Haseeb ASMA, Fazal MA, Jahirul MI, Masjuki HH (2011) Compatibility of automotive materials in biodiesel: a review. Fuel 90:922–931
11. Haseeb ASMA, Masjuki HH, Siang CT, Fazal MA (2010) Compatibility of elastomers in palm biodiesel. Renew Energy 35:2356–2361
12. Sorate KA, Bhale PV (2015) Biodiesel properties and automotive system compatibility issues. Renew Sustain Energy Rev 41:777–798
13. Knothe G (2007) Some aspects of biodiesel oxidative stability. Fuel Process Technol 88:669–677

Clear Sky and Real Sky Solar Radiation Modelling for Locations in India

Jani Das

1 Introduction

It has been estimated that hourly solar energy received on earth is sufficient to supply the yearly energy needs of the world. Electricity can be produced from solar photovoltaic cells or thermal power from solar radiation received by the earth in the form of heat and light. India being a country with adequate solar potential, successful design and implementation of solar installations across the country depends on availability of reliable solar data. Data resolution depends on the user and application. For conducting feasibility studies of PV systems, daily or monthly averaged data is enough. For performance simulation of system components, data with shorter resolution of hours, minutes or seconds are needed [1]. Local solar radiation data is needed for design of energy systems, microgrids and thermal environment of buildings [2–7]. Availability of predictive models for real global solar radiation is relevant for taking decisions related to solar energy applications. Many solar radiation models have been developed throughout these years, which involve solar radiation in different time averages and the components such as global, direct and diffuse.

Models are used to evaluate location wise global, direct and diffuse radiation for analysis, where direct measurement is impossible. This paper presents Bird's Clear Sky Model validation for an Indian location. The real sky values have been validated using Armstrong model using sunshine hours data, instead of real sky models which make use of atmospheric and geographic parameters. This approach is theoretically advanced as directly measured sunshine hours have been used for evaluation of real sky radiation values. The model has been applied to an Indian location and presented in this paper.

J. Das (✉)
Department of Electrical and Electronics Engineering, Muthoot Institute of Technology and Science, Ernakulam, Kerala, India
e-mail: janidas@mgits.ac.in

S. Doolla et al. (eds.), *Advances in Clean Energy and Sustainability*,
Green Energy and Technology, https://doi.org/10.1007/978-981-99-2279-6_2

2 Materials and Methods

One of the earliest models for estimation of global solar radiation is Bird's Clear Sky Model [8]. Diffuse irradiance is considered to be isotopically distributed in the sky in the simplest model [9], diffuse irradiation is considered to be anisotropic in more realistic models. Many complex calculations and fitting procedures are needed for the second case. Advanced models have been presented by Gueymard [10] and Perez et al. [11] based on the concept of "sky categories".

The investigated methodology has been presented in this paper to estimate and validate clear sky radiation for Mumbai, India. It has been understood that sunshine hours are directly proportional to solar radiation. Regression methods can be used to determine the proportionality constants between sunshine duration and solar radiation from scatter plots. Angstrom [12] has established models which have presented this as a linear relationship.

The models have been presented in the following sections.

2.1 *Bird's Clear Sky Model*

One of the initial attempts to review existing clear sky models was done by Bird and Hulstrom [8]. This model is a diffuse radiation models widely used to calculate the hourly clear sky irradiation on a defined horizontal surface. Each model use geographical and atmospheric parameters of a location to model the direct insolation.

The monthly average daily extra-terrestrial solar radiation on the horizontal surface is determined by

$$\overline{H_0} = \frac{24}{\pi} I_{sc}\left(1 + 0.033\cos\frac{360N}{365}\right) \times \left(\frac{\pi\omega_s}{180}\sin\emptyset\sin\delta + \cos\emptyset\cos\delta\sin\omega_s\right)\frac{\text{kwh}}{\text{m}^2}\,\text{day} \quad (1)$$

where I_{sc} is the solar radiation constant given by 1367 W/m^2, N_d is the day of the year, ω_s is the sunshine hour angle of the mean day of the month (°), Ø is the latitude angle in ° and δ is the declination angle (°).

Some basic equations are used by Bird's model to determine the different components of the solar radiation under clear sky conditions is detailed below:

Extra Terrestrial Solar Radiation on a horizontal surface is given by

$$I_o = I_{sc} \times r \times \cos Z \quad (2)$$

r depends on the day of the year.

Bird's Model Base Equations are [8]:

Table 1 Bird's clear sky model input parameters

Specification	Parameter	Unit
Geographical parameters	Latitude and longitude of the location	Degree north positive East positive
	Time zone relative to GMT together with month, day, hour and minutes	Time units (h, east positive)
Atmospheric parameters of the location	Station pressure	millibars
	Ozone thickness of the atmosphere	cm
	Precipitable water in a vertical column from surface	cm
	Aerosol optical depth at 500 nm	
	Aerosol optical depth at 380 nm	
Scattering parameters	Forward scattering of incoming radiation	
	Ground albedo	

$$I_{d,c} = I_O \times 0.9662 \times T_R \times T_0 \times T_{UM} \times T_W \times T_A \tag{3}$$

$$I_{as,c} = I_o \times 0.79 \times T_0 \times T_{UM} \times T_W \times T_{AA} \times \frac{0.5(1 - T_R) + B_a(1 - T_{AS})}{1 - M + M^{1.02}} \tag{4}$$

$$I_{T,c} = \frac{\left(I_{d,c} + I_{as,c}\right)}{1 - r_g r_s} \tag{5}$$

$$I_{diff,c} = I_{T,c} - I_{d,c} \tag{6}$$

where $I_{d,c}$ is the direct clear sky radiation (W/m^2)

$I_{as,c}$ is the solar irradiance on a horizontal surface from atmospheric scattering (W/m^2)

$I_{T,c}$ is the total global solar irradiance on a horizontal surface (W/m^2)

$I_{diff,c}$ is the diffuse solar radiation on a horizontal surface (W/m^2).

The model inputs are summarized in Table 1.

Flow diagram of Bird's clear sky model is shown in Fig. 1.

2.2 *Angstrom Model*

The original Angstrom regression equation related to monthly average daily radiation to clear sky radiation in a given location and average fraction of sunshine hours.

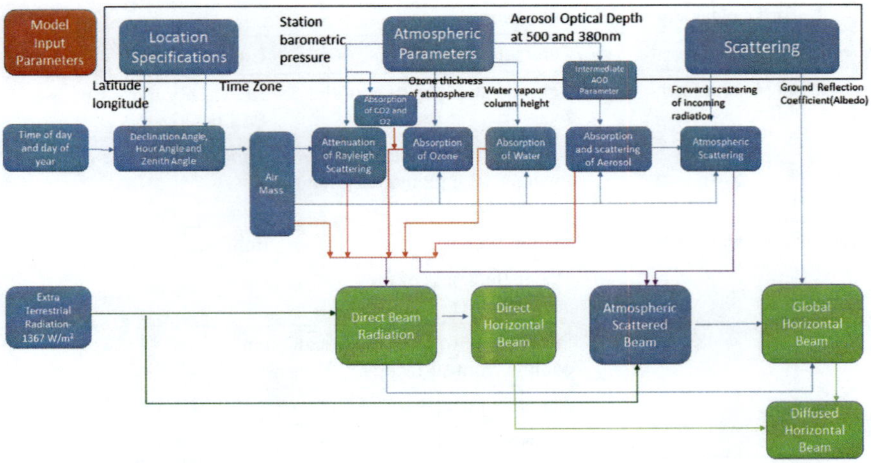

Fig. 1 Flow diagram of Bird's clear sky model

$$\frac{H_o}{H_c} = a + b\frac{s}{S} \tag{7}$$

H_o = monthly average daily global radiation in $Wh/m^2/day$, H_c = monthly average daily clear sky radiation for the location in a given day, s = actual sunshine hours in a day, S = monthly average maximum possible sunshine hours in a day (monthly mean length of the day), a and b are empirical coefficients. These coefficients are location specific.

The Angstrom Model has been used to estimate the first order regression coefficients for Mumbai. In this work, the Angstrom model has been modified on an hourly basis to derive the correlation constants.

Rewriting in terms of hourly values, the equation becomes,

$$\frac{H_{oh}}{H_{ch}} = a + b\left(\frac{s}{S}\right)_h \tag{8}$$

H_{oh} = hourly observed global radiation

H_{ch} = hourly modelled clear sky radiation using Bird's clear sky model

$(s/S)_h$ = hourly sunshine fraction; s/S = 1; perfect cloudless sky and s/S = 0; cloudy sky.

3 Model Validation for Mumbai, India

3.1 Clear Sky Model Validation

Bird's Clear Sky model is used to evaluate the clear sky insolation components for Mumbai. The model input parameters which are considered constant throughout the year for the location have been summarized in Table 2.

Hourly and daily variation of the station pressure has been used for the clear sky model evaluation. Yearly variation of atmospheric pressure for Mumbai is shown in Fig. 2.

Monthly or Seasonal variation of the following parameters are considered for analysis:

- Aerosol Optical Depth at 500 nm
- Angstrom Exponent (for determining AOD at 380 nm)
- Total atmospheric Water vapour column in cm.

The input values for Mumbai are summarized in Table 3.

The model validation has been done for the different climatic zones of the country. Due to lack of year-by-year clear sky data for the location, Typical Meteorological

Table 2 Input parameters for the location

Parameter	Value
Latitude of the location	18.975° N
Longitude of the location	72.825° E
Time zone relative to GMT	+ 5.3
Ozone thickness of the atmosphere (cm)	0.7
Forward scattering of incoming radiation	0.85
Ground albedo	0.2

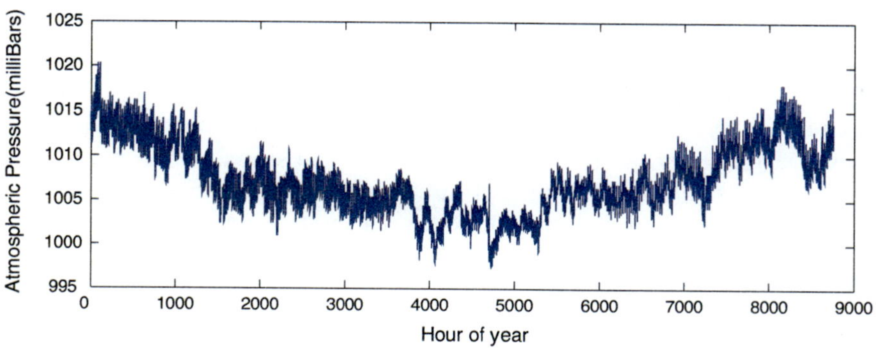

Fig. 2 Yearly variation of atmospheric pressure at Mumbai. *Source* IMD, Pune

Table 3 Monthly variation of input parameters

Month	Water column (cm)	Aerosol optical depth@500 nm	Angstrom exponent	Aerosol optical depth@380 nm	Broadband aerosol optical depth
January	1.25	0.80	1.20	1.25	0.62
February	1.25	0.80	0.90	1.12	0.59
March	1.25	0.80	0.90	1.12	0.59
April	2.25	0.90	0.90	1.26	0.66
May	2.25	0.47	0.95	0.67	0.35
June	1.25	0.80	0.45	0.94	0.54
July	4.00	1.00	0.45	1.18	0.68
August	4.00	0.70	0.45	0.83	0.47
September	4.00	0.70	0.45	0.83	0.47
October	0.50	0.90	1.35	1.48	0.72
November	0.50	0.95	1.19	1.47	0.74
December	0.21	0.95	1.19	1.47	0.74

Year (TMY) clear sky data for Mumbai was used as validation for the four different seasons of the year. The climatic zones of the country are classified as:

(a) Winter (December–February)
(b) Hot or Pre-Monsoon (March–May)
(c) Monsoon or Summer Monsoon (June–September)
(d) Post Monsoon (October–November).

The monthly average clear sky global horizontal radiation value is used for validating the model. IMD do not provide hourly data on a monthly basis for clear sky radiation. The input parameters are tabulated climate wise in Table 4 for Mumbai. The monthly average clear sky radiation comparison between modelled and measured values for the different seasons are shown in Fig. 3a–d. The results show appreciable agreement of the modelled and measured values as can be seen from the statistical analysis from Table 5.

It can be seen from the analysis that the Root Mean Square Errors (RMSE) for the different seasons are in the range of 25.9–67.21 W/m^2 and Mean Bias Error (MBE) which is the average bias of the model is ranging from − 10.11 to 14.8 W/m^2. The Mean Average Percentage Error (MAPE) is for the four seasons are less than 20%, which is a measure of the suitability of Bird's model for determination of clear sky model values for an Indian location. R^2 value is above 0.96 for all the four seasons.

Table 4 Seasonal variation of input parameters for Mumbai

Parameter	Winter (January)	Pre-monsoon (April)	Monsoon (July)	Post monsoon (November)
Water column (cm)	1.25	2.25	4	1.8
Aerosol optical depth@500 nm	0.8	0.9	1	0.95
Angstrom exponent	1.2	0.9	0.45	1.19
Aerosol optical depth@380 nm	1.25	1.26	1.18	1.47

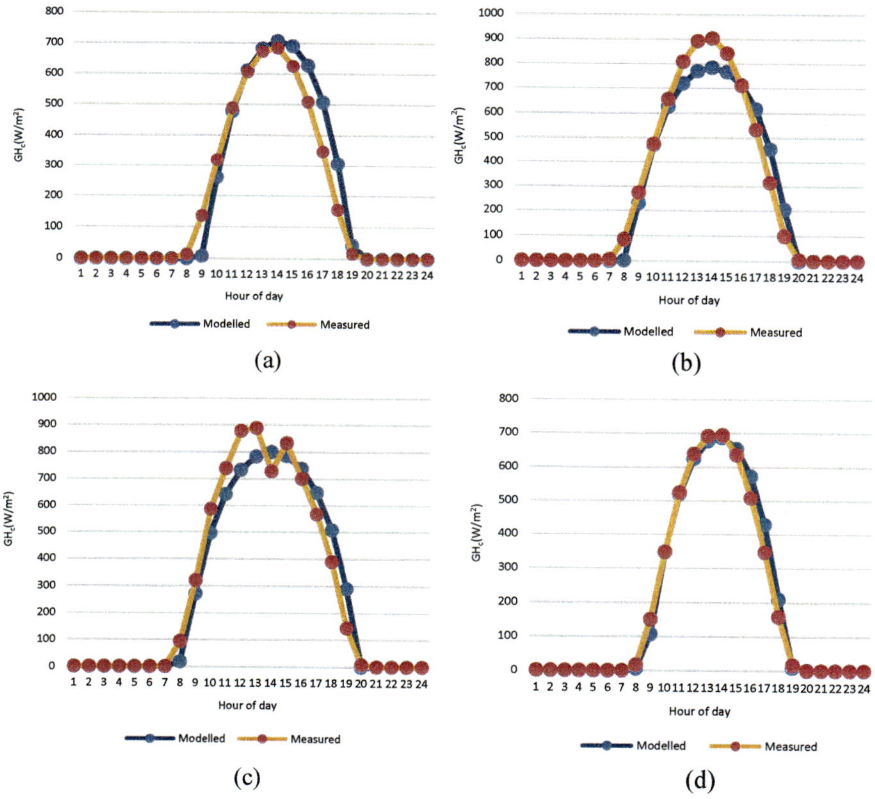

Fig. 3 Monthly average clear sky radiation for or **a** January, **b** April, **c** July, **d** November

Table 5 Statistical analysis of clear sky model validation

Parameter	January	April	July	November
RMSE (W/m^2)	60.8	61.03	67.21	25.9
MBE (W/m^2)	14.8	− 10.11	− 6.66	4.83
MAPE (%)	19	13.7	17.4	8.8
t-statistics	0.1723	0.8	0.5956	0.2765
R^2	0.9583	0.971	0.962	0.99

Table 6 Regression analysis for different months for Mumbai

Month	Degree of correlation	R^2	Regression constants		a + b
			a	b	
January	First	0.76	0.2979	0.5861	0.884
February	First	0.791	0.08	0.80	0.893
March	First	0.842	0.24	0.66	0.909
April	First	0.76	0.41	0.473	0.883
May	First	0.80	0.20	0.693	0.89
June	First	0.88	0.26	0.68	0.94
July	First	0.89	0.19	0.80	1
August	First	0.80	0.28	0.673	0.955
September	First	0.816	0.28	0.62	0.909
October	First	Improper data			
November	First	0.76	0.32	0.598	0.92
December	First	Improper data			

3.2 *Angstrom Model Validation*

As already mentioned, Angstrom model for hourly values have been used to derive the monthly real sky values for the location. The hourly-observed global radiation and the data related to the sunshine fraction were derived from the Indian Meteorological Department, Pune and the hourly clear sky radiation values were derived from the Bird's Clear Sky Model derived above. The monthly Angstrom Relations and their statistics are shown in Table 6.

The regression constants given above are used to determine the real sky values for the location. The comparative results are shown for a typical day in each month in the year (Fig. 4).

The season wise statistical analysis for the calculated and measured real sky values is shown in Table 7.

It can be seen from the analysis that the model best fits for the three seasons except July, which is typically monsoon season in India. The erratic nature of sunshine during that period, shall result in highly unrealistic predictions. But it can be seen that the methodology can be used for other time frames for the location.

4 Conclusion

The present work investigates prediction capability of the Bird's Clear sky model to calculate the clear sky irradiance when applied to a specific location in India. The percentage MAPE is below 20% for all the seasons in Mumbai, India which shows that Bird's model can fit real values within a reasonable accuracy for an Indian location. The information related to sunshine hours is used to deduce the real sky GHI values using Angstrom model. The values have also reasonable regression fit with the measured real sky values.

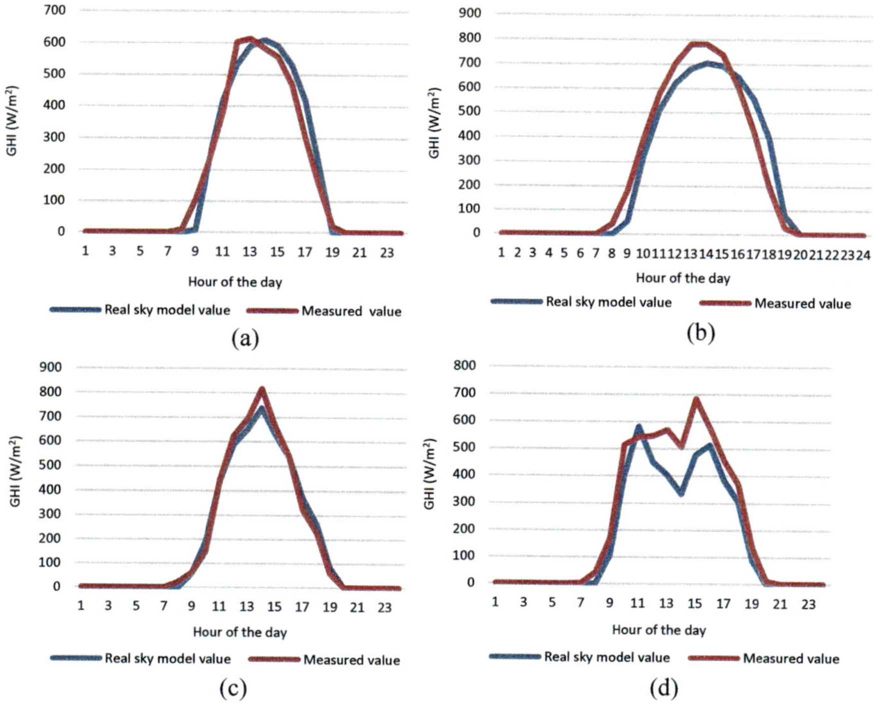

Fig. 4 Real sky values comparison for different months for Mumbai (**a**)–(**j**)

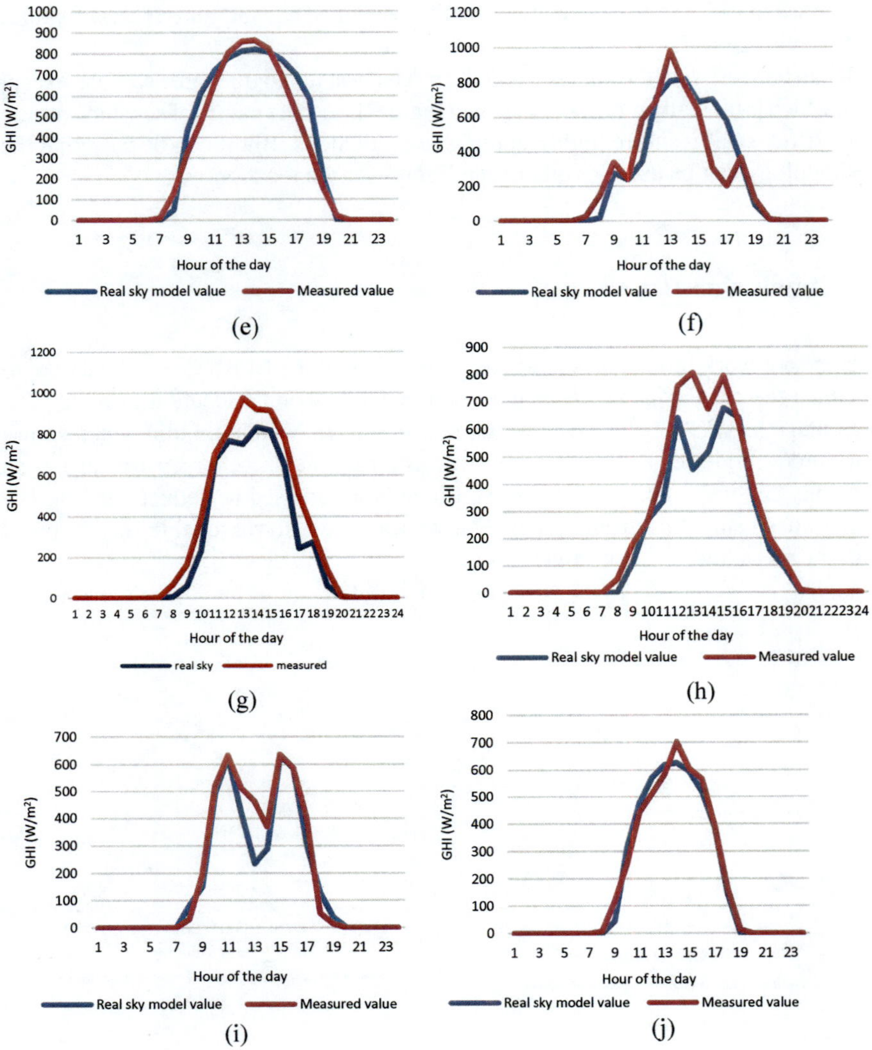

Fig. 4 (continued)

Table 7 Statistical analysis of season wise real sky model validation

Parameter	January	April	July	November
RMSE (W/m^2)	39.14	92.87	91.5	30.3
MAPE (%)	17.1	18.8	23.9	13.7
t-statistics	0.43	0.126	0.499	− 0.09
R^2	0.98	0.984	0.8	0.98

References

1. Viorel B (2008) Modelling solar radiation at the earth's surface recent advances, chap 1. Springer, Berlin, Germany
2. Lu Z, Piedrahita RH, Dos Santos Neto C (1998) Generation of daily and hourly solar radiation values for modeling water quality in aquaculture ponds. Trans Am Soc Agric Eng 41(6):1853–1858
3. Machler MA, Iqbal M (1985) A modification of the ASHRAE clear-sky irradiation model. ASHRAE Trans 91(1):106–115
4. Cartwright TJ (1993) Here comes the sun: solar energy from a flat-plate collector. In: Modelling the world in a spreadsheet—environmental simulation on a microcomputer. The Johns Hopkins University Press, London, pp 121–144
5. Salazar Trujillo JH (1998) Solar performance and shadow behaviour in buildings—case study with computer modelling of a building in Loranca, Spain. Build Environ 33(2–3):117–130
6. ASHRAE (1999) ASHRAE handbook: HVAC applications. ASHRAE, Atlanta, GA
7. Li DHW, Lam JC (2000) Solar heat gain factors and the implications to building designs in subtropical regions. Energy Build 32(1):47–55
8. Bird RE, Hulstrom RL (1981) A simplified clear sky model for direct and diffuse insolation on horizontal surfaces. Seri/Tr-642-761 Uc categories: Uc-S9,61,62,63, Feb 1981
9. Liu BYH, Jordan RC (1961) Daily insulation on surfaces tilted towards the equator. ASRHAE J 3:53–69
10. Gueymard C (1987) An anisotropic solar irradiance model for tilted surfaces and its comparison with selected engineering algorithms. Sol Energy 38:367–386
11. Perez R, Ineichen P, Seals R, Michalsky J, Stewart R (1990) Modeling daylight availability and irradiance components from direct and global irradiance. Sol Energy 44:271–289
12. Angstrom A (1924) Solar terrestrial radiation. Q J R Meteorol Soc 50:121–126

Prediction and Optimization of Thermal Conductivity and Viscosity of Stable Plasmonic TiN Nanofluid Using RSM and ANN Combined Approach for Solar Thermal Applications

Kishor Deshmukh and Suhas Karmare

1 Introduction

The United Nations Department of Economic and Social Affairs' population division reports that global population is growing at an unprecedented rate. The industrial sector is developing continuously to fulfill energy demands. The most frequent energy source is a fossil fuel. Researchers are motivated to find renewable energy sources due to the depletion and shortage of fossil fuels. Solar energy is the only renewable energy source that doesn't harm the environment [1]. Solar thermal collectors harvest energy and transform photons into heat [1–3]. Traditional fluids have poor thermophysical properties. These properties limit solar thermal collector performance [2, 3]. Choi (1995) synthesized a nanofluid by dispersing nanosized (100 nm) particles in a liquid. The next generation of nanofluid is hybrid nanofluid synthesized by suspending two or more nanoparticle species in the base fluid [1, 2, 4]. Choi (1995) synthesized a nanofluid by dispersing nanosized (100 nm) particles in a liquid. The next generation of nanofluid is hybrid nanofluid synthesized by suspending two or more nanoparticle species in the base fluid [2, 3]. Nanomaterials have unique thermal, physical, chemical, and mechanical capabilities due to more atomic atoms on grain boundaries and a greater surface area to volume ratio [2]. Nanofluid with tunable thermophysical characteristics are the next generation. Sarkar et al. [5] highlights the scope and obstacles for study in the nanofluid synthesis, characterization, stability, cost, and application domains. Nanofluid development and use are still in their adolescence. The significant research contributions noticed in recent decades on thermophysical

K. Deshmukh (✉)
Trinity College of Engineering and Research, Pune, India
e-mail: dkishor21@gmail.com

Amrutvahini College of Engineering, Sangamner, India

S. Karmare
Government College of Engineering, Awasari, India

S. Doolla et al. (eds.), *Advances in Clean Energy and Sustainability*,
Green Energy and Technology, https://doi.org/10.1007/978-981-99-2279-6_3

properties [2, 6, 7]. Nanofluid thermophysics depend on base fluid, nanoparticles, temperature, volume concentration, size, and shape. Nanofluid thermal conductivity and viscosity affect heat transfer performance [8, 9]. Nanofluid rheology must be examined with heat transmission properties. Temperature, volume concentration, and stability affect nanofluid thermal conductivity and viscosity [10]. This reliance leads to a complicated interaction between thermophysical qualities and variable factors. Traditional models are incapable of accurately predicting thermophysical characteristics. Soft computing methods including RSM, ANN, NSGA, CFD, particle swarm technology and fuzzy logic save experimental costs and time [2, 7]. Sustainable and cost-effective systems require optimal volume concentration. The ANN technique has inbuilt prediction ability [2]. Seyed et al. suggested a multi-objective optimization model based on ANN and GA. ANN predicts the thermophysical characteristics of a CuO/liquid paraffin nanofluid, while GA predicts the minimal pressure drop and maximum heat transfer coefficient [2, 11]. Maqsood et al. developed a thermal oil-based MWCNT nanofluid for highest thermal conductivity and minimal viscosity [2, 12]. A predictive model was developed using RSM and ANN techniques. In addition to the prediction, optimization was undertaken for four distinct situations based on applications. Ebrahimi-Moghadam and Moghadam [13] used a genetic algorithm to tune geometrical parameters and Al_2O_3 nanofluid flow properties inside a corrugated heat exchanger. The cooling fluid is an Al_2O_3 nanofluid. MATLAB computational code employed for modeling. The following conclusions are drawn from the studied literature:

- Fluid thermal conductivity and viscosity are significant properties in micro convention phenomena and should be considered while exploring nanofluid in heat transfer applications. Nanoparticle volume concentration increases nanofluid thermal conductivity and viscosity [6]. The heat transfer rate increases with thermal conductivity and decreases with viscosity. The pressure drop is proportional to the volume concentration.
- TiN nanofluid surpasses other types of nanofluid in photothermal conversion. Researchers are attempting to embrace modeling approaches due to high testing costs and time consumption. So far, no one has created a model employing soft computing approaches to forecast the thermophysical parameters of TiN nanofluidSimulation model can predict TiN nanofluid thermal conductivity and viscosity at any concentration and temperature without experimentation [14].
- RSM and ANN techniques have their capabilities. RSM and ANN can forecast and optimise TiN nanofluid thermal conductivity and viscosity for heat transfer equipment design and development [15]. An attempt can make to find a more reliable and confident approach to develop thermal conductivity and viscosity correlation.

Table 1 Zeta potential and particle size

% Vol. concentration	0	0.025	0.050	0.075	0.100
Zeta potential (mV)	0	− 61.6	− 61.1	− 37.1	− 30.1
Particle size (nm)	100.0	117.3	141.5	144.3	151.1

2 TiN Nanofluid Preparation, Characterization, and Experimental Investigation of Thermophysical Properties

2.1 *Process Parameters, Nanoparticle, Base Fluid, Surfactant Selection*

According to the studied literature, nanoparticle concentration and temperature are significant factors that influence nanofluid thermophysical properties and system heat transfer performance [6]. The NIMS and MANA research teams found that TiN nanofluid LSPR feature allows it to absorb sunlight well [16]. TiN nanoparticles are explored for this research because of this distinctive LSPR property. Intelligent Materials Pvt. Ltd., Punjab, supplied commercial TiN nanoparticles (India). TiN nanoparticles less than 100 nm size increases stability and heat transmission capability. The experiment's base fluid was distilled and purified utilizing Milli-Q water purification equipment (Pure Lab Flex 3, ELGA UK) [2]. Surfactants stabilize nanofluid. Surfactants make nanoparticles hydrophobic or hydrophilic. Since the base fluid possesses an O–H polar solvent link, LOBA Chemie Pvt. Ltd.-branded, pure, needle-shaped sodium lauryl sulphate (SLS) is used as a surfactant [2, 17]. Nanoparticles start to agglomerate beyond 0.10% volume concentration. For our study, five concentration levels were chosen (0, 0.025, 0.050, 0.075, and 0.1%) [2]. The solar thermal collector's fluid temperature ranged from 30 to 60 °C from 9:00 a.m. to 5:00 p.m. Therefore, five temperature settings (35, 40, 45, 50, and 55 °C) were evaluated for this investigation. Minitab 17 used Taguchi for DOE (L25). Table 1 describes nanoparticles, surfactants, and base fluids. TiN particle morphology is examined using FESEM.

FESEM (FEI Nova NanoSEM 450, SPPU, Pune) photographed TiN nanoparticles. A high-resolution image of nanoparticles at 10 and 500 nm size taken at 100,000× magnification. Sonication is needed for these agglomerated TiN nanoparticles. The average particle size matches the supplier's test results.

2.2 *Nanofluid Preparation and Characterization*

Nanofluid research relies on nanofluid preparation. The two-step approach is preferred to achieve heat transfer efficiency. The mixing law determines how much

TiN nanoparticles are needed to achieve the desired concentration [18]. The mixing law formula determines TiN nanoparticle concentration. Sodium lauryl sulphate is added as a surfactant at one-tenth the nanoparticle weight to stabilise them. A precise balance (Shimadzu AUX220) was used to weigh and add dry titanium nitride powder and sodium lauryl sulphate (SLS) to the base fluid [2]. TiN nanoparticles and sodium lauryl sulphate distributed in 1000 ml of distilled water at volume concentrations of 0.025, 0.050, 0.075, and 0.1%. Van der Waal's attracts TiN nanoparticles. Van der Waal's force of attraction is disrupted by the magnetic stirrer (Remi 1MLH).The magnetic stirrer operates at 1000 rpm for 60 min, followed by a bath-type ultra-sonicator [2]. A probe-type sonicator is utilised at the end to create a homogenous and stable nanofluid. TiN nanofluid is stable in terms of particle size and stability. TiN nanofluid stability is assessed visually and using a zeta potential analyzer. The TiN nanofluid is visibly stable after 15 days. Amrutvahini College of Pharmacy, Sangamner's zeta potential analyzer (SZ-100 HORIBA) measured TiN nanofluid's zeta potential and particle size [2]. The zeta potential value varies with time. The nanofluid is stable because TiN nanoparticles do not aggregate due to their negative zeta potential [2]. Table 1 shows zeta potential and particle size results.

3 Investigation of Thermophysical Properties

Thermal conductivity is a material's ability to conduct heat. The homogeneous dispersion of nanoparticles in base fluids impacts their thermal conductivity. Nanofluid micro convection relies on nano scale Brownian motion. Dr. Sonawane and Prof. Kokate of Sapat College of Engineering in Nashik devised transient hot wire apparatus [2, 19]. The thermal conductivity of TiN nanofluid was tested using his experimental equipment [2]. It has two measuring cylinders connected by a platinum wire. Platinum wire acts as a heating source and a temperature sensor for measuring thermal conductivity. According to the literature, this approach produces accurate findings by excluding the influence of natural convection. The convection effect identified when it presents as a deviation from linearity in the plot of ΔT as a function of ln(t). The possible convection effect minimized using two measuring cylinders [1, 17, 19]. The temperature difference is measured using both cylinder resistances. The transient-hot-wire approach measures nanofluid effective thermal conductivity by comparing platinum wire temperature rise against logarithmic time interval [6, 20]. Linear regression determines linear component slope till unsigned residuals. To calibrate the apparatus, deionized water, air, and toluene thermal conductivity were measured at room temperature [2]. Viscosity inhibits motion. The friction between the contact surface and fluid molecules is affected by fluid viscosity. Nanoparticles enhance viscosity. Nanoparticle concentration, shear stress rate range, base fluid type, agglomeration rate, temperature, purity, size, shape, type, preparation process, and dispersion method affect nanofluid viscosity [2]. A Brookfield Viscometer DV-I Prime (SPPU, Chemistry Department, Pune) assessed TiN-Water nanofluid viscosity at maximum torque and 50 rpm with 0.1 precision [2]. Figure 1 shows thermal

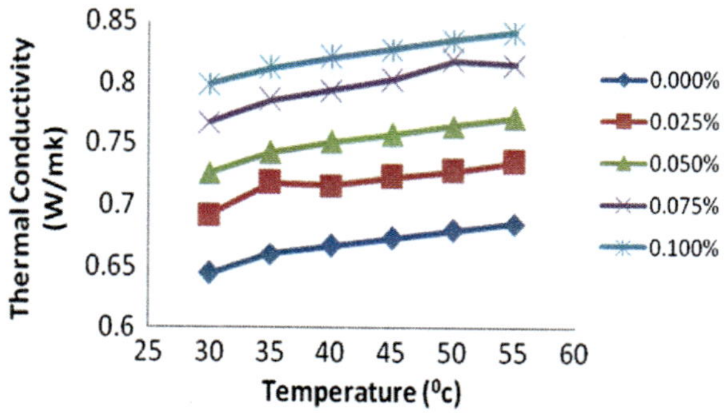

Fig. 1 Thermal conductivity variation as a function of % volume concentration

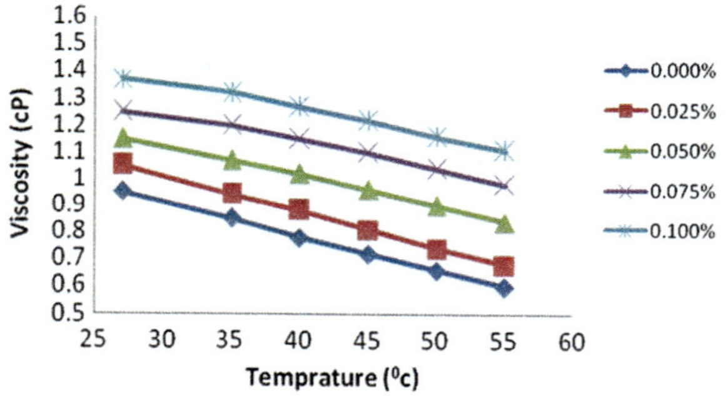

Fig. 2 Viscosity variation as a function of % volume concentration

conductivity variation vs temperature range for varied volume concentrations [2]. Thermal conductivity reached 0.8421% at 55 °C and 0.1% volume concentration. The trend for all nanofluid concentrations is the same as for a 0.1% volume concentration. Micro convection, conduction paths, and Brownian motion promote thermal conductivity. Figure 2 shows how volume concentration affects viscosity. Minimum viscosity was 1.11 cP at 55 °C and 0.1% volume concentration. The trend is the same for all nanofluid concentrations. A maximum enhancement of 19.05% was obtained at a volume concentration of 0.05%. Volume concentration, cohesive forces, and adhesive forces between nanoparticles enhance viscosity [2].

4 Uncertainty Analysis

Uncertainty analysis quantifies input-induced output variability. Sundar et al. analyze experimental data for experimentation errors using uncertainty analysis [21]. The maximum uncertainties in thermal conductivity and viscosity measurement are 12.50% and 8.33%, respectively [2]. The Horiba SZ-1000 has a stability measurement uncertainty of 1.62% and a particle size measurement uncertainty of 1.98%.

5 Mathematical Modelling of TiN Nanofluid Thermal Conductivity and Viscosity

5.1 RSM-Predictive Model

Surface response is the most common statistical and mathematical method [2]. Regression techniques empirically link input and output variables. RSM response variables depend on the least significant input variable [22]. The RSM's primary objective is to forecast thermal conductivity and viscosity. Volume concentration and temperature are input variables [2]. Thermal conductivity and viscosity are response variables. A second-order prediction model considers input-response reciprocity. The prediction model and input variables were tested using an ANOVA. ANOVA calculates the degree of freedom, number of squared deviations, and mean square error for each input. Significant input variables have *p*-values less than 0.05 [2].

$$K = 0.6132 + 1.817A + 0.00157B - 4.693A^2 - 0.000005B^2 + 0.00466\mathrm{AB} \quad (1)$$

Equation (1) predicts TiN nanofluid thermal conductivity using input variables interactions. K denotes thermal conductivity [2]. A, and B are volume concentration and temperature. Thermal conductivity model ANOVA is being considered. F = 732.58, $p = 0.000$. It validates the model. $R^2 = 99.48\%$. Thermal conductivity depends on volume concentration, temperature, and volume concentration squared. Volume concentration and temperature do not affect enhancement.

$$\mu = 1.410 + 3.41A - 0.0173B - 1.6A^2 - 0.000034B^2 + 0.0496\mathrm{AB} \quad (2)$$

Equation (2) shows the input variable interaction viscosity prediction model. A and B denotes volume concentration and temperature, whereas μ represent viscosity. The model's F value is 127.77, and the *p*-value is 0.000. It demonstrates the importance and acceptance of the established model. The R^2 coefficient is 97.11%. Viscosity model ANOVA was considered [2]. The viscosity model states that volume concentration and temperature affect viscosity. Volume concentration and temperature interactions do not effect enhancement. Each input variable's percentage contribution, square, and interactions are calculated in the ANOVA table. For thermal conductivity enhancement, volume concentration (A) is the most important input variable, followed by temperature (B). For viscosity enhancement, temperature (B) is followed by volume concentration (A). Thermal conductivity and viscosity results are plotted in Figs. 3 and 4. Both models have R^2 values between 97 and 98%. Predicted models with maximum R^2 are reliable and efficient. The adequacy and outlier plot tests were done following model construction [2].

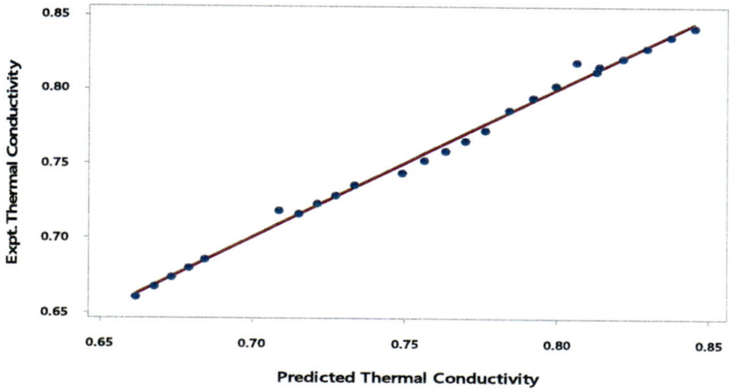

Fig. 3 Experimental and predicted TiN nanofluid thermal conductivity scatterplot

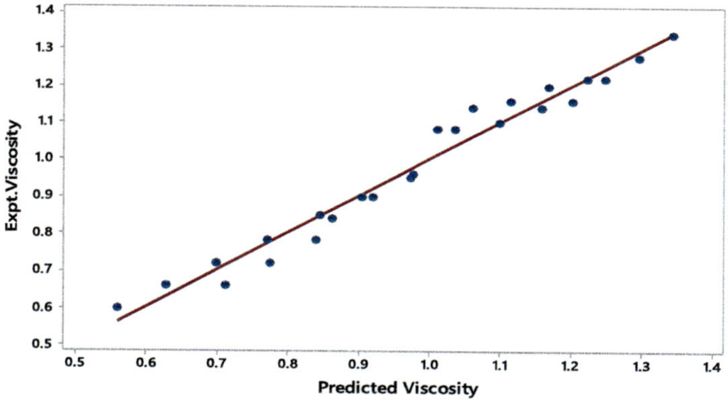

Fig. 4 Experimental and RSM predicted TiN nanofluid viscosity scatterplot

Thermal conductivity and viscosity residual plots are shown in Figs. 5 and 6, respectively. The results are near to the straightened line, randomly distributed, and pattern less. It indicates its suitability for optimization and prediction. Data points that deviate from the trend are called outliers. Predictive model development requires outlier detection and elimination. Figures 7 and 8 exhibit thermal conductivity and viscosity outlier plots. The prediction model allows all data points. Outlier charts show thermal conductivity and viscosity forecasting models are accurate [2]. The contour surface response plot provides the influence of each parameter on thermal conductivity enhancement [2].

Figure 9 shows that volume concentration and temperature improve TiN nanofluid thermal conductivity. Thermal conductivity reached 0.84 W/mK at 0.1% volume concentration and 55 °C [2]. In Fig. 10, TiN nanofluid viscosity increases with volume concentration and decreases with temperature. The minimum viscosity was 0.6 cP at 55 °C and 0% volume concentration [2].

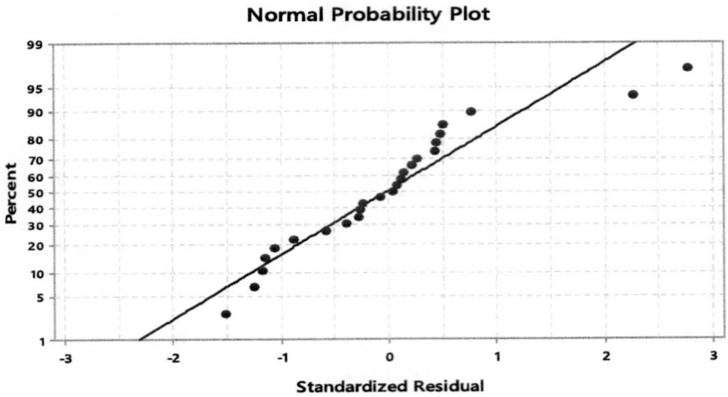

Fig. 5 Normal probability plot of predicted thermal conductivity

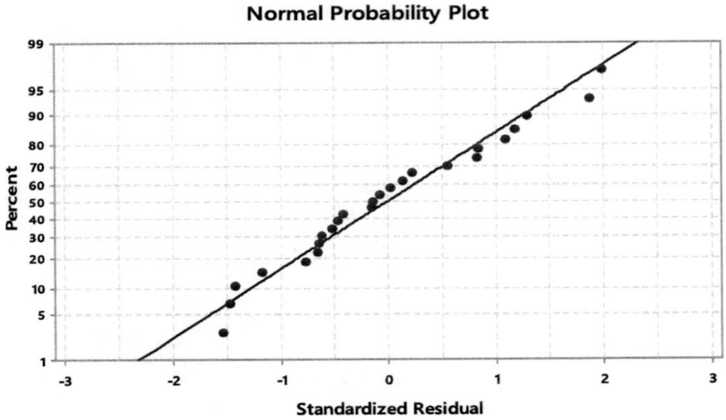

Fig. 6 Normal probability plot of predicted viscosity of TiN nanofluid

5.2 ANN Predictive Model

ANN provides the most efficient computational approach for determining the optimum solution from training data. ANN is inspired by the human brain's biological neuron system [23]. ANN is used to develop an accurate inherent relationship between volume concentration, temperature, and thermophysical properties. ANN uses the multilayer perceptron technique (MLP) to obtain solutions for nonlinear problems. The input layer assigns input variables, while the output layer assigns response variables. The neural network comprises two neurons in the input layer for volume concentration and temperature and two in the output layer for thermal conductivity and viscosity. The appropriate number of neurons in the hidden layer is determined by the greatest R^2 and MSE. Each artificial neuron has a weight (w), bias (b), activation function (y), input signal (x), and output signal (y). All 25 experimental data sets were employed for ANN modelling. 70% of the data was deployed

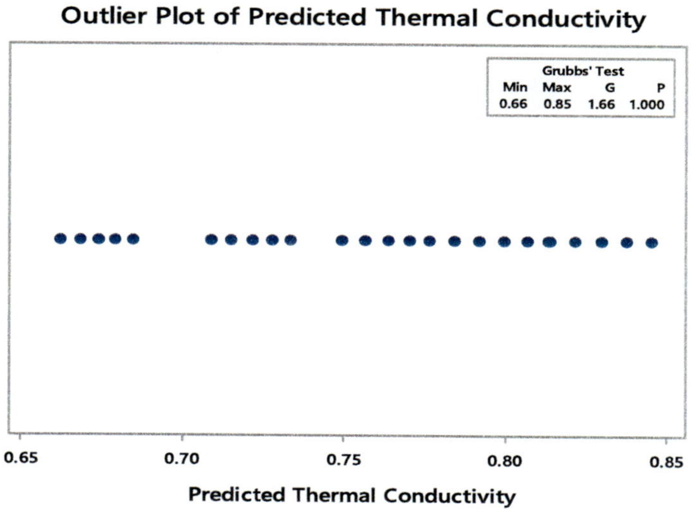

Fig. 7 Outlier plot of predicted thermal conductivity

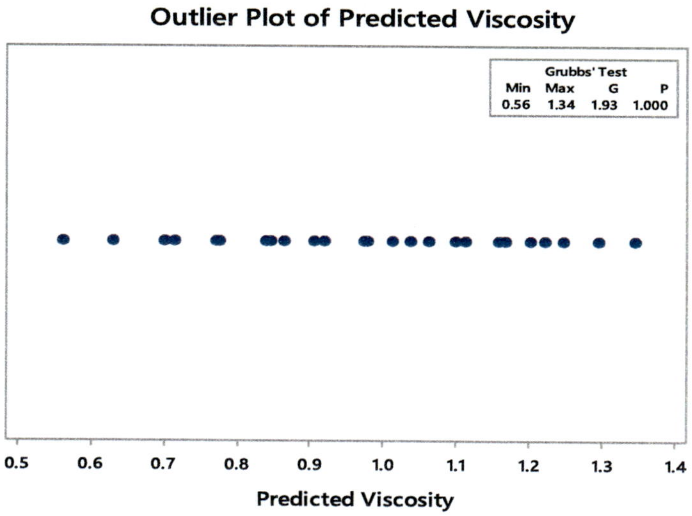

Fig. 8 Outlier plot of predicted viscosity of TiN nanofluid

to train neurons, with the remaining 30% used to validate and evaluate the data. ANN operate in three phases:

- **Experimental data normalization**: Normalization is the preprocessing of experimental data.

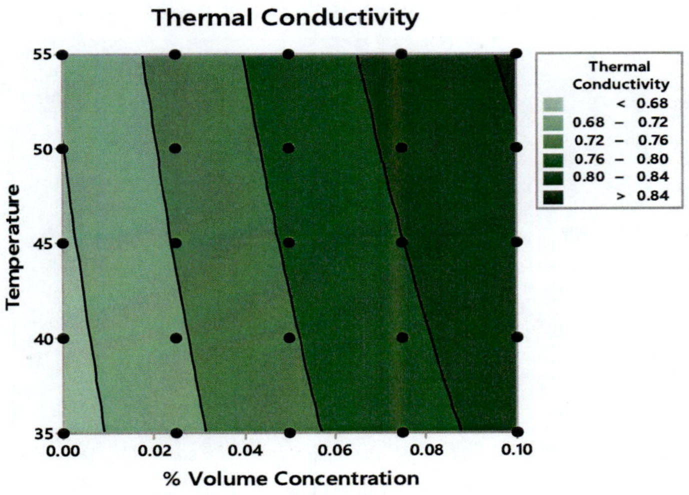

Fig. 9 3D contour plots for experimental and RSM predicted thermal conductivity

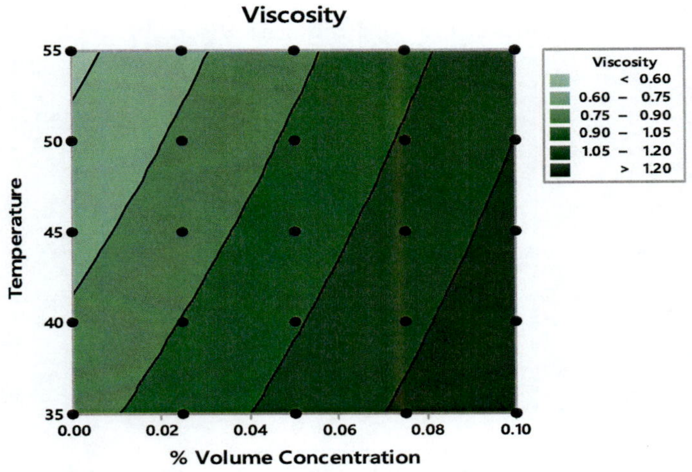

Fig. 10 3D contour plots for experimental and corresponding RSM predicted viscosity

- **Determine optimal neuron number and hidden layers**: 4 neurons in the hidden layer gives minimum MSE of 0.000338, hence 4 neurons are considered to develop the ANN model.
- **Creation, training, testing, and validation of neural network**: In order to avoid over fit or under fit, input data is split into training, validation, and testing.

6 RSM-ANN Model Comparative Error Analysis and Optimization

RSM and ANN models are compared using correlation coefficient (R^2), root mean square error (MSE), standard error of prediction (SEP), and average absolute deviation (AAD) [2]. The regression line maximum R^2 indicates data point proximity. The minimal MSE shows how distant the data points are from the regression line [2]. The result shows good agreement with a minimal error rate of 5%. RSM and ANN thermal conductivity models had 99.48 and 96.97% R^2 values. RSM thermal conductivity model predicts more accurate data than the ANN model. RSM and ANN thermal conductivity models have R^2 97.81% and 99.20% respectively. The ANN viscosity model predicts better data than the RSM model. RSM thermal conductivity and the ANN viscosity model forecast reliable and efficient results [2]. The art of finding the optimum solution within the available constraints is known as optimization. The hurdle for commercializing nanofluid is their fluctuating thermophysical characteristics. Fluctuating nanofluid thermophysical characteristics limit heat transfer enhancement [2]. Solar thermal applications can use TiN nanofluid. Solar thermal applications require nanofluid with high heat conductivity and low viscosity [2]. Optimization uses the ANN-predicted dataset, which is more accurate than the RSM tool. 55 °C and 0.0525% volume concentration yield maximum thermal conductivity of 0.7806 W/mK and minimum viscosity of 0.8766 cP [2].

7 Conclusion

This study covers the stable TiN nanofluid synthesis, characterization, experimental examination, prediction, and optimization of thermal conductivity and viscosity. TiN nanofluid is stable and thermophysically superior. Thermal conductivity and viscosity experiments are costly. RSM and ANN statistical methods use volume concentration and temperature to forecast. Predictive models optimize process parameters. Surface response and artificial neural network thermal conductivity and viscosity forecasting models depend on volume concentration and temperature. RSM thermal conductivity and ANN viscosity models produce correct findings.

$$\begin{aligned} K = {} & 0.6132 + 1.817\varphi + 0.00157T \\ & - 4.693\varphi^2 - 0.000005T^2 + 0.00466\varphi T \end{aligned}$$

$$\begin{aligned} \mu = {} & 1.410 + 3.41\varphi - 0.0173T - 1.6\varphi^2 \\ & - 0.000034T^2 + 0.0496\varphi T \end{aligned}$$

The study focuses on the multi-objective optimization of thermal conductivity and viscosity. 0.052% volume concentration and 55 °C temperature maximize thermal conductivity (0.7806 W/mK) and minimize viscosity (0.8766 cP). Thermophysical characteristics prediction is possible using different computational systems. The best parameters could help you save money and reduce the size of your system. The stable TIN nanofluid found worthiness in solar applications.

References

1. Deshmukh K, Karmare S, Raut D (2022) Preparation, characterization and experimental investigation of thermophysical properties of stable TiN nanofluid for solar thermal application. J Braz Soc Mech Sci Eng 44(10). https://doi.org/10.1007/s40430-022-03733-2
2. Deshmukh K, Karmare S (2022) Prediction and optimization of thermal conductivity and viscosity of stable plasmonic TiN nano fluid using response surface method for solar thermal application. Europe PMC. [Online]. Available: https://europepmc.org/article/ppr/ppr545451
3. Sabiha MA, Saidur R, Hassani S, Said Z, Mekhilef S (2015) Energy performance of an evacuated tube solar collector using single walled carbon nanotubes nanofluids. Energy Convers Manag 105:1377–1388. https://doi.org/10.1016/j.enconman.2015.09.009
4. Ali ARI, Salam B (2020) A review on nanofluid: preparation, stability, thermophysical properties, heat transfer characteristics and application. SN Appl Sci 2(10). https://doi.org/10.1007/s42452-020-03427-1
5. Sarkar J, Ghosh P, Adil A (2015) A review on hybrid nanofluids: recent research, development and applications. Renew Sustain Energy Rev 43:164–177. https://doi.org/10.1016/j.rser.2014.11.023
6. Cakmak NK, Said Z, Sundar LS, Ali ZM, Tiwari AK (2020) Preparation, characterization, stability, and thermal conductivity of rGO-Fe_3O_4-TiO_2 hybrid nanofluid: an experimental study. Powder Technol 372:235–245. https://doi.org/10.1016/j.powtec.2020.06.012
7. Deshmukh KB, Karmare SV (2019) A review on augmentation of convective heat transfer techniques in solar water heating. J Therm Energy Syst 4(3):29–40. https://doi.org/10.5281/zenodo.3542729
8. Deshmukh KB, Karmare SV (2021) A review on convective heat augmentation techniques in solar thermal collector using nanofluid. J Therm Eng 7(5):1257–1266. https://doi.org/10.18186/thermal.978064
9. Cao Y, Khan A, Abdi A, Ghadiri M (2021) Combination of RSM and NSGA-II algorithm for optimization and prediction of thermal conductivity and viscosity of bioglycol/water mixture containing SiO_2 nanoparticles. Arab J Chem 14(7):103204. https://doi.org/10.1016/j.arabjc.2021.103204
10. Sonawane SS, Juwar V (2016) Optimization of conditions for an enhancement of thermal conductivity and minimization of viscosity of ethylene glycol based Fe_3O_4 nanofluid. Appl Therm Eng 109:121–129. https://doi.org/10.1016/j.applthermaleng.2016.08.066
11. Bagherzadeh SA, Sulgani MT, Nikkhah V, Bahrami M, Karimipour A, Jiang Y (2019) Minimize pressure drop and maximize heat transfer coefficient by the new proposed multi-objective optimization/statistical model composed of 'ANN + genetic algorithm' based on empirical data of CuO/paraffin nanofluid in a pipe. Phys A Stat Mech Appl 527:121056. https://doi.org/10.1016/j.physa.2019.121056
12. Maqsood K et al (2022) Multi-objective optimization of thermophysical properties of multi-walled carbon nanotubes based nanofluids. Chemosphere 286:131690. https://doi.org/10.1016/J.CHEMOSPHERE.2021.131690
13. Ebrahimi-Moghadam A, Moghadam AJ (2019) Optimal design of geometrical parameters and flow characteristics for Al_2O_3/water nanofluid inside corrugated heat exchangers by using entropy generation minimization and genetic algorithm methods. Appl Therm Eng 889–898. https://doi.org/10.1016/j.applthermaleng.2018.12.068

14. Deshmukh K, Karmare S, Patil P (2023) Experimental investigation of convective heat transfer inside tube with stable plasmonic TiN nanofluid and twisted tape combination for solar thermal applications. Heat Mass Transf 0123456789. https://doi.org/10.1007/s00231-023-03344-0
15. Azmi WH, Sharma KV, Mamat R, Alias ABS, Misnon II (2012) Correlations for thermal conductivity and viscosity of water based nanofluids. IOP Conf Ser Mater Sci Eng 36(1). https://doi.org/10.1088/1757-899X/36/1/012029
16. Ishii S, Sugavaneshwar RP, Nagao T (2016) Titanium nitride nanoparticles as plasmonic solar heat transducers. J Phys Chem C 120(4):2343–2348. https://doi.org/10.1021/acs.jpcc.5b09604
17. Sonawane S, Patankar K, Fogla A, Puranik B, Bhandarkar U, Sunil Kumar S (2011) An experimental investigation of thermo-physical properties and heat transfer performance of Al_2O_3-aviation turbine fuel nanofluids. Appl Therm Eng 31(14–15):2841–2849. https://doi.org/10.1016/j.applthermaleng.2011.05.009
18. Kumar S, Chander N, Gupta VK, Kukreja R (2021) Progress, challenges and future prospects of plasmonic nanofluid based direct absorption solar collectors—a state-of-the-art review. Sol Energy 227:365–425. https://doi.org/10.1016/j.solener.2021.09.008
19. Kokate Y, Sonawane S (2019) Investigation of particle size effect on thermal conductivity enhancement of distilled water-Al_2O_3 nano fluids. Fluid Mech Res Int J 3(2):3–6. https://doi.org/10.15406/fmrij.2019.03.00052
20. Jannot Y, Degiovanni A, Schick V, Meulemans J (2021) Apparent thermal conductivity measurement of anisotropic insulating materials at high temperature by the parallel hot-wire method. Int J Therm Sci 160:106672. https://doi.org/10.1016/j.ijthermalsci.2020.106672
21. Sundar LS, Sharma KV (2010) Turbulent heat transfer and friction factor of Al_2O_3 nanofluid in circular tube with twisted tape inserts. Int J Heat Mass Transf 53(7–8):1409–1416. https://doi.org/10.1016/j.ijheatmasstransfer.2009.12.016
22. Maqsood K (2021) Multiobjective optimization of thermophysical properties of Indonesian fly-ash nanofluid. Mater Today Proc 49:1255–1262. https://doi.org/10.1016/j.matpr.2021.06.304
23. Kamble LV, Pangavhane DR, Singh TP (2014) Heat transfer studies using artificial neural network—a review. Int Energy J 14(1):25–42

From Slum to Slum Rehabilitation: Comparing the Factors Affecting Energy Consumption and Environmental Satisfaction Among the Low-Income Housing in Mumbai

Ahana Sarkar and Arnab Jana

1 Introduction

Unprecedented urbanization has increased energy consumption growth on one hand [1] while negatively impacting environment on the other hand [2]. It is hypothesized that apart from climate characteristics, varying factors such as market accessibility, affordability, household characteristics, wealth and income patterns might influence the behaviour impacting energy choices. Particularly, daily household activities and occupant behavior shape their need for built-environment space and appliances which further impacts the energy consumption [3]. Consequently, the interaction between energy, built-form and household practices impacts the well-being and health of the occupants.

However, this scenario turns challenging for low-income sector, where it is assumed that households belonging to lower socio-economic group own less energy appliances, thereby belonging to the lowest strata of the energy ladder. However, informal settlements such as slums as well as social housing have now been observed to own various energy appliances [3]. Related literature has found that product cost, quality, brand, and discounts turn to be the major factors impacting appliance characteristics. However, these characteristics might also vary with socio-architectural need of the society and built-environment setup. Therefore, understanding the drivers of energy demand, and environmental satisfaction becomes necessary particularly in disadvantaged areas where space constraints couple with financial restrains, resulting in occupants to struggle for availing essential energy services.

A. Sarkar
KRVIA, University of Mumbai, Mumbai, India

A. Jana (✉)
Centre for Urban Science and Engineering, Indian Institute of Technology Bombay, Mumbai, India
e-mail: arnab.jana@iitb.ac.in

S. Doolla et al. (eds.), *Advances in Clean Energy and Sustainability*,
Green Energy and Technology, https://doi.org/10.1007/978-981-99-2279-6_4

In this study, we investigate whether difference in built-environment setting between horizontal slums and vertical apartments within the slum rehabilitation housing (SRH) affect the appliance ownership decisions and residential energy consumption levels and environmental satisfaction. The novelty of this work lies in the fact of emerging at a system where a linkage between built-environment characteristics, household practice, occupant behavior, and appliance characteristics is established and are determined as essential drivers of environmental satisfaction.

Similar studies conducted in Mumbai SRHs have found disadvantaged locations [4], hyper-density and environment-insensitive built-environment design [5] leading to poor indoor environmental quality and deteriorated health conditions [6]. However, none of these studies investigated the further impact of built-environment characteristics and occupant behavior on appliance purchase decisions and environmental satisfaction, additionally comparing the effects for slums and SRHs. The results and policy implications of this study could pave way towards designing better energy efficiency policies for such low-income communities.

2 Background

Among the major household level determinants, household size, building age and house type are the significant parameters affecting appliance ownership [7]. While factors such as more female members in house, and higher education levels resulted in more appliance ownership in China, quality of electricity supply and household characteristics impacted appliance purchase decisions in India [8, 9]. Other vital non-income drivers of electricity consumption are family composition, age of earning member of the household, housing tenure type etc. The major built-environment elements that influence energy consumption are the number of rooms, bedrooms, floors and floor area, along with space for electric gadgets [9, 10].

While factors such as building design and contextual factors, building type, level of accessibility, occupancy level, space design, work position etc., impact indoor environmental quality and productivity on one hand, residential satisfaction and household energy consumption is hugely impacted by built-environment design parameters such as no of rooms, corridor type, fenestration design impacting indoor ventilation and natural lighting levels, odor and mold on the other hand. Other household characteristics that can be potential drivers of energy choices and appliance ownership levels include household income, education, age and gender, infrastructure and services such as electricity supply, corridor type, lift, accessible road, and other psychological parameters such as indoor comfort and privacy, livelihood opportunity etc. [11].

3 Methodology

3.1 Selection of Housing Communities

A transverse methodology was adopted, where on-field observation coupled with detailed questionnaire survey were conducted across six low-income housing communities of Mumbai. For a representative slum housing configuration, a neighborhood in Dharavi, and in Narayan Nagar, Ghatkopar were selected. Both these case-studies, are characterized by mixed-use community with informal tin and concrete makeshift hutments spanning one to two levels (refer to Fig. 1).

Four representative resettlement configurations, Kohinoor Colony, Lallubhai Compound, Kanjurmarg and Sangharshnagar SRH were selected as sites housing project affected persons (PAP) relocated by major infrastructure projects and slum redevelopment campaigns. Kohinoor SRH colony, containing around 32 13-floored huge building blocks and accommodating approximately 16,000 households across 53 acres (0.213 km^2) has a housing density of 932 DU/Ha (Dwelling Unit/Hectare). Lallubhai Compound contains 9300 apartments across 65 building blocks with units accessed via 2 m wide double-loaded corridors, whose kitchens and bathing rooms are located against the external wall. Kanjur Marg SRH, near Kanjur Marg railway station in the Mumbai suburbs, comprises 12 eight-storied buildings accommodating 2232 households. Sangharshnagar SRH, consisting of 18,362 households, each eight floored building contains 6 dwelling units per floor accessed by central staircase and lift, with toilets and kitchen abutting external wall.

Fig. 1 Photographs of selected housing communities

3.2 Survey Design and Process

The built-environment characteristics included building-to-apartment-level information like age of building, floor-level, apartment design and size, kitchen and bathing room location and furniture details, with information on fenestration design and their operating schedule. Further information was reckoned on the ownership of various energy appliances and their frequency of operation. Five-point perception and satisfaction scales were used to gauge the environmental metrics like air quality and ventilation, daylight, humidity built-environment setup, with a higher value indicating more satisfaction. Lifestyle was measured by collecting information on frequency and overall time spent in cooking and other subsistence activities such as sleeping, taking meals etc. The energy time use of the respondent (i.e., time spent in cooking, washing clothes, using laptop/phone, watching TV, charging electronic devices etc.) were also collected.

It was carried out in the six neighbourhoods of Dharavi slum (109), Ghatkopar slum (109), Kohinoor SRH (122), Kanjurmarg SRH (103), Lallubhai SRH (107) and Sangharshnagar SRH (93) during February–March 2020. The households were identified and the surveys were administered for the eligible and available household members. The families who owned the tenements were included. Since the aim of the research was to explore the relationship between built-environment design, occupant behavior and energy decisions and lifestyle of occupants with a focus on environmental satisfaction, the survey focused on the female members above 18 years, who tend to spend most of the time indoors. The participants completed the computer-aided personal interview (CAPI) based survey questionnaire through personal face-to-face interviews, set in English and Hindi language.

3.3 Hypothetical Model Development

Based on the above literature on built-environment setting, energy time-use and household practices and its effect on ownership of appliances and environmental satisfaction especially in low-income settlements, six hypotheses were formulated, as shown in Fig. 2.

Socio-economic characteristics

In total, 643 slum and resettlement occupants were selected to participate the study, with 100% response rates from female members of the household. While all the respondents were female, the mean age of the respondents was 40.2 years. Most of the respondents (82%) had maximum high-school level education, with only 11% engaged in a part-time job, and remaining housewife. The average household size in Dharavi and Ghatkopar slum was reckoned 4.09 and 3.96 respectively with only 19% households having more than 5 family members. While the average household size in the SRHs was 4, with around one-third (29%) respondent families having a

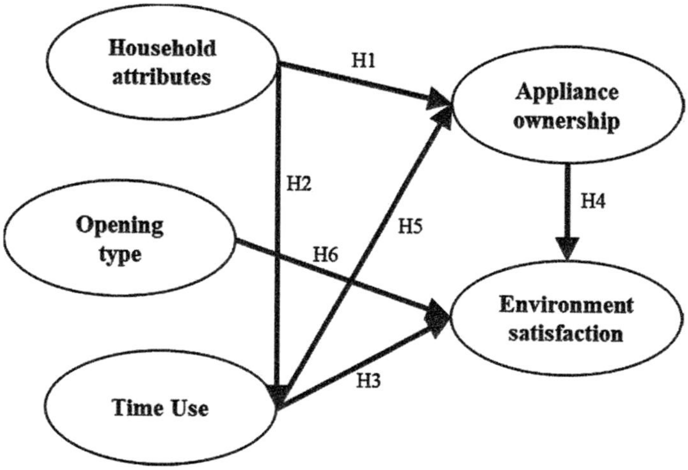

Fig. 2 Hypotheses considered in the study

household size of 5 and above. One-third of the households (30%) of households reported average household monthly income of 9000–18,000 INR.

Built-environment design and lifestyle characteristics

While the age of all the building typologies was reckoned 10–15 years, striking differences were observed in the slums and rehabs. Both Dharavi and Ghatkopar slums had 140 ft^2 tenements raising up to three levels, whereas the SRH families occupied single multipurpose tenements of 200–225 ft^2 area within 7–12 floor high buildings. While Kohinoor and Sangharshnagar SRH had tenements with separate bedroom and living space, the other two SRHs were characterised by single multipurpose tenement.

Concerning their lifestyle, while 60.9% households use clean fuel i.e., Liquefied Petroleum Gas (LPG) as their cooking fuel, 2.5% households still use kerosene for preparing meals. 2.5% households were also observed to use both. In 11 households, male members used to prepare meals. The average age of individual preparing meals was 36.6 years.

4 Results and Discussion

4.1 Appliance Ownership and Occupant Behavior Characteristics

Despite the socio-economic conditions remained similar in slums and SRHs, the effect of built-environment shift from horizontal slums to vertical apartments

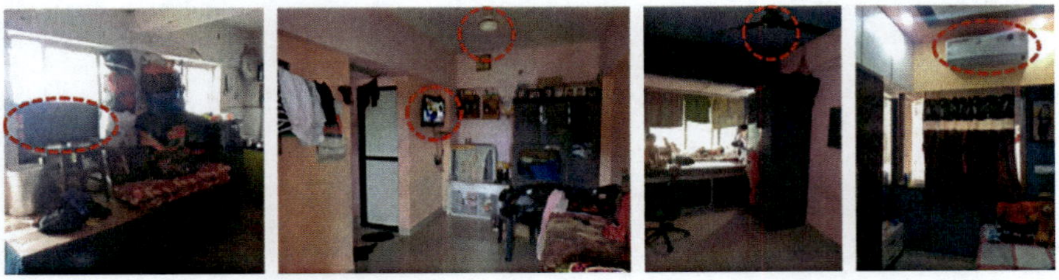

Fig. 3 Household appliances in SRHs

profoundly influenced their household electricity demand through a change of household practices. While 95–100% of the households in both slums and SRHs had mobile phones and ceiling fans, ownership and usage of space-intensive electrical appliances such as computers, television, washing machine, refrigerator and mixture grinder significantly increased in SRHs. In horizontal informal slums, although there might be a provision for access to electricity, extreme space constraints within a tenement (approx. 140 ft^2) restricted higher acceptance of appliance. Moreover, slum neighborhoods having higher social cohesion levels and communal practices, common activities such as washing, cleaning, and cooking, were observed to be performed in outdoor community spaces, thereby reducing the burden of electricity. However, the SRHs with higher tenement floor area (approx. 225–250 ft^2) on one hand, and emphasizing indoor and private living on the other hand might have an impact on purchasing decisions of household appliances. Also, proper housing with permanent tenure position being a social-status indicator in Mumbai, the aspirations of SRH inhabitants in order to climb up the social-ladder might have also influenced the appliance ownership patterns [3]. Figure 3 showcases the appliances found in the slums and SRH apartments.

The current status of appliance ownership as shown in Fig. 4 indicates that approx. 12%, 55% and 70% of the slum households owned washing machine, refrigerator and mixer blender respectively. However, the appliance ownership patterns increased to 32%, 75% and 92% respectively for the SRHs. The rise of appliance ownership after shifting to SRHs was found to be significantly high for washing machine, refrigerators and mixture blender. Furthermore, ownership of air-conditioner (AC) was also found to be higher in SRHs (1.86% in slums whereas 7.76% in SRHs), thereby, indicating increased thermal discomfort in SRHs.

Figure 5 highlights the time spent in household activities in the slums and SRHs. The time spent in daily household activities in slums and SRHs exhibit prominent differences. The number of hours spent and time-use pattern for basic household activities such as cooking and washing clothes were found to remain same for both slums and SRHs, thereby exhibiting that household size-dependent social practices remain unchanged. However, a contrast difference with a sharp rise has been identified in time spent in activities such as listening to music, watching television, and using laptop or mobile phones. Formal housing with improved infrastructure services

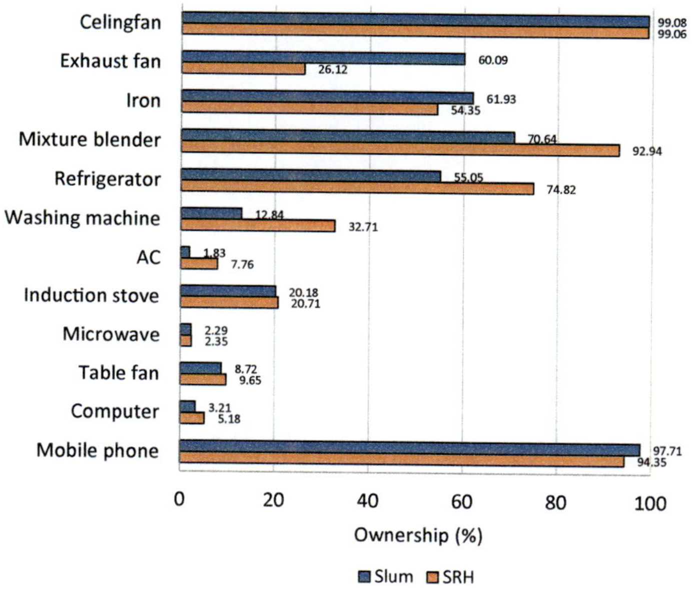

Fig. 4 Ownership of household appliances in the sample size (n = 643)

and electricity connection in SRHs has increased the average time spent in aforementioned activities that involves use of electrical appliances. Whereas in slums, the inhabitants majorly focused on the necessary activities with limited resources and service infrastructure. The results also corroborate with a similar study conducted where the effect of private and indoor living attributed to increased time spent in watching TV and other related ICT usage activities in the SRHs compared to the horizontal slums [3].

Figure 6 shows the monthly average electricity consumed in Gigajoules for both slums and SRHs for a period of six months (May 2019–Jan 2020). The results highlighted that the monthly average electricity consumed in slums were recorded higher than that of SRHs, despite having a fewer number of appliances in comparison to that of the SRHs. This can be attributed to the fact that the slum pockets of Mumbai, also act as a hive of multiple small-scale home-based but energy-intensive industries (such as recycling, potteries, leather industries, soap making, garments etc.) [12], ultimately, leading to higher household energy consumption. Whereas in the SRHs, engaging in such economic activities turn difficult due to common space restrictions as well as are often strictly prohibited due to residential tenure status. Hence, relatively lower electricity costs were recorded in SRH tenements.

Built-environment characteristics coupled with ownership of household appliances and occupant behavior significantly impact the environmental satisfaction. Figure 7 demonstrates that the share of SRH households satisfied with environmental metrices were found to be considerably higher than that of the slum households.

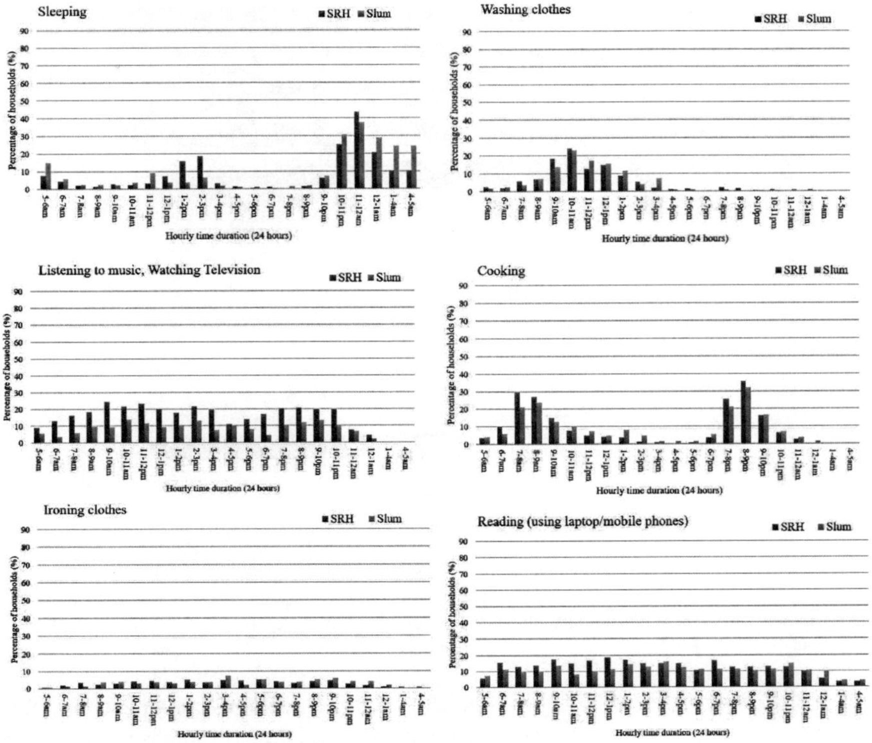

Fig. 5 Time spent in household activities in slums and slum rehabilitation housing

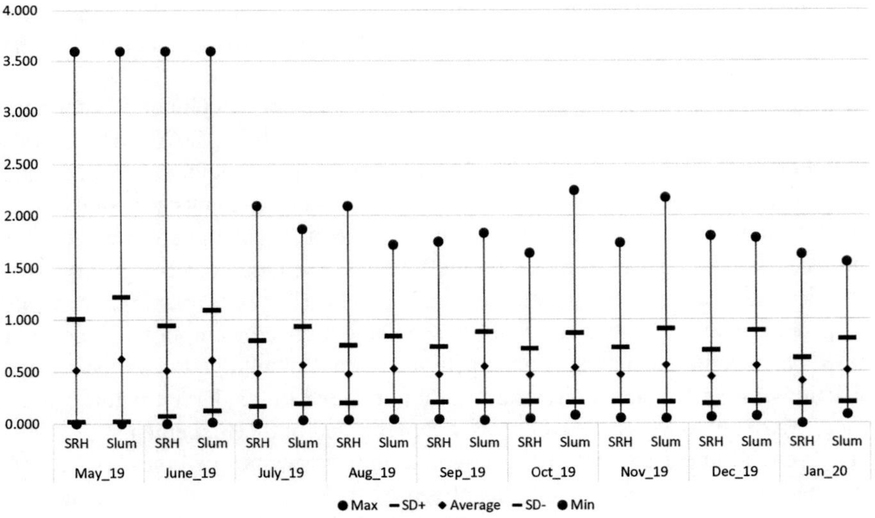

Fig. 6 Monthly average electricity consumption in slums and SRHs from May 19 to Jan 20

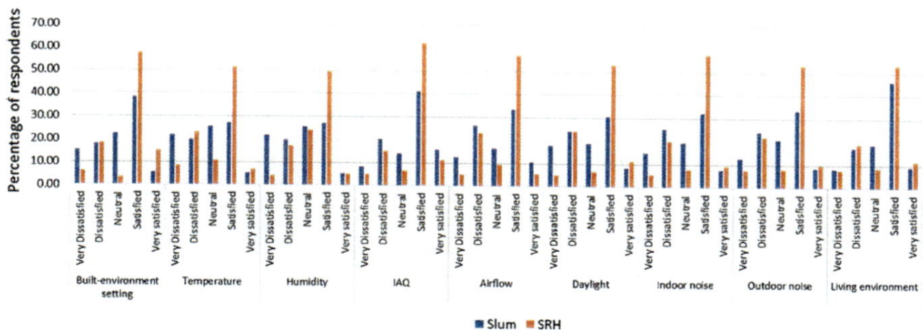

Fig. 7 Satisfaction with environmental metrics in slums and SRHs

4.2 Model Estimation

The final SEM models for both the slums and SRHs are shown in Figs. 8 and 9 respectively, indicating that both the models fit well. The models demonstrate quantitatively that built-environment setting, occupant behavior and energy practices influence appliance ownership, energy decisions, electricity consumption and environmental satisfaction in low-income settlements.

The hypothesis **H1** elucidates the relationship between the built-environment and ownership level of household appliances. The model estimates show that change in

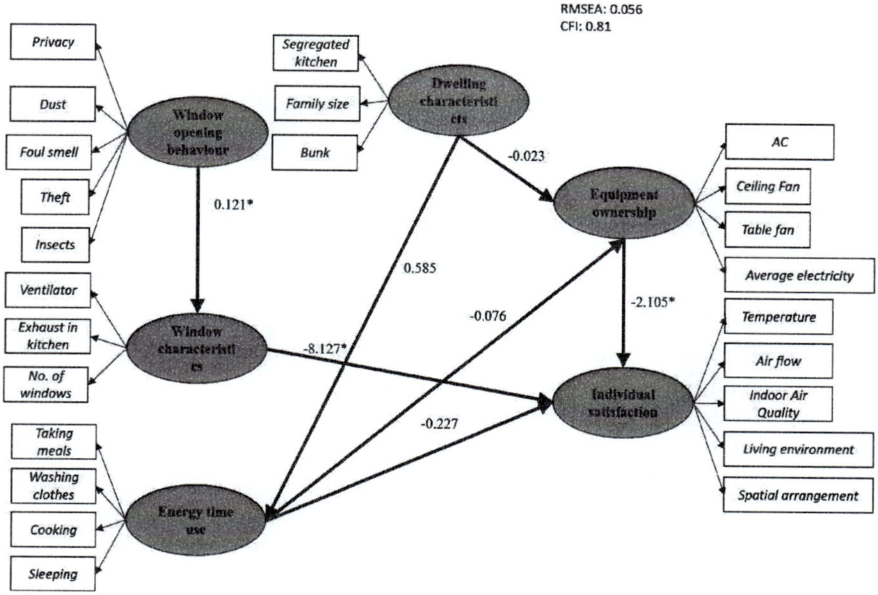

Fig. 8 Structural equation model (SEM) estimation for slum households

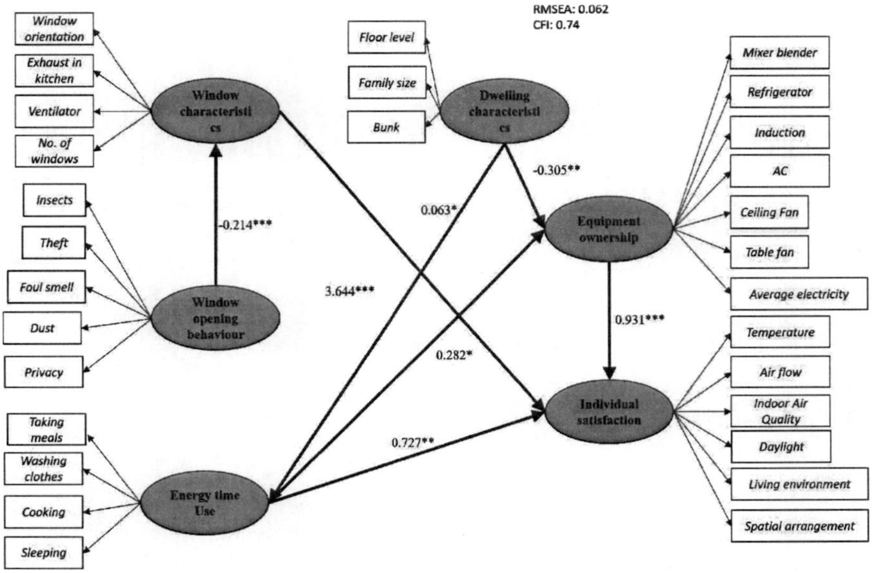

Fig. 9 Structural equation model (SEM) estimation for slum rehabilitation households

built-environment between horizontal slums and vertical SRHs with varying household characteristics has differential impact on appliance ownership levels. The slum households with segregated kitchen, high household size and bunkbeds had a negative and insignificant relationship with appliance ownership patterns. This phenomenon can be attributed to the fact that slum tenements with high household size on one hand and extreme space constraint on the other hand often have restricted use of AC, ceiling fan and table fan due to shortage of clear height, thereby reducing their electricity bills. For the SRHs, as demonstrated in Fig. 9, household attributes exhibited a negative but statistically significant relationship with appliances. This indicates that the SRH households with higher household size and bunkbeds own lesser appliances due to space restriction. On the other hand, the households on the upper floor of SRHs experience better daylight and natural ventilation in comparison to that of lower floors [6]. Better indoor natural environment enables the upper floor dwellers to experience better thermal comfort, thereby reducing load on active appliances such as table fan and AC usage.

Higher household size had a positive but non-significant impact on time use for slums, whereas the relationship turned significant for the SRHs, thereby supporting hypothesis **H2**. This indicates that for both the low-income housing typologies occupant behavior and household practices remain similar. The estimates also showcase that the SRH households which spend more time in mandatory and subsistence activities have a higher probability of owning energy intensive appliances. The results commemorate with another similar study conducted [3], which demonstrated that increase in indoor housing activities upon moving to SRH significantly influenced higher appliance ownership. The SEM results also shows that while the slum households are mainly restricted in owning ceiling fan, table fan, the SRH inhabitants, with a permanent house tend to own more energy intensive appliances such as refrigerator, washing machine, mixer, AC in order to step up the social ladder.

The change in built environment influences changes in relationship between time use and environment satisfaction, and it hence supports the hypothesis **H3**. The insignificant and negative factor loading for slums whereas significant and positive factor loading for SRH can be interpreted as certain aspects of change in built-environment and occupant behavior in SRHs such as indoor and private cooking, avoiding manual washing of clothes, using AC and cooling devices during sleeping positively impacts satisfaction with overall living environment and space arrangement. Whereas increased time outdoor cooking in exposed environment or indoor cooking within constrained space or without ventilation, and sleeping in space-constrained slum tenements without appropriate mechanical ventilation or windows have a negative impact on environmental satisfaction.

Specific appliance ownership has a differential impact on environment satisfaction, i.e., negative and statistically significant for slums, whereas positive and statistically significant for SRHs, thereby supporting hypothesis **H4**. In small slum tenements (approx. 140 ft^2), uptake of appliances leads to further space constraint, thereby negatively impacting satisfaction with space arrangement and living environment especially. Whereas, for the SRH tenements of approx. 250 ft^2, uptake of cooling devices such as AC, ceiling fans positively impact environmental satisfaction.

Household practices such as hours spent in subsistence activities such as taking meals, sleeping, and in performing mandatory activities such as cleaning, washing and cooking has differential impact on appliance ownership, thereby satisfying hypothesis **H5**. While communal, outdoor and manual activities such as cooking, cleaning in slums do not impact uptake of appliances, for SRHs, indoor cooking, washing might enable requirement of energy-intensive appliances.

Lastly, a difference is observed in the relationship between opening characteristics and environmental satisfaction among slums and SRHs, supporting **H6**. The slum rehab households with presence of exhaust in kitchen, ventilator and higher number of windows had experienced higher probability of environmental satisfaction, thereby exhibiting a positive and significant relationship. However, the relationship turned negative and significant for slums. In slums, the concept of windows turn inexistent as the tenements share external walls. Furthermore, even if they have exhaust fans, they are often not in use due to lack of maintenance or high financial burden. Therefore, this study inferred that shift from slums to vertical apartments in SRHs changes occupant behavior and household practices, which further significantly changes energy time-use, energy choices, leading to change in appliance purchase decisions and finally environmental satisfaction.

5 Conclusion

The interrelationship between the drivers of appliance ownership patterns and environmental satisfaction in two major typologies of low-income settlements of Mumbai were explored. With the shift from horizontal setting to vertical apartments, the household activities turned more indoor and private, thereafter demanding uptake of appliances. Despite having similar socio-economic conditions, higher appliance ownership is found in SRHs, majorly due to permanent tenure as well as aspiration for higher social-ladder rise. Higher appliance ownership among SRH inhabitants have a positive impact on satisfaction levels, thereby improving their quality of life. The study emphasizes on the need of designing holistic energy policies by considering the effects of housing characteristics, household practices and energy choices of the people living in low-income social houses.

References

1. Liu X (2022) Impact of urbanization on energy consumption and haze in China—a review. Energy Sources Part A Recov Util Environ Effects 44(1):1959–1976
2. Hao Y, Liu YM (2015) The influential factors of urban PM2.5, concentrations in China: a spatial econometric analysis. J Clean Prod 112:1443–1453
3. Debnath R, Bardhan R, Sunikka-Blank M (2019) How does slum rehabilitation influence appliance ownership? A structural model of non-income drivers. Energy Policy 132:418–428
4. Jana A, Sarkar A, Bardhan R (2020) Analysing outdoor airflow and pollution as a parameter to assess the compatibility of mass-scale low-cost residential development. Land Use Policy 99:105052
5. Sarkar A, Bardhan R (2020) Socio-physical liveability through socio-spatiality in low-income resettlement archetypes—a case of slum rehabilitation housing in Mumbai, India. Cities 105:102840
6. Sarkar A, Kumar N, Jana A, Bardhan R (2021) Association between built-environment and livability: case of Mumbai slum rehabs. In: Urban science and engineering. Springer, Singapore, pp 63–74
7. Leahy E, Lyons S (2010) Energy use and appliance ownership in Ireland. Energy Policy 38(8):4265–4279
8. Rong Z, Yao Y (2003) Public service provision and the demand for electric appliances in rural China. China Econ Rev 14(2):131–141
9. Kemmler A (2007) Factors influencing household access to electricity in India. Energy Sustain Dev 11(4):13–20
10. Filippini M, Pachauri S (2004) Elasticities of electricity demand in urban Indian households. Energy Policy 32(3):429–436
11. Wyatt P (2013) A dwelling-level investigation into the physical and socio-economic drivers of domestic energy consumption in England. Energy Policy 60:540–549
12. Dyson P (2012) Slum tourism: representing and interpreting 'reality' in Dharavi, Mumbai. Tour Geogr 14(2):254–274

Experimental Investigation of Cooling Photovoltaic Panel Using Turbo-Ventilator

Anurag Dixit and Ajoy Debbarma

1 Introduction

Increase in global energy demand and continuous variation in crude oil prices, issue climate change researchers are transitioning toward the usage of renewable energy sources. They are clean and non-polluting. The use of renewable energy became mainstream after the 1973 oil embargo. India ranks 5th in terms of energy generation from solar PV. India has the potential of receiving 5000 trillion kWh energy per year and has sunshine hours between 2600 and 3200 h/year [1]. India has a total installed capacity of 50.303 GW on 31/01/2022 [2]. As of today, the world has 709.674 GW of solar installed capacity [3]. Solar photovoltaics (PV) is one of the famous power generation methods. It directly converts solar insolation into electricity which can be directly used to power household appliances. The energy conversion process in solar cells is only for the visible spectrum of 380–700 nm, except that the radiation is not available for power generation by PV cells [4]. Out of available solar radiation, only 15% part is convertible into electricity and the remaining part of solar energy is turned into heat. As a result of this, the surface temperature of the module rises. The efficiency, power production, and cell life of solar PV panels degrade as their module surface temperatures rise. The efficiency, life span of the PV cell, and power produced by the panel are all negatively affected by a rise in module temperature, as the I_{sc} increases somewhat but the V_{oc} and FF fall significantly [5].

Therefore, cooling is essential to improve efficiency and prolong the life of the panels. The two most common methods of cooling are now in use namely passive and active. The solar panel must be cooled by an external power source in active methods. However, natural air convection is used to remove heat from the system in passive cooling systems. Numerous studies have been using alternative cooling methods

A. Dixit · A. Debbarma (✉)
Department of Mechanical Engineering, NIT Hamirpur, Hamirpur, Himachal Pradesh 177005, India
e-mail: adebbarma@nith.ac.in

S. Doolla et al. (eds.), *Advances in Clean Energy and Sustainability*,
Green Energy and Technology, https://doi.org/10.1007/978-981-99-2279-6_5

for solar panels to be lowering the surface temperature and improve efficiency and durability of solar panels.

Nižetić et al. [6] experimented with water spray applied to both sides of a monocrystalline PV panel. The power output and electrical efficiency of the panel were found to rise by a maximum value of 16.3% and 14.1% respectively. Haidar et al. [7] showed that by using evaporative cooling on solar PV panels, the panel temperature was reduced by 10 °C, while power output was increased by 5%. Teo et al. [8] did experimental and numerical research using a parallel array of ducts with a new design of input and outlet manifold to minimize the panel's temperature. The results showed an increase in efficiency of 4–5% and the optimal air flow rate was determined to be 0.055 kg/s to extract a large amount of heat from the panel. Moharram et al. [9] fabricated an experimental setup to cool the panel by using the least amount of water. The water is sprayed by using a nozzle and it was observed that at 45 °C (MAT) the least amount of water is used. Wu and Xiong [10] conducted the simulation study for using rainwater as coolant and a gas diverging device is used to spread the coolant which reduced the cell temperature up to 19 °C and the average electric yield is improved by 8.3%. Borkar et al. [11] designed a real model of a hybrid PV panel with a combination of thermoelectric. Results revealed that the highest and lowest electrical efficiency of 13.27 and 12.26% were obtained. Popovici et al. [12] conducted a numerical study to analyze the effect of the heat sink with copper ribs. These ribs are connected to the back side of the PV panel at various inclines. At 45° inclination of copper ribs, the rate of heat transfer is enhanced, and the surface temperature of the PV panel is reduced by 10 °C with the maximum power produced is 90% of the nominal one. Kim et al. [13] conducted a numerical study where a reduction of module temperature is recorded by 15.13 °C and by using fins. El Mays et al. [14] did an experimental investigation using a plate with integrated fins stick beneath the PV panel. Results revealed that the minimum power output and the efficiency rose by 1.87 W and 1.77% respectively, with average temperature dropped by 6.1 °C.

Active cooling methods are far better than passive ones in terms of efficiency, performance enhancement, and panel temperature reduction. However, they are useful for large-scale solar power facilities. For the simple reason that the plant's excess energy may be used to power the active cooling system. However, if the solar installation is connected to the local power grid the payback time for an active cooling system will lengthen. Hence, passive cooling methods are most suited for this application, while being less efficient but more cost-effective for a long period of operation. This experimental effort employs the use of a turbo ventilator to extract heat from the rear side of the solar PV panel. It does not drain any power from the grid, but it generates its own.

2 Experimental Facility and Methodology

The present experimental work was performed, and data were recorded for 10 days, from November 17th to 26th 2021. The time for data collection starts from 11:00 a.m. to 4:00 p.m., and data were recorded every 15 min. The location was the NIT Hamirpur campus in Himachal Pradesh, India with a global position of 31.70840 N and 76.52740 E. A Turbo ventilator with a 12-inch. hub diameter made of stainless steel is used for the present experimental work. The technical specification of solar PV panels is listed in Table 1.

2.1 Instrumentation

Instruments are used for the measurement of various parameters and their operating range with accuracy is listed in Table 2. A solar power meter (TENMARS TM-207) is used to measure the total instantaneous solar irradiance. The wind velocity is measured by Anemometer-I (Mastech Ms 6250) and below the turbo ventilator, it is measured by Anemometer-II (Kusam Meco 909). The electrical parameters such as current and voltage are measured by a digital multimeter (FLUKE 106). The module surface temperature is measured by a k-type thermocouple using a data logger (HUATO S220-T8). The speed of the turbo ventilator is measured by the tachometer (Digital Photo Tachometer).

2.2 Experimental Procedure and Calculation

Plywood with a thickness of 6 mm is used to construct a coolant channel that is 1200 mm in length, 720 mm in width, and 280 mm in height. Both sides of the duct are sealed correctly using wood dust. The ambient temperature is assumed to be 29 °C with a wind speed of 1 m/s during the whole experiment. The maximum solar radiation is measured at an angle of 35°, due to this reason the PV panel is put on the wooden duct at this angle of inclination. The turbo ventilator is placed above

Table 1 Technical specifications of PV panel

Specifications	Symbol	Details
Type of PV module	–	Polycrystalline
Maximum power	P_m	50 W
Open circuit voltage	V_{oc}	22 V
Maximum current	I_m	2.78 A
Short circuit current	I_{sc}	3.28 A
Area of module	S	0.277 m^2

Table 2 Instruments and their operating range and accuracy

Instruments	Accuracy	Range	Model
Anemometer-I	± 5%	0.1–30 m/s	MASTECH-MS6250
Anemometer-II	± 3%	0–30 m/s	KUSAM-MECO 909
Solar meter	± 5.1%	0–1599 W/m^2	TENMARS TM-207
Tachometer	± 0.051%	2.51–999.9 RPM	Digital tachometer
Datalogger	± 1%	– 200 to 1800 °C	HUATO S220-T8
Multimeter	± 0.51% (V) ± 1.51% (I)	6–600 V 0.011–10 A	FLUKE-106

the coolant channel in a perpendicular direction which allows the prevailing wind to turn its blades. The turbo-ventilator is designed in such a way, when the ambient air is flowing inside the turbo ventilator it starts to rotate the turbo ventilator. Because of this, a flow of air is generated within the coolant duct channel, and heat is extracted by the flowing air from the rear side of the PV panel. Because of the principle of buoyancy, the hot air is moved upward and cold air flows continuously to cool the back surface of the PV panel. To measure the surface temperature of the rear PV panel, thermocouples are installed at both ends of the panel, as indicated in Fig. 1b. This study measures the relationship between generated current (I_m), short circuit current (I_{sc}), and open circuit voltage (V_{oc}), the temperature of the panel depends on solar radiation (W/m^2), environmental temperature, ambient air velocity, and suction velocity of a turbo ventilator. The formula that links the output voltage and current together may be used to determine the optimum output power.

$$P = V_{oc} \times I_m \tag{1}$$

To calculate how much heat is being expelled by the turbo ventilator, Eq. (2) is used to calculate the rate of mass flow in m^3/s.

$$\mathrm{Q} = \mathrm{A} \times \mathrm{v} \tag{2}$$

where V_{oc} is open circuit voltage in volts, I_m is the maximum current in amp, Q is the mass flow rate in m^3/s, v is the velocity in m/s, and A is the cross-sectional area under the turbo ventilator.

3 Fluctuation of Radiation Intensity Throughout the Day

Radiation and background wind patterns are anticipated during the experimentation procedure. The present research shows that the optimal panel angle for radiation is about 35°. At this suggested angle of inclination, the radiation intensity starts rising from 11:00 a.m. to 12:00 p.m. The maximum intensity was observed at the

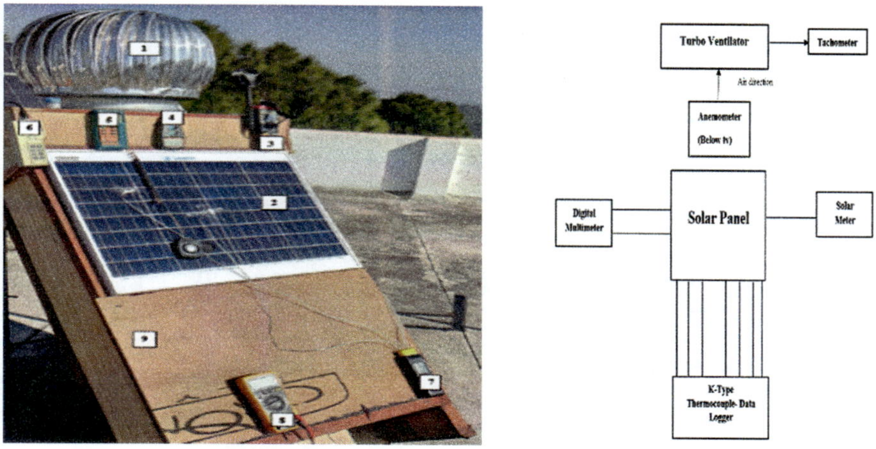

(a) (b)

Fig. 1 Experimental setup **a** schematic diagram of the experimental setup, **b** line diagram

time of noon i.e., from 12:00 to 2:00 p.m., and it starts descending from 3:00 to 4:00 p.m. Initially the observations were taken for two weeks from Nov 17th to Nov 26th without and with turbo ventilator. The first week of observations starts from Nov 17th to Nov 21st and data were calculated without a turbo ventilator, while the second week of observation starts from Nov 22nd to 26th using a turbo-ventilator. The variation of radiation intensity (W/m^2) corresponds to time (h) as shown in Fig. 2a, b, without the turbo ventilator and with the turbo ventilator respectively.

Present experimental observation reveals that the maximum solar intensity was obtained at about 1252 W/m^2 on Nov 23rd and a minimum of about 951 W/m^2 on Nov 17th at peak time in the noon. The maximum and minimum radiation intensity was obtained at about 1213 W/m^2 on Nov 23rd and 848 W/m^2 on Nov 17th respectively for the rising time in the morning, while for the evening time this value of radiation intensity was obtained maximum (830 W/m^2) and minimum (479 W/m^2) on Nov 19th and Nov 18th respectively. The average insolation is 5% more in week 2nd at the rising time, while in week 1st the mean insolation is higher at about 3.89% in the evening. For the time of peak radiation i.e., noon the insolation shows minimum variation for both weeks.

4 Fluctuation of Ambient Wind Velocity Throughout the Day

The vanes of a turbo ventilator are fashioned like aerofoils, and their rotation is prompted by the surrounding airflow. Because of this, the duct is kept at a constant flow rate and the air is constantly being sucked out. The exhaust of access heat

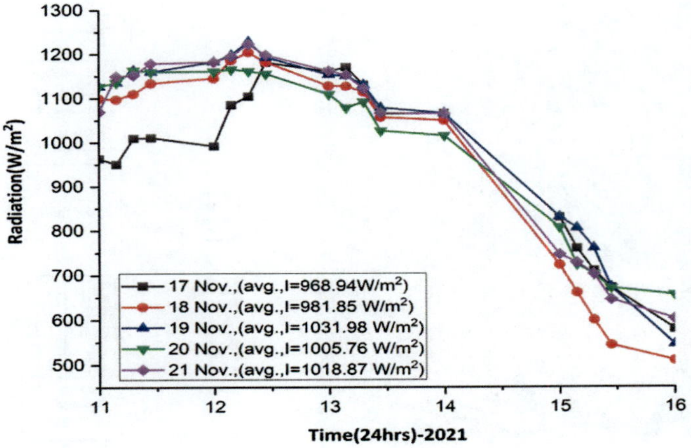

(a) with turbo ventilator (week 1st)

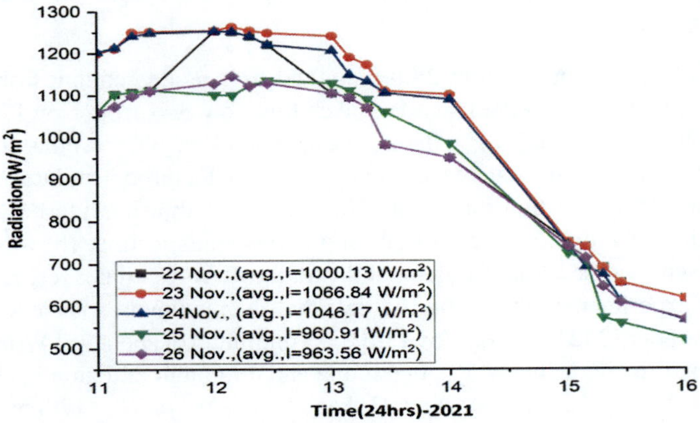

(b) with turbo ventilator (week 2nd)

Fig. 2 Variation of radiation intensity with time

from the rear side of the panel takes place due to this continuous flow of air. The anemometer placed beneath the turbo ventilator measures the air velocity within the chamber. In each interval of 15 min, the data was recorded and collected. To calculate the speed of the induced turbo ventilator in RPM, the number of observations was done every 2 min. The average value of these RPMs that are recorded every 2 min is used for the calculations to minimize the effect of any fluctuations in wind speed. The data for airflow (m/s) and rotational speed (RPM) of the induced turbo-ventilator on the specified date and time are displayed in Fig. 3, and there is a clear correlation between wind velocity and rotational speed in RPM.

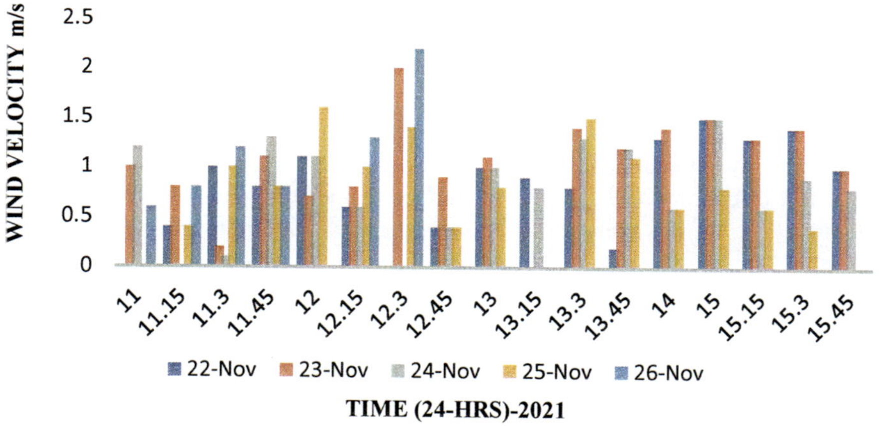

(a) Ambient wind velocity with time

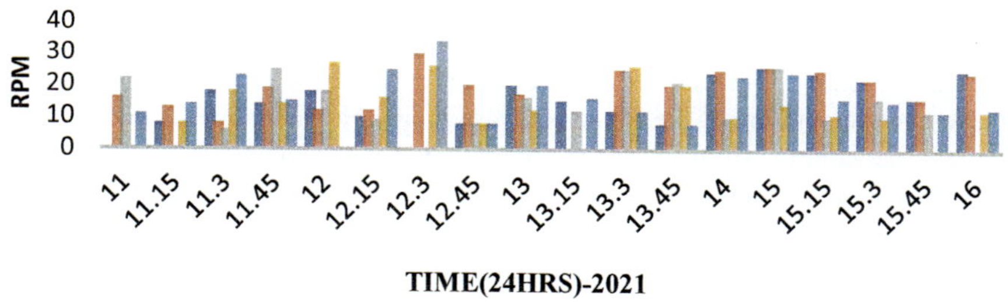

(b) Rotational speed in RPM with time

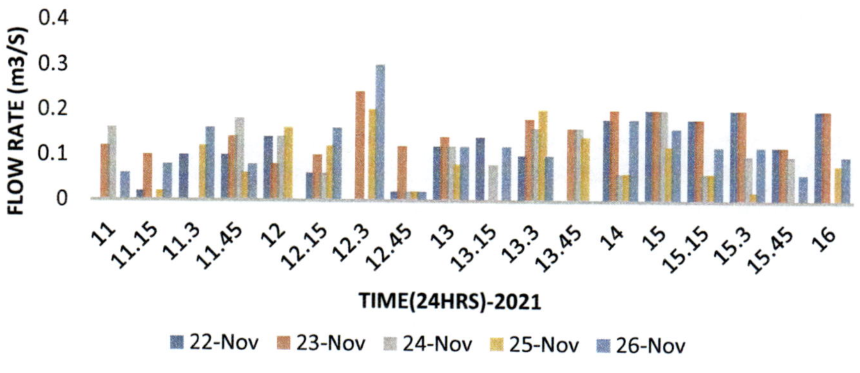

(c) Induced flow by turbo ventilator with time

Fig. 3 Variation of wind velocity on turbo ventilator corresponds to the time

5 Results and Discussions

The experimental investigation of cooling of solar photovoltaic (PV) panel was done using a turbo ventilator. This turbo ventilator acts as a self-induced fan, which does not require any external power source to suck the ambient air for cooling purposes. The effect of cooling of PV panels on surface temperature, electric power output, and output current was analyzed for two weeks from Nov 17th to 26th, 2021 at NIT, Hamirpur campus. The obtained results are discussed as follows.

5.1 Effect on the Surface Temperature of PV Panel

The details are collected without using a turbo ventilator throughout the first week of the trial starts from Nov 17th to 21st. The result was obtained that the highest temperature this week recorded is about 70.8 °C. While the average temperature throughout the week reached 58.87 °C. The second week of the trial starts from Nov 22nd to 26th and involves the use of a turbo ventilator. Over the entire week, the average temperature of the PV panel was recorded at about 54 °C. Figure 4a, b illustrate the variation of the surface temperature of the PV panel corresponding to the time. The plots show the greater quantity of heat in the system during the 1st week, because of the low rate of heat removal without a turbo ventilator. Figure 4a shows a large and erratic temperature profile, while Fig. 4b shows a narrower and less erratic profile. According to the findings, week 2nd has more radiation intensity as compared to week 1st, but still, the surface temperature of the back side of the PV panel is lower. This reduction of surface temperature was obtained due to the application of a turbo ventilator.

Therefore, the heat has been removed from the system consistently and constantly using the turbo ventilator arrangement. After the comparison of these plots, it was observed that the induced rotational speed of the turbo ventilator was significantly influenced by the velocity of ambient air. On Nov 20th, the panel surface temperature reached its highest value during the time of sunrise and sunset, when wind speed was quite low. When using a turbo-ventilator, the average cell temperature in the morning dropped by 15.55%, the temperature in the noon or peak time dropped by 12.47%, and in the evening, the temperature dropped by 13.6%.

5.2 Effect on Electric Power Output (P)

As was previously said, the fluctuation of radiation intensity is proportional to the amount of current that is generated. However, when the temperature of the panel rises, the current increases somewhat but the voltage falls dramatically. The previously expressed Eq. (1) is used to determine the power output, which describes the

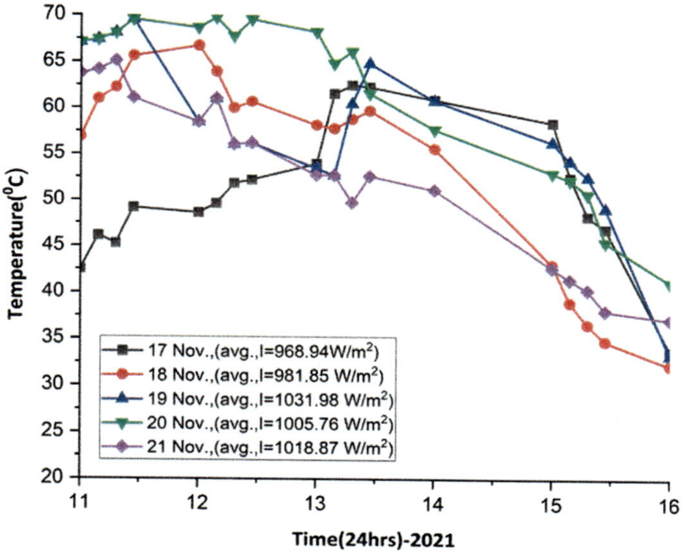

(a) Natural convection without turbo ventilator

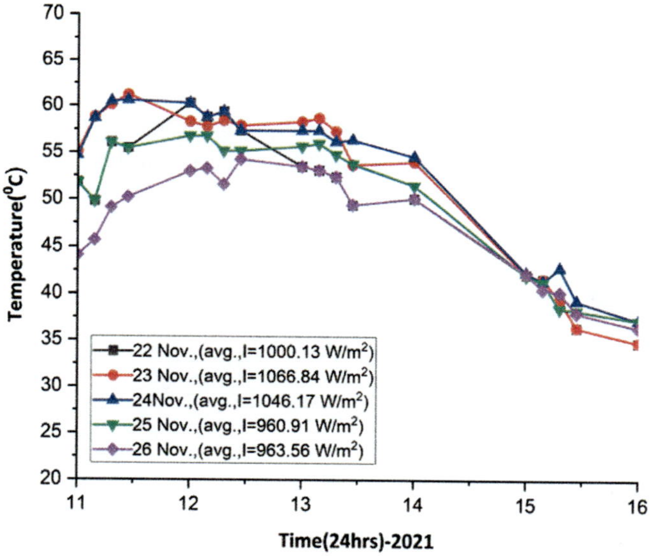

(b) Convection with turbo ventilator

Fig. 4 Variation in surface temperature of PV panel corresponds to radiation intensity with time relationship between voltage and current. The fluctuation of power output concerning the radiation intensity throughout the day using convection cooling without a turbo ventilator and convection cooling with a turbo-ventilator is shown in Fig. 5a, b respectively.

The intermittent nature of ambient wind also has an impact on electricity production by the solar PV panel. Low power output was observed during the natural convection cooling according to the experimental results. The high surface temperature of

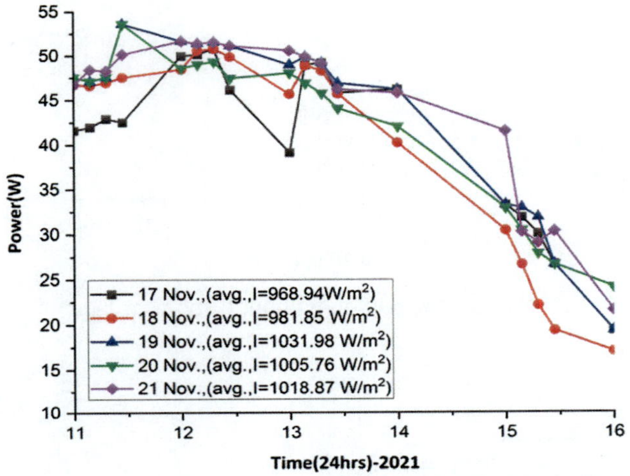

(a) Power generation without turbo ventilator

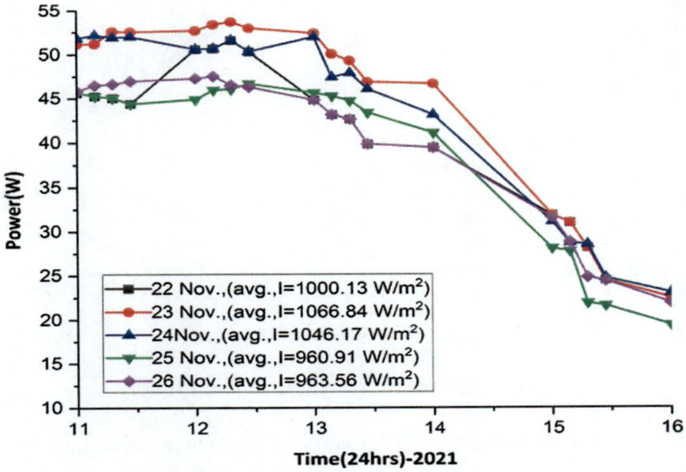

(b) Power generation with turbo ventilator

Fig. 5 Variation of power output corresponds to the radiation intensity with time

PV panels causes low power generation, and the high temperature is due to fluctuation in the flow of ambient air in natural convection cooling. There are times when less electric power is generated than would be expected given the available radiation and wind conditions. The cooling of the PV panel with a turbo-ventilator produces approximately stable power output with less variation as shown in Fig. 5b. It was observed that the power output using the turbo ventilator arrangement increases by 2.094% at peak time i.e., in the noon time from 12:00 a.m. to 2:00 p.m. Because the panel saw its highest temperature and highest solar insolation levels during this peak time of the day. The results also revealed that the greatest voltage drop was observed during the morning and evening time of observation, this phenomenon is accountable for the decreased power production. When compared to the power generated with the natural cooling system, an increase in power production was achieved by about 2.56% more using the turbo ventilator cooling system.

5.3 Effect on Output Current (I)

The output current generated by the solar PV panel directly depends upon the amount of incident solar radiation on the PV panel. The variation of output current corresponding to the radiation intensity with time is presented in Fig. 6a, b without and with turbo ventilator respectively. From the comparison between the cooling of the PV panel without a turbo ventilator and with a turbo ventilator it was observed that the output current is fluctuating in the case of cooling without a turbo ventilator during the morning and evening time. More current is generated during the rising time in week 2nd due to the increased insulation at this time of day using the turbo ventilator. Figure 6b demonstrates that the power generated is less variable, which increases the durability of the panel, and that consistent power generation is achieved.

6 Conclusions

The present experimental work is performed to reduce the surface temperature of solar PV panels and to improve the power output and the life span of the panel. For this purpose, a wooden duct is fabricated, and the heat is transferred upward via the principle of buoyancy and then vented by the turbo ventilator. The following conclusions have been made based on experimental results:

- There is a 5% higher radiation intensity was observed in week 2nd as compared to week 1st during the morning time from 10:00 to 11:00 a.m., while the average radiation intensity is 3.89% higher in week 1st as compared to week 2nd.
- The rotational speed of the turbo ventilator in RPM and the flow rate of air beneath the turbo ventilator directly depend on ambient wind velocity.

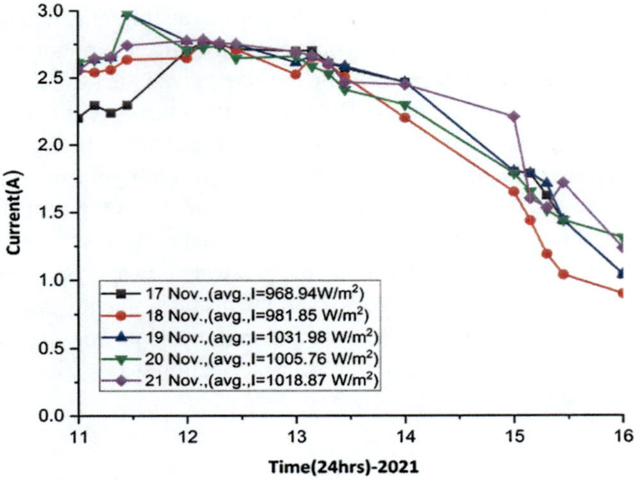

(a) Cooling with natural convection without a turbo ventilator

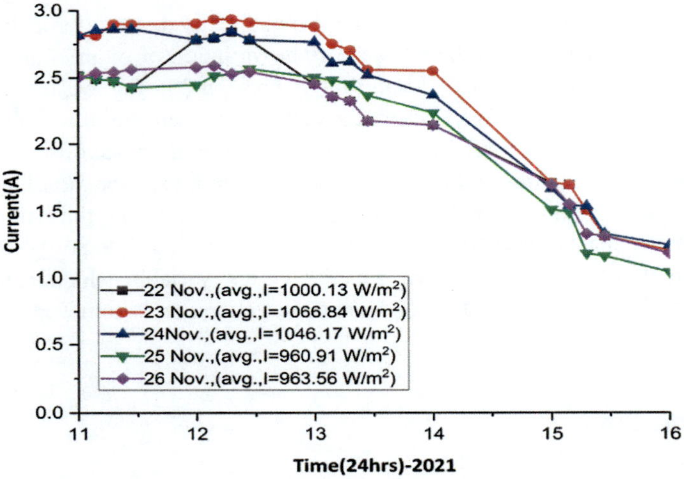

(b) Cooling with turbo ventilator

Fig. 6 Variation of output current corresponds to the radiation intensity with time

- From the results, it is observed that the output power, current, and surface temperature of the PV panel are more uniform in the case of cooling a PV panel with a turbo ventilator as compared to cooling with natural convection.
- The average improvement in the power generated by solar PV panels reached about 2.56% with a turbo ventilator cooling system.
- In addition, adopting the turbo ventilator arrangement increases power production by 1.71%, 2.094%, and 3.09% during the morning period (10:00–11:00 a.m.),

peak or noon time (12:00–2:00 p.m.), and in the evening time (3:00–4:00 p.m.) respectively.

- With a turbo ventilator in place, the present properties fluctuate much less.
- When employing a turbo ventilator cooling arrangement, the surface temperature of the PV module is found to decrease by 15.55%.
- The average reduction in surface temperature of PV panel was obtained about 20.6%, 12.47%, and 13.6% during the morning time (10:00–11:00 a.m.), peak or noon time (12:00–2:00 p.m.), and in the evening time (3:00–4:00 p.m.) respectively.

References

1. Khan BH (2006) Non-conventional energy resources. Tata McGraw-Hill Education, New Delhi
2. Installed RE capacity. https://mnre.gov.in/the-ministry/physical-progress. Accessed 31 Jan 2022
3. Solar energy data. https://www.irena.org/solar. Accessed 31 Jan 2022
4. Dwivedi P, Sudhakar K, Soni A, Solomin E, Kirpichnikova I (2020) Advanced cooling techniques of PV modules: a state of art. Case Stud Therm Eng 21:100674. https://doi.org/10.1016/j.csite.2020.100674
5. Idoko L, Anaya-Lara O, McDonald A (2018) Enhancing PV modules efficiency and power output using multi-concept cooling technique. Energy Rep 4:357–369. https://doi.org/10.1016/j.egyr.2018.05.004
6. Nižetić S, Čoko D, Yadav A, Grubišić-Čabo F (2016) Water spray cooling technique applied on a photovoltaic panel: the performance response. Energy Convers Manage 108:287–296. https://doi.org/10.1016/j.enconman.2015.10.079
7. Haidar ZA, Orfi J, Kaneesamkandi Z (2018) Experimental investigation of evaporative cooling for enhancing photovoltaic panels efficiency. Results Phys 11:690–697. https://doi.org/10.1016/j.rinp.2018.10.016
8. Teo HG, Lee PS, Hawlader MNA (2012) An active cooling system for photovoltaic modules. Appl Energy 90(1):309–315. https://doi.org/10.1016/j.apenergy.2011.01.017
9. Moharram KA, Abd-Elhady MS, Kandil HA, El-Sherif H (2013) Enhancing the performance of photovoltaic panels by water cooling. Ain Shams Eng J 4(4):869–877. https://doi.org/10.1016/j.asej.2013.03.005
10. Wu S, Xiong C (2014) Passive cooling technology for photovoltaic panels for domestic houses. Int J Low-Carbon Technol 9(2):118–126. https://doi.org/10.1093/ijlct/ctu013
11. Borkar DS, Prayagi SV, Gotmare J (2014) Performance evaluation of photovoltaic solar panel using thermoelectric cooling. Int J Eng Res 3(9):536–539
12. Popovici CG, Hudişteanu SV, Mateescu TD, Cherecheş NC (2016) Efficiency improvement of photovoltaic panels by using air cooled heat sinks. Energy Procedia 85:425–432. https://doi.org/10.1016/j.egypro.2015.12.223
13. Kim J, Bae S, Yu Y, Nam Y (2019) Experimental and numerical study on the cooling performance of fins and metal mesh attached on a photovoltaic module. Energies 13(1):85. https://doi.org/10.3390/en13010085
14. El Mays A, Ammar R, Hawa M, Abou Akroush M, Hachem F, Khaled M, Ramadan M (2017) Improving photovoltaic panel using finned plate of aluminum. Energy Procedia 119:812–817. https://doi.org/10.1016/j.egypro.2017.07.103

Thermal Performance of Multiple Tube Sensible Energy Storage with Coil Inserts

Ravi Kumar, Anil Kumar Patil, and Manoj Kumar

1 Introduction

Solar energy can fulfill all energy needs during the daytime. To use it during the off-sunshine hours, thermal energy storage offers a viable solution. The thermal energy storage system stores energy derived from the solar radiations by using a high-temperature heat transfer fluid and the stored energy can be extracted by circulating the low-temperature fluid through the storage system. The thermal energy storage systems can store the energy in either a sensible form or the latent form. In sensible energy storage, a high-temperature heat transfer fluid charges the system by increasing its temperature in a single phase. When the heat transfer fluid comes in contact with a material that absorbs heat energy by changing its phase from solid to liquid, it is said to be a latent energy storage system. Sensible energy storage systems are popular due to their simple operation, availability of material, and low initial and operating costs.

Thermal energy storage has applications in space heating, power plants, drying, desalination, water heating, etc. Hasnain [1] mentioned that the two major techniques for thermal energy storage considered for different applications are latent and sensible heat storage. The integration of thermal energy storage systems with above said solar thermal technologies are well-proven techniques. Prasad and Muthukumar [2] presented the thermal characteristics of concrete, cast iron, and cast steel sensible energy storage with cylindrical finned cavities. The study unfolds the effect of finned

R. Kumar · A. K. Patil (✉) · M. Kumar
Department of Mechanical Engineering, DIT University, Dehradun, Uttarakhand, India
e-mail: dr.anilpatil@dituniversity.edu.in

R. Kumar
e-mail: ravikumar@dituniversity.edu.in

M. Kumar
e-mail: drmanoj.kumar@dituniversity.edu.in

S. Doolla et al. (eds.), *Advances in Clean Energy and Sustainability*,
Green Energy and Technology, https://doi.org/10.1007/978-981-99-2279-6_6

tubes in the concrete bed. A 6 fin tube arrangement yields the optimum charging time.

Muthukumar et al. [3] explored the concrete and cast steel-based high-temperature energy storage system integrated with the solar power plant. The multiple channels of heat transfer fluid were optimized using the numerical approach. Xu et al. [4] discovered that sand can store a large quantity of energy compared to concrete storage when used with a heat transfer fluid of higher conductivity. Özrahat and Ünalan [5] conceptualized the energy storage in a concrete building column using a steel passage for airflow that can fulfill all requirements of energy in a residential flat. Kunwar et al. [6] investigated the sensible energy storage with perforated cylindrical blocks. The perforations in the cylindrical blocks improve the energy diffusion in the system and boost the overall performance.

Rao et al. [7] evaluated the transient characteristics of sensible energy storage with finned tubes. The concrete storage with copper tubes yielded higher thermal performance than that of the concrete system with MS tubes due to higher effective thermal conductivity. Agrawal et al. [8] analyzed the heat transfer and pressure drop in a packed bed consisting of grooved elements. The higher entry air temperature enhances the Nusselt number and stored energy. The empirical correlations for the Nusselt number and friction factor were also proposed. Kumar et al. [9] recently discovered in a study that sensible energy storage with a number of cylindrical passages with wire coil inserts can yield an energy efficiency of 85.9% at a pitch ratio of 0.5.

The novelty of the present work lies in the simultaneous implementation of two potential approaches for thermal performance enhancement: (i) Multiple cylindrical tube passages embedded in a cylindrical SESS configuration, and (ii) Wire coil inserts fitted in the cylindrical tube passages. The experimental data are collected using SESS with and without wire coil inserts for charging and discharging cycles.

2 Lab Facility

In the present work, a SESS is designed to store thermal energy by using air as a heat transfer fluid in the temperature range of 45–75 °C. The multiple cylindrical tube passages of SESS for heat transfer enhancement are inserted with 19 wire coil inserts with a different pitch-to-diameter ratio. The thermal performance of SESS with wire coil inserts is compared with the smooth passage SESS by using wire coil inserts of the different pitch-to-diameter ratios at different mass flow rates and entry temperatures of the air. SESS having a length of 0.95 m and an external diameter (D) of 0.29 m is molded in the laboratory by pouring the concrete mixture. The cylindrical passages of internal diameter (d) of 0.019 m are embedded in concrete storage. The optimal combination is found to be 0.95 m bed length having a total of 19 tubular passages. Proper setting and hardening of the concrete bed were ensured before the experimentation.

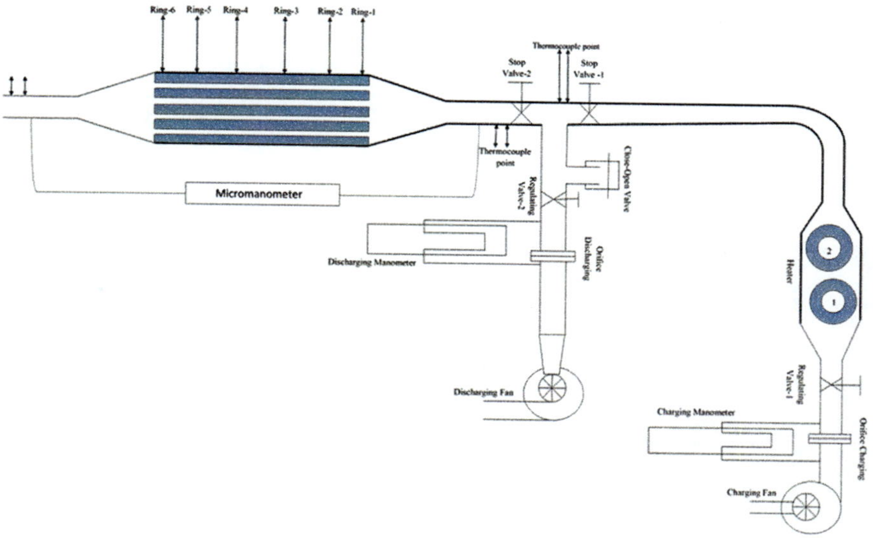

Fig. 1 Schematic of the test rig

The schematic arrangement of the test rig can be seen in Fig. 1. Figure 2 depicts the geometrical features of the concrete energy storage with multiple tubes. The GI wire of 2 mm diameter is used to prepare wire coils of inner and outer diameters of 16 and 19 mm, respectively. The pitch ratio (p/d) of the wire coil varies from 0.25 to 0.75. Figure 3 shows the geometry of wire coil inserts of different pitch ratios. The uncertainty band in the energy stored is found to be ± 3.6%.

The energy stored during the charging process at a time (t) is given as,

$$E_c = \rho_s V_s C_{p_s}[T(t) - T_i] \tag{1}$$

And the energy released during discharging is given as,

$$E_d = \rho_s V_s C_{p_s}[T_i - T(t)] \tag{2}$$

3 Results and Discussions

The variation of energy stored as a function of time at 0.022 kg/s is shown in Fig. 4. The plot reveals that the increase in the entry temperature of the fluid increases the energy storage in SESS in all cases. Also, the energy storage increases with the reduction in the pitch ratio of the wire coils. It is evident that the internal energy of SESS is raised by the use of wire coil inserts, promoting a more uniform distribution of heat energy. The energy stored by the wire coil fitted system varies from 1.7 to

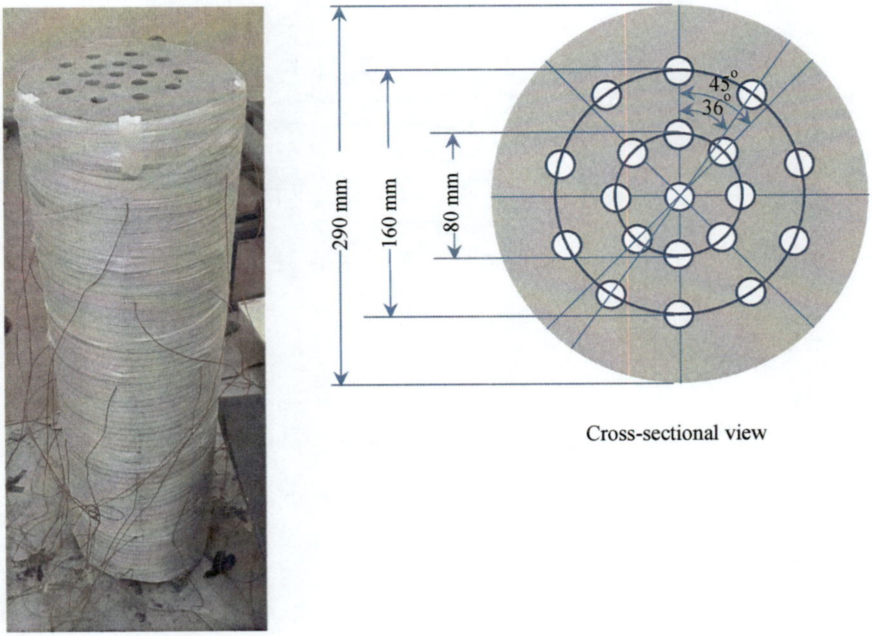

Fig. 2 Geometry of sensible energy storage system (SESS)

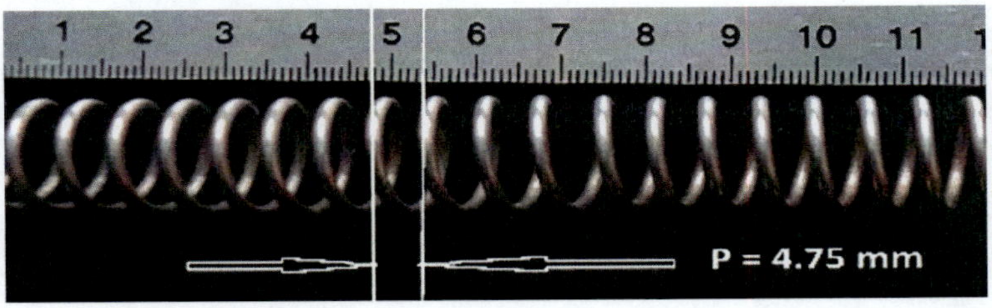

Fig. 3 Wire coil insert

2.38 MJ, 2.58 to 3.64 MJ, 3.48 to 4.14 MJ, and 4.14 to 5.08 MJ corresponding to 45 °C, 55 °C, 65 °C, and 75 °C, respectively. It is observed that the energy stored in wire coil fitted SESS (p/d = 0.25) is 14% higher than smooth SESS at 75 °C entry air temperature after 60 min of charging. However, the energy stored in SESS is increased by 31% after 360 min of charging. The energy stored in the system fitted with the wire coil is increased by 60% and 76% after 60 min and 360 min of charging with respect to the smooth passage, respectively, at 45 °C entry air temperature and (p/d) ratio of 0.25.

The transient energy released from SESS during the discharging process is shown in Fig. 5 at a mass flow rate of 0.022 kg/s. The energy released from the system fitted with the wire coil is increased by 55% after 60 min and 41% after 240 min of

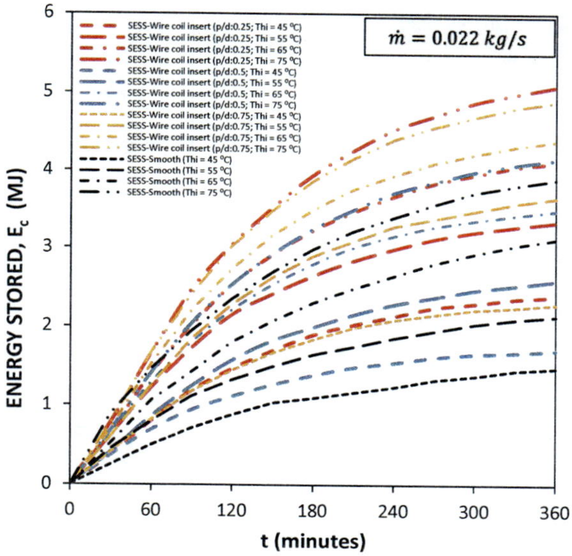

Fig. 4 Energy stored in SESS

discharging with respect to the smooth passage, at 75 °C entry air temperature and (p/d) ratio of 0.25. Whereas, the energy released from the wire coil-fitted system is increased by 106% after 60 min and 89% after 240 min of discharge. It can be noted that the maximum energy discharge is observed nearly the same for SESS with wire coil inserts having (p/d) ratio of 0.25 and 0.75 at an entry air temperature of 75 °C. The energy released from SESS fitted with wire coil inserts varies in the range of 1.41–2.13 MJ, 2.27–3.09 MJ, 3.08–3.74 MJ, and 3.76–4.43 MJ corresponding to 45 °C, 55 °C, 65 °C, and 75 °C, respectively.

4 Conclusions

The rise in internal energy of a sensible energy storage system (SESS) during charging in the presence of wire coil inserts is attributed to the improved spatial distribution of heat energy. The energy stored in the system fitted with the wire coil is increased by 76% after 360 min of charging with respect to the smooth passage at 45 °C entry air temperature and (p/d) ratio of 0.25. Whereas, the energy released from the wire coil-fitted system is increased by 89% after 240 min of discharge with respect to the smooth passage.

The energy stored by the wire coil fitted system varies from 1.7 to 2.38 MJ, 2.58 to 3.64 MJ, 3.48 to 4.14 MJ, and 4.14 to 5.08 MJ corresponding to 45 °C, 55 °C, 65 °C, and 75 °C, respectively. The energy released from the wire coil fitted system varies from 1.41 to 2.13 MJ, 2.27 to 3.09 MJ, 3.08 to 3.74 MJ, and 3.76 to 4.43 MJ corresponding to 45 °C, 55 °C, 65 °C, and 75 °C, respectively.

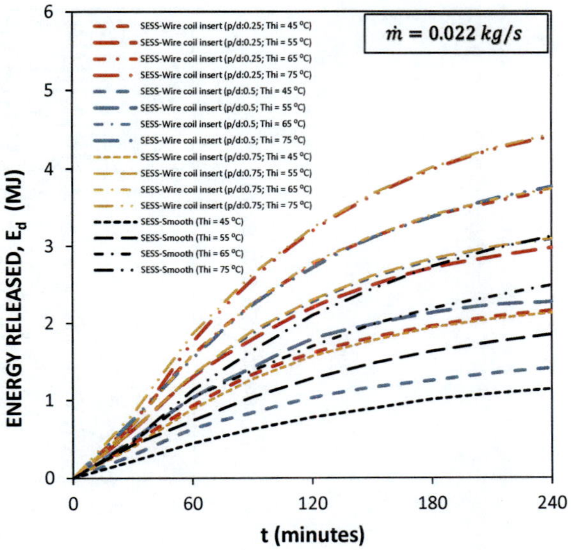

Fig. 5 Energy released from SESS

References

1. Hasnain SM (1998) Review of sustainable thermal energy storage technologies, part I: heat storage material and techniques. Energy Convers Manage 39(5–6):1127–1138
2. Prasad L, Muthukumar P (2013) Design and optimization of lab-scale sensible heat storage prototype for solar thermal power plant application. Sol Energy 97:217–229
3. Muthukumar P, Niyas H, Prasad L (2014) Performance investigation of high-temperature sensible heat thermal energy storage system during charging and discharging cycles. Clean Technol Environ Policy 17(2):501–513
4. Xu B, Han J, Kumar A, Li P, Yang Y (2017) Thermal storage using sand saturated by thermal-conductive fluid and comparison with the use of concrete. J Energy Storage 13:85–95
5. Özrahat E, Ünalan S (2017) Thermal performance of a concrete column as a sensible thermal energy storage medium and a heater. Renew Energy 111:561–579
6. Kunwar A, Kumar M, Gupta A (2019) Experimental investigation of a packed-bed thermal energy storage system fitted with perforated cylindrical elements. Heat Mass Transfer 55:2723–2737
7. Rao CRC, Niyas H, Muthukumar P (2018) Performance tests on lab–scale sensible heat storage prototypes. Appl Therm Eng 129:953–967
8. Agrawal P, Gautam A, Kunwar A, Kumar M, Chamoli S (2018) Performance assessment of heat transfer and friction characteristics of a packed bed heat storage system embedded with internal grooved cylinders. Sol Energy 161:148–158
9. Kumar R, Pathak AK, Kumar M, Patil AK (2021) Experimental study of multi tubular sensible heat storage system fitted with wire coil inserts. Renew Energy 164:1244–1253

Modeling and Comparative Thermal Performance Analysis of a Biomass Gasifier Using Different Gasifying Agents

S. Chowdhury, P. Mondal, and S. Ghosh

1 Introduction

The persistent increase in demand for energy has led to eventual degradation of conventional sources of energy. So the focus has been shifted to explore some renewable and alternate sources of energy. Biomass is one such renewable and alternate source of energy. By performing biomass gasification, syngas is obtained as end product which can be used in many scientific and industrial applications. The composition of syngas derived from biomass gasification depends on biomass content, type of gasifier and gasifiying agent. In biomass gasification, both air and steam can be used as gasifying agents. From air gasification, a low-energy gas is produced due to the dilution of nitrogen in the air. This gas is mainly used for heating purposes and/or for power production. However, air gasification leads to production of hydrogen-lean syngas and thus syngas possesses a lower calorific value. In steam gasification, syngas rich in hydrogen can be obtained which yields in higher syngas calorific value compared to that of air gasification, due to absence of nitrogen in the gasifying agent. Thus, in many aspects steam gasification is preferred to air gasification. Many research works with regard to biomass gasification have been reported in the literature. Shayan et al. [1] developed a model of a biomass gasifier using different gasifying agents: air, oxygen-enriched air, oxygen and steam. Two feedstocks were considered for the theoretical model, viz. wood and paper. From the results, it was noted that the system better energetic performance with air gasification but better exergetic performance while using steam as gasifying agent. Faraji and Saidi [2] developed a biomass gasification model in ASPEN Plus software with groundnut shell as feedstock and air as gasifying agent. The impact of different operating parameters such as the gasification temperature and pressure and the equivalence ratio (ER),

S. Chowdhury · P. Mondal · S. Ghosh (✉)
Department of Mechanical Engineering, IIEST Shibpur, Howrah, West Bengal 711103, India
e-mail: sudipghosh.becollege@gmail.com

S. Doolla et al. (eds.), *Advances in Clean Energy and Sustainability*,
Green Energy and Technology, https://doi.org/10.1007/978-981-99-2279-6_7

on the syngas composition and H_2/CO ratio were investigated using sensitivity analysis. From the sensitivity analysis it was noted that on increasing the gasification temperature, the amount of carbon monoxide and hydrogen increased. With increase in the equivalence ratio, the concentration of hydrogen decreased. Shayan et al. [3] performed thermo-economic analysis of a woody biomass gasifier collaborated with a solid oxide fuel cell (SOFC) for power generation. From the results, it had been noted that the power output deteriorated from 15.86 to 11.69% and the exergetic efficiency increased from 17.34 to 27.51% with the rise in the fuel utilization ratio and the payback period for the plant was 7.5 at cost optimal design (COD) case was found to be 8.5 years. Din and Zainal [4] presented a comprehensive review based on gasification of biomass. The review encompassed the different types of biomass that can be used and different types of gasifier for implementing in the gasification process. The authors also mentioned that apart from the type of feedstock and gasifier used, gasifying agent also have a considerable effect on the composition of the producer. de Sales et al. [5] investigated gasification of biomass using a two-stage downdraft gasifier with the help of gasifying agents such as: air, air along with saturated steam and oxygen with saturated steam. The impact of different gasifying agents had been assessed based on CO, H_2 and CH_4 concentrations and lower heating value, the syngas power and the cold gas efficiency. Zainal et al. [6] simulated the performance of a downdraft gasifier using equilibrium model for different biomass feedstocks. The authors also emphasized on the fact that tar is produced at temperatures higher than 200 °C and other than the drying process.

Although literature reports that research works have been carried out on gasification of sawdust with different gasifying agents but no work has been reported so far by considering the effects of different gasifying agents together on sawdust gasification. In this study, the authors have predicted the comparative performance of a gasifier in terms of hydrogen concentration, LHV of syngas, cold gas efficiency and exergetic efficiency by considering three different modes of gasification: air, steam and air and steam. Sawdust is used as feedstock and operational parameters like gasification temperature, steam-to-biomass ratio and equivalence ratio are considered as input parameters for the analysis.

2 System Modeling and Assumptions

In biomass gasification, biomass undergoes thermo-chemical conversion to produce combustible gas mixture consisting of hydrogen, carbon monoxide, carbon dioxide, water vapour and some traceable amount of methane. For performing biomass gasification process, a model has been established in software Cycle Tempo 5.0. The schematic diagram of the model is shown in Fig. 1.

In the present model, sawdust is considered as feedstock. The ultimate analysis of sawdust is mentioned in Table 1 of the literature. There are some assumptions involved and input parameters considered in Table 2 with the modeling in order to make the analysis more feasible and compatible.

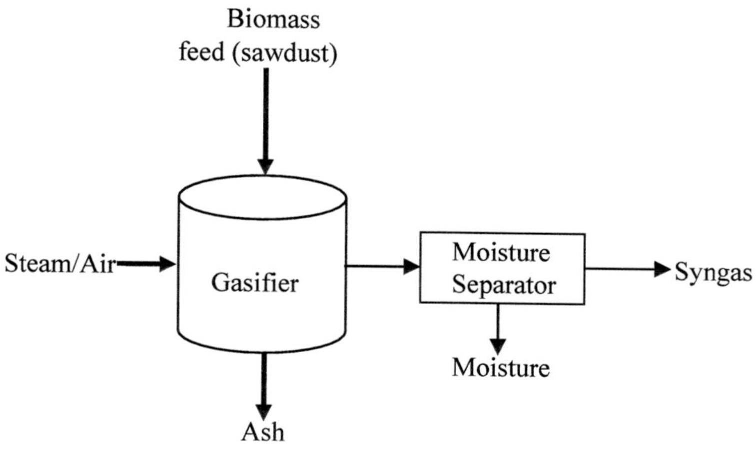

Fig. 1 Schematic of the gasification model

Table 1 Ultimate analysis of sawdust on dry basis [7]

C	H	O	N	Ash
52	6.07	41.55	0.28	0.1

Table 2 Input parameters at the base case level

Input parameters	Value(s)	Units
Biomass feed rate	1	kg/s
Steam inlet condition	150	°C
LHV of biomass	18,508	kJ/kg
Moisture content in biomass	10	%
Gasifier temperature	800	°C

Additionally some more general assumptions regarding the operation of the gasifier are made as follows:

i. The entire system is operating in steady-state configuration.
ii. All the gases are assumed to be perfect gases.
iii. The ambient pressure and temperature are assumed to be 1.013 bar and 25 °C respectively.
iv. Equilibrium model is considered.
v. Temperature and pressure drop across the gasifier is neglected.

2.1 Energy Analysis

A gasifier has been selected for gasification process with both air and/or steam as gasifying agents and sawdust as feedstock. The molecular formula of biomass (in dry form) is written as $CH_xO_yN_z$, where x, y and z is constants obtained from the main gasification equation by equating the weight of hydrogen and oxygen respectively.

The equation for steam gasification of biomass is written as follows:

$$CH_xO_yN_z + wH_2O + \gamma H_2O \rightarrow aH_2 + bCO + cCO_2 + dCH_4 + eH_2O + fN_2 \tag{1}$$

The equation for air gasification of biomass is written as follows:

$$CH_xO_yN_z + wH_2O + m(O_2 + 3.76N_2) \rightarrow aH_2 + bCO + cCO_2 + dCH_4 + eH_2O + fN_2 \tag{2}$$

The Boudouard reaction is written as follows:

$$C + CO_2 \rightarrow 2CO \tag{3}$$

The methanation reaction is as follows:

$$C + 2H_2 \rightarrow CH_4 \tag{4}$$

The water gas shift reaction is mentioned as follows:

$$CO + H_2O \rightarrow CO_2 + H_2 \tag{5}$$

where the letters a, b, c, d, e and f represent the number of moles of hydrogen, carbon monoxide, carbon dioxide, methane, steam and nitrogen respectively. Here, w denotes the amount of moisture content present in biomass and γ denotes the amount of steam that is supplied and m denotes the amount of air.

The amount of moisture content, w can be evaluated as:

$$w = \frac{M_{biomass} \times MC}{M_{H_2O} \times (1 - MC)} \tag{6}$$

Here, $M_{biomass}$ is the molecular weight of biomass and MC is the percentage of moisture present in the biomass. The steam to biomass ratio in terms of m can be written as:

$$STBM = \frac{M_{H_2O} \times m}{M_{biomass} + M_{H_2O} \times w} \tag{7}$$

The lower heating value (LHV) and higher heating value (HHV) of the woody biomass are computed with the help of following correlations as obtained from the work of Zaman et al. [8]:

$$LHV_{biomass} = HHV_{biomass} - h_{fg}(9 \times H_B + MC) \tag{8}$$

$$\begin{aligned} HHV_{biomass} = {} & 349.1 \times C_B + 1178.3 \times H_B - 103.4 \times O_B \\ & - 15.1 \times N_B - 21.1 \times Ash \end{aligned} \tag{9}$$

where h_{fg} represents latent heat of vaporization of water, H_B represents mass fraction of hydrogen, C_B represents mass fraction of carbon, O_B represents mass fraction of oxygen, N_B represents mass fraction of nitrogen, Ash represents the amount of ash present in biomass and MC is representative of moisture content of biomass.

The expression for cold gas efficiency is as follows:

$$\eta_{cg} = \frac{\dot{m}_{gas} \times LHV_{gas}}{\dot{m}_{bio} \times LHV_{bio} + \dot{m}_{agent} \times H_{agent} + \dot{Q}_{in}} \tag{10}$$

where, $\dot{m}_{bio}$ and $\dot{m}_{agent}$ are the mass flow rates of biomass and gasifying agents (air and/or steam) respectively, $\dot{m}_{gas}$ is the mass flow rate of product gas obtained.

2.2 Exergy Analysis

Exergy analysis is generally carried out for finding the best possible thermodynamic performance of a system. Mathematically, the exergy is equal to the sum total of physical exergy and chemical exergy. Physical exergy recognizes the change in enthalpy of a particular gas from standard state to the given pressure and temperature. Chemical exergy can be defined as the standard chemical exergy generated by mixing all the constituents and the loss in entropy due to mixing of different species of gases [9]. The physical exergy at any state point is written as follows:

$$Ex_{phy} = \dot{m} \times (h - h_0 - T_0 \times (s - s_0)) \tag{11}$$

where the parameters with subscript '0' represents the value of the thermodynamic properties at dead state conditions and those without subscript represents the thermodynamic properties at any state point. The dead state conditions are assumed to be 1.013 bar and 25 °C.

The chemical exergy at any state point is evaluated from the given equation:

$$Ex_{che} = \sum y_i ex_i^{ch} + RT_0\Big(\sum y_i \ln(y_i)\Big) \tag{12}$$

where ex_i^{ch} represents standard chemical exergy of any particular gas and y_i represents the mole fraction of that particular gas.

The exergetic efficiency of the plant is defined as the ratio of the total output exergy to the amount of input exergy. The expression for exergetic efficiency is mentioned as follows [2]:

$$\eta_{ex} = \frac{Ex_{syngas}}{Ex_{biomass} + Ex_{agent}} \tag{13}$$

where Ex_{syngas} is the output exergy of the syngas, $Ex_{biomass}$ is the input exergy of the biomass and Ex_{agent} is the input exergy of the gasifying agent.

3 Validation of the Model

In case of steam gasification, the gas composition obtained from the present model has been validated with the reported experimental work of Loha et al. [10] and shown in Table 3. While carrying out the validation work, the input parameters for the model are considered to be same as that of the experimental work of Loha et al. [10]. Simulated results reveal that the modeled gas data obtained are in synchronization with that of Loha et al. [10] with average root mean square (RMS) error of approximately 7.86%.

In case of air gasification, the gas composition has been validated with the reported work of Mondal et al. [11] shown in the form of Table 4. The model has been validated at a gasification temperature of 680 °C and equivalence ratio of 0.35. The composition of syngas obtained has been assessed by considering all the input parameters to be same as that of Mondal et al. [11]. From the simulated results, it is observed that the results obtained from the modeled gas data are similar to that of the reported experimental work. The average root mean square error is found to be 3.51%.

Table 3 Comparison of H_2 yield from the model with Loha et al. [10] considering steam gasification

STBM	Simulated results of H_2 conc (%)	Loha et al. [10]	RMS error (%)
0.75	51.32	48.5	3.02
1	53.57	49.6	5.49
1.25	55.12	50.25	7.81
1.5	56.26	50.75	9.59
1.75	57.12	51.1	11.1
2	57.81	52	10.1

Table 4 Comparison of the gas data composition with Mondal et al. [11] considering air gasification

Gas composition (%)	Simulated results	Mondal et al. [11]	RMS error (%)
H_2	20.23	21.44	3.6
CO	20.68	22.14	4.9
CO_2	10.13	10.57	1.8
N_2	42.43	39.09	6.2
H_2O	5.58	5.76	1.04

4 Results and Discussions

In this study, the performance of a biomass gasifier is assessed using sawdust as feedstock and steam and air as gasifying agents. Effects of several input parameters like steam-to-biomass ratio, equivalence ratio, gasification temperature on the output parameters like LHV of syngas, yield of hydrogen, cold gas efficiency and exergetic efficiency are shown in the following sections.

4.1 Effect on Hydrogen Production

Figure 2 shows the variation of STBM over concentration of hydrogen at different gasification temperatures. It is observed that the concentration of hydrogen increases with STBM. This is so because with increase in STBM, the amount of steam input increases for every unit mass of biomass added. This in turn increases the concentration of hydrogen. It is also noted that the yield of hydrogen is maximum at STBM of 2 and gasification temperature of 800 °C. Figure 3 shows the influence of equivalence ratio over concentration of hydrogen. It is prominent from the figure that there is less yield of hydrogen at higher equivalence ratio. The maximum concentration of hydrogen is found at an equivalence ratio of 0.2 and gasification temperature of 800 °C.

Figure 4 reveals the performance of different gasifying agents over the production of hydrogen at different gasification temperatures. Three different gasifying agents are incorporated: air, steam and both using air and steam. It is evident from the figure that the maximum production of hydrogen is obtained in case of steam gasification followed by air and steam gasification and then air gasification. So for more hydrogen production steam gasification route is more suitable than air gasification.

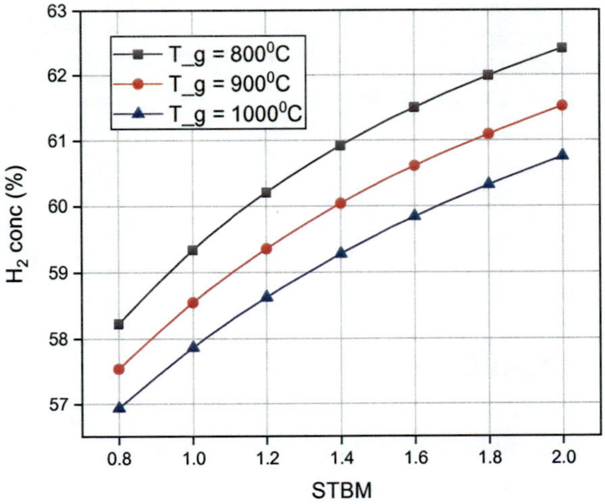

Fig. 2 Effect of STBM on concentration of hydrogen

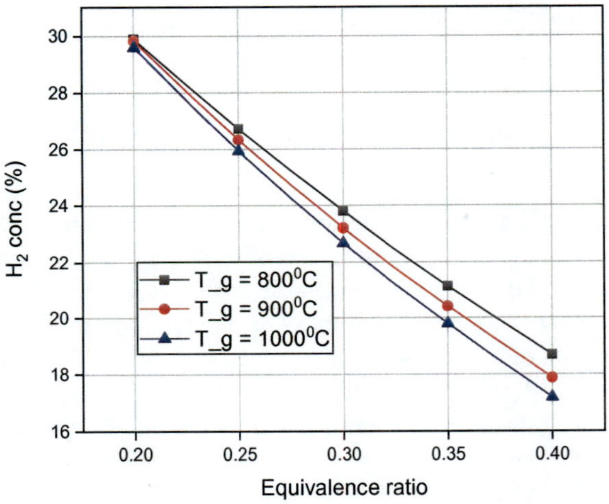

Fig. 3 Impact of equivalence ratio on concentration of hydrogen

4.2 Effect on Cold Gas Efficiency

The impact of STBM on cold gas efficiency is depicted in Fig. 5 of the literature. It has been noted that with increase in STBM, there is reduction in cold gas efficiency. It happens so because with increase in STBM, the energy consumed during steam generation increases which in turn decrease the cold gas efficiency. It is also noticed that the maximum cold gas efficiency is obtained when the STBM is equal to 0.8 and the gasification temperature equal to 800 °C. The impact of equivalence ratio over cold gas efficiency is plotted in Fig. 6 of the literature. From the figure, it is noted that there as a decrement in the value of the cold gas efficiency with increase in the value of equivalence ratio. Figure 7 shows the influence of different gasifying agents

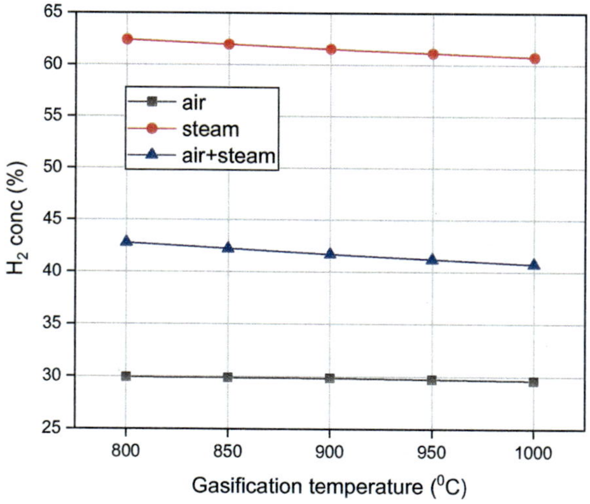

Fig. 4 Effect of different gasifying agents on concentration of hydrogen

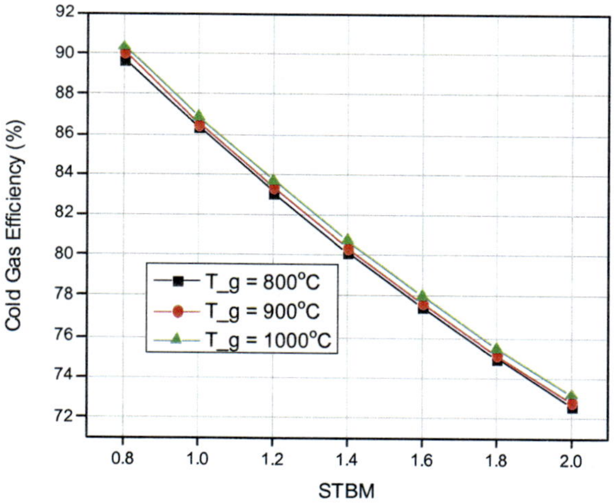

Fig. 5 Effect of STBM on cold gas efficiency

over cold gas efficiency. It is inferred from the results that with air gasification better cold gas efficiency is obtained.

4.3 *Effect on LHV of Syngas*

The effect of STBM and equivalence ratio on LHV of syngas has been plotted in Figs. 8 and 9. From the figures, it is evident that the LHV of the gas decreases with both STBM and equivalence ratio. The largest reduction in LHV with increase in the

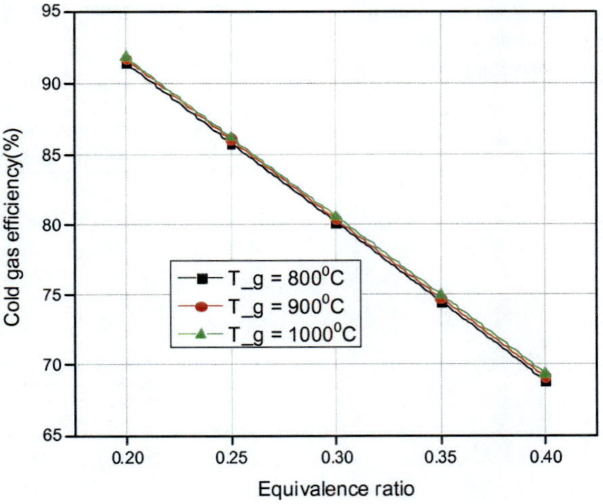

Fig. 6 Effect of equivalence ratio on cold gas efficiency

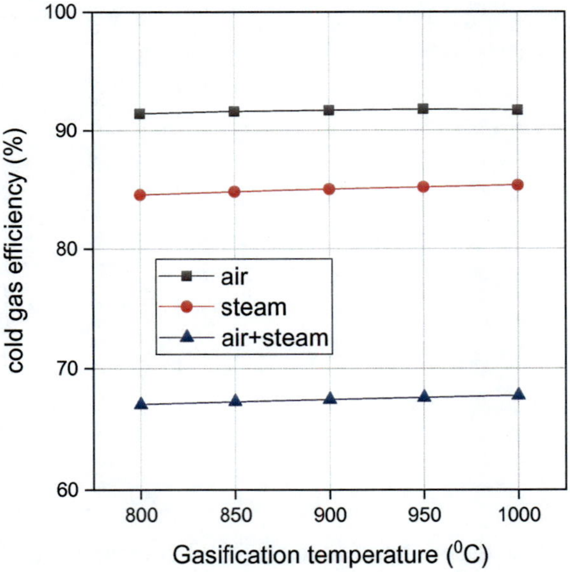

Fig. 7 Effect of different gasifying agents on cold gas efficiency

gasification temperature occurs in case of air gasification. However, at same gasification temperature, the LHV of gas is higher while going through steam gasification route due to the additional heat imparted by steam. From the plots, it is also revealed that the maximum LHV value is found at a gasification temperature of 1000 °C (Fig. 10).

Fig. 8 Effect of STBM on LHV of syngas

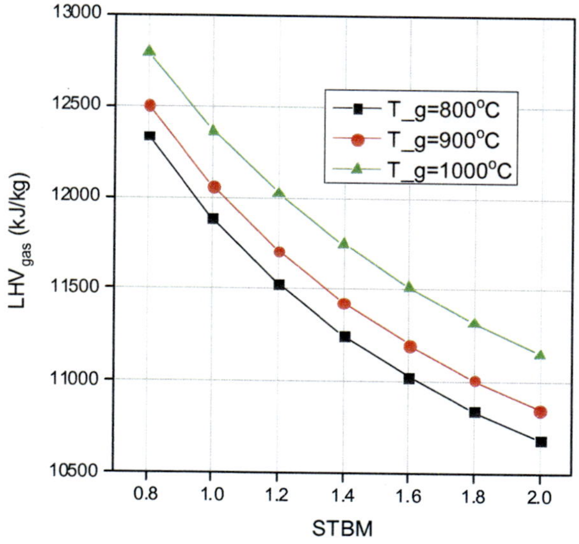

Fig. 9 Influence of equivalence ratio on LHV of syngas

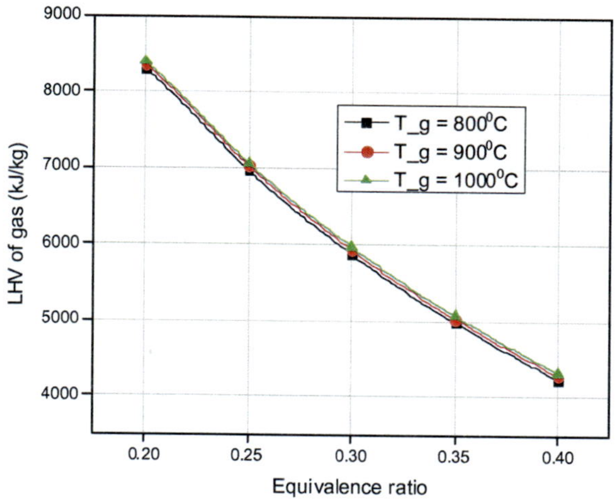

4.4 *Effect on Exergetic Efficiency*

Figures 11 and 12 shows the variation of STBM and equivalence ratio on exergetic efficiency. It is observed that the exergetic efficiency deteriorates with increase in STBM as well as equivalence ratio. This happens so because with increase in STBM content, the amount of steam input for every unit mass of biomass fed increases which in turn increases the input fuel exergy to the system. From the figure, it is also evident that the maximum exergetic efficiency is found at an STBM of 0.8 and equivalence ratio of 0.2 at a gasification temperature of 1000 °C. However, in

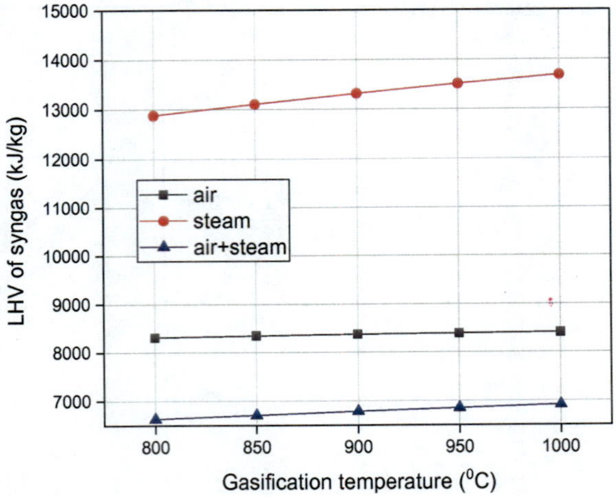

Fig. 10 Effect of different gasifying agents on LHV of syngas

case of air gasification, the exergetic efficiency is coinciding at all the gasification temperatures at the considered range of equivalence ratio. The effect of different gasifying agents on exergetic efficiency has been plotted in Fig. 13 which reveals that more exergetic efficiency is obtained while performing steam gasification.

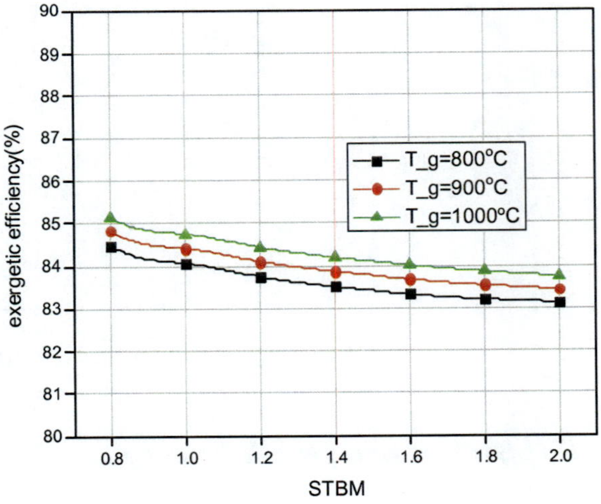

Fig. 11 Effect of STBM on exergetic efficiency

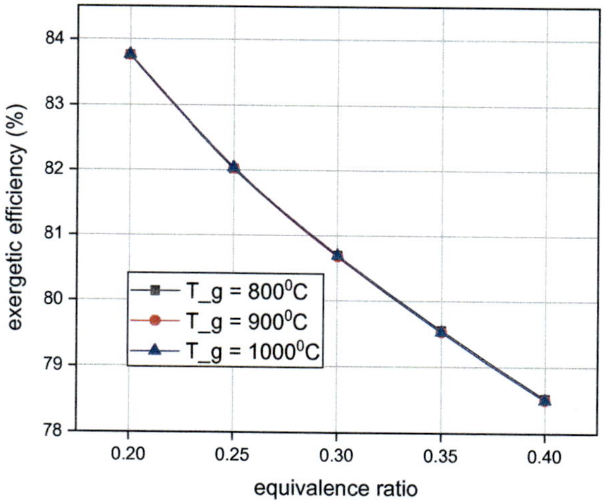

Fig. 12 Effect of equivalence ratio on exergetic efficiency

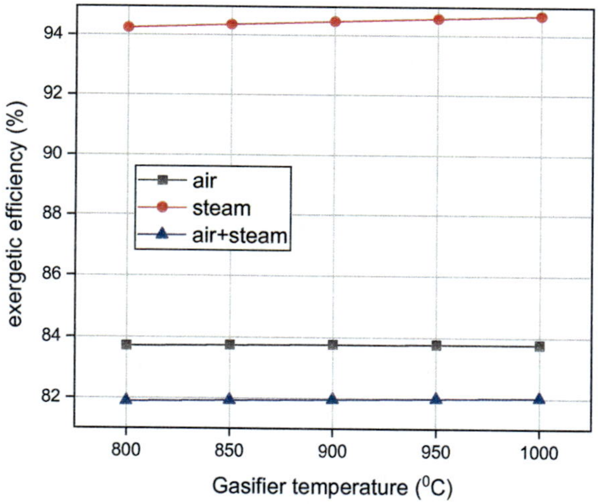

Fig. 13 Effect of different gasifying agents on exergetic efficiency

5 Conclusions

This paper presents the modeling and comparative performance analysis of a fixed bed biomass gasifier with sawdust as feedstock implementing three modes of gasification: air, steam and air and steam both. From the analysis, certain conclusions are drawn:

- With increase in STBM, the yield of hydrogen increases and the maximum concentration of hydrogen are found by steam gasification route.
- With increase in STBM as well as equivalence ratio, the cold gas efficiency as well as the exergetic efficiency of the plant decreases.

- With increase in both STBM and equivalence ratio, the lower heating value (LHV) of syngas decreases and maximum LHV is found to be 12,800 kJ/kg at STBM of 0.8 and 8372 kJ/kg at equivalence ratio of 0.2.

Thus it can be imparted from the analysis that maximum hydrogen percent of the model can be obtained when sawdust is gasified and steam is used as gasifying agent.

References

1. Shayan E, Zare V, Mirzaee I (2018) Hydrogen production from biomass gasification; a theoretical comparison of using different gasification agents. Energy Convers Manage 159:30–41
2. Faraji M, Saidi M (2022) Process simulation and optimization of groundnut shell biomass air gasification for hydrogen enriched syngas production. Int J Hydrogen Energy 47:13579–13591
3. Shayan E, Zare V, Mirzaee I (2019) On the use of different gasification agents in a biomass fueled SOFC by integrated gasifier: a comparative exergo-economic evaluation and optimization. Energy 171:1126–1138
4. Din ZU, Zainal ZA (2016) Biomass integrated gasification–SOFC systems: technology overview. Renew Sustain Energy Rev 53:1356–1376
5. de Sales CAVB, Maya DMY, Lora EES, Jaen RL, Reyes AMM, Gonzalez AM, Martinez JD (2017) Experimental study on biomass (*eucalyptus* spp.) gasification in a two-stage downdraft reactor by using mixtures of air, saturated steam and oxygen as gasifying agents. Energy Convers Manage 145:314–323
6. Zainal ZA, Ali R, Lean CH, Seetharamu KN (2001) Prediction of performance of a downdraft gasifier using equilibrium modeling for different biomass materials. Energy Convers Manage 42(12):1499–1515
7. Roy PC, Datta A, Chakraborty N (2010) Assessment of cow dung as a supplementary fuel in a downdraft biomass gasifier. Renew Energy 35(2):379–386
8. Zaman SA, Roy D, Ghosh S (2020) Process modeling and optimization for biomass steam-gasification employing response surface methodology. Biomass Bioenergy 143:1–13
9. Kumar S, Dassappa S (2014) First and second law thermodynamic analysis of air and oxy-steam biomass gasification. Int J Hydrogen Energy 39(34):19474–19484
10. Loha C, Chatterjee PK, Chattopadhyay H (2011) Performance of fluidized bed steam gasification of biomass—modeling and experiment. Energy Convers Manage 52:1583–1588
11. Mondal P, Mondal K, Ghosh S (2015) Bio-gasification based distributed power generation system employing indirectly heated GT and supercritical ORC: energetic and exergetic performance assessment. Int J Renew Energy Res 5:773–781

The Corrosion Analysis of Diesel Engine Parts on Application of Dual Biodiesel Blend

Sajan K. Chourasia, Absar M. Lakdawala, and Rajesh N. Patel

1 Introduction

Since the ongoing crisis of fossil fuels, there have been many attempts to replace these non-renewable energy fuels with renewable, environment-friendly and reliable alternative fuels. The objective of renewable fuels is to obtain optimum efficiency compared to the existing petrodiesel fuels. Along with other renewable energy sources, biomass is the one that can be altered chemically to have a similar structure as that of petroleum which can be used for transportation engines. A few of the advantages of biomass fuels are that they do not emit sulphur during combustion, there are no dangerous emissions from them, and they can be reused as feedstock in some cases [1]. The fuel produced by biomass is ethanol and biodiesel. The fuel produced by the transesterification of triglycerides present in animal fat or vegetable oils is called biodiesel. The most commonly used oils to produce biodiesel are soya bean oil, palm oil, Castor oil, Rapeseed oil, etc., depending upon the availability of the crop in the area [2–4]. There are around 300 vegetable species that can be used to produce biodiesel; even cooking oil can be used to make biodiesel.

Numerous research has been conducted and is still going on, on the corrosion effect of biodiesels like Palm, Jatropha, and Karanja on the components of the fuel line system of the CI engine [1]. Yet, the current research is still limited to every few crop oils and a few varieties of metals. The composition of the standard fuel and biodiesel has also not been altered even slightly in most recent research. The blends usually considered for this composition are 20% biodiesel–80% Diesel, whereas the permissible amount of use biodiesel can be extended up to 25–30% in the entire composition. There has not been any attempt to blend two or more different biodiesel types to form a different kind of composition with the petrodiesel until now [5]. Our

S. K. Chourasia (✉) · A. M. Lakdawala · R. N. Patel
Mechanical Engineering Department, School of Engineering, Institute of Technology, Nirma University, Ahmedabad, Gujarat 382481, India
e-mail: sajan1chourasia@gmail.com

S. Doolla et al. (eds.), *Advances in Clean Energy and Sustainability*,
Green Energy and Technology, https://doi.org/10.1007/978-981-99-2279-6_8

aim with this research is to attempt a few such things that are yet to be studied for the renewable energy source of biodiesel. To bridge the gap, find an optimum biodiesel blend with petrodiesel for less corrosion rate and higher efficiency.

Many researchers have established that biodiesel with Diesel as a fuel in CI engines can only be around 20%, more than this amount leads to higher corrosion in metals. Which degrades the life of the engine, and optimum power output cannot be achieved. So, this research is based on finding the blends of two different biodiesel with Diesel which will be best suitable for CI engines with the least corrosion rate [2, 3]. Biodiesel's benefits compared to Diesel are the relatively high flash point and lubricating tendency, whereas its major disadvantage is its oxidization property, leading to high corrosion [1–3, 5]. Due to this reason also, until now, the blend of biodiesel and Diesel is taken as 20–30% (B100) and 70–80% (B0) [6, 7].

Several research works on the corrosion rates of materials used for CI engine fuel line components, i.e., Stainless steel, Aluminium, and Copper, for a duration of 600/1200 h with exposure to Diesel and biodiesel at 80 °C. They found that the Copper corrosion rate in biodiesel appears to increase with time, while the Aluminum corrosion rate seems to decrease marginally. The Copper and Leaded Bronze corrosion rates increased at room and elevated temperatures for both metals. Increased corrosion rates are found with the increased concentration of biodiesel in the blended fuel [8]. According to Sarin et al. [2], even when a small area of the above-said metal is exposed to biodiesel, it acts as a catalyst for oxidization. During the oxidization processes, the FAME forms a radical next to a double bond and quickly bonds with the air's oxygen. Niczke et al. [7] discovered acids, aldehydes, ketones, lactones, alkylofurances, and other volatile products after oxidation of Rapeseed biodiesel at 200 °C for 25 h. The above reaction can cause an increase in the water content of the fuel, which can lead to unwanted microbial growth and fuel line corrosion.

Knothe and Steidley [9] have experimented to find out the effect of 26 metals and 29 metal oxides on biodiesel oxidation. They found that Cu and Ru show oxide promoting effect, and Mo shows inhibiting effect.

From the literature, it can be clearly understood that biodiesel has a higher amount of saturated and unsaturated fatty acid methyl/ethyl ester. Due to the presence of fatty acid, the biodiesels have higher corrosive properties, which ultimately results in a higher amount of corrosion with metals. It can also be understood from the literature that the corrosive behavior of biodiesel at elevated temperatures increases. The current experiment aims to reduce the corrosion properties of these test fuels using a dual combination blend of biodiesel and to identify the corrosion behavior of dual biodiesel. Static immersion tests were performed as per ASTM G-31. To evaluate the effect the corrosion rate, weight loss, surface roughness and microscopic analysis were performed.

2 Experiment and Methodology

2.1 Selection of Biodiesel

The selection of biodiesel was made according to the availability, physical–chemical properties of the biodiesel, and the popularity of certain biodiesel worldwide based on many research works.

Castor: Castor biodiesels have the highest and most stable viscosity of any vegetable biodiesel, with a higher molecular weight of 298, a low melting point of 5 °C, and a very low solidification point of – 15 °C. The fuel density should be maintained within an allowable limit to maintain an optimal air–fuel ratio of the fuel for proper combustion. The density of Castor biodiesel is 962 kg/m^3, slightly more than the permissible density in biodiesel. The calorific value of Castor 38.5 MJ/kg, which is slightly lower than the conventional Diesel, the flashpoint of Castor is 239 °C that is higher than Diesel. As shown in Fig. 1 GC–MS test shows Methyl 12-hydroxy-9-octadecenoate (52.2%), cis-Methyl 11-eicosenoate (11.79%), cis-Methyl 11-eicosenoate (10.49%) respectively. India had around 4000 kg per hector yield of Castor seeds last year, Gujarat and Rajasthan being the highest producers of Castor.

Rapeseed: The Rapeseed biodiesel esters are formed due to the chemical reaction between the raw Rapeseed oil and the methyl or ethyl alcohol in the presence of the catalyst. The physical–chemical properties of the rapeseed biodiesel and its ester vary according to the conditions of the raw Rapeseed and producing the rapeseed biodiesel. The density of Rapeseed biodiesel prepared by raw Rapeseed oil is 910 kg/m^3, the flashpoint is 175 °C and the calorific value of Rapeseed biodiesel is 37.2 MJ/kg, almost like Castor. 9-Octadecenoic acid, methyl ester (56.36%), 9,12-Octadecadienoic acid, methyl ester (21.57%), Hexadecanoic acid, methyl ester (14.14%), Methyl stearate (6.60%) these are the significant composition of Rapeseed biodiesel as shown in Fig. 1. Rapeseeds are generally not cultivated in India, but around 40,000 acres of rapeseed plantation are seen in northern India, with about 44% of it being oil yield.

Neem: Neem biodiesel mainly consists of triglycerides and triterpenoid compounds. Its physical–chemical properties, density, viscosity, flash point, and molecular weight are all higher than conventional Diesel. It has four major saturated fatty acids, two of which are palmitic acid and stearic acid. Polyunsaturated fatty acids like oleic and linoleic acids are also present. The density of the Neem biodiesel is 946 kg/m^3, whereas the calorific value is 38.9 MJ/kg, and the flashpoint is around 103 °C. 9-Octadecenoic acid (Z)-, methyl ester (43.66%), Hexadecanoic acid, methyl ester (30.19%), Methyl stearate (15.30%), 1-Cyclohexyldimethylsilyloxybutane (3.68%), Methyl 12-hydroxy-9-octadecenoate (2.40%) are the chief components of Neem biodiesel concluded by the GC–MS tests as shown in Fig. 1. Around 14 million Neem trees are planted all over India, with each tree giving up to 31–51 kg of fruit yield per year under ideal conditions.

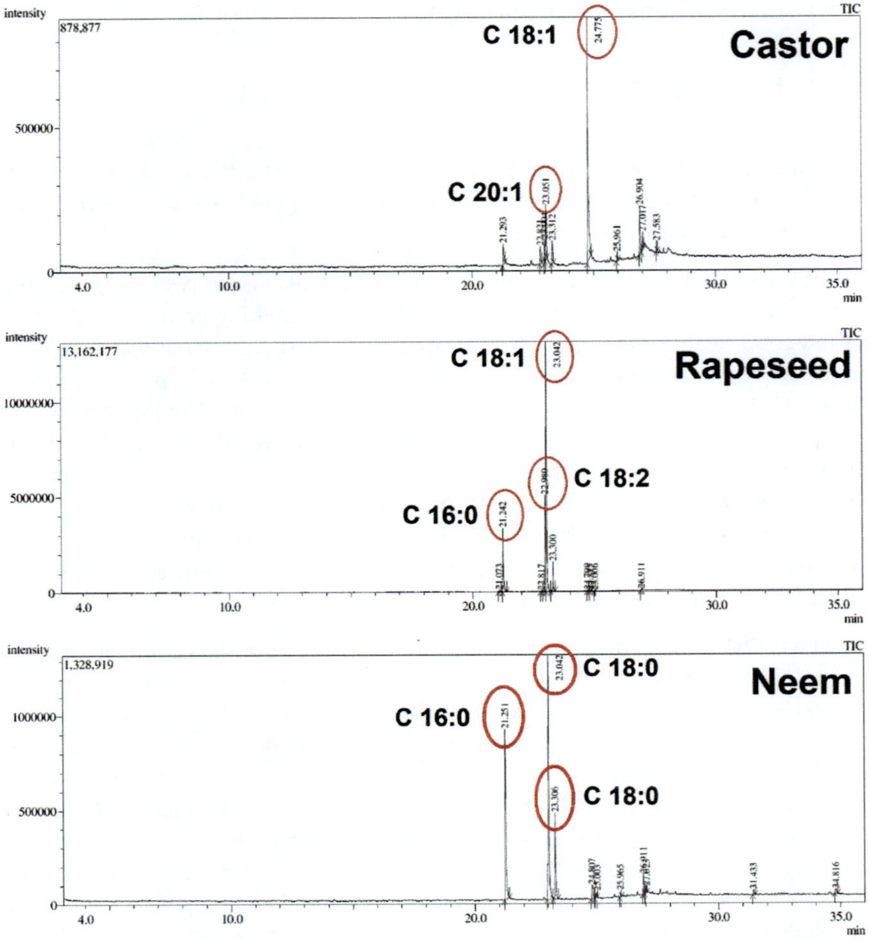

Fig. 1 GC–MS analysis of the Castor, Rapeseed and Neem biodiesel

As it's already stated, the use of only biodiesel as fuel in CI engines cannot be taken due to many drawbacks stated earlier. In the current experiment, it was decided that 70% of the blend will be Diesel, and the other 30% will be the mixture of two different biodiesel in different concentrations. This composition of biodiesel blends was determined by the Taguchi method. The specific composition of blended biodiesel and notation that we are using are: 70% Diesel 25% Castor 5% Rapeseed (70D25CA5R), 70% Diesel 15% Castor 15% Rapeseed (70D15CA15R), 70% Diesel 5% Castor 25% Rapeseed (70D5CA25R), 70% Diesel 25% Castor 5% Neem (70D25CA5N), 70% Diesel 15% Castor 15% Neem (70D15CA15N), 70% Diesel 5% Castor 25% Neem (70D5CA25N).

Table 1 The material composition of CI engine components

Material composition							
CI engine parts	Material percentage in a CI engine parts						
	Al	Fe	C	Cu	S	Z	Others
Piston	83	0.9	–	1.58	12	0.6	–
Piston ring	0.0016	69	0.08	0.018	0.8	–	0.09
Inlet valve	–	96	3	–	1.6	–	0.5
Exhaust valve	–	96	3	–	1.6	–	0.5
Cylinder liner	0.023	67.5	–	0.013	–	0.015	4.5

2.2 *Selection of Metals*

A thorough study of fuel intake and return in a CI engine was conducted before selecting metals to be gone for a static immersion test. The components' materials continuously in contact with the fuel were selected. A CI engine's main fuel contacting components are the engine cylinder, piston, cylinder casing, crank, crankcase, fuel tank, fuel supply system, fuel intake and return lines, fuel filters, fuel pumps, and pressure regulators. The material used in the above setup is both metals and non-metals. The major non-metal used are the elastomers. The nonferrous metals used are Aluminum, Copper and the ferrous metal is Iron. Table 1 shows the composition of each material by weight. Copper, Iron, Brass, Aluminium and Bronze were considered the ideal choice for conducting further experiments to obtain the effects of the biodiesel blends on the engine component material. Small specimen of these materials was obtained from large sheets of the respective metals; the specimen men were cut into identical coin-like/coupon shapes with exact sizes.

2.3 *Weight, Microscopic and Surface Roughness Analysis*

All the selected specimens were prepared and cleaned according to the ASTM standards as given in G1-90, and then weight analysis of these coins was conducted for the initial stage. After immersing these coins for the first 30 days in the above-mentioned dual biodiesel blends, the weight analysis will again be conducted. This will be done after every interval of 30 days for 150 days in total.

In microscopic analysis, the specimen analysis will be done at 200× on a bi-fringe metallurgical microscope to see the specimen's structure at a micro-level and changes that occurred by dipping the coupons in test fuels. For 150 days, initial, intermediate and final images were taken, and a comparison will be made based on the initial reading. For further qualitative results of the specimen, detailed microscopic analysis was performed through SEM images (Make: Hitachi at 30 kV (Mode: Super vacuum)). These detailed images will help determine the types of corrosion, intensity and depths of the pits on the surface of the coupons.

In surface analysis, surface roughness tester model SJ 210, of ISO 1997 Standard, stylus tip radius is of 2 μm, Maker: Mitutoyo is used to check the surface roughness of the metal on the selected length of the coin specimen before and after dipping the coupons into biodiesel. The surface roughness readings were taken initially, as intermediate and final readings. Through this analysis, we get to know about the surface roughness value of the coupons through the values of Ra, Rq, Rz values in μm of the surface of the coupon.

2.4 Corrosion Analysis

The corrosion analysis is conducted to evaluate the results and determine the corrosion rate of the materials immersed in the dual biodiesel blend to determine the corrosion rate (CR) in MMY. The input data of the specimen are mentioned as Density (g/cm^3)—Copper (8.96), Iron (7.86), Brass (8.55), Aluminium (2.7) and Bronze (8.45), Specimen diameter (cm)—1.7, Exposed area (cm^2)—4.5396 and Exposure time (h)—3600. The standard formula according to ASTM G1-90 is:

$$\text{Corrosion rate (CR)} = \frac{\text{Weight loss (g)} \times \text{K}}{\text{Alloy density}\left(\text{g/cm}^3\right) \times \text{Exposed area}\left(\text{cm}^2\right) \times \text{Exposure time (h)}} \quad (1)$$

2.5 Experimental Setup

All the biodiesel blends were collected volumetrically and poured into the beakers. These beakers were previously labeled and partitioned into four different segments to put in four other specimens of selected metals. 80 ml of the total blended biodiesels are kept under observation in each of the beakers for about 150 days. These specimens were cut in the shape of coins and immersed about 2 cm below the free surface of biodiesels with minimal surface contact. These samples are covered with Aluminium foil to restrict the test fuels from any interaction with the atmosphere as it can contaminate the test fuels and hinder the results.

3 Result and Discussion

3.1 Qualitative Analysis

The visual inspection of the specimen is observed after every 30 days intervals. The visible changes in the specimen are noted, photographed and recorded. The naked eyes can note these changes on the surface of the specimen, there are several residues on the coins, whilst some wear has been noted in the material of the coins. The images below can be used to justify these changes. After every 30 days, there have been visual changes in each of the coins, which can be noted. Many literatures have mentioned that biodiesel tends to degrade when in contact with the atmosphere due to oxidation, moisture absorption, and microorganism when kept in storage. The contact of metal can also degrade the fuel properties of biodiesel. The reaction of the biodiesel to exposure to metal can differ from metal to metal. This degradation of biodiesel leads to the increase of oxygen on the surface of the metal, which results in the corrosive nature of the biodiesel. Due to the presence of free fatty acids, water particles and oxidizing agents increase the corrosive nature of the fuel. Figure 2a, b shows the results of day 0, day 150 before and after cleaning images; there are a noticeable amount of deposit on the coin surface; these deposits are of oxides and carbides. In Fig. 2a, b on day 0, a fresh specimen can be seen before the immersion into the biodiesel. These specimens were cleaned, prepared and polished according to ATSM G1-90 standards.

Figure 2a, b contains data on the surface roughness from day 0, and day 150 before and after cleaning. The figure contains the specimen's Ra, Rq, Rz values after the surface roughness test. It can be observed that the surface roughness profile graph has gone through drastic changes thought out each interval due to the corrosive behavior of biodiesel. Even the Ra, Rq, Rz values in μm of the respective specimen have been changed, as we can see the copper immersed in the 70D15CA15R shows roughness in the range of 0.212–0.523 μm, not just copper but such trend is seen in every specimen after 150 days which were immersed in it which is found to be lowest followed by 70D25CA5R.

3.2 Metallurgical Microscopic and SEM Analysis

All the specimens were observed under the microscope to understand better biodiesel's effects on the coins on a structural basis. Each observation was taken after 30 days intervals. Figure 3 shows the state of the specimen on day 0 before the immersion test, and day 150 before and after cleaning. In most literature, four kinds of corrosion are seen on metals due to biodiesel. These are pitting corrosion, cervices corrosion, intergranular corrosion and selective leaching corrosion. The images in Fig. 3 show the specimen under the metallurgical microscope from day 0 day, 150 before and after cleaning. These images show the changes in the specimen's surface

(a)

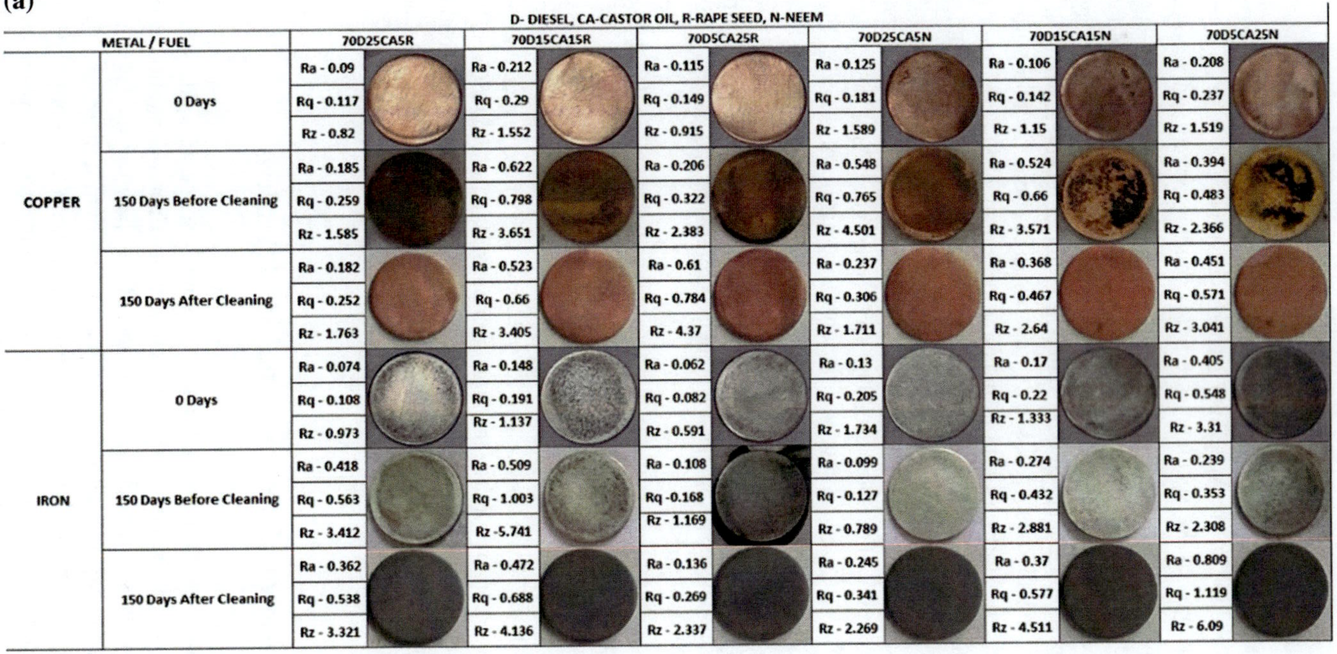

D- DIESEL, CA-CASTOR OIL, R-RAPE SEED, N-NEEM							
METAL / FUEL		70D25CA5R	70D15CA15R	70D5CA25R	70D25CA5N	70D15CA15N	70D5CA25N
COPPER	0 Days	Ra - 0.09	Ra - 0.212	Ra - 0.115	Ra - 0.125	Ra - 0.106	Ra - 0.208
		Rq - 0.117	Rq - 0.29	Rq - 0.149	Rq - 0.181	Rq - 0.142	Rq - 0.237
		Rz - 0.82	Rz - 1.552	Rz - 0.915	Rz - 1.589	Rz - 1.15	Rz - 1.519
	150 Days Before Cleaning	Ra - 0.185	Ra - 0.622	Ra - 0.206	Ra - 0.548	Ra - 0.524	Ra - 0.394
		Rq - 0.259	Rq - 0.798	Rq - 0.322	Rq - 0.765	Rq - 0.66	Rq - 0.483
		Rz - 1.585	Rz - 3.651	Rz - 2.383	Rz - 4.501	Rz - 3.571	Rz - 2.366
	150 Days After Cleaning	Ra - 0.182	Ra - 0.523	Ra - 0.61	Ra - 0.237	Ra - 0.368	Ra - 0.451
		Rq - 0.252	Rq - 0.66	Rq - 0.784	Rq - 0.306	Rq - 0.467	Rq - 0.571
		Rz - 1.763	Rz - 3.405	Rz - 4.37	Rz - 1.711	Rz - 2.64	Rz - 3.041
IRON	0 Days	Ra - 0.074	Ra - 0.148	Ra - 0.062	Ra - 0.13	Ra - 0.17	Ra - 0.405
		Rq - 0.108	Rq - 0.191	Rq - 0.082	Rq - 0.205	Rq - 0.22	Rq - 0.548
		Rz - 0.973	Rz - 1.137	Rz - 0.591	Rz - 1.734	Rz - 1.333	Rz - 3.31
	150 Days Before Cleaning	Ra - 0.418	Ra - 0.509	Ra - 0.108	Ra - 0.099	Ra - 0.274	Ra - 0.239
		Rq - 0.563	Rq - 1.003	Rq -0.168	Rq - 0.127	Rq - 0.432	Rq - 0.353
		Rz - 3.412	Rz -5.741	Rz - 1.169	Rz - 0.789	Rz - 2.881	Rz - 2.308
	150 Days After Cleaning	Ra - 0.362	Ra - 0.472	Ra - 0.136	Ra - 0.245	Ra - 0.37	Ra - 0.809
		Rq - 0.538	Rq - 0.688	Rq - 0.269	Rq - 0.341	Rq - 0.577	Rq - 1.119
		Rz - 3.321	Rz - 4.136	Rz - 2.337	Rz - 2.269	Rz - 4.511	Rz - 6.09

Fig. 2 Quantitative representation of coins at 0 day, 150 days before and after cleaning with Ra, Rq, Rz values

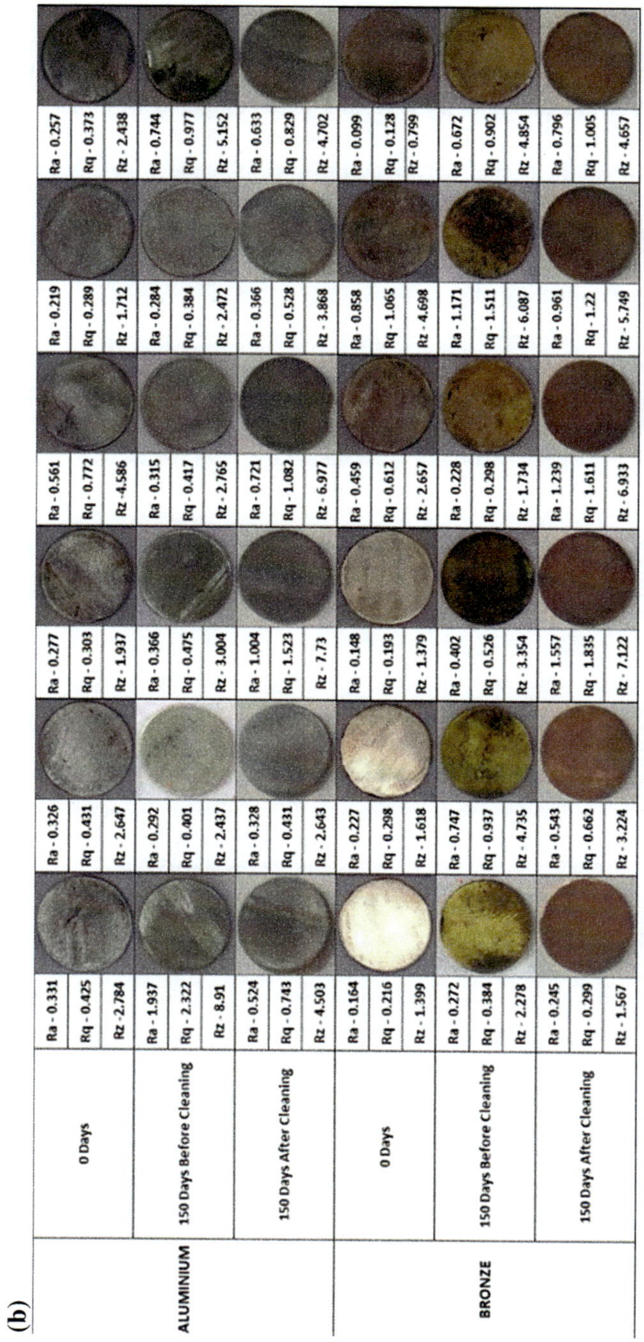

Fig. 2 (continued)

due to corrosion at a microscopic level after conducting all the tests for 150 days. After a deep study of these image results, it is observed that the 70D15CA15R dual biodiesel blend shows the least corroded surface compared to the rest of the blends. For further analysis of metal specimens immersed in this blend under SEM analysis was conducted, and then these results were compared with SEM results of the metal specimen which were immersed in Diesel. From the microscopic results, Fig. 4 shows SEM results of the specimen immersed in the test fuel. With the blend of 70D15CA15R, it has been observed that it is showing the minimal effect of corrosion with biodiesel on the metals coupons/samples compared to Diesel. In both Neem and Rapeseed biodiesel, acids like 9-octadecenoic acid, octanoic acid, nonanoic acid, Hexadecenoic acid, and 9-octadecanoic acid are found with other esters, aldehydes and ketones, which not only degrades the fuel properties but also enhances the corrosive properties of the fuel, while the biodiesel is heated. Esters react with oxygen in the air to produce various chemical species such as aldehydes, ketones, carboxylic acids, etc. [3]. The presence of copper causes biodiesel to become heavily oxidized, as measured by FTIR spectroscopy in accordance with ASTM D7418.

3.3 Quantitative and Corrosion Rate Analysis

The quantitative analysis consists of the weight of each specimen in milligrams from day 0 to day 150 and then the data of the specimen after cleaning them with acid. Figure 5 shows the weight loss in each cleaned specimen after 150 days; these were cleaned according to the ASTM G1-90. The difference of weight at day 0 and weight at day 150 (after cleaning) of the specimen is taken and composed into a graph as seen in Fig. 5. Here the x-axis indicates the test fuels used for the immersion test where 1—70D25CA5R, 2—70D15CA15R, 3—70D5CA15R, 4—70D25CA5N, 5—70D15CA15N, 6—70D5CA25N. It is noted that weightlessness in each metal is seen less except Iron and Brass in 70D15CA15R. In copper, the weight loss in each solution is around 195 mg; it's to be noted that the most weight loss in copper, Bronze and Iron is seen in the solution 70D5CA25N followed by 70D25CA5N, these three metals being few of the most common metals in the C.I engine components.

Weight loss in rubber couldn't be taken as all the specimens were dissolved in each test fuel. Many researchers have observed that the major weight loss of rubber/metals in biodiesel is due to the biodiesel's unsaturated fatty acid content. A few of the major contributors to the corrosion rate are the metal density, immersion time and the surface area of the specimen. In Fig. 5, the corrosion rate in each metal due to the biodiesel is calculated in mills/years, where the exposure time is kept in h, the density in g/cm^3 and K value as 8.75×10^4. As we see in weight loss, the least weight loss is seen in the 70D15CA15R dual biodiesel blend. The likely cause may be more saturated fatty acids and the least unsaturated ones present in the dual blend of Castor and Rapeseed. These corrosion rates are calculated according to the formula mentioned in Eq. 1.

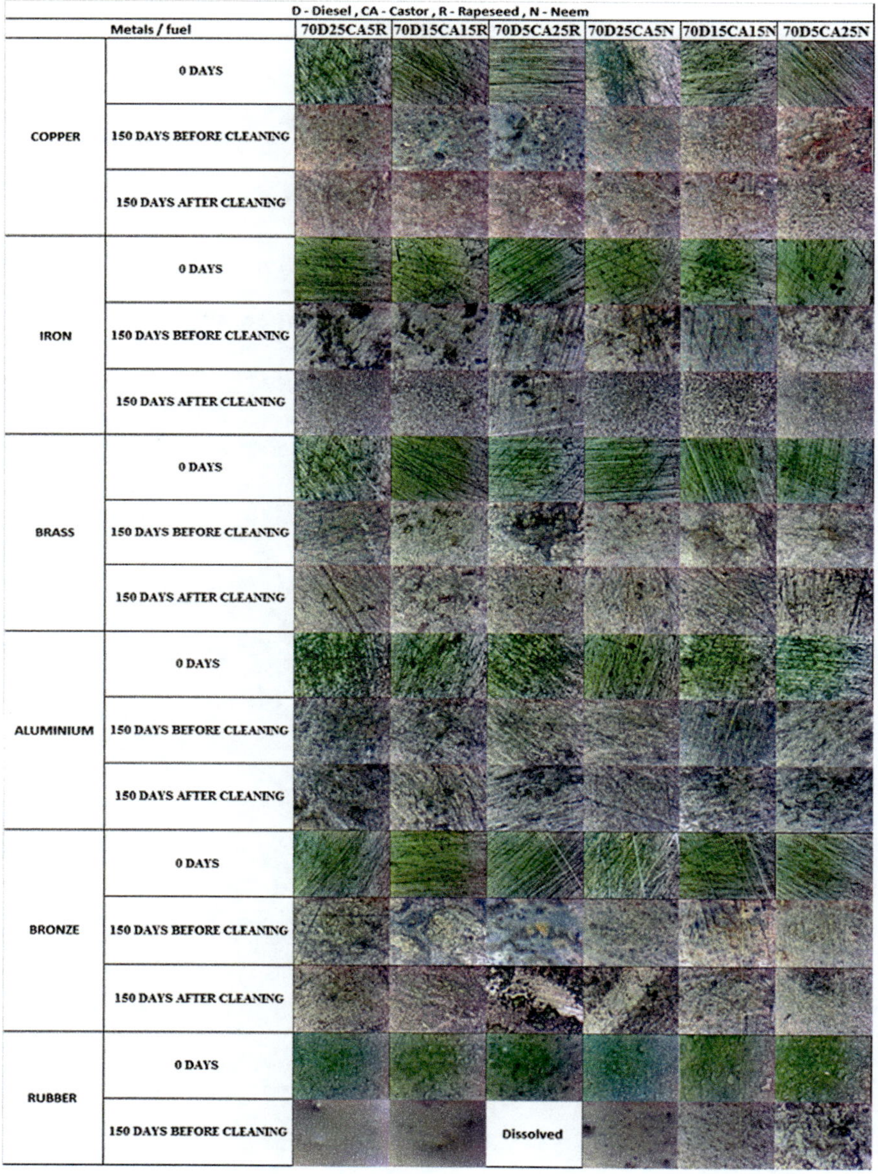

Fig. 3 Microscopic analysis of specimen at 0 day, 150 days and after cleaning

4 Conclusion

The outcome of the current experiment leads to the following conclusion: As the corrosion rate and weight loss graph shows for copper, the dual blend of 15%

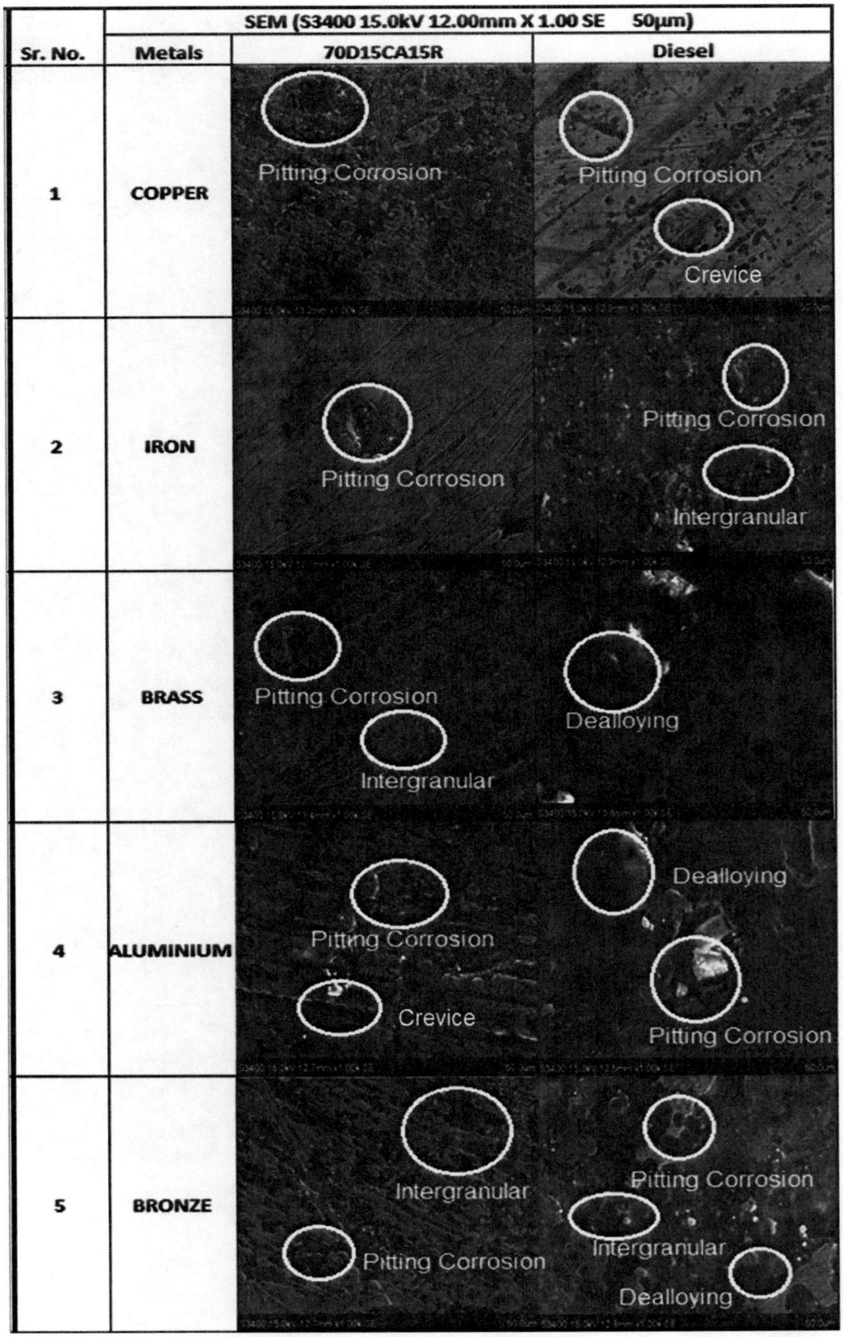

Fig. 4 SEM analysis and comparison between 70D15CA15R with Diesel

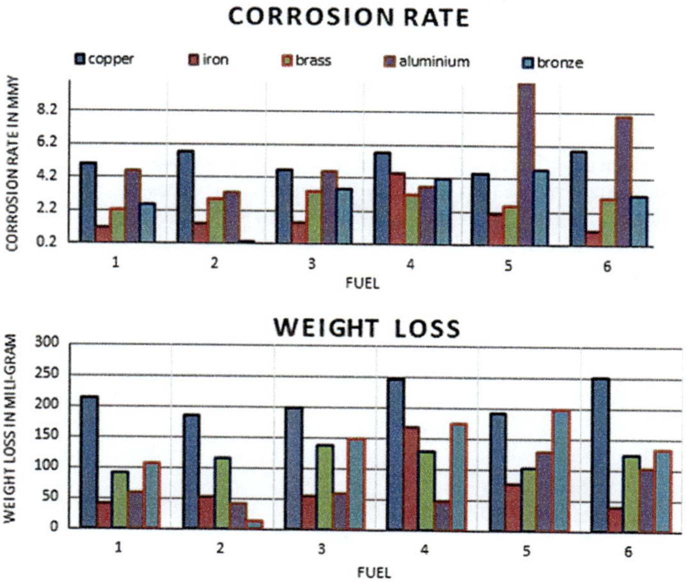

Fig. 5 Weight loss analysis and corrosion rate analysis

Castor and 15% Rapeseed (70D15CA15R) has the least value compared to all the six selected test fuels and is considered the optimum blend. According to the SEM results conducted on specimens, it was found that other than Brass, all other metals in Diesel shows pitting, de-alloying and intergranular type of corrosion and which are seen much less in 70D15CA15R. Bronze, Brass and Iron specimens show the least corrosion rate in 70D15CA15R, followed by 70D25CA5R. GC–MS test conducted on the biodiesels result shows that Castor and Rapeseed consist mainly of Hexadecane alkane, which is known to prevent corrosion. Due to this, the 70D15CA15R resulting has the lowest corrosion rate compared to other test fuels; hence the blend 70D15CA15R can be suggested as the best-optimum fuel.

References

1. Chandran D, Ng HK, Lau HLN, Gan S, Choo YM (2016) Investigation of the effects of palm biodiesel dissolved oxygen and conductivity on metal corrosion and elastomer degradation under novel immersion method. Appl Therm Eng 104:294–308
2. Fazal MA, Haseeb ASMA, Masjuki HH (2010) Comparative corrosive characteristics of petroleum diesel and palm biodiesel for automotive materials. Fuel Process Technol 10(91):1308–1315
3. Fazal MA, Haseeb ASMA, Masjuki HH (2011) Effect of different corrosion inhibitors on the corrosion of cast iron in palm biodiesel. Fuel Process Technol 11(92):2154–2159
4. Fazal MA, Haseeb ASMA, Masjuki HH (2011) Biodiesel feasibility study: an evaluation of material compatibility; performance; emission and engine durability. Renew Sustain Energy

Rev 2(15):1314–1324
5. Singh B, Korstad J, Sharma YC (2012) A critical review on corrosion of compression ignition (CI) engine parts by biodiesel and biodiesel blends and its inhibition. Renew Sustain Energy Rev 5(16):3401–3408
6. Fazal MA, Haseeb ASMA, Masjuki HH (2011) Effect of temperature on the corrosion behavior of mild steel upon exposure to palm biodiesel. Energy 5(36):3328–3334
7. Gis W, Żółtowski A, Bocheńska A (2011) Properties of the rapeseed oil methyl esters and comparing them with the diesel oil properties. J KONES 4(18):121–127
8. Chew KV, Haseeb ASMA, Masjuki HH, Fazal MA, Gupta M (2013) Corrosion of magnesium and aluminum in palm biodiesel: a comparative evaluation. Energy 57:478–483
9. Knothe G, Steidley KR (2018) The effect of metals and metal oxides on biodiesel oxidative stability from promotion to inhibition. Fuel Process Technol 177:75–80

Thermal Analysis of Multi Reflector Compound Parabolic Collector (MRCPC)

Shubhranshu Mishra, Tangellapalli Srinivas, Parmvir Singh, and Ajay Trehan

1 Introduction

According to an estimate, the solar energy that reaches the earth's surface from 20 days of sunshine is equivalent to the entire fossil fuel resources [1]. The conventional fuel combustion system gives more energy conversion efficiency than solar collectors. Different types of solar collectors are available that converts solar energy into useful thermal energy. Flat plate collector (FPC) is a known name used for so long. The efficiency of the same is 10–20%, which is far less than the conventional fossil fuel system, giving almost 90–95% efficiency [2, 3]. There is a need for a non-conventional energy system that can achieve higher efficiency than a conventional system. For the last 3–4 decades, research has been going on for the efficient use of solar energy. Concentrating collectors, such as compound parabolic collectors (CPC) and parabolic trough collectors (PTC), are developed and tested with different sizes and configurations. Though they are more efficient than the FPC, they still have limitations. CPC is suitable for medium-range temperatures with a low concentration ratio. PTC, however, is the most matured concentrating collector, still has some limitations [4]. First, the focal line of the collector is over the concentrator, which casts the shadow on the concentrator's surface. Second, high precision tracking is required for PTC to track the sun, as non-reflecting radiation is useless. Third, the receiver in the concentrator is installed above the concentrator, exposing it directly to the surroundings and wind and heat loss are higher in such conditions. To overcome the disadvantage and using the advantages of both PTC and CPC, a new solar collector is designed by the researchers [4, 5].

S. Mishra (✉) · T. Srinivas · P. Singh · A. Trehan
Department of Mechanical Engineering, Dr B R Ambedkar National Institute of Technology, Jalandhar, Punjab 144011, India
e-mail: shubhranshum.mt.20@nitj.ac.in

S. Doolla et al. (eds.), *Advances in Clean Energy and Sustainability*,
Green Energy and Technology, https://doi.org/10.1007/978-981-99-2279-6_9

1.1 Literature Review

Tao et al. [5] has numerically given the method of designing and working principle of a new trough solar collector. As per his findings, width of the collector is an important parameter in determining the performance of the solar collector. Riffat and Mayere [4] have performed experimental work to study the design and working principle for the desalination of water. At the operating temperature of 100 °C, they get up to 38% thermal efficiency. Natarajan and Srinivas [2] have conceptualized, developed and tested the MRCPC. They experimentally tested the MRCPC to minimize the tracking load error. Zheng et al. [6] use three different shaped receivers for the experimental test of heating of air with a multi-surface trough collector. Chen et al. [7] has compared the solar collector with both flat plate and cylindrical receiver.

1.2 Objective of the Work

In this paper, a multi reflector compound parabolic collector is studied and a mathematical model is developed to get the efficiency and the fluid outlet temperature from the receiver. The objective of the presented work is to find the optimum value of certain parameters based on the properties of working fluid.

2 Working Principle

Figure 1 shows the schematic design of the multi reflector compound parabolic collector. It consists of the 1—upper compound parabolic concentrator, 2—plane mirror for secondary reflection, 3—lower parabolic trough concentrator, 4—path showing parallel lights, 5—symmetric axis of the collector, 6—vacuum glass tube, 7—solar receiver circular tube, 8—connecting flange, which helps in connecting the upper and lower portion of the MRCPC.

The operating principle of the MRCPC is described as follows. The light ray, as shown by 4, is incident parallel to the symmetric axis enters the MRCPC. Surface 1 of the upper CPC reflects light 4 to the plane mirror, which is then reflected to the outer vacuum glass tube of the receiver and ultimately to the receiver. The vacuum glass tube lessens the heat loss in the receiver. The lower PTC, which is connected with the upper CPC by a flange, also reflects solar radiation coming straight to it. In this way, the receiver of MRCPC can accept the solar radiation falling directly to it and reflect solar radiation from upward and downward, enhancing the receiver's efficiency. All these solar radiations are absorbed in the receiver by the receiver's fluid (usually water, ethylene glycol, etc.). The heat from the receiver's fluid is extracted and utilized as per the requirement and the cycle continues as shown in Fig. 2.

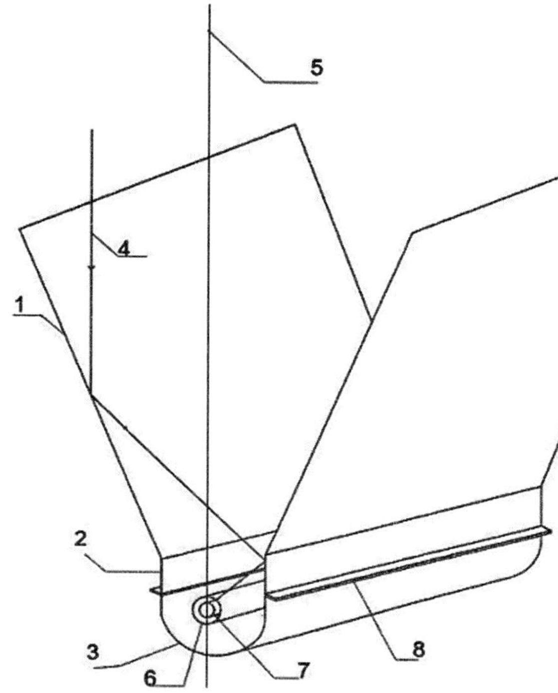

Fig. 1 Schematic design of multi reflector compound parabolic collector

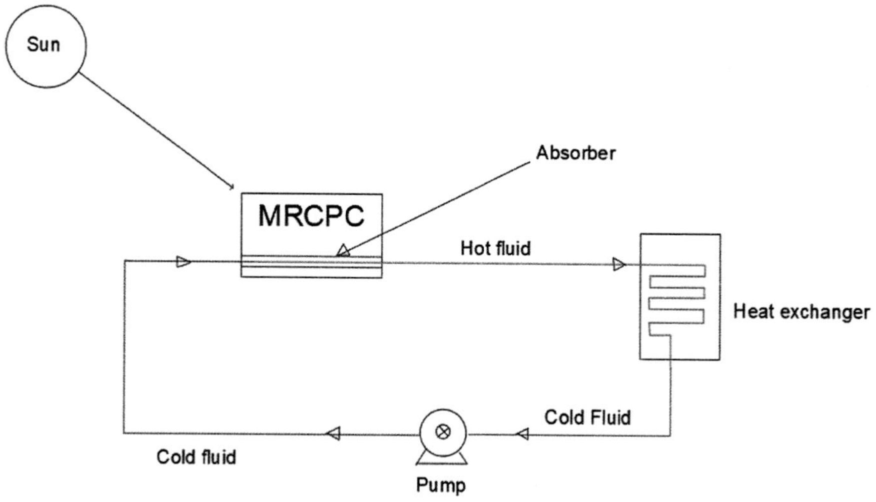

Fig. 2 Working of MRCPC

3 Methodology

See Fig. 3.

4 Numerical Investigation

Mathematical equations for formulation of solar geometry are taken from Sukhatme and Nayak [3] to get the values of certain parameters, such as, declination angle, zenith angle, local apparent time, etc. which is to be used to predict the solar radiation availability.

Empirical equations for predicting the solar radiation availability are taken from Sukhatme and Nayak [3] to get the parameters, such as hourly beam radiation and tilt factor which is useful in thermal analysis of the collector.

4.1 Mathematical Modeling for Thermal Performance Analysis of the Collector

$$S = I_b r_b \tau\alpha\left(\rho_{ref}\rho_{pm} + \frac{d_o}{w} + \rho_{sr}\frac{v - d_o}{w}\right) \quad (1)$$

$$C = \frac{w}{\pi d_o} \quad (2)$$

$$V = \frac{\dot{m}}{\frac{\pi}{4}d_i^2\rho} \quad (3)$$

$$Re = \frac{V d_i}{\upsilon} \quad (4)$$

$$Pr = \frac{C_p \upsilon\rho}{k} \quad (5)$$

$$Nu = 0.023\left(Re^{0.8}\right)\left(Pr^{0.4}\right) \quad (6)$$

$$h_f = \frac{Nu\,k}{d_i} \quad (7)$$

$$F' = \frac{1}{U_l\left(\frac{1}{U_l} + \frac{d_o}{d_i h_f}\right)} \quad (8)$$

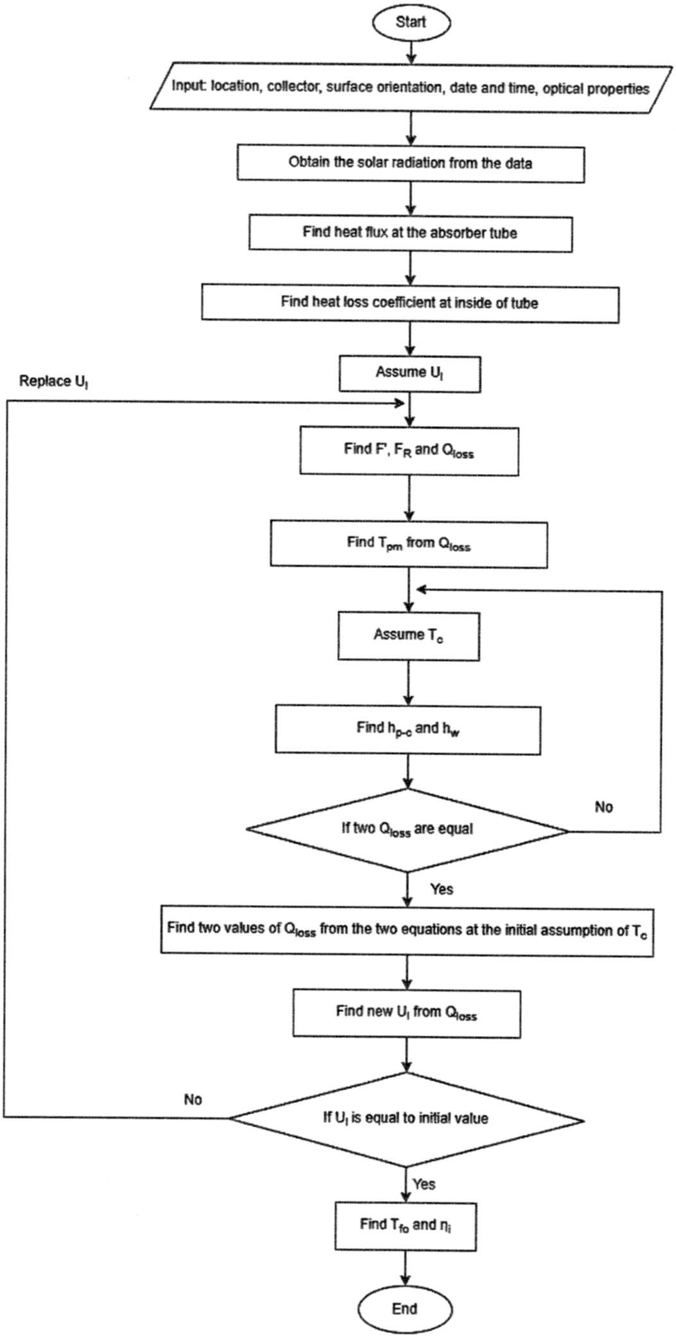

Fig. 3 Methodology flow chart

$$F_R = \frac{\dot{m}C_p}{\pi d_o U_l L}\left[1 - \exp\left(\frac{-F'}{\frac{\dot{m}C_p}{\pi d_o U_l L}}\right)\right] \tag{9}$$

$$q_u = F_R w\, L\left(S - \frac{U_l}{C}\right)\left(T_{fi} - T_{amb}\right) \tag{10}$$

$$q_l = wLS - qu = \pi d_o L U_l\left(T_{pm} - T_{amb}\right) \tag{11}$$

$$\frac{q_l}{L} = h_{p-c}\left(T_{pm} - T_{gc}\right)\pi d_o + \frac{\sigma \pi d_o\left(T_{pm}^4 - T_c^4\right)}{\left\{\frac{1}{\varepsilon_p} + \frac{d_o}{d_{ci}}\left(\frac{1}{\varepsilon_{gc}} - 1\right)\right\}} \tag{12}$$

$$\frac{q_l}{L} = h_w\left(T_{gc} - T_{amb}\right)\pi d_{co} + \sigma \pi d_{co}\varepsilon_{gc}\left(T_{gc}^4 - T_{sky}^4\right) \tag{13}$$

$$U_l = \frac{q_l}{\pi d_o\left(T_{pm} - T_{amb}\right)} \tag{14}$$

$$T_{fo} = \frac{q_u}{\dot{m}C_p} + T_{fi} \tag{15}$$

$$\eta = \frac{q_u}{I_b r_b w L} \tag{16}$$

5 Results and Discussion

Simulation is done using MATLAB software, with the meteorological data of Jalandhar, Punjab, India, on 16 March 2022 from 6 a.m. IST to 6 p.m. IST. The fluid used in the absorber tube is water. The dimension and other parameters of the system are as follows:

L = 1.5 m, w = 0.75 m, v = 0.23 m, d_{co} = 0.058 m, d_{ci} = 0.054 m, d_o = 0.048 m, d_i = 0.042 m, ρ_{ref} = 0.9, ρ_{pm} = 0.8, ρ_{sr} = 0.9, τ_{gc} = 0.9, α_{ab} = 0.95, ε_{gc} = 0.85, ε_{ab} = 0.95.

Conditions to get the optimum value is based on maximum output temperature (≤ 100 °C for water as HTF) with a decent efficiency. Initially 7 different flow rates (viz. 5, 10, 20, 40, 60, 75 and 100 L per hour) has been considered. Based on the above condition and the result obtained in Figs. 4 and 5, 10 L per hour is taken as optimum value of flow rate for present study.

Now using the optimum flow rate as base, length has been varied from 1 to 7 m, keeping other parameters same as previously mentioned. As per the result obtained as shown in Figs. 6 and 7, optimum length has been taken as 2.4 m.

Again, using the previously obtained optimum values, outer dimeter of the absorber glass tube is varied keeping the other diameters in a fixed proportion. For example, if d_{co} is taken as 0.050 m, d_{ci}, d_o and d_i is taken as 0.048 m, 0.044 m and

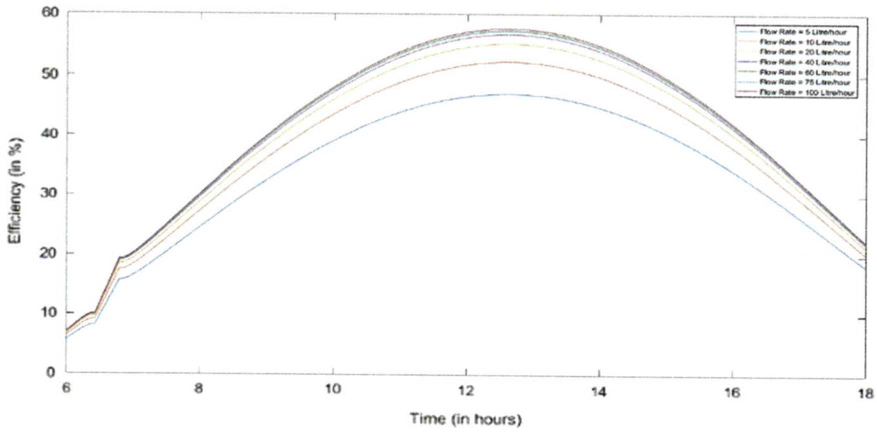

Fig. 4 Efficiency versus time graph for different flow rates

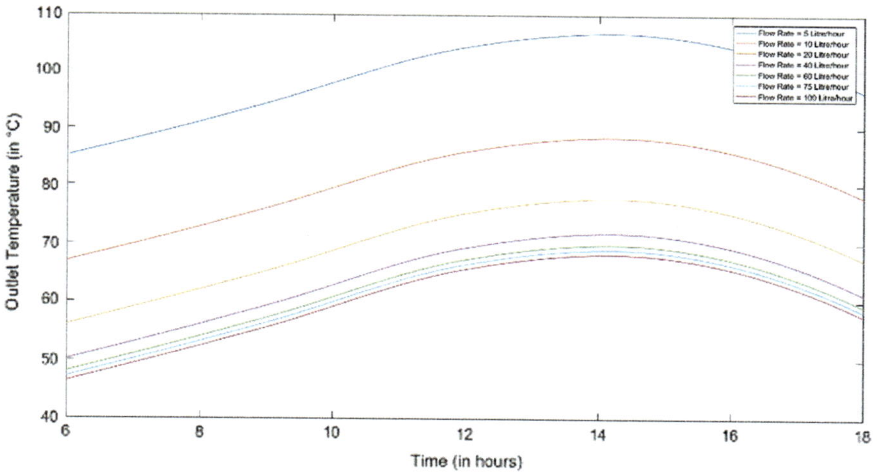

Fig. 5 Outlet temperature versus time graph for different flow rates

0.042 m, respectively. Based on the results obtained in Figs. 8 and 9, the optimum value for d_{co}, d_{ci}, d_o and d_i are taken as 0.060 m, 0.058 m, 0.056 m and 0.052 m respectively (Table 1).

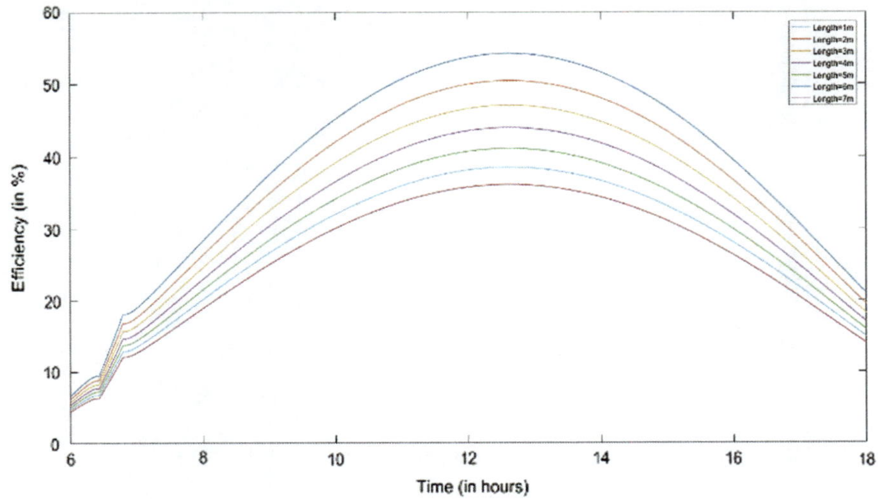

Fig. 6 Efficiency versus time graph for different collector's lengths

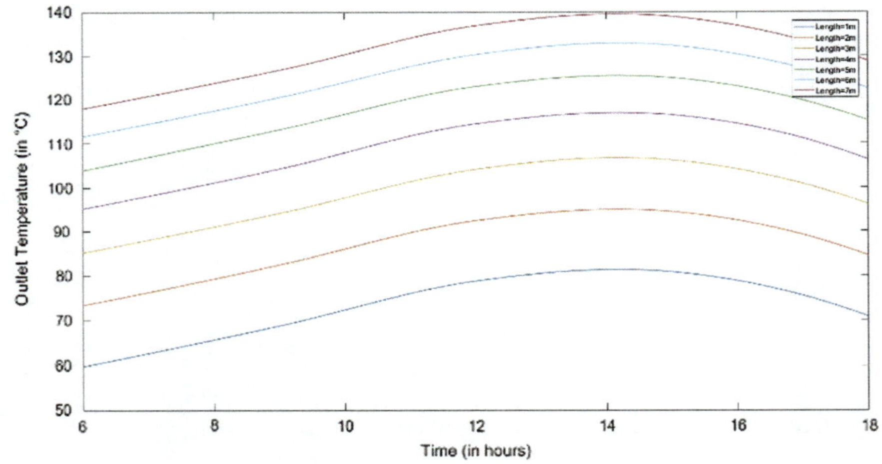

Fig. 7 Outlet temperature versus time graph for different collector's lengths

6 Conclusion

The study was carried out for multi reflection compound parabolic collector. Simulation results are developed to find the optimum values different parameters which are flow rate, length of collector and tube diameter using water as the heat transfer fluid. The maximum outlet temperature below 100 °C was the main criteria for optimum value consideration of collector's parameters. Obtained results shows that 10 LPH flow rate with 2.4 m collector's length and 0.060 m outer tube diameter are

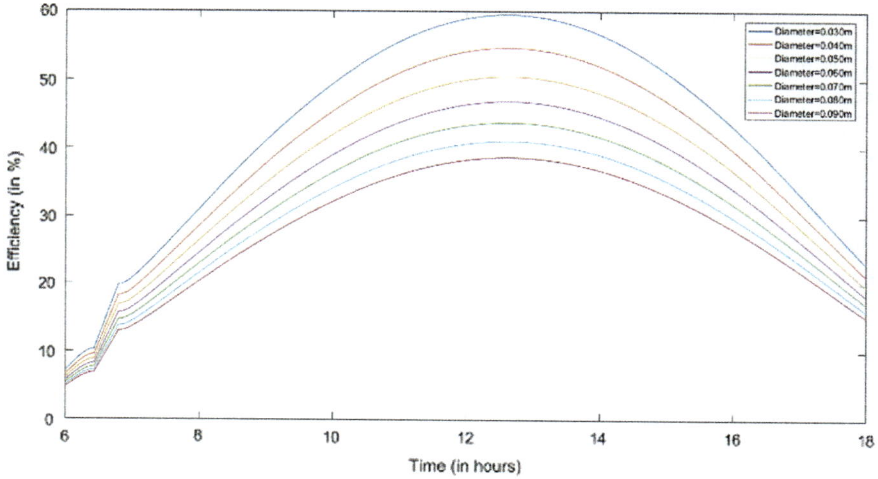

Fig. 8 Efficiency versus time graph for different tube diameter

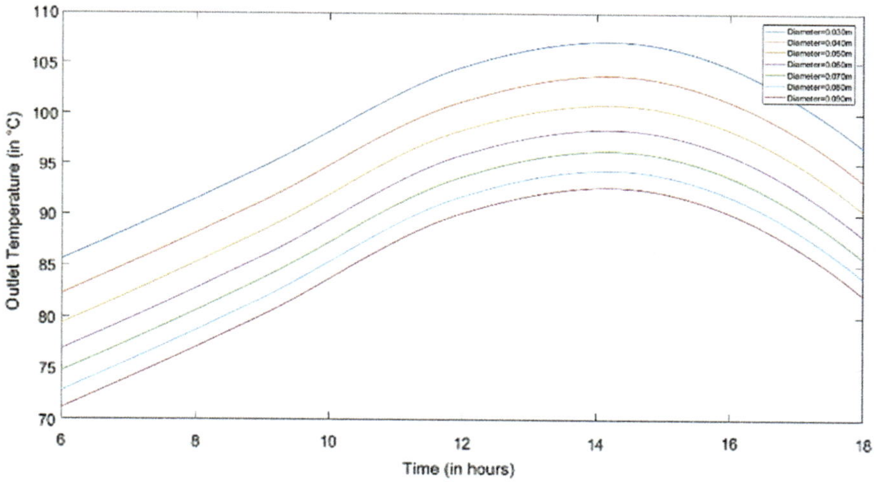

Fig. 9 Outlet temperature versus time graph for different tube diameter

Table 1 Representation of optimum value and respective maximum outlet temperature and efficiency attained

Parameter	Optimum value	Max. outlet temperature (in °C)	Max. efficiency (in %)
Flow rate	10 LPH	88.40	52.26
Length	2.4 m	99.85	49.02
Tube outer diameter	0.060 m	98.40	46.94

the optimum value based on the present condition. Also, the outcomes also show that with an increase in flow rate, efficiency increases and tube outlet temperature decreases, with an increase in collector's length, efficiency decreases and outlet temperature increases and with an increase in absorber tube diameter both efficiency and fluid outlet temperature decreases. The obtained results are in order as previously done research.

Acknowledgements Acknowledge the project grant of the Department of Science and Technology (DST), New Delhi, India (DST/TMD/CERI/RES/2020/38(G)). under Applied Research Solar Stream.

Nomenclature

S	Heat flux at absorber tube
I_b	Hourly beam radiation
r_b	Tilt factor
ρ_{ref}	Reflectivity of upper parabolic collector
ρ_{pm}	Reflectivity of plane mirror
ρ_{sr}	Reflectivity of bottom secondary reflector
τ	Transmissivity of glass cover
d_i	Absorber tube inner diameter
d_o	Absorber tube outer diameter
d_{co}	Glass cover outer diameter
d_{ci}	Glass cover inner diameter
w	Upper aperture width
v	Inner aperture width
L	Length of absorber
C	Concentration ratio
V	Fluid velocity inside the tube
$\dot{m}$	Mass flow rate
ρ	Density of fluid
Re	Reynold's number
Pr	Prandtl number
Nu	Nusselt number
h_f	Heat transfer coefficient inside the absorber tube
ε_p	Emissivity of absorber tube
ε_{gc}	Emissivity of glass cover
q_l	Heat loss
σ	Stefan Boltzmann constant
F'	Collector efficiency factor
F_R	Heat removal factor
q_u	Useful heat gain
υ	Kinematic viscosity

k	Thermal conductivity
U_l	Overall loss coefficient
C_p	Specific heat of fluid
h_{p-c}	Heat transfer coefficient between absorber and glass cover
T_{fi}	Fluid inlet temperature
T_{fo}	Fluid outlet temperature
T_{pm}	Average temperature of absorber tube
T_{gc}	Glass cover temperature
η	Efficiency

References

1. Satyamurty VV (2014) Solar energy technology
2. Natarajan M, Srinivas T (2017) Experimental and simulation studies on a novel gravity based passive tracking system for a linear solar concentrating collector. Renew Energy 105:312–323
3. Sukhatme SP, Nayak JK (2017) Solar energy. McGraw-Hill Education
4. Riffat S, Mayere A (2013) Performance evaluation of v-trough solar concentrator for water desalination applications. Appl Therm Eng 50(1):234–244
5. Tao T, Hongfei Z, Kaiyan H, Mayere A (2011) A new trough solar concentrator and its performance analysis. Sol Energy 85(1):198–207
6. Zheng H, Tao T, Ma M, Kang H, Su Y (2012) Experimental test of a novel multi-surface trough solar concentrator for air heating. Energy Convers Manage 63:123–129
7. Chen M, Li Z, Zheng H, Meng H, Dai J (2015) Performance comparision of a new-type trough solar concentrator thermal system in different installations. Therm Sci 19(suppl. 2):535–545
8. Lei D, Li Q, Wang Z, Li J, Li J (2013) An experimental study of thermal characterization of parabolic trough receivers. Energy Convers Manage 69:107–115
9. Lof GOG et al (1963) Residential heating with solar heated air—the Colorado house. ASHRAE, p 77

Capture and Characterization of Particulates from a Single-Cylinder Diesel Engine Fuelled with Refined Tire Pyrolysis Oil

Akhil Mohan and Vasudeva Madav

1 Introduction

The annual generation of scrap tires is calculated to be 1.5 billion, which results in the accumulation of scrap tires in the form of a heap in landfills [1]. In this scenario, the oil suppliers are ramping up their investment to produce high-quality greener fuels. Tire pile often provides a breeding site for mosquitoes and rodents. There are various methods to address the above issues effectively by rethreading, sports surfacing, furnishing playgrounds, rubber-modified asphalt, civil engineering application, combustion, and pyrolysis. Pyrolysis is found to be one of the interesting options to handle scrap tires effectively by transforming them into value-added products. Several researchers working in the area of tire pyrolysis liquid are trying to recover some value-added products like activated carbon, pyrolysis liquid, and pyro gas. Pyrolysis is a thermochemical conversion process by which the feedstock is subjected to an elevated temperature (400–800 °C) in the absence or presence of oxygen. Tire pyrolysis liquid is a dark black colored liquid with an intense smell, consisting of a complex mixture of aliphatic, aromatic, and polar compounds with a high heating value of 40–44 MJ/kg and found to be comparable with diesel. Studies are shown that crude tire pyrolysis oil (CTPO) can be effectively utilized as an alternative fuel for an internal combustion engine due to its similarities in chemical composition and high energy content in comparison with diesel. Murugan et al. [2] reported that usage of 10–50% of CTPO with diesel fuel in a diesel engine without any modifications results in an increase in brake thermal efficiency (BTE) at constant engine speed but there is a significant increase in the smoke and pollutants to the

A. Mohan (✉)
Department of Energy Science and Engineering, IIT Bombay, Mumbai 400076, India
e-mail: akhilmohan@iitb.ac.in

V. Madav
Department of Mechanical Engineering, NITK, Mangalore 575025, India

S. Doolla et al. (eds.), *Advances in Clean Energy and Sustainability*,
Green Energy and Technology, https://doi.org/10.1007/978-981-99-2279-6_10

environment. This may be due to the long ignition delay and high aromatic composition of CTPO. Then studies have been conducted using CTPO in direct-injected diesel engines produced from the pyrolysis of scrap tires using calcium hydroxide ($Ca(OH)_2$) as a catalyst reported by Ilkilic and Aydin [3]. They observed that the utilization of tire fuel blended with diesel increased the emissions (carbon monoxide, hydrocarbon, and sulfur dioxide) and smoke opacity. The authors attributed the significant rise in emission is due to poor atomization and low cetane number. The high density of fuel caused rich combustion in the combustion chamber, which results in higher smoke opacity of tire fuel blends compared to diesel fuel. Besides, they had also optimized that the utilization of 5% of $Ca(OH)_2$ reduces sulfur content by 35%. They pointed out that 5–35% of tire oil with diesel fuel can be utilized as an alternative fuel for direct-injected diesel engines without any further modifications. In another study, Frigo et al. [4] studied the performance, combustion, emissions, and genotoxicity studies of CTPO from an innovative pilot-scale pyrolyzer (twin-screw extruder with decreasing section facilitates mechanical compression of materials). The performance studies (torque, mechanical efficiency, power) revealed that 20% of tire pyrolysis liquid can be effectively utilized in diesel engines without any engine modifications. However, 20–40% needs modification to compensate for the ignition delay. The usage of a high concentration of CTPO attributes to the high sulfur content and low cetane value of TPO compared to diesel. In addition, they also pointed out that there is no mechanical inconvenience during engine operation and no significant carbon deposits were found in the piston or cylinder surface. Similarly, Tudu et al. [5] observed that the utilization of 10–40% of light fraction pyrolysis oil caused higher emissions (HC, NO, CO) and smoke compared to diesel fuel. This is ascribed due to higher viscosity, high density, poor volatility, and low heating value. According to Vihar et al. [6], the study was conducted in a non-intercooled turbocharged compression ignition multi-cylinder engine using 100% TPO as a fuel source. They observed that CTPO can be fully utilized in the internal combustion engine and found that diesel-like stable combustion was observed without any external modifications. However, the oxides of nitrogen and sulfur are increased with the rise in load, which is due to the high amount of nitrogen and sulfur in the crude tire pyrolysis oil compared to diesel. Kumaravel et al. [7] discussed an overview of the literature that explains an exhaustive idea about the pyrolysis mechanism for designing various types of reactors, performance, and combustion and emission analysis of CTPO in a diesel engine. The review suggests that the CTPO blend of up to 35% can be utilized directly in the diesel engine without any modification but the 50–100% of CTPO blend with diesel decreases the engine performance and increases the emission due to the high aromatic content in CTPO. In short, CTPO can be utilized fully in an internal combustion engine with a reduction in aromatic compounds and viscosity. Sahir et al. [8] investigated the performance and emission characteristics of tire pyrolytic oil using a common rail direct injection diesel engine (CRDI) by varying the CTPO blend with diesel in the ratio of 0 to 50%. They found that the brake thermal efficiency increases with improvement in the blend ratio of tire pyrolysis oil with diesel. In addition, they optimized that 30% crude tire pyrolysis oil with diesel (CTPO 30) exhibits better performance in terms of emission and performance compared to diesel fuel due to higher oxygen content, resulting in lower brake-specific fuel consumption. Another study by Bodisco et al. [9], the study was conducted on a modern commercial light-duty diesel vehicle to measure nitrous oxide emission using blends of tire oil and diesel. They found that the utilization of tire oil blended diesel fuel does not significantly increase nitrous oxide emission in comparison with neat diesel. Similarly, Verma et al. [10] discussed the potential of oil derived from scrap tires as a fuel for internal combustion engines. The majority of studies found that emissions (CO, CO_2, and NO_x) increased with tire-derived fuel. This is ascribed due to the high density and aromatic content of oil derived from scrap tires. Then Mohan et al. [11] conducted extensive studies on the characterization and up-gradation of CTPO from a 10-ton rotating autoclave reactor. They found an innovative up-gradation strategy for CTPO to remove polar acidic fractions using silica gel as an adsorbing agent and petroleum ether as a diluent. The major limitation behind the utilization of CTPO in an internal combustion engine is its low cetane value and high sulfur content. The authors also reduced the sulfur content by 34% and found a cetane number of upgraded tire pyrolysis oil (St-TPO) as 40 (diesel range). To date, no studies have been conducted on the utilization of St-TPO blended fuels in a diesel engine. Most of the studies on tire pyrolysis oil in the scientific literature are focused on the performance, and combustion emission analysis in a conventional diesel engine. In the present study, we have found a novel cost-effective preferential adsorption strategy to upgrade crude tire pyrolysis oil for application in internal combustion engines as an alternate fuel. The study also comprises an extensive investigation of the capture and characterization of particulates from a diesel engine fueled with StTPO as fuel and a comparison with neat diesel. These kinds of studies are not yet reported in the open literature for the analysis of particulate deposition in the piston crown using upgraded tire pyrolysis as a fuel in a diesel engine. The present study focuses on the morphology, chemical content, and structure of carbon deposits formed as a result of the combustion of upgraded tire pyrolysis oil on a piston crown. The study also concentrates on the lubricant oil analysis and understanding of the deposition mechanism of StTPO-derived deposits.

2 Materials and Methods

CTPO (1 L) was mixed with 3 L petroleum ether and the solution was decanted into a 6 L beaker. 1 kg of silica gel was added and the suspension was stirred at 500 rpm by an overhead stirrer for one hour at room temperature. After one hour, the supernatant was decanted and filtered under vacuum through a microfiltration system (pore size 0.2 μm, thickness 140–178 μm, and nylon 6, 6). Then the petroleum ether was evaporated using a rotary evaporator at 45 °C under reduced pressure. The used silica gel was washed with methanol to remove the polar compounds and the used methanol was recycled by distillation. Figure 1 shows the experimental facility for batch scale-up-gradation strategy to upgrade CTPO. Diesel was procured from Indian Oil Corporation Limited. Silica gel (60–120 mesh) and petroleum ether

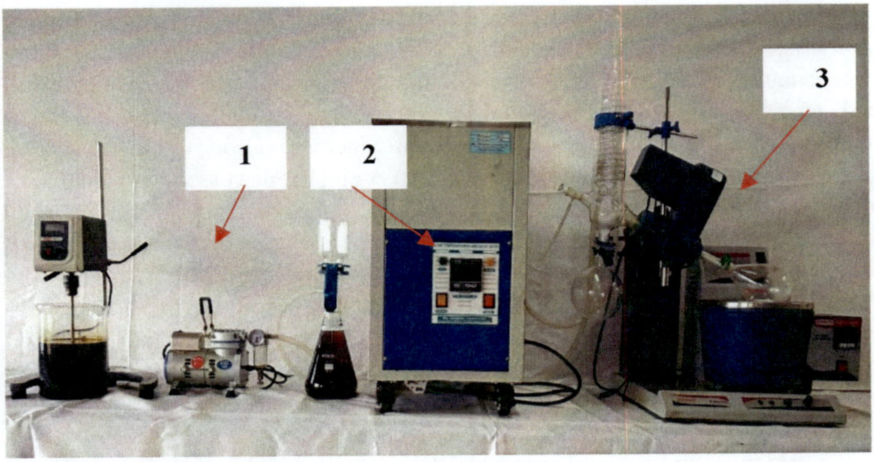

Fig. 1 Schematic of a bench-scale experimental facility for refining CTPO (1—Mixing of CTPO, adsorbent, and solvent, 2—Filtration system, and 3—Evaporation system)

(60–80 °C) was purchased from Loba Chemie Private Limited. The up-gradation process was already discussed in our previous publications. Inductively Coupled Plasma-Atomic Emission Spectroscopy (ICP-AES) was conducted to find out various types of elements in lubricating oil before and after engine experiments. Viscosity analysis was performed using a redwood viscometer and a flash and fire point test was conducted using a Pensky-martens closed cup flash point tester. Microscopic examination of piston deposits was observed using scanning electron microscopy.

The fuel was mixed in a 6 L beaker using a laboratory stirrer operating at 500 rpm, before testing the blended fuels in a single-cylinder diesel engine (make: Kirloskar, model: TV1) for the homogeneity and stability of the samples. The engine set-up consists of a naturally aspirated, direct-injection water-cooled diesel engine system with a rated power of 3.5 kW at 1500 rpm, coupled with an eddy current dynamometer for loading. The lubrication oil and exhaust gas temperature were recorded using a K-type (Cr–Al) thermocouple. Emissions from all fuel samples were measured using a sample probe connected with five gas portable emission analyzers (model: AVL Di-gas 444). Engine experiments were conducted thrice to check repeatability.

3 Results and Discussion

3.1 *Batch-Scale Upgrading Experiments*

The crude tire pyrolysis oil obtained from a rotating autoclave reactor is refined using silica gel as an adsorbent and petroleum ether as a diluent, which is already discussed in our previous publications [12].

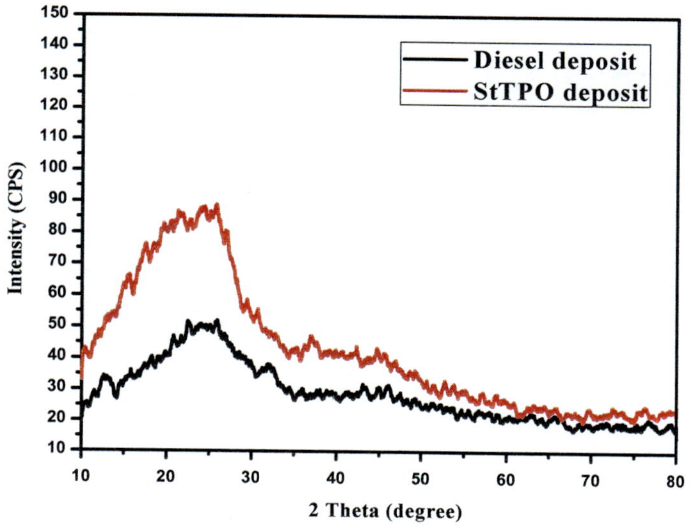

Fig. 2 XRD analysis of particulate deposit from diesel and StTPO

3.2 XRD Analysis

The deposited carbon in the piston crown was analyzed to find out the formation mechanism of carbon deposits. Figure 2 shows the XRD analysis of deposits collected after the engine run using diesel and StTPO. The diffraction pattern of the deposit showed a broad peak at 25°, which can be amorphous carbon and the finding are consistent with the result published by La Rocca et al. [13]. Based on the data obtained from XRD analysis revealed that the carbon deposits from StTPO showed more carbon deposits than diesel.

3.3 SEM/EDAX Analysis

SEM (Scanning electron microscopy) is used to study the microstructure of samples whereas the EDAX helps to identify the various elements present in the samples to identify the deposition mechanism. Figure 3 shows the microstructure of diesel and StTPO deposits. SEM analysis revealed that the particles are randomly dispersed in the carbon matrix in the case of the diesel sample whereas the particles are found to be agglomerated in the case of StTPO. In short, the microstructure of StTPO looks much denser compared to diesel. The high agglomerated mixture in StTPO is due to the mixing of carbon particles with unburned fuel. Figure 5 shows the elemental analysis of diesel and StTPO deposits. Carbon and phosphorous content in the diesel and StTPO deposits are comparatively the same. However, other elements like oxygen, magnesium, aluminum, silicon, sulfur, calcium, iron, and zinc showed a significant

Fig. 3 SEM image of **a** diesel deposits and **b** StTPO deposits

variation. The appearance of sulfur and zinc is due to the combustion of lubrication additives in the combustion chamber. Oxygen content in the piston crown deposits is increased by 4% due to the high oxygenate in StTPO deposits due to insufficient combustion in the combustion chamber. The sulfur content in the StTPO deposits is slightly higher than the diesel deposits due to the high sulfur content in StTPO than in diesel.

3.4 *TEM Analysis*

The TEM image of StTPO looks like a filament-like structure linked by a chain of spherical-shaped particles whereas the diesel particulates look like spheroidal-shaped particles (Fig. 4). The carbon deposits from StTPO form agglomerated fractions and are deposited on the surface of the piston crown and cylinder walls after polymerization and condensation reactions of soot particles. The average particle diameter from diesel and StTPO-derived soot particles was found to be 108.06 nm and 123.78 nm respectively. Soot particles from StTPO are found to be denser than diesel due to the presence of various distillation range fractions in StTPO in comparison with diesel fuel.

The TEM image of StTPO looks like a filament-like structure linked by a chain of spherical-shaped particles whereas the diesel particulates look like spheroidal-shaped particles. The carbon deposits from StTPO form agglomerated fractions and are deposited on the surface of the piston crown and cylinder walls after polymerization and condensation reactions of soot particles. The average particle diameter from diesel and StTPO-derived soot particles was observed to be 108.06 nm and 123.78 nm respectively. Soot particles from StTPO are observed to be denser than diesel due to various distillation range compounds in StTPO in comparison with diesel (Fig. 5).

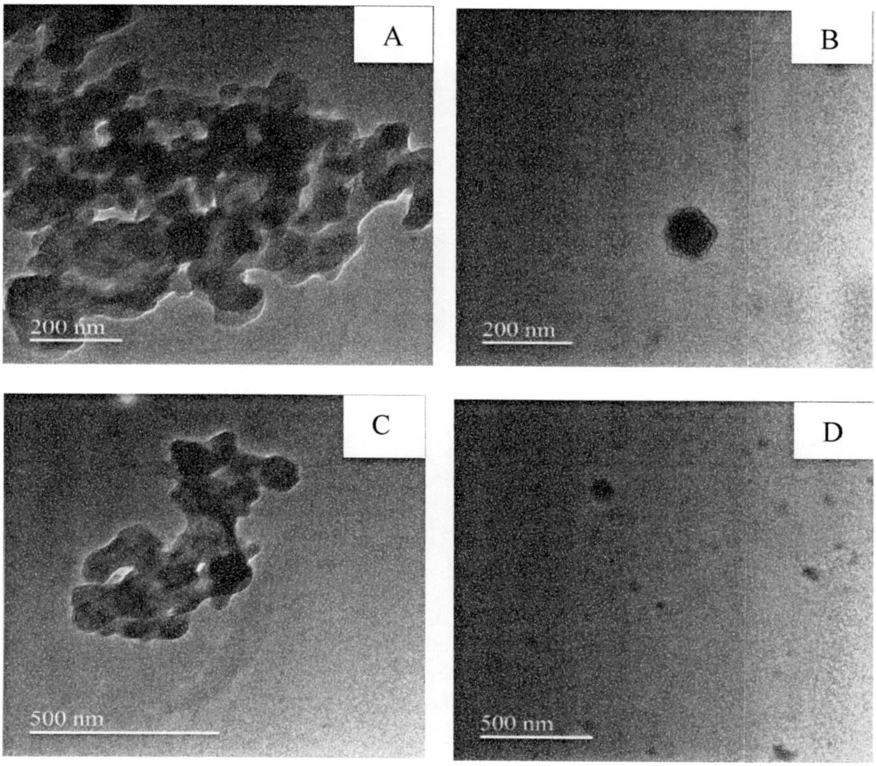

Fig. 4 **a** TEM image of StTPO, **b** diesel

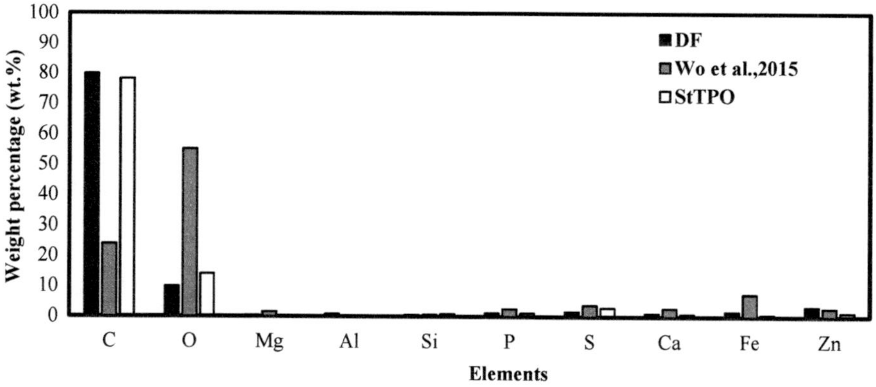

Fig. 5 Elemental analysis of lubricating oil

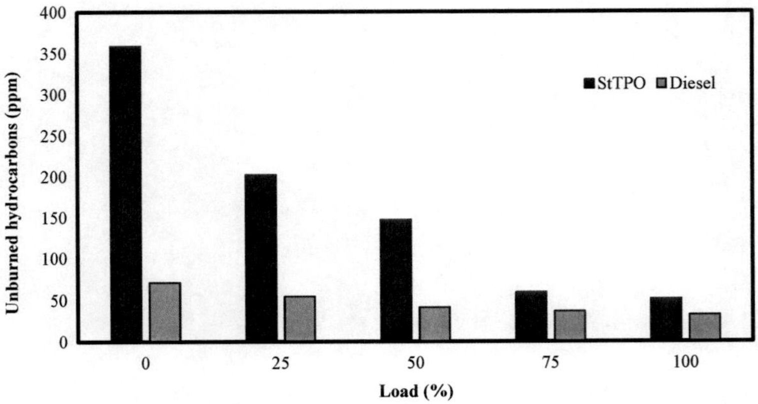

Fig. 6 Variation of unburned hydrocarbons with load

3.5 Discussion on Carbon Deposition from StTPO and Comparison with Diesel

Carbon deposition is formed due to aromatic hydrocarbons (including polynuclear aromatic hydrocarbons) in upgraded tire oil than in diesel [2]. The cetane index of upgraded tire oil and diesel was found to be 40 and 50 respectively. During the combustion of upgraded tire pyrolysis oil in a single-cylinder diesel engine, poly-aromatic compounds such as naphthalene, fluorene, phenanthrene, anthracene, benzopyrene, benzo anthracene, pyrene, chrysene, biphenyl find difficult in dissociating due to high bond dissociation energy and result in the non-homogeneous air–fuel mixture in the combustion chamber [12]. Aromatic compounds are found to be comparatively higher in upgraded tire pyrolysis oil than in diesel. Moreover, the variation of unburned hydrocarbons of StTPO and diesel from the tailpipe of a diesel engine from a gas analyzer with engine load is plotted in a bar diagram as shown in the following bar diagram. The hydrocarbon emission was found to decrease with engine load (Fig. 6). As the load increases, the high temperature and turbulence increase air–fuel mixing and combustion efficiency [14].

3.6 Lubricant Oil Analysis

Lubricating oil contamination can be occurred due to premature decay, dilution into piston and cylinder surfaces leads to cracks and seizure on the surfaces and indication of incomplete combustion. Lubricating oil was collected and sampled at end of the engine test (ca. 100 h). An elemental analysis of fresh and used lubricating oil is reported in Table 1. Used lubricating oil is characterized by a drastic reduction in viscosity and flammability due to fuel leaking into the oil sump and dilution of

Table 1 New and used oil analysis by ICP-AES

Elements (ppm)	New oil	Used oil (StTPO)	Used oil (Diesel)
Chromium	0.341	2.307	0.492
Copper	BDL	1.977	1.722
Iron	BDL	BDL	BDL
Nickel	0.511	BDL	BDL
Zinc	841.34	772.81	804.13
Sulfur	14,522.84	13,264.07	13,062.25

BDL Below detectable limit

lubrication oil with test fuels. Chromium and copper content in the lubricating oil mainly accounts for the additives used in the cylinder liners, piston rings, shaft, antifriction bearing, surface coating, etc. As the usage of StTPO in the engine for 100 h causes the premature degradation of the cylinder and piston rings resulting in an increase in the chromium and copper in used oil in comparison with new oil. The iron content in the lubricating oil mainly accounts for the wear in engine cylinders and piston rings. Interestingly, the used and new oil accounts for no iron content signifies low wear in engine components by the usage of StTPO as an alternate fuel in the engine. Sulfur and zinc are mainly used as the main additive elements used in the lubrication industry. The sulfur and zinc contents in fresh lubricating oil seem to be higher in the comparison with used oil due to the dilution of StTPO and diesel with fresh oil.

Figure 7 shows the variation of lubricating oil temperature with brake power. The lubricating oil temperature of upgraded tire pyrolysis oil is found to be close to diesel at peak brake power compared to diesel. The slight reduction in lubricating oil temperature is due to the presence of oxygen-containing fractions in StTPO in comparison with diesel. During the processing of tires in the industry, the compounds such as zinc oxide, calcium oxide, and stearic acid are used to improve tire properties, which increases oxygen content in StTPO (Table 2).

4 Conclusions

The up-gradation of 1 L crude tire pyrolysis oil yields a 95% yield of refined tire pyrolysis oil with 5% losses. The results from engine experiments revealed that StTPO can be fully applied in single-cylinder diesel engines at all load conditions without any further modifications. The carbon deposits from StTPO were found to be greater than diesel, which is evidenced by SEM and XRD analysis. The high carbon deposition from StTPO is attributed to high amounts of oxygen-containing functional groups and aromatic fractions in StTPO than in diesel.

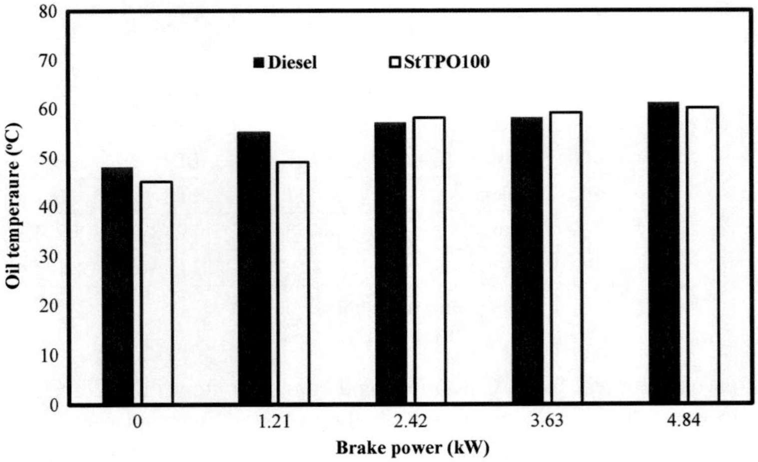

Fig. 7 Variation of lubricating oil temperature with brake power

Table 2 Properties of lubricating oil after an engine run

Properties	New oil	New oil (Ref.)	Used oil (Diesel)	Used oil (Ref.)	Used oil (StTPO)	Methods
Viscosity (cSt)	383.55	104.7	178.75	68.9	234.84	ASTM D455
Flashpoint (°C)	226	220	224	148	216	ASTM D93
Fire point (°C)	246	–	240	–	236	ASTM D92

Acknowledgements The authors would like to thank the staff of the sophisticated analytical laboratory at IIT Bombay, Powai for the data analysis. The authors also like to thank NITK for microscopic examinations such as XRD and SEM. Akhil Mohan thanks IIT Bombay for providing infrastructure and PDF scholarship support.

References

1. Williams PT (2013) Pyrolysis of tires—a review. Waste Manage 33(8):1714–1728
2. Murugan S, Ramaswamy MC, Nagarajan G (2008) The use of tire pyrolysis oil in diesel engine. Waste Manage 28:2743–2749
3. Ilkilic C, Aydin H (2011) Fuel production from waste vehicle tire by catalytic pyrolysis and application in a diesel engine. Fuel Process Technol 92:1129–1135
4. Frigo S, Seggiani S, Puccini M, Vitolo S (2014) Liquid fuel production from waste tire pyrolysis and application in diesel engine. Fuel 116:399–408
5. Tudu K, Murugan S, Patel SK (2014) Light oil fraction from tire pyrolysis plant—an option for energy use. Energy Procedia 54:615–626

6. Vihar R, Seljak T, Opresnik SR, Katrasnik T (2015) Combustion characteristics of tire pyrolysis in turbocharged compression ignition engine. Fuel 150:226–235
7. Kumaravel ST, Murugesan A, Kumaravel A (2016) Tire pyrolysis oil as an alternative fuel for diesel engine—a review. Renew Sustain Energy Rev 60:1678–1685
8. Sahir VK, Jawahar CP, Vinod V, Suresh PR (2018) J King Saud Univ Eng Sci (Article in press)
9. Bodisco TA, Rahman SMA, Hossain FM, Brown RJ (2019) On-road NO_x emission of a modern commercial light-duty diesel vehicle using tire oil and diesel. Energy Rep 5:349–356
10. Verma P, Zare A, Jafari M, Bodisco TA, Rainey T, Ristovski ZD, Brown RJ (2019) Diesel engine performance and emissions with fuel derived from waste tires. Sci Rep 8:2457
11. Mohan A, Dutta S, Madav V (2019) Characterization and up-gradation of crude tire pyrolysis oil obtained from a rotating autoclave reactor. Fuel 250:339–351
12. Mohan A, Dutta S, Balusamy S, Madav V (2021) Liquid fuel from waste tires: novel refining, advanced characterization, and utilization in engine with ethyl levulinate as an additive. RSC Adv 11:9807–9826
13. La Rocca A, Di Liberto G, Shayler P, Parmenter C (2014) A novel diagnostics tool for measuring soot agglomerates size distribution in used automotive lubricating oil. SAE Int J Fuel Lubricants 7(1):301–306
14. Nanthagopal K, Ashok B, Saravanan B, Pathy MR, Sahil G, Ramesh A, Nahi MN, Rasul MG (2019) Study of decanol and Calophyllum Inophyllum biodiesel as ternary blend in compression ignition engine. Fuel 239:862–873
15. Vanderwal RL, Bryg VM, Hays MD (2010) Fingerprinting soot (towards source identification): physical structure and chemical composition. J Aerosol Sci 41(1):108–117
16. Lapuerta M, Fernandez JR, Agudelo JR (2009) Diesel particulate emission from used cooking oil biodiesel. Biores Technol 99(4):731–740
17. Jain A. Report on a compendium of technologies for the recovery of materials or energy from the end-of-life tires. Regional Resource Center for Asia and the Pacific
18. Islam MN (2016) Improvement of waste tire pyrolysis oil and performance test with diesel in compression ignition engine. J Renew Energy 2016, Article ID 5137247
19. Dimitriadis A, Natsios L, Dimaratos A, Katsaunis A, Samaras Z, Bezegianni S, Lehto K (2018) Evaluation of hydrotreated vegetable oil and its effects on passenger car diesel engine. Front Mech Eng 4
20. Zhang G, Chen F, Zhang Y, Zhao L, Chen J, Cao L, Gao J, Xu C (2021) Properties and utilization of waste tire oil: minireview. Fuel Process Technol 211:106–582
21. Banar M, Akyildiz V, Ozkan A, Cokaygil Z, Only O (2012) Characterization of pyrolytic oil obtained from pyrolysis of tire-derived fuel. Energy Convers Manage 62:22–30
22. Diesel fuel technical review. Chevron: https://www.chevron.com/-/media/chevron/operations/documents/diesel-fuel-tech-review.pdf
23. Murugan S, Ramaswamy MC, Nagarajan G (2008) Performance, emission, and combustion of direct-injected diesel engine using distilled tire pyrolysis oil-diesel blend. Fuel Process Technol 89:152–159
24. Murugan S, Ramaswamy MC, Nagarajan G (2008) The use of tire pyrolysis oil in diesel engine. Waste Manag 28:2743–2749
25. Umeki ER, Oliveira CF, Torres RB, Santos RG (2016) Physico-chemical properties of fuel blends composed of diesel and tire pyrolysis oil. Fuel 185:236–242

Estimation of Heat Generation and Thermal Behavior of Cylindrical Lithium-Ion Battery Under Natural Convection

Dinesh Kumar Sharma and Aneesh Prabhakar

1 Introduction

LiBs have become the most preferred battery storage for EVs due to high energy density, high specific power [1], long cycling life and, a relatively low self-discharge rate [2]. However, it is widely acknowledged that the LiBs generate heat during the charging and discharging cycles mainly due to the reversible and irreversible processes. The reversible heat generation is attributed to the entropy change for the electrochemical reaction. On the other hand, irreversible heat generation attributed to active polarization, Ohmic losses, and side reactions [3]. The heat generated in LiBs during discharging cycle, particularly at high discharge rates and high T_{amb}, elevates the temperature of LiBs [4]. The elevated battery temperature degrades the performance of LiBs and decreases its power delivering capacity [5]. Hence, a battery thermal management system (BTMS) is imperative for regulating battery cell/module temperature in the range from 10 to 45 °C [6] for the normal performance, and from 15 to 35 °C for the optimum performance [7]. Hence, estimation of heat generation becomes a necessity for quantifying the heat removal requirement from LiBs battery cell/module during charging and discharging for designing an optimum size BTMS [8].

Recently, various mechanistic mathematical models have been presented to predict the transient electrochemical and thermal behavior of cylindrical LiBs [9, 10]. These models used transient equations for estimation of species, charge, and energy in LiBs. In the detailed models governing equations transformed into the partial differential equations for solving the local transport, electrochemistry and heat generation in LiB [8, 11]. According to the literature, about 40 parameters are needed to establish mechanistic mathematical model [12]. Hence, it becomes

D. K. Sharma (✉) · A. Prabhakar
Centre for Energy and Environment, Malaviya National Institute of Technology Jaipur, Jaipur, Rajasthan 302017, India
e-mail: 2019ren9015@mnit.ac.in

S. Doolla et al. (eds.), *Advances in Clean Energy and Sustainability*,
Green Energy and Technology, https://doi.org/10.1007/978-981-99-2279-6_11

cumbersome to implement this model in real-time applications with limited knowledge about the internal chemistry of LiBs. The general trends in battery modeling were found as, a detailed model for rectilinear geometrical LiBs and a simplified model for cylindrical LiBs, because modeling cylindrical LiBs are more expensive to solve from the numerical and complexity point of view [13].

On the other, Equivalent Circuit Models (ECM) simplicity and acceptable accuracy make them a favorable choice for voltage, SOC, and health predictions [14]. Further ECM is simplified as lumped battery modeling approach for the thermal modeling of LiBs at reduced efforts and less unknown parameters. Lumped parameter model is different from the resistance-capacitor (RC) circuit model; it is solved for the diffusion equation in an idealized spherical particle for solving Ohmic, activation and concentration losses in LiBs during charging/discharge. It is also reported that the detailed structure of cylindrical LiB cells for thermal modeling has little impact on its heat generation. This simplification holds good until as long as the battery is governed by the diffusion process in one of the battery electrodes [12]. Some literature available for lumped parameter thermal modeling of LiB at lower ambient temperature and low discharge rates [15]. Moreover, to the best of the author's knowledge the lumped parameter based thermal model for cylindrical Nickel Cobalt Aluminum Oxide (NCA) spiral-wound LiB at high ambient temperature is not explored yet.

The objective of this work to develop a lumped parameter based thermal model for a cylindrical spiral-wound LiBs for heat generation estimation, and temperature prediction at high ambient temperature. Electrochemical and thermal behavior coupled through the electrochemical heat generation arising from reversible and irreversible processes. The results are discussed here emphasize average heat generation estimation, maximum temperature, and temperature rise for LiB at various discharge rates and T_{amb}.

2 Methodology

2.1 Experimental Study

The experiments current 0.25% at full scale. The temperature data logging was performed using a K-type thermocouple and data acquisition system (Keysight, DAQ973A). Figure 1a shows the experimental setup and Fig. 1b shows the charging current, battery cell voltage, SOC and battery cell temperature during charging at 1C current (CC-CV protocol) at T_{amb} of 30 °C. were performed for a Panasonic LiB (NCR 18650B) to validate the numerical studies. The rated capacity of selected LiB at 25 °C is 3250 mAh and the nominal voltage is 3.6 V. The battery charging was performed at a 1C rate using the constant current constant voltage (CC-CV) protocol while discharging was conducted at 0.5, 1.0, 1.5 and 2.0 C-rates using CADEX C8000 (voltage accuracy of 0.1% and.

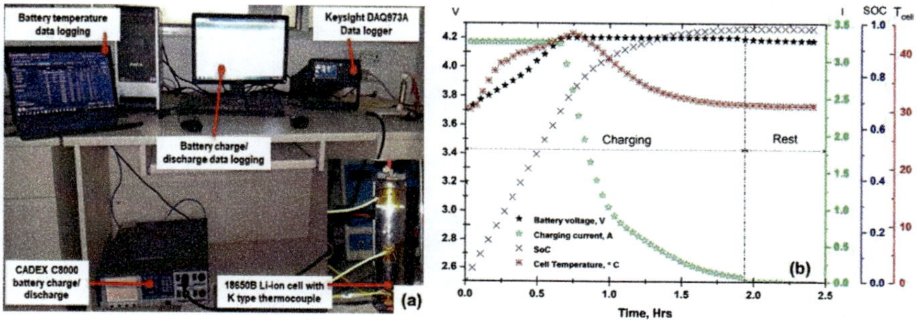

Fig. 1 **a** Experimental setup for single battery cell validation and **b** battery cell charging profile for 1C rate at T_{amb} 30 °C

The OCV-SOC mapping at the T_{amb} of 30 °C was used for the numerical modeling shown in Fig. 2b, and effect of T_{amb} on OCV-SOC mapping is overlooked in the present study due to less impact [16]. Three thermocouples were placed over the battery surface, first at the 5 mm, second at 35 mm, and third at 60 mm from the anode of the LiB, respectively as shown in Fig. 2a. The temperature of the anode recorded 3 °C higher compared to the cathode of LiB cell at the end of the discharge cycle, due to high heat generation from the anode current collector of LiB shown in Fig. 2a. The average temperature of all three positions over the battery surface was considered as the battery surface temperature. Three runs of experiments were performed for checking the repeatability of the experimental results.

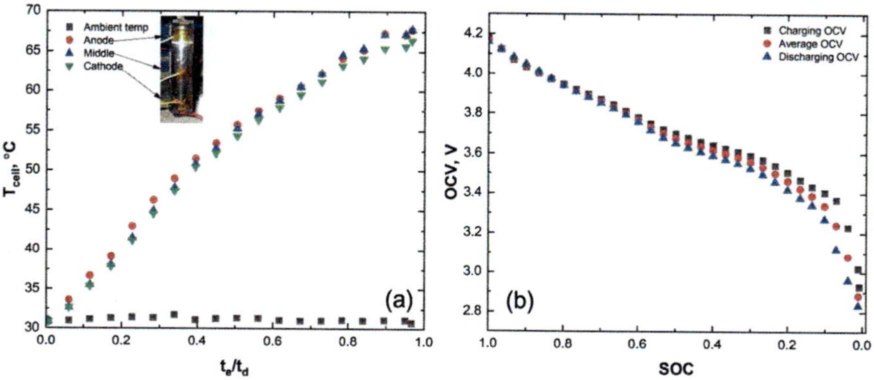

Fig. 2 **a** Surface temperature of the battery cell at different heights at $T_{amb} = 30$ °C for 2C discharge, **b** OCV-SOC mapping for charge/discharge cycle

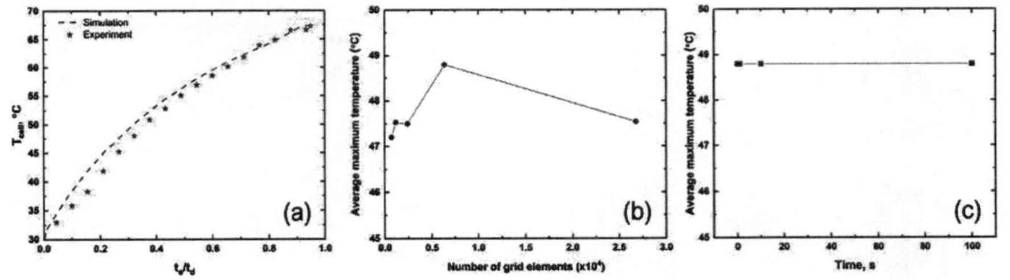

Fig. 3 Plots for **a** model validation at $T_{amb} = 30$ °C at 2C discharge, **b** grid independency, **c** time step, at 1C discharge

2.2 Numerical Study

Computational domain. For the present study a cylindrical battery cell (NCR18650B) of height (H) 60 mm and radius (r) 9 mm were modeled as the only active battery material considered for the simulation. A commercial computational fluid dynamics (CFD) software COMSOL Multiphysics 5.6 has been used for modeling and meshing of the domain. The maximum temperature of the LiBs cell used for model validation, the plot of the temperature from simulation and experiments shown in Fig. 3a. The grid independency test was carried out to investigate the effect of grid size on the temperature of the battery cell to ensure the accuracy of numerical results. The maximum average temperature tends shown in Fig. 3b for the numbers of the mesh grid. Subsequently, a time step dependency test was also performed to check the effect of the time step on temperature results as shown in Fig. 3c and a time step of 10 s was set to the corresponding sampling frequency for temperature recording in the experiments.

Governing equations. The governing equations used to define the energy balance in LiB cell consists of heat generation, storage, and conduction terms as shown in Eq. 1

$$\rho C_p \frac{\partial T}{\partial t} = k_r \frac{1}{r} \frac{\partial}{\partial r}\left(\frac{\partial T}{\partial r}\right) + k_\varphi \frac{1}{r^2} \frac{\partial^2 T}{\partial \varphi^2}\left(r \frac{\partial T}{\partial \varphi}\right) + k_z \frac{\partial^2 T}{\partial r^2} + Q_{gen} \tag{1}$$

The heat generation term (Q_{gen}) taken form Bernadi et al. [3] as given by Eq. 2

$$Q_{gen} = \left(\eta_{IR} + \eta_{IR_{act}} + T \frac{\partial E_{OCV}(SOC_{X=1}, T)}{\partial T}\right) I_{cell} + Q_{mix} \tag{2}$$

where η_{IR} Ohmic over-potential, $\eta_{IR_{act}}$ activation over-potential, X radius of spherical particle and Q_{mix} heat loss from side reactions given as Eq. 3

$$Q_{mix} = \frac{3Q_{cell,0}}{\tau} \int_0^1 \frac{\partial E_{OCV,therm}}{\partial SOC} \frac{\partial SOC}{\partial X} \frac{\partial SOC}{\partial X} X^2 \partial X \tag{3}$$

The thermoneutral voltage, $E_{OCV,therm}$ used in Eq. 3 given as

$$E_{OCV,therm} = E_{OCV,ref}(SOC) - T_{ref} \frac{\partial E_{OCV,ref}(SOC)}{\partial T} \tag{4}$$

Boundary conditions. The geometry considered here is a vertical cylinder and the natural convective heat transfer is takes place from domain boundary to the surrounding. The heat transfer coefficient for natural convection is computed as per Eq. 5 [17].

$$h = \frac{k_a}{H} \left\{ \frac{4}{3} \left(\frac{7Ra_H Pr}{5(20 + 21Pr)} \right)^{1/4} + \frac{4(272 + 315Pr)H}{35(64 + 63Pr)D} \right\} \tag{5}$$

where, k_a thermal conductivity of air, Ra_H Rayleigh number, Pr Prandtl number, and D diameter of battery cell.

Lumped assumption. Due to the small mass and moderate thermal conductivity of cylindrical LiBs, it is reasonable to represent the battery cell using lumped model [18]. The following assumptions are considered for the lumped parameter modeling of LiB cells.

(1) The materials of each layers of the battery are isotropic and with constant physical properties.
(2) Uniform heat generation within the cell.
(3) The internal heat transfer is conduction dominant, ignoring heat convection and radiation. These assumptions can also be justified with the dimensionless Biot number (Bi), which is the ratio of internal heat conduction resistance to surface convective resistance as shown in Eq. 6. A small value of Biot number means any small length, or high thermal conductivity, here battery radius in small [19]. Further, in this study natural convection is considered and at lower cooling rates, the cell behaves as a lumped system with uniform temperature [20].

$$Bi = \frac{h_{avg} L_c}{k_{eff}} = 0.008 \ll 0.1 \tag{6}$$

where L_c is the characteristics length (ratio of battery volume to surface area of heat convective boundaries). h_{avg} average heat transfer coefficient, and k_{eff} battery effective thermal conductivity. Here, thermal conductivity in radial direction considered as effective thermal conductivity of LiB.

Thermophysical properties. The thermophysical properties of LiB are different in the axial and radial direction due to the different materials used to make its layers.

Hence density, specific heat capacity, and thermal conductivity of LiB are calculated as the weighted average of different layers. The thermal conductivity of the LiB used in this study was estimated using Eqs. 7 and 8, and the values of thermal conductivity of each layer were taken from the literature [20], and the thickness was measured by dismantling a LiB cell into different layers. Different parameters used for lumped modeling of LiB are shown in Table 1.

$$k_{T,r} = \frac{\sum_i L_i}{\sum_i L_i / k_{T,i}} \tag{7}$$

$$k_{T,a} = \frac{\sum_i L_i k_{T,i}}{\sum_i L_i} \tag{8}$$

where $k_{T,r}$ radial thermal conductivity of battery, $k_{T,a}$ axial thermal conductivity of battery, L_i thickness of battery i_{th} layer, and $k_{T,i}$ thermal conductivity of battery i_{th} layer.

Voltage losses. The lumped battery model uses a small set of lumped parameters to calculate the contribution of all voltage losses in the LiB, stemming from Ohmic resistances, activation polarization, diffusion processes, and side reactions. The SOC dependent OCV for the modeling was measured from in-house experiments. The temperature effects on OCV-SOC were not considered as the variation is negligible in the temperature range from 30 to 45 °C [21]. The entropy coefficient (∂OCV/∂T) for different C-rates and temperatures are adopted from the literature [15], and extrapolated for unknown values shown in the Table 2.

Coupling of Electrochemical Heating. The Arrhenius equation is used for coupling of the lumped and thermal models considering ηIR, 1C, J0 and τ of the cell as a function of temperature. The OCV at reference temperature from experiments and the temperature derivative of OCV is given as input to the model and based on these inputs the cell equilibrium potential is estimated. The equilibrium potential is updated

Table 1 Values of various parameters used for lumped battery modeling

Parameter	Symbol	Value	Unit	Source
Battery density	ρ	2919		Estimated
Battery heat capacity	C_p	830		[4]
Reference exchange current	$J_{0,ref}$	0.85		[4]
Diffusion time constant	τ	1000		[4]
Thermal conductivity, radial (cross plane)	$k_{T,r}$	1.1		Estimated
Thermal conductivity, axial (in plane)	$k_{T,a}$	3.143		Estimated
Activation energy for temperature sensitivity of $\eta_{IR,c}$	$E_{\eta IR}$	24,000		[4]
Activation energy for temperature sensitivity of J_0	E_{J0}	– 59,000		[4]
Activation energy for temperature sensitivity of τ	E_τ	24,000		[4]

Table 2 Entropy coefficient at different C-rates and T_{amb} [19]

C-rate	Ambient temperature			
	30 °C	35 °C	40 °C	45 °C
0.5C	0.34	0.30	0.28	0.27
1.0C	0.37	0.35	0.33	0.30
1.5C	0.49	0.47	0.45	0.43
2.0C	0.60	0.57	0.56	0.53

in the electrochemical heat coupling, which updates the temperature in the Arrhenius equation to evaluate temperature effect on the voltage losses.

3 Results and Discussion

The evolution of temperate of LiB cell as a function of dimensionless time (elapsed discharge time to total discharge duration, te/td) for different discharge C-rates and T_{amb} are shown in Fig. 4. Heat generation contribution from Ohmic losses, over-potential, activation polarization and side reaction are computed for different discharge C-rates and ambient temperature.

The temperature of the LiB cell surface was used to validate the numerical results at T_{amb} of 30 °C and different C-rates as shown in Fig. 4a. The Mean absolute errors (MAE) in numerical simulation 12.22%, 6.62% 4.36%, and 4.64% for 0.5, 1.0 1.5 and 2.0 C-rates, respectively. Figure 5a shows the average heat generation as a function of te/td to show the effect of different C-rates and T_{amb} on the heat generation. Average heat generation of 0.19 W, 0.5 W, 0.8 W and 1.5 W were recorded for 0.5,1.0 1.5 and 2.0 C-rates, respectively. On other hand, Fig. 5b shows the effect of C-rates and T_{amb} on LiB maximum temperature. The maximum temperature rises by 23.4% when T_{amb} increases from 30 to 45 °C, due to the rise in T_{amb} heat transfer rate to surrounding decreases. The total heat generation inside the LiB during charge/discharge mainly depends on the C-rate (current). The current at which the battery fully charges/discharges in 1 h is defined as 1 C-rate. Higher C-rate leads to higher Ohmic losses, which increases total (irreversible and reversible) heat generation inside LiB as shown in Fig. 6a–d. However, due to the increase in C-rate the irreversible heat generation increases, but reversible heat generation decreases due to reduced elapsed discharge time.

The irreversible heat generation remains almost constant during the discharge as the current remains constant throughout the discharging for corresponding C-rates. This nature of irreversible heat is attributed to the rate of local electrochemical reactions in the LiBs. The reversible heat generation dominates at a low discharge C-rate due to the longer duration of the electrochemical reactions in the LiBs. On the other hand, the maximum temperature of LiB cell increases with an increase in

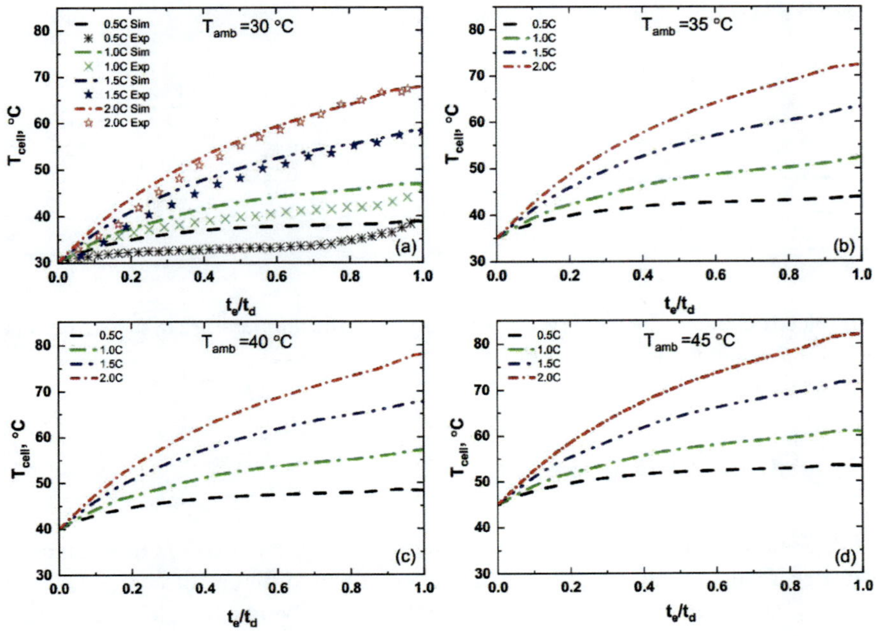

Fig. 4 The evolution of battery cell temperature for 0.5, 1.0, 1.5 and 2.0 discharge C-rates. **a** Model validation and numerical results at $T_{amb} = 30$ °C, **b** numerical results at 35 °C, **c** 40 °C and **d** 45 °C, respectively

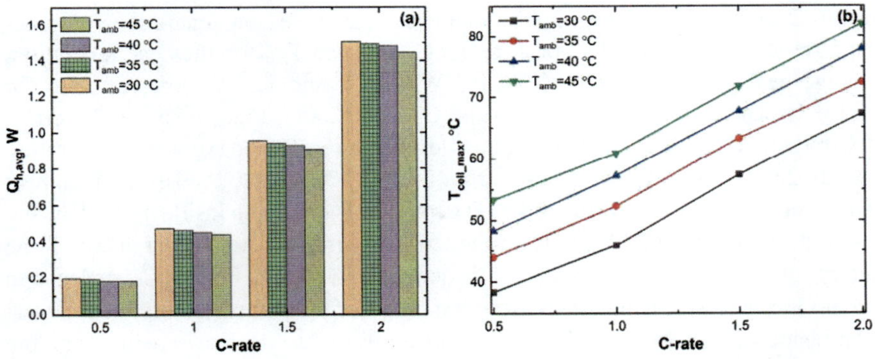

Fig. 5 **a** Average heat generation, **b** maximum cell surface temperature of LiB, for 0.5, 1.0 1.50 and 2.0 discharge C-rate at $T_{amb} = 30$ °C, 35 °C, 40 °C and 45 °C, respectively

discharge current shown as in Fig. 4a–d. The temperature rise is attributed to the high heat accumulation at high discharge rates.

Heat accumulation in the LiB cell at high discharge attributed to high instantaneous heat generation as compared to the heat conduction through LiB cell. The maximum temperature of 38.3 °C, 45.9 °C, 57.4 °C, and 67.3 °C recorded at T_{amb} of 30 °C for

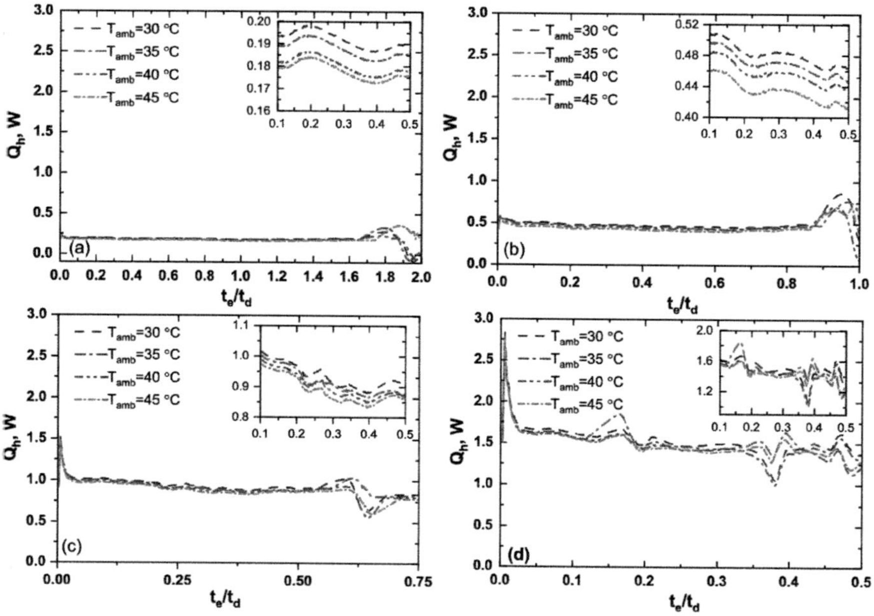

Fig. 6 Heat generation from LiB during discharging at **a** 0.5, **b** 1.0, **c** 1.5, **d** 2.0 C-rate

0.5, 1.0, 1.5 and 2.0 C-rates respectively and temperature rise of 8.3 °C, 15.9 °C, 27.4 °C, and 37.3 °C has been recorded, respectively. The heat generation slightly decreases with an increase in the T_{amb} as shown in the inset plots from Fig. 6a–d. A decrease in heat generation about of 10% noted when the T_{amb} increased from 30 to 45 °C for 1C discharge rate. The decrease in heat generation may be attributed to the decrease in the internal resistance of LiB cell at the high T_{amb} [5], due to a negative correlation with T_{amb} [22]. The heat generation is relatively high at the end of the discharge as shown in Fig. 6a–d, the increased heat generation is ascribed to the increased internal resistance of LiB at low SOC or high depth of discharge [23]. On the other hand, the maximum temperature of the LiB cell at the end of 1C discharge rate increased from 48.8 to 62.1 °C, when T_{amb} increased from 30 to 45 °C. While, the rise in the maximum temperature decreased from 18.8 to 17.1 °C, the decrease in the maximum temperature rise can be ascribed to the low heat generation at the high temperature.

In Fig. 6d some spikes noted in heat generation curves due to abrupt variation in OCV of LiB at 2C discharge rate during experiments. About the end of the discharge cycle heat generation drops sharply due to the experimental OCV-SOC data not being available, because the experimental discharge elapsed time is less than the ideal elapsed discharge time for the corresponding C-rate.

4 Conclusions

In this study, the heat generation estimation for single LiB cell for different C-rates and T_{amb} is carried out. The numerical investigation used for the heat generation estimation is done by considering LiB cell as lumped system. The proposed modeling approach gives good estimation of heat generation and battery surface temperature for single LiB cell, and may be used by battery thermal management designers. The following conclusion are drawn:

(a) The average heat generation at T_{amb} of 30 °C recorded 0.19 W, 0.47 W, 0.95 W and 1.5 W for 0.5, 1.0 1.5 and 2.0 C-rates, respectively.
(b) When the discharge current increases, the heat generation also increases due to the high Ohmic losses. The heat generation is decreased at high $T_{amb,}$ due to the decreased internal resistance of the LiB.
(c) At lower discharge rates the reversible losses dominate, while at the high discharge rates the irreversible losses dominate.
(d) The maximum temperature of 38.3 °C, 45.9 °C, 57.4 °C, and 67.3 °C were recorded at T_{amb} of 30 °C for 0.5, 1.0, 1.5 and 2.0 C-rates respectively.

The future scope of the present study, may include the experimental determination of entropy-coefficient for more accurate modeling using lumped battery approach.

Acknowledgements The authors express their sincere gratitude to the Start-up Research Grant (SRG/2020/001171) awarded to AP from the Science and Engineering Research Board, Department of Science and Technology, Government of India and TEQIP-III, MNIT Jaipur for the financial support of the battery thermal management studies at Centre for Energy and Environment, MNIT Jaipur. DKS would also like to acknowledge the Ministry of Education, Government of India for the doctoral research scholarship. The authors would also like to acknowledge to Dr. Kapil Pareek, MNIT Jaipur for providing access to the experimental setup for this study.

References

1. Ryu H-H, Sun HH, Myung S-T, Yoon CS, Sun Y-K (2021) Reducing cobalt from lithium-ion batteries for the electric vehicle era. Energy Environ Sci 14(2):844–852
2. Li W, Pang Y, Zhu T, Wang Y, Xia Y (2018) A gel polymer electrolyte based lithium-sulfur battery with low self-discharge. Solid State Ionics 318:82–87
3. Bernardi D, Pawlikowski E, Newman J (1985) A general energy balance for battery systems. J Electrochem Soc 132(1):5–12
4. Leng F, Tan CM, Pecht M (2015) Effect of temperature on the aging rate of Li ion battery operating above room temperature. Sci Rep 5(1):12967
5. Ouyang D, Weng J, Chen M, Liu J, Wang J (2020) Experimental analysis on the degradation behavior of overdischarged lithium-ion battery combined with the effect of high-temperature environment. Int J Energy Res 44(1):229–241
6. Guo Y, Luo M, Zou J, Liu Y, Kang J (2016) Temperature characteristics of ternary-material lithium-ion battery for vehicle applications. In: SAE technical papers, vol 2016

7. Sharma DK, Prabhakar A (2021) A review on air cooled and air centric hybrid thermal management techniques for Li-ion battery packs in electric vehicles. J Energy Storage 41:102885
8. Zhang D, Dey S, Perez HE, Moura SJ (2020) Real-time capacity estimation of lithium-ion batteries utilizing thermal dynamics. IEEE Trans Control Syst Technol 28(3):992–1000
9. Jeon DH, Baek SM (2011) Thermal modeling of cylindrical lithium ion battery during discharge cycle. Energy Convers Manag 52(8):2973–2981
10. Kim GH, Pesaran A, Spotnitz R (2007) A three-dimensional thermal abuse model for lithium-ion cells. J Power Sources 170(2):476–489
11. Srinivasan V, Wang CY (2002) Analysis of electrochemical and thermal behavior of Li-ion cells. J Electrochem Soc 150(1):A98–A106
12. Ekström H, Fridholm B, Lindbergh G (2018) Comparison of lumped diffusion models for voltage prediction of a lithium-ion battery cell during dynamic loads. J Power Sources 402:296–300
13. Somasundaram K, Birgersson E, Mujumdar AS (2012) Thermal-electrochemical model for passive thermal management of a spiral-wound lithium-ion battery. J Power Sources 203:84–96
14. Saqli K, Bouchareb H, M'sirdi KN, Naamane A, Oudghiri M (2020) Electric and thermal model of Li-ion battery pack with cylindrical components. In: 2020 5th international conference on renewable energies for developing countries (REDEC), pp 1–6
15. Özdemir T, Amini A, Ekici Ö, Köksal M (2022) Experimental assessment of the lumped lithium ion battery model at different operating conditions. Heat Transf Eng 43(3–5):314–325
16. Zheng C, Ge Y, Chen Z, Huang D, Liu J, Zhou S (2017) Diagnosis method for Li-ion battery fault based on an adaptive unscented Kalman filter. Energies 10(11)
17. Kang G-U, Chung B-J, Kim H-J (2014) Natural convection heat transfer on a vertical cylinder submerged in fluids having high Prandtl number. Int J Heat Mass Transf 79:4–11
18. Jilte RD, Kumar R (2018) Numerical investigation on cooling performance of Li-ion battery thermal management system at high galvanostatic discharge. Eng Sci Technol Int J 21(5):957–969
19. Kirillov PL, Ninokata H (2017) Heat transfer in nuclear thermal hydraulics. In: D'Auria F (ed) Thermal-hydraulics of water cooled nuclear reactors. Woodhead Publishing, pp 357–492
20. Al Hallaj S, Maleki H, Hong JS, Selman JR (1999) Thermal modeling and design considerations of lithium-ion batteries. J Power Sources 83(1):1–8
21. Baccouche I, Jemmali S, Manai B, Omar N, Ben Amara NE (2017) Improved OCV model of a Li-ion NMC battery for online SOC estimation using the extended Kalman filter. Energies 10(6)
22. Chen L et al (2021) Estimation the internal resistance of lithium-ion-battery using a multi-factor dynamic internal resistance model with an error compensation strategy. Energy Rep 7:3050–3059
23. Inui Y, Hirayama S, Tanaka T (2019) Detailed estimation method of heat generation during charge/discharge in lithium-ion battery using equivalent circuit. Electron Commun Japan 102(12):3–14

Numerical Investigation of the Thermal and Emissions Performance of a Hybrid Draft Biomass Cookstove

Suraj Ghiwe, Vilas Kalamkar, and Pravin Sawarkar

1 Introduction

Poor combustion of solid biomass yields toxic emissions, notably carbon monoxide (CO) and particulate matter ($PM_{2.5}$), resulting in indoor air pollution, which causes heart disease, strokes, lung cancer, and chronic diseases, culminating in 4 million deaths each year [1]. As a result, researchers emphasize the necessity of decreasing hazardous emissions rather than boosting the thermal efficiency of a biomass cookstove, leading to a variety of biomass cookstove advancements [2]. The introduction of a forced draft in biomass cookstoves was a pivotal event in cookstove development. Forced draft employs a fan to deliver air, creating turbulence that improves mixing and hence combustion. Large-scale industrial furnaces and boilers [3], as well as residential biomass heating equipment [4], all use forced draft. Forced draft has been widely used in gasifier stoves for both primary and secondary air, which are batch-fed stoves that require pre-processed fuel [5–7]. Direct-combustion stoves are ideal for consuming biomass in its natural state without considerable fuel preparation. However, natural draft direct-combustion stoves, on the other hand, perform poorly in terms of combustion and emissions [8]. Thus, attaining good combustion performance in a direct-combustion biomass cookstove necessitates the use of a forced-air supply. Hence, a new mechanism known as "Hybrid Draft" which combines natural and forced draft, has been devised to alleviate the drawbacks of forced draft and can be applied to small-capacity biomass combustion devices. Kshirsagar and Kalamkar [9] recently developed a hybrid draft biomass cookstove (HDBC) by investigating the effects of various parameters on emissions and efficiency performance, using the same hybrid draft mechanism. To reduce emissions, they have adopted a cross-flow arrangement for forced draft secondary air supply. For the experimental investigation,

S. Ghiwe (✉) · V. Kalamkar · P. Sawarkar
Visveswaraya National Institute of Technology, Nagpur, India
e-mail: surajghiwe@gmail.com

S. Doolla et al. (eds.), *Advances in Clean Energy and Sustainability*,
Green Energy and Technology, https://doi.org/10.1007/978-981-99-2279-6_12

a water boiling test (WBT) was performed, and a mathematical model was created using VBA code. They reported tier 4 emissions and tier 2 efficiency.

Since experiments cannot explain the underlying physics inside the cookstove, computational fluid dynamics (CFD) could overcome these constraints. CFD has recently been employed as an efficient tool to model the thermal and flow behaviour, as well as gas combustion, in various cookstoves. Pande et al. [10, 11] used an Ansys-fluent to perform a 2D CFD simulation of a natural draft biomass cookstove to study the effect of the inlet area ratio on its performance. Combustion was modelled as non-premixed combustion using a presumed probability density function (PDF) model, ignoring reaction kinetics. The effects of radiation and soot production were also overlooked. Pundle et al. [12] investigated the flow and temperature behaviour of a 2D axisymmetric natural draught cookstove using a laminar finite rate combustion model. Radiative heat transfer was modeled using a discrete ordinate model, but soot generation was neglected. Husain et al. investigated homogeneous [13] and heterogeneous [14] combustion in forced-draft biomass cookstoves. Homogeneous combustion was modelled using a single combustion reaction of wood volatile and air, while heterogeneous combustion was modelled using a different composition of wood volatile with multiple reaction kinetics. An eddy dissipation model was used to model the turbulent-chemistry interaction. Soot was not included in the model. Núñez et al. [15] and Medina et al. [16] investigated the heat transfer, fluid flow, and combustion mechanisms in a natural draft plancha-type cookstove. Solid wood combustion was modelled as wood volatiles reacting with an oxidizer without considering turbulent-chemistry interactions. Radiation and soot were not included in the simulation.

As per advances in cookstove modelling involving the use of CFD, the current numerical study employs the Ansys-Fluent to analyses the thermal and emission (CO and $PM_{2.5}$) performance of a hybrid draft biomass cookstove (HDBC). A 3D numerical analysis was carried out, which included combustion and soot modelling, as well as the effect of secondary air on the overall performance of HDBC. The numerical results were validated by comparing them to the experimental data of the water boiling test.

2 Numerical Modelling

2.1 *Domain Configuration and Meshing Strategy*

Figures 1 and 2 depicts the schematic and actual model of HDBC, which is the subject of the current investigation. It consists of a natural draft primary air supply and a fan-assisted forced draft secondary air supply in a cross-flow configuration. Secondary air is delivered into the combustion chamber through an annular cavity and holes on the combustion chamber's surface.

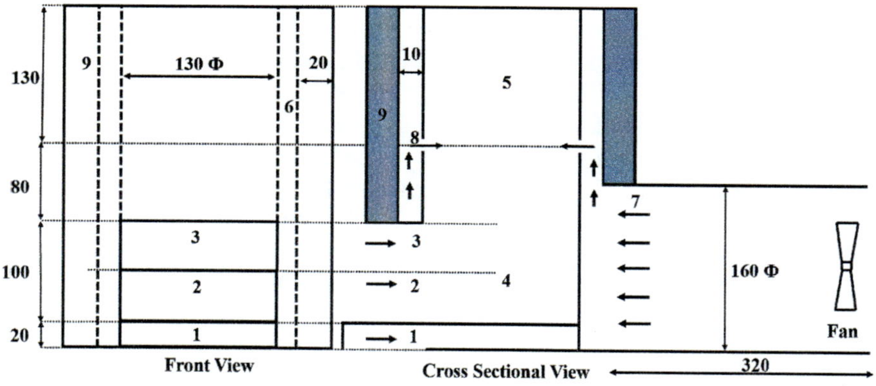

1. Fixed inlet area (Grate) for primary air (natural convection); 2. Biomass fuel supply; 3. primary air supply through natural convection (variable area); 4. Primary combustion chamber; 5. Secondary combustion Chamber; 6. Annular cavity; 7. Secondary air supply; 8. Holes for secondary air; 9. Insulation (All dimensions are in mm)

Fig. 1 Schematic view of HDBC

Fig. 2 Working model HDBC

The HDBC computational domain used for CFD analysis is shown in Fig. 3. The computational domain is composed of the following: the combustion chamber volume, the pot (represented by the pot bottom), the fuel inlet, and the primary and secondary air inlets (represented by flat boundaries). The overall height of the domain is 340 mm, with a 330 mm actual stove height and a 10 mm pot gap represented by the outlet. The combustion volume measures 130 mm in diameter, while the pot base is 260 mm. For secondary air supply, 36 holes of 4 mm in diameter are provided.

The domain was discretized into fine 3D tetrahedron grids to capture the flow dynamics, with finer grids developed in crucial regions like secondary air supply. To simulate the HDBC, a mesh configuration of 784,585 elements was used. This mesh configuration was chosen after a mesh independency test that was conducted on four different grid configurations (M1, M2, M3, and M4), as shown in Fig. 4. Where, M1: 382,315 elements with a minimum grid size of 5 mm, M2: 495,853 elements with a minimum grid size of 4 mm, M3: 784,585 elements with a minimum grid size of 3 mm, and M4: 1,111,097 elements with a minimum grid size of 2.5 mm.

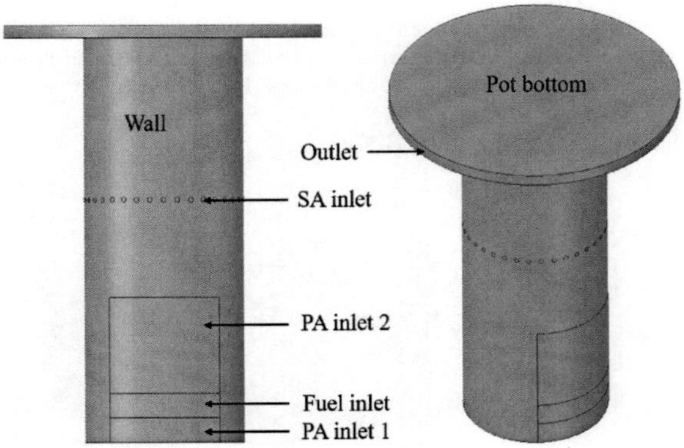

Fig. 3 Computational domain of HDBC

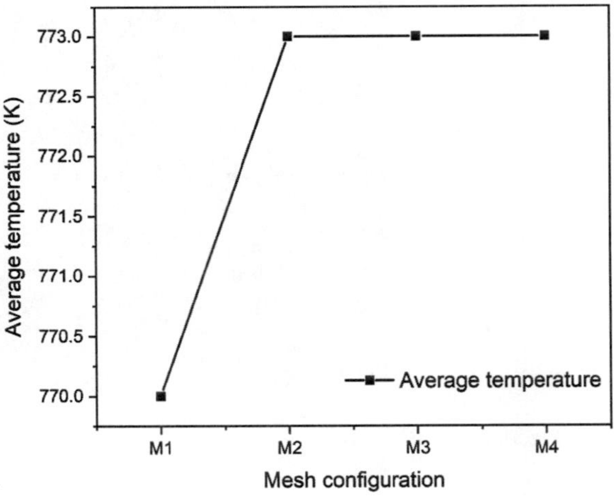

Fig. 4 Mesh independency analysis

2.2 Simulation Approach

The analysis was carried out using a steady-state and pressure-based solver. Governing equations such as continuity, momentum, and energy balance are solved for each grid volume. Turbulence was modelled using the realizable k-ε turbulence model with enhancement wall treatment for thermal and pressure effects [11, 12, 16]. A discrete ordinate model was used to model radiative heat transfer in a HDBC. The pressure–velocity coupling was addressed using the SIMPLE scheme with a second-order upwind scheme to discretize the governing equation and species.

Combustion Modelling

Combustion of biomass involves moisture evaporation, volatile combustion, and char combustion. The moisture evaporation and char generation phases involved in actual biomass combustion are not taken into account in this study. Following moisture evaporation, volatile matter, primarily CO, CO_2, H_2O, CH_4, H_2, higher hydrocarbons, and tar, is released from solid biomass at higher temperatures, reacting with oxygen and causing combustion [14, 17]. Since the exact compositions of volatiles are unknown, the compositions of volatiles and their mass fractions are drawn from experimental work by Husain et al. [18] based on the total volatile matter of biomass fuel from the proximate analysis. The volatiles comprised H_2, CO, CH_4, CO_2, C_2H_4, C_2H_6, and C_3H_8. Combustion was modelled using a 7-step reaction mechanism of volatiles with oxygen, as shown below, in conjunction with Magnussen and Hjertager's eddy dissipation model for turbulence chemistry interaction.

$$H_2 + 0.5O_2 \rightarrow H_2O \tag{1}$$

$$CO + 0.5O_2 \rightarrow CO_2 \tag{2}$$

$$CH_4 + 2O_2 \rightarrow CO_2 + 2H_2O \tag{3}$$

$$2CO_2 \rightarrow 2CO + O_2 \tag{4}$$

$$C_2H_4 + 3O_2 \rightarrow 2CO_2 + 2H_2O \tag{5}$$

$$C_2H_6 + 7O_2 \rightarrow 4CO_2 + 6H_2O \tag{6}$$

$$C_3H_8 + 5O_2 \rightarrow 3CO_2 + 4H_2O \tag{7}$$

One of the harmful pollutants released from biomass combustion is soot (particulate matter). Hence, when modelling biomass combustion, predicting soot formation is essential. In the present study, a one-step soot model was employed to predict soot generation in HDBC due to its great applicability to a broad spectrum of hydrocarbons. C_3H_8 (propane) as a fuel and O_2 as an oxidant were defined in the species definition.

Defining appropriate boundary conditions that approximate experimental data is crucial for getting good results in CFD analysis. In the present analysis, primary air inlets were specified as pressure inlets as a boundary condition since they were fed through natural convection. Experiment-derived mass flow rate conditions were provided at the fuel and secondary air inlets. Each gaseous species' mass fraction was defined in the fuel inlet. At the outlet, a pressure outlet boundary condition was defined. The bottom of the pot was described as a wall at 100 °C, and the combustion chamber wall was believed to be perfectly insulated.

3 Result and Discussion

HDBC, as earlier mentioned, uses secondary air that is delivered through a fan. Hence, the optimum mass flow rate of secondary air was first determined experimentally using WBT, and further investigation was then conducted at this optimum mass flow rate. In numerical analysis, the same approach was reproduced.

3.1 *Optimum Secondary Air Mass Flow Rate* (m_{sa})

Emission performance of HDBC was assessed at four different mass flow rates to identify the optimum mass flow rate: 0, 0.011, 0.017, and 0.027 kg/s. Based on CO and $PM_{2.5}$ emission performance, the optimum mass flow rate was determined. Figure 5 depicts the emission performance of HDBC for experiments and CFD at various secondary air mass flow rates.

CO emissions follow similar trends in both experiments and CFD, as seen in Fig. 5a. Although the reduction in CO emissions is marginal in CFD, the initial trend shows that increasing the mass flow rate from 0 to 0.017 kg/s reduces emissions. Increased mass flow rates enhance turbulence and the mixing of combustion gases with excess oxygen, promoting combustion and lowering CO emissions. Increased CO emissions at higher mass flow rates imply an influx of cooler into the combustion chamber, quenching the flame and cooling the combustion zone, which impedes combustion and raises CO emissions. Therefore, too high or low a secondary air mass flow rate reduces combustion efficiency and, as a result, increases CO emissions.

$PM_{2.5}$ (particulate matter) emissions follow the same pattern as CO emissions: as flow rates rise to 0.017 kg/s, $PM_{2.5}$ emissions fall (see Fig. 5b), as small jets of secondary air trim the flame in the combustion chamber, causing fine particulate matter to settle to the bottom. The growth and formation of particulate matter is caused by the cooling and crystallization of combustible gases in the exhaust into solid particles [19]. The additional secondary air in the combustion chamber cools

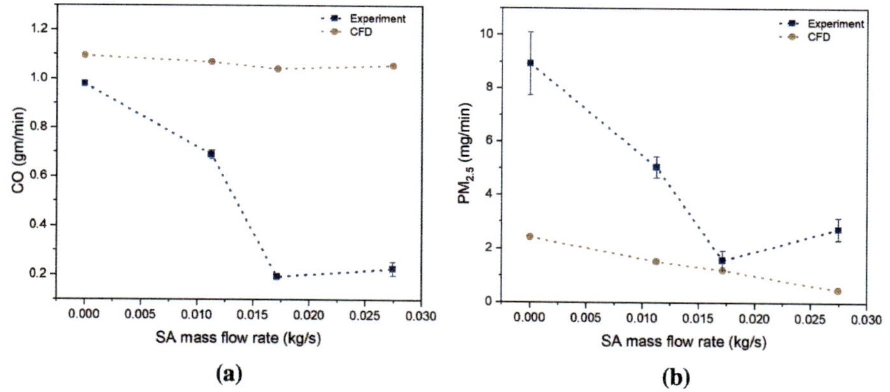

Fig. 5 HDBC emission performance with SA mass flow rate: **a** CO versus m_{sa}; **b** $PM_{2.5}$ versus m_{sa}

the temperature in the combustion zone at a higher mass flow rate, boosting particle condensation and crystallization. However, the CFD result for $PM_{2.5}$ emissions at a higher mass flow rate is conflicting. This is because the temperature of secondary air entering the combustion chamber is assumed to remain constant in simulation, but in practice, as mass flow rate increases, velocity increases, and therefore lower temperature air enters the combustion chamber.

Hence, the optimum secondary air mass flow rate was determined to be 0.017 kg/s, and further investigation of HDBC was carried out at this rate to establish the thermal and emission performance.

3.2 *Emission Performance*

The experimental and computational CO and $PM_{2.5}$ emission performance of the HDBC at the optimum secondary air mass flow rate are shown in Fig. 6a, b. At the optimum secondary air flow rate, HDBC emitted significantly less CO and $PM_{2.5}$, implying that the purpose of having forced secondary air in HDBC was justified. When compared to the experimental result, the combustion model used in the present investigation completely overpredicts CO emissions (see Fig. 6a). In contrast, $PM_{2.5}$ emissions from CFD agree well with experimental data, as shown in Fig. 6b. According to the IWA performance tiers [20], the CO emission obtained from the experiment was tier 4, whereas the CFD result was tier 1 performance. On the other hand, the $PM_{2.5}$ emissions from both analyses have achieved tier 4 performance. Hence, with propane as a fuel and oxygen as an oxidizer, the one-step soot model employed for soot prediction works effectively.

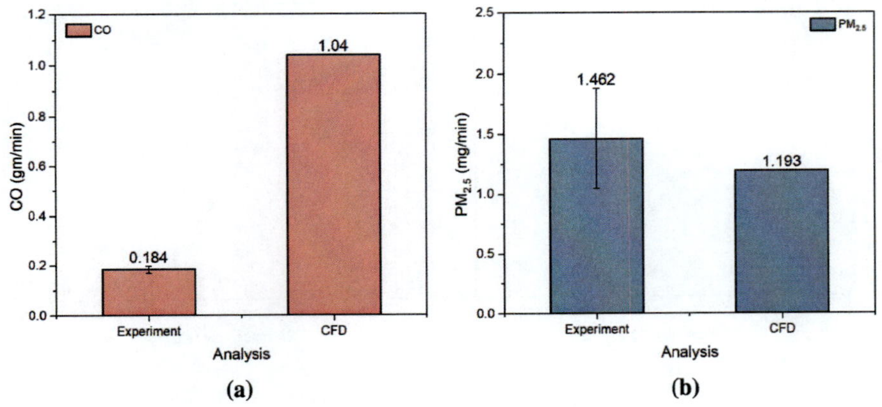

Fig. 6 Emission performance: **a** CO emissions and **b** $PM_{2.5}$ emissions

3.3 Thermal Performance

To validate the combustion model employed in this study, efficiency and temperatures at various locations inside the combustion chamber and at the outlet were compared to experimental results. The temperatures were recorded using thermocouples at two separate radial locations and two different axial distances from the stove's bottom: 205 mm inside the combustion chamber and 337 mm at the outlet. The thermocouple configurations and their locations are illustrated in Fig. 7.

The CFD model's predicted efficiency at the optimum mass flow rate, as computed by using Eq. (8), accords well with the WBT-derived efficiency (see Fig. 8), and both attained tier-2 performance. All of the experiments in this study were conducted at constant firepower of 4.8 kW, which is used in Eq. (8). According to Eq. (8), for constant firepower, efficiency is only influenced by heat flux to the pot, which is

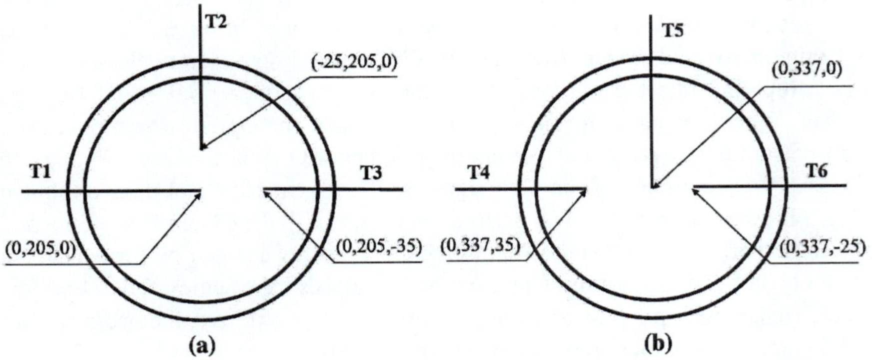

Fig. 7 Thermocouple's location: **a** 205 mm from bottom of HDBC and **b** 337 mm from bottom of HDBC (T1 to T6 represents thermocouples)

governed by combustion quality. Furthermore, the average temperatures predicted by the combustion model at 205 mm and 337 mm from the bottom of the stove are in good agreement with the temperatures measured during the water boiling test (WBT), see Fig. 9.

$$\eta = \frac{Q}{Firepower} \tag{8}$$

where, Q is average heat flux through the pot bottom.

Fig. 8 Efficiency performance of HDBC

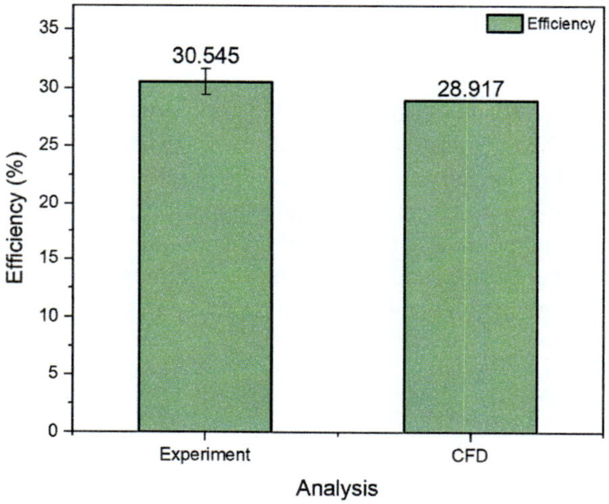

Fig. 9 Temperature at different locations

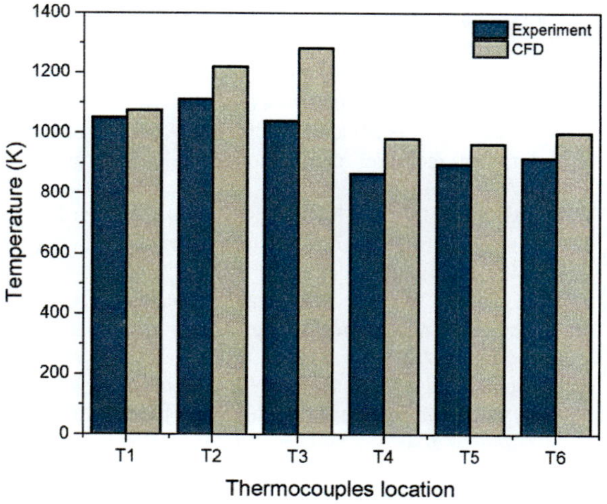

4 Conclusion

The thermal and emission performance of HDBC was assessed using CFD analysis, which included combustion modelling and soot modelling. The WBT experimental data was used to validate the CFD results. This is one of the HDBC studies that has produced some notable results.

- The use of a cross-flow arrangement of forced drafts secondary air to reduce emissions has proven to be very effective. However, higher mass flow rates might yield contradictory outcomes, so optimizing the secondary air mass flow rate is necessary. Therefore, when designing HDBC, secondary air is an important parameter to address.
- The combustion model employed was found to successfully predict the thermal performance, which exhibited good agreement with the experimental data, except for CO emission, which was overpredicted.
- The one-step soot model used to predict soot generation was effective and agreed well with experimental PM emissions. It can thus be used to anticipate the generation of soot in a wide range of biomass combustion systems.
- The use of computational fluid dynamics (CFD) tools to investigate the complex combustion phenomenon has proven to be a cost-effective and efficient method.

References

1. World Health Organization. Household air pollution and health 2018. https://www.who.int/news-room/fact-sheets/detail/household-air-pollution-and-health.
2. Sambandam S, Balakrishnan K, Ghosh S, Sadasivam A, Madhav S, Ramasamy R et al (2015) Can Currently available advanced combustion biomass cook-stoves provide health relevant exposure reductions? Results from initial assessment of select commercial models in India. EcoHealth 12:25–41. https://doi.org/10.1007/s10393-014-0976-1
3. Nussbaumer T (2003) Combustion and co-combustion of biomass: fundamentals, technologies, and primary measures for emission reduction. Energy Fuels 17:1510–1521. https://doi.org/10.1021/ef030031q
4. Šyc M, Horák J, Hopan F, Krpec K, Tomšej T, Ocelka T et al (2011) Effect of fuels and domestic heating appliance types on emission factors of selected organic pollutants. Environ Sci Technol 45:9427–9434. https://doi.org/10.1021/es2017945
5. Ndindeng SA, Wopereis M, Sanyang S, Futakuchi K (2019) Evaluation of fan-assisted rice husk fuelled gasifier cookstoves for application in sub-Sahara Africa. Renew Energy 139:924–935. https://doi.org/10.1016/j.renene.2019.02.132
6. Hosseini Rahdar M, Nasiri F, Lee B (2019) A review of numerical modeling and experimental analysis of combustion in moving grate biomass combustors. Energy Fuels 33:9367–9402. https://doi.org/10.1021/acs.energyfuels.9b02073
7. Panwar NL, Rathore NS (2008) Design and performance evaluation of a 5 kW producer gas stove. Biomass Bioenerg 32:1349–1352. https://doi.org/10.1016/j.biombioe.2008.04.007
8. Kshirsagar MP, Kalamkar VR (2014) A comprehensive review on biomass cookstoves and a systematic approach for modern cookstove design. Renew Sustain Energy Rev 30:580–603. https://doi.org/10.1016/j.rser.2013.10.039

9. Kshirsagar MP, Kalamkar VR (2022) Hybrid draft direct-combustion with secondary air jets in cross-flow for reducing CO and PM2.5 emissions in biomass cookstoves. Sustain Energy Technol Assessments 51:101913. http://doi.org/10.1016/j.seta.2021.101913
10. Pande RR, Kalamkar VR, Kshirsagar MP (2019) The effect of inlet area ratio on the performance of multi-pot natural draft biomass cookstove. Proc Natl Acad Sci India Sect A Phys Sci. https://doi.org/10.1007/s40010-019-00650-3
11. Pande RR, Kshirsagar MP, Kalamkar VR (2018) Experimental and CFD analysis to study the effect of inlet area ratio in a natural draft biomass cookstove. Environ Dev Sustain. https://doi.org/10.1007/s10668-018-0269-x
12. Pundle A, Sullivan B, Means P, Posner JD, Kramlich JC (2019) Predicting and analyzing the performance of biomass-burning natural draft rocket cookstoves using computational fluid dynamics. Biomass Bioenerg 131:105402. https://doi.org/10.1016/j.biombioe.2019.105402
13. Husain Z, Tiwari SS, Pandit AB, Joshi JB (2019) Computational fluid dynamics study of biomass cook stove—part 1: hydrodynamics and homogeneous combustion. Ind Eng Chem Res. https://doi.org/10.1021/acs.iecr.9b03181
14. Husain Z, Tiwari SS, Kataria A, Mathpati CS, Pandit AB, Joshi JB (2020) Computational fluid dynamic study of biomass cook stove—part 2: devolatilization and heterogeneous combustion. Ind Eng Chem Res 59:14507–14521. https://doi.org/10.1021/acs.iecr.9b07109
15. Núñez J, Moctezuma-Sánchez MF, Fisher EM, Berrueta VM, Masera OR, Beltrán A (2020) Natural-draft flow and heat transfer in a plancha-type biomass cookstove. Renew Energy 146:727–736. https://doi.org/10.1016/j.renene.2019.07.007
16. Medina P, Núñez J, Ruiz-García VM, Beltrán A (2021) Experimental and numerical comparison of CO_2 mass flow rate emissions, combustion and thermal performance for a biomass plancha-type cookstove. Energy Sustain Dev 63:153–159. https://doi.org/10.1016/j.esd.2021.07.001
17. Porteiro J, Collazo J, Patiño D, Granada E, Gonzalez JCM, Míguez JL (2009) Numerical modeling of a biomass pellet domestic boiler. Energy Fuels 23:1067–1075. https://doi.org/10.1021/ef8008458
18. Husain Z, Ansari KB, Chatake VS, Urunkar Y, Pandit AB, Joshi JB (2020) Valorisation of biomass pellets to renewable fuel and chemicals using pyrolysis: characterisation of pyrolysis products and its application. Indian Chem Eng 62:78–91. https://doi.org/10.1080/00194506.2019.1635047
19. Torvela T, Tissari J, Sippula O, Kaivosoja T, Leskinen J, Virén A et al (2014) Effect of wood combustion conditions on the morphology of freshly emitted fine particles. Atmos Environ 87:65–76. https://doi.org/10.1016/j.atmosenv.2014.01.028
20. EPA, PCIA A. The water boiling test, version 4.2.3. vol 4.2.2 (2014)

Numerical Investigation of Melting of Layered PCM in Squared Cavity

Devendra Raut, B. Kalyan Raj, N. Pavan Satyanarayana, P. Bhaskar Reddy, and V. R. Kalamkar

1 Introduction

The heat addition and removal are the fundamental block of any thermal system. In many applications, the heat is the unwanted and it is desired to absorb the heat without any temperature rise of the sink. Such addition or removal of heat at constant temperature involves the latent heat storage (LHS) technology, which employs phase change material as energy storing media. The energy stored in the LHS is the combination of sensible and latent heat and it is given by the equation as mentioned in Ref. [1]. Moreover, the rate of energy storage depends on the thermophysical properties of the PCM. The major properties are like thermal conductivity, latent heat, viscosity. They affect the melting of PCM, which can be affected by altering these properties.

In the available literature, there are many works on the performance improvement techniques like addition of fins, encapsulation, use of additives and configuration optimization. Other techniques which are adopted to realize higher energy storage efficiency is multiple PCM modules cascaded in a storage tank, such arrangement is also called multi-layered PCM [2]. Such arrangement offers modified heat transfer characteristics and energy storage rates with minimal efforts. Selection of suitable PCM and their arrangement in correct sequence can result in desired performance. Generally, the PCM with higher melting points are placed next to the heat source so that latent heat can be utilized completely to absorb the heat from source [3].

Multi-layered or cascaded PCM arrangement has wide application in space thermal regulation, electronic component cooling, composite wall of refrigerated unit and cooling of PV panel also find it useful.

For instance, Reddy et al. [4] investigated the single and double layer PCM in roof structure. The base layer in the double PCM layer roof is found to be hotter than the

D. Raut (✉) · B. Kalyan Raj · N. Pavan Satyanarayana · P. Bhaskar Reddy · V. R. Kalamkar
Department of Mechanical Engineering, Visvesvaraya National Institute of Technology, Nagpur, India
e-mail: devendra.raut123@gmail.com

S. Doolla et al. (eds.), *Advances in Clean Energy and Sustainability*,
Green Energy and Technology, https://doi.org/10.1007/978-981-99-2279-6_13

base layer in the single PCM layer roof. This is due to the fact that the additional PCM layer remains in the melting phase for the majority of the time, preventing complete solidification of the PCM. In a similar study by Rehman et al. [5], the simulation of a novel dual layered phase change material brick wall for human comfort in hot and cold climatic conditions is performed. The results show that using two PCMs together provides better human comfort with lower energy requirements throughout the year than using a single PCM (29 °C) for summer or 13 °C for winter alone.

The use of layered PCM in energy storage technology has been adopted, Ahmed et al. [6] used 3 layered PCM to investigate the influence of combined sensible and latent heat storage material on thermocline characteristics. They reported that PCM within admissible temperature, sequence of PCM arrangement alters the rate of charging and discharging cycle.

The application of PCM layer for cooling of electronic components is in practise. Moraga et al. [7] investigated the thermal cooling of a solar car numerically. They used 7 arrays of 4 PCMs at the outer shell of the battery. The optimal arrangement is use of 3 layers of PCM in which the PCM with the highest thermal conductivity is placed next to the battery body while the PCM with the lowest thermal conductivity is placed in the outer wall. Based on the transient results, it was concluded that the battery cooling is mainly due to the conduction. Other applications of PCM layers are mentioned in the references [8–10]. The benefit of the layered PCM over the single PCM is that the layered PCM offers modified thermal conductivity, which in turns affects the heat transfer rate.

Based on the literature review, it urged authors to investigate the melting characteristics of the 3 layered PCM in a square cavity with constant heat source at the bottom wall. The arrangement of the PCM is done in decreasing melting point, the melting parameters and heat transfer characteristics are analysed to provide the detailed thermal insight on the obtained thermal performance of the presented setup.

2 Numerical Simulation

2.1 Geometry and Boundary Conditions

The system studied in this report is a two-dimensional square cavity in dimension of 100 mm (H) × 100 mm (W) with bottom wall being thermally active and rest walls are insulated. The cavity is filled with PCM and their thermophysical properties are listed in Table 1. The configuration of the layered PCM is illustrated in Fig. 1. The initial temperature (T_i) of paraffin wax is 27 °C while the hot wall temperature is 90 °C. The PCM used for the study are Tritriacontane [16], RT58 [17] and P116 [18], the thermophysical properties are taken from the reference.

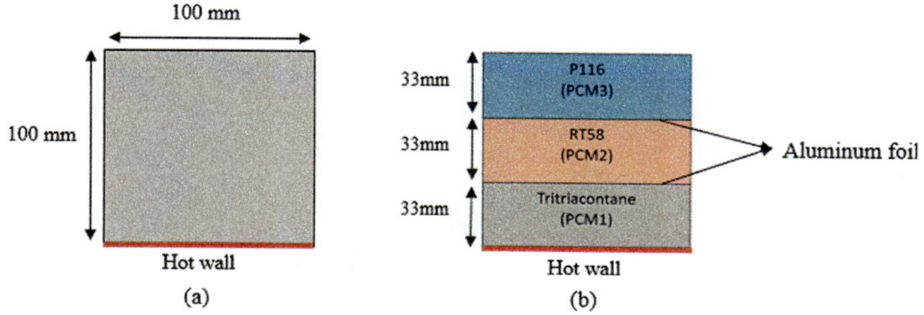

Fig. 1 Square geometry for single and layered PCM

2.2 *Governing Equations and Numerical Method*

In the current study, the enthalpy-porosity method was used. The thermophysical properties of the PCMs were assumed constant during the melting process, except for density, which was modelled using the Boussinesq approximation with initial value as average of solid and liquid density. Raut et al. [11] provided the governing equations for the model's conservation of mass, momentum, and energy.

2.3 *Grid Independence Test*

Independence test of grid size 100 × 100, 90 × 90, 110 × 110 and time step of 0.1 s is examined for the PCM Tritriacontane ($C_{33}H_{68}$). From Fig. 2 the grid 100 × 100 is optimum for the current mode and further increase of grid number has insignificant variation and nearly same as 100 × 100 grid.

2.4 *Numerical Model Validation*

The numerical model is validated with the experimental results of the Mahdi et al. [12], in which PCM lauric acid is heated from side wall in a rectangular cavity. It can be observed from Fig. 3 that a good agreement between the present numerical and experimental work of [12]. Hence, the presented numerical model and boundary conditions are adopted for further study.

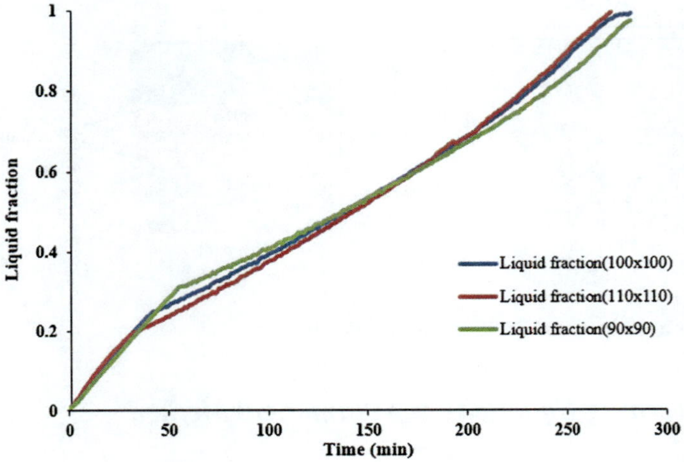

Fig. 2 Grid independent test

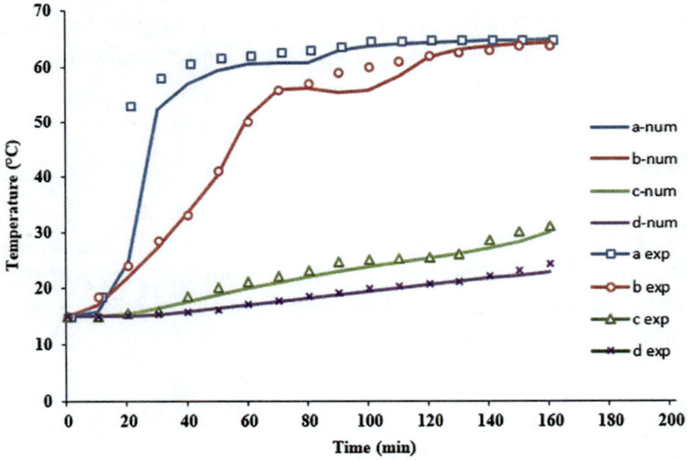

Fig. 3 Comparison of numerical and experimental work [4] temperatures with time during melting

3 Result and Discussion

3.1 Single Layered PCM

Melting Fraction

The simulation work of single layered PCM is divided in 3 cases. In case1 Tritriacontane was taken as PCM, in case2 RT58, and in case3 p116 were taken as PCMs. The results of melting behavior and temperature distribution for three cases are discussed below.

The liquid fraction contours of the three PCMs at melting are depicted in Fig. 4. Because conduction is the primary mode of heat transfer and buoyancy forces cannot overcome the resistance imposed by viscous forces, the liquid–solid interface is flat at the start of the melting process. Following that, convection occurs, and as the fluid motion becomes stronger, the liquid–solid interface gradually becomes distorted.

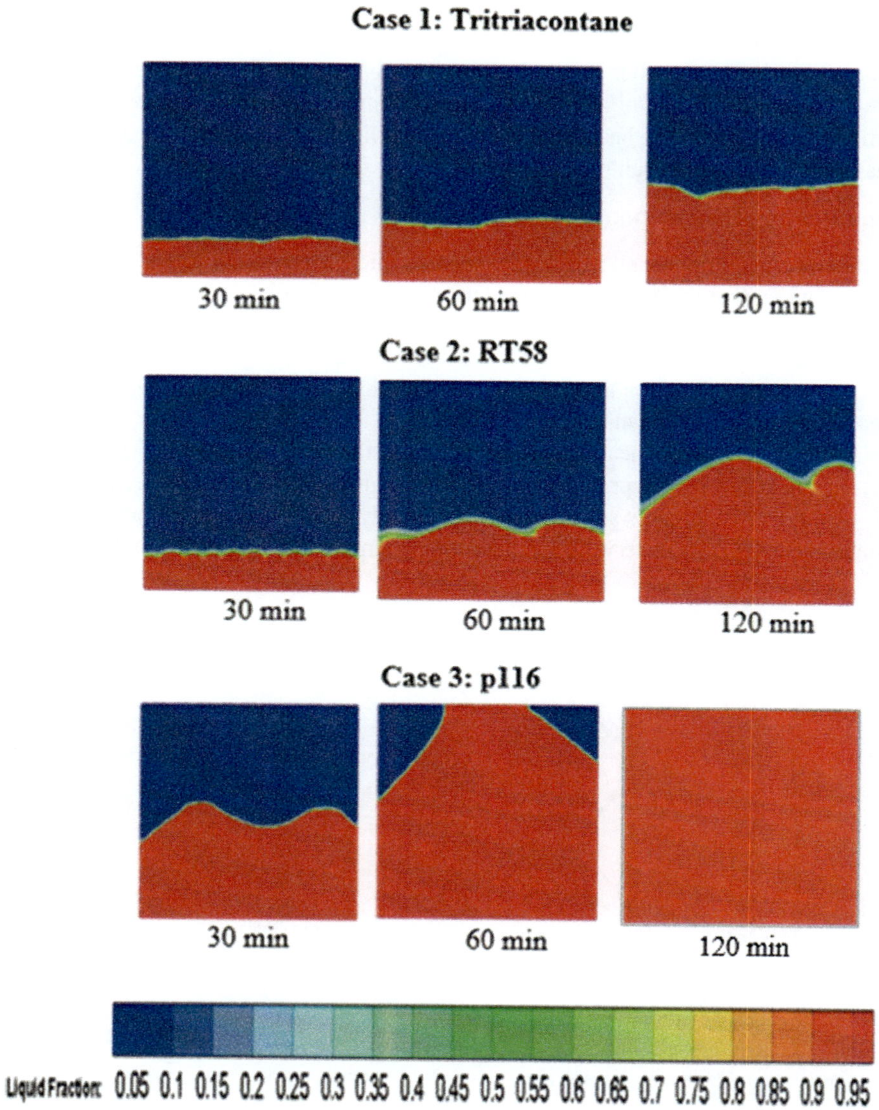

Fig. 4 Liquid fraction contour of the 3 PCM at melting time

As time escapes, the solid–liquid interface shows a crest near the center of the storage unit, it is due to the rise of melted PCM due to buoyancy. As the density of solid is more compared to that of liquid PCM it tends to move downward. Hence the semi-solid PCM at the interface tends to slip down into the liquid from the sides and forms lumps of solid PCM above the liquid one. In numerical setup, average value of solid liquid density is taken and Boussinesq approximation is considered for density variation.

Figure 5 shows the graph of liquid fraction values during melting time for 3 PCM. The graph depicts the rise in liquid fraction over time as a result of PCM melting. It can be seen that in the case of p116 have a higher liquid fraction than that of Tritriacontane after about 30 min. This is because the lower temperatures are achieved faster than higher ones. There is no much difference between melting time for RT58 and p116 because the solidus and liquidous temperature are of these materials are close to each other.

Temperature Distribution

Temperature contours of all 3 PCMs are sketched at 3 different times are the results are shown in Fig. 6. It is observed that as time proceeds volume of yellow zone increases while blue zone decreases. Since, heat is being provided from the bottom wall the temperature of molecules in contact to the thermally active surface are heated first by the process of conduction and later on heat is transferred to surrounding molecules above the surface by means of convection.

As the distance from heating surface increases, the temperature of PCM decreases because heat is being supplied from the lower surface. As time proceeds, volume of melted PCM increases substantially. As already mentioned in the liquid fraction contours, the PCM is observed to achieve higher temperatures when the melting point

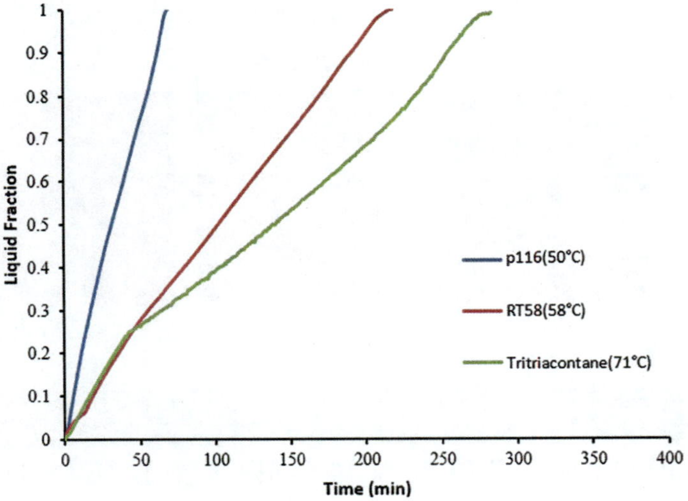

Fig. 5 Liquid fraction for single PCM arrangement

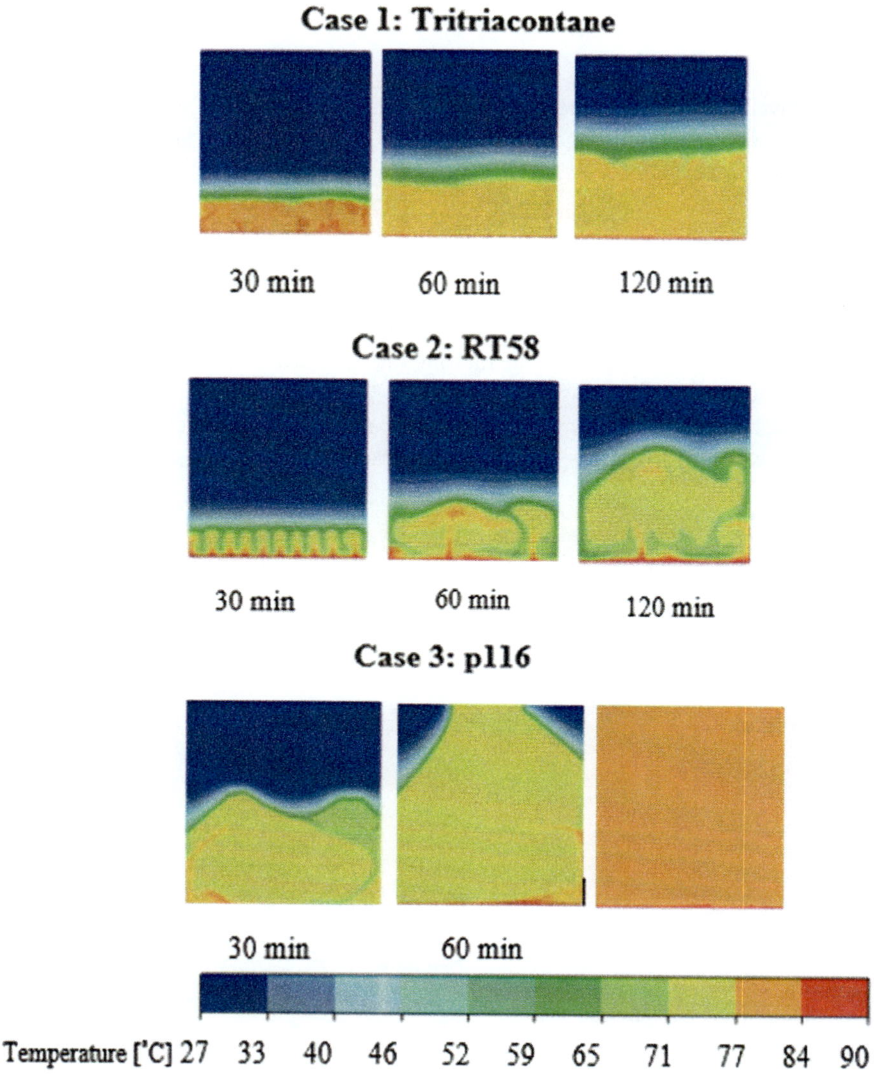

Fig. 6 Temperature distribution of PCMs at different time points

of PCM is low. It is because of convection process is achieved faster in p116. Since, convection heat transfer is faster than conduction, the temperature rises rapidly.

3.2 Multilayered PCM

A multi layered PCM of geometry as shown in Fig. 7 is simulated numerically using the same operating conditions as in single layer PCM simulation and the following observations were made.

In the first 30 min, PCM1 heated up to melting point as a result, Most of the PCM1 gets melted. PCM1 transfer its heat content to the PCM2 which in turn transfers to PCM3. The temperature map shows that the stratification of temperature in vertical direction. PCM2 takes major part of the melting duration as a result of which the slope of liquid fraction and temperature becomes steady as shown in Figs. 8 and 9. The effect of layered PCM is that the melting rate is retarded and it can be compared with individual PCM's melting rate.

The clear insight of melting process can be made from the melt fraction color map as shown in Fig. 10. It is shown in melt fraction that the shape of solid liquid

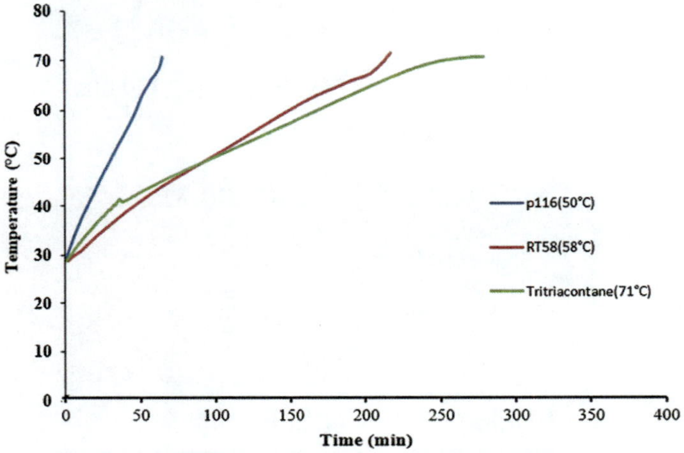

Fig. 7 Temperature for individual PCM

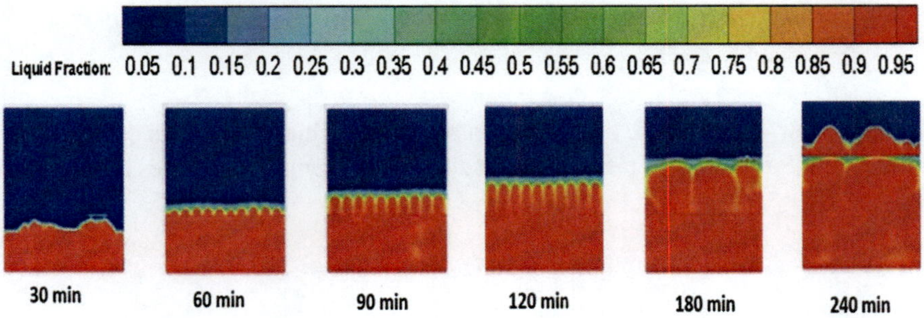

Fig. 8 Liquid fraction contours at different time instant when base wall temperature is 90 °C

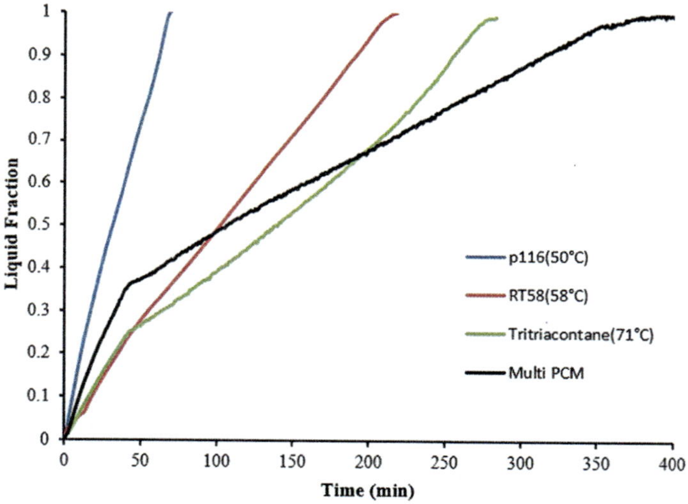

Fig. 9 Liquid fraction curve for multi and single layered PCM

interface varies with the PCM. In PCM1, it irregular while in PCM2 it shows wavy shape with smaller peaks while PCM3 shows wavy nature with bigger peaks. The cause for such variation in solid liquid interface can be understand from temperature contours as the temperature layer is narrow and respective solid–liquid interface has smaller peaks.

Thus, it can be concluded that the combined effect of heat and melt fraction distribution is that the melting rate gets delay even the source wall temperature remains same. This finding enables the use of layered PCM for the application where heat extraction with lesser temperature rise is required (Fig. 11).

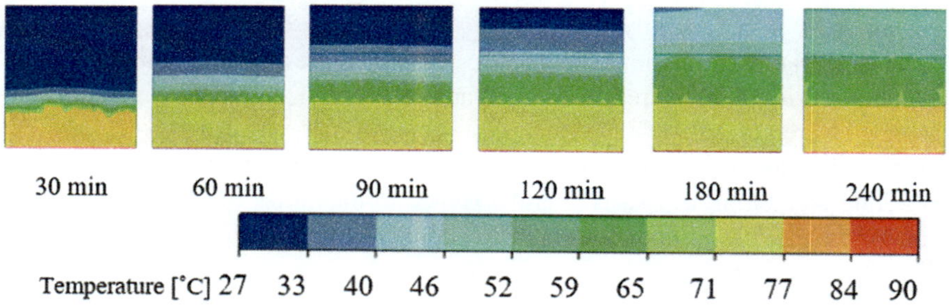

Fig. 10 Temperature contours at different time points

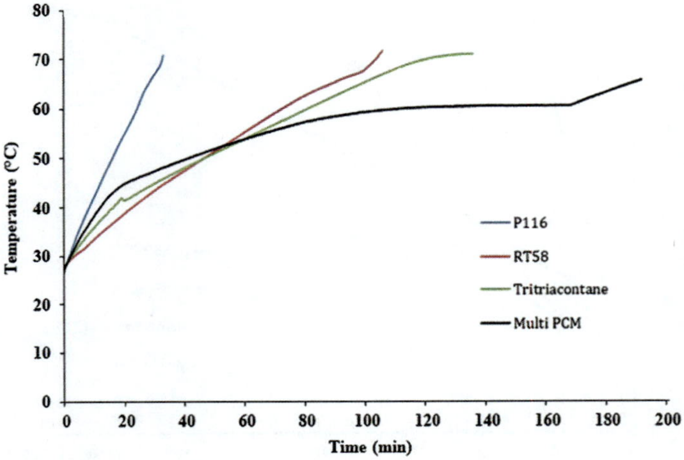

Fig. 11 Temperature for multi layered PCM and single layered PCMs

4 Conclusion

The new model aims to improve the natural convection of PCM storage by developing a multi layered model instead of conventional single layered PCM model. Based on the numerical results obtained following conclusions can be drawn:

- The simulation results show that the average heat storage capacity of multi layered PCM is higher than that of single layered PCM. The multi-layered PCM enhanced the heat storage capacity by 30%, 46.75%, 82.5% for single layered PCM of Tritriacontane, RT58, P116.
- The natural convection and heat storage capacity of a thermal energy storage system can be improved by adapting the system of multi layered PCM developed using PCM of same family.
- The temperature simulations show that a steady increase of temperature of system is obtained in multi layered PCM compared to rapid changes of temperatures in single layered PCM.
- Hot wall temperature study shows that with the increase in inlet wall temperature the rate of melting increases slightly. Hence to maximize the capacity of heat intake the melting point of bottom PCM should be established as high as the inlet wall temperature.

References

1. Sarbu I, Sebarchievici C (2018) A comprehensive review of thermal energy storage. Sustainability 10. http://doi.org/10.3390/su10010191
2. Chirino H, Xu B, Xu X (2019) Parametric study of cascade latent heat thermal energy storage (CLHTES) system in concentrated solar power (CSP) plants. J Energy Inst 92:653–664. https://doi.org/10.1016/j.joei.2018.03.007
3. Izquierdo-Barrientos MA, Sobrino C, Almendros-Ibáñez JA (2015) Energy storage with PCM in fluidized beds: modeling and experiments. Chem Eng J 264:497–505. https://doi.org/10.1016/j.cej.2014.11.107
4. Reddy KS, Mudgal V, Mallick TK (2017) Thermal performance analysis of multi-phase change material layer-integrated building roofs for energy efficiency in built-environment. Energies 10. http://doi.org/10.3390/en10091367
5. Rehman AU, Sheikh SR, Kausar Z, McCormack SJ (2021) Numerical simulation of a novel dual layered phase change. Energies 14
6. Ahmed N, Elfeky KE, Lu L, Wang QW (2020) Thermal performance analysis of thermocline combined sensible-latent heat storage system using cascaded-layered PCM designs for medium temperature applications. Renew Energy 152:684–697. https://doi.org/10.1016/j.renene.2020.01.073
7. Moraga NO, Xamán JP, Araya RH (2016) Cooling Li-ion batteries of racing solar car by using multiple phase change materials. Appl Therm Eng 108:1041–1054. https://doi.org/10.1016/j.applthermaleng.2016.07.183
8. Farid MM, Khudhair AM, Razack SAK, Al-Hallaj S (2004) A review on phase change energy storage: materials and applications
9. Fumo N, Bortone V, Zambrano JC (2011) Comparative analysis of solar thermal cooling and solar photovoltaic cooling systems. In: ASME 2011 5th international conference on energy sustainability, ES 2011, pp 85–90. http://doi.org/10.1115/ES2011-54162
10. Trelles JP, Dufly JJ (2003) Numerical simulation of porous latent heat thermal energy storage for thermoelectric cooling. Appl Therm Eng 23:1647–1664. https://doi.org/10.1016/S1359-4311(03)00108-X
11. Raut D, Lanjewar S, Kalamkar VR (2022) Effect of geometrical and operational parameters on paraffin's melting performance in helical coiled latent heat storage for solar application: a numerical study. Int J Therm Sci 176:107509. https://doi.org/10.1016/j.ijthermalsci.2022.107509
12. Mahdi MS, Mahood HB, Campbell AN, Khadom AA (2021) Natural convection improvement of PCM melting in partition latent heat energy storage: numerical study with experimental validation. Int Commun Heat Mass Transf 126:105463. https://doi.org/10.1016/j.icheatmasstransfer.2021.105463

Electrochemical Hydrogen Storage Within a Modified Reversible PEM Fuel Cell and Its Performance Analysis with Interdigitated and Spiral Micro Flow Channels

Gurwinder Singh, Amandeep Singh Oberoi, Harmesh K. Kansal, and Amrinder Pal Singh

1 Introduction

Fuel cell is a device which convert chemical energy into electrical energy. Fuel cells are known to be the most effective energy conversion technique. Because of their efficiency, fuel cells are seen as important sources of power for the future and is better than that of internal combustion engines, and also the fact that their byproducts are water and electricity [1]. Among all forms of fuel cell, proton exchange membrane (PEM) fuel cell has emerged as a potential device with wide acceptability across various fields. It offers ease of implementation, long life, minimum emission, quick starting and durability compared to its counterparts. PEMFC found wide applications in the fields of auto industry and stationary power supply [2, 3]. In conventional system, hydrogen is extracted from water dissociation using a typical electrolyser powered by solar photovoltaics. The produced hydrogen stored in a pressurized cylinder at around 350 bars. The day time electric load demand is met directly by solar photovoltaics and any excess power is used to drive the electrolyser and store hydrogen. During night, the hydrogen which was stored, is fed to fuel cell

G. Singh (✉) · H. K. Kansal · A. P. Singh
Mechanical Engineering Department, University Institute of Engineering and Technology, Panjab University, Sector 25, South Campus, Chandigarh 160014, India
e-mail: er.gurwindersingh14@gmail.com

H. K. Kansal
e-mail: harmesh@pu.ac.in

A. P. Singh
e-mail: apsingh@pu.ac.in

A. S. Oberoi
Mechanical Engineering Department, Thapar Institute of Engineering and Technology, Patiala, Punjab 147004, India
e-mail: oberoi@thapar.edu

S. Doolla et al. (eds.), *Advances in Clean Energy and Sustainability*,
Green Energy and Technology, https://doi.org/10.1007/978-981-99-2279-6_14

in order to give out electricity and water. The system's round-trip efficiency is low due to number of stages involved and associated losses. To overcome this limitation, reversible unitized regenerative fuel cell (URFC) was introduced. However, the system still dependent upon compressor and pressurized gas cylinder that require energy expenditure [4]. To enhance round-the-trip efficiency and work ability of the system, an integrated hydrogen storage medium is introduced at cathode that eliminated the requirement of the external storage and the system is named as a modified reversible PEMFC. It works on the principle of battery and stores energy in hydrogen form. While running as electrolyser (Charging mode), it splits water into H_2 and O_2 through applied voltage as well as catalyst. The generated oxygen allowed to exit the cell and hydrogen gas further gets split into ions. The protons pass through electrolyte membrane whereas, e^- travels towards cathode via electric circuit. The H^+ and e^- forms a stable hydrogen atom at the cathode side that gets adsorbed in a porous storage electrode instead of forming hydrogen gas. While running as fuel cell (Discharging mode), the stored hydrogen is only the power source. The entire unit is subjected to an electrical load which creates a potential difference across the cell. The stored hydrogen splits into H^+ and e^- due to potential difference. The hydrogen proton follows the obvious path of membrane and e^- satisfies the applied electric load. At other end, supplied oxygen reacts with hydrogen protons and electrons to reform water [5]. The reactant gases are made to enter the cell through the bi-polar end plates that consists of micro-flow channels. The main purpose of the micro flow channels is to spread the reactants even across the cell. Therefore, cell's performance, is affected by design of micro-flow channels [6]. Several studies have been conducted on fuel cells. Many factors influence the performance of a cell like catalyst loading density, type of gas diffusion layers, oxidation rate, flow channels design etc. However, testing, and comparative analysis of interdigitated and spiral channel and their effect on an experimental modified reversible PEMFC have not been found in the published literature. The current research examines the effect of interdigitated and spiral channel on a storage medium's ability to adsorb hydrogen when integrated in an experimental modified reversible proton exchange membrane fuel cell.

2 Experimentation

2.1 Fabrication of the Carbon Electrode and Cell

A weighed amount of carbon powder is mixed with polytetrafluoroethylene (PTFE) to hold the particles of carbon together. The slurry mixture has been stirred for 120 min at NTP conditions. With the help of a mould, the mixture has been casted into a required shape. After that, mould baked in electric oven at around 100 °C for 120 min with purpose to eliminate moisture from mixture. In this way, carbon electrode was fabricated. The actual photograph of electrode is presented in Fig. 1.

Fig. 1 Carbon electrode

The fabricated cell is also given in Fig. 2. In this cell, hydrogen is stored in the form of H^+ ions. The bi-polar end plates with interdigitated and spiral design have been experimentally examined and used in the cell. These end plates are electrically insulated and serve two purposes i.e., distributing reactants in the cell and allowing gases to exit the cell. The gas diffusion layers have been utilized on both sides of the cell. Membrane electrode assembly is a combination of nafion-117 and catalyst. The membrane is supplied with oxygen side catalyst to assist the reaction of water reduction in the charging mode to discharge O_2, H^+ and e^- ions due to applied voltage, whereas catalyst on hydrogen side is utilized to assist reaction of hydrogen reduction during discharging cycle to discharge H^+ and e^- ions. The prepared electrode has been employed in the cell on the cathode-side. In the current research work, stainless steel 316 which has a high resistance of corrosion has been used to fabricate end plates with interdigitated and spiral design. The end plates were of same length and breadth but not same thickness. 9 mm thick plate has been used as hydrogen side plate and 7 mm thick plate has been used as oxygen side plate. The actual photographs of two flow channels are also provided in Fig. 3.

2.2 *Testing of the Fabricated Cell with Interdigitated and Spiral Flow Fields*

The cell has been assembled in two ways i.e., charging and discharging cycle. The fabricated cell has been set to run in both cycles at normal temperature and pressure. In E-mode operation, the cell has been connected to H_2O, direct current supply, O_2 cylinder at anode side and H_2 cylinder at cathode side. In this operation, the protons are developed and adsorbed in an equipped carbon electrode. In FC-mode operation, direct current power supply, H_2O cylinder and H_2 cylinder, have been disconnected from setup. Therefore, powering source left in this mode, is the stored hydrogen i.e., stored protons and O_2 cylinder have been essential to generate power. In this operation, the stored protons are recovered from carbon electrode and pass back

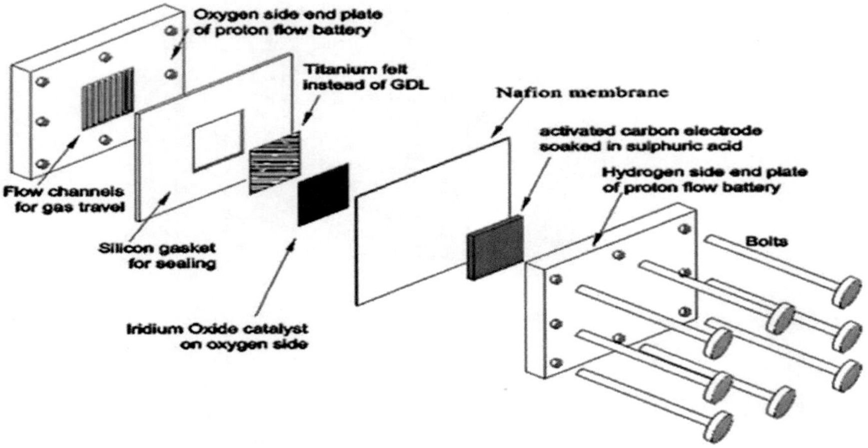

Fig. 2 Schematic of an experimental modified reversible PEMFC [4]

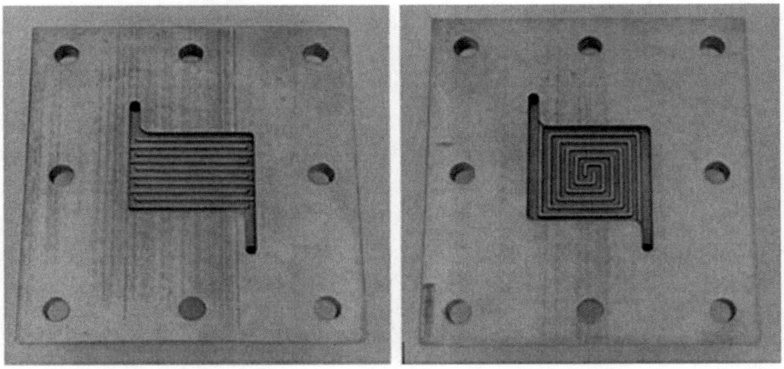

(a) (b)

Fig. 3 Flow channel design: **a** interdigitated; and **b** spiral

through the membrane of the cell to generate electricity. An experimental setup is given in Fig. 4.

In the E-mode operation, voltage, current, time, H_2O level, O_2 level and H_2 level, have been considered as operating parameters in this experimental work. To start cycle, Direct current voltage had been given through five ampere power set. The voltage has been increased and readings of the parameters have also been noted. Therefore, H_2O had been splits into O_2, H^+, and e^- at anode side. At the reading of 1.2 V, O_2 gas bubbles have been noted in O_2 cylinder which shows water breakdown, but no bubbles of gas have been seen in H_2 cylinder. The produced O_2 flows across the channel in anode side i.e., oxygen side plate and exit the cell. After this, oxygen gas allowed to exit the cell and same has been collected, but H^+ ions attracted towards cathode side and travels via nafion-117 to other area of cell and the electrons moved

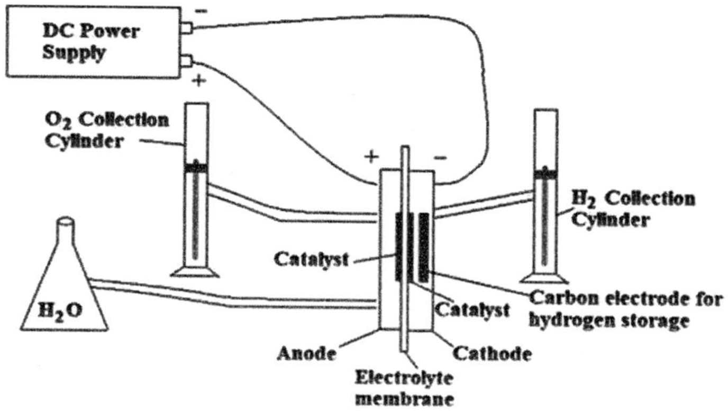

Fig. 4 Experimental setup

via power circuit. After couple of hours, few gas bubbles of H_2 have been seen in the H_2 cylinder which means all produced hydrogen ions i.e., H^+ ions, are not adsorbed electrochemically in the employed electrode, but few combine to form H_2 gas and received in the H_2 tank [12]. The operation was stopped thereafter. The mass of total generated hydrogen has been found out with the help of Eq. 1.

$$M_{TH} = \frac{It}{1000F} \tag{1}$$

where, M_{TH} is mass of total hydrogen (grams), I is current (mA), t is time (seconds), and F is Faraday constant.

Volume of received hydrogen had been used to determine mass of gas (collected) with the help of following Eq. 2.

$$PV = mRT \tag{2}$$

where, P is gas's pressure (bar), V is gas's volume (cm^3), m is gas's mass (grams), R is specific gas constant and T is temperature (K).

The charging cycle has been performed by changing plates with interdigitated and spiral design in the cell step-by-step. The cell performance with above said channels in E-mode operation was measured. In FC-mode, electrical load has been connected to setup to gain current from it. The stored oxygen gas was supplied back to the oxygen side plate. Due to electric potential, the inadequate surface chemical connection splits and adsorbed protons i.e., hydrogen atom, exits the storage material. Here, a reaction of hydrogen reduction occurs wherein platinum catalyst dissociate atom of hydrogen into protons and electrons. These protons and electrons then return to the anode side through membrane and electrical system and interact with O_2 and electrons on other side to reform H_2O and produce power. The voltage of cell has been dropped as per assembled load. The parameters viz. current, voltage, discharge time, and O_2 levels have been noted. The hydrogen weight percent was measured. The cycle has been performed by changing said plates in the cell. The cell performance in FC-mode operation was measured and all the readings have been recorded.

3 Results and Discussions

3.1 Electrochemical Hydrogen Storage

The two cells have been fabricated with interdigitated and spiral design. The one electrode has been placed in each cell. The cells have been tested via charging and discharging cycle. The operating parameters have also been noted. Accordingly, the graphs i.e., current versus time, H_2 versus time, and H_2O versus time, are presented in Fig. 5a–c. In Fig. 5a, the increase in the current is visible with respected to the ramping up of the voltage across the cell. The applied voltage resulted in the water disassociation into oxygen and hydrogen. The current value increase at twelve hundred seconds of operation signifies that storage point has reached. The line between time 400–600 s showed electrical and chemical adsorption of hydrogen in the electrode. Figure 5b represents rate of hydrogen generation during charging mode of the cell with respect to time. The water disassociation has been given O_2, H^+ and e^- with the help of voltage and catalyst. Since all produced hydrogen ions i.e., H^+ ions, have not been adsorbed in the electrode but few of them merge to generate H_2. The sudden rise on the hydrogen production rate showed that storage is currently full and all remaining hydrogen atoms would evolute as H_2 gas only.

Figure 5c shows that cell start drawing current from it as indicated in downfall curve. The adsorbed hydrogen in weight percent in the electrode in charging cycle is consolidated in Table 1. The H_2 weight percent calculated in discharging cycle is also presented in Table 2.

3.2 Result Analysis

The spiral channel is allowed the electrode to store 1.67 wt% of hydrogen in electrolyser cycle, whereas 1.21 wt% of hydrogen has been obtained in fuel cell mode. The tested cell with interdigitated channel stored 1.47 wt% of hydrogen during charging and corresponding discharging result was found to be 1.14 wt% of hydrogen. The comparative analysis conducted for two tested flow channel designs revealed that spiral channel performs better than another channel. This type of flow channel is the highest performer and interdigitated flow channel design is the lowest performer.

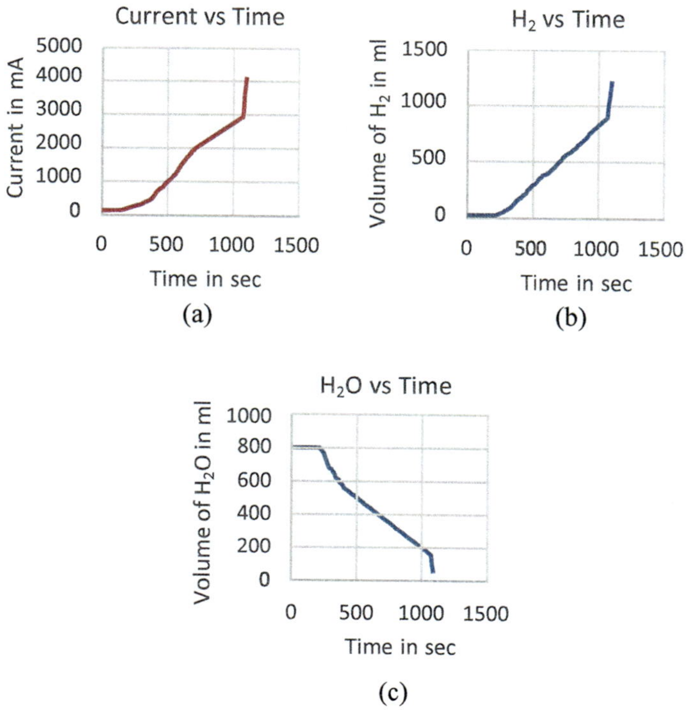

Fig. 5 **a** Current versus time; **b** H_2 versus time and **c** H_2O versus time

Table 1 The cell performance in charging cycle

Channel	Mass of total hydrogen as determined from Eq. 1 'b' in g	Volume of received hydrogen 'a' in ml	Hydrogen's mass enters the storage 'c' = b − a in g	Hydrogen accumulates in electrode c/mass of carbon * 100 in weight percent
Interdigitated	0.10	850	0.0300	1.47
Spiral	0.10	800	0.0342	1.67

Table 2 The cell performance in discharging cycle

Channel	Hydrogen weight percent
Interdigitated	1.14
Spiral	1.21

4 Conclusion and Future Scope

Stainless steel 316 was selected for the fabrication of two end plates with varied channel designs. The fabricated carbon electrodes one for each set of the cell were integrated in the cell. The cell was fabricated using interdigitated and spiral channel. The experimental cell equipped with carbon electrode has been set to run in E-mode (charging mode or electrolyser mode) and FC-mode (discharging mode or fuel cell mode) subsequently at normal temperature and pressure. The performance of the fabricated cell was measured, and all the operational parameters were also recorded. The interdigitated channel helped the carbon electrode to adsorb 1.47 wt% of hydrogen in charging mode out of which 1.14 wt% was recovered in discharging mode. The fabricated cell with spiral channel orientation stored 1.67 wt% of hydrogen during E-mode and gave out 1.21 wt% of hydrogen during discharging mode. The results of two channels in terms of hydrogen weight percent are compared. The obtained results are comparable with hydrogen adsorption in a porous graphene-based electrode reported by Jindal et al. [12]. The comparative analysis conducted for two tested flow channel designs revealed that spiral channel design leads to better dispersion of the reactants. The obtained results revealed the optimum design of flow channels that could help to achieve the higher hydrogen adsorption within the cell compared to previous design. It is a way forward towards developing a fuel-cell based proton battery that does not emit harmful fumes like lithium batteries and therefore, is environmentally friendly. Alternate catalysts with different catalyst loading densities could be tested in the cell for higher efficiency.

References

1. Fahim KH, Alfayydh EM, Dhahad HA (2017) Effect of geometric design of the flow fields plat on the performance of a PEM fuel cell: a review. Int J Sci Eng Res 8(7):25–34
2. Chen YS (2009) A segmented model for studying water transport in a PEM fuel cell. Ph.D. dissertation in Mechanical Engineering Department, University of Michigan, USA
3. Maher AR, Sadiq AB (2008) CFD models for analysis and design of PEM fuel cells. Nova SciencePublisher, Inc., New York
4. Oberoi AS (2015) Reversible electrochemical storage of hydrogen in activated carbons from Victorian brown coal and other precursors. Doctor of Philosophy (Ph.D.), Aerospace, Mechanical and Manufacturing Engineering, RMIT University
5. Kapoor D, Oberoi AS, Nijhawan P (2019) Hydrogen production and subsequent adsorption/desorption process within a modified unitized regenerative fuel cell. Processes 7(238):1–18
6. Fesser JP, Prasad AK, Advani SG (2007) Particle image velocimetry measurements in a model proton exchange membrane fuel cell. J Fuel Cell Sci Technol 4:328–335
7. Udoetok ES (2008) An investigation into fuel cells and flow cytometers optimal design. Ph.D. dissertation in the Department of Mechanical Engineering, Louisiana State University, USA
8. Hamilton PJ, Pollet BJ (2010) Polymer electrolyte membrane fuel cell (PEMFC) flow field plate: design, materials, and characterization. Fuel Cells 10(4):489–509
9. Yuan XZ, Wang H, Zhang J, Wilkinson DP (2005) Bipolar plates for PEM fuel cells: from materials to processing. J New Mater Electrochem Syst 8:257–267
10. Turan C, Cora ON, Koc M (2011) Effect of manufacturing processes on contact resistance characteristics of metallic bipolar plates in PEM fuel cells. Int J Hydrogen Energy 36:12370–12380

11. Park S, Lee JW, Popov BN (2012) A review of gas diffusion layer in PEM fuel cells: materials and designs. Int J Hydrogen Energy 37:5850–5865
12. Jindal H, Oberoi AS, Sandhu IS, Chitkara M, Singh B (2020) Graphene for hydrogen energy storage—a comparative study on GO and rGO employed in a modified reversible PEM fuel cell. Int J Energy Res 1–12

Effects of Ambient Condition on the Performance of Ammonia Based Loop Heat Pipe

Shail N. Shah, Fagun A. Pithadiya, and Sanjay V. Jain

1 Introduction

Loop heat pipe (LHP) is a heat transfer device used for thermal management in applications like spacecraft, mobile, laptop, electronics cooling etc. Packages used in spacecrafts generate heat which needs to be controlled within the threshold limit. Loop heat pipe takes heat from the heated surface and working fluid gets converted into vapor form. Wick structure is used within LHP to allow the fluid, flow back to evaporator under different orientations. LHP was developed in the year 1972 by Maydanik et al. [1] at the Ural Polytechnical Institute of Thermal Physics. LHP of 1.2 m length with a capacity of approximately 1 kW was tested by them. Conventional heat pipe (HP) consists of single wick whereas LHP has secondary wick as shown in Fig. 1. The major structural difference between HP and LHP is addition of secondary wick and C.C. in LHP. Heat source and heat sink are separated by transport lines in LHP that leads to more flexibility in LHP. Dry out condition occurs when primary wick is not having sufficient liquid. During such condition, due to pressure difference between evaporator and compensation chamber (C.C.), liquid will flow from C.C. through secondary wick to evaporator section and subsequently to primary wick and LHP will be able to take higher heat loads from the heat source. Conventional heat pipe has major limitations of dry out and length whereas LHP can be operated at higher heat loads and for longer distance between evaporator and condenser section. LHP for long distance heat transport up to 10 m and heat load of 1000 W was studied by Nakamura et al. [2].

S. N. Shah (✉) · F. A. Pithadiya · S. V. Jain
Nirma University, Ahmedabad, India
e-mail: shailthermal@gmail.com

S. V. Jain
e-mail: sanjay.jain@nirmauni.ac.in

S. Doolla et al. (eds.), *Advances in Clean Energy and Sustainability*,
Green Energy and Technology, https://doi.org/10.1007/978-981-99-2279-6_15

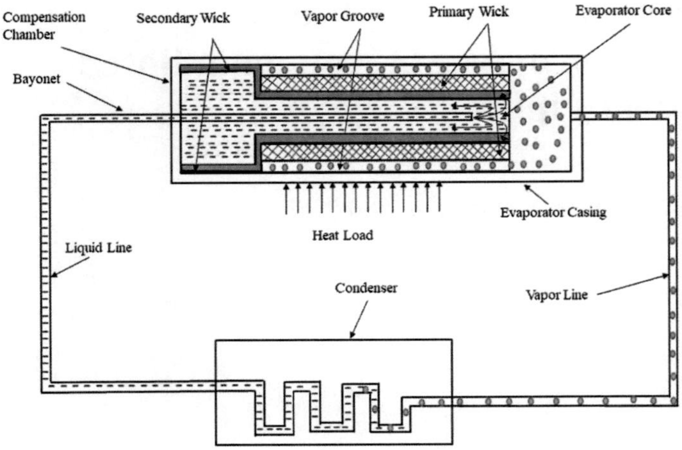

Fig. 1 Schematic diagram of LHP

Working fluid rejects the heat in condenser section or to cold surface/ambient in actual application. Separate liquid and vapor line are available in LHP which helps in using it for longer distance between heat source and heat sink.

First mathematical model of LHP was developed by Kaya et al. [3]. Single phase pressure drop was considered for modeling. Kaya et al. [4] later advanced model with two phase analysis and explained significance of two-phase pressure drop at higher heat loads. It was observed that with increase in heat load, condenser two phase pressure drop increases which leads to increased overall pressure drop across LHP. However, at lower heat loads, this effect was found to be insignificant. Capillary pressure should always be higher than the total pressure drop for the passive working of LHP. Two phase pressure drop analysis was found to be significant to decide passive heat load carrying capacity of the LHP. Effect of ambient condition on steady state operating temperature (SSOT) was studied. Mathematical model with single layer and two-layer compound wick was developed by Bai et al. [5]. Algorithm of LHP modeling was explained in depth by Weng and Leu [6]. Energy balance was applied at each section and two-phase pressure drop was considered. Two phase pressure drop in condenser section has significant effect on the working of LHP at higher heat loads. Few models are developed for two phase pressure drop over the years [4–8].

Ambient effect was not considered in initial models. Hamdan and Elnajjar [9] studied thermodynamic behavior of LHP and effect of different parameters on the performance of LHP. Adoni et al. [10] developed steady state model by applying energy and mass conservation equations and explained two phase heat transfer coefficient and pressure drop in depth. To overcome limitation in distance between heat source and sink, Zhang et al. [11] analyzed pump assisted LHP for longer distance application. In recent years, LHP is widely used in applications like space, mobile, laptop, electronic cooling [12–14] for thermal management. Miniature and micro LHP are under immense investigation for terrestrial applications [15–20].

Ambient condition has significant effect on the performance of LHP. From the literature review, it was found that very few researchers have exhaustively studied the combined effects of ambient condition on different parameters of LHP. The novelty of this paper is to analyze the effects of ambient conditions on the parameters like steady state operating temperature, condenser outlet temperature, liquid line outlet temperature, pressure drops across liquid line, vapor line, wick and condenser two-phase, heat exchange from compensation chamber and liquid line to ambient. The results obtained with the basic model are validated with the results published in the standard literature.

2 Modeling of LHP

Modeling of LHP was done for ammonia as a working fluid. Structural dimensions were considered as per literature for validation purpose [6] (Table 1).

The total applied heat load (Q_l) is distributed among the evaporator (Q_{evp}), heat leak from evaporator compensation chamber (Q_{HL}) and heat rejection from C.C. and liquid line to ambient ($Q_{c.c.\text{-}amb}$, $Q_{ll\text{-}amb}$) as per Eq. (1):

$$Q_l = Q_{evp} + Q_{HL} + Q_{ll\text{-}amb} + Q_{c.c.\text{-}amb} \tag{1}$$

Evaporator wall temperature (T_{evp}) was calculated using Eq. (2) developed by Chuang [21]

$$q'' = \sum_{n=0}^{n=4} C_n (T_{evp} - T_{c.c.})^n \tag{2}$$

Vapor temperature was calculated using Eq. (3) [9] and by applying Clausius Clapeyron equation for saturation temperature.

Table 1 Structural parameters of LHP

Components	Dimensions (OD × ID × Length)	Unit
Evaporator	Φ 24.1/19.1 × 610	mm
Compensation chamber	Φ 69.3/68.3 × 160	mm
Vapor line	Φ 6.4/5.3 × 740	mm
Liquid line	Φ 6.4/5.3 × 970	mm
Condenser	Φ 6.4/4.6 × 4650	mm
Wick	Φ 19.1/9.5 × 610	mm
Wick porosity	60%	–
Pore radius	1.6	μm

$$T_h = T_{evp} - \left(T_{evp} - T_{evpsat}\right)e^{\left(\frac{h_{evp}A\Delta L}{mC_pL}\right)} \tag{3}$$

Ambient heat exchange was considered in liquid line sand Eq. (4) for flow through pipe and with ambient heat exchange was applied for T_{lo}.

$$T_{lo} = T_{amb} + (T_{co} - T_{amb})e^{\left(\frac{-L}{mRC_p}\right)} \tag{4}$$

Equation (5) for pressure drop across wick was developed analytically by Kaya et al. [4].

$$\Delta P_w = \frac{\mu m ln\left(\frac{d_{o,w}}{d_{i,w}}\right)}{2\pi\rho L_{wick}K_{wick}} \tag{5}$$

Lockhart model [21] was used for two phase analysis across condenser section and Eqs. (6–10) were applied.

$$\Delta P_{2\phi} = \int_{x_{in}}^{x_{out}} \phi_l^2 \frac{f_l}{D}\frac{(1-x)\rho_l u^2}{2}\left[\frac{dz}{dx}\right]dx \tag{6}$$

$$\phi_l = \left[1 + \left[\frac{C}{X}\right] + \left[\frac{1}{X^2}\right]\right] \tag{7}$$

$$X = \left[\frac{f_l}{f_g}\right]^{0.5}\left[\frac{\rho_g}{\rho_l}\right]^{0.5}\left[\frac{1-x}{x}\right] \tag{8}$$

$$\alpha = \left[1 + 0.28X^{0.71}\right]^{-1} \tag{9}$$

$$L_{2\phi} = \int_{x_{in}}^{x_{out}} \frac{Q_l}{T_{sat} - T_{sink}} * \left[\left[\frac{UA}{L}\right] + \left[\frac{UA}{L}\right]\left[\frac{T_{sat} - T_{amb}}{T_{sat} - T_{sink}}\right]\right]^{-1} \tag{10}$$

where, q″ = Heat flux, ϕ_l = two phase multiplier, α = void fraction, $d_{o,w}$ = wick outer diameter, $d_{i,w}$ = wick inner diameter, X = martinelli multiplier, T_{sink} = sink temperature, x = vapor quality, K_{wick} = thermal conductivity of the wick material, $\frac{UA}{L}$ = overall heat transfer co-efficient, f = darcy friction factor, l = liquid phase, g = vapor phase

Heat exchange and heat leak were calculated by applying Eqs. (11–13).

$$Q_{c.c.\text{-}amb} = \frac{(T_{c.c.} - T_{amb})}{R_{c.c.}} \tag{11}$$

$$Q_{sc} = mC_p(T_{c.c.} - T_{lo}) \tag{12}$$

$$Q_{HL} = \frac{2 * \pi * \text{Keff} * \text{Lwick} * \Delta\text{Twick}}{\ln\left(\frac{D_{o,w}}{D_{i,w}}\right)} \tag{13}$$

2.1 *Flowchart of Modeling*

Modeling was done in MATLAB software as per flowchart presented in Fig. 2 to find Steady State Operating Temperature (SSOT) for any applied heat load. Fifth order polynomial equations given by Faghri [22] were used to find thermo-physical properties like k, C_p, C_v, h_{fg}, μ, σ, ρ for both liquid and vapor phase. Iterations were run till convergence criteria $Q_{HL} - Q_{sc} - Q_{c.c.\text{-}amb} < 0.001$ was achieved at the specified heat load. Total pressure drop across the loop should always be lower than the capillary pressure for the passive working of LHP. Steady state is achieved when both above conditions are satisfied.

3 Validation and Effects of Ambient Condition

3.1 *Validation of Two-Phase Modeling of LHP*

Two phase modelling of LHP was done as per the flow chart shown in Fig. 2 using MATLAB software and was validated with different models available in open literature viz., Kaya [4], Weng [6], Mishkinis [23], Chuang [21], and Cavillini [24] at different heat loads as shown in Fig. 3. Validation of SSOT and $\Delta P_{2\phi}$ with varying heat load from 20 to 600 W is shown in Fig. 3. Present model was found to be in line with above models and was found best suited with Weng [6] model. SSOT reduced to 17.26 °C at 150 W and increased to 26.08 °C at 600 W. Two-phase pressure drop model was validated with different models available in open literature as shown in Fig. 3b and was found to be within 3.76% with Chuang model [21]. It was observed that $\Delta P_{2\phi}$ does not have significant effect on pressure drop at lower heat loads up to 150 W but beyond that two-phase pressure drop needs to be analyzed to secure the passive flow of fluid in the LHP.

3.2 *Effects of Ambient Condition on the Performance of LHP*

In the present study, effects of ambient conditions (i.e., 5 °C, 19 °C and 25 °C) on the parameters like i.e., SSOT, T_{co}, T_{lo}, ΔP_{ll}, ΔP_{vl}, ΔP_w, $\Delta P_{2\phi}$, $Q_{c.c.\text{-}amb}$, $Q_{ll\text{-}amb}$ were

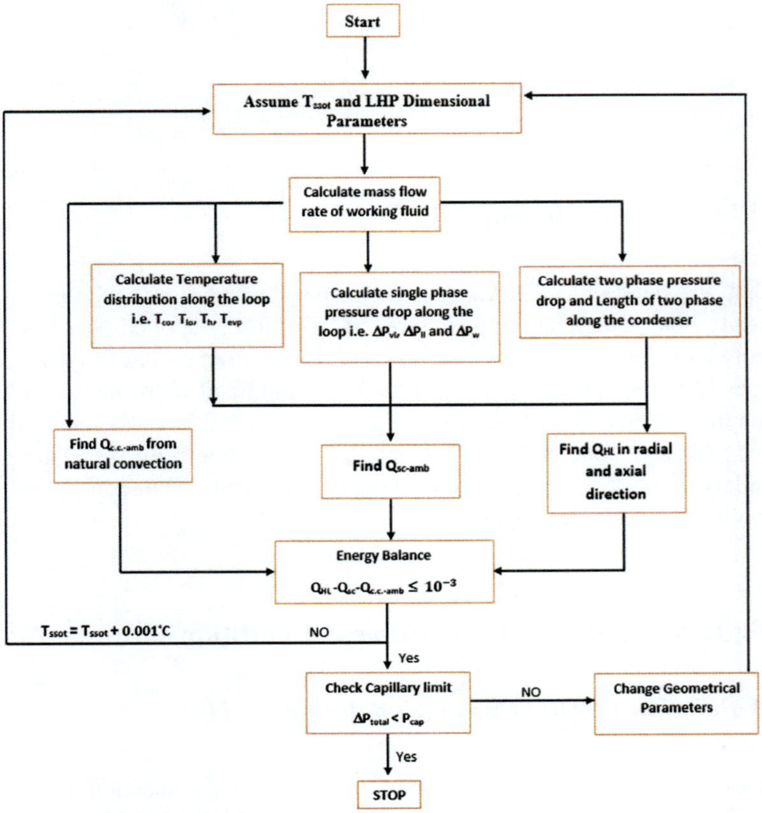

Fig. 2 Flowchart of modeling

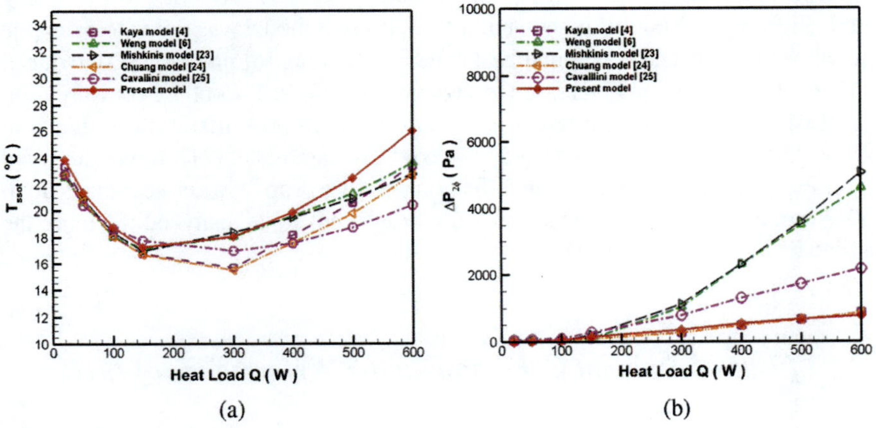

Fig. 3 **a** SSOT and **b** $\Delta P_{2\phi}$ validation

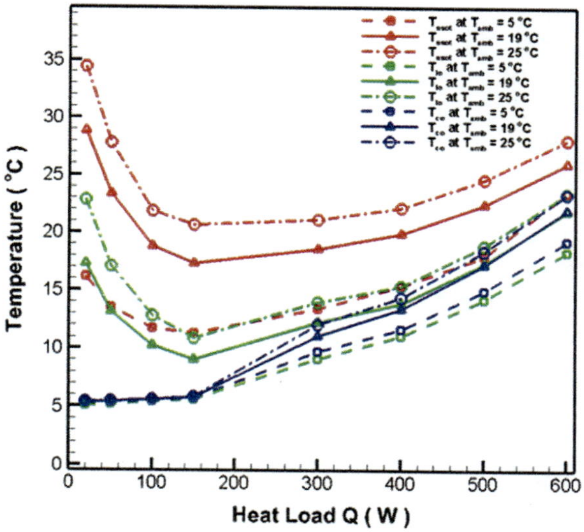

Fig. 4 (SSOT, T_{co}, T_{lo}) – Q (W) at T_{amb} = **a** 5 °C, **b** 19 °C, **c** 25 °C

analyzed. Effect of T_{amb} (5 °C, 19 °C and 25 °C) on SSOT, T_{co} and T_{lo} is shown in Fig. 4.

Increase in loop temperatures was observed with increase in T_{amb} due to transfer of heat from ambient to the system. Increase in SSOT, T_{co} and T_{lo} was seen respectively as 4.59 °C, 4.06 °C and 5.35 °C at 600 W when T_{amb} was increased from 5 to 25 °C. Same trend was observed by Chuang [21]. T_{co} was found to be higher than T_{lo} at lower ambient temperature due to heat transfer from liquid line to ambient while T_{lo} was found to be higher than T_{co} at higher T_{amb} due to heat transfer from ambient to liquid line.

Effect of ambient condition on pressure drop across the loop at heat load 50 W and 600 W at T_{amb} = 5 °C, 19 °C and 25 °C are shown in Figs. 5 and 6. It was observed from Fig. 5 that pressure drop across the wick is having dominant effect on overall pressure drop. This was due to the fact that loop temperature was lower at lower heat load which resulted in lesser pressure drop in the condenser section. ΔP_w at 50 W was found to be 91.5%, 90% and 89% of overall pressure drop at T_{amb} = 5 °C, 19 °C and 25 °C respectively at 50 W. It can be seen from Fig. 5 that ambient condition does not have significant effect on overall pressure drop at lower heat loads.

$\Delta P_{2\phi}$ was found to be increasing from 5 to 31.5%, 7 to 39% and 8 to 42% at T_{amb} = 5 °C, 19 °C and 25 °C respectively when heat load was varied from 50 to 600 W. At the same time ΔP_w reduced not in terms of value but in terms of proportion in overall pressure drop across the LHP. $\Delta P_{2\phi}$ at 600 W was found to be 31.5%, 39% and 42% of overall pressure drop at T_{amb} = 5 °C, 19 °C and 25 °C respectively. Loop temperature should increase with increase in the ambient temperature due to transfer of heat from surrounding to the system. Increased loop temperature leads to higher condenser inlet temperature and thus higher two-phase pressure drop. Same

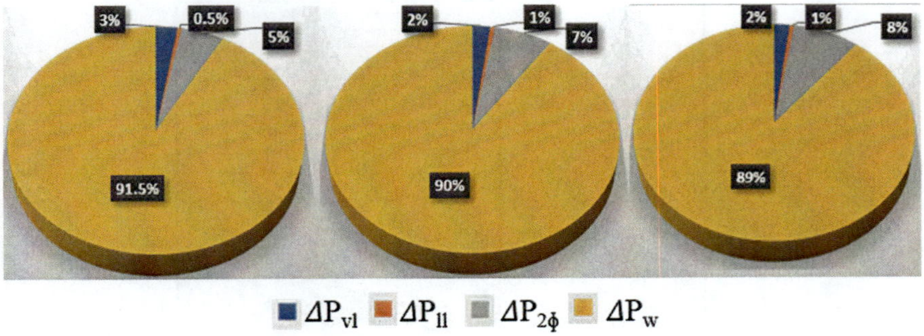

Fig. 5 $(\Delta P_{ll} + \Delta P_{vl} + \Delta P_w + \Delta P_{2\phi})$ at Q = 50 W and T_{amb} = **a** 5 °C, **b** 19 °C, **c** 25 °C

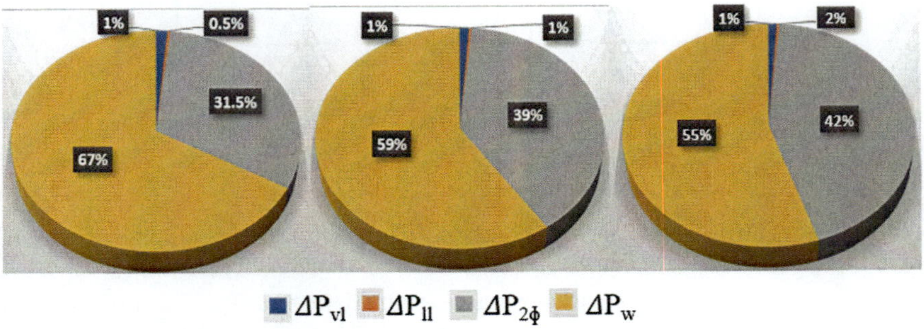

Fig. 6 $(\Delta P_{ll} + \Delta P_{vl} + \Delta P_w + \Delta P_{2\phi})$ at Q = 600 W and T_{amb} = **a** 5 °C, **b** 19 °C, **c** 25 °C

trend was observed in Fig. 6. It can be understood that, condenser $\Delta P_{2\phi}$ and ΔP_w both have significant effect on ΔP_{total} at higher heat load. Same observation about significance of $\Delta P_{2\phi}$ and ΔP_w at higher heat loads was made by Hoang [4].

Two phase pressure drop in condenser section $(\Delta P_{2\phi})$ was observed to be increasing with increase in heat load Q (W) and T_{amb} as shown in Fig. 7. Increase in heat load leads to increased evaporator temperature that leads to increased condenser two phase length as well as two phase pressure drop. Increase of 359.80 W (63.84%) in $\Delta P_{2\phi}$ was observed at 600 W when T_{amb} was increased from 5 to 25 °C. It can be understood that for passive flow of LHP, $\Delta P_{2\phi}$ plays a significant role at higher heat loads and ambient condition has significant effect on it. Present results are compared in percentage form as well which helps in predicting results for LHP with different structural parameters.

This was due to the fact that at higher heat loads, higher operating temperature was seen and that leads to larger $L_{2\phi}$ in condenser section which increases the $\Delta P_{2\phi}$.

SSOT increased with increase in heat load from 20 to 600 W as shown in Fig. 8a due to increased evaporator temperature. Heat exchange between C.C. and ambient is dependent on SSOT and T_{amb} values at a specific heat load. Absolute value of $Q_{c.c.\text{-}amb}$ is shown in Fig. 8a on Y-2 axis. Very less ambient temperature should lead to increased heat exchange from C.C. to ambient due to the higher difference in

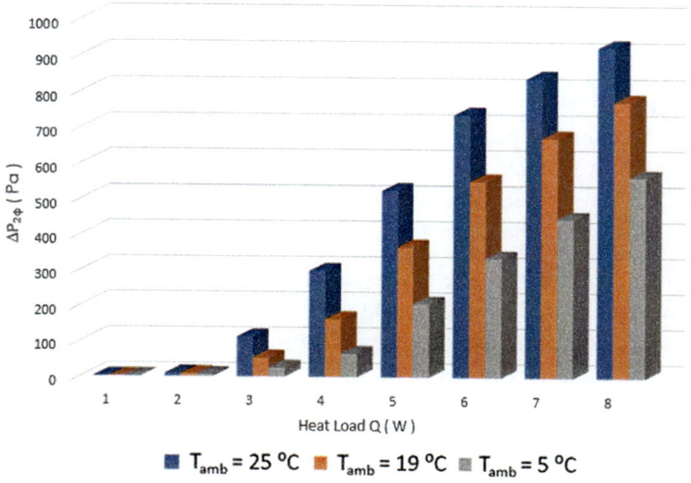

Fig. 7 $\Delta P_{2\phi}$ − Q (W) at T_{amb} = **a** 5 °C, **b** 19 °C, **c** 25 °C

temperature. At 600 W and T_{amb} = 25 °C, SSOT was found to be 28.08 °C, which is nearer to ambient temperature and leads to lesser $Q_{c.c.\text{-}amb}$ at higher T_{amb} as can be seen from Fig. 8a. Heat exchange between ambient and liquid line depends on the ΔT between T_{amb} and T_{co} as shown in Fig. 8b. T_{co} was observed to be increasing with increase in heat load due to increased evaporator temperature which leads to increased ΔT at lower ambient temperature and ΔT keeps on reducing with increase in ambient temperature as shown on Y-2 axis. It indicates ambient condition has significant effect on $Q_{ll\text{-}amb}$.

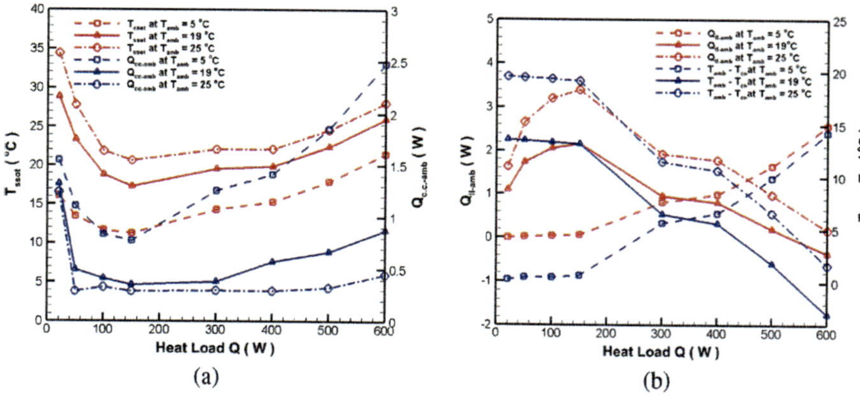

Fig. 8 Effect of T_{amb} on **a** $Q_{c.c.\text{-}amb}$, **b** $Q_{ll\text{-}amb}$

4 Conclusions

In the present study, mathematical model was developed to understand the effect of ambient condition on the performance of LHP. The two-phase model developed in the current study was found to be in close proximity with the model presented by Chuang [21] and the deviation between these two models was found to be 3.76% in terms of pressure drop. The major conclusions drawn from the work are as under:

- When an ambient temperature (T_{amb}) was increased from 5 to 25 °C, SSOT, T_{co} and T_{lo} were increased by 4.59 °C, 4.06 °C and 5.35 °C respectively at 600 W.
- It was found that the pressure drop across the transmission line has very less effect on the overall pressure drop of LHP.
- With an increase in T_{amb} from 5 to 25 °C, two phase pressure drop in condenser section ($\Delta P_{2\phi}$) increased by 11.5% and pressure drop in wick (ΔP_w) reduced by 12% at 600 W in terms of proportion in overall pressure drop.
- The heat exchanged from compensation chamber and liquid line to ambient ($Q_{c.c.\text{-}amb}$ and $Q_{ll\text{-}amb}$) reduced by 2.03 W and 2.38 W respectively when T_{amb} was increased from 5 to 25 °C at 600 W.

References

1. Maydanik Y et al (1985) Heat transfer apparatus. U.S. Patent no. 4515209
2. Nakamura K, Odagiri K, Nagano H (2016) Study on a loop heat pipe for a long-distance heat transport under anti-gravity condition. Appl Therm Eng 107:167–174
3. Kaya T, Hoang T, Ku J, Cheung M (1999) Mathematical modeling of loop heat pipes. In: 37th aerospace sciences meeting and exhibit 13
4. Hoang T (1999) AIAA 99-3448 mathematical modeling of loop heat pipes with two-phase pressure drop 33rd thermophysics conference 28 June–1 July, 1999, Norfolk. VA for permission to copy or republish, contact the American Institute of Aeronautics and Astronautics
5. Bai L, Lin G, Zhang H, Wen D (2009) Mathematical modeling of steady-state operation of a loop heat pipe. Appl Therm Eng 29:2643–2654
6. Weng C, Leu T (2016) Two-phase flow pattern based theoretical study of loop heat pipes. Appl Therm Eng 98:228–237
7. Zhu L, Yu J (2016) Simulation of steady-state operation of an ejector-assisted loop heat pipe with a flat evaporator for application in electronic cooling. Appl Therm Eng 95:236–246
8. Kim J, Golliher E (2002) Steady state model of a micro loop heat pipe. In: Annual IEEE semiconductor thermal measurement and management symposium, pp 137–144
9. Hamdan M, Elnajjar E (2009) Thermodynamic analytical model of a loop heat pipe. Heat Mass Transf 46:167–173
10. Adoni A, Ambirajan A, Jasvanth V et al (2007) Thermohydraulic modeling of capillary pumped loop and loop heat pipe. J Thermophys Heat Transf 21:410–421
11. Zhang H, Jiang C, Zhang Z et al (2020) A study on thermal performance of a pump-assisted loop heat pipe with ammonia as working fluid. Appl Therm Eng 175:115342
12. He S, Zhou P, Liu W, Liu Z (2020) Experimental study on thermal performance of loop heat pipe with a composite-material evaporator for cooling of electronics. Appl Therm Eng 168:114897
13. Vasiliev L, Lossouarn D, Romestant C et al (2009) Loop heat pipe for cooling of high-power electronic components. Int J Heat Mass Transf 52:301–308

14. Adoni A, Ambirajan A, Jasvanth V, Kumar D (2009) Effects of mass of charge on loop heat pipe operational characteristics. J Thermophys Heat Transf 23:346–355
15. Yu M, Chen F, Zheng S et al (2019) Experimental investigation of a novel solar micro-channel loop-heat-pipe photovoltaic/thermal (MC-LHP-PV/T) system for heat and power generation. Appl Energy 256:113929
16. Shioga T, Mizuno Y (2015) Micro loop heat pipe for mobile electronics applications. In: Annual IEEE semiconductor thermal measurement and management symposium 2015-April, pp 50–55
17. Choi J, Sano W, Zhang W et al (2013) Experimental investigation on sintered porous wicks for miniature loop heat pipe applications. Exp Therm Fluid Sci 51:271–278
18. Maydanik Y, Vershinin S, Pastukhov V, Fried S (2010) Loop heat pipes for cooling systems of servers. IEEE Trans Compon Packag Technol 33:416–423
19. Anand A, Jaiswal A, Ambirajan A, Dutta P (2018) Experimental studies on a miniature loop heat pipe with flat evaporator with various working fluids. Appl Therm Eng 144:495–503
20. Chu C, Wu S, Chen P, Chen Y (2004) Design of miniature loop heat pipe. Heat Transf Asian Res 33:42–52
21. Chuang P (2003) An improved steady-state model of loop heat pipes based on experimental and theoretical analyses. Ph.D. thesis, Pennsylvania State University
22. Faghri A. Heat pipe science and technology. Taylor & Francis
23. Mishkinis D, Ochterbeck J (2003) Analysis of tube side condensation in microgravity and earth-normal gravity. In: Proceedings of the fifth Minsk international seminar heat pipes, heat pumps, refrigerators, Minsk, Belarus
24. Cavallini A, Censi G et al (2002) Condensation of halogenated refrigerants inside smooth tubes. HVAC&R Res 8:429–451
25. Faghri A, Zhang Y (2006) Transport phenomena in multiphase system. Elsevier, Burlington, MA

Design and Development of Artificial Neural Network-Based Prediction Model for Hemispherical Solar Still

Badduru Chinna Savaraiah, Siddharth Ramachandran, and Naveen Kumar

1 Introduction

Water is the one of the main things in human life. Without water, human beings can't even think of surviving on earth. But now days due to industries and some other causes, pollution of water is increasing. On earth over 97% of water is in oceans and which can't able to use because of salt nature of sea water. Solar still is the device which converts the sea water into potable water. Solar distillation is one of the best methods to convert sea water into potable water. For enhancing the water productivity of solar still, researchers have tried various methods such as modifications to absorber plate, use of dyes, sponges, condenser units, external and internal reflectors, nano particles etc. In these enhancing methods modifications to glass cover is one method. In this shape of glass cover will change such as single slope, double slope, pyramid shape and dome shape etc. In dome shape, based on dome cap angle classifications are there. If dome cap angle is 90-degrees, then it is called as hemispherical.

A daily water productivity output from the hemispherical solar still from 2.8 to 5.7 L/m^2 day was achieved by Ismail [1]. Arunkumar et al. [2] have done experimental study on hemispherical solar still with top cover cooling effect. The results showed efficiency of solar still was improved from 34 to 42% with top cover cooling effect. Development of a predictive based models can give estimation of the water productivity from a solar still before experiments. This can save a lot of time and money [3]. Several heat and mass transfer models have been proposed by researchers to predict the solar still water productivity, such as Dunkle [4], Clark [5], Kumar and

B. Chinna Savaraiah (✉) · S. Ramachandran
Department of Mechanical Engineering, Indian Institute of Information Technology, Design and Manufacturing Kancheepuram, Chennai 600127, India
e-mail: mds20m003@iiitdm.ac.in

N. Kumar
Department of Science and Humanities, Indian Institute of Information Technology, Design and Manufacturing Kancheepuram, Chennai 600127, India

S. Doolla et al. (eds.), *Advances in Clean Energy and Sustainability*,
Green Energy and Technology, https://doi.org/10.1007/978-981-99-2279-6_16

Tiwari [6] etc. But while predicting the water productivity from these models the major problem faced is complexity involved in these heat and mass transfer models [7]. But, by using ANN for predicting the solar still water productivity, it can predict much easily and reliably compared to other heat and mass transfer models [3].

From the literature survey, it is found that water productivity prediction model for hemispherical solar still has not been developed. So, this paper presents a novel ANN performance prediction model of hemispherical solar still water productivity.

2 Design of Solar Still

This hemispherical solar still model was developed by modifying the single slope solar still into hemispherical solar still. The single slope solar still model was adopted from Nazari et al. [8]. Modifications was done in a such a way, changing the rectangular basin of single slope solar still into hemispherical solar still circular basin with same basin area. The basin area of single slope solar still is 0.5 m^2. After converting that area into circular area, radius of hemispherical solar still basin will be 0.3989 m. With equal radius of circular basin, hemispherical glass cover was modelled. All properties and other specifications of solar still was adopted from Nazari et al. [8] to use in theoretical analysis of hemispherical solar still, except from the hemispherical collector area, which was considered as 1 m^2 (Fig. 1).

In this theoretical study, optimization of developed hemispherical solar still was done by changing the dome cap angle, here two cap angles were taken i.e., 75° and 60°. Here, Area and mass of glass will change when cap angle of dome changes. Because change in cap angle will leads to change in radius and height of glass. For

(a) (b)

Fig. 1 **a** Single slope solar still [8], **b** basic structure of hemispherical solar still

(a) (b)

Fig. 2 **a** 75° cap angle dome shaped solar still, **b** 60° cap angle dome shaped solar still

75 and 60° cap angle dome shaped solar stills, area and mass of glass cover are 0.715 m^2, 0.433 m^2, 7.15 kg and 4.33 kg respectively (Fig. 2).

3 Theoretical Analyses

3.1 Mathematical Modelling for Solar Radiation on Hemispherical Cover

Solar radiation falling on hemispherical cover was found by dividing the total surface area into in small flat surfaces. Consider the element on hemispherical surface and angular orientations of this element are i = angle of incidence, β = tilt angle and γ = surface azimuth angle. β and γ can be relate to ϕ and θ from Fig. 3 [9]:

$$\beta = 90 - \phi \tag{1}$$

$$\gamma = \theta \tag{2}$$

The total solar radiation on hemispherical can be calculated by using following surface integral [9]:

$$I_t = \iint (I_d + I_b)dA = \int_0^{2\pi} \int_0^{\frac{\pi}{2}} (I_d + I_b) r^2 \cos\phi d\phi d\theta \tag{3}$$

Here I_b and I_d are beam and diffuse radiations falling on one element. I_b can be found by following equation [9].

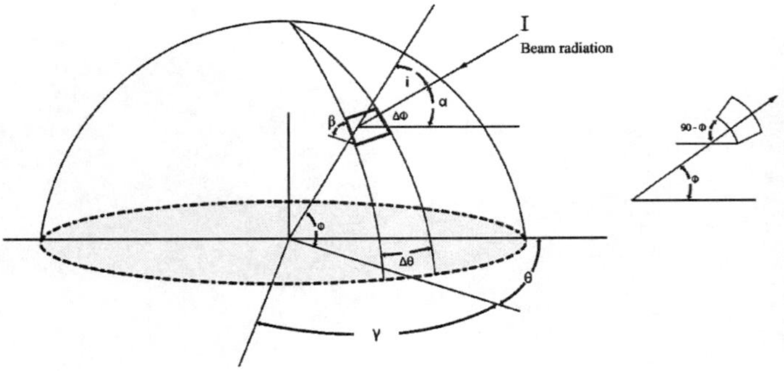

Fig. 3 Solar radiation on hemisphere [9]

$$I_b = \frac{I_h}{\cos\theta_z}\cos i \tag{4}$$

Here, I_h is beam radiation falling on horizontal surface.

The angle of incidence for any tilted surface is given by [9]:

$$\begin{aligned}\cos i &= \sin\delta(\sin L\cos\beta - \cos L\sin\beta\cos\gamma)\\ &+ \cos\delta\cos\omega(\cos L\cos + \sin L\sin\beta\cos\gamma)\\ &+ \cos\delta\sin\beta\sin\gamma\sin\omega\end{aligned} \tag{5}$$

By Eqs. 1, 2 and 5, Eq. 5 can be rewritten as [9]:

$$\begin{aligned}\cos i &= \sin\delta\sin L\sin\Phi - \sin\delta\cos L\cos\Phi\cos\theta\\ &+ \cos\delta\cos L\sin\Phi\cos\omega + \cos\delta\sin L\cos\Phi\cos\theta\cos\omega\\ &+ \cos\delta\cos\Phi\sin\theta\sin\omega\end{aligned} \tag{6}$$

To calculate the I_h and I_d, same methodology has been adopted which Arunkumar et al. [2] used.

3.2 *Mathematical Modelling for Thermal Analysis*

Mathematical modelling for solar still was performed by taking energy balance equations in absorber plate, saline water (feed water) and glass cover to attain the basin plate, water and glass cover temperatures.

Energy Balance Equations

Energy balance equation for glass cover, water and absorber plate can be carried out such that the sum of energy received by these components is equals to sum of energy leaving from glass cover, water and absorber plater. Energy balance equations for

Table 1 Parameters in thermal analysis

Parameter	Equation
$h_{c,w-g}$	$0.884\left[T_w - T_g + \frac{(P_w-P_g)(T_w+273)}{268.9\times 10^3-P_w}\right]^{1/3}$
$h_{e,w-g}$	$\frac{0.016237\times h_{c,w-g}\times(P_w-P_g)}{(T_w-T_g)}$
$h_{r,w-g}$	$\varepsilon_{eff}\,\sigma\left[(T_w+273)^2+(T_g+273)^2\right]\left[(T_w+T_g+546)\right]$
$h_{c,b-w}$	Nu. k/L_C
P_w	exp (25.317 – (5144/T_w))
P_g	exp (25.317 – (5144/T_g))

glass cover, water and basin plates are expressed in Eqs. 7, 8, and 9 respectively [8] (Table 1).

$$\alpha_g A_g I_s(t) + h_{t,w-g} A_w (T_w - T_g) = m_g (C_p)_g \cdot \frac{dT_g}{dt} + h_{t,g-a} A_g (T_g - T_a) \tag{7}$$

$$\begin{aligned}\alpha_w \tau_g A_w I_s(t) + h_{c,b-w} A_w (T_b - T_w) = (\rho_w A_w d_w)(C_p)_w \frac{dT_w}{dt} \\ + h_{t,w-g}(T_w - T_g)\end{aligned} \tag{8}$$

$$\begin{aligned}\alpha_b \tau_g \tau_w A_b I_s(t) = m_b (C_p)_b \cdot \frac{dT_b}{dt} + h_{c,b-w} A_b (T_b - T_w) \\ + h_b (A_b + A_{sides})(T_b - T_a)\end{aligned} \tag{9}$$

where,

$$h_{t,g-a} = 5.7 + 3.8V \tag{10}$$

$$h_{t,w-g} = h_{c,w-g} + h_{e,w-g} + h_{r,w-g} \tag{11}$$

Here, $h_{c,w-g}$, $h_{e,w-g}$ and $h_{r,w-g}$ are convective, evaporative and radiative heat transfer coefficients between glass cover and water surface respectively. $h_{c,b-w}$ is convective heat transfer coefficient between absorber plate and water [8].

$$\varepsilon_{eff} = \left(\frac{1}{\varepsilon_w} + \frac{1}{\varepsilon_g} - 1\right)^{-1} \tag{12}$$

$$\begin{cases} Nu = 0.54 Ra^{1/4} & (10^4 \le Ra \le 10^7) \\ Nu = 0.15 Ra^{1/3} & (10^7 \le Ra \le 10^{11}) \end{cases} \tag{13}$$

Here, Ra = Rayleigh number, which can be obtained by [8]:

$$Ra = \frac{g\beta^{-1}(T_b - T_w)\rho^2 L_c^3 (C_P)}{\mu.k} \tag{14}$$

$$h_b = \left(\frac{L_i}{K_i} + \frac{1}{h_{t,b-a}} \right)^{-1} \tag{15}$$

$$h_{t,b-a} = 5.7 + 3.8V \tag{16}$$

Water productivity of solar still is [8]:

$$\dot{m}_{ev} = \frac{h_{e,w-g} \times (T_w - T_g)}{h_{fg}} \times 3600 \tag{17}$$

h_{fg} is latent heat of water in J/kg, which can be obtained by following equation:

$$h_{fg} = 3{,}161{,}500 - 2407.41 \times T_w \tag{18}$$

where, ε_{eff} is effective emissivity and P_w, P_g are partial pressures of water and glass respectively and V is wind velocity, I is solar radiation. L_C denotes the ratio of surface area to perimeter of the basin. T_w, T_g and T_b represents water, glass and basin temperatures respectively.

Methodology of Thermal Analysis

A Mathematical program developed in MATLAB R2020b for solving energy balance equations of glass cover, water and basin plate to obtain the temperatures of glass cover, water and basin plate. The Runge–Kutta 4th order method was implemented to solve energy balance equations. Calculation period is from 9:00 AM to 4:00 PM. The time interval for calculating the temperatures is considered as 1 s. At every time step i.e., 1-h, initial values temperatures are equal to final temperature of previous step. For initial step calculations, initial temperatures of glass cover, water and basin plate are assumed as ambient temperature. Input parameters are solar radiation of location (i.e., Kermanshah, Iran), wind speed, ambient temperature, constant parameters like solar still specifications and design parameters and thermophysical properties of water shown in Table 2. Similarly thermal analysis has been done to dome shaped solar stills i.e., 75° and 60° cap angle dome shape solar stills.

4 Artificial Neural Network

The neural networks are inspired by human brain. These neural networks are designed to perform like human brain. It consists of minimum three layers i.e., input, output and hidden layer. Two different learning mechanisms (i.e., supervised and unsupervised) will be used to adjust the weights and biases during the training of neural

Table 2 Thermophysical equations for water [8]

Physical properties	Water
Density (kg/m^3), ρ	$1000 \times \left[1.0 - \frac{(T-4.0)^2}{119000+1365\times T-4\times T^2}\right] \pm 0.07\%$
Viscosity (kg/m s), μ	$0.0015 - 3.16325 \times 10^{-5} \times T + 3.04789 \times 10^{-7} \times T^2 - 1.1104 \times 10^{-9} \times T^3 \pm 2.75\%$
Specific heat (J/kg K), C_P	$4217.629 - 3.20888 \times T + 0.09503 \times T^2 - 0.00132 \times T^3 + 9.415 \times 10^{-6} \times T^4 - 2.5479 \times 10^{-8} \times T^5 \pm 2.46\%$
Thermal conductivity (W/m K), k	$0.55994 + 0.00216 \times T - 1.02749 \times 10^{-5} \times T^2 + 6.7279 \times 10^{-9} \times T^3 \pm 1.4\%$
Thermal expansion (K^{-1}), β^{-1}	21×10^{-5}

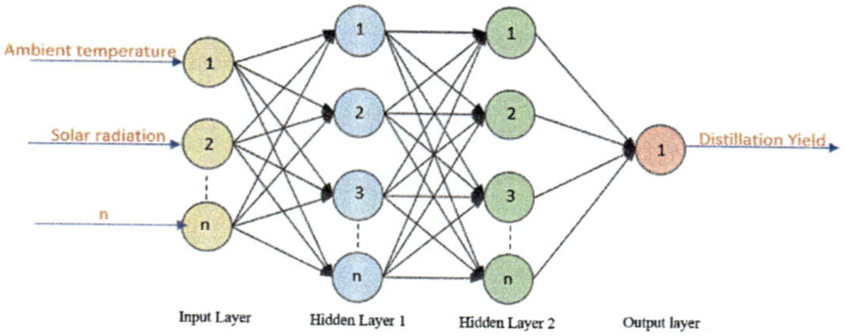

Fig. 4 Schematic view of artificial neural network

network. In supervised learning mechanism, weights and biases are adjusted based on difference between neural network output and actual output. In unsupervised learning mechanism, actual output is unknown so weights and biases are adjusted based on input variables. In this research, supervised learning algorithm has been used (Fig. 4).

4.1 Methodology of ANN

For predicting the water productivity of hemispherical solar still an artificial neural network has been developed in MATLAB R2020b. The feedforward neural network has been adopted in this study. The network consists of three layers i.e., one input layer and one output layer and one hidden layer. The number of nodes in hidden layer has been selected based trial and error method. The Levenberg Marquardt (LM) algorithm has been used to train the model. Total available data is 48. From the available data 70, 15 and 15 percentage of data has been assigned to training,

Table 3 Performance parameters

Performance parameter	Equation
RMSE (Root mean square error)	$\sqrt{\left(\sum_{i=1}^{n}(y' - y)^2\right)/n}$
MAE (Mean absolute error)	$\frac{1}{n}\sum_{i=1}^{n}\lvert y' - y\rvert$
R^2 (Coefficient of determination)	$1 - \frac{\sum(y-y')^2}{\sum(y-\overline{y})^2}$

validation and testing of neural network respectively. Six variables (i.e., solar radiation, time, ambient temperature, glass cover temperature, water temperature and basin temperature) are considered as input parameters for input layer of feedforward neural network. Performance evaluation of training algorithm judged on the basis of the performance parameters mentioned in Table 3.

Here, y is actual output value and y′ is predicted value and n is total number of observations (i.e., total number of available data). Lower RMSE and MAE value means that the model is predicting accurately. If R^2 values is very near to 1, which represents a perfect fit between actual value and predicted value.

5 Results and Discussions

In order to obtain the water productivity of hemispherical solar still, mathematical modellings have been developed. For solving these mathematical modellings, different methodologies have been adopted as discussed in previous chapters. Nazari et al. [8], have done experimental and theoretical analysis for single slope solar still on June 16, 18, 19, 20, 23 and 24 from 9:00 AM to 4:00 PM at Razi university in Kermanshah city, Iran (34.32° N, 47.07° E). For validation, they have compared daily accumulated water for six days of theoretical and experimental values and the comparison was shown in Fig. 5a. By comparison, it is clearly knowing there is good agreement between theoretical and experimental values. The same theoretical analysis which used in Nazari et al. [8] has been adopted in this study for hemispherical solar still thermal analysis because their results are quietly acceptable with experimental values. In hemispherical solar still thermal analysis, area and mass of glass will change due to change shape of glass cover and the remaining properties and specifications are taken same for theoretical analysis because hemispherical solar still was developed by modifying the single slope solar still.

The solar radiation on hemispherical solar still from 9:00 AM to 4:00 PM for 6 days, is shown in Fig. 6a and compared with solar radiation falling on single slope solar still. Later, thermal analysis has done to get glass, water and basin temperatures and water productivity of hemispherical solar still. In thermal analysis, ambient temperature was taken as constant value i.e., 27 °C and initial water depth in solar still was taken by 2 cm. All the constant values used in theoretical analysis of solar still are tabulated in Table 1. After solving energy balance equations of glass cover,

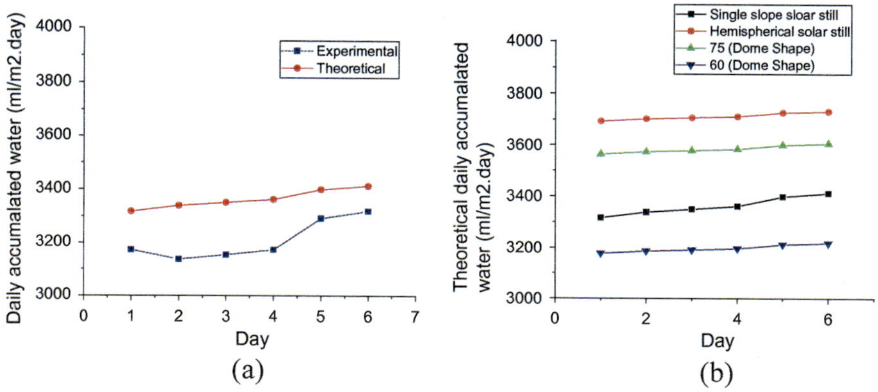

Fig. 5 **a** Experimental and theoretical daily accumulated water for single slope solar still, **b** theoretical daily accumulated water for hemispherical solar still and single slope solar still

water and basin plate, the values of glass, water and basin temperatures are as shown in Fig. 6b. In Figs. 6, 7 and 8, data number in X-axis represents hour of the day i.e. in this study 6 days were taken and 8 h per day, total number of data is 48.

Later, theoretical analysis for dome shaped solar stills has been completed. Daily accumulated water for dome shape solar stills for total six days compared with hemispherical solar still and single slope solar still. The comparison of water productivity is shown in Fig. 5b. The water productivity of hemispherical solar still is producing an average of 350 ml/day compared to single slope under same conditions like ambient temperature, insulation material, basin area, same water depth in basin etc. Solar radiation falling on hemispherical solar still is more compared to single slope solar still at particular time. Due to this, water productivity of hemispherical solar still

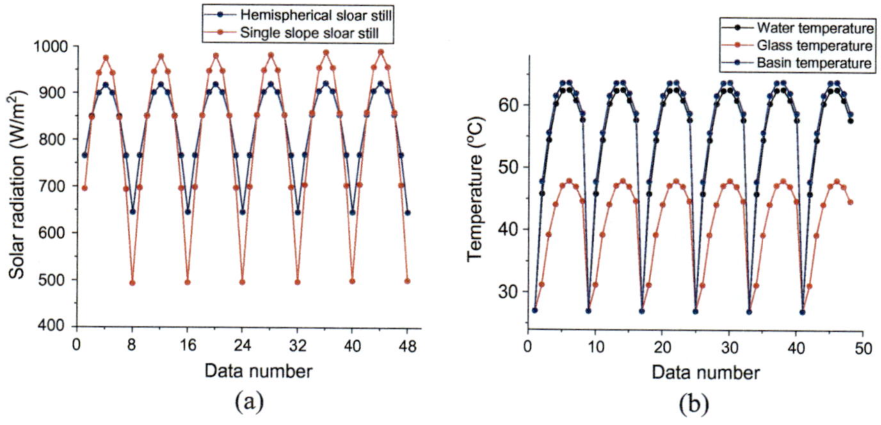

Fig. 6 **a** Solar radiation on single slope and hemispherical solar stills, **b** hemispherical solar still theoretical water, glass and basin temperatures

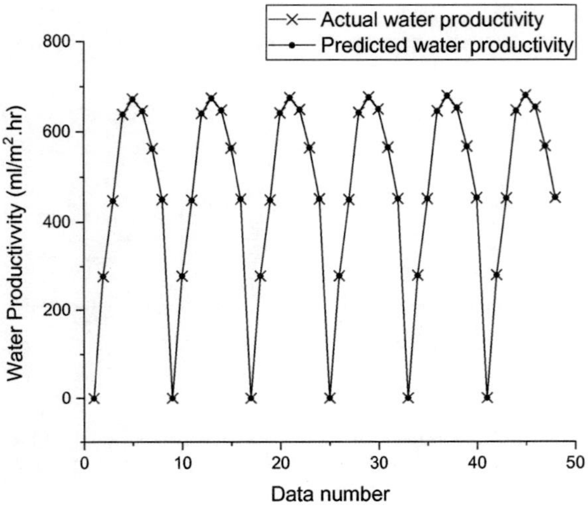

Fig. 7 Comparison of actual and predicted water productivity of hemispherical solar still

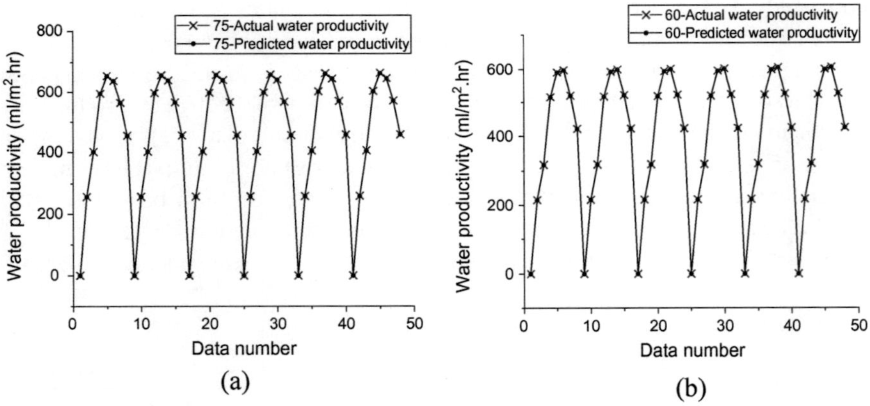

Fig. 8 **a** Comparison of actual and predicted water productivity of 75° cap angle dome solar still, **b** comparison of actual and predicted water productivity of 60° cap angle dome solar still

is more compared to single slope solar still under same conditions. Water productivity for hemispherical solar still is more compared to 75° and 60°. If cap angle is decreasing the radius and height of glass cover is decreasing but radius of basin is still same. So, the remaining portion i.e., difference between basin radius and glass cover will be filled with insulation material at top. As the remaining portion filled with insulation material, the losses from sides of solar still will increase so water productivity of solar still unit will decrease.

These six days hemispherical solar still theoretical data was used to predict the water productivity with help of ANN. Three performance parameters have taken

Table 4 Performance parameter values for ANN analysis

	Training	Validation	Testing
RMSE	0.5443	2.3701	3.5015
MAE	0.3741	1.9739	2.9430
R^2	0.9999	0.9999	0.9993

to check the performance of ANN and those performance parameter values for hemispherical solar still data are given in Table 4.

Here R^2 values of training, validation and testing sets are very close to 1. It means provided data to developed ANN is perfectly fitted. RMSE and MAE values for hemispherical solar still are 1.7390, 2.9660, 3.7181 and 1.4056, 2.3194, 3.2362 for training, validation and testing sets respectively. RMSE and MAE values are near to 0, so developed ANN model performance is good. Similarly in case of 75° and 60° cap angle cases, R^2 values are near to 1 and RMSE and MAE values are near to 0. So developed ANN models for hemispherical solar still and 75°, 60° cap angle solar stills (i.e., dome shaped solar stills) have good performance and predicting the water productivity with high accuracy. The correlation between actual output and predicted output for hemispherical solar still and dome shaped solar stills are shown in Figs. 7 and 8.

Finally, due to excellent performance of ANN while predicting the water productivity, it was decided to develop a mathematical formula to predict the water productivity of hemispherical solar still. To develop Eq. 19, optimized connecting weights and biases of ANN were extracted and values are given in Table 5. For hemispherical solar still prediction 1 hidden layer with 5 neurons in it has been taken. Note that, each one terms $C_1, C_2 \ldots C_6$ represents the response of the hidden neurons and are defined in Table 5. X1, X2 ... X6 represents the input variable, W is weightage and b is bias.

$$\begin{aligned}\text{Water Productivity} = & -1.0504 + (C_1 \times 1.9270) + (C_2 \times 5.9113) \\ & + (C_3 \times 6.0001) + (C_4 \times 6.3017) + (C_5 \times 5.5207) \end{aligned} \quad (19)$$

Table 5 The optimized weights and biases of ANN model

Neuron (i)	$C_i = \text{logsig}(W_{i1} \times X1 + W_{i2} \times X2 + \cdots + W_{i6} \times X6 + b_i)$					
	W_{i1}	W_{i2}	W_{i3}	W_{i4}	W_{i5}	W_{i6}	b_i
1	−6.44138	−4.37998	0.787886	9.298062	−15.3766	22.49386	−9.9829
2	−4.16823	−2.61418	−1.36891	0.30308	−15.6785	23.29095	−9.3625
3	−4.00065	−2.52503	−0.56433	0.69823	−15.3001	23.67019	−8.8312
4	−0.8269	−11.2831	−2.32776	−5.35089	−18.1066	21.06888	−9.7814
5	−13.1711	−0.90888	0.076465	4.446781	−10.5815	14.89479	−9.3548

6 Conclusion

In this study, prediction of hemispherical solar still water productivity with help of ANN have done. Hemispherical solar still was developed from single slop solar still by modifying it keeping the same basin area. Thermal analysis has been done to hemispherical solar still and compared the water productivity of hemispherical solar still with single slope solar still. Hemispherical solar still is producing daily 350 ml an average of quantity more water compared to single slope solar still, under same conditions. With theoretical data as input to ANN, water prediction of hemispherical solar still has been done. From the results, the developed ANN is predicting with good performance and with accuracy of 98%. So, this developed ANN can be used for predicting the water productivity of hemispherical solar still with given specifications i.e. developed hemispherical solar still with help of developed Eq. 19.

References

1. Ismail BI (2009) Design and performance of a transportable hemispherical solar still. Renew Energy 34:145–150
2. Arunkumar T, Jayaprakash R, Denkenberger D, Ahsan A, Okundamiya MS, Kumar S, Tanaka H, Aybar HŞ (2012) An experimental study on a hemispherical solar still. Desalination 286:342–348
3. Chauhan R, Dumka P, Mishra DR (2019) Modeling conventional and solar earth still by using LM algorithm based artificial neural network. Int J Ambient Energy 43(1):1389–1396
4. Dunkle R (1961) Solar water distillation: the roof type still and a multiple effect diffusion still. In: International developments in heat transfer, ASME, Proceeding of international heat transfer, Part V, University of Colorado, pp 895–902
5. Clark JA (1990) The steady-state performance of a solar still. Sol Energy 44(1):43–49
6. Kumar S, Tiwari GN (1996) Estimation of convective mass transfer in solar distillation systems. Sol Energy 57(6):459–464
7. Tripathy PP, Kumar S (2009) Neural network approach for food temperature prediction during solar drying. Int J Therm Sci 48(7):1452–1459
8. Nazari S, Safarzadeh H, Bahiraei M (2019) Experimental and analytical investigations of productivity, energy and exergy efficiency of a single slope solar still enhanced with thermoelectric channel and nanofluid. Renew Energy 135:729–744
9. Sabzevari A, Golneshan AA (1990) Solar radiation intensity on domed roofs. Sol Wind Technol 7(6):625–647

Design and Development of Artificial Neural Network-Based Prediction Model for Truncated Pyramid Type Solar Cooker

Koppisetty Sri Sai and Siddharth Ramachandran

Nomenclature

Q	Heat flow
A	Area (m^2)
h	Heat transfer coefficient (W/m^2 °C)
T	Temperature (°C)
m	Mass (kg)
t	Time (s)
σ	Stefan Boltzmann constant (5.67×10^{-8} W/m^2 °C4)
ρ	Reflectivity
ε	Emittance
α	Absorptance
θ	Angle of side wall with horizontal
τ	Transmissivity

Subindex

g	Glass
r	Recipient
amb	Ambient
f	Fluid
t	Lid
m	Wet

K. S. Sai (✉) · S. Ramachandran
Indian Institute of Information Technology, Design and Manufacturing Kancheepuram, Chennai, India
e-mail: mds20m008@iiitdm.ac.in

S. Doolla et al. (eds.), *Advances in Clean Energy and Sustainability*,
Green Energy and Technology, https://doi.org/10.1007/978-981-99-2279-6_17

int-1 Inner 1
int-2 Inner 2

1 Introduction

Energy is the significant of all development and improvement. The energy consumption in our country is increasing day by day. To meet the needs, our country majorly depends in the thermal power plants, that consume coal as fuel. Depletion of fossil fuels and global warming are the biggest concerns. So, a clean and pollution free energy is needed to be developed and improved. Solar is one among the renewable energies and considered as one of the promising choices. The sunlight can be harnessed into proper form of energy with low cost and less labour. Solar equipments like solar collectors, solar water heaters, solar cookers etc., are most widely used.

The use of solar cookers is widely increasing across the globe. Many different types of cookers are developed and being tested and utilized. The performance of hot box solar type cooker was investigated by Nahar [1]. It was found that the stagnation temperature in the cooker was retained for long time in cooker with storage when compared without storage and made attempts to cook the food at the evening hours (17.30–20.00 h). Srivastava et al. [2] performed the energy and exergy analysis of solar box type cooker by varying the energy storage mediums like Mustard oil and Sunflower oil. It was observed that Mustard oil is 25% more efficient as a storage medium than Sunflower oil. There were some studies got the detailed thermal and mathematical analysis of the solar cookers. The energy balance equations were computed by Ozkaymak [3] for a hot solar box type cooker with three reflectors hinged at the top of the cooker. Cramer's rule was applied to solve theoretical equations and observed the results of theoretical and experimental are in a good agreement. Similarly, Terrés-Pena and Quinto-Diez [4] developed a mathematical method for the solar box type cooker with multiple reflectors. Runge Kutta method is adopted for the numerical solution. A detailed temperature analysis of each component is observed.

Pyramid type solar cooker is similar to box type cooker which small variations in the design like truncations from the base part to the top of the cooker. Kumar et al. [5] designed a and constructed a truncated pyramid type solar cooker. It was designed in such a way, the reflected rays hit the absorber tray with better intensity. In continuation of work, Kumar et al. [6] designed and fabricated a truncated pyramid type solar cooker, a multipurpose appliance, which could be used for both cooking and water heating. After the briefing about the pyramid type cooker, the below mentioned objectives were made.

Ghritlahre and Prasad [7] carried out an experiment on porous solar air bed heater and developed a computational model using artificial neural networks (ANN) and trained with Levenberg Marquardt (trainLM) algorithm. The trained model was used to predict the thermal efficiency of the solar heater is compared with the actual experimental values and found the accuracy is better. Mukaro [8] conducted an experiment for energy and exergy analysis of a hot box type solar cooker. A feed forward neural network based on LM algorithm was developed with five input variables and four output parameters. With the use of correlations comparison of ANN predicted values and experimental results was made. An ANN model for the prediction of output parameters of solar cooker. The value of coefficient of determination was found to be 0.994 and 0.9964 for unidirectional and cross flow type of collectors respectively.

From the previous investigations, it was observed that a research gap of neural networks for pyramid cooker is identified. Therefore, the present work focus on the development of a mathematical model for the truncated pyramid cooker. Later, an ANN model should be developed and predict the outcomes of the cooker. Comparison of theoretical and predicted data is observed, using the performance parameters.

2 Description of Truncated Pyramid Type Solar Cooker

Kumar et al. [6] designed and fabricated a truncated type solar cooker as shown in Fig. (i). The truncated style of design helps to collect solar energy and retain high temperatures for considerable amount of time. As the sun rays hit the side walls of the cooker, the rays get reflected down towards the absorbing plate. Thus, the loss of solar radiation is minimized and the solar concentration is enhanced compared to the conventional box type cookers. However there is a restriction of angle θ and it can be solved by using following equation.

$$2(n + 1)\theta + \alpha \leq 90^\circ \qquad (1)$$

where n refers to number of reflections, α is angle made by the sunrays with the glazing surface. At particular period of time in a day, the maximum value of α would be 52°. By using these values, the θ should be approximately equal to 10°. Here θ refers to angle of the sidewalls with respect to vertical axis as shown in Fig. 1.

The glass (glazing surface) was made horizontal with respect to the surface but the side walls of the cooker are tilted with an angle approximately 10° from the vertical (80° to the horizontal). The size of the glazing surface was 50 cm × 50 cm the top most component, whereas the base of the cooker i.e., the absorbing plate size was 32.6 cm × 32.6 cm. The absorbing plate was painted with black for better absorptivity as shown in Fig. 2a, b. The depth of the cooker is 49.2 cm.

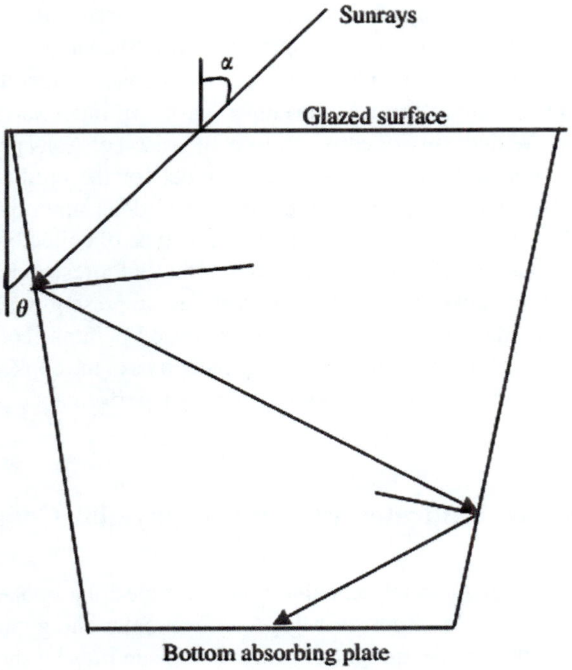

Fig. 1 Schematic diagram

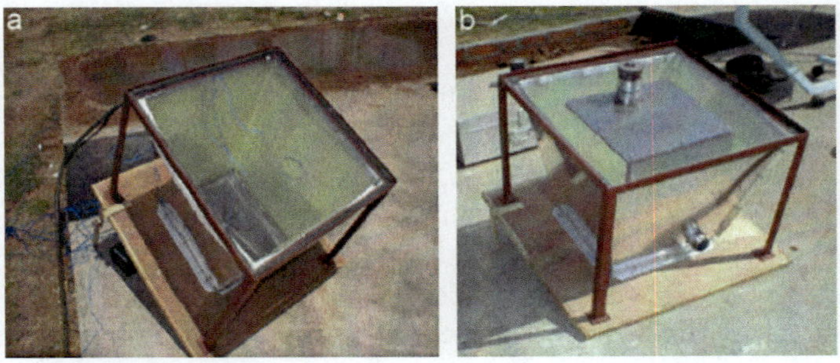

Fig. 2 **a**, **b** Truncated type pyramid cooker

3 Mathematical Modelling

The mathematical modelling equations of the truncated pyramid cooker were developed with reference to the work of Terrés-Pena [9]. The following are the assumptions required to be made to derive the model for the different heat flows as shown in Fig. 3a, b.

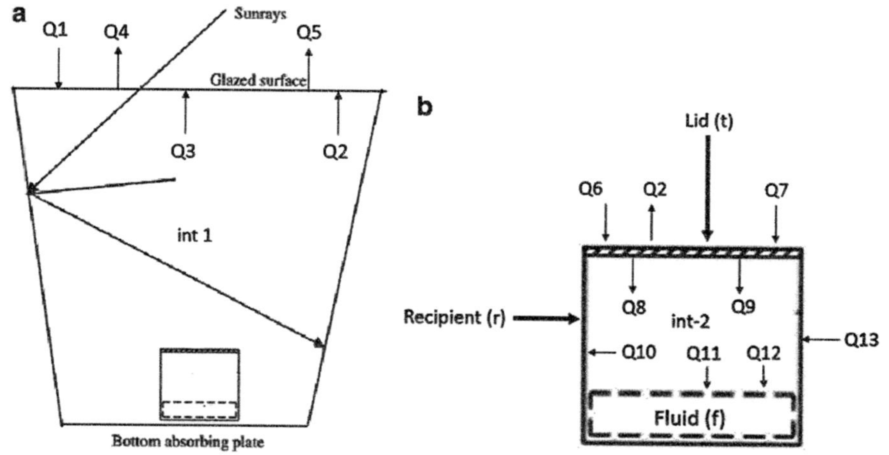

Fig. 3 **a** Heat flow considered for truncated pyramid cooker, **b** heat flow considered through cooking pot inside the cooker

1. The heat exchange takes place only in one direction
2. All the properties of materials like glass, lid, recipient remain constant throughout the investigation
3. The temperature is uniform for each component in the cooker
4. Heat transfer coefficients are constant throughout the process.

Energy balance equation for glass:

$$(m_g c_g)\frac{dTg}{dt} = Q1 + Q2 + Q3 - Q4 - Q5$$

$$\begin{aligned}(m_g c_g)\frac{dTg}{dt} &= (A_g G\alpha_g) + A_t\sigma\varepsilon_t(T_t^4 - T_g^4) + A_g h_{g-int1}(T_{int1} - T_g) \\ &\quad + Ar\sigma\varepsilon_t(T_r^4 - T_g^4) - A_g\sigma\varepsilon_g(T_g^4 - T_c^4) - A_g h_{g-amb}(T_g T_{amb})\end{aligned} \quad (2)$$

Energy balance equation for lid:

$$(m_t c_t)\frac{dTt}{dt} = Q6 + Q7 - Q2 - Q8 - Q9$$

$$\begin{aligned}(m_t c_t)\frac{dTt}{dt} &= A_t h_{t-int1}(T_t - T_{int1}) + \left(A_g G\tau_g^2\alpha_g\right) - A_t\sigma\varepsilon_t(T_t^4 - T_g^4) \\ &\quad - A_t h_{t-int2}(T_t - T_f) - A_t\sigma\varepsilon_t(T_t^4 - T_f^4)\end{aligned} \quad (3)$$

Energy balance equation for recipient:

$$m_r c_r \frac{dTr}{dt} = Q10 - Q11 - Q12 - Q13$$

$$m_r c_r \frac{dTr}{dt} = A_r h_{r-int1}(T_{int1} - T_r) - A_r \sigma \varepsilon_r (T_r^4 - T_f^4) - A_m h_{r-Am}(T_r - T_f) + 4 * \rho A_{ref} G \tau_g^2 \cos(90 - \theta_{ref}) \quad (4)$$

Energy balance equation for fluid:

$$m_f c_f \frac{dTf}{dt} = Q8 + Q9 + Q11 + Q12$$

$$m_f c_f \frac{dTf}{dt} = A_t h_{t-int2}(T_t - T_f) + A_t \sigma \varepsilon_t (T_t^4 - T_f^4) + A_r \sigma \varepsilon_r (T_r^4 - T_f^4) + A_m h_{r-Am}(T_r - T_f) \quad (5)$$

In the above equation, heat flow through glass is explained. Q1 is the radiation heat flow of glass and Q2 refers to heat radiation flow of lid through glass. Q3 is explained as convection flow of heat from inner-1 to glass. The radiation flow of glass towards sky and Q5 is convection flow from glass to ambient. The term Q6 refers to heat convection flow from glass to lid and Q7 refers to radiation of sun absorbed by the lid. Other terms like Q8 is convection flow of heat from lid towards fluid and Q9 is radiation flow from lid to fluid. Q10 is the convectional heat received by the recipient from inner-1 and Q11 is radiation flow of recipient towards fluid. The term Q12 refers to the convectional flow of heat of recipient towards fluid.

The Temperature T_c is calculated by means of correlation by Swinbank [8]

$$T_c = 0.0552 T_{amb}^{1.5} \quad (6)$$

The temperature T_{int} is considered as function of following temperatures [9]

$$T_{int} = (T_g + T_r + T_t)/3 \quad (7)$$

All the equations mentioned above are solved using Runge Kutta order 4 method, using the boundary conditions. A mathematical model was developed using MATLAB for the numerical solution. The initial conditions i.e., the initial temperature of each component is assumed to be ambient temperature (25 °C). A 2 L load of water is considered for the analysis. The data was derived for every 15 min from the timings 9 am to 4 pm on 15th August, at the location of Chennai, using theoretical calculations mentioned above as shown in Fig. 4.

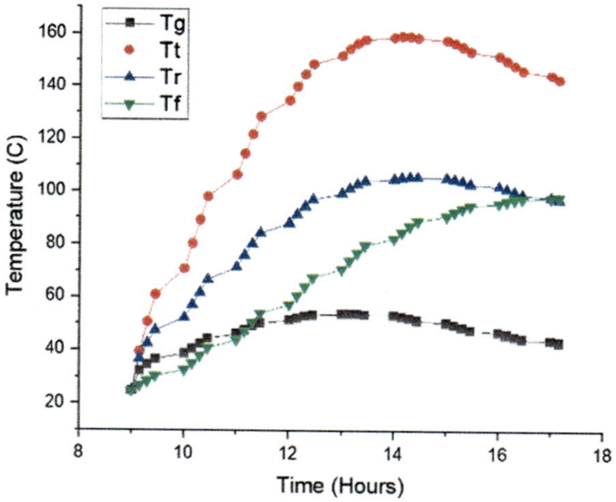

Fig. 4 Temperatures of each component with respect to time

4 Energy and Exergy Analysis

Mathematical equations are required to compute the thermal performance parameters manually and they are also required to be predicted by the ANN model.

The mathematical equations for the energy and exergy are derived from the work of Mukaro [11].

The rate of energy input to the cooker is known as Power Input P_i

$$P_i = A_p H \tag{8}$$

where H is Solar Radiation, Aa is aperture area

The rate at which energy is transferred or gained by the water is known as Power Output. This is regarded as cooking power

$$P_o = \frac{m_w C_w (T_{wf} - T_{wi})}{\Delta t} \tag{9}$$

where m_w is mass of water, c_w is specific heat of water, T_{wi} and T_{wf} are initial and final temperatures of water, Δt is time in secs.

The portion of energy supplied which can be converted into useful work by means of reversible mechanism is known as Exergy (or) Input Exergy [10]

$$\xi_i = A_p H \left(1 + \frac{1}{3}\left(\frac{T_a}{T_s}\right)^4 - \frac{4}{3}\frac{T_a}{T_s}\right) \tag{10}$$

where T_s is sun temperature 5800 K, T_a is ambient temperature.

The rate of exergy transferred or gained by water is known as Exergy Output.

$$\xi_o = m_w c_w \left(\frac{(T_{wf} - T_{wi}) - T_a \ln\left(\frac{T_{wf}}{T_{wi}}\right)}{\Delta t} \right) \tag{11}$$

Energy efficiency is defined as the ratio of output power to input power

$$\eta_{ef} = \frac{P_o}{P_i} = \frac{m_w C_w (T_{wf} - T_{wi})}{\Delta t * A_p H} \tag{12}$$

Exergy efficiency is the ratio of exergy output to exergy input

$$\eta_{ex} = \frac{\xi_o}{\xi_i} \tag{13}$$

5 Artificial Neural Networks and Its Architecture

ANN stands for Artificial Neural Networks, named because of its structure and functioning is similar to the neurons in the human brain. It is simply a mathematical computing technology, helps to solve nonlinear engineering and complex problems in our day-to-day life. It helps to predict the outcomes of an experiment without the need of apparatus and setup. The ANN mainly consists of 3 layers namely input, hidden and output layers. The input layer takes in the input give. The function of output is to predict the outcomes. The hidden layer acts as an intermediate between input and output layers. The hidden layer consists of activation function that helps to find relationship between input and output layers.

The Architecture of ANN comprises of input, hidden and output layers. There are four parameters namely, glass temperature, lid temperature, recipient and fluid temperature considered to be the four input layer neurons. Similarly, other four output predicted values namely, power out, exergy out, efficiency and exergy efficiency are the output layer neurons. The input data given to neural networks will be by default divided into 3 categories, training, testing and validation in the ration of 70:15:15. Levenberg Marquardt algorithm is adopted. A feed forward backpropagation network is been used in this scenario. After different configurations, it was found that a single hidden layer with 13 neurons is observed to be the best network.

The activation function adjusts the weights of the neurons by calculating the weighted sum and further adding bias to the model. This method helps to solve the non-linearity of the problem. The weighted sum of input layer is fed to the hidden layer or a number of hidden layers one by one. Further similar process is adopted where the weighted sum is fed to the output layer. And the weighted sum keeps on updating based on the loss function Mean square error (MSE). So, this is way how activation function generates an equation for the input and output layers. The activation function considered are tansig from input to hidden layer and purelin from hidden layer to output layer.

6 Performance Parameters

T = Theoretical values, t = ANN predicted values

Correlation coefficient

$$R = \sqrt{1 - \frac{\sum(T(i) - t(i))2}{\sum(T(i))2}} \tag{14}$$

Root Mean Square Error

$$RMSE = \sqrt{\left(\frac{1}{N}\right)*\left(\sum(T(i) - t(i)\right)2} \tag{15}$$

Mean Absolute Error,

$$MAE = \left(\frac{1}{N}\right) * \left|(\sum(T(i) - t(i))\right| \tag{16}$$

Mean Absolute Percentage Error

$$MAPE = \left(\frac{1}{N}\right) * \frac{\left|(\sum(T(i) - t(i))\right|}{|(\sum(T(i)|} * 100 \tag{17}$$

7 Results and Discussions

The ANN analysis was performed based on the data of one sunny day. From Figs. 5, 6, 7, and 8 it is clear that there is a good agreement between the output and predicted values.

Tables 1 and 2 describes the statistical analysis of each parameter like correlation coefficient, RMSE, MAE, MAPE, MSE.

The following is the optimized weights and bias distribution table describes the mathematical equation being developed to solve the problem of non-linearity. ANN structure used consists of 4 input, 13 hidden layers and 4 output layers. Levenberg Marquardt algorithm is used.

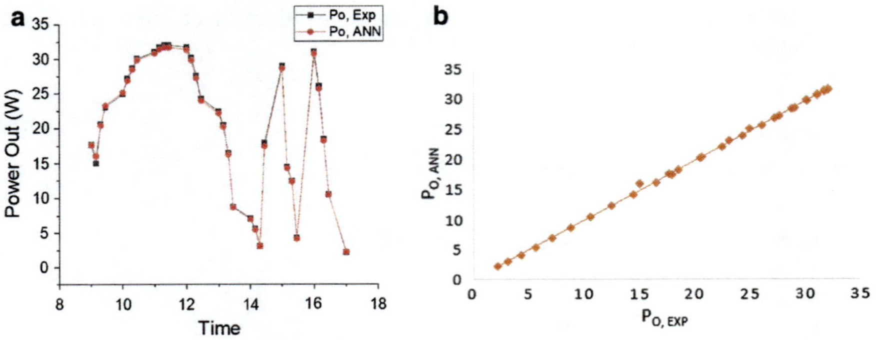

Fig. 5 Comparison of experimented and ANN predicted values of power out with respect to time and corresponding correlation

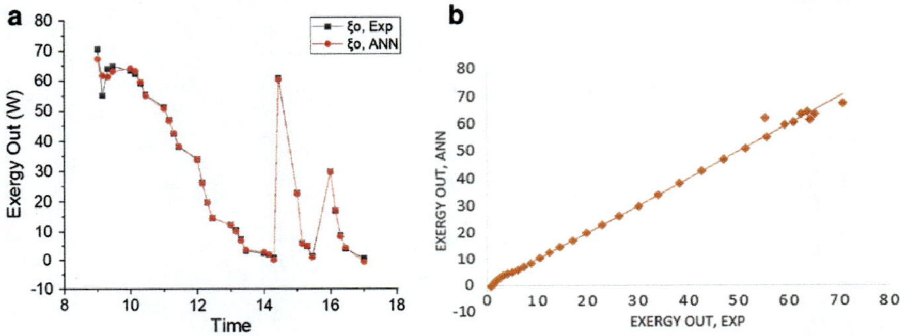

Fig. 6 Comparison of experimented and ANN predicted values of exergy out with respect to time and corresponding correlation

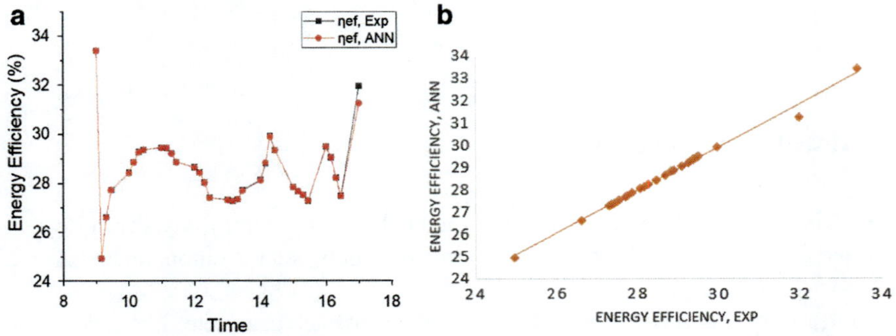

Fig. 7 Comparison of experimented and ANN predicted values of energy efficiency with respect to time and corresponding correlation

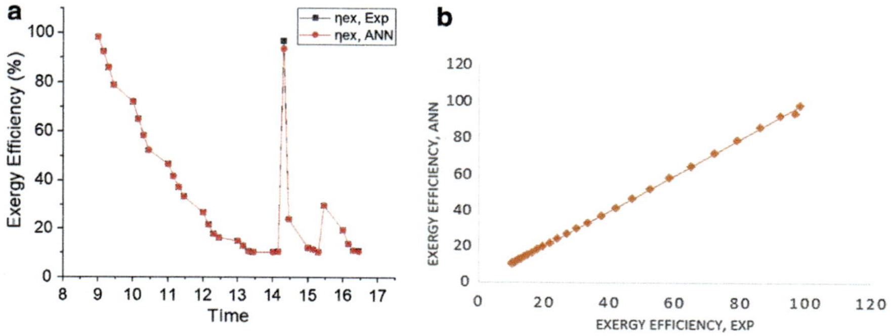

Fig. 8 Comparison of experimented and ANN predicted values of exergy efficiency with respect to time and corresponding correlation

Table 1 Comparison of statistical parameters of cooker parameters

Cooker parameter	R	RMSE	MAE	MAPE (%)	MSE
Power out	0.999	0.2923	0.1364	0.7371	0.0855
Exergy out	0.9992	1.4629	0.0354	0.1328	2.1401
Energy efficiency	1.00	0.0012	0.000215	0.0843	0.0001
Exergy efficiency	0.9999	0.0057	0.0011	0.3585	0.0003

Table 2 The optimized weights and bias

Neuron (i)	$C_i = \text{tansig}(W_{i1} \times X_1) + (W_{i2} \times X_2) + (W_{i3} \times X_3) + (W_{i4} \times X_4) + b_i$				
	W_{i1}	W_{i2}	W_{i3}	W_{i4}	b_i
1	−1.9679	−0.3335	−1.3051	1.16421	2.6436
2	−1.8961	0.7161	1.2137	−1.2093	2.2385
3	1.5515	0.8969	−1.0471	1.6549	−1.8108
4	1.4299	1.9354	0.0843	−1.0406	−1.2666
5	−1.0002	−1.2749	1.1528	−1.7435	0.8294
6	1.5311	0.6634	1.5544	−0.7696	−0.2969
7	−1.7692	0.5861	1.8620	0.4626	−0.2176
8	−0.5201	−2.5304	0.5273	0.2959	−1.1083
9	−0.4441	−1.7234	−1.6961	−0.8918	−0.4997
10	0.9389	−0.1842	−2.0159	1.8879	1.1422
11	0.9153	1.6648	−1.2869	0.3126	1.6366
12	−1.7194	−0.9028	1.8005	0.2294	−2.2148
13	−0.0659	0.4836	−1.3044	2.2953	−2.6214

$$\begin{aligned}\text{Power out} = &-0.8477 - (C_1 \times 0.4059) + (C_2 \times 0.6216) + (C_3 \times 0.5611) \\ &+ (C_4 \times 0.2729) + (C_5 \times 0.3149) - (C_6 \times 0.0180) - (C_7 \times 0.12508) \\ &- (C_8 \times 0.8850) + (C_{10} \times 1.3210) - (C_{11} \times 0.1146) - (C_{13} \times 0.3172)\end{aligned} \quad (18)$$

The above mentioned describes the mathematical equation helps to find the final output that is predicted by ANN. Here C_i (where $i = 1, 2, \ldots 13$) refers to the constants that are being solved by the correlation explained in the tabular column 2. Similarly, the equations are developed for the other output parameters.

8 Conclusion

The developed ANN model using Levenberg Marquardt algorithm and feed forward backpropagation network is very effective. The values of correlation coefficient close to one in every case. The values of RMSE and MSE are less than 1.5 and 2.2 respectively and the other statistical parameters also very less. Therefore, it can be concluded that the above designed model is successful in estimating the output parameters power out, exergy out, energy efficiency and exergy efficiency with good accuracy.

References

1. Nahar NM (2003) Performance and testing of a hot box storage solar cooker. Energy Convers Manage 44:1323–1331
2. Srivastava SK, Shukla A, Tiwari R, Rai AK (2020) Studies on solar cooker with thermal storage. Int J Adv Res Eng Technol (IJARET) 11(10):410–417, Article ID: IJARET_11_10_044
3. Ozkaymak M. Theoretical and experimental investigation of a hot box-type solar cooker performance. http://doi.org/10.1243/09576509JPE326
4. Terrés-Pena H, Quinto-Diez P (2003) Applications of numerical simulation of solar cooker type box with multiple inner reflector. In: Proceedings of ISEC 2003, 2003 international solar energy conference, Hawaii, USA, ISEC2003-44060
5. Kumar N, Chavda T, Mistry HN (2010) A truncated pyramid non-tracking type multipurpose domestic solar cooker/hot water system. Appl Energy 87:471–477
6. Ghritlahre HK, Prasad RK (2019) Modelling of back propagation neural network to predict the thermal performance of porous bed solar air heater. Arch Thermodyn 40(4):103–128
7. Mukaro R (2021) Comparison of experimental and artificial neural network-estimated thermal performance parameters for a hot-box solar cooker, vol 32, no 3. ISSN: 0868-4952
8. Swinbank WC (1963) Long-wave radiation from clear skies. Q J Roy Meteorol Soc 89
9. Kumar N, Agravat S, Chavda T, Mistry HN (2008) Design and development of efficient multipurpose domestic solar cookers/dryers. Renew Energy 33:2207–2211
10. Petela R (2003) Exergy of undiluted thermal radiation. Sol Energy 74:469–488

Effect of Pseudopotentials on the Electronic Structures for Hydrogen Adsorption on Titanium (Ti) Doped B_{40} Fullerene—A First Principle Study

Harshavardhan Thodupunoori and Paramita Haldar

1 Introduction

Hydrogen is considered an alternative energy source because of its abundance, environmental friendliness, and cost-effectiveness [1]. Due to low volumetric density, it is very difficult to store hydrogen in liquid form [2]. Developing an appropriate hydrogen storage medium is very crucial for wide applications. During physical adsorption of hydrogen, metal doping amplifies the storage capacity due to the strong binding between hydrogen and host material [3]. Boron fullerene-based materials are considered promising hydrogen storage compounds because of their lightweight and good binding capability with metal adatoms [4]. Ti decorated nano-structures have exhibited an outstanding performance for hydrogen storage. Ti adatoms can be doped on the surface of B_{40} fullerene to overcome the poor reversibility and thermodynamic limitations of the fullerene molecule [5]. It is been observed that during exchange-correlation (XC) calculations, the interaction between valence and core electrons changes significantly in presence of different pseudopotentials. As a result, the cohesive and electronic properties of the molecules also vary. In this work we have studied hydrogen storage in Ti decorated B_{40} fullerene in presence of four different pseudopotentials based on local density approximations (LDA) [6, 7] and generalized gradient approximations (GGA) [8–11]. We have performed calculations for structural analysis, Mulliken charge analysis [12], and average adsorption energy in presence of different pseudopotentials. From the charge population density of electrons, it is evident that the pseudopotentials play a crucial role in the study of hydrogen adsorption on Ti decorated B_{40} fullerene.

H. Thodupunoori · P. Haldar (✉)
Department of Chemical Engineering, BITS Pilani, Goa Campus, Sancoale, Goa 403726, India
e-mail: paramitah@goa.bits-pilani.ac.in

S. Doolla et al. (eds.), *Advances in Clean Energy and Sustainability*,
Green Energy and Technology, https://doi.org/10.1007/978-981-99-2279-6_18

The entire paper is planned as follows. Section 2 is outlined for describing the computational methodology. Results from structural analysis, binding energy calculations, Mulliken charge analysis, and hydrogen adsorption are discussed in Sect. 3. In the end, we have presented the conclusion in Sect. 4.

2 Methodology

All the first principle DFT [13] calculations are performed using the plane wave-based Quantum-ESPRESSO package. For geometric optimization and exchange-correlation energy calculations, LDA [6] and GGA [8] functionals are used. Perdew-Burke-Ernzerhof (PBE) exchange-correlation [9] and Perdew-Wang 91 (PW91) gradient-corrected functionals [8] are considered for GGA. PBE with ultrasoft Vanderbilt pseudopotentials (USPP) [10] and Projector-Augmented-Wave (PAW) functional [9] based pseudopotentials, Perdew-Zunger exchange-correlation (pz) [7] with ultrasoft Vanderbilt pseudopotentials (LDA-pz-USPP) and Perdew-Wang 91 (PW91) gradient-corrected functionals with ultrasoft Vanderbilt pseudopotentials (PW91-Van-USPP) are considered for Ti, boron (B) and hydrogen atoms. A K-point grid, centered on the gamma point of the Brillouin zone, is considered for sampling. A cubic cell of 15 × 15 × 15 dimension and kinetic energy cutoff of 18, 60, and 70 Ry for B_{40}, Ti_6B_{40}, and $(Ti\text{-}nH_2)_6B_{40}$ respectively are selected to perform energy minimization and self-consistency field (SCF) [14] calculations. All the calculations are carried out at a convergence threshold of 10^{-3} Ry and a mixing factor of 0.3. Visual molecular dynamics (VMD) [15] and XCRYSDEN [16] software are used for visualizing different structures. Mulliken charge analysis calculations are carried out using the B3LYP/6-311++G(d,p) level of theory by GAUSSIAN [17] program.

3 Results and Discussion

3.1 *Structural Analysis and Binding Energy Calculation at Different Pseudopotential*

B_{40}. The surfaces of B_{40} fullerene consist of four non-planar heptagonal (S1, S2, S3, and S4) cavities and two planar hexagonal holes (S5, S6) embedded with 48 triangles. Before relaxation, the diameter of B_{40} fullerene is 3.37 and 3.77 Å for hexagonal and heptagonal holes respectively. After relaxation, the diameter of the optimized structures varies with different pseudopotentials. We have observed that the diameter of hexagonal and heptagonal cavities changes to 3.30 and 3.72 Å for LDA, 3.38 and 3.70 Å for PBE-USPP, 3.36 and 3.71 Å for PBE-PAW, and 3.35 and 3.76 Å for PW91 respectively. Before relaxation, the average bond length of heptagonal and hexagonal cavities is 1.65 and 1.70 Å. After relaxation, the average bond length changes to 1.65

and 1.70 Å for LDA, 1.66 and 1.72 Å for PBE-USPP, 1.66 and 1.72 Å for PBE-PAW and 1.66 and 1.71 Å for PW91 respectively. Table 1 provides the diameter (Å) and average bond lengths (Å) of B–B atoms calculated before and after relaxation of B_{40} at different pseudopotentials for all surfaces (S1–S6). The top view of the heptagonal (S1) surface of optimized structures of B_{40} at different pseudopotentials (a) LDA, (b) PBE-USPP, (c) PBE-PAW, and (d) PW91 is shown in Fig. 1. The pink color (color code 09) represents the boron atoms. From our study, it is evident that in presence of different pseudopotentials, there is no significant structural difference in B_{40} even after relaxation and all the structures are stable.

Ti-doped B_{40}. Ti metal atoms are placed on the centers of heptagonal and hexagonal cavities of B_{40} fullerene because they are the most stable and energy favorable sites in comparison with other sites on the surface of B_{40} [18]. To avoid cluster formation, we have doped a single Ti atom on each surface. Optimized structures of B_{40}, obtained for different pseudopotentials, are considered initial structures for Ti doping. Before structural relaxation, the diameter of heptagonal and hexagonal surfaces of Ti_6B_{40} varies between 3.72 and 3.30 Å for LDA, 3.70 and 3.38 Å for PBE-USPP, 3.71 and 3.36 Å for PBE-PAW and 3.76 and 3.35 Å for PW91 respectively. After relaxation the heptagonal and hexagonal diameter changes to 3.56 and 3.42 Å for LDA, 3.68 and 3.58 Å for PBE-USPP, 3.66 and 3.48 Å for PBE-PAW, and 3.60 and 3.47 Å for PW91 respectively. Though before relaxation, the average bond length of the Ti-B atoms of the heptagonal and hexagonal surface is 2.17 Å, it changes significantly after relaxation calculation in presence of different pseudopotential. After relaxation calculations, the average bond length of Ti-B atoms on heptagonal and hexagonal surfaces varies from 2.18 and 2.19 Å for LDA, 2.30 and 2.31 Å for PBE-USPP, 2.28 and 2.24 Å for PBE-PAW and 2.21 and 2.22 Å for PW91 respectively. Table 2 gives detailed information regarding the change in diameter and average bond length of the heptagonal and hexagonal surfaces of Ti-B atoms of Ti_6B_{40} at different pseudopotentials. The optimized structures of S1 surface Ti_6B_{40} at different pseudopotentials (a) LDA, (b) PBE-USPP, (c) PBE-PAW, and (d) PW91 are illustrated in Fig. 2. The pink color (Color code 09) represents B atoms and the cyan color (Color code 10) indicates the Ti atoms. It is observed from Table 2 and Fig. 2 that the optimized structures of Ti_6B_{40} change significantly in presence of different pseudopotentials.

Binding Energy Calculation at different pseudopotential. Pseudopotentials play a very important role in the interaction between ionic cores and electrons. The total energy is a sum of the functional of valence charge density, the interaction of valence electrons with the nucleus, the interaction of valence electrons with core electrons (Pauli repulsion), electrostatic interaction, and XC interaction. As a result, the total binding energy of the molecules undergoes a prominent variation under the influence of different pseudopotentials. We have calculated the binding energy ($E_{Binding}$) of B_{40} and Ti_6B_{40} at different pseudopotentials using the following equation

$$E_{Binding} = \frac{[E_{B_{40}} + n \times E_M - E_{M_6B_{40}}]}{n} \quad (1)$$

Table 1 Diameter (Å) and average bond lengths (Å) of B–B atoms were calculated before and after the relaxation of B_{40} at different pseudopotentials

Structure	Surface	Before relaxation		After relaxation							
		Diameter (Å)	Average bond length (Å)	Diameter (Å)				Average bond length (Å)			
				LDA	PBE-USPP	PBE-PAW	PW91	LDA	PBE-USPP	PBE-PAW	PW91
B_{40}	S1	3.77	1.65	3.72	3.70	3.71	3.76	1.65	1.66	1.66	1.66
	S2	3.77	1.65	3.72	3.70	3.71	3.76	1.65	1.66	1.66	1.66
	S3	3.77	1.65	3.72	3.70	3.71	3.76	1.65	1.66	1.66	1.66
	S4	3.77	1.65	3.72	3.70	3.71	3.76	1.65	1.66	1.66	1.66
	S5	3.37	1.70	3.30	3.38	3.36	3.35	1.70	1.72	1.72	1.71
	S6	3.37	1.70	3.30	3.38	3.36	3.35	1.70	1.72	1.72	1.71

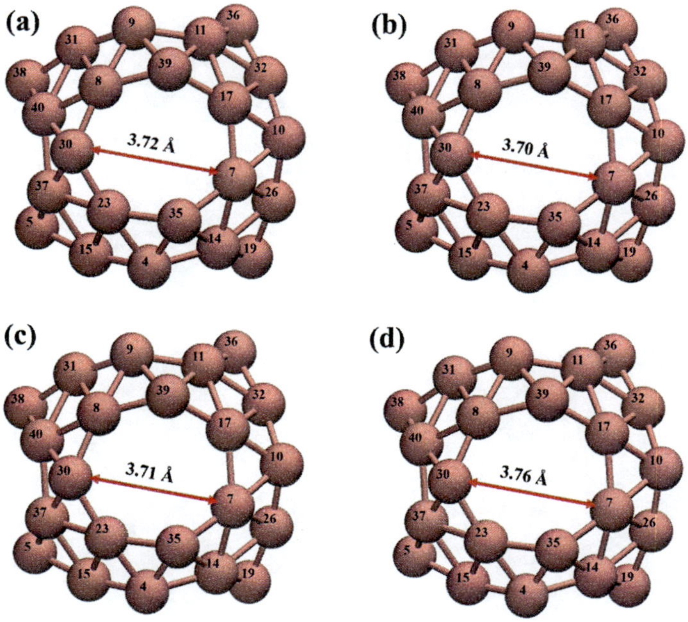

Fig. 1 Optimized structures of B_{40} fullerene at different pseudopotentials **a** LDA, **b** PBE-USPP, **c** PBE-PAW, and **d** PW91 (S1 Surface)

Table 2 Diameter (Å) and average bond lengths (Å) of Ti-B atoms calculated before and after relaxation of Ti_6B_{40} at different pseudopotentials

Structure	Surface	Diameter (Å)							
		LDA		PBE-USPP		PBE-PAW		PW91	
		Before	After	Before	After	Before	After	Before	After
Ti_6B_{40}	S1	3.72	3.56	3.70	3.68	3.71	3.66	3.76	3.60
	S2	3.72	3.56	3.70	3.68	3.71	3.66	3.76	3.60
	S3	3.72	3.56	3.70	3.68	3.71	3.66	3.76	3.60
	S4	3.72	3.56	3.70	3.68	3.71	3.66	3.76	3.60
	S5	3.30	3.42	3.38	3.58	3.36	3.48	3.35	3.47
	S6	3.30	3.42	3.38	3.58	3.36	3.48	3.35	3.47
Ti-B average bond length (Å)									
Ti_6B_{40}	S1	2.17	2.18	2.17	2.30	2.17	2.28	2.17	2.21
	S2	2.17	2.18	2.17	2.30	2.17	2.28	2.17	2.21
	S3	2.17	2.18	2.17	2.30	2.17	2.28	2.17	2.21
	S4	2.17	2.18	2.17	2.30	2.17	2.28	2.17	2.21
	S5	2.17	2.19	2.17	2.31	2.17	2.24	2.17	2.22
	S6	2.17	2.19	2.17	2.31	2.17	2.24	2.17	2.22

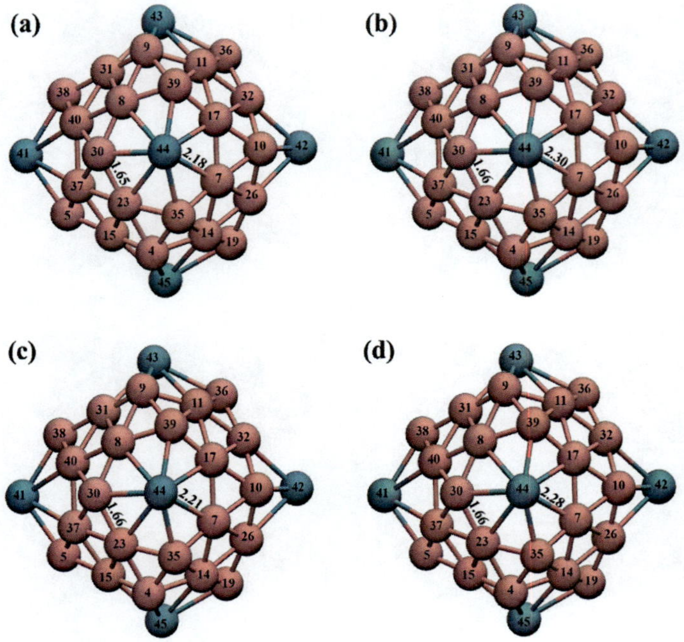

Fig. 2 Optimized structures of Ti_6B_{40} fullerene at different pseudopotential **a** LDA, **b** PBE-USPP, **c** PBE-PAW, and **d** PW91.The pink color (color code 09) represents B atoms and the cyan color (color code 10) indicates the Ti atoms (S1 surface)

where $E_{B_{40}}$, E_M and $E_{M_6B_{40}}$ denote the energy B_{40} fullerene, single metal adatom, and metal adatom doped B_{40} fullerene respectively. n signifies a total number of doped metal adatoms. From Table 3 it is evident that the binding energies of the B_{40} molecule vary moderately with LDA (6.5329 eV), PBE-USPP (6.0299 eV), PBE-PAW (5.9671 eV), and PW91 (5.916 eV). In the case of $Ti_6B_{40,}$ though there is a little change in binding energies for LDA (7.6820 eV), PBE-USPP (7.8010 eV), and PBE-PAW (7.5601 eV), PW91 pseudopotential have a huge impact on electrons interaction resulting in the binding energy of 11.6764 eV.

Table 3 Binding energy ($E_{Binding}$) (eV) of B_{40} and Ti_6B_{40} calculated at different pseudopotentials

Structure	$E_{Binding}$ (eV)			
	LDA	PBE-USPP	PBE-PAW	PW91
B_{40}	6.5329	6.0299	5.9671	5.916
Ti_6B_{40}	7.6820	7.8010	7.5601	11.6764

3.2 Mulliken Charge Analysis

Mulliken charge analysis [12] is used to calculate the electronic charge distribution in a molecule and the behavior of the molecular orbitals (bonding, non-bonding, and anti-bonding) can be predicted for different pairs of atoms. Pseudopotentials play a crucial role in calculating electron density matrix, Eigenvalues, and atomic orbital populations. Table 4 lists the average charge of B and Ti for B_{40} and Ti_6B_{40} molecules respectively calculated at different pseudopotentials. The average charge of boron atoms is calculated as + 0.0072e for S1–S4 surfaces and – 0.0168e for S5 and S6 surfaces due to LDA. PBE-USPP provides + 0.0048e of charge for S1–S4 surfaces and – 0.0113e for S5 and S6 surfaces. In presence of PBE-PAW, it is calculated as + 0.0061e for S1–S4 surfaces and – 0.0143e for S5 and S6 surfaces. PW91 impacts the average charge to change in + 0.0064e for S1–S4 surfaces and – 0.0150e for S5 and S6 surfaces. The charge distribution of boron atoms for B_{40} and boron and Ti atoms for Ti_6B_{40} is shown in Fig. 3a, b respectively. In Fig. 3, green and black represent the charges for LDA and PBE-USPP respectively, whereas charges due to PBE-PAW and PW91 pseudopotentials are exhibited by the yellow and red colors respectively. The average charge of B atoms for S1–S4 surfaces and S5–S6 surfaces changes from – 0.043e and – 0.072e for LDA, – 0.041e and – 0.083e for PBE-USPP and – 0.042e and – 0.072e for PW91 respectively. For PBE-PAW, the average charge of B atoms varies for different surfaces like – 0.250e for S1 and S3, – 0.237e for S2 and S4, and – 0.367e for S5 and – 0.354e for S6 surfaces respectively. For Ti atoms, the Mulliken charge is + 0.313e for S1–S4 surfaces and + 0.41e for S5–S6 surfaces respectively for LDA, + 0.327e for S1, S3 surfaces, + 0.328e for S2, S4 surfaces and + 0.42e for S5 and S6 surfaces for PBE-USPP, + 1.87e for S1–S4 surfaces and + 1.82e for S5 and + 1.83 for S6 surfaces respectively for PBE-PAW and + 0.312e for S1–S4 surfaces and + 0.40e for S5–S6 surfaces for PW91. In Ti_6B_{40}, the boron atoms of all the surfaces exhibit negative charges, which causes anti-bonding leading to the doping of Ti atoms. Due to the negative charge, Ti atoms get easily attached to the hexagonal and heptagonal cavities and contribute to the charge transfer process. We can observe a strong population of positive charges on Ti atoms for all the pseudopotentials, causing the Ti atoms to free their electrons during doping and become suitable for accepting electrons making the Ti_6B_{40} molecule suitable for hydrogen storage.

3.3 Hydrogen Adsorption on Ti-Doped B_{40} at Different Pseudopotential

According to the 18-electron rule, a maximum number (N_{max}) of 6 hydrogen molecules can be adsorbed on each surface of Ti_6B_{40}. The 18-electron rule can be written as

Table 4 Mulliken charge distribution on B and Ti atoms over S1 and S5 surfaces of Ti_6B_{40}

Structure	Surface	Mulliken charges (e)							
		Average charge on B				Charge on Ti			
		LDA	PBE-USPP	PBE-PAW	PW91	LDA	PBE-USPP	PBE-PAW	PW91
B_{40}	S1	0.0072	0.0048	0.0061	0.0064	–	–	–	–
	S2	0.0072	0.0048	0.0061	0.0064	–	–	–	–
	S3	0.0072	0.0048	0.0061	0.0064	–	–	–	–
	S4	0.0072	0.0048	0.0061	0.0064	–	–	–	–
	S5	− 0.0168	− 0.0113	− 0.0143	− 0.0150	–	–	–	–
	S6	− 0.0168	− 0.0113	− 0.0143	− 0.0150	–	–	–	–
Ti_6B_{40}	S1	− 0.043	− 0.041	− 0.250	− 0.042	0.313	0.327	1.87	0.312
	S2	− 0.043	− 0.041	− 0.237	− 0.042	0.313	0.328	1.87	0.312
	S3	− 0.043	− 0.041	− 0.250	− 0.042	0.313	0.327	1.87	0.312
	S4	− 0.043	− 0.041	− 0.237	− 0.042	0.313	0.328	1.87	0.312
	S5	− 0.072	− 0.083	− 0.367	− 0.072	0.41	0.42	1.82	0.40
	S6	− 0.072	− 0.083	− 0.354	− 0.072	0.41	0.42	1.83	0.40

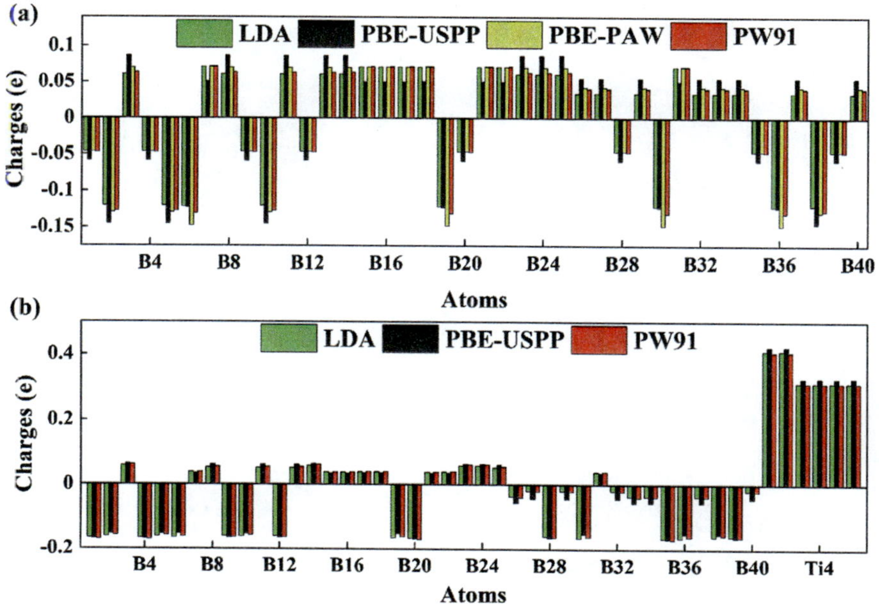

Fig. 3 Mulliken charge distribution on **a** B_{40} and **b** Ti_6B_{40} molecule at different pseudopotentials

$$2N_{max} = 18 - n_v(M) - n_v(B_{40}) \tag{2}$$

where $n_v(M)$ and $n_v(B_{40})$ denote valence electron number of metal electron and electrons contributed by B_{40} respectively. The maximum number of valance electron contributions of Ti atom for both hexagon and heptagon surfaces is 3 and B_{40} contributes 4 electrons for hexagonal and 3 electrons for the heptagonal surface. So N_{max} can be evaluated as 5.5 and 6 for hexagonal and heptagonal surfaces respectively following the above equation. We have incorporated 1–6 hydrogen molecules on each surface of the Ti_6B_{40} molecule. The optimized structures of $(Ti\text{-}6H_2)_6B_{40}$ (S5 surface) for different pseudopotentials are presented in Fig. 4. In the figure, the pink color (Color code 09) represents B atoms and cyan color (Color code 10) indicates the Ti atoms and the grey color (Color code 02) denotes the hydrogen atoms. These figures clearly show that all the $(Ti\text{-}6H_2)_6B_{40}$ structures are stable after relaxation calculations performed with different pseudopotentials. In presence of different pseudopotentials, we have calculated the average adsorption energy ($E_{Adsorption}$) to understand the adsorption capacity of the B_{40} molecule. The adsorption energies are calculated using the following equation

$$E_{Adsorption} = \frac{[E_{M_6B_{40}} + n \times E_{H_2} - E_{(M-nH_2)_6B_{40}}]}{n} \tag{3}$$

where $E_{M_6B_{40}}$, E_{H_2} and $E_{(M-nH_2)_6B_{40}}$ denote the energy of metal-doped B_{40} fullerene, hydrogen molecule and hydrogen molecule adsorbed metal-doped B_{40} fullerene

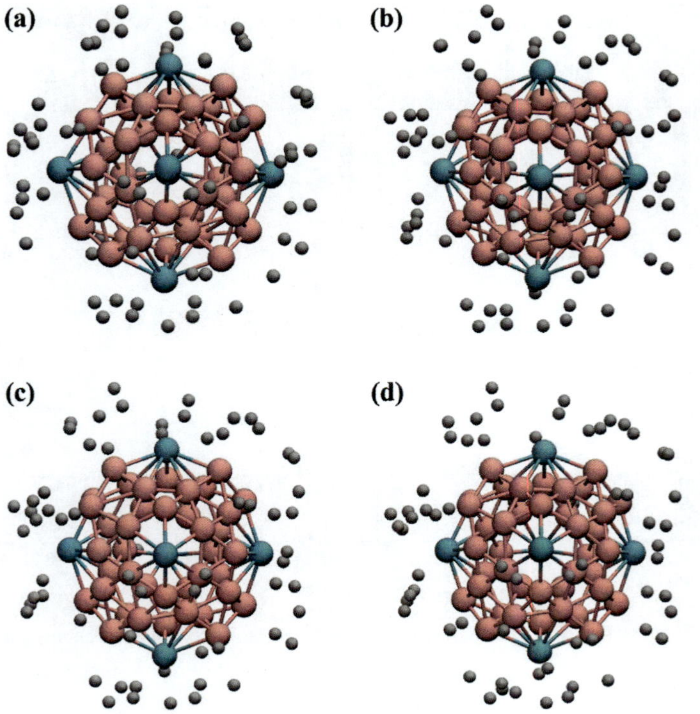

Fig. 4 Optimized structures of (Ti-6H2)$_6$B$_{40}$ fullerene at different pseudopotential **a** LDA, **b** PBE-USPP, **c** PBE-PAW, and **d** PW91. The pink color (color code 09) represents B atoms and cyan color (color code 10) indicates the Ti atoms and the grey color (color code 02) denotes the hydrogen atoms (S5 surface)

respectively. n indicates a total number of hydrogen molecules adsorbed at the surface of metal-doped B_{40} fullerene.

Table 5 represents the adsorption energies for (Ti-nH_2)$_6B_{40}$ where n = 1–6, calculated at different pseudopotentials. From the table, we can observe that the average adsorption energy is 0.2681 eV for 1 hydrogen molecule whereas it changes to − 0.0685 eV, − 0.0238 eV, − 0.1681 eV, − 0.1716 eV, and − 0.2240 eV for 2–6 hydrogen molecules respectively in presence of LDA pseudopotential. For PBA-USPP the average adsorption energy is positive with 1.0722, 0.6931, 0.6305, 0.3224, and 0.2334 eV for 1–5 hydrogen molecules respectively whereas 6 adsorbed hydrogen molecules exhibit average adsorption energy of − 0.0324 eV. In the case of PBE-PAW pseudopotential, all the average adsorption energies are positive with a value of 0.4358, 0.3505, 0.3486, 0.4074, 0.3770, and 0.0632 eV for 1–6 hydrogen molecules respectively. For PW91, the average adsorption energies are all positive for 1–6 hydrogen molecules with values of 0.5512, 0.2937, 0.3300, 0.2770, 0.3068 and 0.3015 eV respectively. In the case of LDA, the XC energy functional depends only on the value of local electron density calculated for each point in space. But the LDA fails when the electron density undergoes faster changes and considering

Table 5 Average adsorption energy ($E_{Adsorption}$) at different pseudopotentials for $(Ti\text{-}nH_2)_6B_{40}$ where n = 1–6

Structure	$E_{Adsorption}$ (eV)			
	LDA	PBE-USPP	PBE-PAW	PW91
$(Ti\text{-}1H_2)_6B_{40}$	0.2681	1.0722	0.4358	0.5512
$(Ti\text{-}2H_2)_6B_{40}$	– 0.0685	0.6931	0.3505	0.2937
$(Ti\text{-}3H_2)_6B_{40}$	– 0.0238	0.6305	0.3486	0.3300
$(Ti\text{-}4H_2)_6B_{40}$	– 0.1681	0.3224	0.4074	0.2770
$(Ti\text{-}5H_2)_6B_{40}$	– 0.1716	0.2334	0.3770	0.3068
$(Ti\text{-}6H_2)_6B_{40}$	– 0.2240	– 0.0324	0.0632	0.3015

the gradient of electron density it can be improved. GGA modifies the XC interaction of the valance electrons as well as the core valence interactions. In the case of PW91 and PBE, the gradient expansions of XC energies are considered. PBE is found to perform more accurately in comparison with PW91 and PAW. As a result, we have obtained more accurate adsorption energies for GGA pseudopotentials in comparison with LDA. That is the reason we have observed a reduction in adsorption energies for LDA in comparison with GGA pseudopotentials. From this table, it is evident that the PBE-USPP pseudopotential contributes the best estimation for the calculation of adsorption energy. The negative adsorption energy for $(Ti\text{-}6H_2)_6B_{40}$ indicates that a strong attraction of hydrogen molecules is occurring on the Ti_6B_{40} surfaces. We can conclude that Ti_6B_{40} can adsorb a maximum of 36 hydrogen molecules. Figure 5 displays the average adsorption energy vs adsorbed hydrogen molecules for $(Ti\text{-}nH_2)_6B_{40}$ obtained at different pseudopotentials. From this figure, it is seen that the total adsorption energy decreases as the number of hydrogen atoms are increasing. This figure reflects that the Ti_6B_{40} molecule exhibits better average adsorption energy (Positive to negative) in presence of PBE-USPP in comparison with other pseudopotentials.

4 Conclusion

We have performed a hydrogen adsorption study of Ti-doped B40 fullerene in presence of different pseudopotentials. The B_{40} fullerene structures are stable after doping of Ti atoms. Though the optimized structures of the B_{40} molecule do not change significantly, the optimized structures of Ti_6B_{40} differ with different pseudopotentials. The binding energies of B_{40} and Ti_6B_{40} also change significantly. As pseudopotential plays a crucial role in the interaction between core and valence electrons, the change of pseudopotential affects the electron charge density and adsorption energy. A strong population of positive energy surrounding the Ti atoms for all the pseudopotentials indicates that Ti atoms can release the electrons easily during

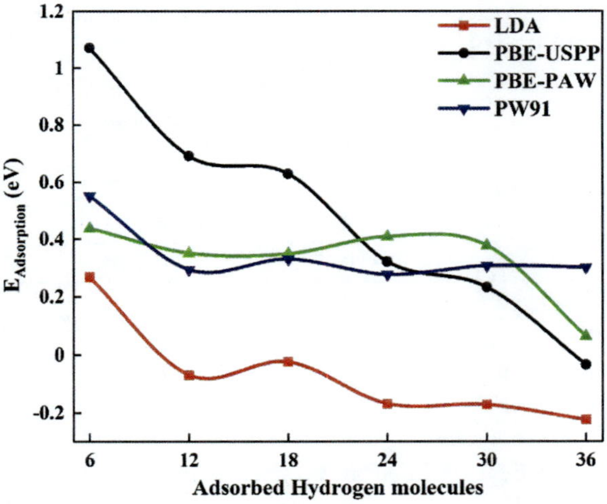

Fig. 5 Average adsorption energy ($E_{Adsorption}$) versus adsorbed hydrogen molecules at different pseudopotentials for $(Ti\text{-}nH_2)_6B_{40}$ where n = 1–6

doping and become suitable to accept electrons during the hydrogen molecules. PBE-USPP provides the best estimations during average adsorption energy calculations of $(Ti\text{-}nH_2)_6$B40 (1.0722 to − 0.0324 eV) in comparison with LDA (0.2681 to − 0.2240 eV), PBE-PAW (0.4358 to 0.0632 eV) and PW91 (0.5512 to 0.3015 eV). A negative average binding energy for $(Ti\text{-}6H_2)_6B_{40}$ indicates the strong attraction of hydrogen molecules on the Ti_6B_{40} surfaces. It also can be concluded that Ti_6B_{40} can adsorb a maximum number of 36 hydrogen molecules.

Acknowledgements The authors are thankful for the Research Initiation Grant (BPGC/RIG/2018-19, Dt.08/11/2018) and Additional Competitive Grant (GOA/ACG/2019-20/Nov/08, Dt. 14/11/2019) by BITS Pilani Goa for carrying out the work. The authors would also like to acknowledge the computational facilities provided by BITS Goa for carrying out the research work.

References

1. Cortright RD, Davda RR, Dumesic JA (2002) Hydrogen from catalytic reforming of biomass-derived hydrocarbons in liquid water. Nature 418:964–967
2. Chen X, Wang L, Zhang W, Zhang J, Yuan Y (2017) Ca-decorated borophenes as potential candidates for hydrogen storage: a first-principle study. Int J Hydrogen Energy 42:20036–20045
3. Wu G, Wang J, Zhang X, Zhu L (2009) Hydrogen storage on metal-coated b80 buckyballs with density functional theory. J Phys Chem C 113:7052–7057
4. Bai H, Bai B, Zhang L, Huang W, Mu YW, Zhai HJ, Li SD (2016) Lithium-decorated borospherene B_{40}: a promising hydrogen storage medium. Sci Rep 6:35518
5. Dong H, Hou T, Lee ST, Li Y (2015) New Ti-decorated B_{40} fullerene as a promising hydrogen storage material. Sci Rep 5:9952
6. Kohn W, Sham LJ (1965) Self-consistent equations including exchange and correlation effects. Phys Rev 140(4A):B864–B871

7. Perdew JP, Zunger A (1981) Self-interaction correction to density-functional approximations for many-electron systems. Phys Rev B 23(10):5048–5079
8. Perdew JP, Burke K, Ernzerhof M (1996) Generalized gradient approximation made simple. Phys Rev Lett 77(18):3865–3868
9. Ernzerhof M, Scuseri GE (1999) Assessment of the Perdew–Burke–Ernzerhof exchange-correlation functional. J Chem Phys 110(11):5029–5036
10. Kresse G, Joubert D (1999) From ultrasoft pseudopotentials to the projector augmented wave method. Phys Rev B 59(3):1758–1775
11. Blöchl PE (1994) Projector augmented-wave method. Phys Rev B 50(24):17953–17979
12. Mulliken RS (1955) Electronic population analysis on LCAO–MO molecular wave functions. I. J Chem Phys 23(10):1833–1840
13. Hohenberg P, Kohn W (1964) Inhomogeneous electron gas. Phys Rev 136(3B):A1133–A1138
14. Hartree DR (1928) The wave mechanics of an atom with a non-coulomb central field. Part I. Theory and methods. Math Proc Camb Philos Soc 24(89):89–110
15. Humphrey W, Dalke A, Schulten K (1996) VMD—visual molecular dynamics. J Mol Graph Model 14:33–38
16. Kokalj A (1999) XCrySDen—a new program for displaying crystalline structures and electron densities. J Mol Graph Model 17:176–179
17. Gaussian 16, Revision C. Gaussian, Inc., Wallingford CT (2016)
18. Harshavardhan T, Akshay G, Paramita H (2020) Titanium atoms-decorated B40 fullerene: first-principle study to predict the structural evolution during hydrogen storage. J Inst Eng (India) Ser E 1–7

Comparative Analysis for Performance Characteristics of a Compression Ignition Engine Running on Microalgae Methyl Ester and Diesel Blends with Base Engine and Coated Engine

K. Sai Babu, B. Sai Rama Krishna, V. Djana Raju, and N. Rama Krisna

1 Introduction

Biodiesel is an organic product so it is sulphur free that's why in emission point of view which gives favourable results. But due to carbohydrate content present in the biodiesel it gives some carbon related emissions but with the help of some techniques like exhaust gas recirculation, emissions may control. So, biodiesel is gives favourable results in case of emission point view that's why this research work is mainly focussed on the performance characterises.

As the part of literature survey under the zone of bio–diesel, many research articles, conference papers, books and some research papers are referred, and the outcome of those research are discussed in detail as given below.

Agarwal et al. [1] confined that B20 gave a optimum performance among all the other. The results indicated that percentage of difference between diesel and the blend in case of BTE at peak condition was 2.5%. Which is a very small variation that's why biodiesel is an alternative fuel.

Avinash et al. [2] has concluded that in diesel engine characteristics wise karanja biodiesel reported the favourable spry and atomization conditions when the fuel was preheated. Tarabet et al. [3] conducted an investigation on a altered DI engine functioned with can produce eucalyptus biodiesel and natural gas under dual fuel manner and the outcome shown that the inconsiderable NO_X emissions.

K. Sai Babu (✉) · B. Sai Rama Krishna · V. Djana Raju · N. Rama Krisna
Department of Mechanical Engineering, Lakireddy Bali Reddy College of Engineering, JNTU-K, Mylavaram, India
e-mail: kolahalamsaibabu332@gmail.com

V. Djana Raju
e-mail: dhanaraju.v@lbrce.ac.in

Department of Mechanical Engineering, Amrita Sai Institute of Science & Technology, JNTU-K, Paritala, India

S. Doolla et al. (eds.), *Advances in Clean Energy and Sustainability*,
Green Energy and Technology, https://doi.org/10.1007/978-981-99-2279-6_19

Demirbas et al. [4] used algae as a biofuel and conducted the experiment on the diesel engine. This experiment has given the suggested results on the biodiesel especially CO_2 were less as compared with the other fuels. Chisti et al. [5] investigated on microalgae as a non-traditional source and it had been described those microalgae gave equivalent amount of biodiesel related to rapeseed or soybean crops using around 49–132 times less land.

Converti et al. [6] stated that, the lipid content of microalgae decreased from 14.7% to 5.90% when the growing temperature increased from 25 to 30 °C. Charyulu et al. [7, 8] investigated a compression ignition engine under the effect of microalgae methyl ester and diesel blends, it has been conveyed that the blend B30 formed well performance and discharge outcomes linked to that of diesel. Kumar et al. (2017) have studied the consequence of insulation with ceramic constituents on the performance, combustion, and discharge features of an engine.

The surveys revealed that the biodiesel blends can be used as substitute fuels for engines. Current study has shown that the practice of biodiesel blends can considerably growth the engine performance and diminish the emissions. So, in the existing study investigate the performance characteristics of microalgae methyl ester (MME) in the ratios 10–30% with pure diesel blends. So, the present work focussed on performance characteristics of an engine powered by the blend of diesel and MME when the piston coated with the zirconium oxide.

2 Materials and Methods

In this study, the research work done on microalgae as the feed stock for producing bio-diesel. As per the literature survey, Microalgae are auspicious substitute oil for engines. Algae can be transformed into Bio-diesel, Bio-ethanol, Bio-hydrogen, Bio-oil and Bio-methane through bio-chemical methods (Fig. 1).

Fig. 1 Microalgae

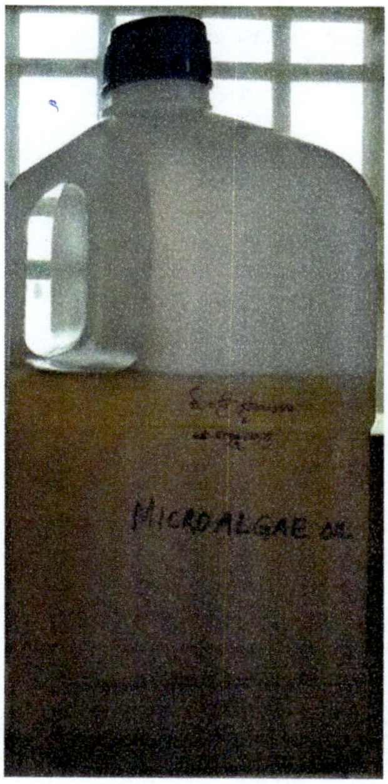

Fig. 2 Microalgae crude oil

The crude oil which is extracted from the algae presented below. The major disadvantage in case of MME is viscosity content. There are different methods to reduce the viscosity but the most adopted methodology is transesterification (Fig. 2).

2.1 *Transesterification*

It is one of the primary mythologies to alter the unwanted fuel properties and fatty acids. This research has done on the catalytic transesterification.

The "Catalytic transesterification" procedure is the response of a triglyceride by an alcohol in the incidence of some catalyst such as sodium hydroxide or potassium hydroxide. The quantity of 1000 ml of crude algae oil is collected in flask. The measure 18 g of KOH is evaluated and 250 ml of methanol is mixed in beaker. KOH is merged with the alcohol and it is stirred while waiting for they are correctly dissolved (Fig. 3).

Micro algae oil has the fatty acid content of palmitic acid is 36.90, palmitoleic acid is11.90 margaric acid is 0.89, myristic acid is 9.10, oleic acid is 6.70 and linolenic acid is 22.30 etc. (Fig. 4).

Fig. 3 Transesterification of microalgae crude oil

Fig. 4 Test samples

2.2 *Fuel Properties*

Bio-diesel is an orgonic product. Usually, organic products contained hydro carbons due this reason it is must to verify the bio diesel properties will match or near the diesel or not. In this research work we measured main fuel properties with standard apparatus (Table 1).

Table 1 Physiochemical properties of diesel–biodiesel blends

S. No.	Fuel property	Diesel	MME	MME10D90	MME20D80	MME30D70
1	Density (kg/m^3)	840	872.6	820.20	825.40	827.4
2	Kinematic viscosity	3.8	5.13	2.75	2.93	3.11
3	Flash point	60	178	56	57	58
4	Cetane number	47	46.69	54.64	54.85	55.06
5	Calorific values	42,500	38,963	42,551	41,921	41,291

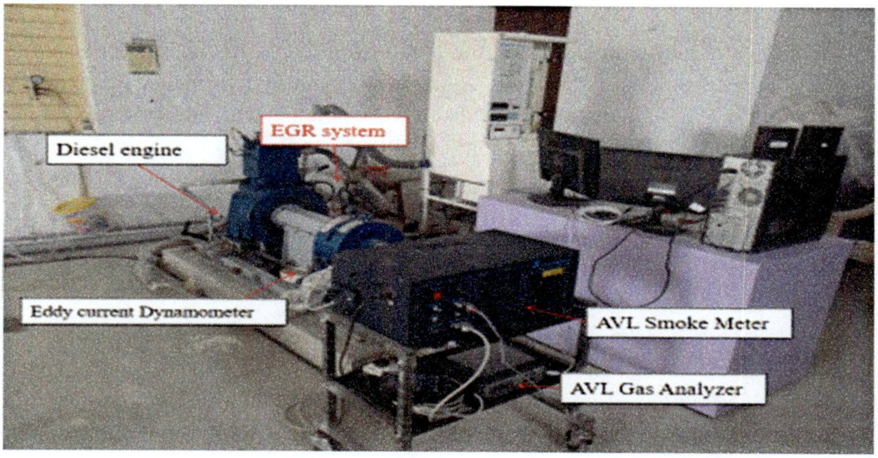

Fig. 5 Photographic view of experimental setup

2.3 *Experimental Engine*

The research was carried out to investigate the effect of microalgae methyl ester blends MME10D90, MME2080, and MME30D70 on performance of a diesel engine and compared to diesel fuel. The experimentation was carried out on Kirloskar make single cylinder four stroke water cooled and direct injection diesel engine (Fig. 5).

2.4 *Preparation of Coated Piston*

In the current study, the piston crown was treated with 0.35 mm thickness of zirconium oxide ceramic material to achieve semi LHR (Low Heat Rejection) engine

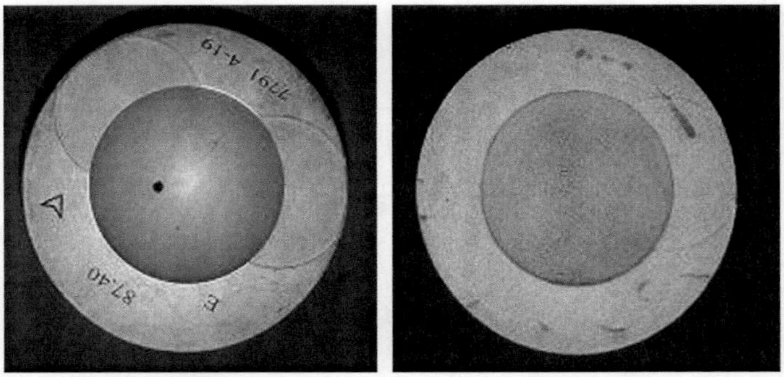

(a) Piston before Coating (b) Piston after Coating

Fig. 6 Preparation of coated piston

which endures higher temperature in the combustion chamber that encourages the combustion. The motto of the work is to progress the performance of a diesel engine with ZrO_2 treated piston by means of MME blends with several proportions fluctuating from 10 to 30% at dissimilar load situations (Fig. 6).

3 Results and Discussions

The diesel fuel was substituted with the mixtures of conventional diesel and biodiesel in the ratios 10–30% i.e., MME10D90, MME20D80, and MME30D70 by fluctuating 0–100% of load on the engine with raise of 25%. Subsequently completion of the test, now the normal piston of the engine is substituted with zirconium oxide treated piston and the above experiment is repeated in same way.

3.1 Relation Between the Brake Power and Brake Specific Fuel Consumption

Figure 7 represents the relation between the brake power (BP) and brake specific fuel consumption (BSFC) of all the test samples. It is evident from the graph that, for all the test fuels the BSFC decreased with increased brake power. Generally, brake specific fuel consumption is the measure of how the supplied fuel energy to the engine is converted into brake power and it can be defined as the amount of fuel consumed in unit time for generating kW power.

The experimental result revealed that, the BSFC of test fuels operated with coated engine was improved compared to pure diesel and base engine. Decrease in BSFC is due to the reduction in the fuel consumption and improved energy conversion rate at all loading conditions in the coated engine. This may be due to the increased

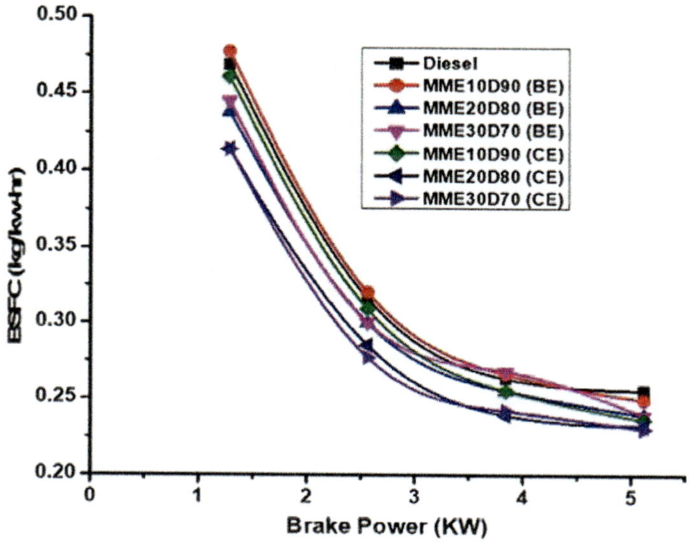

Fig. 7 Variation of BSFC with respect to BP

temperature of the combustion chamber walls, which increase the temperature of the fuel issued from the heated fuel injecting nozzle resulting in the reduced fuel viscosity and better combustion of the fuel. At maximum load condition, MME30D70 with coated engine has registered lowest BSFC amongst all and it was improved by 9.80% compared to pure diesel and by 3.77% compared to base engine data.

3.2 *Variation of Brake Thermal Efficiency with Brake Power*

Figure 8 shows the variation of brake thermal efficiency (BTE) with brake power (BP) for all the test fuels. It can be noticed from the experimental results that, the BTE of all the test samples under coated engine was higher than that of base engine at all load conditions. This is due to the fact that, the thermal resistance of the piston crown which cannot allow the heat energy to coolant there by reducing the fuel consumption for the same amount of power output.

It can be observed that, at maximum load condition, the BTE of MME30D70 with coated engine is higher than that of pure diesel and base engine data. At full load condition, diesel has recorded 33.40% of BTE, MME30D70 with base engine has recorded 36.36% of BTE and with coated engine has recorded 37.92% BTE, therefore for coated engine the BTE of MME30D70 was increased by 13.54% compared to pure diesel and by 4.29% compared to base engine data.

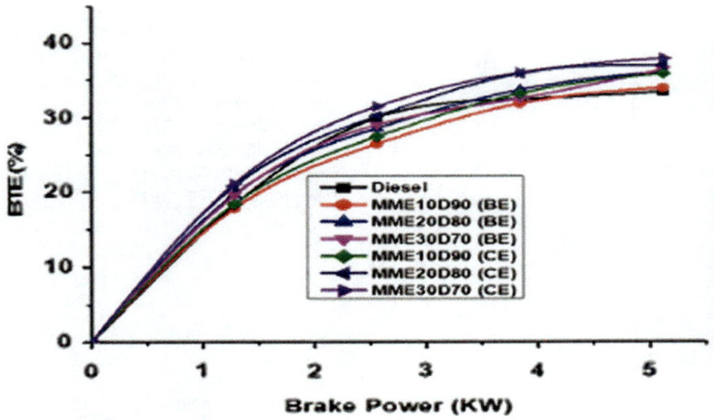

Fig. 8 Variation of BTE with respect to BP

3.3 Variation of Mechanical Efficiency with Brake Power

Figure 9 shows the variation of mechanical efficiency with brake power for all the test samples. Generally, the significance of mechanical efficiency is to measure the effectiveness of a machine in transforming from input to output by overcoming frictional losses.

It can be observed from the trend of the figure that, the mechanical efficiency was gradually increased for all the test samples from zero to maximum. As the biodiesel provides more lubrication and smooth engine operation than that of pure diesel, experimental results showed that the mechanical efficiency of the biodiesel samples operated with base engine is more than that of coated engine and pure diesel. At full load condition, among all the blends MME20D80 with base engine

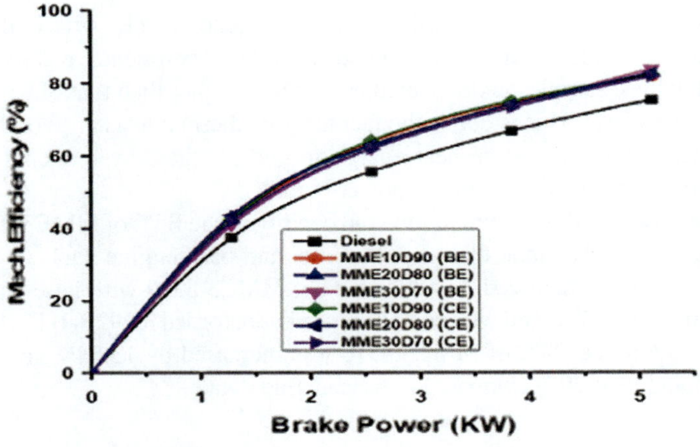

Fig. 9 Variation of mechanical efficiency with respect to BP

has registered 83.69%, MME20D80 with coated engine has recorded 82.61% and diesel has recorded 75.44% so, a little drop was found with MME30D70 with coated engine. Therefore, finally the mechanical efficiency of MME30D70 with coated engine was decreased by 2.39% compared to base engine data, and increased by 8.28% compared to pure diesel.

3.4 Variation of Mechanical Efficiency with Brake Power

Figure 10 revealed the variation of volumetric efficiency with brake power for all the test samples from zero to full load. In general, as far as the engine performance is concerned the volumetric efficiency is one of the important factors. It measures the breathing capacity of the naturally aspirated engines. It is defined as the ratio of volume of air/charge introduced into the engine cylinder during intake to the theoretical displacement volume of all the cylinders at atmospheric pressure. It can be influenced by design parameters such as intake and exhaust restrictions, valve timing, cylinder sealing, inertia of gas, speed etc., and also by some external parameters such as pressure, temperature and humidity of air.

The result shows the volumetric efficiency of each test fuel is continuously decreased from zero to full load as the brake power increases. This is because of the rise in temperature inside the engine cylinder, this was found true in case of base engine results but in case of coated engine results it was found increased instead of decreasing though it was less than that of pure diesel. At full load condition, the volumetric efficiency of MME30D70 with coated engine is higher than that of MME30D70 with base engine by 2.82% and decreased by 1.34% compared to pure diesel.

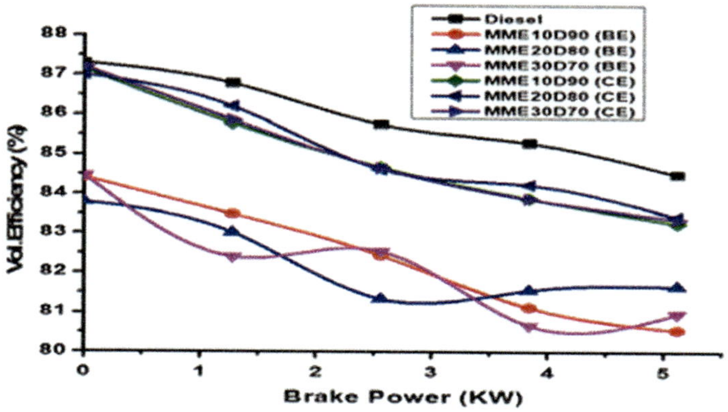

Fig. 10 Variation of volumetric efficiency with respect to BP

4 Conclusions

The performance characteristics of compression ignition engine powered with conventional diesel, diesel and biodiesel blend were investigated with and without piston coating. The conclusions of this investigation at full load condition are as follows,

- In general, the brake specific fuel consumption decreases with increase in load on the engine, among all the blends, MME30D70 with coated engine has registered lowest BSFC amongst all and it was improved by 9.80% compared to pure diesel and by 3.77% compared to base engine data.
- Diesel has recorded 33.40% of BTE, MME30D70 with base engine has recorded 36.36% of BTE and with coated engine has recorded 37.92% BTE, therefore for coated engine the BTE of MME30D70 was increased by 13.54% compared to pure diesel and by 4.29% compared to base engine data.
- Among all the blends, MME20D80 with base engine has registered 83.69% of mechanical efficiency, MME20D80 with coated engine has recorded 82.61% and diesel has recorded 75.44% so, a little drop was found with MME30D70 with coated engine. Therefore, finally the mechanical efficiency of MME30D70 with coated engine was decreased by 2.39% compared to base engine data, and increased by 8.28% compared to pure diesel.
- The TFC of MME30D70 with coated engine was improved by 7.81% compared to MME30D70 with base engine and by 9.92% compared to pure diesel.
- The volumetric efficiency of MME30D70 with coated engine is higher than that of MME30D70 with base engine by 2.82% and decreased by 1.34% compared to pure diesel.

From the above conclusions it can be noticed that, as far as the performance parameters are concerned, out of all the blends MME30D70 with coated piston has showed better results and it was reported as the optimum blend with coated engine.

References

1. Agarwal AK, Das LM (2001) Bio-diesel development and characterization for use as a fuel in compression ignition engines. J Eng Gas Turbine Power Trans ASME 123:440–447. https://doi.org/10.1115/1.1364522
2. Agarwal AK, Dhar A (2010) Karanja oil utilization in a direct-injection engine by preheating. Part 1: Experimental investigations of engine performance, emissions and combustion characteristics. Proc Inst Mech Eng Part D J Automob Eng 224(1):73–84. https://doi.org/10.1243/09544070JAUTO1266
3. Tarabet L, Loubar K, Lounici MS, Khiari K, Belmrabet T, Tazerout M (2014) Experimental investigation of DI diesel engine operating with eucalyptus biodiesel/natural gas under dual fuel mode. Fuel 133:129–138. https://doi.org/10.1016/j.fuel.2014.05.008
4. Demirbas A, Fatih Demirbas M (2011) Importance of algae oil as a source of biodiesel. Energy Convers Manag 52:163–170. https://doi.org/10.1016/j.enconman.2010.06.055

5. Chisti Y (2007) Biodiesel from microalgae. Biotechnol Adv 25(3):294–306. https://doi.org/10.1016/j.biotechadv.2007.02.01
6. Converti A, Casazza AA, Ortiz EY, Perego P, Del Borghi M (2009) Effect of temperature and nitrogen concentration on the growth and lipid content of *Nannochloropsis oculata* and *Chlorella vulgaris* for biodiesel production. Chem Eng Process 48:1146–1151. https://doi.org/10.1016/j.cep.2009.03.006
7. Charyulu TN, Naveenchandran P (2019) The experimental features of a compression ignition engine running on pure diesel and mixtures of diesel and Manilkara zapota methyl ester. Int J Innov Technol Expl Eng 8(11):834. https://doi.org/10.35940/ijitee.K1499.0881119
8. Charyulu TN, Naveenchandran P (2019) The behaviour of a compression ignition engine under the influence of diesel and microalgae biodiesel blends. Int J Mech Prod Eng Res Dev 9(4):447–456. https://doi.org/10.24247/ijmperdaug201944
9. Charyulu TN, Naveenchandran P, Raja E, Babu RN (2020) The behavioural attributes of a compression ignition engine powered with diesel and *Artocarpus heterophyllus* methyl ester blends. Rasayan J Chem 13(2):876–886. https://doi.org/10.31788/RJC.2020.1325560
10. Agarwal D, Agarwal AK (2007) Performance and emission characteristics of Jatropha oil (preheating and blends) in a direct injection compression ignition engine. Appl Therm Eng 27:2314–2323. https://doi.org/10.1016/j.applthermaleng.2007.01.009

Derivation of Hydrodynamic Characteristics of a Medium Speed Francis Turbine Operated Under Various Loading Conditions

Md. Mustafa Kamal, Ali Abbas, Vishnu Prasad, Brijkishore, Prashant Kumar, and Shaurya Varendra Tyagi

1 Introduction

Globally, the power sector is mainly dependent on conventional energy resources such as coal, oil, natural gas or nuclear energy for power generation. Many drawbacks like limited quantity, adverse effects on the environment [1], climate change [2], carbon emissions [3], etc., are associated with the usage of a conventional method for power generation. Recently, the approach has been shifted towards non-conventional resources like solar, wind, hydro, biomass, etc. [4].

From available resources of renewable energy, hydro energy has gained more popularity than other counterparts due to its good predictability and availability [5]. A device that harnesses the available hydro energy from water is called a 'Hydro turbine'. Its classification can be done on the basis of available head and discharge, specific speed, flow direction to the runner, the orientation of the rotational axis, etc. [6]. However, a Francis turbine can be used for the low-head as well as the high-head hydro site [7, 8]. In this paper, a low-head Francis turbine has been chosen for numerical study.

A CFD tool can be treated as a virtual laboratory due to its reliability in turbomachines. A CFD (computational fluid dynamics) is a prediction tool that predicts the flow within a domain by solving the partial differential form of governing equations [9]. These flow variables discretise into the algebraic form of equations, and with the help of computational power, these equations are solved by the iterative method. The solution is always associated with some numerical errors. CFD provides an

Md. Mustafa Kamal (✉) · P. Kumar · S. Varendra Tyagi
Department of Hydropower and Renewable Energy, IIT Roorkee, Roorkee, India
e-mail: itskamal92@gmail.com

A. Abbas
FLOVEL Energy Private Limited, Faridabad, India

V. Prasad · Brijkishore
Maulana Azad National Institute of Technology, Bhopal, India

S. Doolla et al. (eds.), *Advances in Clean Energy and Sustainability*,
Green Energy and Technology, https://doi.org/10.1007/978-981-99-2279-6_20

approximate solution to governing equations, not an exact solution. These errors can be minimized by using appropriate boundary conditions, turbulence modelling, element size in grid generation, convergence criteria, etc. In the present study, a commercially available CFD solver, ANSYS CFX 18.1, has been used to predict the flow behaviour inside the turbine.

Various numerical investigations have been carried out to derive the performance characteristics of a hydro turbine. A few studies have been summarized in tabular form and presented in Table 1.

Table 1 The numerical investigations carried out by various researchers on the hydro turbine

Authors	Solver code	Turbulence model	Objectives	Conclusions
Khare et al. [10]	CFX	SST	To evaluate the flow characteristics of the Francis turbine using CFD	• Rotational speed has less effect on velocity parameters at the inlet than GVO • At the outlet, the relative velocity is more affected by speed than GVO
Goyal et al. [11]	CFX	Std. $k-\varepsilon$ SST	CFD analysis of high-head Francis turbine model at part load condition and validation	• Two types of high amplitude frequencies in the draft tube were observed, i.e., blade passing frequency and vortex breakdown frequency • A vortex rope was captured at the inlet of the draft tube caused by vortex breakdown
Jeon et al. [12]	CFX	SST	Prediction of the hydraulic performance of the turbine with various design parameters of the draft tube	• There was an up-gradation of 1.3% in efficiency and elimination of the vortex region near the wall of the draft tube

(continued)

Table 1 (continued)

Authors	Solver code	Turbulence model	Objectives	Conclusions
Stoessel et al. [13]	Open-FOAM	Std. $k - \varepsilon$ RNG $k - \varepsilon$ Std. $k - \omega$ SST	Steady and transient simulation of the Francis turbine model at various operating conditions	• The steady simulation predicted nearly similar behaviour to experimental ones, but the transient simulation did not give the desired result using open-FOAM as a solver CFD code
Khare et al. [14]	CFX	SST	Effect of a number of blades on the performance of elbow draft tube	• It was observed that the variation in draft tube loss and efficiency is parabolic in nature • The reduction in runner solidity caused the maximum efficiency and minimum loss to shift towards a higher speed factor
Anup et al. [15]	CFX	RNG $k - \varepsilon$	Prediction of vortex shedding in the draft tube using CFD	• There was an occurrence of vortex rope at the outlet of the runner or inlet of the draft tube, and it was observed that the rotational frequency of the vortex was 19% of the rotational frequency of the runner

2 Numerical Methodology

This segment discusses the 3D modelling of the Francis turbine, grid generation, boundary conditions, turbulence modelling and selection of CFD parameters.

2.1 3D Domain and Mesh Generation

A Francis turbine consists of a casing, guide vanes (GV) and stay vanes (SV) assembly, runner and draft tube (DT). To model these complex-shaped components of the turbine, two CAD software viz. 'Solid works' and 'ANSYS design modeller' have been exercised. A complete set-up of the Francis turbine is presented in Fig. 1.

Mesh generation is a process of discretization of the flow domain into a small number of cells. The meshing of different components of the model is presented in Fig. 2. The convergence of a numerical solution is highly dependent on the quality of the mesh. Some influential parameters such as skewness (< 1), orthogonal quality (> 1) and aspect ratio (< 100) [16], which define the quality of the mesh, have been well taken care of.

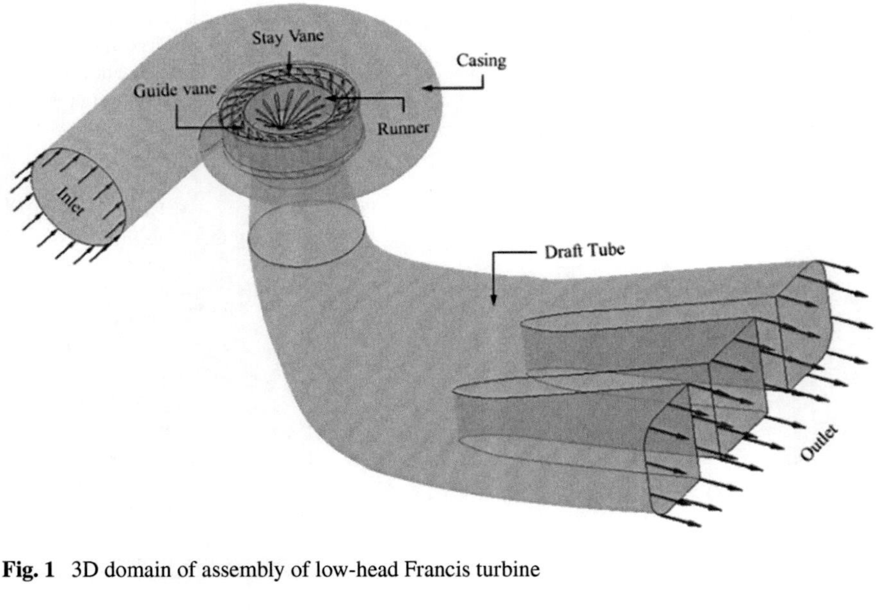

Fig. 1 3D domain of assembly of low-head Francis turbine

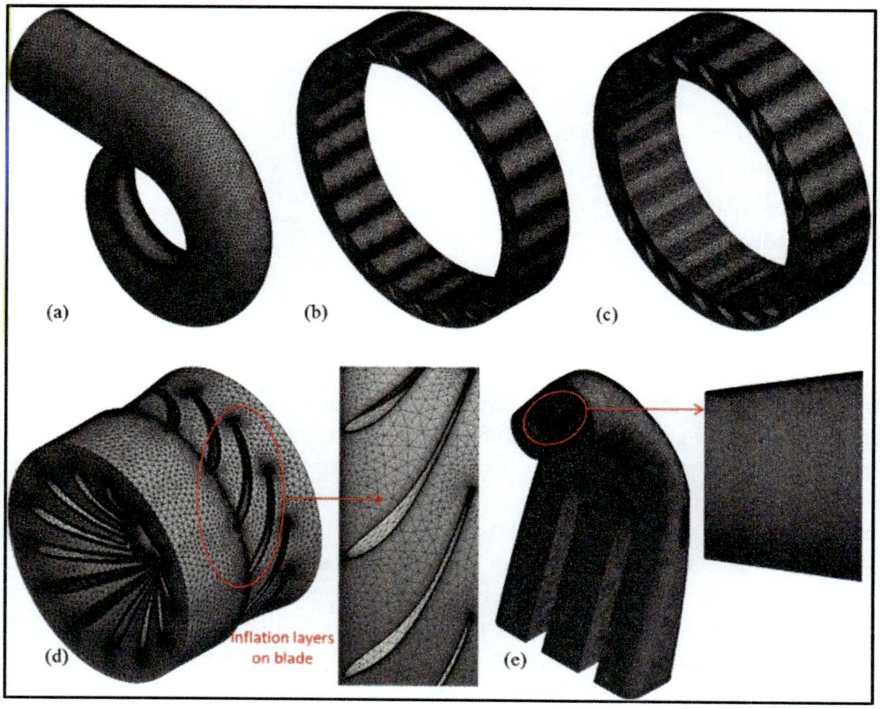

Fig. 2 Meshing of different components of the Francis model; **a** casing, **b** guide vanes assembly, **c** stay vanes assembly, **d** runner, **e** draft tube

2.2 Operating Points and Boundary Conditions

The performance characteristics of the Francis turbine have been driven at three loading conditions, viz. part-load, rated load (BEP) and overload condition. These conditions are varied by guide vanes' closing and opening operation, which control the flow rate inside the turbine.

To solve the partial form of differential equations, boundary conditions need to be provided. At the inlet of the casing, the 'mass flow rate' boundary condition is given, while at the outlet of the draft tube, 'static pressure' is provided. A 'wall' boundary condition with a no-slip condition has been chosen for guide vanes, stay vanes and runner blades. Both stationary and rotating sub-domains are interconnected by interfacing, which conserves the flux while passing from one sub-domain to another. The details of boundary conditions and operating conditions are presented in Table 2.

Table 2 Operating and boundary conditions considered in the present study

Loading condition	Inlet condition (mass flow rate, kg/s)	Outlet condition (average static pressure, Pa)
Part-load (α = 14.01°)	343.46	101,325
BEP (α = 25.47°)	665.49	
Over-load (α = 34°)	840.76	
Reference pressure (0 Pa)		

2.3 Turbulence Modelling and Other Solver Parameters

In order to capture the fluctuating components of the flow variable, an extra set of equations (other than RANS equations) is needed, termed 'turbulence modelling'. Several authors suggested that the two-equations turbulence model offers a good response between computational power and numerical accuracy. Hence, in the present study, the SST turbulence model has been used [17, 18].

Some influential parameters which affect the accuracy of numerical results are the advection scheme, residual convergence criteria and a number of iterations for solving the algebraic form of differential equations. A 'high resolution' has been selected for the advection scheme. In this numerical study, the number of iteration and convergence criteria have been kept as 500 and 10^{-5} [19], respectively.

3 Numerical Results

This section discusses the mesh independency test, hydraulic efficiency of the turbine with validation and flow behaviour of fluid in the turbine vicinity.

3.1 Mesh Independency Limit

To mould a relationship between the number of mesh elements of the domain and hydraulic efficiency, a mesh independency limit (MIL) test has been carried out. A 'mesh refinement method' has been chosen for the MIL test [20]. Four different sets of mesh elements have been considered to accomplish this activity, i.e., 9,474,570, 13,751,672, 14,822,671 and 16,146,536. The minimum numerical error has been observed with 14,822,671 elements. Therefore, this mesh density has been carried forward for all simulations. A graphical representation of the MIL test is shown in Fig. 3.

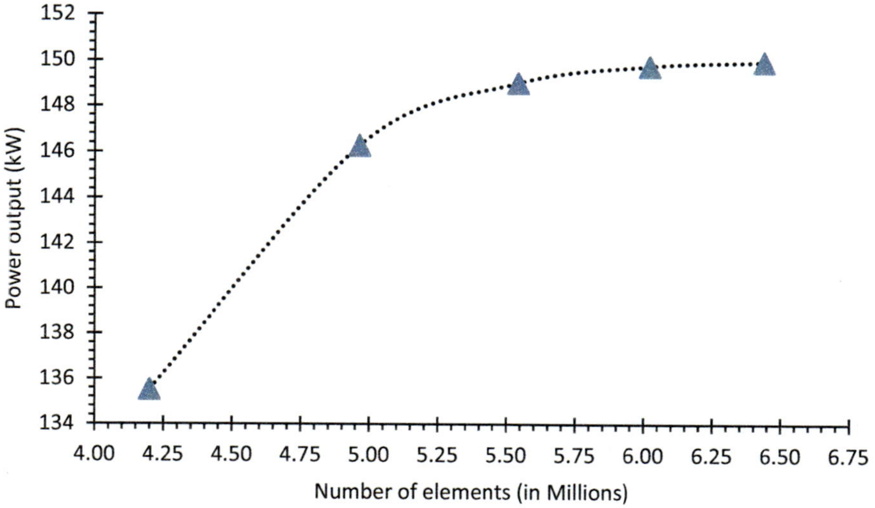

Fig. 3 Details of the mesh independency test

3.2 *Validation of Numerical Results*

A non-dimensional parameter, hydraulic efficiency, has been chosen to validation of the numerical model of the considered hydro turbine. It is defined as the ratio of power extraction by turbine blades to the power available at the inlet of the turbine. It has been found that the difference between the hydraulic efficiency obtained from the numerical and experimental results is about $\pm$ 1.7%. The present validation shows a good agreement of the numerical results with experiments, and computed hydraulic efficiency is similar to the experimental efficiency. The hydraulic efficiency obtained from the numerical study and experimental study is presented in Fig. 4.

3.3 *Hydrodynamic Behaviour of Turbine*

In this segment, the variations in the velocity components at the inlet and the outlet of the turbine runner operated under the different loading conditions have been elaborated. Further, the head loss across different components of the Francis turbine has also been discussed. The flow behaviour of water across the turbine is also assessed.

Variations in Velocity Components Across Francis Runner

Under the present investigation, velocity components have been computed at an inlet and outlet of the runner at different guide vane openings to determine the variations in velocity carried by flowing water. The graphical representation of variations in

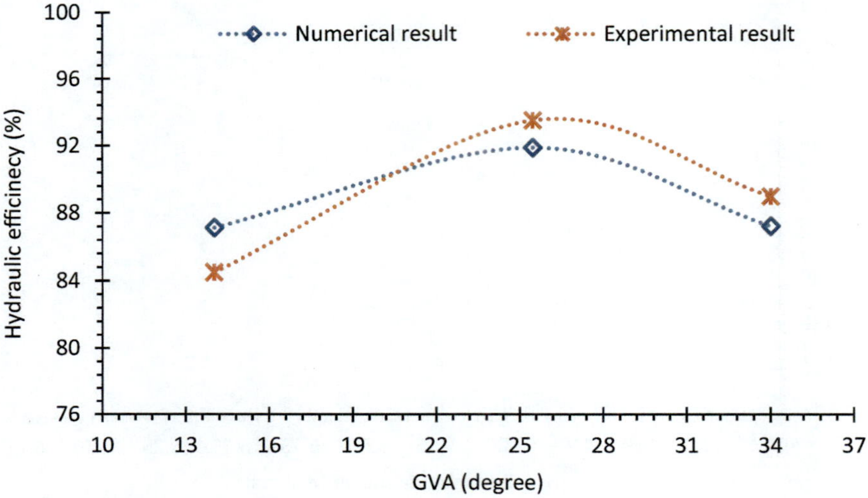

Fig. 4 Hydraulic efficiency at various loading conditions

absolute velocity, relative velocity, whirl velocity and flow velocity of flowing water is presented in Fig. 5a, b, c and d, respectively.

It can be observed from Fig. 5 that the absolute and whirl component of the velocity of water at the inlet of the runner decreases as the guide vane opening increases. In contrast, at the rated condition (GVA = 25.47°), the difference in whirl velocity

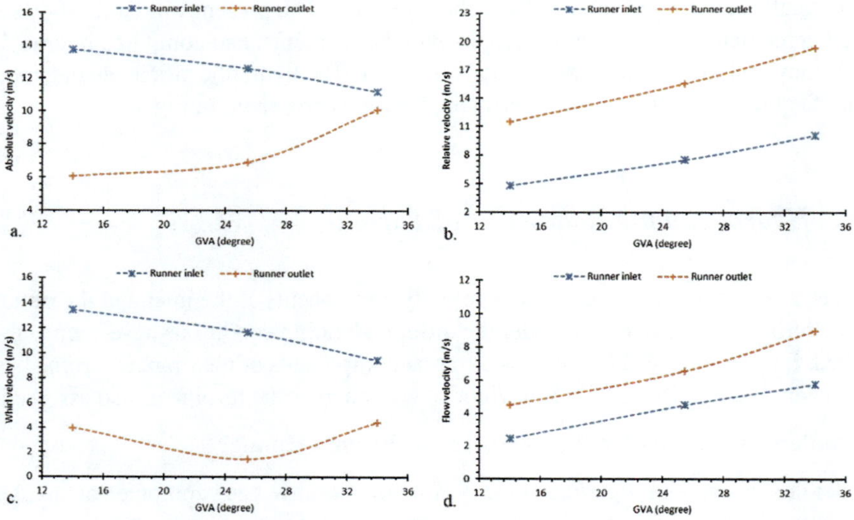

Fig. 5 Variations in water velocity at the inlet and outlet of the runner; **a** absolute velocity, **b** relative velocity, **c** whirl velocity and **d** flow velocity

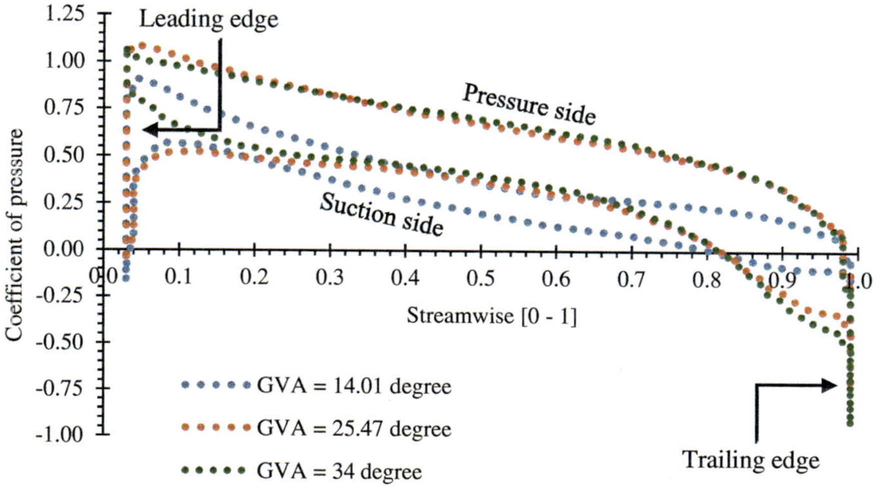

Fig. 6 Blade loading curve for different guide vane openings

was higher than in the other loading conditions, which signifies the higher torque developed by the turbine runner at the rated condition. However, relative velocity and flow velocity shoot up across the runner as the guide vane opening increases.

Variations in Pressure Across Francis Runner Blade

In order to quantify the pressure distribution around the runner blade surface, a blade loading curve has been plotted at different guide openings and presented in Fig. 6. The X-axis of the graph represents the streamwise from 0 (leading edge of the blade) to 1 (trailing edge of the blade), and Y-axis represents the coefficient of pressure.

Based on the blade loading curve, it was perceived that the pressure side of the blade surface experiences higher pressure compared to the suction side, generating a lift force that aids the rotational motion of the runner. However, a sudden drop in pressure was observed at the blade's trailing edge at the guide vane opening of 34°. This is due to the flow detachment from the blade's surface. Therefore, the suction side of the blade is more prone to cavitation.

Head Loss Across Different Components of the Francis Turbine

Under the present study, the head loss across various components of Francis turbine such as casing, stay vane, guide vane, runner and draft tube, has been computed through numerical analysis. The graphical representation of head loss in various components of the Francis turbine at different guide vane openings is depicted in Fig. 7.

It can be observed that the head loss for the casing, S.V. and G.V., was found to be lower than that of runner and D.T., irrespective of the guide vane opening. However, the minimum head loss in runner and D.T. was found when the turbine operated at rated operating conditions.

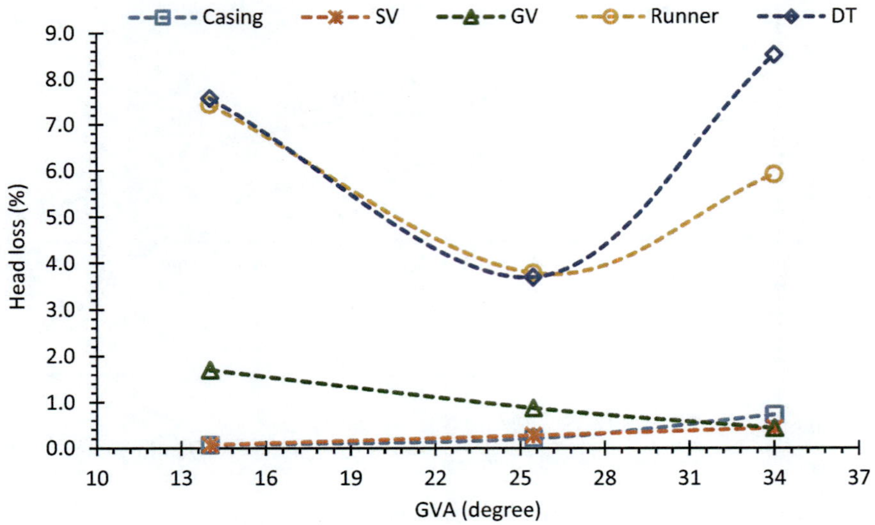

Fig. 7 Head loss in various components of the Francis turbine at different guide vane openings

Pressure Contours and Streamlines Across the Turbine Runner

Based on the pressure contours plots, the pressure gets reduced as it travels from the guide vane to the runner. The maximum pressure was encountered at the pressure side of the guide vanes, while the minimum was at the trailing edge of the suction side of the runner blade. At overload condition (GVA = 34°), a high magnitude of negative pressure was observed at the trailing edge of runner blades which may lead to the occurrence of the cavitation phenomenon. As shown in Fig. 8a, the difference between the pressure at suction and the pressure side of the guide vane is higher in part-load conditions than in other conditions. Hence, this may lead to higher pressure pulsation at the vanless space of the hydro turbine. The pressure at the pressure side of the runner blade was captured higher than the suction side of the runner blade at all loading conditions, which helped in the generation of lift and rotation of the Francis wheel. The blade surface's maximum and minimum pressure were observed at over-load conditions. The variation of pressure around the blade is presented in Fig. 8.

A relative velocity inside the turbine can be visualized with the help of stream tubes across the Francis turbine. It signifies the direction of the velocity vector of flowing fluid. The maximum velocity was observed at the overload condition. In part-load conditions, the velocity carried by water is less, leading to adverse flow conditions and flow becoming more turbulent, which can be conceived in Fig. 9a. The most stable flow condition was observed at the rated load condition (GVA = 25.47°). The variations of velocity inside the turbine are shown in Fig. 9.

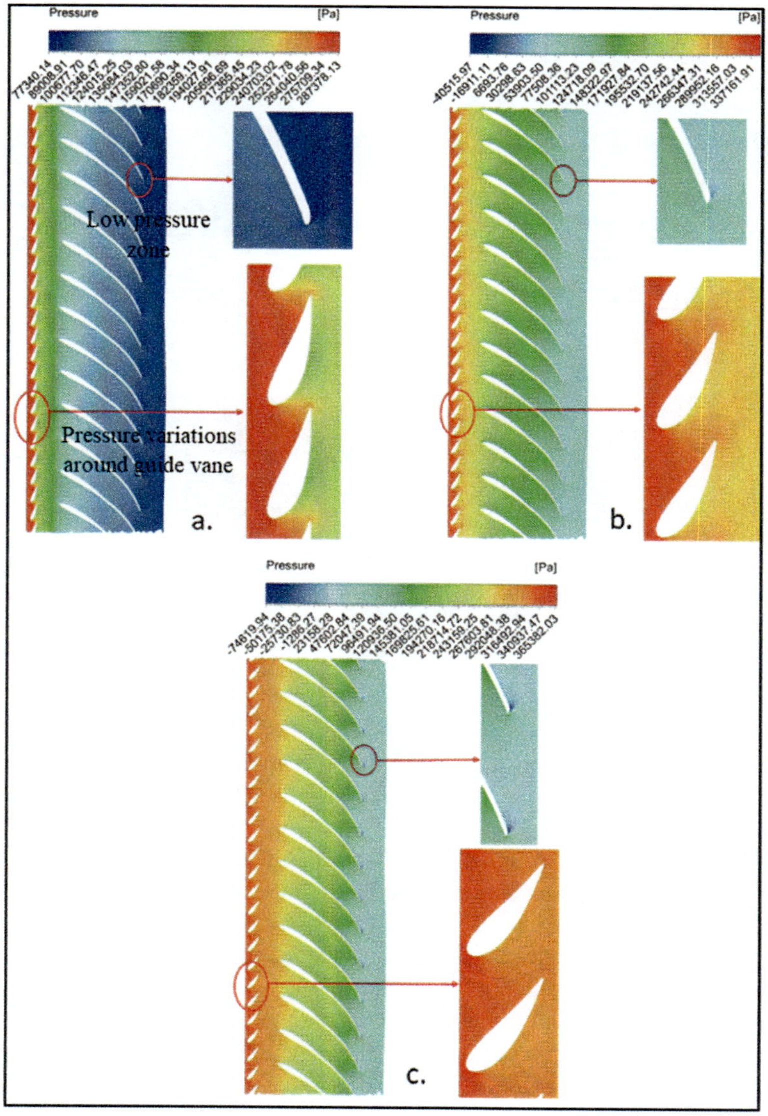

Fig. 8 Blade-to-blade pressure variation at; **a** GVA = 14.01°, **b** GVA = 25.47° and **c** GVA = 34°

4 Conclusions

An extensive numerical study has been carried out on a medium-speed Francis turbine to derive the turbine's hydrodynamic characteristics under different loading conditions, viz. part-load, BEP and over-load conditions. Based on the numerical analysis, the following conclusions are drawn;

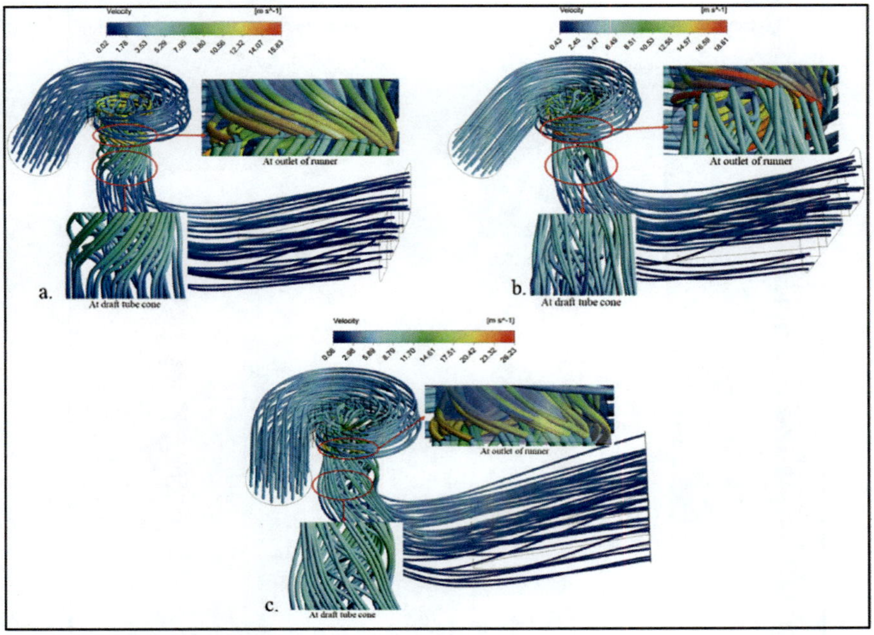

Fig. 9 Velocity variation at; **a** GVA = 14.01°, **b** GVA = 25.47° and **c** GVA = 34°

i. Based on the study, it is concluded that the predictability of hydraulic efficiency of the hydro turbine is more accurate at BEP conditions than off-design conditions.
ii. Guide vane opening significantly affects the different velocity components carried by flowing water across the turbine.
iii. The fluid pressure declined abruptly at the trailing edge of the suction side of the Francis runner blade, which signifies that the trailing edge of runner blades is more prone to cavitation.
iv. Based on the numerical results, it is observed that the minimum head loss was obtained when the turbine is operated at BEP condition.
v. By plotting pressure contours on the blade surface, it is observed that the chances of cavitation are more at over-load conditions due to the negative pressure experienced by the blade. Similarly, it is perceived that the highly stable flow condition occurs at the BEP condition, while a highly adverse flow condition was observed at the part-load condition.

References

1. Kamal M (2017) Scenario of small hydro power projects in India and its environmental aspect. Int Res J Eng Technol 4:228–234
2. Kaunda CS, Kimambo CZ, Nielsen TK (2012) Hydropower in the context of sustainable energy supply: a review of technologies and challenges. ISRN Renew Energy 2012:1–15. https://doi.org/10.5402/2012/730631
3. Sood M, Singal SK (2019) Development of hydrokinetic energy technology: a review. Int J Energy Res 43:5552–5571. https://doi.org/10.1002/er.4529
4. Jawahar CP, Michael PA (2017) A review on turbines for micro hydro power plant. Renew Sustain Energy Rev 72:882–887. https://doi.org/10.1016/j.rser.2017.01.133
5. Kamal MM, Saini RP (2022) A review on modifications and performance assessment techniques in cross-flow hydrokinetic system. Sustain Energy Technol Assess 51:101933. https://doi.org/10.1016/j.seta.2021.101933
6. Brijkishore, Khare R, Prasad V (2021) Prediction of cavitation and its mitigation techniques in hydraulic turbines—a review. Ocean Eng 221:108512. https://doi.org/10.1016/j.oceaneng.2020.108512
7. Trivedi C, Cervantes MJ, Dahlhaug OG (2016) Numerical techniques applied to hydraulic turbines: a perspective review. Appl Mech Rev 68:1–18. https://doi.org/10.1115/1.4032681
8. Mustafa Kamal M, Abbas A, Kumar R, Prasad V (2021) The cause and control of failure of hydraulic turbine due to cavitation: a review, pp 1099–1112. https://doi.org/10.1007/978-981-16-0235-1_85
9. Tiwari G, Kumar J, Prasad V, Patel VK (2020) Utility of CFD in the design and performance analysis of hydraulic turbines—a review. Energy Rep 6:2410–2429. https://doi.org/10.1016/j.egyr.2020.09.004
10. Khare R, Prasad V, Kumar S (2010) CFD approach for flow characteristics of hydraulic Francis turbine. Int J Eng Sci 2:3824–3831
11. Goyal R, Trivedi C, Kumar Gandhi B, Cervantes MJ (2018) Numerical simulation and validation of a high head model Francis turbine at part load operating condition. J Inst Eng Ser C 99:557–570. https://doi.org/10.1007/s40032-017-0380-z
12. Jeon JH, Byeon SS, Kim YJ (2013) Effects of draft tube on the hydraulic performance of a Francis turbine. IOP Conf Ser Mater Sci Eng 52. https://doi.org/10.1088/1757-899X/52/5/052034
13. Stoessel L, Nilsson H (2015) Steady and unsteady numerical simulations of the flow in the Tokke Francis turbine model, at three operating conditions. J Phys Conf Ser 579. https://doi.org/10.1088/1742-6596/579/1/012011
14. Khare R, Prasad V, Verma M (2012) Design optimisation of conical draft tube of hydraulic turbine. Int J Adv Eng Sci Technol 2:21–26. http://www.ijaest.com/docs/IJAEST12-02-01-21.pdf
15. Kc A, Lee YH, Thapa B (2016) CFD study on prediction of vortex shedding in draft tube of Francis turbine and vortex control techniques. Renew Energy 86:1406–1421. https://doi.org/10.1016/j.renene.2015.09.041
16. Kamal M, Saini G, Abbas A, Prasad V (2021) Prediction and analysis of the cavitating performance of a Francis turbine under different loads. Energy Sources Part A Recover Util Environ Eff 1–25.https://doi.org/10.1080/15567036.2021.2009941
17. Kamal M, Abbas A, Prasad V, Kumar R (2021) A numerical study on the performance characteristics of low head Francis turbine with different turbulence models. Mater Today Proc. https://doi.org/10.1016/j.matpr.2021.02.155
18. Kaniecki M, Krzemianowski Z (2016) CFD analysis of high speed Francis hydraulic turbines. Trans Inst Fluid-Flow Mach 131:111–120

19. Kamal M, Saini RP (2022) A numerical investigation on the influence of savonius blade helicity on the performance characteristics of hybrid cross-flow hydrokinetic turbine. Renew Energy 190:788–804. https://doi.org/10.1016/j.renene.2022.03.155
20. Kamal M, Saini RP (2023) Performance investigations of hybrid hydrokinetic turbine rotor with different system and operating parameters. Energy 267:126541. https://doi.org/10.1016/j.energy.2022.126541

Experimental Assessment of Liquid Suction Heat Exchanger Used with Vapour Compression Refrigeration System by the Application of Twisted Strip Inserted Condenser

A. Pratyush, V. Dhana Raju, K. Sai Babu, and M. Oliva

1 Introduction

The global challenges lead our society to employee clean and carbon neutral economy. In this context of heating and cooling sectors accounted for 200–400 kW/s m per year of the total final energy consumption in India. India emits 3 giga-tonnes of greenhouse gasses per year, it is half the world average. Current refrigeration technologies are based on compression refrigeration system, having power consumption maximum [1]. To meet the climatic conditions objectives and reduce pollutions to atmosphere by more research in energy saving and improving methods in HVAC technologies [2].

Wang et al. [3] studied theoretical model was prepared to investigate the performance of the VCR cycle like COP, exergy efficiency and loss. For this study R290 and R600 liquids are used which are low GWP and no ozone depletion potential (ODP) acts as replacement for R134. Two systems are considered like series and parallel 2 stage evaporators. The refrigeration capacity increased by 102% for R600 compared to R134. The low temperature evaporator dehumidifies the outdoor air and used to cool indoor air coming from high temperature evaporator. This improves the performance of VCR cycle. The 2 stage system becomes more complex than that of parallel stage.

Babarinde et al. [4] conducted to replace the pure refrigerant, multi walled carbon nano tubes are used in single evaporator refrigerator. The research work is conducted in the varying concentration of nano lubricants and refrigeration R600a different mass ratios. The temperature in the evaporator is − 8 to − 11 °C at 0.4–0.6 g/l of nano lubricant concentration and 50–60 g of refrigerant. The system uses least electricity at the condition of CNT lubricant 0.4 and 50 g of charge (R600). At these operating

A. Pratyush (✉) · V. Dhana Raju · K. Sai Babu · M. Oliva
Department of Mechanical Engineering, Lakireddy Bali Reddy College of Engineering, Mylavaram, India
e-mail: pratyush0301@gmail.com

S. Doolla et al. (eds.), *Advances in Clean Energy and Sustainability*, Green Energy and Technology, https://doi.org/10.1007/978-981-99-2279-6_21

conditions this situation is best suitable for maximum refrigeration effect at 50, 60, 70 g the pull down time and the power consumption is lower by 25.9%, 20.2% and 13.7% respectively compared with pure lubricant under 0.4–0.6 g L of blended lubricant.

Selvnes et al. [5] reviewed on PCM materials under cold thermal energy storage incorporated in the refrigeration system. The air conditioning application to food freezing by making use of different PCM materials. Wide studies of PCM's for refrigeration of ice/water PCM can be used as cold energy for the food transport, commercial refrigerator and packing of food. This can take share of future growth and vast application of refrigeration.

Prateek et al. [6] reviled that refrigerant are major pollutants of the environment like HCFC, HFC due to vast application of refrigeration units in different fields. The alternative refrigerants may the eco-friendly. The feasibility of alternative refrigerants is identified by assessment of the energy. The Simulink model of multi-stage multi evaporation in the VCR unit to check the thermal performance of 18 refrigerants which are safe, environment friendly. 0 to − 15 °C, − 15 to − 5 °C and 25–33 °C are the temperature variations in the 1-evaporator, 2-evaporator, condenser respectively R245c, R141B, R245F, R123 and R152 exhibits high thermal working properties. The performance equal to conventional refrigerants given by R1234yf and R1234e (E) which are environment safe. These are validated in the Simulink MATLAB software.

Liu et al. [7] examined that refrigeration CO_2 low performance can be solved by involvement of ejector, subcooling in the market refrigeration unit. The feasibility and possibility of the view was conducted in MATLAB. The heat performance, consumption of energy and SEER are the major factors to estimate the system performance in typical environment condition of china cities. The thought of involving ejector and mechanical subcooling reducing pressure of system operation and COP enhanced by 62% (mechanical subcooling), 19% (MEJ) and 20.21% (mechanical sub-cooling + MEJ). The experiment was carried out at an atmospheric temperature condition of 40 °C. The mechanical subcooling and MEJ saves energy up to 30 and 43% increment in energy efficient refrigeration.

Goyal et al. [8] investigated the atomic effect of ozone depletion and acclimation in the greenhouse emissions are damaging environment from past decades. The vapour compression refrigerator is tested with R134a and R152 blends which are environment friendly, inflammable, corrosive resistance and non poisons. Different lengths of expansion valves are implanted in VCR unit to analyse the performance with R134a and blends of R134a-R152a. R134a gave higher compressor work and refrigeration effect that R152a. With first expansion valve more COP at 10/90 blend ratio of R134a and R152a but the reliable COP is at 40/60 blend ratio. High COP is at high evaporator temperature that can be attained by blend rather than pure R134a because of less compressor work. High volumetric efficiency is given by blends by exhibiting low pressure ratios. R134a gives high heat extraction rate. It was concluded that R152a (60/40) is best chosen as refrigerant for its minimum GWP and requirement of high cooling rate. The system operating at medium temperature and high performance given by R134a.

Chen et al. [9] performed to increase the degree of subcooling in the lower temperature circuit CO_2/NH_3 cascade system is proposed to auxiliary refrigeration circuit. In view of the essential standards of thermodynamics, a numerical model is laid out and hypothetical done by simulation to acquire the important parameters that shows impact on the cycle execution. Contrasted and the traditional cascade refrigeration framework (CCRS), the final traced that exists an ideal condensation temperature of low-temperature cycle (TMC.opt) to maximum COP. When subcooling degree is 10 °C CRSS COP is 4.58% more than CCRS. The extreme exergy effectiveness is 0.391, expanding to 4.40%. With the increment of the degree of sub-cooling, both the TMC.opt and the COP of CRSS increments. At the point the degree of subcooling increments from 5 to 15 °C, the increment 2.73–6.00% in its performance. Exergy efficiency and COP decreases when changes in temperature of evaporation. And it is observed that better performance when lower temperature of evaporator at steady temperature of condensation. Moreover, 9.9 °C reduction in the CRSS discharge temperature of ammonia compressor at − 30 °C temperature of evaporation.

Ahmed [10] reviewed new strategy of using nanofluid as an optional liquid in an auxiliary circuit of refrigeration unit is inspected. Nanofluid heat is extracted by refrigerant R134a in evaporator segment which is across the auxiliary circuit. The refrigerant streamed in the coil and nanofluid streamed in the shell in the shell and coil heat exchanger evaporator. The significant parts of the unit are water cooled condenser, expansion valve and compressor. Nanofluid in temperatures (30–40 °C), mass stream rates (40–80 g/s) and different concentrations of volumes of Al_2O_3 nanofluid (0–15%) at these conditions test was conducted. The altered unit displayed predominant execution when Al_2O_3 nanofluid was utilized in the auxiliary circuit when contrasted with base liquid (refined water) while working at a similar mass stream rate and in temperature. At mass stream rate 80 g/s, volume concentration 15% a greatest COP of 6.5 was accomplished for nanofluid with inlet temperature of 40 °C.

Kotu et al. [11] performed convective heat flow is vital in the microelectronics cooling HVAC and refrigeration applications. The heat movement characteristics and thermophysical properties are enhanced by expansion of nanoparticles [12] to the refrigerant outcomes, thereby enhancing the working performance of the refrigerator. The staging of the domestic refrigerator with mineral oil /HFC134a unit was contrasted and nano-refrigerant, mineral oil along with HFC134a-two pipeline heat exchanger-mineral oil. Results shows that unit performance has further developed when HFC134a, DPHE, mineral oil unit is utilized instead of using HFC134a, mineral oil, nano-mineral oil refrigerator. The nano refrigerant and mineral oil and HFC134a, DPHE and mineral oil works ordinarily and securely in the refrigerator. The HFC134a, DPHE and mineral oil unit decreased the energy utilization by 30% and nano refrigerant and mineral oil unit diminished that 26% of energy utilization when contrasted with HFC134a, mineral oil unit. There are additionally improvements in performance coefficient when DPHE was presented in the ordinary unit.

Adelekan et al. [13] studied was led on TiO_2 nanoparticle mixed to R600a as the working liquid medium. In a household fridge with minute system recreation TiO_2-R600a nano-refrigerant was utilized. Pull down test, compressor work-input, refrigeration effect and performance coefficient (COP) under these boundaries' refrigerator performance was researched by steady state examination utilization. In refrigerator nano-refrigerants TiO_2-R600a showed optimum results and performance when compared with R600a. The best refrigeration impact and COP was given by 40 g R600a with 0.1 g TiO_2 and gave un-explicit compressor input work inside the test rig.

For past 20 years, Refrigeration technologies [14] received much attention to cope with high demand due to vast applications. Subcooling, condenser, refrigerants and heat exchangers gave effective solution to improve the refrigerator performance without varying the refrigerator structure. The identified peak demand on the refrigeration especially on domestic vapour compression system [15] is one of the important requirement to make the system successful by implementation of these techniques for the existing VCR unit [16]. Up to date, many work publications on refrigeration aiming on domestic refrigerator, AC systems, storage systems, transportation systems. The system with liquid suction heat exchanger and twisted strip inserted condensers are less studied, these may be essential for commercial, domestic and industrial refrigeration. On the focus of generalised costs and better performance was monitored on these techniques involvement. This work gives overview of thermal performance of the VCR unit can be applied to food transport, food preservation, and commercial refrigeration with small capacity.

2 Material and Methodology

Wide research is done on the effect of liquid suction heat exchanger and twisted strips individually, it is noticed that only few studies are conducted on twisted strips inserted condensers. In this research the pair of LSHE [17] and strip inserted condensers are incorporated in the common compression refrigeration [18] unit.

2.1 Materials

Tetra fluoro ethane (R134a) was used as refrigerant in VCR system [19], as this is non-flammable, eco-friendly, non-toxic and corrosive resistance. The ozone depletion level is zero and the global warming potential is observed to be 1200. Some of the physio-chemical properties are listed in Table 1.

Table 1 Various physio-chemical properties of tetra fluoro ethane

Properties	R134a
Critical temperature (°C)	101.1
Heat of vaporization (kJ/kg)	175.5
Boiling point at 1.013 bar (°C)	− 26.1
Molecular weight	102
Specific heat capacity (liquid) (kJ/kg K)	1.423
Specific heat capacity (vapour) (kJ/kg K)	0.876
Density liquid (kg/m^3)	1208

2.2 Methodology

The refrigerant is made to flow through the circuit as shown in Fig. 1. Two way valves are provided at inlet and outlet of each condenser. That facilitate to operate that particular condenser. The high pressure refrigerant is coming from compressor is made to pass through any one condenser by opening its valve while the valves of remaining condensers remains closed. The condensed refrigerant is expanded iso-enthalpically in the expansion device. This low temperature working medium takes heat load in the evaporator.

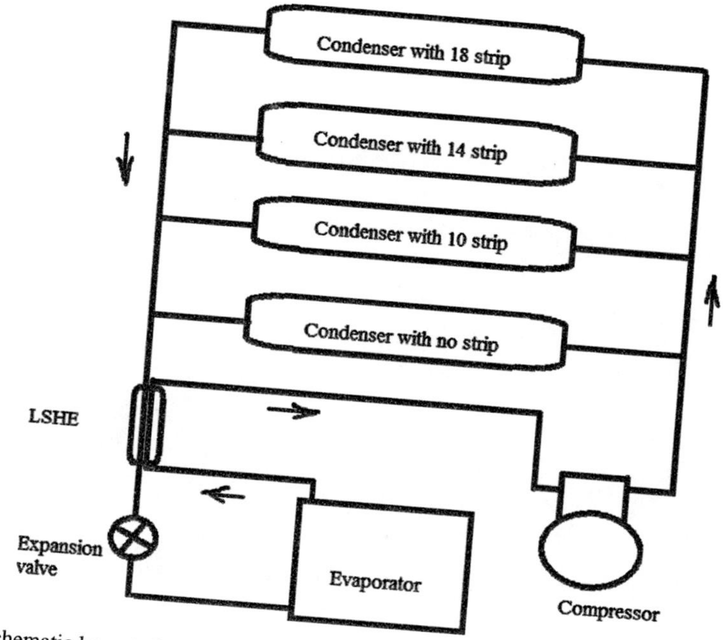

Fig. 1 Schematic layout of the test rig

3 Experimental Setup

The experiment [20] test rig is equipped with a reciprocating semi-hemisphere compressor of capacity 1 tonne of refrigeration (TOR), the condenser made of copper piping which having ID of 10 mm and OD of 12 mm, length of 3 m. This copper pipe is bent into 7 passes that is installed in shell [21] of the condenser. The shell of the condenser [22] is made of tin sheet, in 52 cm length as represented in Fig. 2. In the pipes of the condenser twisted strips are inserted which are made of mild steel of thickness 0.4 mm. The angles of these twisted strips are 10°, 14° and 18°. Four separate condensers are made by inserting these strips in tubes along with plain tube i.e. no strips inserted. To operate condensers 2 valves are placed at inlet and outlet for each. This arrangement facilitates to operate that particular condenser. A blower is provided for forced convection heat transfer in the condenser.

Tube in tube liquid suction heat exchangers [23] is made of copper. Parallel flow was made for the working medium to pass in the LSHE. The length of this heat exchanger is 200 mm. Inner tube carrier's working fluid from condenser and refrigerant from evaporator passes through 12 mm piping network. The valve mechanism is implanted to suction exchanger providing it to involve or exclude from the basic circuit of normal VCR unit as represented in Fig. 3.

Pressure gauge with 0.1 psi accuracy are located at condenser entrance, exit and also at compressor inlet. To make a note of refrigerant temperature at respective locations thermo couples with ± 1 °C accuracy was used. The evaporator chamber was made of stainless steel tank of 15 L capacity which is surrounded with the piping from expansion valve. The Evaporator chamber and the pipelining are insulated with foam. The energy consumed by the compressor, blower is measured with voltmeter and ammeters.

Fig. 2 View of experiment test rig

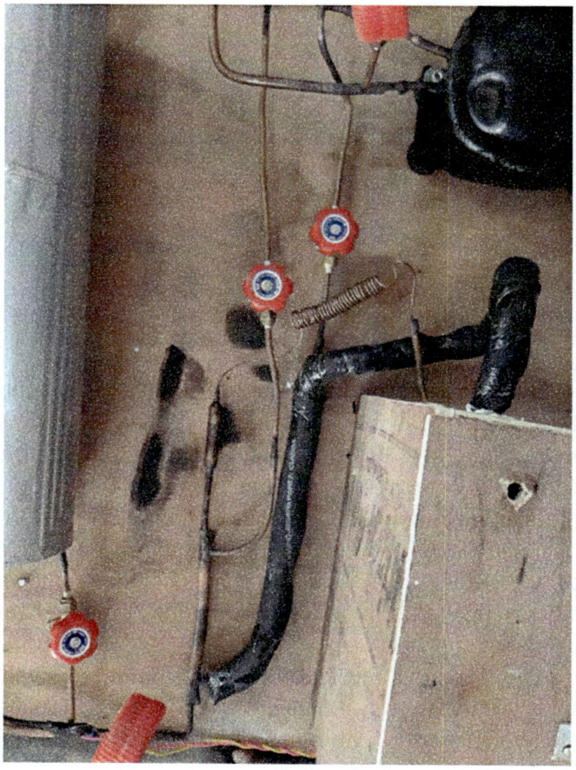

Fig. 3 Construction of LSHE

The refrigerating effect of the system is tested by cooling the water this is given as load to the VCR system. During the experimentation LSHE is involved for sub-cooling and superheating the refrigerant from condenser, evaporator and excluded depending on the requirement during experimentation. Four condensers with 10°, 14°, 18° twisted strip inserted condenser [24] along with no strip condenser involved one by one noting its performance in the cycle. The temperature of the water, refrigerant and pressures are noted at regular intervals of time i.e. 10 min. The overall duration of the experiment is 60 min.

The relation giving refrigeration effect is

$$\mathrm{RE} = \mathrm{mC_p(dT/dt)\ Kw} \quad (1)$$

where

m—water in evaporator by mass,
C_p—water specific heat kJ/kg K,
dT—water temperature change (°C),
dt—time logged corresponding to dT,
Coefficient of performance = RE/W.

4 Results and Discussion

To perform the analysis of LSHE-VCR and VCR a model was created and also employed with twisted strips inserted condensers. To validate the performance parameters like refrigeration effect, COP, refrigeration efficiency, Pressure difference across the condenser and the temperature of the evaporator and load.

4.1 Temperature Variation

The system is performed during subcooling and no-subcooling conditions to see the effect of temperature variation of water in the evaporator chamber. The variation of temperature of water with respect to time represents the COP. Sub-cooling and no sub-cooling conditions are applied by involving and withdrawing of LSHE. During the experimentation all the four condensers (plain tube, strip with 10°, 14°, 18°) operated one by one. It is noticed that in subcooling condition low final temperature of water was attained and also concluded that as the angle of the strip increased the water temperature is low. The variation of temperature with respect to time is as shown in Figs. 4 and 5.

The condenser inlet and outlet temperature vary with respect to the strip angle. It is observed that the temperature difference across the condenser is increased as the strip angle increased. Means with 18° strip inserted condenser the condensation process is better than other condensers. This was shown in the table below at maximum time of experiment i.e. at 60 min of experiment running.

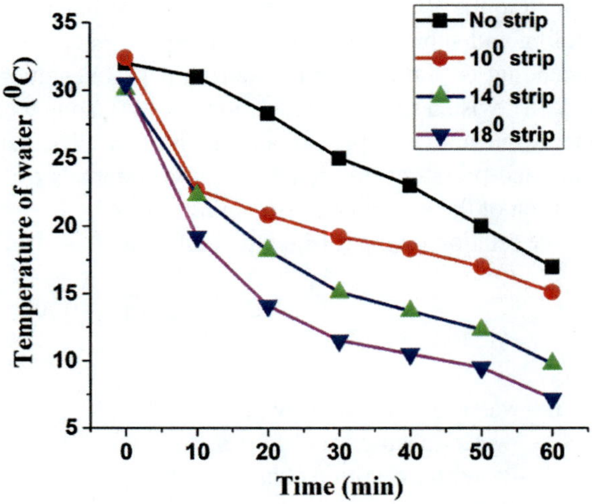

Fig. 4 Variation of temperature in sub-cooling condition

Fig. 5 Variation of temperature in no sub-cooling condition

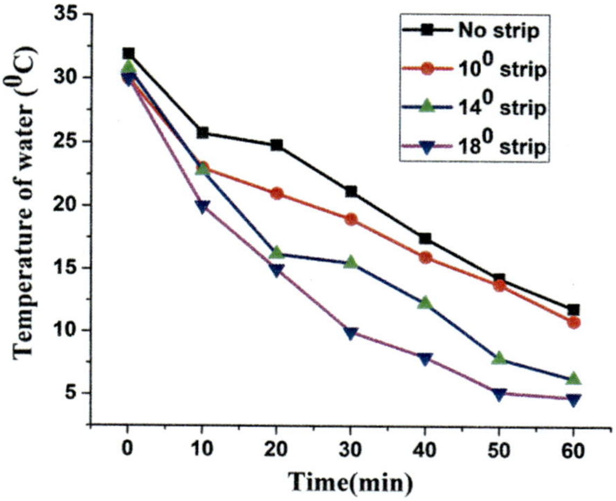

4.2 Coefficient of Performance Variation

The actual COP [25] of the common VCR is best for 18° strip condenser. It is observed that as the strip angle increases the actual COP of the unit also increased. The same was observed in sub-cooling condition by LSHE. Figure 6 showing the variation of actual COP.

Fig. 6 COP for different condensers

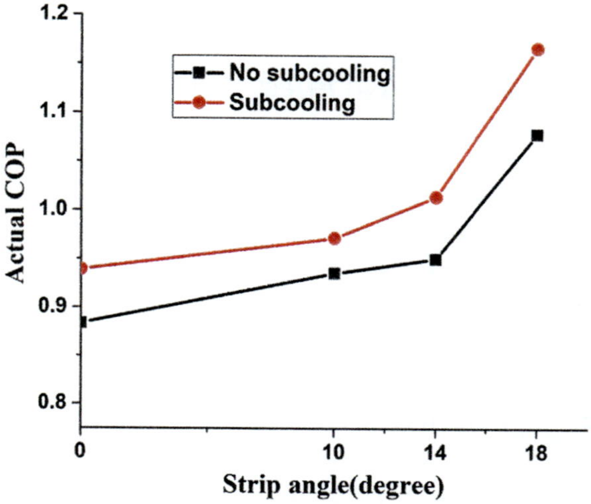

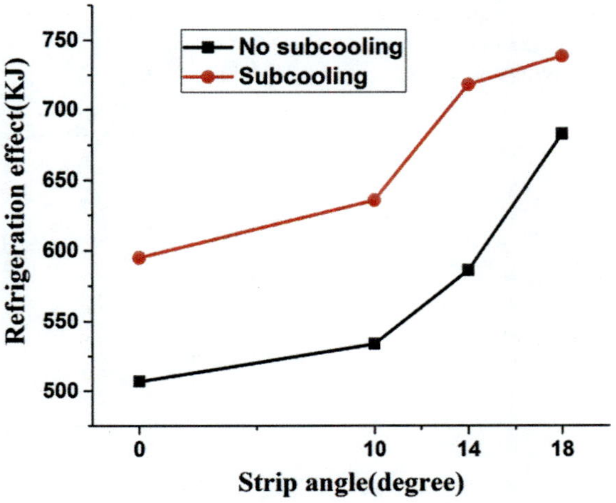

Fig. 7 Refrigeration effect for different condensers

Table 2 Details of performance parameters

Strip angle	COP		Refrigeration effect (kJ)	
	No-sub cooling	Sub cooling	No-sub cooling	Sub cooling
No strip	0.88	0.93	507.04	594.97
10° strip	0.93	0.97	533.80	635.48
14° strip	0.95	1.01	586.18	718.07
18° strip	1.07	1.16	682.89	738.58

4.3 Refrigeration Effect Variation

The initial and final temperature of the evaporator is directly influencing the Refrigeration effect of the unit. It is noticed that the temperature difference is large for 18° strip condenser at sub-cooling condition. The refrigeration effect increases from the range of 513.80 to 682.89 kJ in no sub-cooling and 594.97 to 738.58 kJ in sub-cooling condition (Fig. 7 and Table 2).

5 Conclusion

This paper has carried an assessment of twisted strips inserted condensers in VCR system and LSHE. The calculations are conducted for both sub-cooling and no sub-cooling conditions four different condensers (plain tube, strip with 10°, 14°, 18°) are involved for the analysis. This investigation evaluated the influence of temperature difference of evaporator and condenser on the performance parameters like COP, RE by R134a used as refrigerant.

The COP of unit variations as 0.88, 0.93, 0.95 and 1.07 under no sub-cooling condition for plane, 10°, 14° and 18° strip inserted condensers respectively. The COP of unit variations as 0.93, 0.97, 1.01 and 1.16 under sub-cooling condition for plane, 10°, 14° and 18° strip inserted condensers respectively.

However, the 18° strip condenser shows better performance under sub-cooling than the other condensers.

The RE is maximum at 18° strip with sub-cooling than without sub-cooling condition. The RE increases up to 52% than simple VCR system.

Future Recommendations

It was clear that the implementation of LSHE, twisted strip inserted condensers benefits the thermal performance of VCR system that plays a key role in domestic refrigerator, cold storage etc. The attractive research was proposed on PCM's, nano refrigerants, alternative refrigerants into refrigeration over past decades. Further development of the refrigeration system is in these aspects.

References

1. Carsten H, Petersen S, Huls W, Albers J (2021) Primary energy efficiency potentials of district heat driven refrigeration systems. Energy Rep 7:79–87
2. Ahmad N, Abdul MQ, Islam M, Raza NS, Min S, Lee S, Lee M (2022) Performance enhancement of hydrogen liquefaction process via absorption refrigeration and organic Rankine cycle-assisted liquid air energy system. Energy Convers Manag 254:115200
3. Wang Q, Tailu Li, Jia Y, Zhang Y (2021) Thermodynamic performance comparison of series and parallel two-stage evaporation vapour compression refrigeration cycle. Energy Rep 7:1616–1626
4. Babarinde TO, Madyira DM, Mashinini PM, Marangwanda GT (2022) Performance of multi-walled CNTs suspended with hydrocarbon refrigerant (R600a) and lubricating oil in vapour compression refrigeration system. Fuel Commun 10:100036
5. Selvnes H, Allouche Y, Manescu RI, Hafner A (2021) Review on cold thermal energy storage applied to refrigeration systems using phase change materials. Therm Sci Eng Progress 22:100807
6. Prateek DM, Shaikh J, Gawali BS (2022) Exergy assessment of a multistage multi-evaporator vapour compression refrigeration system using eighteen refrigerants. Energy Rep 8:153–162
7. Liu X, Kaihong Yu, Wana X, Li X (2021) Performance evaluation of CO_2 supermarket refrigeration system with multi-ejector and dedicated mechanical sub cooling. Energy Rep 7:5214–5227
8. Goyal K, Nanditta RV, Dharma Teja P, Malarmannan S, Manikandaraja G (2021) Analysis of vapour compression refrigeration system employing tetrafluoroethene and difluroethane as refrigerants. In: International conference on advances in thermal engineering and applications. https://doi.org/10.1088/1742-6596/2054/1/012054
9. Chen X, Yang Q, Chi W, Zhao Y, Liu G, Li L (2022) Energy and exergy analysis of NH_3/CO_2 cascade refrigeration system with sub cooling in the low-temperature cycle based on an auxiliary loop of NH_3 refrigerants. Energy Rep 8:1757–1767
10. Ahmed F (2021) Experimental investigation of Al_2O_3-water nanofluid as a secondary fluid in a refrigeration system. Case Stud Therm Eng 26:101024
11. Kotu TB, Reji Kumar R (2014) Comparison of heat transfer performance in domestic refrigerator using nano refrigerant and double pipe heat exchanger. Int J Mech Ind Eng (IJMIE) 4(2). ISSN: 2231-6477

12. Jose CA, Bravob L, Roberto CM, Hatakeyamad K, Barrantes E (2021) Knowledge construction and systematization of solar adsorption refrigeration prototypes. Energy Rep 7:428–440
13. Adelekan DS, Ohunakin OS, Gill J, Atiba OE, Okokpujie IP, Atayero AA (2019) Experimental investigation of a vapour compression refrigeration system with 15nm TiO_2-R600a nano-refrigerant as the working fluid. Procedia Manuf 35:1222–1227
14. Franco SS, Henríquez JR, Ochoa AAV, Da Costa JAP, Ferraz KA (2022) Thermal analysis and development of PID control for electronic expansion device of vapour compression refrigeration systems. Appl Therm Eng 206:118130
15. Jeon Y, Lee D, Cho H (2022) Optimization of motive nozzle position in a modified two-phase ejector expansion household refrigeration cycle using an artificial neural network. Energy Rep 8:1114–1123
16. Lu S, Zhang J, Liang R (2022) Experimental research on the vapour injected photovoltaic-thermal heat pump for heating, power generation and refrigeration. Energy Convers Manag 257:115452
17. Navarro J E, Mole's F, Barragan-Cervera A (2013) Experimental analysis of the internal heat exchanger influence on a vapour compression system performance working with R1234yf as a drop-in replacement for R134a. Appl Therm Eng 59:153–161
18. Goyal K, Nanditta RV, Dharma Teja P, Malarmannan S, Manikandaraja G (2021) Analysis of vapour compression refrigeration system employing tetrafluroethane and difluroethane as refrigerants. In: International conference on advances in thermal engineering and applications. https://doi.org/10.1088/1742-6596/2054/1/012054
19. Gill J, Singh J (2017) Performance analysis of vapour compression refrigeration system using an adaptive neuro-fuzzy inference system. Int J Refrig 82:436–446
20. Jian S, Zhang M, Gehl A, Fricke B, Nawaz K, Gluesenkamp K, Shen B, Munk J, Hagerman J, Lapsa M (2022) COVID 19 vaccine distribution solution to the last mile challenge: experimental and simulation studies of ultra-low temperature refrigeration system. Int J Refrig 133:313–325
21. Farahani SD, Farahani M, Ghanbari D (2021) Experimental study of the effect of spiral-star fins and nano-oil-refrigerant mixture on refrigeration cycle characteristics. J Therm Anal Calorim. https://doi.org/10.1007/s10973-021-10921-0
22. Pritam D, Kumar BM (2021) Performance enhancement of a shell-and-tube evaporator using Al_2O_3/R600a nanorefrigerant. Int J Heat Mass Transf 170:121015
23. Kotu BT, Kumar RR (2013) Comparison of heat transfer performance in domestic refrigerator using nano refrigerant and double pipe heat exchanger. Int J Mech Ind Eng 4(2):Article 8. https://doi.org/10.47893/IJMIE.2014.1195
24. Keklikcioglu O, Ozceyhan V (2021) Thermo hydraulic performance evaluation for horizontal tube by using combination of modified coiled wire inserts and grapheme nanoplatelet-water nanofluids. Int Commun Heat Mass Transf 123:105206
25. Manuel JL, Galindo J, Dolz V, Alberto PM (2022) Optimization of the thermal storage system in a solar-driven refrigeration system equipped with an adjustable jet-ejector. J Energy Storage 45:103495

Electricity Production from Different Effluent Using Microbial Fuel Cell

Vandana, N. K. Tiwari, and Rajesh Kumar

1 Introduction

It is well known that depletion of non-renewable resources and climate change drag attention towards the need for alternate sources of energy. Current dependency on fossil fuel is not suitable because it causes pollution and it is limited. In the past, several studies are carried out for finding a wide range of energy solutions. As one solution is not enough to meet the energy demands. Various renewable energy sources are discovered such as solar, wind, hydro-energy etc. One of them is energy derived from microorganisms i.e. bioenergy. Waste products are gathered at a very high rate in highly populated areas. Waste produced if left unrefined would be accumulated to a point that it would pollute the area with microorganisms along with streams and lakes. Microbial fuel cell (MFC) is a biotechnology that is not only capable to produce electricity from the degradation of organic compounds in effluent but also can be used for treating waste [1, 4, 7, 9] These previously developed MFC systems are very adaptable and hold much promise to provide energy sustainably, however, major improvements are required in the systems for their widespread applications. In this study, we point toward improving the performance of microbial fuel cells by using external resistance so that maximum power can be obtained Microbial fuel cells are utilized to generate energy by using various pure compounds including acetate [2], lactate, and glucose [13]. Effluents used in MFCs include domestic effluents [6], papermill effluents [7], corn stover hydrolysates [15] and swine effluent [14].

Vandana (✉)
School of Renewable Energy and Efficiency (SREE), NIT Kurukshetra, Kurukshetra, Haryana, India
e-mail: vandanabrar1709@gmail.com

N. K. Tiwari
Department of Civil Engineering, NIT Kurukshetra, Kurukshetra, Haryana, India

R. Kumar
Department of Mechanical Engineering, NIT Kurukshetra, Kurukshetra, Haryana, India

S. Doolla et al. (eds.), *Advances in Clean Energy and Sustainability*,
Green Energy and Technology, https://doi.org/10.1007/978-981-99-2279-6_22

The power output using these effluents varies depending on the MFC design and is typically greater with pure compounds than with mixed compounds according to previous studies [2, 6, 14, 15]. The revelation that microbial metabolism may supply energy in the form of electrical energy has attracted many researchers to carry out the study on MFC. technology Furthermore, a significant increase in the number of publications on MFC technology can be seen.

The purpose of this study is to examine the electricity production capacity of municipal effluent, dairy effluent and paper mill effluent utilizing microbial fuel cell technology.

2 Materials and Method

2.1 Substrate

Domestic effluent was gathered from an existing sewage treatment plant in NIT, Kurukshetra, Haryana, India. Dairy effluent was collected from Gaurav Yadav Dairy Farm in Sanwala, Haryana, India. Papermill effluent was collected from Akar Shakti Engg. Works, Yamuna Nagar, Haryana, India. The general properties of all the effluents are shown in Table 1. Before usage, all effluent samples were placed in a fridge at 5°C. For all MFC studies, the effluents were utilized as the inoculum without any changes in their properties, such as pH, the accumulation of nutrients, mediators and metals present. All trials were carried out using full-strength effluent at room temperature and in an immobile state.

2.2 Electrodes

Anode Material

Some significant criteria must be set while choosing the right anode material. The material should be

- Low-cost large-scale application
- Highly conductive

Table 1 Properties of effluent

Effluent	pH	BOD (mg/l)	COD (mg/l)	TSS (mg/l)
Domestic Effluent	7.6	234	1235	256
Dairy Effluent	5.5	654	1487	329
Papermill effluent	8.2	269	1591	396

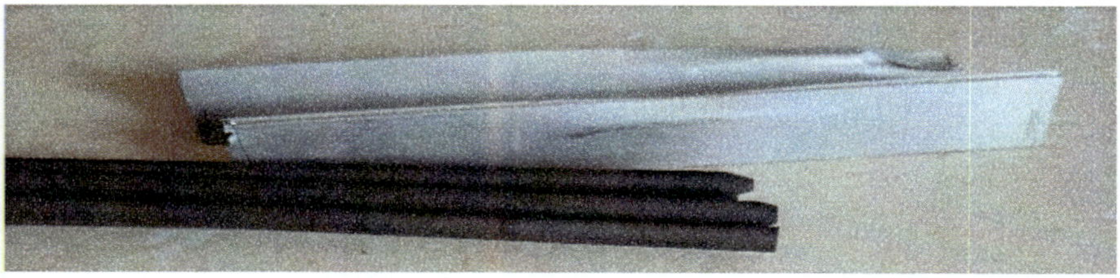

Fig. 1 Aluminium and graphite electrodes

- High Porosity
- Non-corrosive nature
- Optimal for bacterial growth.

These criteria may be met by a variety of materials. Carbon paper or cloth is typically employed as the anode in most studies as it is extremely conductive, non-corrosive, and porous however, it is brittle, which is its principal drawback [5]. Graphite plates or rods, which have high conductivity, are also used in many studies [10]. In the present study, we used aluminium with a surface area of 26 cm^2 as shown in Fig. 1.

Cathode Material

The cathode must have two major properties: high conductivity and non-corrosiveness. These properties are the same as that of an anode, hence the same material has been used as a cathode.

2.3 *MFC Construction*

Salt Bridge

This is an important component of an MFC. The Salt bridge separates the anodic and cathodic chambers so that water from the cathodic chamber cannot travel to the anodic one as it contains dissolved oxygen. Furthermore, it must allow for the normal passage of protons from the anodic chamber to the cathodic chamber [5].

We made a salt bridge by soaking a 1.5 m cotton thread in a saturated salt solution for 2 h at 95 °C, then letting it absorb the fluid overnight (24 h). Then the salt bridge was inserted into the PVC pipe. The ends of the PVC pipe were connected to both the chambers using a glue gun which makes the chambers leakproof.

Design

Here we construct a DCMFC. It consists of an anodic chamber and a cathodic chamber. The anodic chamber is filled with the substrate while the cathodic chamber is filled with water. Both the chambers are connected through a salt bridge. MFCs with different effluents are shown in Table 2.

Table 2 MFCs with different effluents

MFCs	Effluent
MFC1	Municipal effluent
MFC2	Dairy effluent
MFC3	Papermill effluent

2.4 MFCs Operation

The MFC functioned in the batch mode because it provide sufficient time for the growth of microorganisms as well as for the degradation of organic matter. The anodic chamber was filled with different substrates. The MFC functioned at the hydraulic retention time of 7 days. The cathodic compartment was filled with normal water and was continuously aerated with the help of an air pump. After making the microbial fuel cell, it was kept for 24 h without external resistance so that the voltage becomes stable. Then external resistance of 10 Ω was applied for 7 days and daily observations were recorded.

2.5 Monitoring Electricity

Using a Digital Multimeter (MASTECH M830BZ) and a 10 Ω external resistance, voltage measurements were carried out to calculate the current.. COD measurements were carried out by standard methods [3]. Prior to COD measurements, all samples were filtered via a membrane filter with a pore diameter of 0.22 m. The amount of COD removed was calculated using Eq. (1)

$$E_{COD} = \left[COD_{in} - \frac{COD_{out}}{COD_{in}}\right] \times 100\% \quad (1)$$

where COD_{in} = Influent COD,
COD_{out} = Effluent COD.

3 Results and Discussion

3.1 Current Generation

After the construction of DCMFCs, each DCMFC were operated with different effluent samples under different states. During the experiment, the MFCs were continuously monitored, and data was recorded for every 24 h. The MFCs was operated for

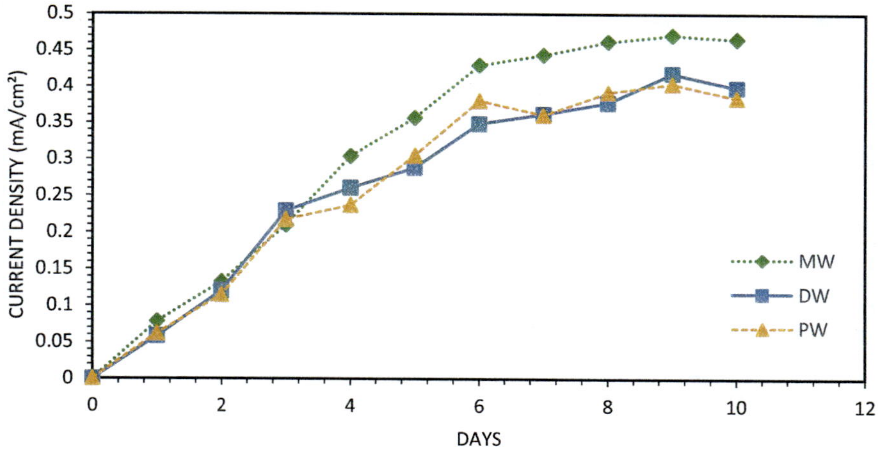

Fig. 2 Impact of agitation on current generation

10 days. Initial trials with MFCs revealed that energy may be produced from various effluents. After 24 h, stable current output was attained.

The presence of compounds that are readily consumed by microorganisms existing in the effluents can be linked to the initial increase in current here. After the depletion of the readily degradable substrates, energy generation began to fall. Meanwhile, complicated compounds were degraded, resulting in a reduced current density.

Impact of Agitation

To test their power production capability, each DCMFC were run with distinct effluent samples for 5 days at a time, first in a stagnant state and then in an agitated state. The experimental results revealed that when a stagnant solution was stirred, there is a rise in the current density of all substrates. Microbial fuel cell with Municipal effluent shows the best result of 0.471 mA/cm^2 current on the 9th day. DW, PW gives 0.42 mA/cm^2, and 0.44 mA/cm^2 current as shown in Fig. 2.

Impact of Substrate Strength

All MFCs functioned with different effluents samples to scrutinize the impact of the strength of substrate on the current density and on power density of MFCs. Firstly, full-strength effluent was utilized as substrate in an anodic chamber however, after 5 days, half of the effluent was substituted with distilled water. Figure 3 depicts the influence of effluent concentrations on current density.

From the experimental result, it is noticed that the power production decreased with the decrease in the strength of the substrate. There is no decrease in the current density of MFC2, this is because of certain microorganisms which experience substrate limitation at higher effluent concentrations. MW, DW, PW achieved a maximum of 0.34 mA/cm^2, 0.29 mA/cm^2 and 0.301 mA/cm^2 current in full strength

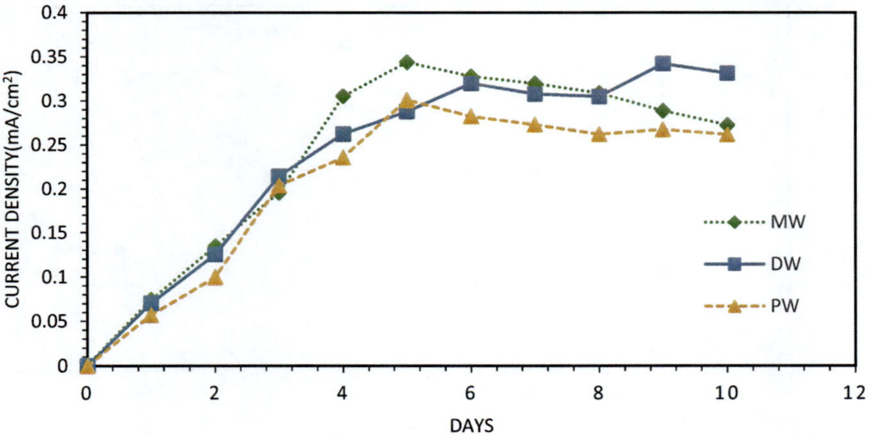

Fig. 3 Impact of substrate concentration on current generation

effluent whereas 0.33 mA/cm^2, 0.34 mA/cm^2 and 0.282 mA/cm^2 respectively in 50% effluent.

This variance in current production might be attached to the existence of fewer oxidizable substrates in half of the effluent samples.

3.2 Chemical Oxygen Demand Removal Efficiency

Throughout functioning, all MFCs were continuously examined for COD elimination to determine the MFCs ability to function as an effluent treatment unit. All effluents samples show COD removal potential, representing the role of bacteria in metabolizing the carbon source as electron suppliers. Experimental findings demonstrate that current generation and COD elimination are relatively compatible. In all MFCs, continuous COD elimination was observed.

Impact of Agitation

Figure 4 demonstrate the COD removal efficiency of different effluent used in both stagnant and agitated state. For the first 5 days, MFCs were operated in stagnant and then for the next 5 days in an agitated state. Results shows high COD removal efficiency in an agitated state. MW, DW, PW gained 48.35%, 50.39%, and 40.15% COD removal in stationary states after 5 days of operation. While after 5 days these MFC shows 85.15%, 81.29%, and 76.535% COD removal in an agitated state. Data showed that the time required for carbon depletion was shorter in agitated states. This might be due to increased dispersion and mixing of substrate and microorganisms, which contribute to the COD removal efficiency effectiveness of the MFCs.

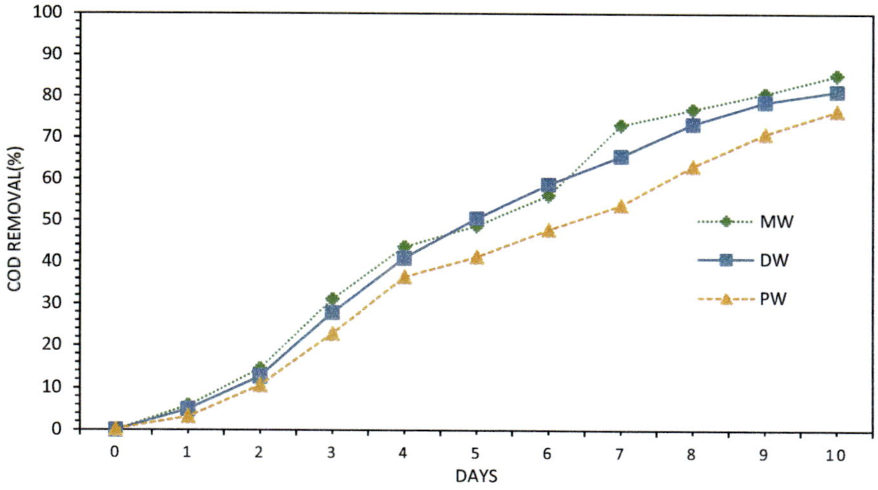

Fig. 4 Impact of agitation on COD removal

Impact of Effluent Concentration

To scrutinize the impact of the strength of substrate on the COD removal of MFCs, all MFCs functioned with different effluent samples. Firstly, full-strength effluent was utilized as substrate in an anodic chamber however, after 5 days, half of the effluent was substituted with distilled water. Figure 5 depicts the influence of effluent concentrations on COD removal.

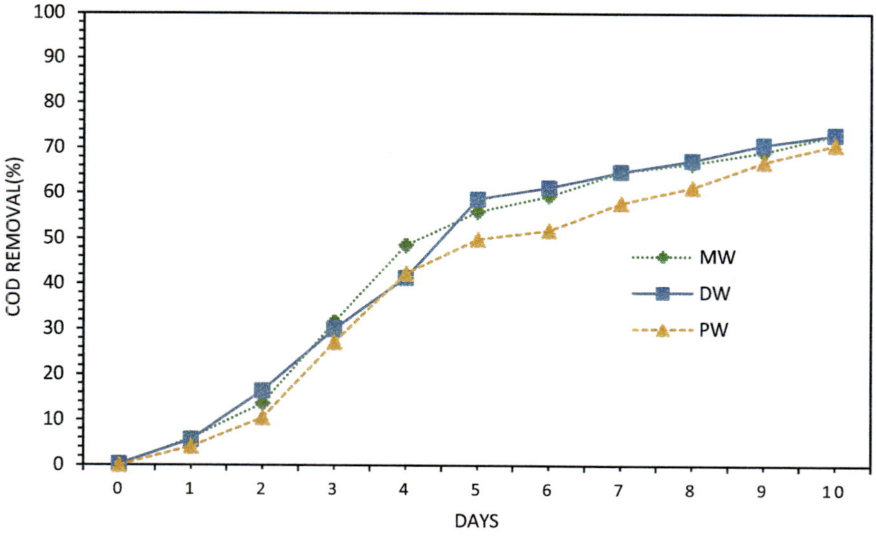

Fig. 5 Impact of effluent concentration on COD removal

MW, DW, and PW obtained 55.08%, 55.83% and 48.92% COD removal in full strength effluent, after 5 days of operation. However, these samples attained 72.95%, 72.95% and 71.75% COD removal in 50% effluent after the 10th day. The low COD removal might be attributed to the lower accessibility of decomposable substrate in 50% effluent samples compared to full strength effluent, resulting in reduced inhibition in microorganisms.

The results shows that the best performance of municipal effluent, which is due to easily degradable mater and generating good current. Papermill effluent had the lowest findings, which might be attributed to the existence of cellulose in the effluent sample. MFCs will be used to generate electricity from effluent, giving a novel technique to treat effluent while also gaining a clean and sustainable energy source [8, 12].

4 Conclusions

The following key points were concluded from the current study:

- The energy was effectively produced with waste removal from dairy effluent, municipal effluent and paper mill effluent utilizing DCMFCs.
- If the electricity output of MFC is enhanced, MFC technology may give an innovative way to reduce effluent treatment plant working costs, making effluent treatment more reasonable for developing countries.
- As a result, the grouping of effluent treatment with power generation may aid in cost savings as effluent treatment is currently expensive.
- The major challenges to be resolved for realistic use are capital cost, high internal resistance and incomplete effluent usage.

References

1. Allen RM, Bennetto HP (1993) Microbial fuel-cells. Appl Biochem Biotechnol 39(1):27–40
2. Bond DR, Lovley DR (2003) Electricity production by Geobacter sulfurreducens attached to electrodes. Appl Environ Microbiol 69(3):1548–1555
3. Greenberg A, Clesceri LS, Eaton AD (1992) Standard methods for the examination of water and effluent, 18th edn. American Public Health Association, Washington, D.C
4. Kim, B. H., Ikeda, T., Park, H. S., Kim, H. J., Hyun, M. S., Kano, K., ... Tatsumi, H. (1999). The electrochemical activity of an Fe (III)-reducing bacterium, Shewanella putrefaciens IR-1, in the presence of alternative electron acceptors. Biotechnology Techniques, 13(7), 475-478
5. Logan BE (2008) Microbial fuel cells. John Wiley & Sons
6. Lui H, Ramnarayanan R, Logan BE (2004) Production of electricity during effluent treatment using a single chamber microbial fuel cell. Environ Sci Technol 38:2281–2285
7. Mathuriya AS, Sharma VN (2009) Bioelectricity production from papermill waste using a microbial fuel cell by Clostridium species. J Biochem Tech 1(2):49–52
8. Min B, Logan BE (2004) Continuous electricity generation from domestic effluent and organic substrates in a flat plate microbial fuel cell. Environ Sci Technol 38:5809–5814

9. Rosenbaum M, Zhao F, Quaas M et al (2007) Evaluation of catalytic properties of tungsten carbide for the anode of microbial fuel cells. Appl Catal B: Environ 74:262–270
10. Park DH, Zeikus JG (2003) Improved fuel cell and electrode designs for producing electricity from microbial degradation. Biotechnol Bioeng 81(3):348–355
11. Rosenbaum M, Zhao F, Quaas M, Wulff H, Schröder U, Scholz F (2007) Evaluation of catalytic properties of tungsten carbide for the anode of microbial fuel cells. Appl Catal B 74(3–4):261–269
12. Ra CS, Lo KV, Shin JS, Oh JS, Hong BJ (2000) Biological nutrient removal with an internal organic carbon source in piggery effluent treatment. Water Res 34(3):965–973
13. Rabaey, K., Lissens, G., Siciliano, S. D., & Verstraete, W. (2003). A microbial fuel cell capable of converting glucose to electricity at high rate and efficiency. Biotechnology --letters, 25(18), 1531–1535.
14. Sevrin Reyssac, J. Biotreatment of swine manure by production of aquatic valuable biomasses.--p. 177–186 (No. HEM). En: Agriculture Ecosystems and Environment (Netherlands).--Vol. 68, no 3 (Apr 1998).
15. Zuo Y, Maness PC, Logan BE (2006) Electricity production from steam-exploded corn stover biomass. Energy Fuels 20(4):1716–1721

Precursor Tuning for Post-treatment Free $MAPbI_3$ Films for Efficient and Stable Perovskite Solar Cells

Ramya Krishna Battula, C. Sudakar, P. Bhyrappa, Ganapathy Veerappan, and Easwaramoorthi Ramasamy

1 Introduction

A new method to address the stability and scalability aspect of perovskite absorber layer has come into the limelight employing single crystals [1, 2]. A unique ink obtained by liquefying single crystals is employed to coat onto the charge transport layers to improve the scalability and stability of PSCs employing low boiling point solvent combination of methylamine and acetonitrile [3]. The ink derived from single crystals was found to be containing the methylamine (MA) ion within the framework on the PbI_6 octahedra as opposed to the conventional precursor where MA is free to escape [4]. This prevents loss of MA and the ink is capable of being coated by any technique including large area deposition such as blade coating, bar coating, slot-die coating, spray coating, etc. [5]. This ink can be coated in the absence of conventionally used high boiling point solvents such as dimethylformamide, dimethylsulfoxide, etc.

This field of research employing low boiling point solvents has tremendous potential in developing PSCs for large area application but only a handful reports are available in this regard [6]. However, the negative impact of annealing is not reported by this method to the best of our knowledge [5]. We have employed the single crystal derived precursors to develop perovskite absorber films by liquefying single crystals with and without annealing. The films were thoroughly characterized, devices were

R. Krishna Battula · G. Veerappan · E. Ramasamy (✉)
Centre for Solar Energy Materials, International Advanced Research Centre for Powder Metallurgy and New Materials, Hyderabad 500005, India
e-mail: easwar@arci.res.in

R. Krishna Battula · C. Sudakar
Multifunctional Materials Lab, Department of Physics, Indian Institute of Technology Madras, Chennai 600036, India

R. Krishna Battula · P. Bhyrappa
Department of Chemistry, Indian Institute of Technology Madras, Chennai 600036, India

S. Doolla et al. (eds.), *Advances in Clean Energy and Sustainability*,
Green Energy and Technology, https://doi.org/10.1007/978-981-99-2279-6_23

fabricated in the FTO/c-TiO_2/m-TiO_2/$MAPbI_3$/Spiro/Au n-i-p device architecture followed by I-V testing and stability studies.

2 Materials and Methods

2.1 Characterization Details

The morphology was characterized using field emission scanning electron microscopy (FESEM) (ZEISS GeminiSEM 500). XRD patterns were obtained from a Rigaku SmartLab 9 kW diffractometer. Photoluminescence (PL) of the samples were recorded on a Horiba Scientific PL spectrophotometer (FL3C-21) with a 450 W ozone free Xe source, excitation wavelength—540 nm. Absorbance studies of the films were carried on a Varian Cary UV-Vis Absorbance Spectrophotometer. Class AAA solar simulator (Oriel instruments) was used for I-V measurements in ambient conditions at 960 mW/cm^2 illumination. Samples were stored in dark and measured periodically at 25±3 °C and 50 ± 10% RH for stability measurements.

Experimental

Single crystals of $MAPbI_3$ are grown by inverse temperature crystallization (ITC) method. The $MAPbI_3$ precursor was prepared by exposing the as grown single crystals to MA vapor until it turns into a liquid phase. This transformation happens as the MA molecule sitting on the $MAPbI_3$ slices the lattice into clusters of $MAPbI_3$ [4, 7]. The disintegrated parts dissolve in the MA vapor turning into a high pure perovskite liquid which is compatible with any form of coating [5]. The resulting solution is diluted in ACN for spin-coating onto m-TiO_2 substrates. The heat-treated films are annealed at a temperature of 100 °C for 5 min. Devices were fabricated in FTO/c-TiO_2/m-TiO_2/$MAPbI_3$/Spiro/Au architecture. The device related information except for the perovskite film deposition is carried out as per our previous works [8–10].

3 Results and Discussion

The films coated on m-TiO_2 substrates were analyzed for their surface and cross-sectional morphology as shown in Fig. 1. FESEM surface morphology in Fig. 1a reveals that the grains are crystalline with well-defined boundaries in the unannealed film. Whereas the grains are fused to result into a rough film in the annealed case as depicted in Fig. 1b. This shows the non-beneficial effect of annealing on the perovskite film. The unannealed film shows large columnar grains in the cross-sectional morphology as shown in Fig. 1c. The same is not reflected in the surface morphology because during spin coating the weak cohesive forces at the liquid-air interface cause multiple nucleation. Whereas in the bulk, the strong chemical

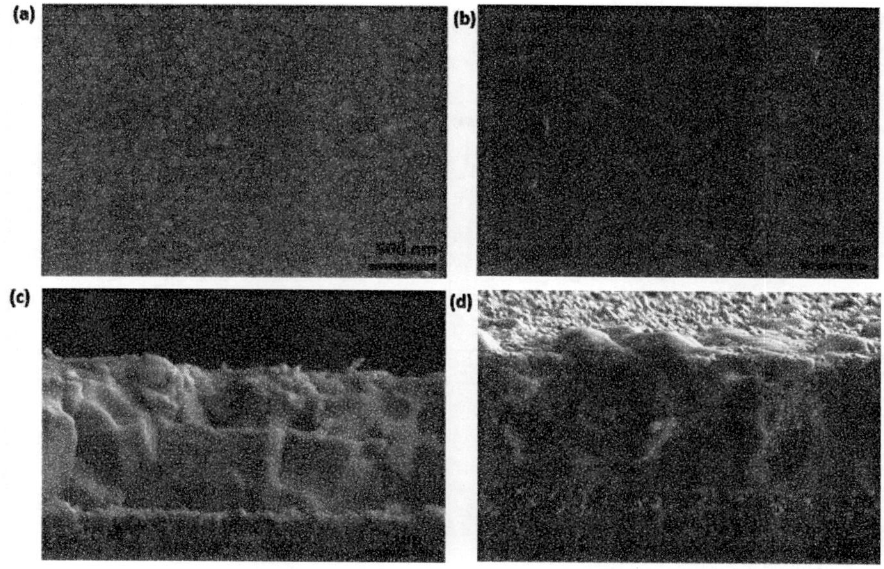

Fig. 1 FESEM surface morphology and cross-sectional images of the (**a**, **c**) unannealed and (**b**, **d**) annealed films respectively

bonding between the precursor molecules and the substrate results in columnar oriented growth. Hence, the film tends to form multiple nucleation resulting in a fine-grained topmost layer in contrast to the bulky cross-sectional morphology. The fusing of grains of the annealed films is also observed in the cross-sectional morphology of the films as shown in Fig. 1d. This is an indicator that the film formed after spin coating is already in the crystalline $MAPbI_3$ phase and annealing it to higher temperature is non-beneficial for its crystallinity as opposed to the conventional films.

XRD patterns shown in Fig. 2a throws light on the crystallinity of the annealed and unannealed films. Intense $MAPbI_3$ phase peaks were observed for the unannealed films compared to the annealed films. An average full width at half maximum (FWHM) of 0.080° was recorded for the unannealed films indicating better crystallinity over the FWHM of 0.306° of the annealed films. It is interesting to note that even when exposed to 100 °C, the $MAPbI_3$ films did not degrade to the PbI_2 phase highlighting the robustness of the film. The films were tested for their absorbance in the visible range as depicted in Fig. 3b. Evidently, the annealed film showed lesser absorbance than the unannealed film due to the fused grain morphology and lower crystallinity of the films. PL of the films shown in Fig. 3c revealed that the annealed film has more defects resulting in a less intense emission compared to the unannealed film. TRPL (Fig. 3d) also follows similar trend with an average higher lifetime of 147ns for the unannealed films compared to only 59 ns for the annealed film.

To support the unwanted effect of annealing, devices were fabricated in the n-i-p architecture and tested for I-V characteristics. We can see that the unannealed devices outperformed the annealed devices by a great margin as shown in Fig. 3a. The box

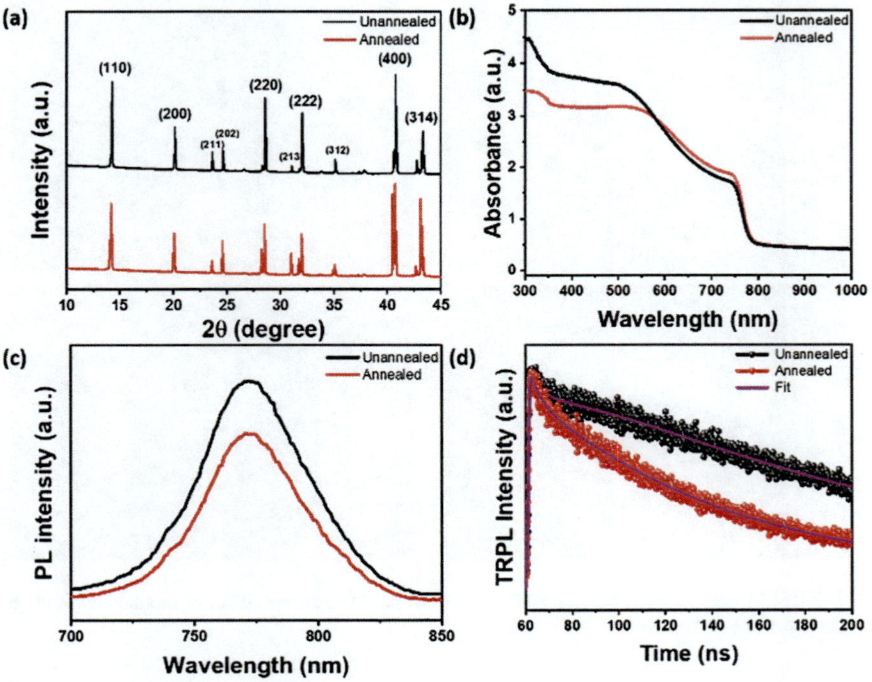

Fig. 2 **a** XRD **b** UV–Vis absorbance **c** PL and **d** TRPL plots of the annealed and unannealed films

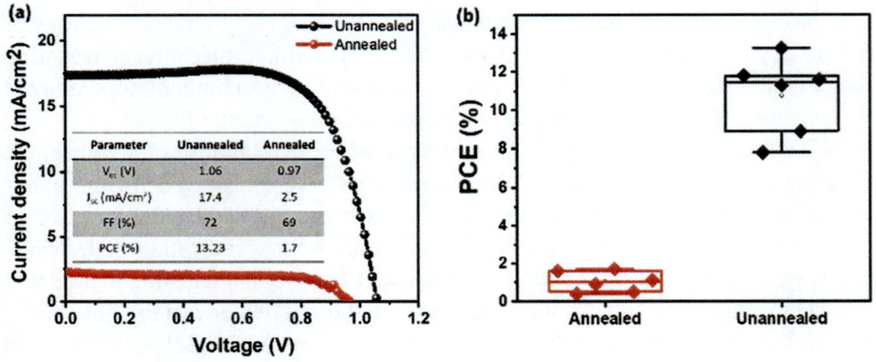

Fig. 3 **a** I-V characteristics of the best performing devices in both cases and **b** corresponding box plots to show the distribution of performance

plot in Fig. 3b shows the stark difference in distribution of PCEs in both the cases. The champion device in the unannealed case, exhibited a high-power conversion efficiency (PCE) of 13.23% with 1.06 V open circuit voltage (V_{OC}), 17.4 mA/cm^2 current density (J_{SC}) and 72% fill factor (FF). On the other hand, the annealed best performing device showed 1.7% PCE with 0.97 V V_{OC}, 2.5 mA/cm^2 J_{SC} and 69% FF.

Fig. 4 Device stability for 1500 h in ambient conditions

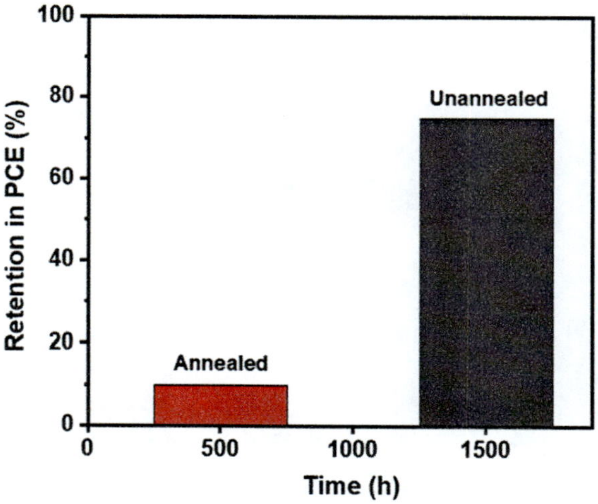

So, it is clear that J_{SC} is a significant factor for the low performance of the annealed device. This is majorly due to the poor absorber quality and crystallinity of $MAPbI_3$ films as evidenced by the films' characterization. We owe this poor performance to disruption in the lattice of the annealed films leading to fusing of grains that has significantly affected its current collection and transport properties.

The annealed devices showed poor stability and discontinued functioning within 500 h. Whereas the unannealed devices showed an impressive retention in PCE of more than 75% when stored in ambient conditions for more than 1500 h as depicted in Fig. 4.

In case of the conventional one-step recipe $MAPbI_3$ case, post treatment of the film by anti-solvent and annealing ensures complete crystallization and removal of solvent as shown in Fig. 5a. However, in the crystal derived films, annealing of films is detrimental severely affecting its performance and stability. We owe the poor morphological, structural, optical quality of the annealed film to the possible disruption of the lattice when exposed to temperature which is reflected in its photovoltaic performance and stability as shown in Fig. 5b. Single crystal when liquefied does not break down into PbI_2 and MAI to form a complex with CAN solvent which is usually the case in the conventional one-step films. Instead, it remains as multiple abridged entities with very low formation energy. When subjected to spin or bar coating, the MA and ACN readily evaporate, facilitating the semi-formed perovskite phase to rapidly crystallize. Thus, the resulting film does not require any post treatment. Thus, any thermal treatment is proven to damage the film quality and consequently the performance. Since these abridged entities have very low formation energy, we hypothesize that when exposed to temperature, the lattice is quickly distorted without degrading the perovskite. Such distorted lattice even if not degraded to its precursors is not of desired photovoltaic quality resulting in deleteriously impacted films.

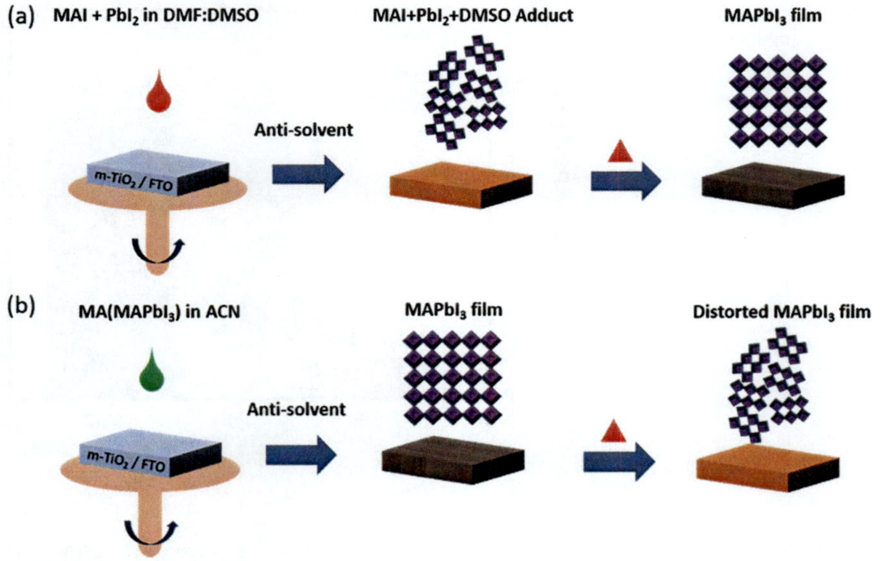

Fig. 5 Effect of thermal annealing on **a** one-step films and **b** crystal derived film

4 Conclusion

Perovskite precursor is developed by liquefying single crystals. The films derived from the precursor ink were found to have better morphological and optical properties without annealing. Annealing of these films led to distortion in the lattice resulting in films unsuitable for photovoltaic applications. The devices fabricated from the unannealed films were found to be superior in terms of PCE and stability. The champion device showed 13.23% PCE with 75% stability retention when exposed to ambient conditions for more than 1500 h. The as developed ink is also compatible with any type of coating. Thus, the potential of this ink might open up a plethora of large-scale applications given its compatibility with any type of coating, absence of post treatment reducing the costs, high performance and stability in ambient conditions.

Acknowledgements The authors are grateful to the Technical Research Centre project (AI/1/65/ARCI/2014) and the Clean Energy Research Initiative project (DST/TMD/CERI/C247(G)), Department of Science and Technology, Government of India, for funding this work.

References

1. Battula RK, Sudakar C, Bhyrappa P, Veerappan G, Ramasamy E (2022) Single-crystal hybrid lead halide perovskites: growth, properties, and device integration for solar cell application. Cryst Growth Des [Internet] 22:6338–6362. Available from: https://doi.org/10.1021/acs.cgd.2c00789
2. Battula RK, Veerappan G, Bhyrappa P, Sudakar C, Ramasamy E (2023) Growth of single-crystalline MAPbI3 perovskite film by a modified space-confined inverse temperature crystallization method. Surfaces and Interfaces [Internet] 36:102475. Available from: https://linkinghub.elsevier.com/retrieve/pii/S2468023022007349
3. Noel NK, Habisreutinger SN, Wenger B, Klug MT, Hörantner MT, Johnston MB et al (2017) A low viscosity, low boiling point, clean solvent system for the rapid crystallisation of highly specular perovskite films. Energy Environ Sci [Internet] 10:145–52. Available from: http://xlink.rsc.org/?DOI=C6EE02373H
4. Jeong D-N, Lee D-K, Seo S, Lim SY, Zhang Y, Shin H et al (2019) Perovskite cluster-containing solution for scalable D-bar coating toward high-throughput perovskite solar cells. ACS Energy Lett [Internet] 4:1189–1195. Available from: https://doi.org/10.1021/acsenergylett.9b00042
5. Wu C, Wang K, Li J, Liang Z, Li J, Li W et al (2021) Volatile solution: the way toward scalable fabrication of perovskite solar cells? Matter [Internet] 4:775–793. Available from: https://linkinghub.elsevier.com/retrieve/pii/S259023852030727X
6. Battula RK, Ramasamy E, Bhyrappa P, Sudakar C, Veerappan G (2022) Oxide free materials for perovskite solar cells. Oxide free nanomater energy storage and conversion applications [Internet]. Elsevier, pp 287–306. Available from: https://linkinghub.elsevier.com/retrieve/pii/B9780128239360000012
7. Wu C, Li H, Yan Y, Chi B, Pu J, Li J et al (2017) Cost-effective sustainable-engineering of $CH_3NH_3PbI_3$ perovskite solar cells through slicing and restacking of 2D layers. Nano Energy [Internet] 36:295–302. Available from: https://linkinghub.elsevier.com/retrieve/pii/S2211285517302367
8. Battula RK, Veerappan G, Bhyrappa P, Sudakar C, Ramasamy E (2022) Dual functional inorganic CuSCN for efficient hole extraction and moisture sealing of MAPbI3 perovskite solar cells. Mater Adv [Internet]. Available from: http://pubs.rsc.org/en/Content/ArticleLanding/2022/MA/D1MA00861G
9. Battula RK, Veerappan G, Bhyrappa P, Sudakar C, Ramasamy E (2020) Stability of MAPbI 3 perovskite grown on planar and mesoporous electron-selective contact by inverse temperature crystallization. RSC Adv [Internet] 10:30767–30775. Available from: http://xlink.rsc.org/?DOI=D0RA05590E
10. Ashina A, Battula RK, Ramasamy E, Chundi N, Sakthivel S, Veerappan G (2021) Dip coated SnO_2 film as electron transport layer for low temperature processed planar perovskite solar cells. Appl Surf Sci Adv [Internet] 4:100066. Available from: https://linkinghub.elsevier.com/retrieve/pii/S266652392100012X

Power Play or Prudent Policy? An Analysis of the Lifeline Electricity Scheme in Delhi

Afsal Najeeb, Satish B. Agnihotri, and Anand B. Rao

1 Introduction

Quality and affordable electric power, specifically for the residential sector is an unavoidable check box in the path to political power in India. The history of electrification policies in India stands testimony to this correlation. India has seen a rapid expansion of residential electricity access in the last decade. The accelerated realization of village electrification targets, conceptualization of the advanced target of household electrification and achievement of almost universal household electrification can be seen as a positive impact of political accountability to the demand of residential electricity. There are instances of electoral pressure leading to policies that have not served the sector well. Perpetual financial woes of the electricity Distribution Companies (DISCOMs) have been attributed to leakages in revenues due from the state, formal and informal subsidies targeted at specific consumer groups etc. [3, 11, 18].

Electricity subsidies to residential consumers have been a common but significant policy lever available to the governments, especially at the State level to directly influence consumer perceptions about governance [11]. Subsidising electricity prices was introduced as a policy tool to make necessary utilities affordable to the public, encourage uptake, modify consumer behaviour and as a method for income redistribution [14]. The opposition to subsidies stem from the lack of quantification, effect on financial health of utilities, inclusion and exclusion errors and undesirable changes in consumer behaviour. Critics point out that subsidies in essential utility sectors like electricity and water are entrenched with no exits as they are directly linked to political acceptance [13]. All states in India cross subsidise the residential sector. In its most common form, industrial and commercial consumers are charged higher to

A. Najeeb (✉) · S. B. Agnihotri · A. B. Rao
Indian Institute of Technology Bombay, Mumbai, Maharashtra 400076, India
e-mail: 204354005@iitb.ac.in

S. Doolla et al. (eds.), *Advances in Clean Energy and Sustainability*,
Green Energy and Technology, https://doi.org/10.1007/978-981-99-2279-6_24

cover the subsidised prices provided to residential consumers. Though cross subsidies are not reported at the national level, different studies estimate the quantum of cross subsidy to be around 70,000 crores [1, 2]. States like Jharkhand, Karnataka, Kerala and Uttar Pradesh provide direct subsidies and preferential tariffs to Below Poverty Line (BPL) households (Sharma et al. 2020).

1.1 Free Lifeline Electricity Consumption Scheme in Delhi

In March 2015, the Government of Delhi announced a 50% subsidy on monthly power consumption of up to 400 units for domestic consumers with connected load less than 5 kW (Goswami 2019) [8]. The scheme was aimed at fulfilling a poll promise and was expected to cover more than 80% of Delhi's residents. Consumers who consumed more than 400 units were excluded from the scheme. In August 2019, the scheme was further extended to introduce "lifeline consumption" electricity units for the first time in India. A subsidy for the entire bill amount (including fixed charges, energy charges, all surcharges and electricity tax) was announced for consumers utilizing up to 200 units per month, with connected load up to 4 kW. Consumers utilizing 201–400 units were provided a subsidy of around 50% subject to a maximum of Rs. 800 per month. No subsidy would be provided for consumption above 400 units. The amount of subsidy available for various categories of connected loads are shown in Fig. 1. The stated objective were two fold. First, it was aimed as a governance measure emanating from the commitment to provide public services at reasonable costs, avoiding mismanagement. Secondly, the policy was expected to discipline consumption to the threshold levels by incentivizing energy saving. The present study deals with the analysis of the free lifeline electricity scheme announced in 2019.

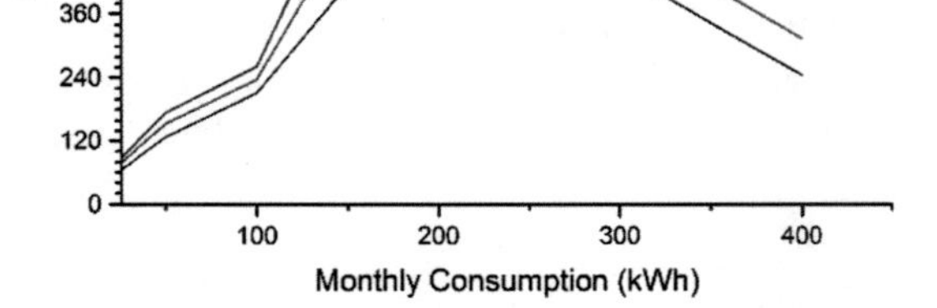

Fig. 1 Monthly subsidy for residential consumers in Delhi

1.2 Electricity, the New Arena of Competitive Populism

The promise of a certain quantum of electricity units as free, dubbed lifeline consumption is becoming a new element of competitive populism in India. Tamil Nadu had implemented a similar scheme in 2016. During the state elections of 2022, free electricity schemes of different hues and shades were on offer in all states. Free electricity for residential consumers was a poll promise in Uttar Pradesh for the 2022 elections [15]. Free electricity of up to three hundred units was a poll promise for the Punjab elections of 2022. Similar schemes are being offered in Chhattisgarh and Goa. It was particularly attractive in Uttar Pradesh which had seen an increase in electricity tariffs two times in the last five years, first by 12.7% and again by 11.7%. Mumbai has created a ministerial committee to look at the feasibility of providing 100 units of power for residential consumers [6]. The trend is likely to continue and the analysis of the scheme where it has been implemented for a considerable period of time becomes important.

2 Objectives, Methodology and Steps of Analysis

2.1 Objectives and Methodology

The overall objective is to understand the effect of the subsidy scheme on the electricity expenditure and appliance ownership using pooled cross section data on household expenditure and appliance ownership. The objectives and methodology are explained in detail in Table 1.

Table 1 Objectives and methodology

Sl. No.	Objective	Methodology
1	To understand the effects of the subsidy on household electricity expenditure and the distribution of benefits across economic classes	Analysis of survey data from 1375 households. The households were divided into quintiles based on annual income and electricity expenditure of each quintile was analysed
2	To understand the effect of the subsidy on appliance ownership, specifically equipment for space conditioning	Analysis of appliance ownership data. Appliances analysed included air conditioners, air coolers, refrigerators and televisions
3	The third objective is to understand the effect of the subsidy on the state finances and DISCOMS	Review of literature, analysis of policy documents about state finances and DISCOM performance

2.2 *Data Sources and Steps of Analysis*

The data from Consumer Pyramids Household Survey conducted by the Centre for Monitoring Indian Economy have been used for the analysis [4]. The sample for Delhi falling into the time period consisted of 2255 households covering rural and urban areas surveyed thrice every year. 1375 samples were used for the analysis after removing non responsive households and univariate outliers identified through z score-based detection. Variables of interest like the monthly expenditure on electricity, total income of the household, appliances in use and power availability were extracted from the overall dataset. As the data on household electricity consumption is not readily available, the expenditure on electricity is used as a proxy for consumption. The households were divided into five income quintiles based on their annual income and the monthly expenditure on electricity of each quintile was analysed using data visualisation and analysis of summary statistics. T tests were conducted to ensure the validity of the results obtained. The change in appliance ownership was understood by considering air conditioners, air coolers, television sets and refrigerators. The third objective relates to the political economy implications of the subsidy scheme was understood through a systematic review of literature considering state and DISCOM finances for the state of Delhi.

3 Results and Discussion

3.1 *Analysis of Household Expenditure on Electricity from 2014–2021*

The violin charts in Fig. 2 indicates the violin chart for average monthly expenditure on electricity by households in Delhi from 2015 to 2021. The width of each section indicates the number of consumers paying a specific amount as monthly electricity bull. We find that there have been an overall leftward shift in the electricity expenditure. The wide portions of the years 2020 and 2021 indicate that a large number of households that had zero or near zero expenditure on electricity.

The mean expenditure on electricity of each quintile are indicated in Fig. 3. Median expenditure shows a similar trend. The mean expenditure shows a sudden dip in the year 2015 which had seen the introduction of subsidy on electricity consumption less than 400 units. But the expenditure gradually increases in the following years before decreasing again in 2019 when the lifeline electricity scheme was announced. The drastic decrease in the year 2020 could be accounted to the disruption in the electricity dues collection system owing to the COVID 19 pandemic.

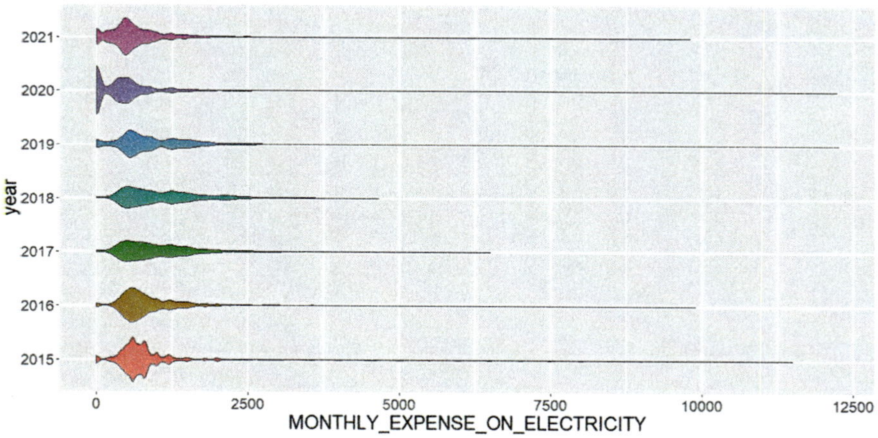

Fig. 2 Distribution of monthly expense on electricity

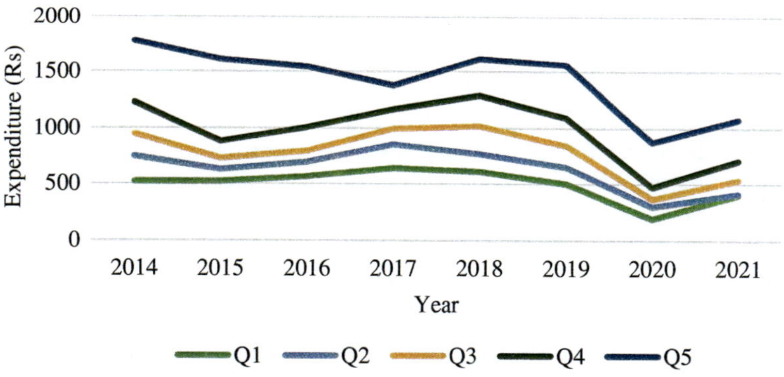

Fig. 3 Mean expenditure on electricity by quintiles

3.2 Analysis of Monthly Expenditure on Electricity from 2019

The analysis of monthly electricity expenditure excluding the months of the pandemic shows that there has been a considerable decease in household (Fig. 4). There has been substantial reduction in monthly expenses on electricity over all months. The savings vary from about 35% in November to more than 60% in February. T test statistics with 95% confidence interval was carried out to check if the scheme had introduced a significant difference in monthly household expenditure on electricity. The results from the pre scheme time period (M = 976, SD = 96.4) and months after implementation of the scheme (M = 502, SD = 86) indicate that the scheme has resulted in a reduction in monthly expenses at confidence level $p = 0.005$.

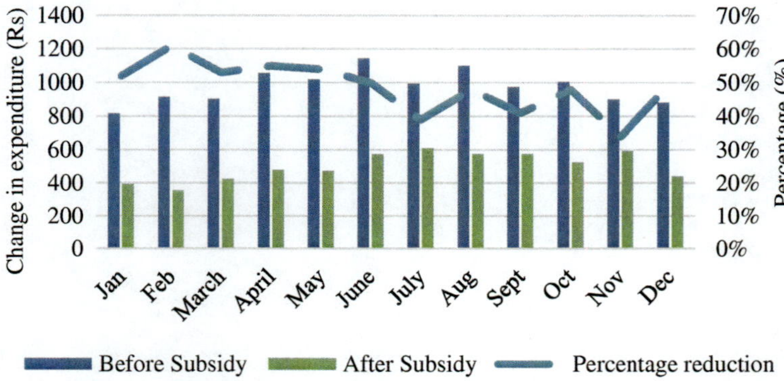

Fig. 4 Mean monthly expenditure on electricity 2019–2021

The period between March 2020 and August 2020 might not be representative of the general trend. The restrictions imposed could have affected the billing and payment of electricity dues. As the monthly expenditure on electricity was collected by a recall method, this could have affected the responses. T test of means comparing the expenses of each quintile for each month from the period January to August of 2019 and 2021. The period January 2019 to August 2019 represents the period before the lifeline scheme and the period January 2021 to August 2021 represents the period when the lifeline scheme was in effect. The results are shown in Table 2.

For understanding the effect of the subsidy on different quintiles, we perform at We find that the savings for the lowest quintile of income vary from 12% to about 50% for various months. The lifeline scheme is regressive and benefits even the highest quintiles of income earners. We see that as consumption increases during the summer months of March to May, more and more households of all categories move away from the subsidy benchmarks of 200 units and 400 units.

Table 2 Percentage change in monthly expenditure on electricity

Quantile	Percentage Change from 2019 to 2021								
	January	February	March	April	May	June	July	August	Average
1	-48%	-45%	-47%	-50%	-12%	-38%	-44%	-39%	-40%
2	-50%	-47%	-44%	-37%	-29%	-44%	-49%	-46%	-43%
3	-52%	-43%	-39%	-38%	-33%	-41%	-44%	-45%	-42%
4	-38%	-30%	-31%	-34%	-32%	-41%	-45%	-41%	-37%
5	-40%	-23%	-19%	-21%	-22%	-32%	-33%	-31%	-28%

3.3 Who Gets What: Equity Implications of the Subsidy

To ensure that subsidies are affordable and redistributive, they must disproportionately benefit groups with lower incomes. It is well acknowledged that quantity based subsidies provided through tariffs are regressive [13]. In any form of subsidy, coverage and metering are critical [13]. The case of residential electricity sector in Delhi, shows high coverage with over 99% and high metering rate. From the results, we find that the consumer expenditure on electricity has reduced considerably over all income quintiles. The variation across months can be attributed to climatic factors that modify electricity consumption [19]. The second and third quintiles of income have shown the highest savings, possibly due to their ability to invest in new appliances. In the case of households that earlier consumed slightly above 400 units, reduction of units below 400 would lead to savings of close to 50%.

Considering the equity implications of the subsidy, we find that the lower quintiles have the maximum percentage reduction in subsidy, though the actual benefits in rupee terms transferred might still be regressive. There is evidence to believe that the largest number of beneficiaries occur in the months winter months when the consumption levels are lower. The temporal analysis of electricity expenditure from 2015 shows that all quintiles of income show considerable decrease in electricity expenditure in the short run, but the expenditure rises close to the pre subsidy levels in the medium term. This could be due to natural increase in appliance ownership or conspicuous consumption or both.

Thus, more evidence is necessary to conclude if the subsidy schemes have led to disciplining of consumption to set thresholds. Not surprisingly, the savings of all income groups are highest in the summer months resulting from base effect of larger consumption.

3.4 Analysis of Appliance Ownership

Electricity consumption is realized through use of electricity consuming appliances. Incentives to reduce or limit electricity consumption would lead to reduction in the number of appliances, reduced time of use or use of energy efficient appliances. We find that the total wattage of appliances considering all the households have decreased with the onset of subsidy. But the trend has not been uniform. The lower three quintiles have reduced appliance ownership while fourth quintile shows marginal increase. The fifth quintile shows considerable increase of about two hundred watts (Fig. 5).

Space conditioning contributes to a significant portion of the electricity demand in Delhi [10, 17]. The subsidy regime has affected the ownership of air conditioning equipment also. The ownership of air conditioners in the three lowest quintiles have remained constant or declined while it has increased for the fourth and fifth quintile (Fig. 6). In the case of air coolers, the trend is almost the opposite (Fig. 6). The three lower quintiles have shown the maximum increase in ownership. As air coolers are

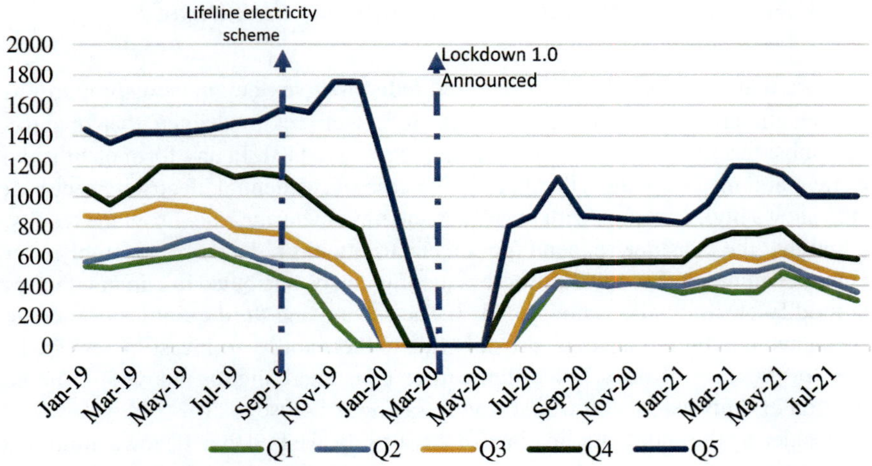

Fig. 5 Median monthly expenditure on electricity 2019–2021

much cheaper and easy to acquire and use, the lower quintiles would have increased the number of air coolers. But with the higher quintiles with more disposable income are likely to have invested in air conditioners. The ownership of refrigerators does not show any particular pattern, while the ownership of televisions have increased across the board by about 1%.

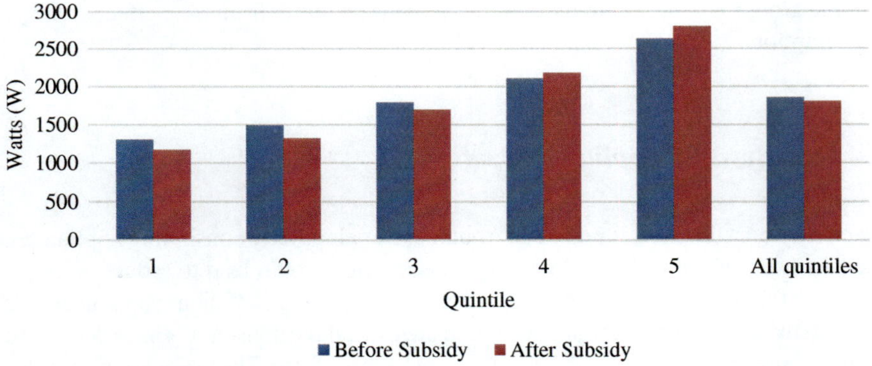

Fig. 6 Connected load of appliances

4 Political Economy of Free Electricity

Flat rate subsidies of the scale seen in Delhi are a novelty, barring a similar attempt in Tamil Nadu with a lower threshold of consumption. It is based on the idea of a lifeline consumption of electricity and provision of "zero bills". Multiple independent studies and reports have found the scheme to have boosted the public perception about the government (Goswami 2019) [21]. A large existing school of thought considers electricity (like water) as a public good meant for non rival consumption. The implementation of the subsidy has its root in the protest for efficiency and transparency in Delhi's electricity sector that was spearheaded by the ruling dispensation. Free electricity has been touted as a governance reform that helps the average person. The thresholds have been set to ensure quasi universality of almost 80%. In addition, the announcement of the scheme in many ways focus on "efficiency", "lack of corruption" and "transfer of public resources back to the public". Thus, the scheme has been explicitly linked with reduction of corruption. The scheme is an electoral plank with open challenges electoral rivals to announce similar schemes (ANI 2021) [16]. Not surprisingly, the scheme has become a poll issue in states where the ruling dispensation of Delhi is an important contestant (Fig. 7).

4.1 The Effect of Subsidy on State Finances

The level of cross subsidies in Delhi was higher than the national average in the year 2015 with certain categories being subsidised by as much as 25% [20]. The free electricity scheme has created a large subsidy commitment that affects 50.98% of all electricity consumption in Delhi. For Delhi, power subsidy estimates for the year 2021 was Rs. 3090 crores with a 15% increase over the past financial year. But it accounts to only 5.2% of the total expenditure of the government with the

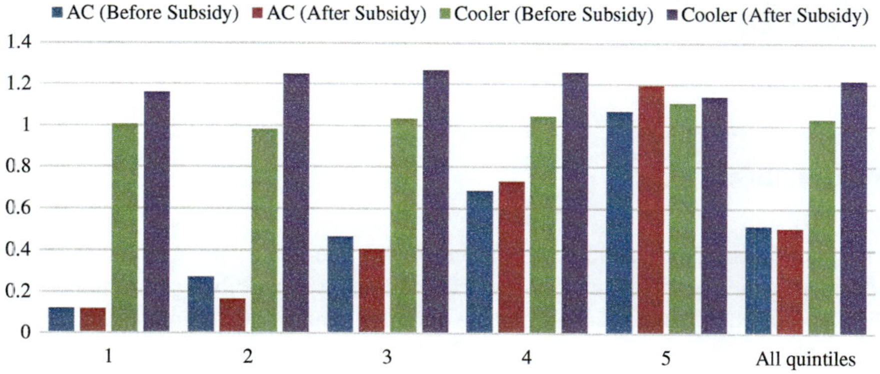

Fig. 7 Ownership of air conditioners and air coolers before and after subsidy

national average of all state governments is 4.2%. It is also considerably lower than the expenditure in other sectors like education (23.4%), health (14.4%), transport (10%) and roughly equal to the expenditure on social welfare and nutrition (6%) [12].

The allocations to health and education are most prominently for continuing revenue expenditure and capital expenditure for infrastructure expansion while the allocation to electricity goes almost exclusively to consumer subsidies through Distribution Companies (DISCOMs) [9, 12]. Thus, we find that the subsidy scheme has limited effect on the financial health of the state but has become the most important expenditure within the electricity sector. It could lead to lower capital expenditure on important aspects like maintenance and replacement expenditure and ideation of new schemes. For instance, there has been a reduction in subsidies to electric vehicles and renewables [5].

4.2 The Effect of Subsidy on Financial Sustainability of the DISCOMs

The direct subsidy to residential consumers would be more acceptable to DISCOMs than the cross subsidy regimes. The entrenched cross subsidies and surcharges increased the price of electricity for industrial and commercial consumers, incentivizing them to opt for captive generation or private power purchase. Subsidies and cross subsidies are theoretically a "zero-sum" game for the DISCOM where the losses from one segment is compensated by another segment or transfers from the state. But non quantification of subsidies, deferred payments and classification as "regulatory assets" have led to mounting losses. DISCOMs in India were losing 74 paise per unit of electricity sold in financial year 2016. Iindependent reports suggest a timely transfer of electricity subsidies by the Government of Delhi to the DISCOMs till the year 2020 [9]. In 2020, the state government of Delhi disbursed 50% of the subsidy commitment while accounting the rest of the subsidy towards dues owed by the DISCOMs to the state run power producers [7]. This can be considered as a form of forceful extraction, albeit with formal justification.

5 Conclusion

The paper has analysed the performance of the free electricity scheme in Delhi considering the change in out of pocket expenditure on electricity by households and the distribution of benefits across income quintiles. Analysing the annual household electricity expenditure from the period 2015 onwards, we find that the immediate effect of the electricity subsidy is a decrease in electricity expenditure. But the expenditure gradually increases, often to pre subsidy levels within a span of two or three years. Thus, subsidy as a stand-alone tool without measures to improve penetration of energy efficient appliances would not lead to disciplined consumption. The analysis of the lifeline electricity scheme shows that all quintiles have benefitted from the subsidy. The threshold for inclusion of 200 units can be revised to make the subsidy less regressive. In terms of total wattage of appliances in use, the lower quintiles show a decrease while the higher two quintiles show an increase. The subsidy scheme accounts for less than 6% of the total expenditure of the Government of Delhi but overshadows all other expenditure of the sector. It should be ensured that the large and benevolent scheme does not strain the financial space of other capital expenditure and support for new and innovative schemes.

References

1. Aggarwal P, Viswamohanan A, Sharma S (2020) Unpacking India's electricity subsidies. International Institute for Sustainable Development
2. Basumatary S, Devi M, Basumatary K (2021) Determinants of household electricity demand in rural India: a case study of the impacts of government subsidies and surcharges. Int J Energy Econ Policy 11(6):243–249. https://doi.org/10.32479/ijeep.11716
3. Chunekar A, Varshney S, Dixit S (2015) Residential electricity consumption in India, p 60. https://www.prayaspune.org/peg/trends-in-india-s-residential-electricity-consumption
4. CMIE (2021) Consumer Pyramids Household Survey (Online). https://consumerpyramidsdx.cmie.com/
5. Delhi govt withdraws subsidies on electric cars. Here's why (2021) Livemint
6. Deshpande T (2020) Minister reiterates 100 units free power to residential consumers. The Hindu. https://www.thehindu.com/news/states/minister-reiterates-100-units-free-power-to-residential-consumers/article30976841.ece
7. FE Bureau (2020) Delhi govt to use pvt discom subsidies to clear dues of state-run power stations. Financial Express. https://www.financialexpress.com/industry/delhi-govt-to-use-pvt-discom-subsidies-to-clear-dues-of-state-run-power-stations/1935729/
8. Government of NCT Delhi (2019) Government order on subsidy for residential electricity consumers
9. Government of NCT Delhi (2021) Annual financial statement. https://openbudgetsindia.org/dataset/delhi-annual-financial-statement-2021-22/resource/6285586d-2f5e-4ea8-8c01-f6e78bfb7189
10. IEA (2021) India energy outlook 2021. https://www.iea.org/reports/india-energy-outlook-2021
11. Kale SS (2014) Electrifying India: regional political economies of development. Stanford University Press. http://www.sup.org/books/title/?id=22441
12. Kaur P (2021) Delhi budget analysis 2021–22. PRS Legislative Research. https://prsindia.org/files/budget/budget_state/delhi/2021/Delhi%20Budget%20Analysis%202021-22.pdf
13. Komives K, Halpern J, Foster V, Wodon Q (2006) The distributional incidence of residential water and electricity subsidies. World Bank Policy Research Working Paper, No. 3878, pp 1–5. http://econ.worldbank.com
14. Kumar A, Chatterjee S (2012) Electricity sector in India: policy and regulation. Oxford University Press. https://doi.org/10.1093/acprof:oso/9780198082279.001.0001
15. Misra A (2021) Why free electricity is a poll issue in UP. India Today. https://www.indiatoday.in/india-today-insight/story/why-free-electricity-is-a-poll-issue-in-up-1855392-2021-09-21
16. PTI (2021) Arvind Kejriwal's dare to Punjab chief minister over electricity bills. NDTV. https://www.ndtv.com/india-news/arvind-kejriwals-dare-to-punjab-chief-minister-over-electricity-bills-2627470

17. Singh J, Mantha SS, Phalle VM (2018) Characterizing domestic electricity consumption in the Indian urban household sector. Energy Build 170:74–82. https://doi.org/10.1016/j.enbuild.2018.04.002
18. Singh MK (2022) Eradicating energy poverty: overcoming "barriers" to decentralized energy systems in India. Springer, Singapore
19. Tewathia N (2014) Determinants of the household electricity consumption: a case study of Delhi. Int J Energy Econ Policy 4(3):337–348
20. Tongia R (2017) Delhi's household electricity subsidies: highly generous but inefficient? Brookings India IMPACT Series No. 042017
21. VP S (2019) Over 47 lakh people in Delhi benefit from the electricity subsidy, but is it really a good welfare scheme? Newslaundry.Com. https://www.newslaundry.com/2019/12/02/delhi-aam-aadmi-party-electricity-subsidy

Performance Emission Vibration Analysis of Petrol Engines Using Alcoholic Fuel Blends

N. Ravi Kumar, U. Sai Srivatsha, R. Roopa Keerthana, and S. K. Nooruddin

1 Introduction

Road transportation currently consumes the most petroleum fuels (25%) and is the primary source of air pollution in metropolitan areas (> 75%). The burning of fossil fuels plays a major role in transportation. At the same time, it is estimated that the amount of readily available fossil fuels such as natural gas and oil will decline rapidly over the next 30 years. Therefore, the search for alternative fuel sources and new technologies to reduce the consumption of fossil fuels is much needed and it can also reduce the cost of oil import which helps to improvise the country's economy. India's oil import bill over the past decade is shown in Fig. 1. Because it is a fuel made primarily from plant resources like sugarcane and corn, ethanol is a desirable alternative to gasoline for reducing reliance on fossil fuels and emissions, given off into the air. Additionally, ethanol burns more efficiently since it has a higher oxygen content due to its higher-octane rating compared to gasoline.

1.1 Literature Review

The concept of alternative fuels is not new, and numerous academic papers have been released on it. We read a number of research publications to gain a sense of how earlier studies have been conducted.

The use of E40 among other mixes was shown to greatly increase thermal efficiency while also reducing the specific fuel consumption to a minimum. As the mixture's ethanol content rises, the emissions are similarly seen to decline [2].

N. Ravi Kumar · U. Sai Srivatsha (✉) · R. Roopa Keerthana · S. K. Nooruddin
Department of Mechanical Engineering, MVGR College of Engineering, Vizianagaram, AP 535005, India
e-mail: uppaladhadiamsaisrivatsa@gmail.com

S. Doolla et al. (eds.), *Advances in Clean Energy and Sustainability*,
Green Energy and Technology, https://doi.org/10.1007/978-981-99-2279-6_25

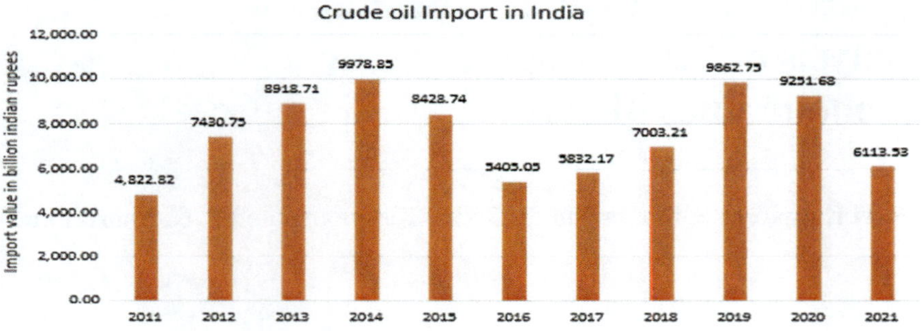

Fig. 1 Import quantity and cost of crude oil by India [1]

At a little lower temperature than gasoline, combustion using an ethanol blend is more complete than combustion using pure gasoline. Less combustion deposits and a longer spark plug life are the results. Because it is an oxidant and contains oxygen in its molecules, ethanol burns efficiently. Under stoichiometric circumstances, it also requires less air because ethanol already contains oxygen atoms [3].

Engine damage might result from an issue called engine vibration. Alcoholic mixes can lessen engine vibration, likely as a result of the higher latent heat of vaporization of ethanol and the oxygen content, as well as the increased compression ratio (DP/DT) and peak in-cylinder pressure values during combustion processes. Because ethanol will have three to five times the heat of association as pure gasoline, the latent heat of an ethanol blend will be higher than that of pure gasoline [4].

Prakash Verma, Alok Choube, and other researchers [5] concentrated on two major topics: the impact of ethanol blends and compression ratio on engine performance. High octane ethanol allows for higher compression ratios and fewer emissions. Because ethanol is also an environmentally beneficial gasoline, it is readily available, less expensive than gasoline, and has a larger supply than fossil fuels.

Aqueous ethanol and a gasoline-ethanol blend (E22) were contrasted in terms of engine performance and exhaust emissions by Costa and Sodré [6]. Due to its lower calorific value compared to gasoline, the use of aqueous ethanol increased specific fuel consumption while also modestly improving engine torque and power production at high engine speeds. Because ethanol burns more quickly than gasoline, the thermal efficiency was better when ethanol was utilized instead of gasoline.

Alcohol typically reacts with brass, copper, and rubber components to jam the fuel line. As a result, fluorocarbon rubber is advised as an alternative to rubber [7].

Alcohols' anti-knock property has been mentioned, and in addition, their higher heat of vaporization at high temperatures and higher flame speed enables improved fuel conversion efficiency when compared to gasoline. Additionally, alcohol combustion produces a larger amount of product, increasing cylinder pressure and exerting more force on the piston.

On the other hand, alcohol's lower calorific value leads to increased specific fuel consumption when compared to gasoline, meaning that more alcohol must be used

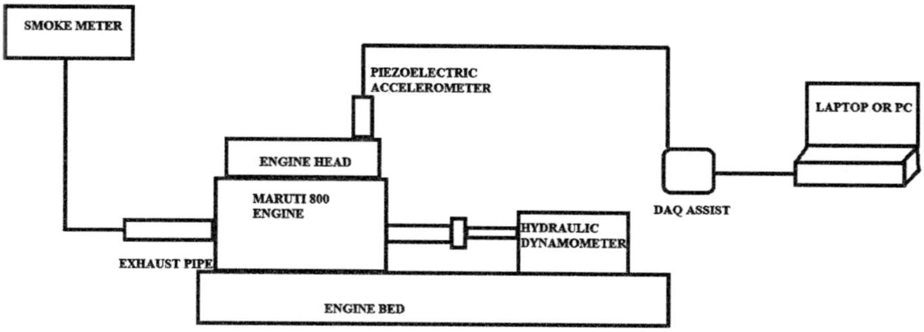

Fig. 2 Schematic diagram of a part of an experimental rig

overall to produce the same amount of power. Due to their low vapor pressure, alcohol fuels also have a problem with cold starts [8].

The engine can be started steadily using E5, E10, E20, and E30, according to research on the effects of ethanol gasoline blended fuel on gasoline engine cold start emissions. Due to the E40's excessively lean air/fuel combination, the engine idle became unsteady [9].

The multi-cylinder internal combustion (SI) engine's response to gasoline containing ethanol and oxygenated additives was studied in the research article. There were two phases to the experimentation. According to experimental findings, the blend raised BTE more than a single fuel, like gasoline, did [10] (Fig. 2).

2 Experimental Setup

2.1 Maruti 800cc

An SI engine with three cylinders and four strokes is subjected to performance, emission, and vibration analyses in this study. The engine's specifications are as in Table 1.

2.2 Digital Bomb Calorimeter

The quantity of heat produced by a substance's unit volume upon full combustion is referred to as its calorific value. The better the fuel efficiency, the higher the calorific value of the fuel. The Calorific Value was calculated using a calorimeter.

Pure petrol and alcoholic mixtures' values (E05, E10, E20, E30, E40) (Table 2).

Table 1 Specifications of Maruti 800

Engine	Maruti 800
BHP	12 HP (8.94 kW)
Speed	1500 rpm
Fuel	Petrol
No. of cylinders	Three
Bore (D)	68.5 mm
Stroke length	72 mm
Arm length	0.32 m
Starting	Self-start
Working cycle	Four stroke
Method of cooling	Water cooled

Table 2 Specifications of bomb calorimeter

Specification	Model CC01/M3	Model CC01/M2
Working principle	Iso-thermal	Iso-thermal
Standards	BS 1016: Part 5:1967 IS 1359-1959 IP 12/63T	BS 1016: Part 5:1967 IS 1359-1959 IP 12/63T
Experiment duration	10–15 min	10–15 min

2.3 Piezoelectric Accelerometer

It operates on the idea that when a mass exerts force on a piezoelectric material (such as a crystal or ceramic), the resulting stress causes the material to produce an electrical charge that is proportionate to the applied force.

Then, it was sent to DAQ help, which converts the form and displays the vibration's amplitude in LABVIEW (Table 3).

Table 3 Specifications of piezo-electric accelerometer

Product	Accelerometer
Model number	352CO3
Sensitivity (− 20 to + 20)	9.95 mV g
Measurement	− 500 to + 500 gpk
Frequency range (− 5 to + 5%)	0.5–10,000 Hz
Resonant frequency	≥ 50 kHz
Weight	5.8 g

2.4 *Smoke Meter*

The working principle is the amount of light blocked in the smoke released by engines from vehicles is detected and measured using opacity meters. The smoke meter readout shows the density of smoke, which is an indicator of combustion efficiency. The Beer-Lambert principle is used for reading. It shows the amount of smoke present in the exhaust (Table 4).

3 Methodology

A certain amount of ethanol is blended with petrol and stirred well to form an ethanol-petrol blend. The blend is taken in a small container and measured the mass of the blend, and volume of the blend are. Then using mass, the volume of the blend is determined. Calorific value is measured for each ethanol-petrol blend using a Digital Bomb Calorimeter, and the obtained calorific values are listed in Table 5. Each blend is poured into the fuel tank and the engine is run by varying the speed by throttle opening and loads using a hydraulic dynamometer. The fuel consumption of the engine is measured using volumetric analysis thereby calculating the performance characteristics. The vibration characteristics are captured using a piezoelectric accelerometer placed over the engine head. The vibrations obtained are recorded at different blends, loads, and speeds using LabVIEW software. The smoke percentage in the exhaust gasses is captured using a smoke meter.

Table 4 Specifications of smoke meter

Smoke intensity	0–100 BSN	+ 1%	− 2BSN to + BSN

Table 5 Properties of different ethanol-petrol blends

Blend	The density of fuel (kg/m^3)	Calorific value (kJ/kg)
E00	784.90	46,670.86
E05	758.22	44,181.00
E10	742.32	41,791.33
E20	730.71	39,366.76
E40	711.85	35,295.26

4 Results and Discussions

Pure petrol without mixing of ethanol has a density of around 784.90 kg/m^3 mixing 5% ethanol will decrease the density value by. Usually, the density of ethanol is slightly lesser than the density of petrol. So, when the proper blending of the mixture leads to a drop in the density of the fuel mixture.

With 40% of ethanol in the mixture, the density is reduced to 711.85 kg/m^3. As, density is a function of temperature, as the temperature increases the density will vary. For, Constant volume of fuel, for higher density more mass is inducted into the engine chamber compared to the lower-density fuel.

Fuel with a high viscosity produces fuel droplets that are larger because density increases also cause fuel viscosity to rise. Large fuel droplets cause incomplete combustion of the fuel. Therefore, efficiency drops.

Calorific value, which is calculated by measuring the heat produced when a specific amount of a substance completely burns, is the amount of energy present in food or fuel. These days, this is typically stated in kilojoules per kilogram.

Ethanol has a lower calorific value than gasoline. Pure gasoline has a calorific value of 46,670.86 kJ/kg, but a blend containing 40% ethanol has a calorific value that is nearly 24% lower.

So, it is observed that the decline in the calorific value with increase the ethanol content in the blend. As less calorific value is observed, so more fuel is burned to get the same power as pure petrol.

4.1 Brake Thermal Efficiency

Internal combustion engine fuel efficiency is gauged by brake thermal efficiency. Fuel usage and greenhouse gas emissions decrease with increasing brake thermal efficiency. Maximum brake thermal efficiency for the engine is often preferred. Figure 3 depicts the brake thermal efficiency for various mixes at various loads at a constant speed (1500 rpm). A 12 kg load results in the highest thermal efficiency. The volumetric efficiency depicted in Fig. 6 indicates that it is lower with lower loads. As a result, it may be noted that at lower loads, the smaller air intake may result in incomplete combustion and a reduction in the thermal efficiency of the brakes. The Brake Thermal Efficiency ranges from a minimum value of 9.27% (Engine running with E05 blended fuel) to a high value of 37.73% at a constant speed of 1500 rpm (Engine running with E20 blended fuel). The thermal efficiency of the E00 blend at 3 kg is approximately 14.38%, whereas the addition of 40% ethanol increases the thermal efficiency of the engine brakes by roughly 14.4%. This may be a result of the reason why adding ethanol to gasoline raises its octane rating. By accelerating the flame, the increase in octane number promotes improved combustion. Ethanol has a larger oxygen content than gasoline, complete combustion, and higher thermal efficiency in the brake system. It has been observed that as the amount of ethanol in the gasoline blend increases, fuel consumption decreases. Consequently, there is less fuel combustion increasing in the mixes causes a reduction in duration, which lowers heat transfer losses. This may also be the cause of blended fuels' higher thermal efficiency after they have carried 3/4 of the load. Figures 3, 4 and 5 show that whereas the thermal efficiency of pure gasoline starts to decline after about 3/4 of the load, the thermal efficiency of blended fuels continues to increase as the load increases. Compared to blended fuels, pure gasoline has a higher mass flow rate. Since thermal efficiency is inversely related to the mass flowrate of fuel, the calorific value of the fuel, higher mass flow rate, and calorific value of pure gasoline might lead to poorer thermal efficiency in the case of fixed brakepower.

With an increase in engine speed, the brake power will also increase. Since speed directly proportions to thermal efficiency. The thermal efficiency rises quickly when the mass flow rate and calorific value are constant (comparison of a single blend). Figure 4 displays the BTE vs. Speed graph for various blends with a constant load of

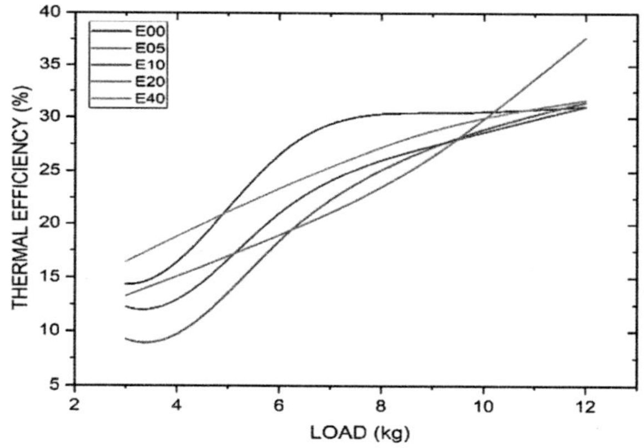

Fig. 3 Variation of brake thermal efficiency concerning load at a constant speed of 1500 rpm for different blends

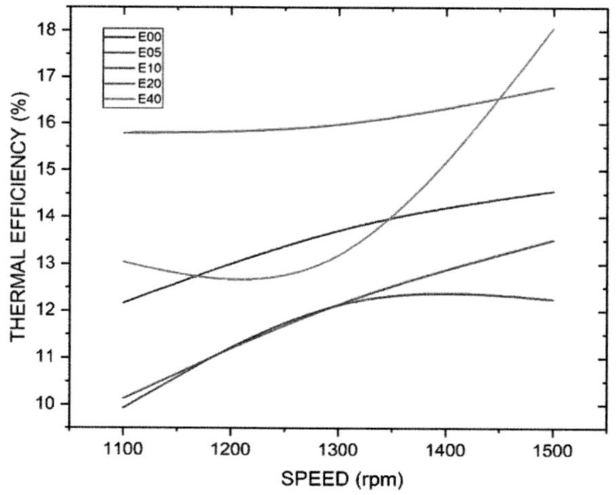

Fig. 4 Variation of brake thermal efficiency for speed at a constant load of 3 kg for different blends

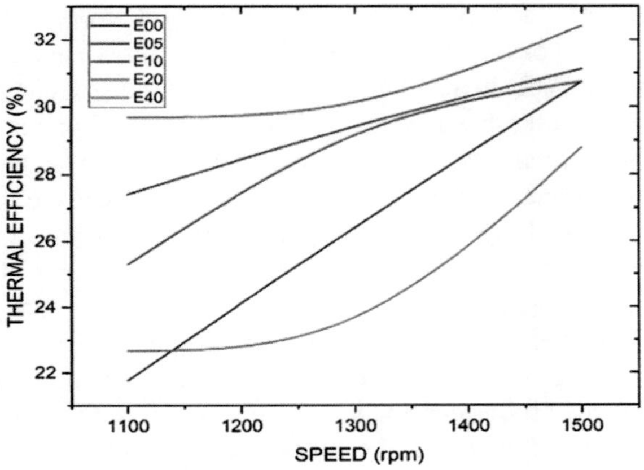

Fig. 5 Variation brake thermal efficiency for speed at a constant load of 9 kg for different blends

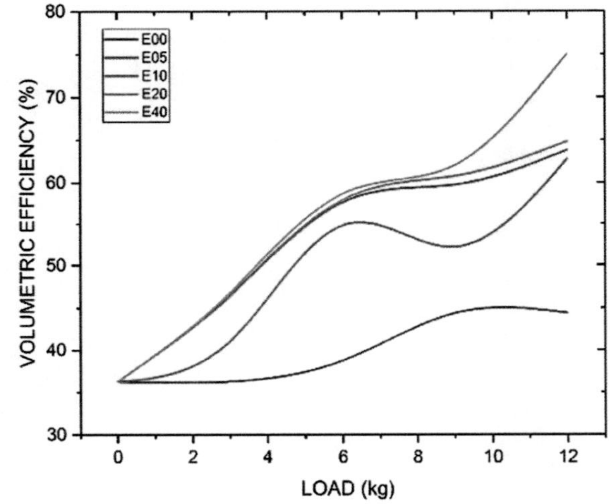

Fig. 6 Variation of Volumetric efficiency for load at a constant speed of 1500 rpm for different blends

3 kg. Ethanol may be added to increase flame speed, which improves combustion. For speed, the BTE rises for all blends. E20 blend is discovered to have a high thermal efficiency of 15.75% at 1100 rpm, while pure petrol is observed to have a thermal efficiency of 12.16% at the same speed. So it has been observed that almost 3.59% of the increase in the BTE at 1100 rpm for engine running with E20 blend comparing with the pure petrol.

The graph of BTE versus speed for various blends is shown in Fig. 5 with a constant load of 9 kg. Ethanol may be added to increase flame speed, which improves combustion. However, the engine must be modified to take into consideration the different combustion characteristics of ethanol concentrations beyond 15–20%, which might result in engine damage. Beyond the E20 blend, the engine we used in the experiment has a constraint. For speed, the BTE rises for all blends. E20 blend has a high thermal efficiency of 29.79% at 1300 rpm, compared to pure gasoline's thermal efficiency of

26.36% at the same speed. So it has been observed that almost 3.43% of the increase in the BTE at 1300 rpm for engine running with E20 blend comparing with the pure petrol.

Due to perfect combustion at all loads and speeds, the blend's increased ethanol percentage increases thermal efficiency. More over 20% of the ethanol in the blend cannot be burned by the engine. Therefore, a decrease in brake thermal efficiency is shown following the E20 blend.

4.2 *Volumetric Efficiency*

It is the ratio of the total displacement of all the cylinders at atmospheric pressure to the volume of air or charge pulled into them during the suction stroke. The engine should generally operate at its highest volumetric efficiency. Figure 6 illustrates the volumetric efficiency for various blends at zero to maximum load situations at a constant speed of 1500 rpm, the highest.

When using an E40 fuel blend, volumetric efficiency was measured at full load. About 75.04% of it. In comparison to E40 blended fuel, the volumetric efficiency at a full load of pure gasoline was raised by 30.65%. Generally speaking, ethanol contains more oxygen than pure gasoline. Therefore, it was anticipated that the volumetric efficiency would decrease as the ethanol level in the fuel rose. However, because of the low stoichiometric ratio of ethanol, blended fuel has a lower stoichiometric ratio and becomes leaner. Due to ethanol's lower stoichiometric ratio, more fuel may burn simultaneously with the same amount of air. However, when the ethanol concentration increases, the lower vapor pressure and higher heat of evaporation cause the volumetric efficiency of the engine to increase. As a result, it causes the same volume of cylinder to be filled with more F-A mixture, which is then burned effectively by the fast flame. Additionally, lower emissions may be extremely important to the engine's ability to breathe. There is a reduced risk of valve seizure and exhaust gas recirculation as emissions decline with an increase.

It has been noted that all blends' volumetric efficiency has increased with regard to loads. The reason for this was that more Air:Fuel mixture had to be introduced into the same volume of the cylinder for heavier weights. As a result, the volumetric efficiency with regard to load increases. Higher speeds produce a greater vacuum at the port, which results in a greater air flow rate, which increases volumetric efficiency. A further rise in engine speed brings volumetric efficiency to its highest level. Figure 7 illustrates difference in the volumetric efficiency for the engine's speed for all fuel blends with a 3 kg load. At 1100 rpm, an engine running on pure gasoline has a volumetric efficiency of 36.8%; however, when the engine speed is increased to 1500 rpm and the amount of ethanol in the blend is increased to 40%, the volumetric efficiency rises to 68.96%.

More torque and power are produced as a result of an increase in volumetric efficiency. A mixture in the cylinder has a higher thermal capacity when the volumetric efficiency is higher. Therefore, it is assumed that when the same quantity of heat is

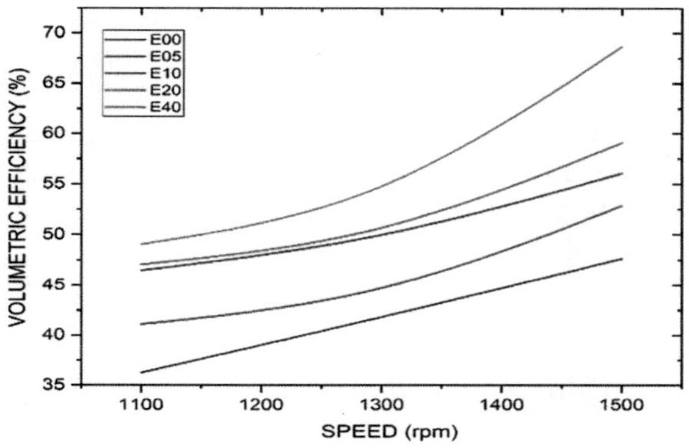

Fig. 7 Variation of volumetric efficiency for speed at a constant load of 3 kg for the different blends

released in the cylinder, a lesser increase in in-cylinder temperature is obtained with a higher volumetric efficiency.

Brake-Specific Fuel Consumption

Any engine that consumes gasoline and generates rotational power output has a BSFC value, which represents how efficiently it uses fuel. How effectively the engine turns the fuel given into useful work is shown by the BSFC value. The best mileage is indicated by a lower BSFC number.

Figure 9 displayed the BSFC versus load for various loads at 1500 rpm for various mixes. It should be remembered that for all fuel blends, BSFC falls as engine load increases (E00, E05, E10, E20, E40) Higher loads cause the engine's turbulence and in-cylinder temperature to increase, which improves combustion efficiency and increases BP through optimal fuel atomization and mixing in the cylinder.

The ethanol component is seen to improve the volumetric efficiency. The improvement in volumetric efficiency with regard to speed and ethanol content in the blend is shown in Fig. 8. As the volume proportion of ethanol fuel in the combination rises, the volumetric efficiency rises as well.

The heat of ethanol evaporation is roughly 2.5 times that of gasoline, which lowers intake manifold temperature and improves volumetric efficiency. It should be noted that the volumetric efficiency of an engine running on pure gasoline is measured at a minimum of 1100 rpm, whereas the volumetric efficiency of an engine running on an E40 blend is measured at a maximum of 1500 rpm.

Figure 9 shows that the maximum brake-specific fuel consumption is recorded at 3 kg load when the engine is operated with E05 blended fuel, and the minimum specific fuel consumption is recorded at 12 kg, even though the fuel consumption increased with the load. This was because the increase in the BP was greater than the increase in the fuel consumption (E20 blended fuel). It should be noted that the BSFC for pure gasoline decreases at lower loads by up to 3/4th, but thereafter increases approaching full load conditions. Due to the mixed fuels' improved combustion efficiency at higher loads, there is a greater gain in BP than there is in fuel consumption, which results in a decrease in BSFC at through loads.

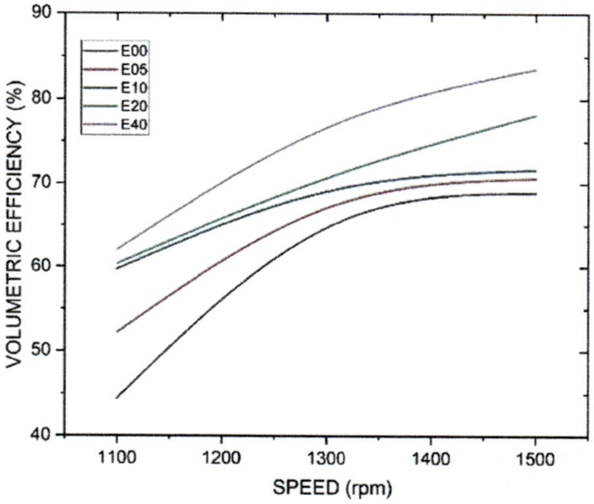

Fig. 8 Variation of Volumetric efficiency for Speed at a constant load of 9 kg for different blends

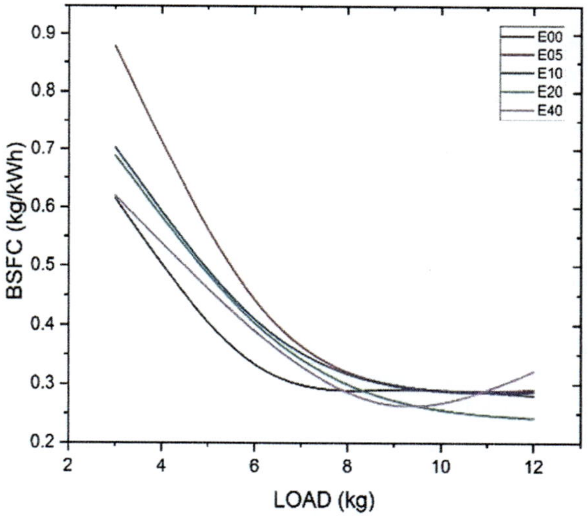

Fig. 9 Variation of brake specific fuel consumption for load at a constant speed of 1500 rpm for different blends

The fuel's conversion efficiency has been found to cause the BSFC to continuously decline as load increases. The combustion chambers' internal turbulence and temperature will be high at greater loads, which aids in the optimum fuel mixing since better atomization leads to increased combustion efficiency. It is possible to rephrase BSFC as "Work per Fuel Consumed". In other words, because less of the fuel entering the engine is wasted, an engine is more effective at converting fuel to work at a higher load for the same speed utilized to pump air in and out, heat the engine, turn the engine, or exit into the exhaust.

The engine running on the E20 mix with 0.242 kg/kWh is determined to have the lowest BSFC at 1500 rpm. With an increase in load, it is discovered that the efficiency of E20 mix decreases exponentially. The Speed vs. BSFC for each blend is shown in Fig. 10 at a constant load of 3 kg, with E20 and E10 blends having the lowest and greatest BSFC values at 1100 rpm, respectively. The reason for the anonymous behaviour is because pure petrol engines first burn more BSFC, which decreases with speed.

Volumetric flow has been observed to become more efficient quickly. E40 has been found to have better volumetric efficiency at full load (12 kg) and 1500 rpm. The largest BTE and the least amount of brake-specific fuel consumption were recorded with the engine operating on the E20 blend. It became clear that there is a mechanical restriction on the engine's ability to properly burn fuel that goes beyond the blend's 20% ethanol level (Fig. 11).

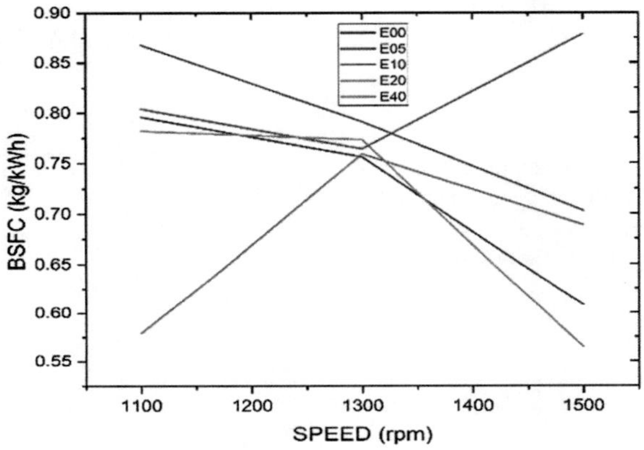

Fig. 10 Variation of brake specific fuel consumption for speed at a constant load of 3 kg for different blends

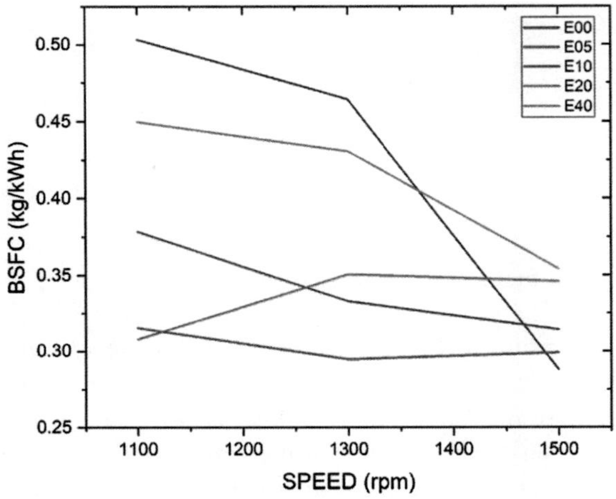

Fig. 11 Variation of brake specific fuel consumption at a constant load of 3 kg for different blends

4.3 Smoke Opacity

Figure 12 displays the proportion of smoke produced by various loads and ethanol and gasoline blends when the engine is running at a constant speed of 1500 rpm. When the mixes' proportion of smoke was investigated. More aggressive than gasoline, ethanol can wear down or damage materials and components, cause elastomers to expand, and cause metals to corrode.

The type of engine used and its substance both contribute to the level of smoke. The aggregate findings indicate that compared to the other blends, E05 and E10 release a higher amount of smoke. For E20 and E40 mixes, proper combustion takes place with an increase in oxygen concentration and results in a decreased amount of smoke. The largest amount of smoke was measured under no-load situations at 38%, which is decreased to 9% with the E40 combination. At a weight of 3 kg, the E05 mixture's observed percentage of smoke is approximately 33%, while the E40 mixture's is decreased to 10%. The maximum recorded smoke percentage for the E10 mix at 6 kg load circumstances is over 40%, while the lowest recorded smoke percentages for the E20 and E40 blends are roughly 11%. The proportion of smoke was steadily decreased from 40 to 14% with a load of 9 kg by adding ethanol to gasoline. With an E05 mixture, the percentage of smoke is higher at a weight of 12 kg and is lower with an E20 combination at a load of 10%. Each blend's smoke percentage rises with an increase in load. It ranged from 4 to 10% for the E00 blend. It varied from 34 to 42% for the E05 blend. It ranged from 38 to 40% for the E10 mixture. It ranged from 10 to 14% for the E20 mixture. It varied from 9 to 12% for the E40 combination. Because ethanol fuel includes more oxygen than regular gasoline does, the fuel burns completely and the amount of pollutants in the exhaust gas is reduced as the percentage of ethanol in gasoline rises. This phenomenon has been observed. Due to the quick load variation during testing and the impact of the fuel's water content, the results show that the percentage of smoke in the E10 mix is higher than that in the E05 blend.

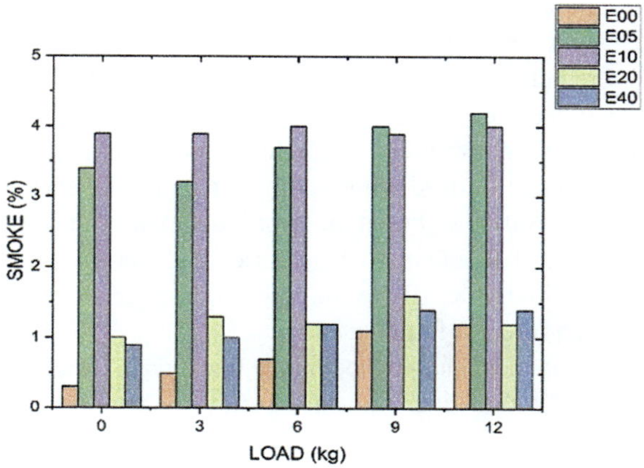

Fig. 12 Variation of smoke percentage for load at different blends at a constant speed (1500 rpm)

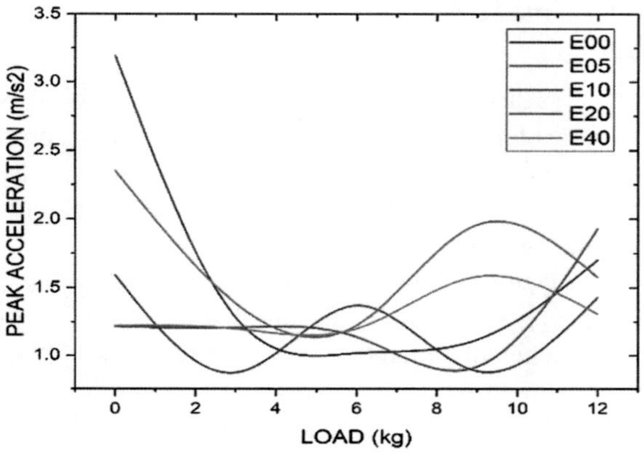

Fig. 13 Variation of acceleration concerning load at a constant speed of 1500 rpm for different blends

4.4 Vibration Analysis

Figure 13 shows the relationship between peak acceleration values and load for different ethanol-gasoline blends. Frictional resistance is higher when the motor is started than later, resulting in higher power consumption at start-up and thus higher RMS values at lower loads.

Higher peaks in the graphs indicate large changes in in-cylinder pressures during fuel combustion due to sudden fluctuations in load and engine speed. Pre-ignition and ignition delay can also be the cause of maximum RMS values.

5 Prediction Model

By looking for patterns in a set of input data, predictive modelling uses mathematics to foretell future events or results. It is a crucial part of predictive analytics, a sort of data analysis that makes use of both recent and old data to forecast patterns in behaviour, activity, and trends.

Predictive model types:

Choice Trees: Data (mined, public, or internal) is plotted in branches using decision tree algorithms to highlight the potential results of various choices. Decision trees can be utilised with incomplete data sets, classify response variables, predict response variables based on prior judgements, and are simple to understand and use for rookie data scientists. Time series analysis is a method for projecting future events using a time series. By examining historical tendencies and extrapolating from them, you can forecast future events.

A statistical analytic technique that aids in data preparation is logistic regression. As more data is given into the algorithm, it becomes better at sorting and categorising it, which allows predictions to be produced.

Neural Networks: This method looks at a lot of labelled data to see if there are any correlations between the variables. Many modern applications of artificial intelligence (AI), such as image recognition, intelligent assistants, and natural language synthesis, are based on neural networks.

For this prediction, a non-linear regression analysis was performed in origin pro software and here the R2 value is 1, which means that the values are almost the same with the given data and from the equation obtained, the error is calculated using the equation and the known parameters.

X1 = Compression ratio

X2 = Blend (% of ethanol content)

X3 = Load (kg).

Brake thermal efficiency:

$$(51625.1485)X_1^{-1.73688} + \left(0.08227X_2^{0.38309}\right) + (0.06342)X_3^{0.6623}$$

Volumetric efficiency:

$$(0.33671)X_2^{0.2106} + (0.33597)X_3^{0.29659}$$

BSFC:

$$\left(1.40384 * 10^{12}\right)X_1^{-4.02736} + \left(1.9943X_2^{-0.59444}\right) + (1.79499X_3^{-0.75006}).$$

6 Conclusion

In the current research, the performance, emission, and vibration characteristics of petrol engines were studied using different fuel mixtures at different loads and speeds.

BTE increases with load for all mixes and decreases for E00 mix at full load.

The increase in BTE of an E40 blend when increasing from 3/4 to full load is approximately 10%, but the increase in BTE for an E20 blended fuel engine when increasing from 3/4 to full load is approximately 40%.

While the drop in BSFC for the E20 mixture when increasing the load from 3/4 to full is around 30%, while for the E40 mixture it was around 10%.

Vibrations decrease at a similar rate for both the E40 and E20 compounds, with the E40 experiencing a lower amplitude at full load.

In terms of performance, emissions, and vibrations, the E20 has good performance at higher loads with significant vibrations and emissions.

Nomenclature

SI ENGINE	Spark ignition engine
E00	0 ml of ethanol in 100 ml of a mixture
E05	50 ml of ethanol in 1000 ml of the mixture

E10	100 ml of ethanol in 1000 ml of the mixture
E20	200 ml of ethanol for 1000 ml of the mixture
E40	400 ml of ethanol for 1000 ml of a mixture
BSFC	Brake specific fuel consumption (kg/kWh)
BTE	Brake thermal efficiency
B.P	Brake power

References

1. Source: Statista 2022
2. Mohammed MK, Balla HH, Al-Dulaimi ZM, Kareem ZS, Al-Zuhairy MS (2021) Effect of ethanol-gasoline blends on SI engine performance and emissions
3. Chen R, Okazumi R, Nishida K, Ogata Y (2015) Effect of ethanol ratio on ignition and combustion of ethanol-gasoline blend spray in DISI engine-like condition. SAE Int J Fuels Lubr 8(2)
4. Erdiwansyah, Sani MSM, Mamat R, Khoerunnisa F, Rajkumar AR, Razak NFD, Sardjono RE (2018) Vibration analysis of the engine using biofuel blends: a review. MATEC Web Conf 225:01010
5. Verma AP, Choube A (2012) Ethanol as alternative fuel for SI engine—a review. Int J Curr Res Rev (IJCRR)
6. Costa RC, Sodré JR (2011) Compression ratio effects on an ethanol/gasoline fuelled engine performance. Appl Therm Eng 31
7. Naegeli DW, Lacey PI, Alger MJ, Endicott DL. Surface corrosion in ethanol fuel pumps. SAE paper 971648
8. Owen K, Coley T (1995) Automotive fuels reference book, 2nd edn. Society of Automotive Engineers, USA
9. Chen R-H, Chiang L-B, Chen C-N, Lin T-H (2021) Cold-start emissions of an SI engine using ethanol-gasoline blended fuels
10. Ananda Srinivasan C, Saravanan CG (2010) Study of combustion characteristics of an SI engine fuelled with ethanol and oxygenated fuel additives. J Sustain Energy Environ 1

Effect of Alcoholic Fuel Blends on Performance, Combustion, Emission and Vibration Analysis of a Variable Compression Ratio Diesel Engine

N. Ravi Kumar, G. Aswin, B. Sanyasi Naidu, and A. Harika

1 Introduction

The world is heading towards a global energy crisis mostly due to running out of conventional energy sources; decreasing the dependency on fossil fuels is recommended. Reduced emissions from gasoline and diesel engines are the primary driver driving the development of alternative fuels for IC engines. Nowadays, alcohol is used as an alternative fuel which is used for blending with diesel and petrol. There are many studies on the use of ethanol in spark ignition (SI) engines. Research on the utilization of alcohols in compression ignition (CI) engines is less compared to spark-ignition engines. The first studies on the use of ethanol in diesel engines were conducted in South Africa in the 1970s [1], and subsequent studies were conducted in Germany and the US in the 1980s [2, 3]. The most popular engine is the diesel one, which accounts for more than 80% of the world's transportation market and makes a priceless contribution to societal advancement [4]. However, because the diesel engine produces so much particulate matter (PM), hydrocarbon (HC), nitrogen oxides (NOx), sulphur oxides (SO_x), and other poisonous or dangerous compounds [5]. The goal of this research was to examine the performance of diesel engines running on ethanol–diesel blend fuels and determine the maximum and ideal amount of diesel fuel that could be replaced with ethanol [6]. Phase separation is the main obstacle to ethanol–diesel blends, especially at low temperatures. To maintain the mixture as a homogenous emulsion or solution known as diesohol/E-diesel, an emulsifier or co-solvent is necessary [7–9]. The fuels used by internal combustion engines are fossil-based. Internal combustion engine users may be able to lessen their reliance on fossil-based fuels by switching to alternative fuels. Additionally, it is becoming

N. Ravi Kumar · G. Aswin · B. Sanyasi Naidu (✉) · A. Harika
Department of Mechanical Engineering, MVGR College of Engineering, Vizianagaram, AP 535005, India
e-mail: sanyasinaidu0326@gmail.com

S. Doolla et al. (eds.), *Advances in Clean Energy and Sustainability*,
Green Energy and Technology, https://doi.org/10.1007/978-981-99-2279-6_26

Table 1 Properties of ethanol and diesel

Properties	Ethanol	Diesel
Molecular weight	46	170–198
Calorific value (kJ/kg)	26,700	42,600
Octane number	105–110	–
Cetane number	0–5	50
Density (kg/m^3)	789	846
Latent heat of vapourization (kJ/kg)	904	700
Viscosity (mPa s)	1.074	3.546
Flash point (°C)	13	70
Oxygen content (mass%)	34.78	0

more and more crucial to develop efficient and environmentally friendly combustion systems and alternative fuels. Alcohol has been proposed as a substitute for diesel fuel [10]. Researchers have looked into the ethanol–diesel blend fuel utilized in CI engines in recent years as a result of technological advancements [11]. The altering of fuel the spray characteristics, combustion, performance, and emissions of the engine are altered by its physicochemical qualities. The use of E-Diesel in CI engines has been investigated using a number of different strategies. They are dual injection, alcohol fumigation, alcohol-diesel fuel mixes, and alcohol-diesel fuel emulsions [12].

2 Properties of Ethanol and Diesel

See Table 1.

3 Experimental Setup

An eddy current dynamometer is connected to a single-cylinder, four-stroke, water-cooled, direct injection diesel engine with a variable compression ratio (VCR) engine. A strain gauge is attached to the eddy current dynamometer's output shaft to measure the load. A 10 cc burette and stopwatch are used to measure the fuel flow rate. The air box is utilized for airflow measurement and modification. For measuring combustion pressure, an engine head-mounted piezoelectric pressure sensor with a sensitivity of 1 mV/PSI and a range of 5000 PSI is installed. For measuring pressure, an average of 10 consecutive combustion cycles are employed. RTD, PT100, and K-type thermocouples are used to measure temperature. PCB single-axis accelerometer sensor is mounted on the cylinder head in the vertical direction for measuring engine vibrations. This non-intrusive type of accelerometer operates according to the piezoelectric

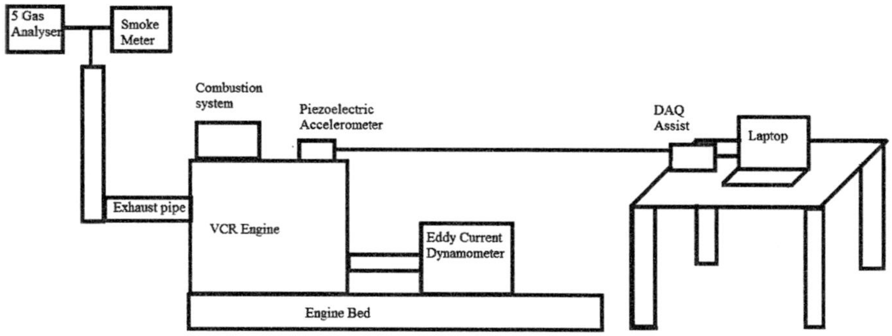

Fig. 1 Schematic of engine setup

effect's basic tenets. The vibrations are measured in the time domain by the PCB single-axis accelerometer sensor. Data from the time-domain is transformed into the utilizing a quick Fourier transform in the frequency domain (FFT). For combustion diagnosis, the vertical axis vibration sensor is more sensitive. With the aid of Loctite adhesive, the accelerometer is fixed to the engine head (Fig. 1).

The NI9234 (A/D converter) data acquisition card, which offers a connection to an input channel with an input range of 5 V, is attached to this accelerometer. The accelerometer's input channel is connected to the A/D conversion's AI0. The gathered information is then sent to the laptop for recording. The engine vibrations are measured using a programme created in the NI LabVIEW software. The graphs of acceleration versus time and versus frequency are created using this data. For the evaluation of exhaust gases CO, HC, carbon dioxide (CO_2), NO_x, and oxygen (O_2), an AVL Di gas 444, 5-gas analyzer is employed. In Hartridge Smoke Units, smoke is measured using an AVL 437C smoke meter (HSU). The constant speed used for every trials was 1500 rpm. The compression ratios used in this experiment are 15, 16, 17, 17.5.

4 Specifications of Engine Setup and Methodology

Digital Bomb Calorimeter, When a substance completely burns, its calorific value measures how much heat it produces per unit of its volume. Fuel efficiency improves with increasing calorific value. A calorimeter was used to determine the Calorific Values of Pure Diesel and Alcoholic blends (E05, E10, E15, E20). The piezoelectric accelerometer operates on the idea that when a mass exerts force on a piezoelectric material (such as a crystal or ceramic), the resulting stress causes the material to produce an electrical charge that is proportionate to the applied force. After that, it was sent to DAQ help, which converted the form and provided the vibration's amplitude in LAB VIEW. Measurements of carbon monoxide, hydrocarbons, carbon dioxide, nitrogen oxide, and oxygen are made using the 5 gas analyzer (AVL Di-gas

Table 2 Specifications of engine setup

Engine type	Single cylinder, four-stroke, VCR, water-cooled, direct injection, diesel engine
Stroke	110 mm
Bore	87.5 mm
Speed	1500 rpm
Range of compression ratio	12–18
Injection timing	23° before top dead center
Injection pressure	200 bar
Power	3.5 kW

444). When infrared light of all wavelengths is shone on a thin layer of a solid, liquid, or thicker layer of gaseous substance, the wavelengths that correspond to the molecules' vibrational frequencies are absorbed, while all other wavelengths are transmitted. The premise of a smoke meter is that opacity meters are used to detect and measure how much light is obstructed in smoke produced by vehicle engines. The density of the smoke, a measure of combustion effectiveness, is displayed on the smoke meter readout. For reading, the Beer-Lambert principle is applied. It displays how much smoke is in the exhaust. A Suitable and required percentage of ethanol is blended with Diesel with a stirrer for homogeneous mixing. The Calorific value is measured for each ethanol–Diesel blend using a Digital Bomb Calorimeter and the obtained calorific values are listed down in Table 2. Now, using a hydraulic dynamometer, run the engine at various loads and CRs (15, 16, 17, 17.5) while utilizing each mix as fuel. Volumetric analysis is used to measure the engine's fuel consumption and calculate the performance curves.

The vibration characteristics are captured using a piezoelectric accelerometer placed over the engine head. The vibrations obtained are recorded at different blends, loads, and speeds using LabVIEW software. A smoke meter is used to measure the amount of smoke present in the exhaust gases. The exhaust gases are measured using a 5-gas analyzer (HC, CO, CO_2, NO, O_2).

5 Results

See Table 3.

6 Performance Curves

From Fig. 2, which depicts how the load and brake thermal efficiency change at CR17.5, it can be seen that the load causes an increase in brake thermal efficiency.

Table 3 Properties of ethanol–diesel blends

Blended fuel	Calorific value (kJ/kg)	Density (kg/m^3)
E0	45,728.11	835
E5	44,162.38	900
E10	43,837.38	920
E15	41,768.20	942
E20	40,847.28	974

This is due to the fact that more suction pressure will develop as the load increases, which could have led to efficient combustion. The ethanol blends show increased thermal efficiency compared with diesel. Despite having a lower calorific value than diesel due to the presence of a larger oxygen concentration, ethanol blends burn more effectively (i.e., completely convert their energy) than diesel. When compared to E0, the BTHE with load increases for E5, E10, E15, and E20 by 1.05%, 3.75%, 6.39%, and 13.33%, respectively. Figure 3 illustrates the relationship between brake thermal efficiency and compression ratio. It can be seen that as compression ratio rises, brake thermal efficiency also rises. As the higher compression ratio results in higher brake power, and so the brake thermal efficiency rises. Ethanol blends shows increased brake thermal efficiency at the corresponding compression ratio and the E20 blend shows the higher brake thermal efficiency at every compression ratio and at CR17.5 it 29.1%. At CR15 E20, E15, E10, E5 and E0 has brake thermal efficiencies 27.8%, 27.63%, 27.6%, 27.4% and 27.2% respectively.

From Fig. 4, brake specific energy consumption for Diesel and Ethanol Blends at CR17.5 under various loads; it is noted that the brake specific energy consumption reduces as the loads rise. The quantity of heat input necessary to produce one unit of power is referred to as brake specific energy consumption. Difference in the Ethanol blends is because of the low calorific value of ethanol blends as the blending increases. At full load condition the decrease in brake specific energy consumption for E5, E10, E15 and E20 are 3.42%, 4.13%, 5.61% and 7.69% compared to E0.

The fluctuation of brake specific energy consumption and the compression ratio can be seen in Fig. 5. As the compression ratio increases, the brake specific energy

Fig. 2 Brake thermal efficiency varies with load

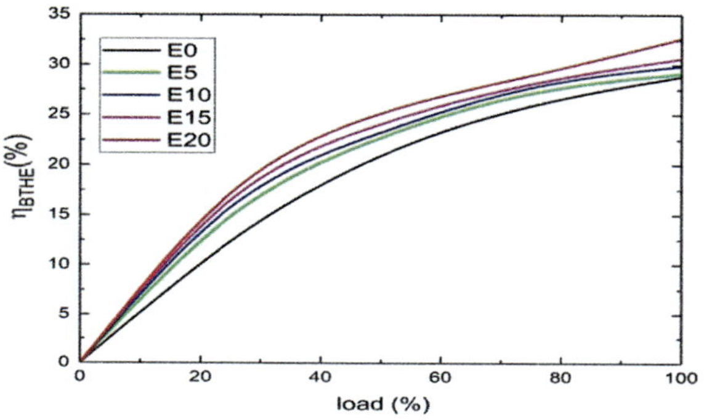

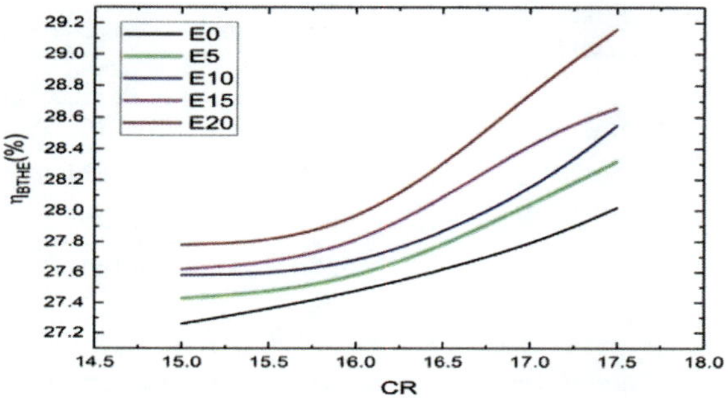

Fig. 3 Brake thermal efficiency varies with CR

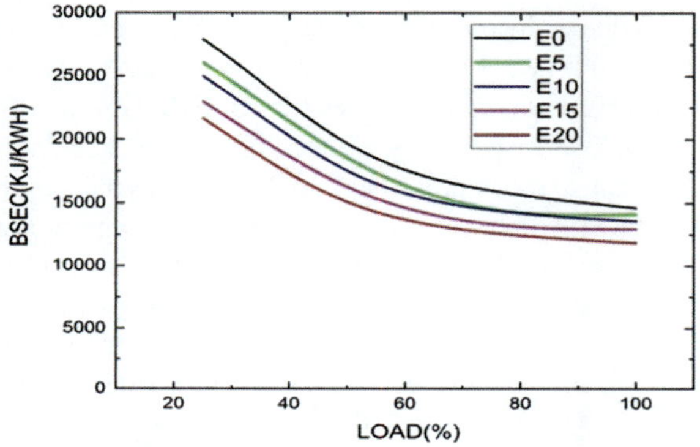

Fig. 4 Brake specific energy consumption varies with load

consumption is slightly increasing and almost similar in some cases. It is evident that the brake specific energy consumption reduces as blending ratio of ethanol gets increases, because of the lower calorific value of blends compared with pure diesel. At CR15, brake specific energy consumption (kJ/kWh) for E20, E15, E10, E5 and E0 are 12,254.18, 12,948.14, 13,589.58, 13,690.33 and 14175.71.

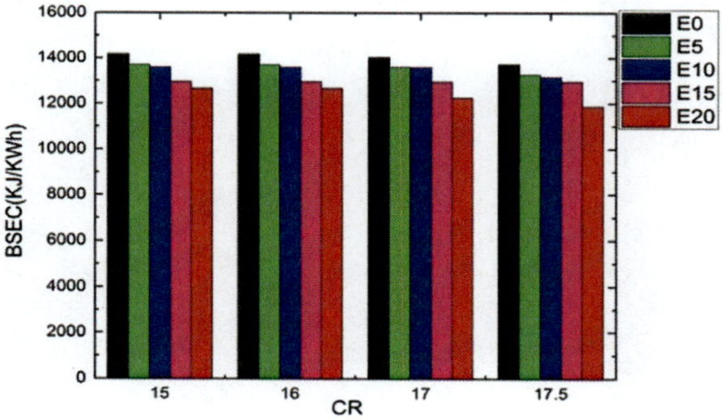

Fig. 5 Brake specific energy consumption varies with CR

7 Combustion Characteristics

From Fig. 6, it shows the variation of maximum rate of pressure rise and different fuels at CR17.5 full load condition. It is evident that the rate of pressure rise is decreasing, when the ethanol content is getting increased and as the blend ratio is increasing the rate of pressure rise increases in small trends when compared to starting blends. An engine research has to analyse rate of pressure rise because it can tell how smoothly the combustion process moves along in the cylinder. In the graph, it is evident that the rate of pressure rise (bar) for E0, E5, E10, E15 and E20 are 5.4, 5.2, 5.0, 4.9 and 4.9 respectively. From Fig. 7 it shows the variation of maximum rate of pressure rise and compression ratio for all blends at full load condition. From the graph, it resembles that the rate of pressure rise is increasing as compression ratio increases, and for the blends it shows that the rate of pressure rise is decreasing at any particular compression ratio. The maximum rate of pressure rise (bar) for the blends E0, E5, E10, E15 and E20 are 5.2, 5.1, 5, 5, 5 at CR15, 5.4, 5.3, 5.2, 5.2, 5.1 at CR16, 5.5, 5.4, 5.3, 5.2, 5.2 at CR17 respectively. For improved engine lifespan and noise reduction, the maximum rate of pressure rise must be reduced.

Figure 8 shows net heat release rate and the blends for pure Diesel and Ethanol blends in full load condition at CR17.5. Net heat release rate increases in the blends with compared to diesel. This is because as the blending is increased the chemical delay is reduced and the combustion phase occurs more effectively. Hence the heat released per degree of crank angle increases as blending increases. The NHRR for E5, E10, E15 and E20 are 1.02, 14.55, 23.9, 25.98% higher compared to Pure Diesel. The net heat value is higher for E20 blend and lower for E0 blend. From Fig. 9, it displays how the compression ratio affects the net heat release rate. Net heat release rate is getting decreased as the compression ratio is getting increased. Net heat release rate at the particular compression ratio is getting increase as the blending of the ethanol content is increasing. The net heat release rate depends on the chemical delay and combustion process. Here the higher chemical delay takes place for higher blends and so net heat release rate is getting higher as the combustion phase occurs more

Fig. 6 Rate of pressure rise varies with blends

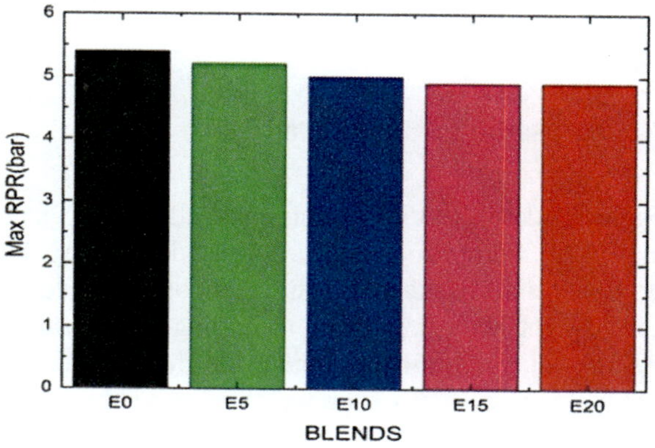

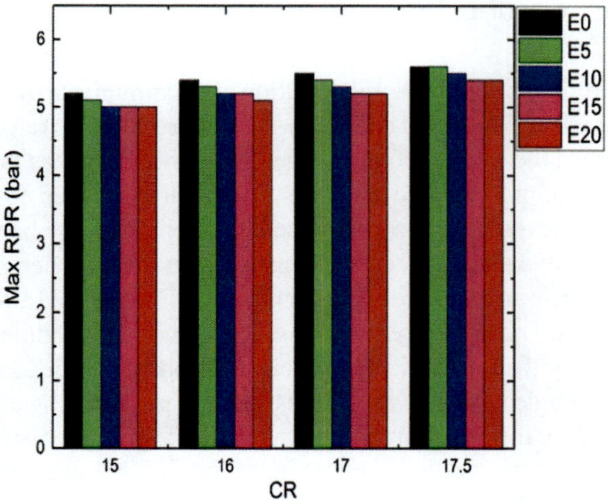

Fig. 7 Rate of pressure rise varies with CR

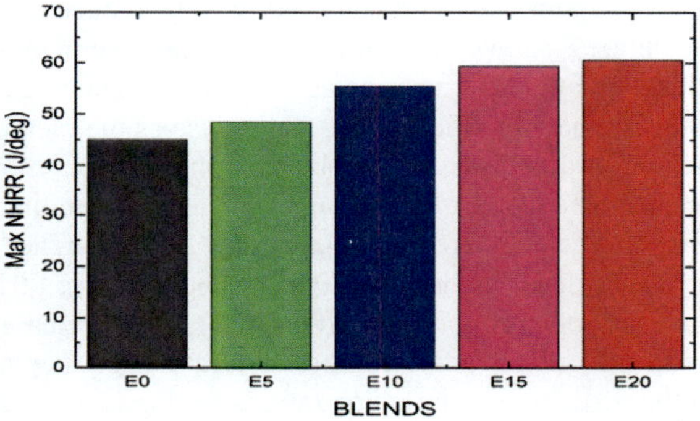

Fig. 8 Net heat release rate varies with blends

effectively. As shown in the above chart Net heat release rate (J/deg) for E0, E5, E10, E15, E20 are 58, 59.4, 61.3, 65.4, 65.4 at CR15, 51.6, 54.6, 58.6, 58.8, 63.3 at CR16, 49.1, 49.6, 55.1, 60.1, 61.7 at CR17 respectively.

8 Emission Characteristics

Figure 10 describes the CO emissions and the load for the various blends (E0, E5, E10, E15, E20). Carbon monoxide is formed during the combustion process with rich-fuel mixtures and when there is insufficient oxygen to fully burn all the carbon in the fuel to CO_2. The above trend with increasing in the load of CO emissions may be due to rising temperatures in the combustion chamber, physical and chemical properties of the fuel, air–fuel ratio, shortage of oxygen at high speed, and lesser amount of time available for complete combustion. The effects of fuel viscosity on fuel spray quality

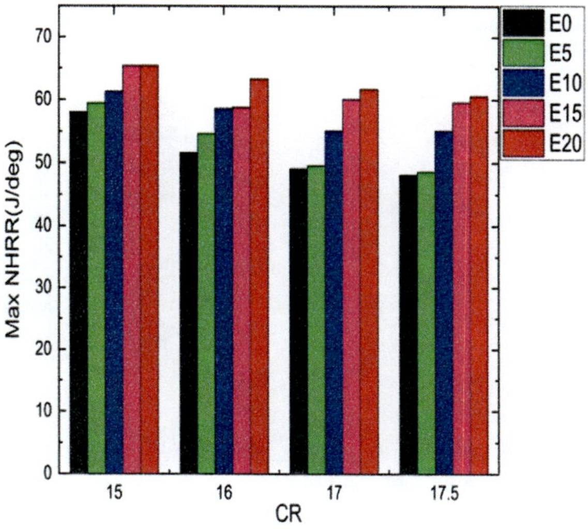

Fig. 9 Net heat release rate varies with CR

would be expected to make CO decrease with increase in higher oxygen content. As the load increases the CO emissions is decreased for the blends compared to Diesel as increase in oxygen content. From Fig. 11, it shows the variation of CO emissions and the compression ratio at the full load condition, as the compression ratio is increasing the CO emissions is getting decreased and at the higher compression ratio the CO emissions are almost equal to that of the blends in the full load condition. The CO (% Volume) emissions for E0, E5, E10, E15 and E20 are 0.03, 0.02, 0.02, 0.02 and 0.02 at CR15, 0.02, 0.01, 0.01, 0.01 and 0.01 at CR16, At CR17 and CR17.5 the CO emissions are almost similar as shown in the chart.

From Fig. 12, it shows variation of CO_2 and load at CR17.5 (i.e. higher compression ratio). The CO_2 decreases as the blending of the ethanol get increases as shown

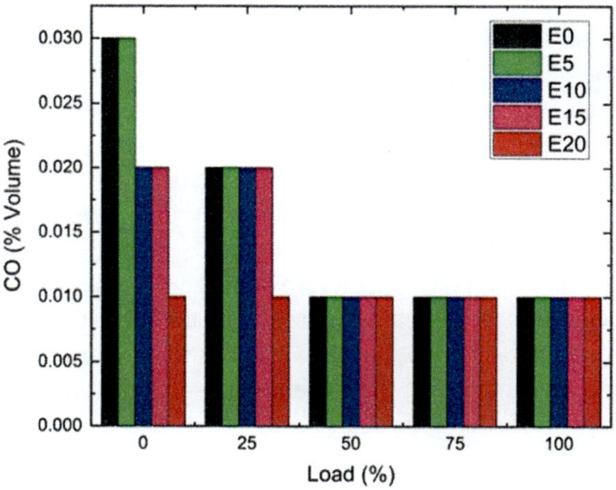

Fig. 10 Variation of CO with load

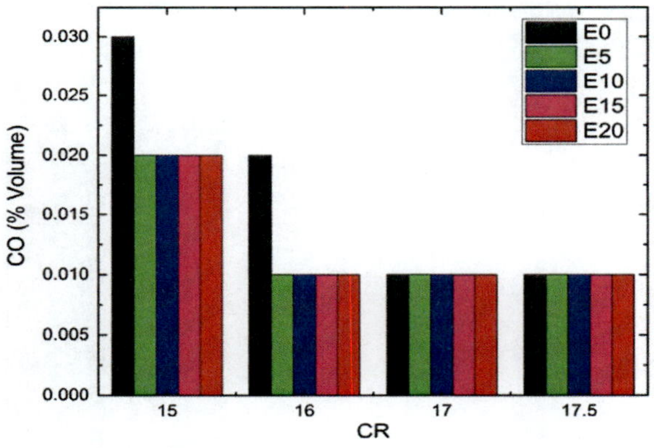

Fig. 11 Variation of CO with CR

in above graph. Carbon Dioxide is formed during the combustion process with rich-fuel mixtures regions and when there is insufficient oxygen to fully burn all the carbon in the fuel to CO_2. The above trend with increasing in the load of CO_2 emissions may be due to rising temperatures in the combustion chamber, physical and chemical properties of the fuel, air–fuel ratio, shortage of oxygen at high speed, and lesser amount of time available for complete combustion. At full load condition CO_2 emissions for E0, E5, E10, E15 and E20 0.5, 0.5, 0.4, 0.4 and 0.4 respectively. From Fig. 13, it shows the variation of CO_2 and the compression ratio as the compression ratio is getting increase CO_2 emissions are getting decreased and the CO_2 emissions at any particular compression ratio for various blends getting decreases compared to E0 as there is higher oxygen content in the ethanol and so results in rich air–fuel mixture and efficient amount of oxygen required for the burning of fuel and so the CO_2 emissions gets decreased. At CR15 the carbon dioxide emissions are 0.53, 0.51, 0.5, 0.5 and 0.5 for E0, E5, E10, E15 and E20 respectively and the values are almost similar for higher blends.

From Fig. 14, it shows the variation of Hydrocarbons emissions and the load at CR17.5, it is observed that HC emissions are increasing as the ethanol content is getting increased due to higher temperatures that is attained and the incomplete combustion of the hydrocarbons after the After burning stage in the combustion chamber. The one of the major disadvantage of using ethanol as an alternative fuel is HC emissions are drastically increasing as the ethanol content gets increases. In full load condition the HC (ppm) emissions for E0, E5, E10, E15 and E20 are 1, 5, 5, 7 and 9 respectively. Figure 15 shows the comparison of unburnt HC with compression ratio for blends and Diesel at full load condition. The amount of HC emissions mainly depends upon the temperature of the gas. As the temperature of combustion chamber walls and gas increases, it accelerate the formation of mixture gas and promotes the combustion of fuel, but after combustion unburnt hydrocarbons would be represent and thus increasing the HC emissions. All the blends shows HC emissions are more than the diesel. This apart from above reasons other reason is due to oxygen content in the Blends which increase with increase in blending. At CR17 the HC emissions are 5, 6, 6, 7 and 8 for E0, E5, E10, E15 and E20 respectively.

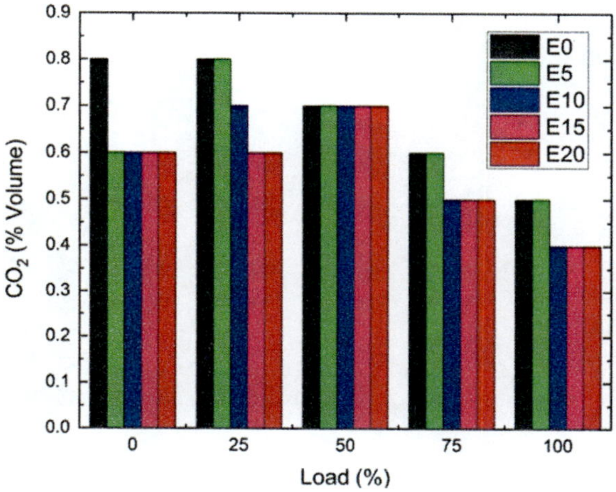

Fig. 12 Variation of CO_2 with load

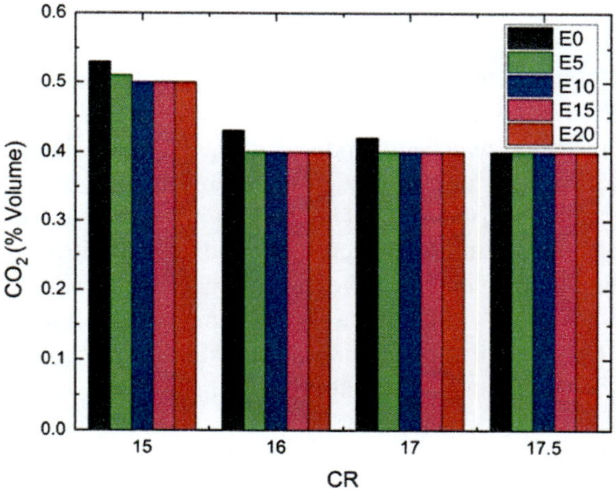

Fig. 13 Variation of CO_2 with CR

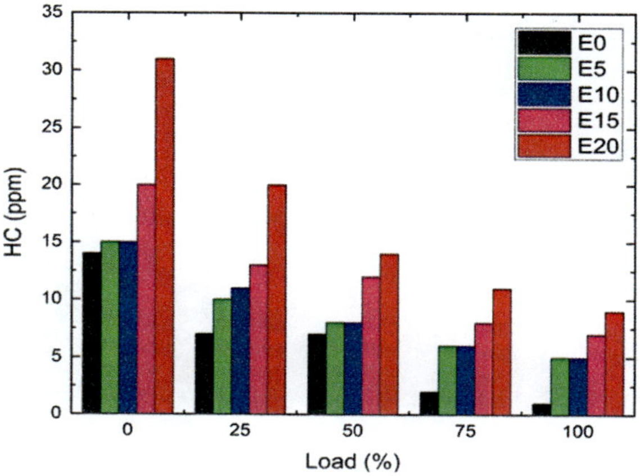

Fig. 14 Variation of HC with load

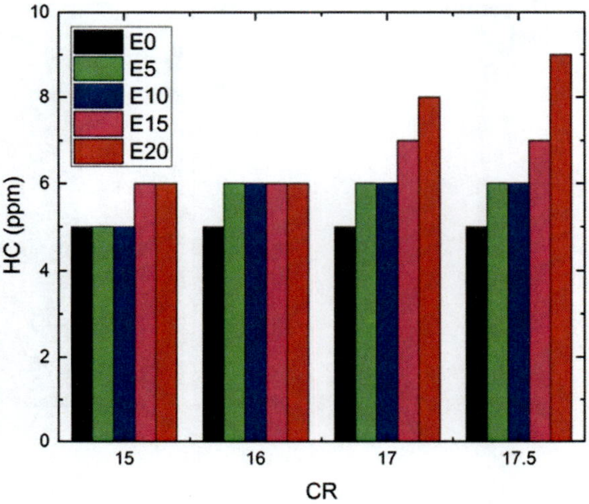

Fig. 15 Variation of HC with CR

Figure 16 describes the emissions of oxides of nitrogen and the load for E0, E5, E10, E15 and E20. Oxides of nitrogen emissions are very dangerous pollutant emissions, which are produced when the fuel is unburnt at high temperature causing dissociation of nitrogen. From the results of test oxides of nitrogen is very low at lower loads, this amount increases with increase in load from E0 to E20. The ethanol blending with diesel decreases the oxides of nitrogen value as shown in the graph. Engine temperature increases at higher loads, which is responsible for raising the level of NO_x Exhaust gases. Blending of ethanol to pure diesel leads to have higher heat of vaporization value and so oxides of nitrogen get decreased for the blends as shown in the chart. From Fig. 17, it shows the variation of oxides of nitrogen with the compression ratio for the various blends in the full load condition. At CR15 NO_x emissions for E0, E5, E10, E15 and E20 are 41, 35, 31, 31 and 30 respectively.

The NO_x emissions at CR17.5 for E0, E5, E10, E15 and E20 are 31.7%, 42.85%, 51.61%, 45.61%, 43.33% higher compared to CR15. As the increase in compression ratio decreases the clearance volume and so the volume left for combustion is small

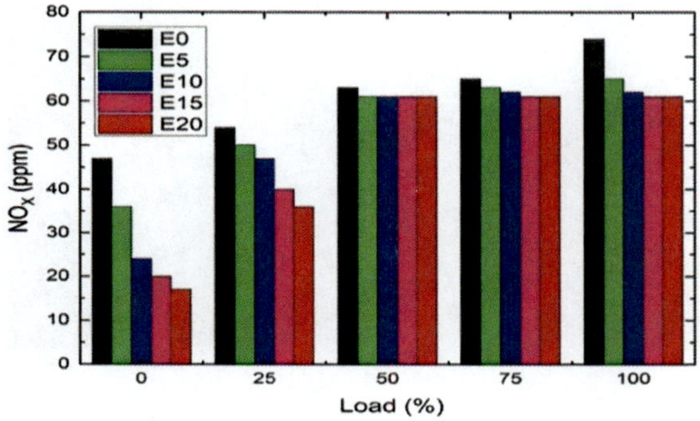

Fig. 16 Variation of NO_x with load

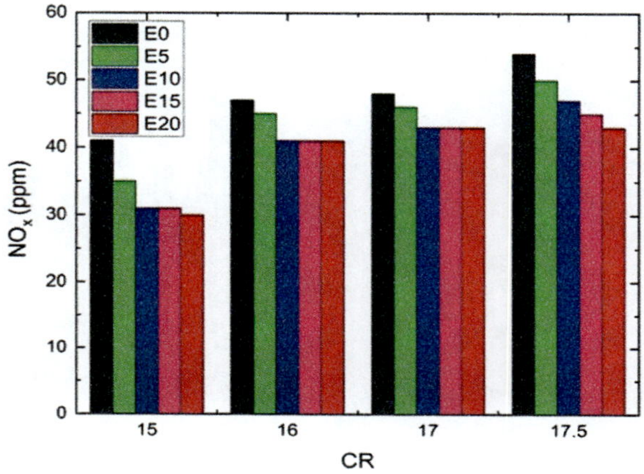

Fig. 17 Variation of NO_x with CR

and so the oxides of nitrogen emissions gets increases as increase in temperature but for the blends it is having higher heat of vaporization value and so the oxides of nitrogen emissions decreases for blends. The NO_x emissions might also depend on viscosity, calorific value of blends.

9 Vibration Analysis

From Fig. 18 it shows the superimposed time waves, variation of acceleration and the time at the compression ratio 17.5 and full load condition for E0, E20, it is observed that the max acceleration peaks obtained are less for E20 blend compared to E0 blend, this is because as it is mentioned that rate of pressure rise is decreasing as blending of ethanol increases in fuel. Also, the main reason is that the ethanol blended fuels are having the lower cetane number compared to pure diesel, and thus results in higher ignition delay for the blended fuels, leads to decrease in the vibrations. The peaks in the time domain signals represents the peak pressures in the cylinder during combustion process. Figure 19 shows the superimposed curves of frequency spectrums for E0, E20 at CR17.5 and full load condition. The peak amplitudes of the Diesel fuel are 2.62 and 2.52 m/s^2 at the frequencies of 63.32 and 163.49 Hz. The peak amplitudes of the E20 fuel are 2.37 and 2.13 m/s^2 at the frequencies of 63.67 and 164 Hz. The decrease in the peaks of the E20 compared to E0, is due to the reason that ethanol having lower cetane number and so the blended fuels have lower cetane number compared to Diesel and so their would increase in the higher ignition delay and so the peaks obtained are lesser compared to E0 fuel, but in the graph in some frequencies the peaks are more than Pure Diesel it may be due to that the vibrations occurs mainly due to the rotational parts and reciprocating parts, as the rotational speed of the crank gets increased for blended fuel, so in some frequencies the peaks are somewhat higher for E20 compared to E0.

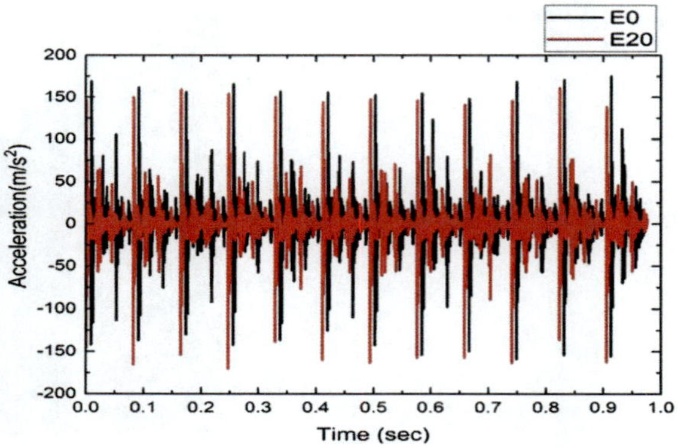

Fig. 18 Variation of acceleration with time

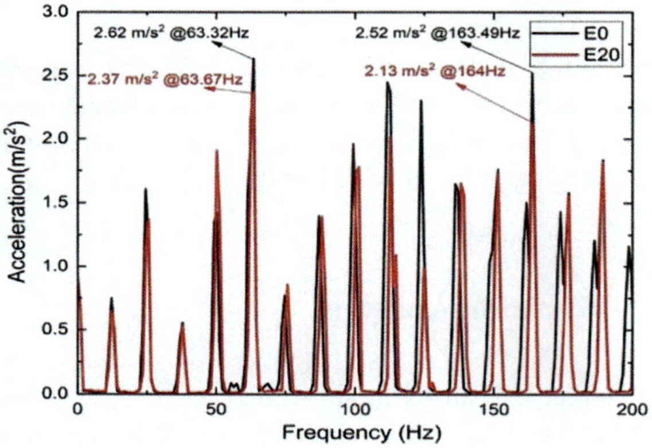

Fig. 19 Variation of acceleration with frequency

Figure 20 shows the Maximum acceleration values and the load variation for all blends at higher compression ratio. As the load is increases the peaks of the acceleration values gets increased as due increase in load leads to increase in fuel consumption so as to maintain the constant speed and so the vibrations gets increases. At full load condition the max acceleration values are lower by 5.37, 27.7, 29.03, 30.10% for E5, E10, E15 and E20 compared to E0. Figure 21 shows the max acceleration and the compression ratio at the full load condition for all the blended fuels. The vibrations in the diesel engine occurs due to the unidirectional combustion forces caused by the changes in the gas pressures inside the cylinder, structural resonances and alternating inertia forces concentrated on engine parts. From graph as the noise and the vibration of engine reduced with increment of the compression ratio. Ethanol blended fuels shows a decrease in the vibration as the compression ratio increases. The decrease in the blends for E5, E10, E15 and E20 are 1.89, 9.76, 12.44 and 14.74% compared to E0 at compression ratio 16. At compression ratio 17 the decrease in blends for E5, E10, E15 and E20 are 2.43, 5.37, 7.77 and 9.94% compared to E0. Comparison between the CR15 and CR17.5 for E0, E5, E10, E15 and E20 are 34.5, 34.8, 48.3, 47.2 and 45.83%.

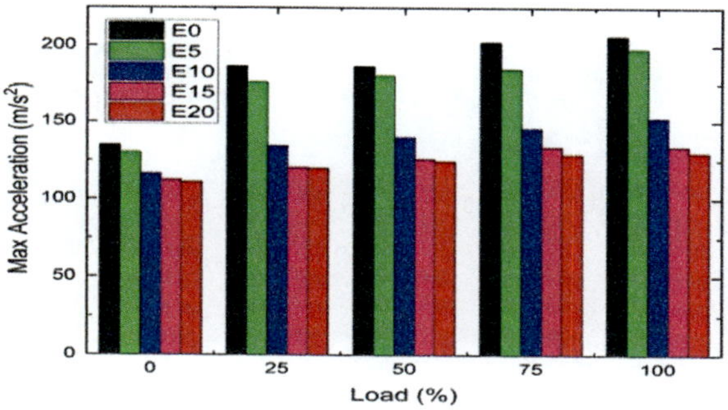

Fig. 20 Variation of acceleration with load

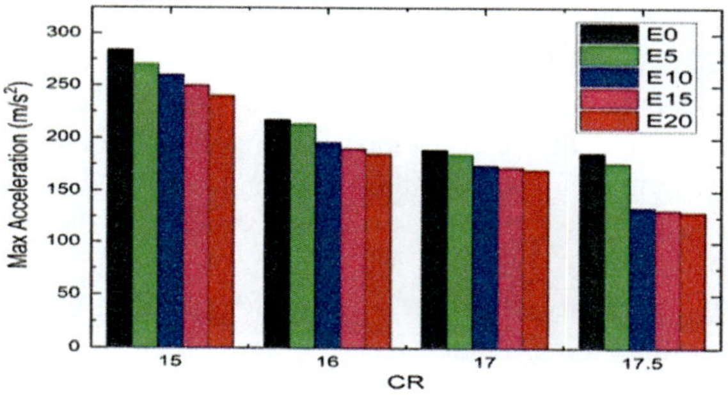

Fig. 21 Variation of acceleration with CR

10 Prediction Model

Predictive modeling is a mathematical process used to predict future events or outcomes by analyzing patterns in a given set of input data. It is a crucial component of predictive analytics, a type of data analytics which uses current and historical data to forecast activity, behavior and trends.

For this prediction the non-linear regression analysis has been done in origin pro software and here the R^2 value is 1, it represents the values are almost equal to given data and the equation obtained the error is been calculated with the equation and the known parameters.

X_1 = Compression Ratio,
X_2 = Blend (% of ethanol content),
X_3 = Load (kg).

$$\text{Brake thermal efficiency} = 19.52546 * (X_3)^{-0.0218} + 10.05458 * (X_2)^{0.42484} - 13.34415 * (X_1)^{0.11487}$$

$$\text{Brake Specific fuel consumption} = 0.9489 * (X_3)^{-0.47888} + 0.29337 * (X_2)^{0.02744} - 0.42204 * (X_1)^{-0.10263}$$

$$\text{Maximum Net Heat Release Rate} = 21.42148 * (X_3)^{0.38965} + 41.2126 * (X_2)^{-0.00677} - 11.15867 * (X_1)^{0.45957}$$

$$NO_X \text{ Emission} = 44.21589 * (X_3)^{0.05274} + 35.21269 * (X_2)^{0.05645} - 0.04906 * (X_1)^{2.42738}$$

11 Conclusion

This study is done to find out the blend at which the performance parameters, combustion, emissions, and vibrations are optimum. The thermal efficiency increases with the load for all blends would get increases and for the E20 blend, it is higher. Increase in thermal efficiency of the E20 blend is 13.33% higher compared to pure Diesel at CR17.5. The rate of pressure rise is lower for E15 compared to pure diesel at full load conditions. NHRR is higher for blended fuel compared to pure diesel. CO and smoke density emissions are lower for all blends compared to pure diesel. Vibrations are lower for E5 and E10 are lower and emissions are also lower. So, the blends that are recommended for the efficient working of the engine by considering the performance, combustion, emissions, and vibrations of the engine are E5 and E10 blends.

Nomenclature

E0	100% diesel
E5	Diesel 95% ethanol 5%
E10	Diesel 90% ethanol 10%
E15	Diesel 85% ethanol 15%
E20	Diesel 80% ethanol 20%
BSEC	Brake specific energy consumption
BSFC	Brake specific fuel consumption
CR	Compression ratio
BTE	Brake thermal efficiency
CI	(Compression ignition) engine

References

1. CSIR (Council for Scientific and Industry Research) (1980) Reports nos. ME1584 and ME1651. National Mechanical Engineering Research Institute, Pretoria (Republic of South Africa)
2. Wrage KE, Goering CE (1980) Additives with diesel-ethanol blends for use of diesel engines. Trans ASAE 1338

3. Weidmann K, Menrad H (1985) Society of automotive engineers. SAE Tech Paper 5(841441):800
4. Cai T, Zhao D, Li X, Shi B, Li J (2021) Mitigating NO_x emissions from an ammonia-fueled micro-power system with a perforated plate implemented. J Hazard Mater 401:123848
5. Jiaqiang E, Pham M, Zhao D, Deng Y, Le D, Zuo W, Zhu H, Liu T, Peng Q, Zhang Z (2017) Effect of different technologies on combustion and emissions of the diesel engine fueled with biodiesel: a review. Renew Sustain Energy Rev 80:620–647
6. Schuetzle D, Han W, Srithammavong P, Akarapanjavit N, Norbeck JM, Cornwell K (2002) The evaluation of diesel/ethanol for diesel, pp 12–15
7. Ahmed I. Oxygenated diesel: emissions and performance characteristics of ethanol-diesel blends in CI engines. SAE Paper No. 2001-01-2475
8. Jackson M, Cornwell CK, Degroote CC. Study of diesel and ethanol blends stability. SAE Paper No. 2003-01-3191
9. Environmental Protection Agency (2008) Diesel-barriers and benefits. EPA STRIVE Programme 2007–2013, STRIVE Report
10. Singh AP, Agarwal AK (2011) Combustion characteristics of diesel HCCI engine: an experimental investigation using external mixture formation technique. Appl Energy 88:1169–1180
11. Mayura RK, Agarwal AK (2011) Experimental study of combustion and emission characteristics ethanol fuelled port-injected homogeneous charge compression ignition (HCCI) combustion engine. Appl Energy 88:1169–1180
12. Can Ö, Celikten I, Usta N (2004) Effects of ethanol addition on performance and emissions of a turbocharged indirect injection diesel engine running at different injection pressures. Energy Convers Manag 45(15):2429–2440
13. Budharaju MV, Naradasu RK, Guvvala P (2019) Effect of oxygenated fuels on performance, combustion, emission, and vibration characteristics of a compression ignition engine. Biofuels 10:453–461. ISSN: 1759-7269

Influence of Negative Overlap Ratio on the Performance of Semicircular Savonius Rotor with Straight Edge Extension on Overlap Region

Jaykumar S. Patel, Vimal K. Patel, and Vikram P. Rathod

1 Introduction

Hydrokinetic turbine utilized the kinetic energy of the flow to generate electricity. It does not require heavy construction of dam and transmission of water from reservoir to turbine. There are mainly two type of hydrokinetic turbine one is Savonius and other is Darrius. Savonius is drag force driven turbine and Darrius is lift force driven turbine. The main advantages of the Savonius turbine over Darrius turbine is its good starting characteristic. The coefficient of power of the Savonius turbine is lower compare to Darrius turbine. For the same flow velocity rotational speed of the Darrius turbine is higher compare to Savonius turbine.

In three open channels, a small laboratory channel, a large laboratory channel, and a real-world irrigation channel, Patel et al. [1] investigated the impact of overlap ratio and aspect ratio on the performance of the Savonius hydrokinetic turbine. Additionally, they had looked at how the Savonius turbine's end plate affected its performance. The effectiveness of the Savonius rotor was studied by Kamoji et al. [2] in relation to geometrical parameters such the overlap ratio, blade shape factor, aspect ratio, blade arc angle, and Reynolds number. Additionally, they had looked at the effect of the shaft between the rotor's two end plates. The influence of an auxiliary blade in the overlap region of a Savonius turbine with a semicircular blade was numerically examined by Abdelaziz et al. [3]. They had used three configurations out of that two configurations with different shape auxiliary blade one is arc and other one is straight plate at 90° and 3rd configuration they changed the profile position of the conventional Savonius turbine. Salleh et al. [4] experimentally investigated the influence of deflector plate on advancing and returning side of the rotor for different angular positions on the performance of the Savonius turbine by using wind tunnel at wind velocity 7 m/s (0.4 m/w equivalent velocity of water) for advancing plate

J. S. Patel (✉) · V. K. Patel · V. P. Rathod
Sardar Vallabhbhai National Institute of Technology, Surat, India
e-mail: jayspatel51@gmail.com

S. Doolla et al. (eds.), *Advances in Clean Energy and Sustainability*,
Green Energy and Technology, https://doi.org/10.1007/978-981-99-2279-6_27

angle ranging from 15° to 75° and returning blade angle ranging from 30° to 105°. Thiyagaraj et al. [5] had studied the effect of number of blade ranging from 2 to 6 on the performance of the Savonius turbine. They also studied the influence of different position of the flip on the performance of the Savonius turbine. Patel et al. [6] optimized the angle between deflector plate for convergent and divergent orientation to enhance the performance of the Savonius hydrokinetic turbine with duel rotor configurations.

2 Data Reduction

2.1 *Input Power*

Power is available in the flowing water at a section near to the turbine or power imparted by flowing water on the blade of the turbine.

$$P_{in} = \frac{1}{2}\rho A V^3 \tag{1}$$

2.2 *Coefficient of Power*

It is the ratio of output power to input power.

$$C_p = \frac{P_{out}}{P_{in}} \tag{2}$$

$$C_p = \lambda C_t \tag{3}$$

2.3 *Coefficient of Torque*

It is the ration of output torque to input torque.

$$C_t = \frac{T_{out}}{T_{in}} \tag{4}$$

2.4 Tip Speed Ratio

It is the ratio of blade tip speed to free stream velocity of water

$$\lambda = \frac{u}{V} \quad (5)$$

3 Conceptual Discussion

Savonius turbine is made up of two semicircular blades arranged in such a way that it form an S shape profile as shown in Fig. 1 [4]. Each blade have two side, one is advancing side and other is retarding side. Shape of the advancing side is concave and retarding side is convex as shown in figure. In Savonius turbine can be made more than two blades. Based on the differential in drag force between its advancing blade and returning blade, the Savonius turbine operates [4]. When the Savonius turbine is placed in the flow of water, water will impart the force on both blade at a same time means on advancing side of the one blade and retarding side of the other blade. The advancing blade of concave surface will experience more drag force compare to returning blade of convex shape [4]. The difference in this drag force will produce a positive net torque on the turbine, which is what causes the turbine to rotate. While negative torque produced by the returning blade's drag force prevents the turbine from rotating [4].

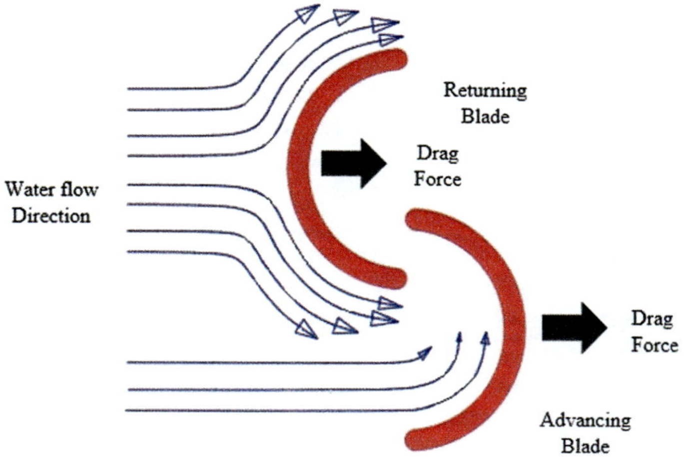

Fig. 1 Savonius turbine

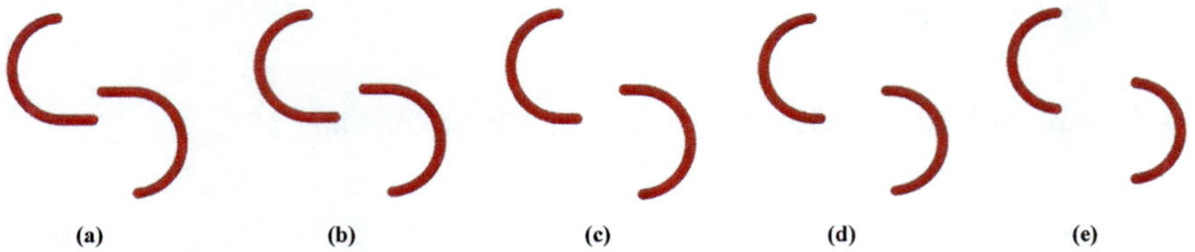

Fig. 2 **a** 0 OR rotor. **b** − 0.1 OR rotor. **c** − 0.2 OR rotor. **d** − 0.3 OR rotor. **e** − 0.4 OR rotor

4 Parameter Investigated

The present work aim is to numerically study the effect of negative overlap ratio on the performance of the Savonius rotor with semicircular blade and straight edge while the diameter of the rotor remains constant so the kinetic energy imparted by water on the rotor remain same in all configurations as shown in Fig. 2. Study is carried out for the gap between two blades of 0 mm (OR 0), 10 mm (OR − 0.1), 20 mm (OR − 0.2), 30 mm (OR − 0.3), and 40 (OR − 0.4) mm as shown in Fig. 2.

5 Numerical Study

5.1 Grid Independents Study

Computational fluid dynamics is a numerical method in which problem can be solved by solving governing differential equations. In this approach the process of grid generation plays most critical role. The purpose of the grid generation is to divide the domain in to number of small elements and for each cell value of different parameters can be found by solving governing equations with finite volume approach. The output of the numerical method is depending on the size of the element and quality of the mesh (Figs. 3 and 4).

During the study, it is observed that the value of coefficient of torque become stagnant after 79,000 elements, means value of the coefficient of torque become independents of the no of elements and size of elements. So more than 79,000 number of elements is selected for present study.

In this study simulations were carried out with orthogonal quality 0.9701 and skewness 0.06435 which is in the excellent range of skewness mesh metrics spectrum. Maximum value of skewness is 0.6839 which is lower than the default value for the skewness threshold of 0.98 (Fig. 5).

So, grid independents test is carried out to choose the optimum density of the mesh in the domain so that the output of the simulation become independents of elements size and also simulation time can be minimized. With increase in the density of the mesh beyond the optimum density then also output of the method remain stagnant.

Fig. 3 Grid shape near rotor wall

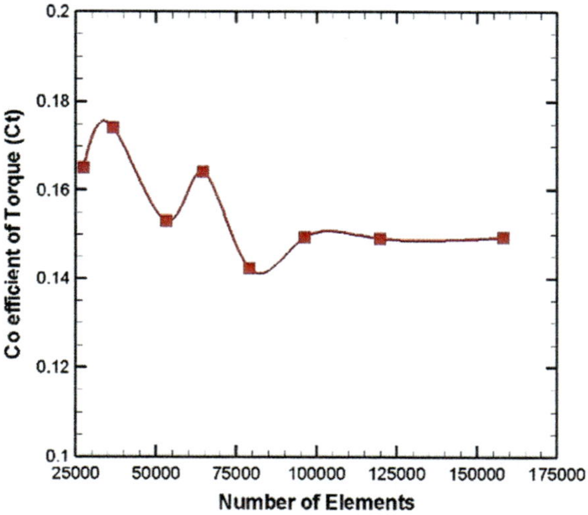

Fig. 4 Variation in torque coefficient in respect to the number of elements

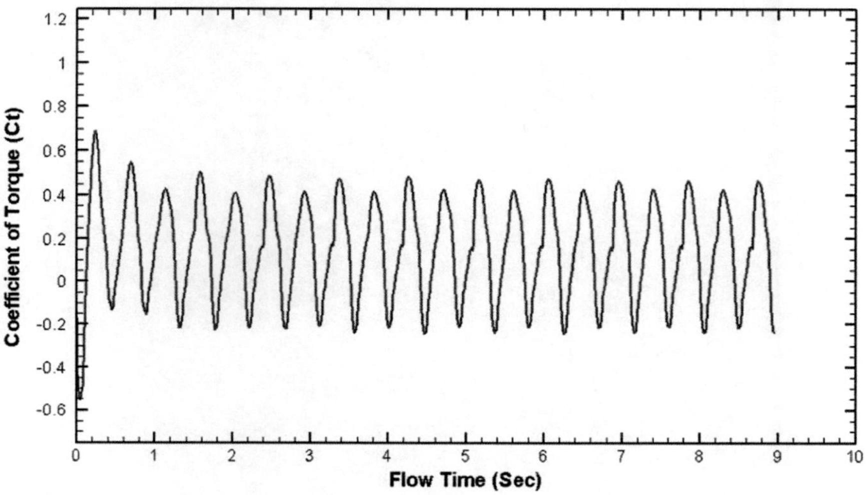

Fig. 5 Coefficient of torque with flow time

During the study it was found that the value of average coefficient of torque for one revolution isn't stagnant but varying with the number of revolutions as shown in figure and become stagnant after 5 s. So, 5 s was considered for the simulation at which steady condition is achieved.

5.2 Validations

Numerical simulations were carried out for different value of angular velocity to fine the coefficient of torque and from that coefficient of power. The numerical result were validated with the experimental result of Patel et al. [6] for different value of tip speed ratio as shown in Fig. 6.

5.3 Rotor Design and Geometry Parameter

The study was carried out on the rotor whose dimensions are as per Table 1. The dimensions of the rotor are directly taken from the research paper of Patel et al. [6].

Here Geometry of the vane is symmetrical means do not change along the height of the blades and that's why 2D Simulations were carried out in Ansys fluent. The size of the domain selects such that it does not affect the performance of the rotor. Which is 20 D front side, 100 D back side and 20 D on top and bottom side as shown in Fig. 7. The computational domain is divided in to two sub domain, rotational

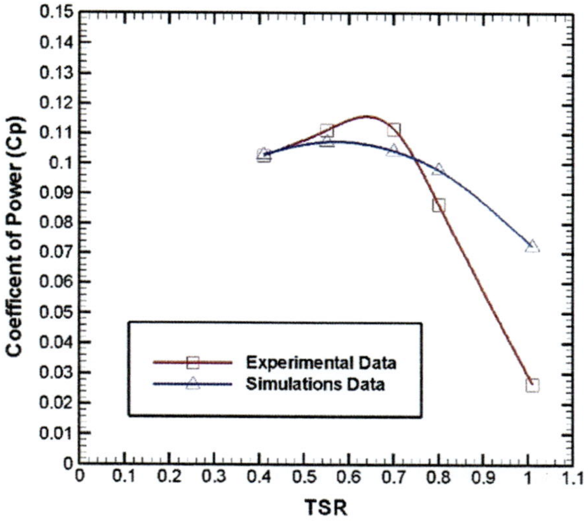

Fig. 6 Validation of present simulation method

Table 1 Rotor detail used in present investigation

Sr. No.	Details	Dimensions (mm)
1	Rotor diameter (D)	100
2	Vane diameter (d)	50
3	Vane thickness (t)	5
4	Gap between blades (e_y)	10
5	Height of blades (H)	350
6	Gap between blades (e_x)	0, 10, 20, 30, 40

domain and stationary domain. These two sub zones are in contact with each other through interface.

5.4 Turbulence Modeling

A two-equation eddy simulation model called k-ω Shear Stress Transport (k-ω SST) was used to simulate the turbulent flow. It is a hybrid model that combines the k-ω and k-ε models. It imitates k-ω model near the wall which allowing the flow separation to be accurately predicted and k-ε model at region far away from the wall to accurately predict wake, free shear effects and recirculation. In other word this model can utilized the strength of both k-ω and k-ε model to ensure that the wake profile and rotor performance characteristics that are produced are true.

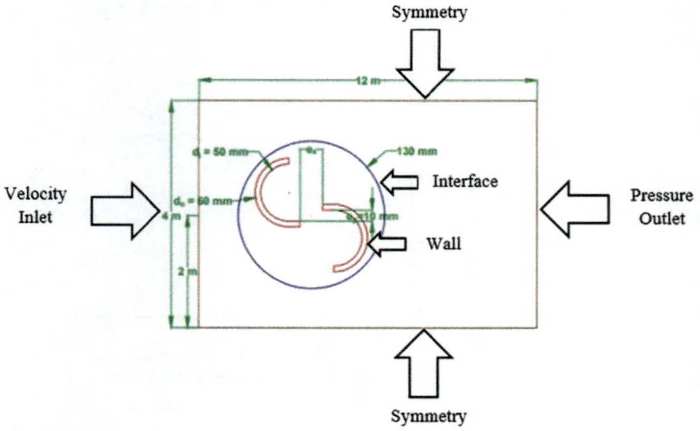

Fig. 7 Conceptual diagram for domain with boundary condition

Table 2 Boundary conditions

Boundary type	Boundary condition
Inlet	Velocity inlet (0.5 m/s)
Outlet	Pressure outlet (0 GPa Gauge pressure)
Rotor surface	Wall (no slip)
Rotating and stationary domain surface	Interface
Domain water surface	Interface
Wall 1 and Wall 2	Symmetry

5.5 Boundary Condition

Initial conditions for the present work is directly taken from the experimental work done by Patel et al. the boundary conditions used for present numerical study are shown in figure and table (Table 2).

6 Result and Discussion

6.1 Simulation Result

The numerical simulation was carried out for different value of negative overlap ratio of 0 to 0.4 with flow velocity 0.5 m/s and TSR ranging from 0 to 1.0 for thirteen complete revolution of the rotor for time step corresponding to 1° rotation of rotor. The pressure and velocity contour near the rotor region is as shown in Fig. 8. Pressure

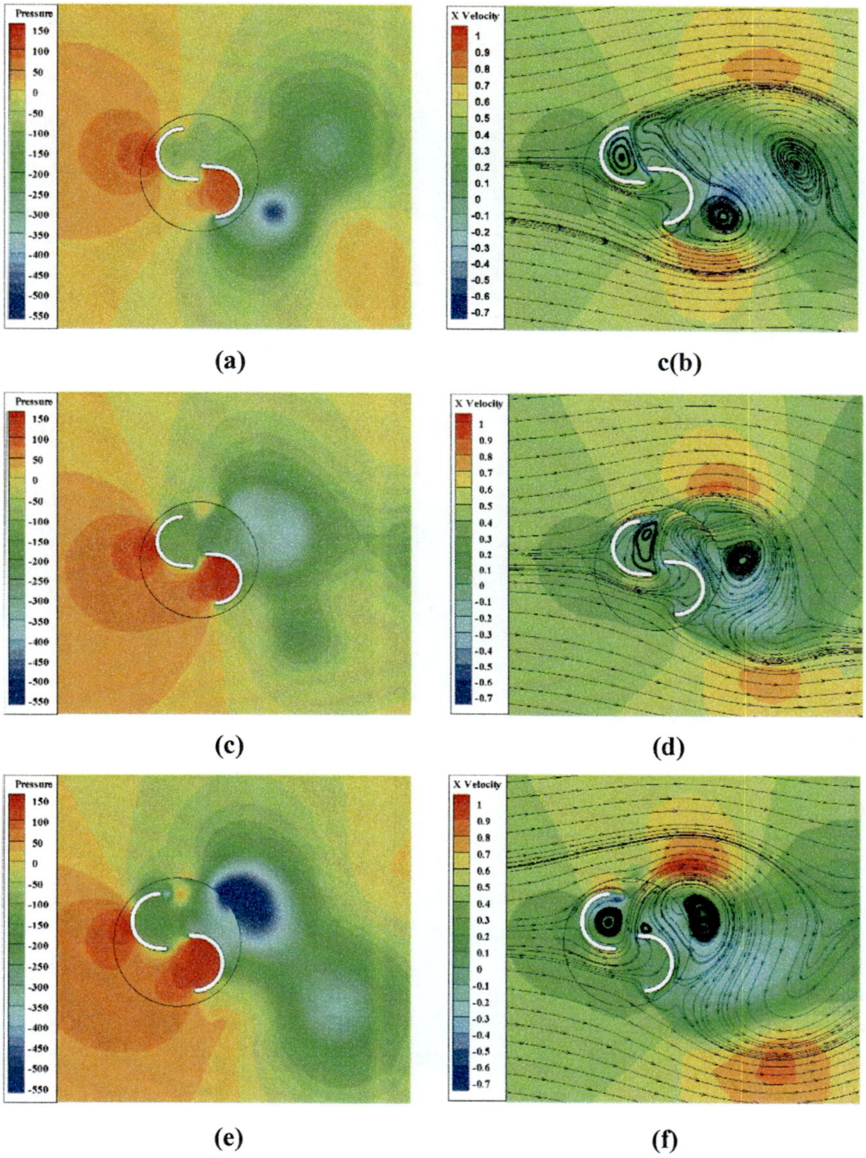

Fig. 8 **a** Pressure contour for 0 OR and 0.1 TSR. **b** Velocity contour for 0 OR and 0.1 TSR. **c** Pressure contour for −0.1OR and 0.1 TSR. **d** Velocity contour for − 0.1 OR and 0.1 TSR. **e** Pressure contour for − 0.2 OR and 0.1 TSR. **f** Velocity contour for − 0.2 OR and 0.1 TSR. **g** Pressure contour for − 0.3 OR and 0.1 TSR. **h** Velocity contour for − 0.3 OR and 0.1 TSR. **i** Pressure contour for − 0.4 OR and 0.1 TSR. **j** Velocity contour for − 0.4 OR and 0.1 TSR

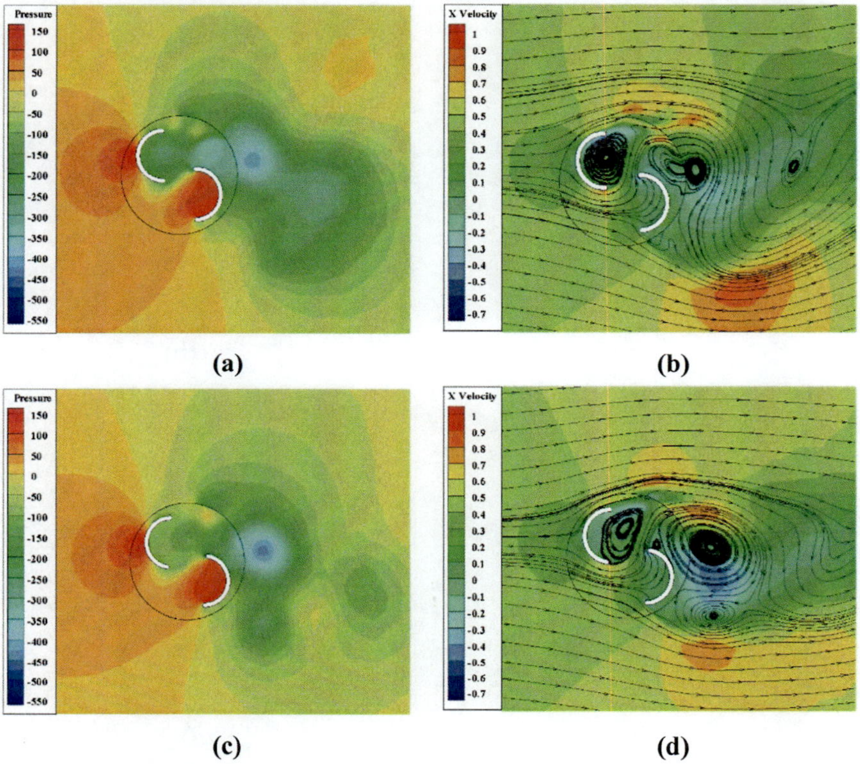

Fig. 8 (continued)

on the leading blade increases with increase in the negative OR up to 0.2 and then decreases with increase in negative OR while trailing side pressure increases with increase in negative OR. With increase in Negative OR, Gap between two blade increases and velocity of the flow passing through gap between blades increases because of the diversion of the flow as shown in Fig. 8. Because of that effective mass flow rate striking on the leading blade will reduce which leads to decrease in performance of the rotor.

Coefficient of torque and coefficient of drag of the rotor was calculated by taking an average of the instantaneous coefficient of torque for one complete revolution of the rotor with time step size corresponding to 1° angle of rotation. Figure 9 shows the instantaneous coefficient of torque and instantaneous coefficient of drag for one complete revolution or the rotor for different negative overlap ratios and tip speed rations.

Figure 10 shows the variation in coefficient of torque, coefficient of power and coefficient of drag with TSR for different negative overlap ratios. Maximum coefficient of torque is obtained for overlap ratio of − 0.1 corresponding to 0.1 TSR While maximum coefficient of power is obtained for 0 Overlap ratio corresponding to 0.6

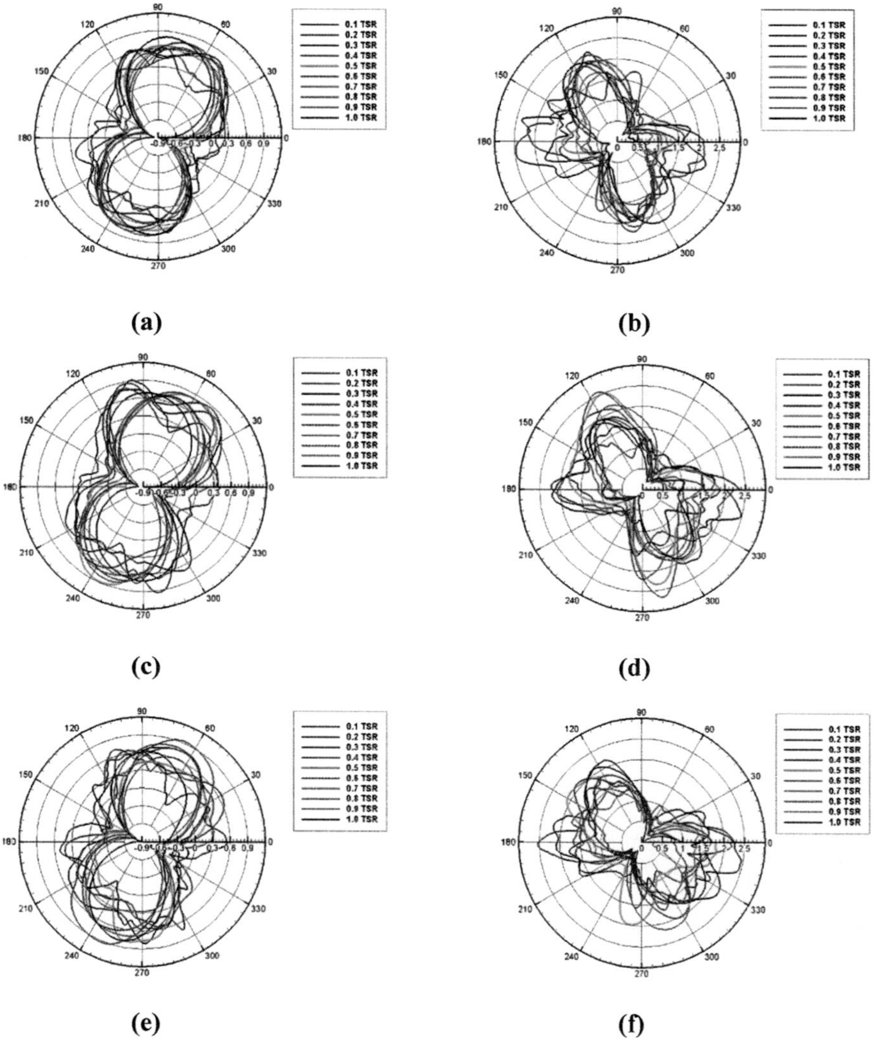

Fig. 9 **a** Instantaneous coefficient of torque for 0 OR. **b** Instantaneous coefficient of Drag for 0 OR. **c** Instantaneous coefficient of torque for − 0.1 OR. **d** Instantaneous coefficient of torque for − 0.1 OR. **e** Instantaneous coefficient of torque for − 0.2 OR. **f** Instantaneous coefficient of Drag for − 0.2 OR. **g** Instantaneous coefficient of torque for − 0.3 OR. **h** Instantaneous coefficient of Drag for − 0.3 OR. **i** Instantaneous coefficient of torque for − 0.4 OR. **j** Instantaneous coefficient of Drag for − 0.4 OR

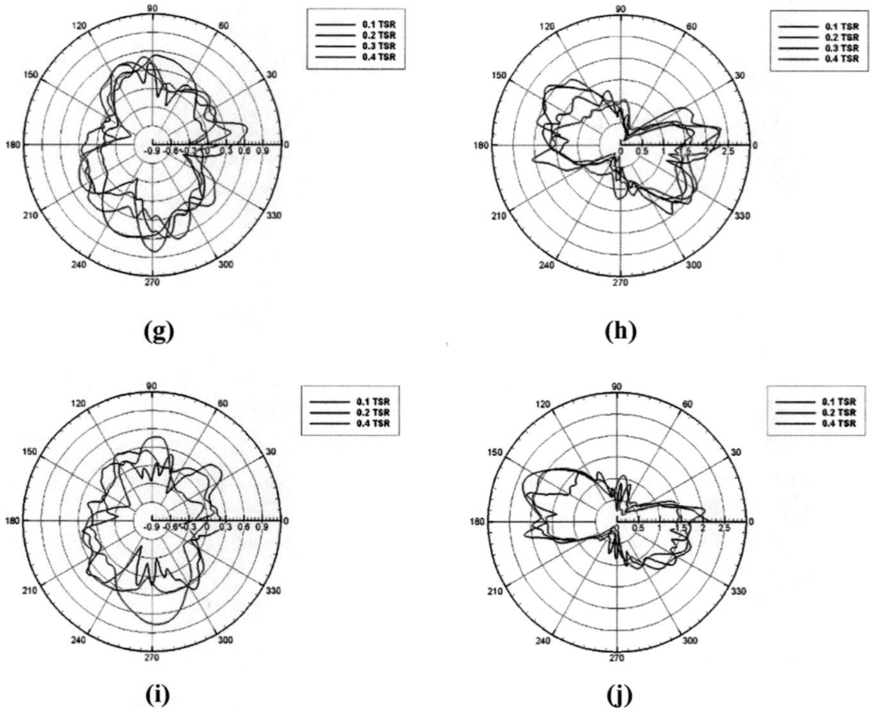

Fig. 9 (continued)

TSR and maximum coefficient of drag is obtained for overlap ratio corresponding to 0.2 TSR. Reduction in coefficient of power with OR is due to diversion of flow through the gap between blades.

7 Conclusion

The purpose of the present study is to investigate the influence of negative overlap ratio on the performance of the Savonius rotor with straight edge extension on overlap region. The simulations were carried out for the negative overlap ratio ranging from 0 to 0.4 and constant flow velocity of 0.5 m/s. The maximum coefficient of power of 0.0953 was obtain with 0 overlap ratio corresponding to 0.6 TSR. As the negative overlap ratio increase, the gap between two blade increases which leads to increase in velocity of the flow passing through the gap between blades because of the diversion of the flow. As a result, there will be a decrease in the mass flow rate of water impacting the blade, which leads to reduction in the performance of the rotor (Fig. 11).

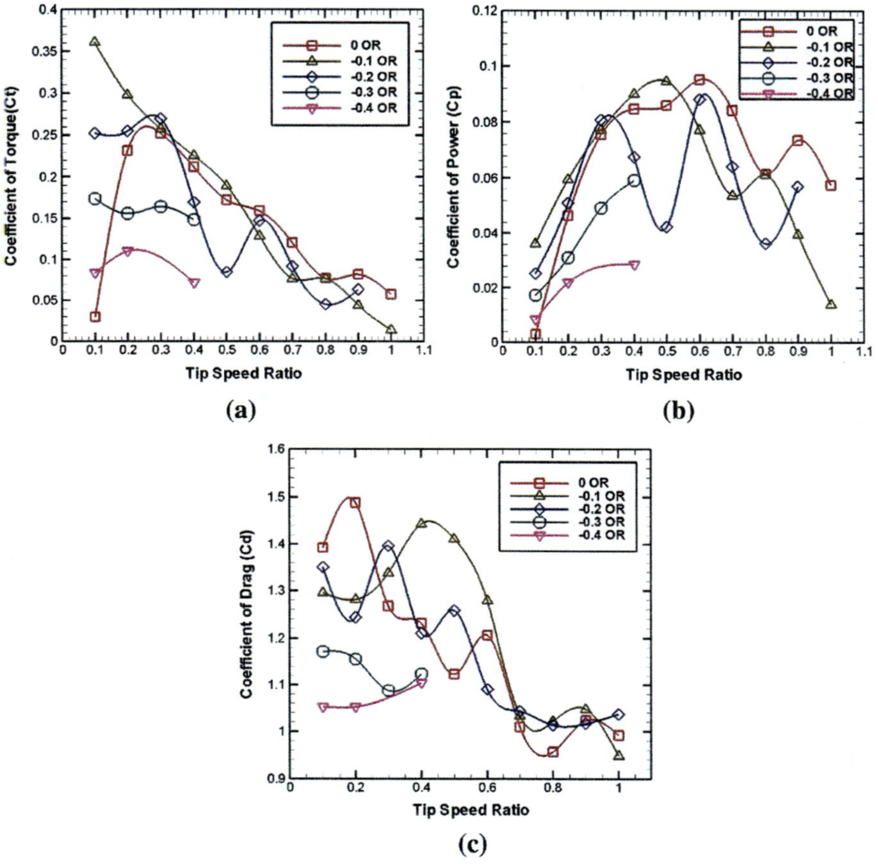

Fig. 10 **a** Coefficient of torque at different negative overlap ratio. **b** Coefficient of power at different negative overlap ratio. **c** Coefficient of drag at different negative overlap ratio

Abbreviation

A	Projected area of the rotor (m)
C_d	Coefficient of drag (–)
C_p	Coefficient of power (–)
C_t	Coefficient of torque (–)
D	Rotor diameter (m)
d_i	Inner diameter of the blade (m)
D_o	Outer diameter of the blade (m)
e	Gap between blades (m)
H	Rotor height (m)
P_{out}	Output power of the turbine (W)
T	Thickness of the blade (m)
u	Tangential velocity of the rotor (m/s)

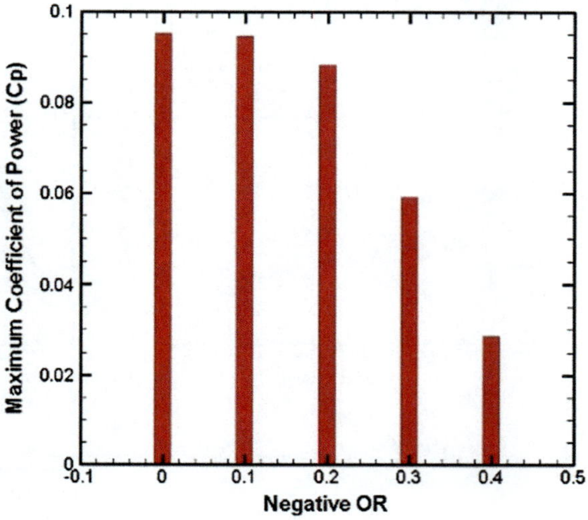

Fig. 11 Maximum coefficient of power at different negative overlap ratio

V Free stream velocity of flowing water (m/s)

Greek Symbol

ρ Density (kg/m^3)
λ Tip speed ratio

Abbreviation

OR Overlap ratio
TSR Tip speed ratio

References

1. Patel V, Bhat G, Eldho TI, Prabhu SV (2017) Influence of overlap ratio and aspect ratio on the performance of Savonius hydrokinetic turbine. Int J Energy Res 41(6):829–844
2. Salleh MB, Kamaruddin NM, Mohamed-Kassim Z (2022) Experimental investigation on the effects of deflector angles on the power performance of a Savonius turbine for hydrokinetic applications in small rivers. Energy 247:123432
3. Thiyagaraj J, Rahamathullah I, Anbuchezhiyan G, Barathiraja R, Ponshanmugakumar A (2021) Influence of blade numbers, overlap ratio and modified blades on performance characteristics of the Savonius hydro-kinetic turbine. Mater Today: Proc 46:4047–4053
4. Kamoji MA, Kedare SB, Prabhu SV (2009) Experimental investigations on single stage modified Savonius rotor. Appl Energy 86(7–8):1064–1073
5. Abdelaziz KR, Nawar MA, Ramadan A, Attai YA, Mohamed MH (2022) Performance improvement of a Savonius turbine by using auxiliary blades. Energy 244:122575
6. Patel VK, Patel RS (2022) Optimization of an angle between the deflector plates and its orientation to enhance the energy efficiency of Savonius hydrokinetic turbine for dual rotor configuration. Int J Green Energy 19(5):476–489

Refuse Derived Fuel (RDF) Co-processing in Kiln Main Burner in a Cement Plant: A Case Study

Prateek Sharma, Kapil Kukreja, K. P. K. Reddy, Ankur Mittal, D. K. Panda, and Bibekananda Mohapatra

1 Introduction

Global warming is a major threat in near future as temperature the world over are going to rise by 2.1–3.5 °C according to latest sixth IPCC report. It is well above the limit by 1.5–2.0 °C as laid down in Paris agreement. So the efforts to contain greenhouse gas emissions has been increased manifold globally. Cement industry is accountable for approx. 7% GHG emissions contributing significantly in this direction. One of the identified levers to achieve this target as per the low carbon technology roadmap devised by the industry is utilization of waste derived fuels replacing fossil fuels in cement plants [1, 2]. Cement plants have worldwide attained high TSR by utilization of alternative fuels with the world average TSR of approx. 18% in 2017. India is playing its part with an average TSR of 4% [3] and striving to achieve 25% TSR by 2025 [4, 5]. Refuse Derived Fuel (RDF) has been identified as one of the most promising alternative fuel. But RDF utilization in large volume in cement production is still not achieved thus far in India. Most of the cement plants in India are using RDF in calciner only due to the low heat value of RDF. Recently, RDF grading (SCF, grade I, grade II, grade III) carried out by Ministry of Housing and Urban Affairs in consultation with different stakeholders is based upon the quality to maximize RDF utilization [6]. This will support cement plants to achieve TSR through kiln main burner as well. In this regard, one cement plants in southern parts of India targeted to replace 25% of their fuel requirement by RDF available in the vicinity of the plant as alternative fuel. These efforts towards Municipal Solid Waste will support the cement plant in two ways: Substitute the main fuel i.e. a blend of coal & petcoke and secondly consume the waste generated, which otherwise would consume additional resources for its. The objective of the paper is to check the technical feasibility for achieving 25% TSR through kiln main burner in a cement plant.

P. Sharma (✉) · K. Kukreja · K. P. K. Reddy · A. Mittal · D. K. Panda · B. Mohapatra
National Council for Cement and Building Materials, Ballabgarh, Haryana, India
e-mail: ncbprateek@gmail.com

S. Doolla et al. (eds.), *Advances in Clean Energy and Sustainability*,
Green Energy and Technology, https://doi.org/10.1007/978-981-99-2279-6_28

2 Material and Methods

2.1 Methodology

A technical assessment was conducted to evaluate the feasibility of 25% TSR through RDF co-processing in kiln main burner of the cement plant. Raw materials and current fuel mix were analyzed and mass and energy balance was established by the process measurements during the site visit. Impact assessment for process, environment and clinker quality parameters are conducted using excel spreadsheets and plant data. System was designing done based upon the alternative fuel quantity and quality and outcomes of the impact assessment. Experiences shared by some European countries with RDF coprocessing in kiln main burner were also referred.

2.2 Fuel Availability and Specification

The system requirement to achieve 25% TSR through RDF is 87 tpd which will be made available to the plant by waste generators. The specification of RDF Grade I and Grade II is found most suitable to achieve high TSR through the main burner and is given in Table 1.

Presently, the plant is operating by using a mix of petcoke and coal. The analysis of fuel mix and proposed RDF Grade-I is given in Tables 2 and 3.

Table 1 RDF specification [6]

S. No.	Parameters	RDF-Grade II	RDF-Grade I
1	Particle size, mm	< 50 mm or < 20 mm depending upon use in ILC or SLC, respectively*	
2	Net calorific value (NCV)—in kcal/kg	> 3750	> 4500
3	Ash—maximum	< 10%	< 10%
4	Moisture—maximum	< 15%	< 10%
5	Chlorine—maximum	< 0.7%	< 0.5%
6	Sulphur—maximum	< 1.5%	< 1.5%

Table 2 Main fuel & proposed RDF (Grade-I) specification

Fuel	Proximate analysis (% ad basis)				Ultimate analysis (% daf basis)					NCV
	M	VM	FC	Ash	C	H	N	S	O	kcal/kg
Coal and petcoke mix	2.56	24.7	49.48	23.26	84.63	5.08	1.59	3.09	5.61	5586
Proposed RDF (Grade-I)	10	72.5	7.5	10	50	10	1.87	0.63	37.5	4500

Table 3 System requirement

Parameters	Unit	Value
Target AF consumption	tpd	87
Days of plant operation	tpa	330
Plant running hrs per day	hrs	24
AF quantity for 25% TSR	tph	3.60
Shredder running hrs per day	hrs	15
Shredder capacity @ 15 h per day operation	tph	6
Design margin in material handling equipment	%	20
Effective storage capacity	days	2

3 Feasibility for RDF Firing in Kiln Main Burner

3.1 Evaluation of Existing Plant Operation to Establish % TSR

The cement plant has a rated capacity of ~ 15 tpd clinker and producing approx. 1450 tpd. There were reducing conditions prevailing at kiln inlet. Specific heat consumption works out to 960 kcal/kg clinker based upon the heat balance, operating parameters and plant data for entire pyro processing section.

3.2 Process and System Design Limitations

It was observed that reducing conditions were prevailing in kiln system due to high CO peaks. The plant is having conventional mono-channel burner which is also contributing to high specific heat consumption. In the existing coal mill operation, the mill fan volume which is handled by coal mill bag filter fan is being split to bag filter and primary air fan due to insufficient bag filter capacity. Coal dosing system is old generation type (venturi type pneumatic conveying). Recuperation zone of grate cooler is not having static grate leading to low recuperation efficiency and high

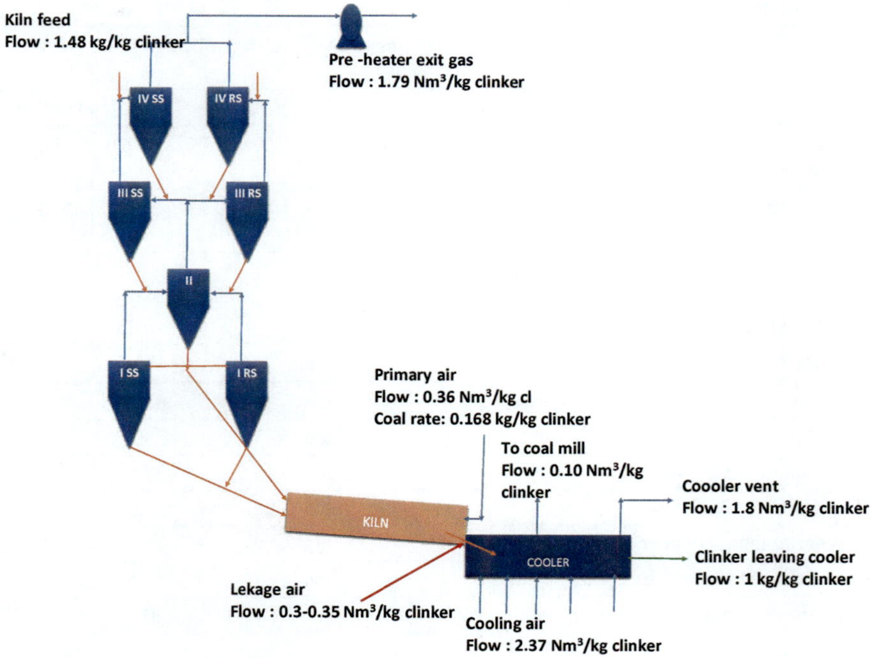

Fig. 1 Kiln-Cooler and preheater air/gas balance

preheater fan volume. Kiln outlet seal was damaged leading to high false air ingress in the system. Figure 1 shown below is indicating the air/gas flow rate measured. The total cooling air is 2.37 Nm^3/kg clinker, the cooler recuperated air is only 0.47 Nm^3/kg clinker and primary air is 0.36 Nm^3/kg clinker. The estimated air required for complete combustion with respect to the coal rate of 0.168 kg/kg clinker is 1.13 Nm^3/kg clinker. During the visit it was observed that the kiln outlet seal was not present and a lot of atmospheric air estimated at around 0.3–0.35 Nm^3/kg clinker was entering.

3.3 *Firing Location*

Kiln inlet and main kiln burner are two firing locations available for co-firing. Residence time of gas in kiln riser is less than 1 s which is not sufficient for combustion of pre-processed alternate fuel at this location. Moreover, varying CO was observed at kiln inlet with temperature lying in the range of 850–900 °C. For complete combustion and destruction of dioxins and furans generated due to the feeding of chlorinated waste, it requires residence time of gas in kiln riser should be at least 3–4 s at ~ 900 °C [7], which was not available. Hence, it is not recommended to feed alternative fuels like RDF at kiln inlet location. Therefore, the only option left to feed AF is in the kiln main burner.

3.4 Proposed System Design for RDF Firing in Kiln Main Burner

RDF is proposed to be received at plant through trucks in bales (2 m × 1 m)/in bags (50 kg cement empty bags)/in bulk (Self-tipping type trucks). A covered shed [20 m (Width) × 35 m (Length) × 19.2 m (Height)] is considered for complete pre-processing station where trucks will dump the received RDF. Material is proposed to be extracted by multipurpose fully automated grab crane from the over ground storage area and will be fed to fine shredder for material sizing. The input material will have maximum bale size 2 m × 1 m, hence fine shredder is proposed to get the required output size of < 10 mm. Shredded material shall be fed to underground pits (2 nos.) with the help of belt conveyor. Shredded material from the pit shall be extracted with the help of grab crane and will be fed to box feeder. Flexible conveyor shall lift the material vertically and discharge it on to the trough belt conveyor further to be transported which will be further transported to the material to the storage bin located at the cooler top level floor. Belt conveyor will enter into clinker cooler top floor and will discharge the material to hopper of screw weigh feeder; further into screw weigh feeder and after weighing, material shall be fed to new multichannel burner through IDMS blow through rotary valve. The flowsheet for the proposed system is shown in Fig. 2. It is proposed to replace the single channel burner with multi-channel multi fuel burner and Coal dosing system with highly accurate Pfister Pump/Rota Scale.

3.5 Impact Assessment

Chatterjee and Sui [8] reported the impacts of co-processing of alternative fuels in larger quantity on clinker production along with specific heat consumption. Parameters like clinker production, specific heat consumption, clinker quality along with environmental emissions were assessed at 25% TSR with the use of waste / alternative fuels. Impact assessment of RDF utilisation indicated the increase in specific heat consumption and reduction in clinker production. The introduction of moisture of ~ 15% in the process results in increase of specific heat consumption by ~ 5 kcal/kg clinker along with net reduction in flame temperature of ~ 30 °C. The reason for decrease in clinker production is mainly related to the high moisture of RDF. When the moisture in the system increases after introduction of RDF, part of the thermal energy input will be utilised to remove the fuel moisture. Thus, actual heat input for clinkerisation gets reduced. Hence more fuel needs to be fired in the system which

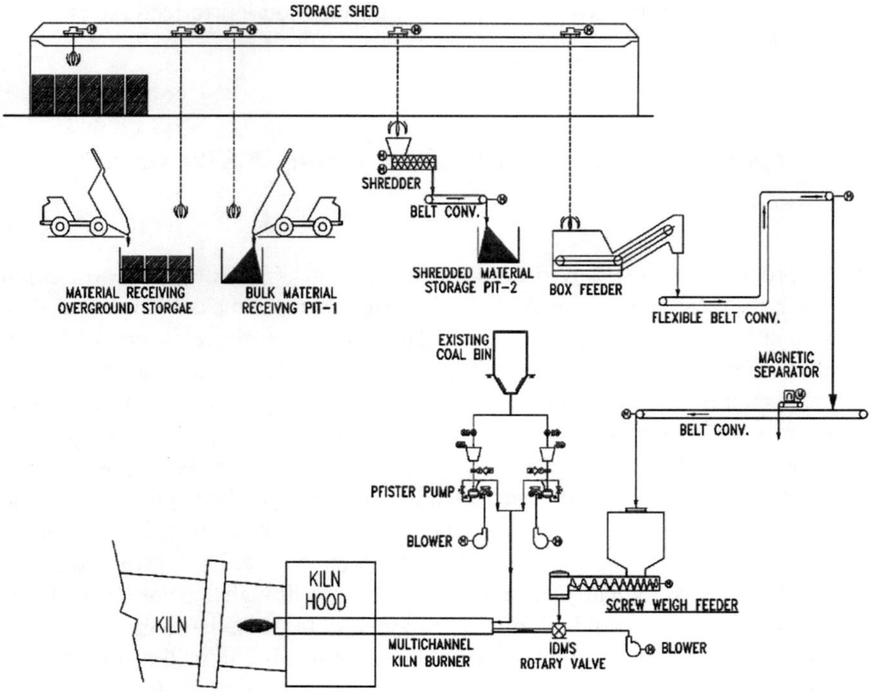

Fig. 2 System design for proposed alternative fuel handling system

requires more combustion air. But the existing preheater fan which extracts the flue gases will not be able to take the extra volume due to the design limitation. Therefore, fuel quantity cannot be increased, thereby reducing the kiln feed and clinker production to maintain the clinker quality. It is anticipated that there shall be 3.5% reduction in gas volume in preheater fan with the similar production levels. Increase in overall fuel moisture and decrease in overall ash content also results in net production loss of ~ 25 tpd of clinker after introduction of RDF.

High Cl content in RDF has several operational issues as reported in literature [9]. Cl content in hot meal and clinker is 0.04% and 0.007% which is within the normal limits of 0.8% and 0.1% respectively. As per the quality of RDF, chlorine content shall be < 0.7% which leads to rise in Cl content of total input to 0.043% which is higher than normal operating norms of 0.023%. This limits the TSR to 10% thus meeting the upper limit of %Cl. In the current scenario, SO_3 input from the fuel is 0.95 kg/kg clinker. After utilization of RDF (1.5% S content) at 25% TSR, the SO_3 input from the total fuel will decrease to 0.73 kg /kg clinker. Thus, SO_3 will further decrease which will reduce the coating/build up formation. Ash content in RDF would be ~ 10%. Considering 25% TSR by RDF, overall ash content shall reduce to ~ 18% in the fuel mix as compared to the existing ash content of 23%. Accordingly, raw mix design needs to be optimized to meet the LSF (also SM and AM) target in the clinker.

Table 4 The total project cost for implementing 10% TSR through RDF

S. No.	Description	Total cost (In Rs lakhs)
1	Shredder	450
2	New multichannel burner	200
3	Static grate & cooler drive	300
4	Grab crane, screw weigh feeder	200
5	Pfister pump	120
6	Other auxiliaries (ducts, hoppers, bag filters, blowers, outlet sealing etc.)	1200
	Total project cost	2470
7	Thermal energy savings annually @ 55 kcal/kg clinker	329
8	Fuel cost saving by replacing RDF	281
9	Payback period	~ 4 years

3.6 CO_2 Mitigation Potential

At 25% TSR through RDF, it is estimated that there shall be savings of ~ 0.135 t CO_2/t clinker which corresponds to ~ 30% CO_2 savings annually considering 62% carbon content in present fuel mix and 17% specific fuel consumption.

3.7 Project Cost and Payback Period

The total project cost for implementing 10% TSR through RDF is given in Table 4. The project cost includes new burner, cooler drive, RDF feeding system etc. The fuel cost considered is Rs 7000/tonne and RDF cost is Rs 3500/tonne. The payback period for 10% TSR through RDF with the above suggested modification in the plant will be ~ 4 years.

4 Results and Discussions

Plant must modify the existing kiln system to operate at high TSR using RDF as alternative fuel. The operation can be improved by arresting the false air entry from the kiln outlet and increasing the high temperature recuperated air from the cooler. This will further result in reduction of preheater exit fan volume. It is recommended to have lamella or pneumatic type kiln outlet seal which can further reduce the specific heat consumption to the level of ~ 10 kcal/kg clinker. If the plant implements multichannel low NOx burner, it can further reduce the specific heat consumption to ~ 20–25 kcal/kg clinker. During installation of the multi-channel burner, plant has to consider a new bag filter for handling the additional coal mill fan volume going to Primary Air (PA) fan. After installation of multi-channel burner with new PA fan and arresting kiln outlet leakages, the secondary air volume will increase. However, the secondary air temperature reduces due to conventional grate cooler with movable grate. It is recommended to install static grate from reputed technology suppliers in cooler along with up-gradation of cooler drive (for trouble free operation) which will reduce cooler losses leading to energy savings of 15–20 kcal/kg clinker and improve recuperation efficiency. Further, it will improve the kiln operation and facilitate higher TSR. Thus, a comprehensive study was conducted and a suitable system design catering to above aspects was proposed.

5 Conclusion

RDF utilization in kiln main burner necessitates stringent fuel quality parameters. Variation in fuel input parameters will affect kiln operation and clinker quality. Thus, pre-processing is the key to high TSR which has been considered in the proposed system design concept. Further, impact assessment has been done which concludes that there will be a loss of clinker production of the tune of 25 MT per day along with the increase in specific energy consumption of 5 kcal/kg clinker. Incapability of mono channel burner to handle AF, substantial moisture in the AF (RDF), low secondary air flow & temperature in kiln, present leakages through kiln outlet seal, high primary air requirement and incapability of coal mill bag filter demands certain modifications in the plant which are modification in cooler, replacement of kiln outlet seal, installation of multi-channel burner and accurate coal dosing system to achieve high TSR. Chloride content in RDF is major bottleneck to achieve high TSR of ~ 25%. With Grade I and Grade II RDF having maximum chloride content of 0.7%, the maximum achievable TSR is 10%. If plant wants to operate at high TSR with same chloride content in RDF, then plant must install kiln bypass system.

References

1. World Business Council for Sustainable Development: Low Carbon Technology Roadmap for the Indian Cement Sector: Status Review (2018)
2. International energy agency: Technology Roadmap Low-Carbon Transition in the cement industry. IEA (2018). pp. 5, 29. https://www.wbcsd.org/Sector-Projects/Cement-Sustainability-Initiative/Resources/Technology-Roadmap-Low-Carbon-Transition-in-the-Cement-Industry. Accessed 11 Jan 2022
3. Mohapatra B, Chaturvedi S, Saxena A, Sharma P, Bohra A, Naidu G (2019) Use of alternative fuels and raw materials in cement industry in India—prospects & challenges in conserve green & sustainable resources. New Delhi
4. Confederation of Indian Industry: Approach paper for achieving 25% thermal substitution rate in Indian cement industry by 2025 (2016), pp 1–8. http://www.ciiwasteexchange.org/doc/annexure_6.pdf. Accessed 20 April 2022

5. Confederation of Indian industry: Status paper on alternate fuel usage in Indian cement industry (2018), pp 1–2. http://www.ciiwasteexchange.org/doc/afr2018.pdf
6. Ministry of housing and urban affairs Government of India: Guidelines on usage of refuse derived fuel in various industries (2018). pp 13–14. http://cpheeo.gov.in/upload/5bda791e5afb3SBMRDFBook.pdf. Accessed 15 May 2022
7. Karstensen DKH (2017) Formation and control of dioxins in dry pre-heater/pre calciner kilns co-processing wastes. In: 13th NCB International seminar on cement, concrete and building materials, New Delhi
8. Chatterjee A, Sui T (2019) Alternative fuels—effects on clinker process and properties. Cem Concr Res 123:1–19
9. Mora N, Martinez J, Ayala C (2019) Clinker production with high chlorine alternative fuels. In: 15th International congress on the chemistry of cement, Prague, Czech Republic

Analysis of Several Parabolic Trough Collector Structures Using Finite Element Analysis and Multicriteria Decision-Making Method

Punit V. Gharat, Snehal S. Bhalekar, Vishwanath H. Dalvi, Sudhir V. Panse, Suresh P. Deshmukh, and Jyeshtharaj B. Joshi

Abbreviations

PTC	Parabolic trough collector
HTE	Heat transfer element
HTF	Heat Transfer Fluid
FEM	Finite Element Method
FEA	Finite Element Analysis
MS	Mild steel
CAD	Computer-Aided Design
TOPSIS	Technique for order performance by similarity to ideal solution
MCDM	Multicriteria Decision Making

1 Introduction

1.1 Parabolic Trough Collector

Parabolic Trough Collector (PTC) technology is the emerging solar thermal power technology for low and medium-temperature applications. The PTC system consists of a collector structure, receiver, receiver support, pylon, and reflector as shown in

P. V. Gharat (✉) · S. S. Bhalekar · V. H. Dalvi · S. V. Panse · S. P. Deshmukh · J. B. Joshi
Institute of Chemical Technology, Matunga, Mumbai 400019, India
e-mail: mec18pv.gharat@pg.ictmumbai.edu.in

P. V. Gharat · S. V. Panse · J. B. Joshi
Marathi Vidnyan Parishad, Chunabhatti, Mumbai 400022, India

J. B. Joshi
Homi Bhabha National Institute, Anushaktinagar, Mumbai 400085, India

S. Doolla et al. (eds.), *Advances in Clean Energy and Sustainability*,
Green Energy and Technology, https://doi.org/10.1007/978-981-99-2279-6_29

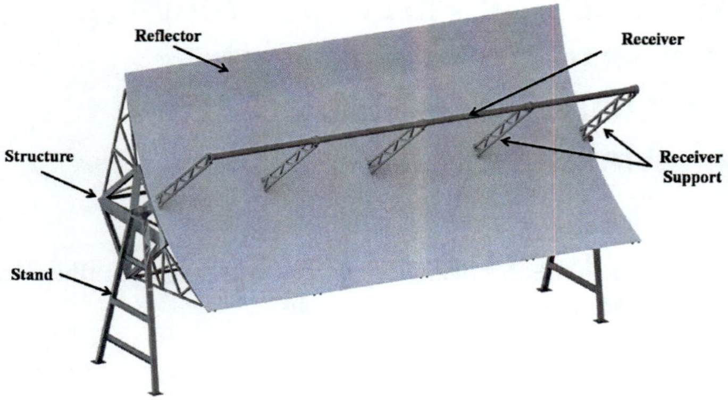

Fig. 1 Components of parabolic trough collector

Fig. 1. The reflector is parabolic in shape therefore, the solar beam radiation that falls on the reflector is reflected at the focal point of the parabola where the receiver is fixed, containing an HTF [1, 2]. These radiations are converted into heat which goes to heating HTF, flowing within the receiver pipe [3]. As radiation is reflected from a large area onto the receiver's small area, a Concentration ratio between 40 and 90 can be achieved using In PTC [4]. The generated thermal energy using a solar thermal system can be used in food, chemical, and dye industries for several processes demanding process heat. The solar thermal system can provide process heat to 40–50% of the industrial processes which require a temperature less than 250 °C. [5] The thermal energy generated can also be used for generating electricity using steam turbines in solar thermal power plants [6].

1.2 Studies Based on FEA and Optimization Are Given Below

The cost-effective and efficient solar collectors are necessary for the current era of energy crises for producing sustainable and renewable energy. The evolution in PTC structure from first to latest was reviewed by several researchers [7]. It gave us the idea of the available PTC structure, its evolution, the best PTC structure, and its adopted techniques [8]. Also, the researchers are working on making a cost-effective, highly efficient, and optimized PTC system with the use of various principles like FEA, ray tracing, and Optimization, few of them are mentioned in this section.

The structural investigation of the EuroTrough using various parameters like dead load, wind loads, different angles of the collector, and wind directions were carried out by Geyer et al. by using FEM [9]. The model analysis of PTC was performed to understand the natural vibration characteristics of the structure by Zou et al. By using dynamic structural analysis, the natural frequencies and mode shapes of the collector structure were measured at different pitch angles and these results can help to improve the accuracy of the PTC structure [10]. The optimized design of the collector structure was developed using FEA and the Genetic Optimal algorithm. At different wind velocities and different angles of pitching of the collector, the stress and deformation in the collector were obtained. It was found that the maximum stress and deformation were obtained under 10th level gale at 90° pitching [11]. Liu et al. implemented an optimization, for the reduction of weight of the structure without compromising stiffness. The use of different sizes of the bracket leads to optimized structure. Optimization led to a 10.68% reduction in the weight of the system which also reduced the manufacturing cost [12]. In 2018, the US Department of Energy set a benchmark of $100/m^2 for the cost of the PTC. It was achieved by various analyses including structure analysis, efficiency analysis, changing material, labor rates, advanced manufacturing, assembly methods, and world supply chain logistics. Every component of the PTC system was optimized and analyzed for cost and performance [13].

From the literature, it was clear that the four major PTC structures have been developed to date namely *Torque tube, Torque box, Space Frame, and Space Tube.* Also, the cost breaks up of the parabolic trough collector system show, that the solar field contributes around 60% of the cost, i.e. in the area of the solar field, there is more scope for cost reduction, which can be achieved by the use of various optimization techniques. Therefore, here we have considered four major types of collector structures for our analysis study.

1.3 Objectives of This Study

The main objective of this study was a comparative analysis of optimized designs of four major types of PTC based on the parameters like mass, mass balancing, and ease of manufacturing using MCDM and finding the better collector among all.

1.4 Methodology

The comparative study of major parabolic collector structures starts with the study of the development of the structural design of the collector [7]. Where we found the four major structures namely *Space Frame, Torque Tube, Torque Box, and Space Tube.* The CAD model of every design was made with dimensions and then FEA was performed to find out the deformation and stresses in the structure at predefined loading conditions.

The effect of deformation in the collector structure on the radiation loss was measured with help of Ray tracing and the maximum limit of the deformation was set for the entire analysis. With this deformation limit, the optimized structure of each type was selected with considering the minimum torque value of the individual structure. Finally, with the help of TOPSIS, a multicriteria decision-making method,

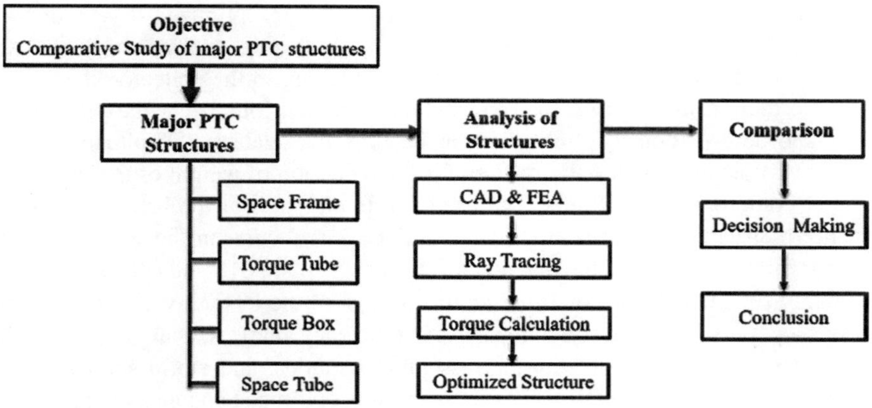

Fig. 2 Methodology of the comparative study of collector structures

a comparative analysis was made for selecting a better structural model out of four models which can be used for further research. The methodology of the study is shown in Fig. 2.

2 PTC Structure Designs

From literature, we found the four major collector structures namely *Torque Tube, Space Frame, Torque Box, and Space* evolved until a date, and CAD models of collectors are shown in Fig. 3. The basic principle involved in the evolution of structure is to transmit torque from one end of the collector to another end of the collector.

In this study, these four major structures were considered for the comparative analysis using FEA and optimization.

2.1 Parameter for Analysis

As study involves the analysis of four major collector structures as listed in the earlier chapter. The comparison of all optimized models has to carry out for the factors like Mass balancing, the mass of the structure, and ease of manufacturing. However, while performing the analysis the few parameters listed below has to keep constant in every design, as they are common in every structure.

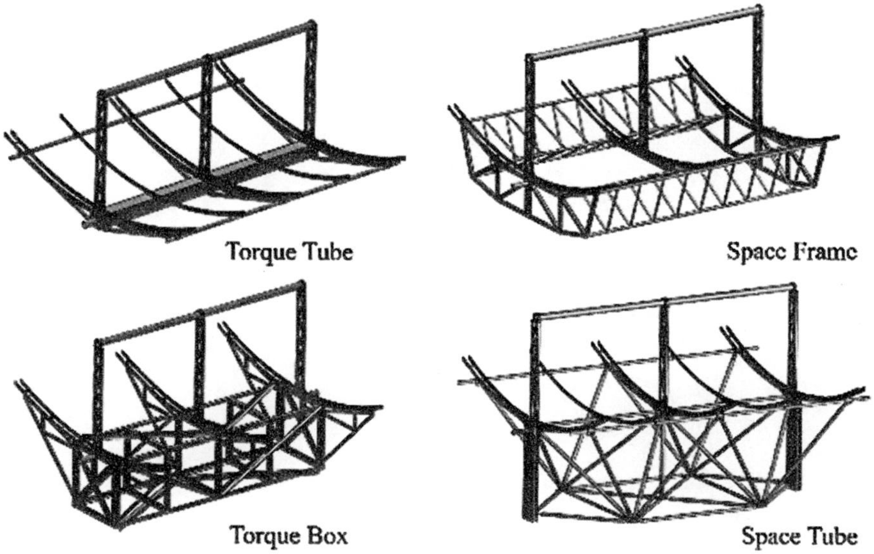

Fig. 3 CAD of major types of collector structure

System Specifications

Aperture	3 m
Focal length	1.3 m
Mass of reflector	6.67 kg/m^2
Mass of receiver support	5.5 kg/component
Mass of receiver	12 kg/m (including HTF)
Material of construction	Mild steel
Yield stress of MS	3.52×10^8 N/m^2
Density of MS	7900 kg/m^3
Study-1—aperture area	3 m × 3 m = 9 m^2
Study-2—aperture area	3 m × 6 m = 18 m^2

2.2 Structural Deformation at a Different Tilt Angle of Collector

The PTC rotates from the East to West direction in a day for tracking the sun perfectly for obtaining maximum solar energy. Normally PTC rotates by 180°, the 90° on the east side and 90° on the west side, measured w.r.t. vertical. Therefore, before starting an analysis of receiver support, it was necessary to find out the position (Angle w.r.t.

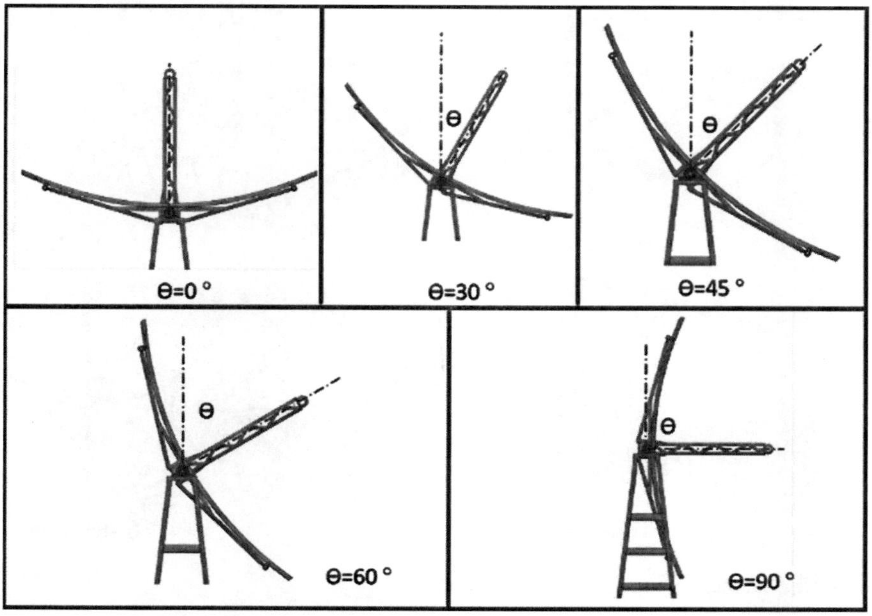

Fig. 4 Collector at a different tilt angle

vertical) of the collector, at which it gives maximum deformation in the support structure.

The static structural analysis of the collector structure was carried out on several angles of the collector with vertical as shown in Fig. 4.

It was observed that the maximum deformation of 2.091 mm and maximum stress of 4.56×10^7 N/m^2 in the collector structure were found at the position of 90°. Therefore, further analysis/static simulations were performed considering the collector tilted at 90° to the vertical.

2.3 *Radiation Loss Due to Deformation Measured by Ray Tracing*

In an earlier chapter, we analyzed several models of collector structure to find out the minimum deformation and minimum mass under a given load. The only minimum deformation can't be a constraint for a given analysis, but finding out optimum mass with less deformation which won't turn in loss of radiation was a major concern. Therefore, with the help of ray tracing, we co-relate the deformation with the loss of radiation. Here loss of radiation means concentrated rays going off the receiver pipe due to excessive deformation. The in-house ray-tracing program was created with python language, in which we can create any shape by equation, can decide the

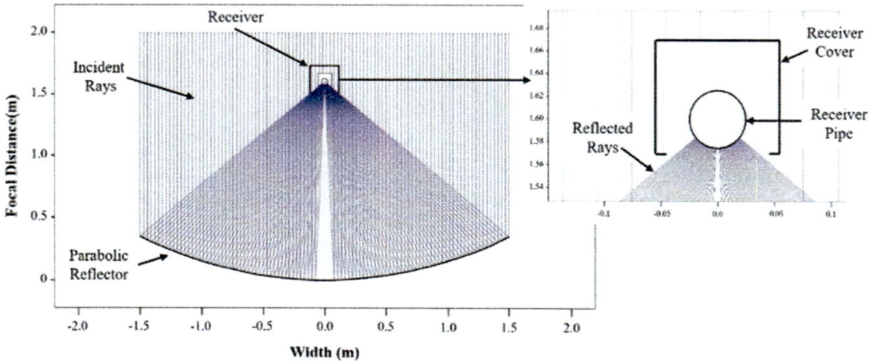

Fig. 5 Ray tracing for a parabolic reflector with receiver housing

number of incident rays and the program will give the result of reflected rays falling on the focus of the parabola as shown in Fig. 5. For analysis, we have considered a receiver pipe of 50 mm dia. We have considered the number of rays as 100. Four deformations were considered for analysis, 5, 10, 15, and 20 mm. The amount of deformation was equal to the offset of the pipe from focus of the parabola.

It shows that there was no radiation loss up to 10 mm deformation in the PTC system, but there were 10 rays out of 100 rays going out of the pipe at 15 mm deformation i.e. 10% loss in radiation. While in the case of 20 mm deformation in the system, 25% loss in radiation. From Ray tracing, we have set the structural deformation limit to 10 mm (Maximum) for further analysis.

Note: Here we have considered a half-insulated pipe as the receiver of PTC, therefore any radiation fall in the upper part of the pipe was considered a loss in radiation, and deformation was overall deformation in PTC structure.

3 Analysis of Four Major Types of PTC Structure

In this study, we have analyzed four major collector structures as explained earlier. The CAD and FEA of all models were performed and then optimization is done for minimum mass for the given deformation limit of the structure. The comparison of all optimized models was carried out for the factors like Mass. Torque generation (mass balancing), cost, and ease of manufacturing using TOPSIS.

3.1 Study-1: Analysis of 3 m × 3 m Collector Structure

In this section, the four kinds of collectors with an aperture of 3 m, and length of 3 m were analyzed. For all structures, several models with different dimensions were

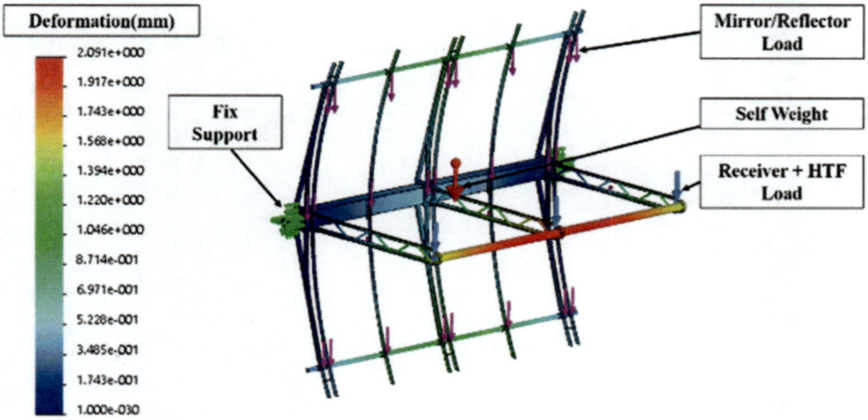

Fig. 6 Deformation analysis (FEA) of the torque tube structure

created and analyzed for deformation and stress under given loading conditions. The loading values were: Reflector Load: 120 kg, and receiver load 36 kg. Later the model with the optimum mass and optimum deformation was selected as an optimized structure and considered for comparison.

3.2 Analysis of Structure (Torque Tube Structure)

All four kinds of collector structures were analyzed thoroughly and here we have presented the analysis of Torque Tube structure in brief. After analyzing several models for torque tube structure, the optimized model was selected with 99.6 kg (mass of receiver and receiver support is not included). The deformation and stresses in the structure at given loading conditions were 2.091 mm and 4.5×10^7 N/m^2 respectively. The deformation analysis of torque Tube structure is shown in Fig. 6.

As the collector size was bigger and the assembly was having several parts, the combination of tetrahedron mesh and the hexahedral mesh was used for analysis. The number of nodes and elements were 13,987,413, and 7,446,650 respectively for the element size of 2 mm.

3.3 The Torque Acting on the Driving Shaft

The torque acting on the driving torque is important for the selection of a driving system for PTC. The total torque generated at the shaft is the summation of torque generated at the point of rotation by all components. The basic formula of torque (moment) is given by,

Table 1 The torque generated by components of the torque tube structure

Component	Mass (kg)	Distance between C.G. and point of rotation (cm)	Torque (kg-cm)
Central Tube, & Ribs	99.6	9.01	897.4
Reflector	55	17.5	962.5
Receiver support	11	61.4	675.4
Receiver (with HTF)	18	132.5	2385
		Total	**4920.296**

Note Here, torque is calculated when PTC is inclined at 90° with vertical. (Refer to Fig. 4)

$$\text{Torque} = (\text{Perpendicular distance between the point of rotation and C.G of component}) \times (\text{Mass of component}).$$

Table 1 shows that the torque generated in the Torque Tube structure is around 4920 kg-cm which was very high in value, therefore it required a driving system with such a high torque handling capacity, ultimately increasing the cost of the system. This value also shows that the structure is not mass-balanced; therefore, it requires the counterweights for mass balancing otherwise needs to move the point of rotation above the current point of the rotation this may lead to the discontinuity in the reflector while PTC joined in the row.

Similarly, the deformation, stress, and torque calculations were performed for the other three types of collector structures, and details are given in Table 2.

It can be seen that the higher torque is generated in the Space Frame structure with a magnitude of 5397 kg-cm. After the space frame, the Torque tube with the value of 4920 kg-cm came into line. The torque generated in the torque box and space frame structure were 2737 kg-cm and 2289 kg-cm respectively, which shows a 49 and 57% reduction torque value as compared to the highest torque generation in the space frame structure.

The reduction in the torque or lower value of torque in Torque Box and Space Tube structure as compared to Torque Tube and Space Frame structure were due to the good distribution of mass throughout the structure. The good distribution of mass led to the move of the C.G of the structure near the point of rotation. We have also analyzed structures by increasing the length to study the effect of increasing length on the various parameters of structure.

Table 2 The analysis of four-collector structures (3 m × 3 m)

Collector	Mass (kg)	Torque (kg-cm)	Deformation (mm)	Stress (N/m^2)
Space frame	116	5397	2.78	8.08×10^7
Torque tube	99.6	4920	2.091	4.5×10^7
Torque box	117	2737	3.32	1.01×10^8
Space tube	116	2289	2.86	7.8×10^7

Note Structure mass does not include the mass of the receiver and receiver support

3.4 Analysis of 3 m × 6 m Collector Structure

In this section, the four kinds of collectors with an aperture of 3 m, and a length of 6 m were analyzed. For all four types of collector structure, several models with different dimensions were created and analyzed for deformation and stress under given loading conditions. The loading values were: Reflector Load: 240 kg, and receiver load 72 kg. The analysis of structures for parameters like mass, deformation under load, and torque generation values are given in Table 3.

The mass and the torque value for 3 m length and 6 m length collector structures are shown in plot Fig. 7. It was observed that, along with the increasing length of the structure, the torque generation increases in Space Frame and Torque Tube structures, and due to the better distribution of mas in the Torque Box and Space tube type structure the torque value change by a small amount even after increasing the length. Also, in these two structures, we can achieve zero torque generation at the point of rotation just by moving the point of rotation by a few mm distance. Therefore, these two well-mass-balanced collector structures can be very useful in the PTC system development.

Table 3 The analysis of 6 m long structures

Collector	Mass (kg)	Torque (kg-cm)	Deformation (mm)
Space frame	238	11,216	9.8
Torque tube	229	8990	7.76
Torque box	234	2171	7.02
Space tube	232	1753	8.2

Note The torque value of the Torque Box and Space tube structure can be made zero, but for calculation purposes, we have made it positive

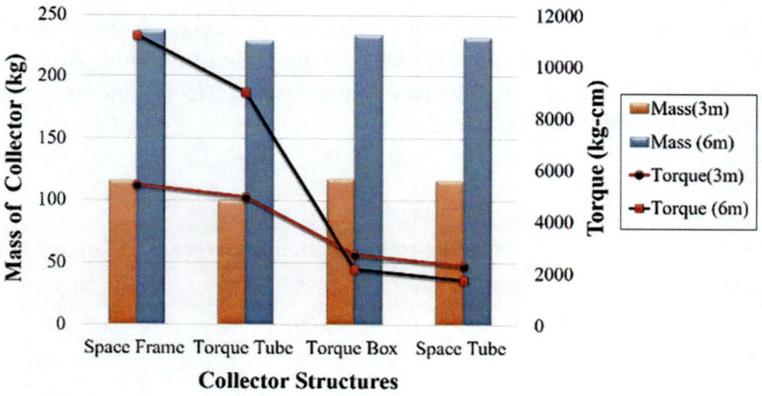

Fig. 7 Mass and torque value for 3 and 6 m length collector

4 Multicriteria Decision-Making Method

Multi-criteria Decision Making is the most well-known method of decision making. It deals with the decision problem when there are several decision criteria [14]. Also, to avoid any uncertainties due to the human-based decision-making process, MCDM-based decision-making was used [15]. When there are more than three alternatives and criteria, the MCDM like TOPSIS can be used for solving decision-making problems [16]. This method is widely used to decision making due to the efficient computation, and simple and easy-to-understand concept [17, 18].

In our study, the decision criteria were the mass of the collector, torque generation, ease of manufacturing, and deformation in the system. Here, we have used TOPSIS as an MCDM method to find out the best structure among the four major structures considering all four listed criteria. The Table 4 gives the details (or Matrix) of criteria and alternatives, and Table 5 gives a final solution.

The final solution of TOPSIS shows that the Torque Box collector structure stands at 1st rank with a performance score of 0.64. Therefore the Torque Box type collector structure is better among all four structures.

While Space Tube structure was at 2nd rank with a performance score of 0.61. The difference in the performance score of the Space tube and Torque Tube collector was very less. Hence, we can say that along with the Torque Box structure, the Space Tube structure can be considered for the development of the PTC system.

Table 4 Details of criteria and alternatives for TOPSIS problem

Alternatives	Criteria			
	Mass (kg)	Torque (kg-cm)	Deformation (mm)	Ease of manufacturing
Space frame	238	11,216	9.8	8
Torque tube	229	8890	7.76	10
Torque box	234	2171	7.02	9
Space tube	232	1753	8.2	8

Table 5 Performance score and rank

Collector	Space frame	Torque tube	Torque box	Space frame
Score	0.1139	0.2780	0.6429	0.6111
Rank	4	3	1	2

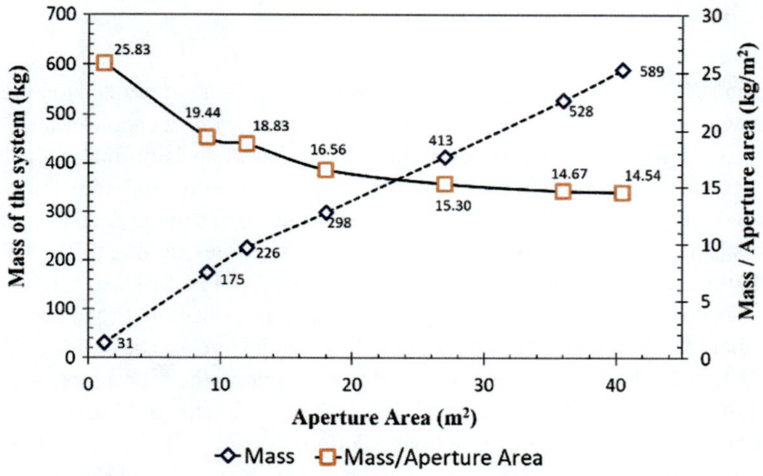

Fig. 8 Effect of increasing aperture area on mass per unit area

5 Effect of Increasing Aperture Area on a Mass Per Unit Area

To find out the effect of increasing aperture area on the mass per unit area of the collector we made several collectors with different aperture widths, and lengths. As MCDM shows the Torque Box collector as a better structure among all, we have considered the same for this analysis.

- Aperture was ranging from 1 to 6 m, and length ranged from 1.2 m to 9 m. The smallest, and largest area of the collector was 1.2 m^2 & 40.5 m^2 respectively.
- The analysis shows that as the area of the collector was increasing the mass per unit area decreased as shown in Fig. 8. This means, that the more aperture area per unit module will be better for cost reduction of the system, and The aperture area can be increased by increasing the aperture of the collector or/and increasing the length of the collector module.

6 Conclusion

The comparative analysis of four major structures of parabolic trough collector namely *Torque Tube, Space Frame, Torque Box, and Space Tube* was performed using finite element analysis, and structural optimization. Finally, using MCDM with criteria like the mass of the structure, Torque Generation, ease of manufacturing, and deformation the best structure was selected.

The Pros and Cons of all structures are as follow:

- *Torque Tube:* The fewer number of components compares to other structures makes it easy to assemble, and also, leads to lesser installation time which reduced installation cost. The main disadvantage of this structure is an unbalanced structure, which required counterweights for mas balancing.
- *Space Frame:* Here we can use a larger aperture but deformation in the structure is more, and also it may require a jig while assembly.
- *Torque Box:* The central box design made the structure stiff, and good distribution mass lead to the mass balanced structure. The use of nut and bolts make it weld less assembly. But more number of components may lead to more installation time.
- *Space Tube:* The helical truss design gives better strength to the structure, and like the Torque box structure due to better distribution of mass the structure is mass balanced. The only complexity in Space tube collectors is the designing and manufacturing of pipe connectors. In addition to this, a more number of components may lead to an increase in installation time.

6.1 Other Key Points

- The torque generation in the torque tube and space frame collector structure was high as compared to the other structures.
- Along with increasing the length of the structure, the torque generation was increased in Torque Tube and Space Frame structure, while in the case of the Torque Box and Space Tube structure it remains unchanged (or very small change) due to the better distribution of mass in the structure.
- The Torque Box and Space Tube structures are well mass balanced, therefore these structures are more cost-effective than other structures as there is no need for extra counterweights.
- The results of the MCDM show that Torque Box is better in all four collector structures. Therefore Torque Box structure can be considered for the current and future development of the trough collector system.
- In the development of the Space Tube structure, more focus must be given to developing simple tube connectors for easy assembly.
- Instead of a nut and bolt arrangement for fastening, other techniques like press-fit arrangement can be considered which can help in fast assembly.
- Finally, we conclude that further research in the design and development of *Torque Box,* and *Space Tube* structures can be led to the innovative, cost-effective, and simpler design of trough collector structures.
- In the development of the collector, the more aperture area per unit module will be better for cost reduction of the system. The aperture area can be increased by increasing the aperture of the collector or/and increasing the length of the collector module.
- It is also recommended that, along with the aperture of the collector, the length of the collector must be increased for better cost reduction.

References

1. Fernandez A, Zarza E, Valenzuela L, Perez M (2010) Parabolic-trough solar collectors and their applications 14:1695–1721. https://doi.org/10.1016/j.rser.2010.03.012
2. Belyakov N (2019) Solar energy. Sustain Power Gener Curr Status, Futur Challenges Perspect, pp 417–438. https://doi.org/10.1126/science.192.4236.202
3. Joardder MUH, Halder PK, Rahim MA, Masud MH (2017) Solar pyrolysis: Converting waste into asset using solar energy. Clean energy sustain dev comp contrasts new approaches. Elsevier Inc., pp 213–235. https://doi.org/10.1016/B978-0-12-805423-9.00008-9
4. Kodama T (2003) High-temperature solar chemistry for converting solar heat to chemical fuels. Prog Energy Combust Sci 29:567–597. https://doi.org/10.1016/S0360-1285(03)00059-5
5. Thomas A, Guven HM (1993) Parabolic trough concentrators-design, constructions and evaluation. Energy Convers Manag 34:401–416
6. Mohammad ST, Al-Kayiem HH, Assadi MK, Sabir O, Khlief AK (2018) An integrated program of a stand-alone parabolic trough solar thermal power plant: code description and test. Case Stud Therm Eng 12:26–37. https://doi.org/10.1016/j.csite.2018.02.006
7. Gharat PV, Bhalekar SS, Dalvi VH, Panse SV, Deshmukh SP, Joshi JB (2021) Chronological development of innovations in reflector systems of parabolic trough solar collector (PTC)—a review. Renew Sustain Energy Rev 145:111002. https://doi.org/10.1016/j.rser.2021.111002
8. Bellos E, Tzivanidis C (2019) Alternative designs of parabolic trough solar collectors. Prog Energy Combust Sci 71:81–117. https://doi.org/10.1016/j.pecs.2018.11.001
9. Geyer M, Solar A, Osuna R, Esteban A, Schiel W (2003) Eurotrough—parabolic trough collector developed for cost efficient solar power generation EURO TROUGH—parabolic trough collector developed for cost efficient solar power generation
10. Zou Q, Li Z, Wu H (2017) Modal analysis of trough solar collector. Sol Energy 141:81–90. https://doi.org/10.1016/j.solener.2016.11.026
11. Tao L, Ling X, Zhu Y (2008) Design of solar parabolic trough collector by FEM. Int Des Eng Tech Conf Comput Inf Eng Conf, pp 1–6
12. Liu J, Ye B, Guan X, Shi Y (2014) Mechanical analysis and structural optimization of the parabolic trough solar collector. Key Eng Mater 580:916–923. https://doi.org/10.4028/www.scientific.net/KEM.579-580.916
13. Schuknecht N, Mcdaniel J, Filas H, Schuknecht N, Mcdaniel J, Filas H (2018) Achievement of the $100/m^2 parabolic trough. AIP Conf Proc 2033, pp 030019-1–030019-10. https://doi.org/10.1063/1.5067035
14. Velasquez M, Hester P (2013) An analysis of multi-criteria decision making methods. Int J Oper Res 10:56–66
15. Maheshwari N, Choudhary J, Rath A, Shinde D, Kalita K (2021) Finite element analysis and multi-criteria decision-making (MCDM)-based optimal design parameter selection of solid ventilated brake disc. J Inst Eng Ser C 102:349–359. https://doi.org/10.1007/s40032-020-00650-y
16. Zhongyou X (2012) Study on the application of TOPSIS method to the introduction of foreign players in CBA games. Phys Procedia 33:2034–2039. https://doi.org/10.1016/j.phpro.2012.05.320
17. Rahim R, Supiyandi S, Siahaan APU, Listyorini T, Utomo AP, Triyanto WA, Irawan Y, Aisyah S, Khairani M, Sundari S, Khairunnisa K (2018) TOPSIS method application for decision support system in internal control for selecting best employees. J Phys Conf Ser 1028. https://doi.org/10.1088/1742-6596/1028/1/012052
18. Balioti V, Tzimopoulos C, Evangelides C (2018) Multi-criteria decision making using TOPSIS method under fuzzy environment. Application in spillway selection 637. https://doi.org/10.3390/proceedings2110637

Effect of Induction Heating in Minimizing Cold Start Emissions in Catalytic Converter

Sumana Dey, Ankan Man, Kamlesh Sahu, and Bijan Kumar Mandal

1 Introduction

The advent of civilization has resulted in an increase of toxic gaseous pollutants due to rampant use of fossil fuels. As a result, today's fuels are subjected to stringent emissions norms to comply with the standards set by different regulatory bodies throughout the world. Hence, reducing emissions of harmful gases from the IC engines is an important step towards a greener environment as the emissions from the IC engines contribute significantly to environmental pollution and are a threat for human health. As a result, the progressive strictness in the emission norms for reducing carbon monoxide (CO), hydrocarbons (HC), nitrogen oxides (NO_x) and particulate matter forces car manufacturers to develop exhaust after treatment systems to meet these emission standards. One of the many ways to deal with this problem would be to reduce emissions by addressing the issues caused by Cold Start.

Cold start occurs when the vehicle starts after being non-operational for a period of time and the engine is colder than its normal operating temperature. During cold start air becomes denser and low intake air temperature causes autoignition difficulties in CI engine and reduces flame speeds in SI engine. This affects the air fuel ratio, which in turn affects the flammability of the mixture as mentioned by Yusuf and Inambao [1]. Catalytic converters are used to convert the toxic by-products of fuel into less hazardous substances by mitigating carbon monoxide (CO), hydrocarbon (HC), oxides of nitrogen (NO_x) and particulate matter emissions from internal combustion engines into carbon dioxide (CO_2), water vapour (H_2O) and nitrogen gas (N_2). To achieve efficient toxic exhaust gas conversion, catalytic converters must operate at temperatures greater than 300 °C with conventional catalysts using platinum group metal catalysts such as Pt, Pd, Rh, which does not happen during cold starts. Liu et al.

S. Dey (✉) · A. Man · K. Sahu · B. K. Mandal
Indian Institute of Engineering Science and Technology, Shibpur, Howrah 711103, India
e-mail: sumanadey492000@gmail.com

S. Doolla et al. (eds.), *Advances in Clean Energy and Sustainability*,
Green Energy and Technology, https://doi.org/10.1007/978-981-99-2279-6_30

[2] identified cold start is responsible for about 45–65% of the net urban emissions which represents the most toxic period of the engine operating cycle.

Miao et al. [3] proposed the change in the position of the SCR (selective catalytic reduction) unit which can change the light-off significantly. Bakatwar et al. [4] observed that the improvement in the thermal management system using electric heaters controlled by a thermostat showed a considerable reduction in the emissions. One of the new methods to reduce cold start is the preheating of the catalyst to reach light off faster. Aktas et al. [5] carried out an experimental investigation of cold start emission using a preheating system on the exhaust line in a spark ignition engine. The experimental and numerical analysis of Kurzydym et al. [6] showed that the flow distribution through the monolith structure affected the catalyst utilization, conversion efficiency and faster light off during cold start. Dey et al. [7] reviewed the effectiveness of manganese oxide catalysts for catalytic oxidation of carbon monoxide at ambient conditions. Manganese oxide catalyst in ambient conditions are very active for the overall oxidation of carbon. Williamson et al. [8]) noted that the use of an upstream resistance heater with the conventional catalytic converter unit, had shown incredible results in solving the issues caused by cold start but compromised the compactness and increased heat losses in the after treatment system.

The idea of induction heated catalyst in a catalytic converter is fairly new and not much research and experimental work has been done on its effects on light off temperature under variation of various physical parameters. The effect of different sizes and porosities on the thermal response of the converter for reaching the light off temperature have been addressed in this work.

2 System Description

Figure 1a, b show the corresponding CAD model and the schematic diagram of the physical model of the catalytic converter respectively. The catalytic converter model consists of an outer steel casing, inlet and outlet ducts, an insulation and shock absorbing mat, an induction coil (copper) housing, a ceramic (cordierite) honeycomb core and magnetic heating elements. The core of the catalytic converter is made up of cordierite ceramic material which is covered by a steel casing that accommodates the induction coils responsible for heating the cordierite. The core is divided into 3 sections longitudinally having a honeycomb structure with alternate open and closed gas flow passages. The honeycomb structure provides high catalyst surface area which maximizes the contact between the catalyst and the pollutants in the exhaust. The design of alternating open and closed channels forces the gasses to pass through the porous walls and the residence time of the gasses inside the converter is expected to increase, which will ensure effective heat distribution.

In conventional catalytic converters the internal surface of the whole monolithic core is coated with PGM (Pt, Pd and Rh) catalysts. Here, the core is divided into three chambers. The first section consists of Mn coated Fe catalyst that mainly converts CO to CO_2 according to the following equation:

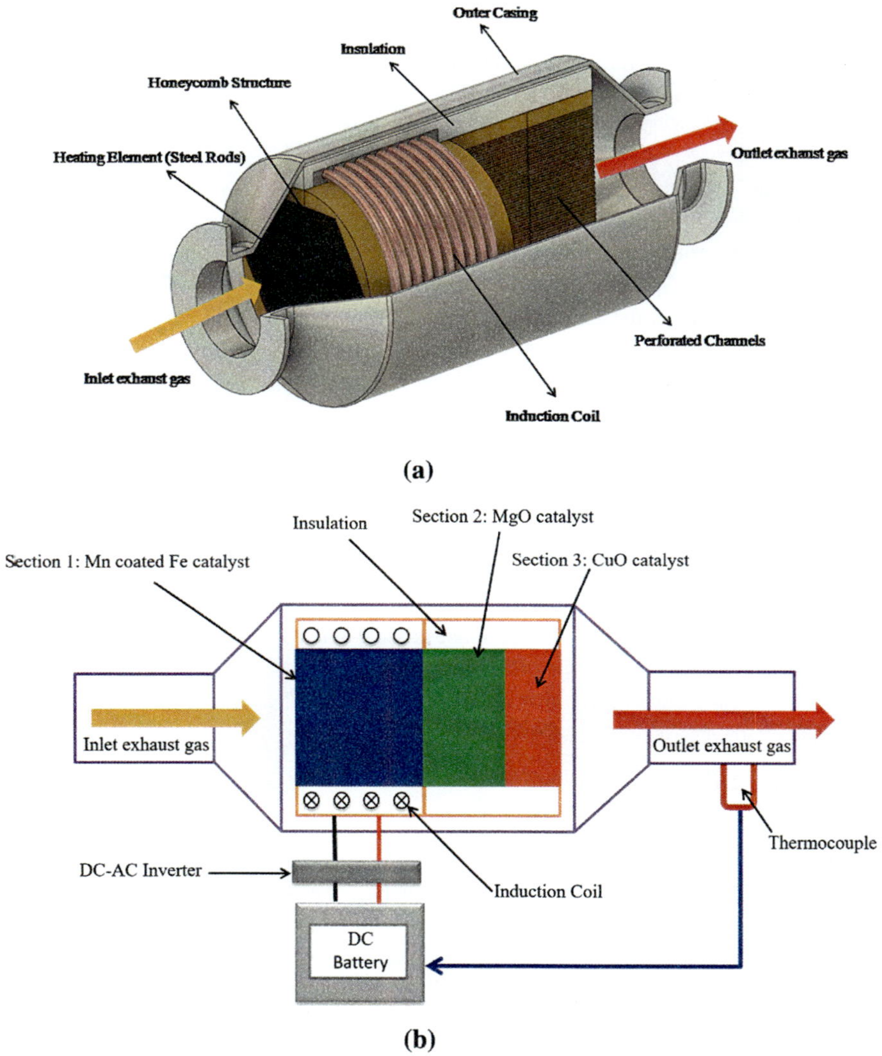

Fig. 1 **a** CAD model of catalytic converter, **b** schematic diagram

$$CO + O_2 \rightarrow CO_2 \tag{1}$$

The next section consists MgO as catalyst that mainly converts the NO_x to N_2 following the reaction:

$$NO_x + CO + O_2 \rightarrow N_2 + CO_2 \tag{2}$$

The third section contains CuO as catalyst that primarily converts the hydrocarbons to CO_2 and water vapour as shown below:

$$C_xH_x + O_2 \rightarrow CO_2 + H_2O \tag{3}$$

The monolithic substrate is perforated which entraps the particulate matter in them that would undergo passive regeneration due to high temperature of the core.

The light off temperature of catalysts used in the model is lower than PGM catalysts. The light off temperature of Fe coated Mn is about 153 °C. The sections are made keeping in mind the amount of corresponding reactants present in the engine out gasses and the light off temperature of the catalysts in ascending order for quicker light off.

Heating of the ceramic core and gas is accomplished using induction heating. Induction heating is a non-contact energy transfer process through an induced magnetic field in a ferromagnetic heating material. The heating of the ferromagnetic elements occurs due to two phenomena namely, eddy current heat dissipation and hysteresis effect. Alternating current applied to the coil induces an alternating electromagnetic field which causes eddy current losses and hysteresis losses in the steel rods. Steel rods are only introduced to the first section from which heat would be transferred to the monolith and exhaust gasses. The remaining two sections will be heated via conduction and convection.

3 Mathematical Model

The main objective is to observe the temperature rise of the catalyst and of the gas flowing in the monolith with time. For this purpose, transient heat transfer and fluid flow analysis in the catalytic converter have been carried out using commercially available software ANSYS Fluent. The simulation is governed by mass conservation of the working fluid, the momentum equation of the porous media and the energy equation for both the solid and fluid region. The heat source term in the energy equation is obtained by Maxwell's equation of electrodynamics.

3.1 Assumptions

The walls of the monolithic core are perfectly insulated. The porous media is assumed to be homogeneous. The heat generated due to induction heating is believed to be uniform. The combustion in the engine is considered to be uniform and hence the inlet velocity and the temperature are assumed to be constant. Effect of gravity is neglected since its effect would be negligible at such high fluid velocities. In k- ε model the effect of the porous medium on the turbulence field is only approximated.

3.2 Governing Equations

The governing equations considered and used in the ANSYS fluent for the present analysis are as follows:

Mass conservation equation (Continuity equation):

$$\frac{\partial \rho}{\partial t} + \nabla.(\rho U) = 0 \tag{4}$$

Momentum equation for porous media:

$$\frac{\partial(\rho U)}{\partial t} + \nabla.(\rho U_i U_j) = -\nabla p + \nabla.(\mu((\nabla_j U_i) + (\nabla_i U_j))) - \frac{2}{3}\mu(\nabla.U)I + \rho g + S \tag{5}$$

For simple homogeneous porous medium the source term in the standard Navier-strokes equation for momentum conservation is modified and written as

$$S = \frac{\Delta p}{\Delta L_i} = -\left(\frac{\mu}{\alpha}U_i + C_2\frac{1}{2}\rho|U|U_i\right) \tag{6}$$

In the above expression for S the first part represents viscous resistance, with proportionality coefficient $1/\alpha$ and the second term represents inertial resistance with inertial resistance coefficient C_2. The porosity and dimensions of the catalytic converter help in determining the values of $1/\alpha$ and C_2. For determining the viscous and inertial resistance coefficients for the porous media, an experimentally determined pressure drop correlation is considered following Nowak et al. [9] and presented below.

$$\Delta p = aU_i^2 + bU_i \tag{7}$$

Now comparing Eqs. (6) and (7) the viscous and inertial resistance coefficients can be determined by the following relations:

$$\frac{1}{\alpha} = \frac{b}{\mu \Delta L_i} \text{ and } C_2 = \frac{2a}{\Delta L_i} \tag{8}$$

Energy Equations:

For non-equilibrium simulation of porous media, a dual cell approach is used. A solid zone that is spatially coincident with the porous fluid zone and the solid and fluid zone interact only with respect to heat transfer. Heat transfer takes place between the induction heated cordierite and the exhaust gas. The energy conservation equation for fluid zone is given by

$$\frac{\partial}{\partial t}\left(\rho_f E_f\right) + \nabla.\left(U\left(\rho_f E_f + p\right)\right) = \nabla.\left(\gamma k_f \nabla T_f - \sum_i h_i J_i + \tau.U\right) + S_f^h + h_{fs} A_{fs}\left(T_s - T_f\right) \tag{9}$$

And the conservation equation for the solid zone is given by

$$\frac{\partial}{\partial t}((1-\gamma)\rho_s E_s) = \nabla.((1-\gamma)k_s \nabla T_s) + S_s^h + h_{fs} A_{fs}\left(T_f - T_s\right) \tag{10}$$

The source terms in Eqs. (9) and (10) are represented by $h_{fs} A_{fs}\left(T_s - T_f\right)$ and $h_{fs} A_{fs}\left(T_f - T_s\right)$ respectively.

3.3 Boundary Conditions

The velocity is taken to be uniform and constant at the inlet section, i.e., at $z = 0$. The boundary conditions at the inlet section can be written mathematically as

$$\frac{1}{r}\frac{\partial v}{\partial \theta} = 0, \quad \frac{\partial v}{\partial r} = 0, \quad \frac{\partial v}{\partial t} = 0 \tag{11}$$

The walls of the catalytic converter are insulated, so there is no heat transfer across the walls. Therefore, at $r = R$, $\frac{\partial T}{\partial r} = 0$. The exhaust gases are released to the atmosphere. So, at outlet, $p = p_a$.

4 Simulations

First a simplified 3D model of the catalytic converter was created using Design modeler using the standard dimensions available in the literature. However, the geometry was simplified by taking only the monolithic core of the converter; all of the insignificant details such as the filets, grooves, edges were omitted from the design. The core was divided into three sections, and the first section was sliced into an internal solid core and an outer annular region. Models of three different sizes have been used for simulation categorized by the power output of the engine to be used in the vehicle as shown in Table 1, 2 and 3.

The accuracy of the simulation depends on the quality and number of mesh elements generated. Since the mesh generated by ANSYS Workbench is unstructured, the size of each cell is not the same. Table 1 also shows the number of control volumes and nodes in each model.

The viscous model used in the simulation is the realizable k-ε model, which is an improved version of the standard k-ε model. It provides an alternative formulation for the calculation of turbulent viscosity. The dissipation rate has been derived from

Table 1 Model parameters

Model	Length of monolith (in.)	Dia. of monolith (in.)	Inlet diameter (in.)	Outlet diameter (in.)	No. of elements	No. of nodes
1	16	8.46	3	3	1,874,049	434,452
2	18	8.46	3	3	2,413,205	553,455
3	20	8.46	3	3	2,676,296	613,792

Table 2 Viscous and inertial resistance for different models

Model	C_2			$1/\alpha$		
	X	Y	Z	X	Y	Z
Model 1	7	11	11	300,000	600,000	600,000
Model 2	6	11	11	280,000	600,000	600,000
Model 3	5.5	11	11	270,000	600,000	600,000

Table 3 Viscous and inertial resistance for different porosities for Model 1

Porosity	C_2			$1/\alpha$		
	X	Y	Z	X	Y	Z
0.60	7	11	11	300,000	600,000	600,000
0.45	7.5	12	12	320,000	660,000	660,000
0.35	7.7	13	13	350,000	700,000	700,000

an exact equation for the transport of the mean-square vorticity fluctuation. The near wall treatment criteria have been set to enhance wall treatment which allows non-linear variation of the flow parameter and thermal heating close to the wall. The irreversibility between the fluid and its adjacent layer causes heat to develop, termed as viscous heating, and has been taken into account under the k-ε model.

The working fluid has been considered to be air. The values of the air parameters have been assumed constant over the temperature range of 100–250 °C. An average value has been considered for the density, viscosity, specific heat and thermal conductivity of air. The porous solid material is made of cordierite.

A non-equilibrium model is assigned to the fluid cell zone which is present in the porous part of the catalytic converter. The porosity of the cordierite is taken as 45% for standard simulation. The porosity is taken as 35, 45 and 60% for Model 1 for understanding its effect on light off. The inertial and viscous resistances are calculated using Eq. (8) and using the pressure versus velocity graph presented by Kurzydym et al. [6]. The solid core is given a uniform heat source term of 4.5×10^6, 3.4×10^6 and 3×10^6 W/m^3 for Model 1, Model 2 and Model 3 respectively, corresponding to a power of 15 kW provided via induction heating.

For the CFD simulation a SIMPLE pressure–velocity coupling scheme is used and flux type estimation is done using Rhie-Chow interpolation. The spatial discretization of the convection terms in the solution equations is done by setting the method for computing the gradient as least square based. The momentum is calculated using second order upwind method. Pressure equation is solved by a second order solver and for turbulent kinetic energy and turbulent dissipation rate and energy equation second order upwind method is used. The transient formulation is set to first order implicit type. It was initialized with inlet velocities of 5 m/s, 0 m/s and 0 m/s in the x, y and z directions respectively.

5 Results and Discussion

The scaled residuals of various parameters (momentum, velocity, energy, etc.) after each iteration have been plotted and tested. By default, in ANSYS Fluent the residuals must be less than 10^{-2} for the solution to be convergent and that has been strictly obeyed for all the parameters.

Figure 2 shows the velocity contour at 60 s after the key start which has a close resemblance with that observed by Kurzydym et al. [6]. As the gases enter the porous monolith, due to high inertial and viscous resistance, some of the pressure and kinetic energy gets converted to heat energy giving a negative pressure and velocity gradient over the complete monolith. As a result, highest pressure is observed just before the gases enter the porous media. Due to the high resistances offered by the porous substrate recirculation of flow occurs in the diverging inlet section which results in eddy losses.

Figure 3 shows the temperature contours at different times. One can observe from the figures that the rate of rise in volume averaged temperature of the air (exhaust gas) and solid substrate reduces as the lengthwise dimension of converter is increased with constant input power of the induction heater (15 kW). The light-off temperature

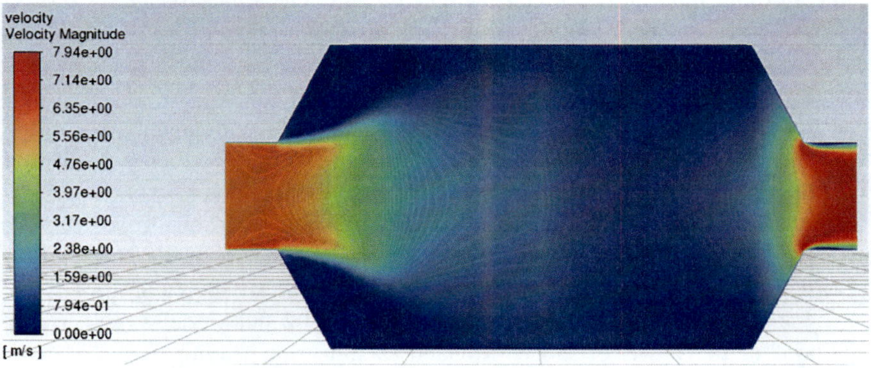

Fig. 2 Velocity contour at 60 s after key start

for the catalyst used in heating section is 153 °C and this temperature is reached within the Model-1 in 45–50 s, but with same power input the light off temperature will be reached after 60 s for the larger converters as shown in Fig. 4. Since the length of heating section increases for Model 2 and 3, heat source value (W/m^3) decreases with the same input power of 15 kW. This decrease in volumetric heat generation results in reduction in temperature rise. This can be attributed to the increase in volume, decrease in inertial resistance which in turn increase advection heat transfer reducing temperature rise. The viscous resistance also decreases with increase in size, as a result heat generation due to viscous heating reduces. The values of input power values for the larger converters for achieving light-off temperature in 40–50 s are estimated to be 20 kW for model-2 and 22.5 kW for Model-3 respectively.

The snapshots of the temperature contour at 30, 60, 90 and 120 s are taken for studying the heat distribution inside the catalytic converter. With the initiation of air flow, the heat from the heating section is convected away to the later sections. Most of the heat is concentrated in the first two sections. This is due to the lower values of thermal conductivity of cordierite substrate and air and also due to the high values of flow resistances offered by the porous substrate medium. As in the present catalytic converter model, utilizing the contactless power transfer nature in induction heating, the heat can be generated in an inner core which in turn reduces excess heat flow towards the walls, hence reduces heat losses from the walls. Thus, induction heating reduces losses, eradicates the use of costly insulation materials and safeguards wall materials from reaching failure temperatures.

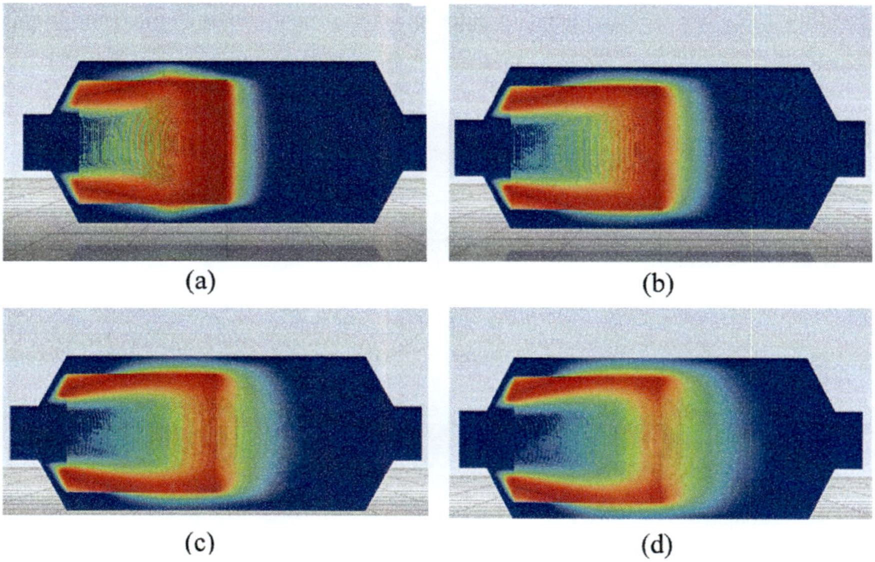

Fig. 3 Temperature contours at **a** 30 s **b** 60 s **c** 90 s **d** 120 s

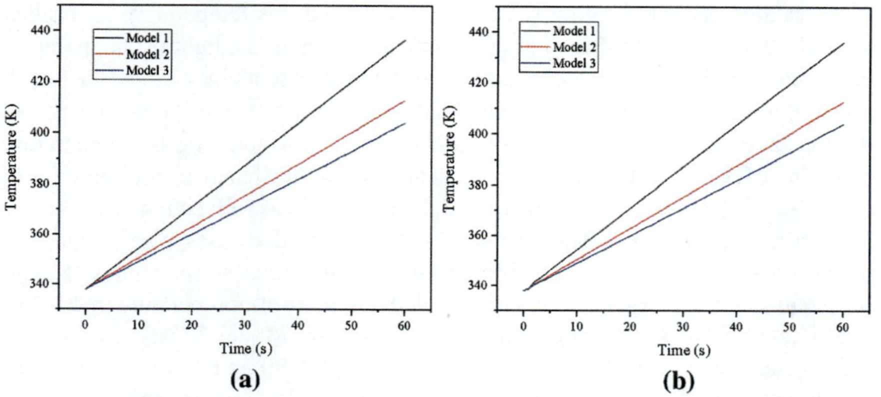

Fig. 4 Temperature versus time for **a** exhaust gas **b** for solid core For different Models

Variation in porosity can be caused by various reasons. For the present model this can be because of the presence of heating elements inside the core, or due to concentrated high heat loads which may cause sintering of the substrate material.

It is observed from Fig. 5a, b that the temperature of gases flowing inside the substrate increases at a faster rate when the porosity of the medium is increased. The same trend is observed for the solid medium also. With the increase in porosity the volume of solid portion decreases and the temperature rise will be faster for solid. On the other hand, for the gases, with increase in porosity, flow resistances decrease, mass flow rate increases and hence greater amount of heat is being convected to the gases resulting in faster temperature rise.

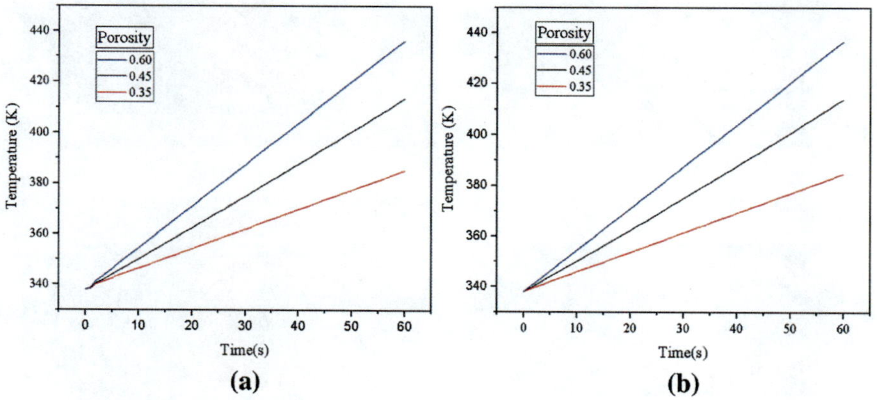

Fig. 5 Temperature versus time for **a** exhaust gas **b** solid core for different porosities

6 Conclusion

A 15-kW induction heater helps the converter (Model 1) to reach its light-off temperature within 45–50 s after key start. Increase in size showed decrease in rate of temperature rise in all three converter models. Through the study of flow and heat distribution it is found that heat losses at walls are greatly reduced, while re-circulation losses at inlet section still exist. These losses can be reduced with proper inlet section design of the converter by reducing the divergence angle. It was also observed that the decrease in porosity will lead to reduction in the rate of temperature rise for both solid and gas inside the core. With porosity of 0.6, the light-off is reached in about 45 s for Model 1. When porosity is decreased to 0.45 and 0.35, it is observed that it will take more than 60 s to reach light-off. Hence, porosity of 0.6 is desirable but a too high porous medium will make the substrate fragile. In summary, the simulation results show that the induction heated catalytic converter can be a useful technology for decreasing cold-start emissions in both small and large IC based vehicles.

References

1. Yusuf A, Inambao F (2019) Effect of cold start emissions from gasoline-fueled engines of light-duty vehicles at low and high ambient temperatures: recent trends. Case Stud Therm Eng 14:100417
2. Liu B, Zimmerman N (2020) Fleet-based vehicle emission factors using low-cost sensors: case study in parking garages. Transp Res Part D: Transp Environ 91:102635
3. Miao Y, Chen LD, He Y, Kuo TW (2009) Study of SCR cold-start by energy method. Chem Eng J 155:260–265
4. Bakatwar R, Nesamani K, Bhargava A, Jain R (2018) Performance analysis & optimization of engine cooling system by using electronically controlled thermostat for improving thermal efficiency. SAE Technical Paper 01-0053
5. Aktas F, Dinler N, Karaaslan S, Turker A, Yucel N (2021) Experimental investigation of cold start emission using preheating system on the exhaust line at the idle conditions on a spark ignition engine. Sadhana 46:136
6. Kurzydym D, Klimanek A, Żmudka Z (2018) Experimental and numerical analysis of flow through catalytic converters for original part and WALKER's replacement using reverse engineering and CFD. IOP Conf Ser Mater Sci Eng 421:042044
7. Dey S, Mehta N (2020) Selection of Manganese oxide catalysts for catalytic oxidation of carbon monoxide at ambient conditions. Resour Environ Sustain 1:100003
8. Williamson WS, Gonze EV (2008) Diesel particulate filter (DPF) regeneration by electrical heating of resistive coatings. U.S. Patent No. 7, 469, 532. 30 Dec
9. Nowak R (2016) Estimation of viscous and inertial resistance coefficients for various heat sink configurations. Procedia Eng 157:122–130

Thermal Properties of Nano-SiO_2/Paraffin Composite Phase Change Material for Thermal Energy Storage

Neetu Bora, Jaspreet Singh Aulakh, and Deepika P. Joshi

1 Introduction

In current time, the deficiency of energy resources and rapid using up of fossil fuels for its production has been increasing serious environmental issues [1]. So, to maintain a harmonious relationship between energy and the environment, it is needed to look for alternatives such as clean energy resources and a large number of energy techniques. Renewable energy sources (RES) and thermal energy storage (TES) techniques have attracted the attention of lots of researchers [2, 3]. TES includes sensible heat storage (SHS) and latent heat storage (LHS). LHS has better energy strength with invariable temperature throughout the phase change procedure. However, the intermittent nature of RES limits its applications. Therefore, TES solves the intermittency issues of RES. In TES, PCM has been recognized as a better candidate for thermal energy storage material because of its larger energy density which helps to stock up a huge amount of thermal energy in the form of latent heat [4]. This means PCM can successfully decrease the gap and sustain the regularity among the availability and utilize of energy. It can reduces energy wastage and thus, increase the energy consumption appreciably because PCM can soak up/release a vast amount of thermal energy contained by a narrow temperature range when a change of form occurs between solid to liquid form [2]. Because Organic PCM doesn't change over time, hence these PCM has been proven to have a good potential for usage in solar heating systems, water heating and building materials. It also exhibits outstanding electrical insulator properties with a resistivity of between 10^{12} and $10^{17} \Omega$ meters. However, some paraffin PCM characteristics, including as form instability and lw heat conductivity, need to be enhanced for the thermal storage application [5–8]. The SiO_2/paraffin PCM composite has been subjected for TES in the current work. In which silicon dioxide

N. Bora (✉) · J. S. Aulakh · D. P. Joshi
Department of Physics, Govind Ballabh Pant University of Agriculture and Technology, Pantnagar, Uttrakhand 263145, India
e-mail: neetubora95@gmail.com

S. Doolla et al. (eds.), *Advances in Clean Energy and Sustainability*, Green Energy and Technology, https://doi.org/10.1007/978-981-99-2279-6_31

(SiO_2) NP_S is used to improve the shape stability of PCM. SiO_2 NPs is an easily available and eco-friendly compound [9]. The role of SiO_2 NPs is to confine paraffin in its nature to form the shape stable PCM. In the present work, an effort has been prepared to study the effect of SiO_2 NPs on the melting temperature, shape-stability, latent heat capacity, and thermal stability of the PCM composite.

1.1 Innovativeness of Work

The proposed SiO_2/Paraffin composite PCM has completely eliminated the paraffin leakage by impregnating nano-silica in paraffin, which can absorb a significant quantity of leaking paraffin. Silica has increased the thermal stability of paraffin. This makes it a potential candidate in TES and specialized fields like waste heat recovery systems, Photovoltaic electricity systems, Heat pump systems, Thermal Protection of electricity, etc.

1.2 Materials Used

The chemical used in this experiment is Tetraethylortho silicate (TEOS, 99%), purchased from Sigma Aldrich; Ethanol (C_2H_2OH, 99.9%), Ammonium hydroxide (NH_2OH, 29% in water), and Paraffin wax obtained from Hi-media. All the chemicals are used as received without further purification. Double distilled water is used throughout the experiment.

1.3 Tailoring of Samples

The Stöber method has been used to synthesized silica (SiO_2) NPs. Ethanol (290 ml), water (30 ml), and ammonium hydroxide (8 ml) are mixed and stirred for half an hour at room temperature using a magnetic stirrer. TEOS (Tetraethylortho silicate) is introduced dropwise and then stirred magnetically at room temperature for three hours. After three-hour, a white precipitate solution is observed and centrifuged at 12,000 rpm for 10 min [9]. The obtained white precipitate is washed with ethanol 3 to 4 times and dried at room temperature. Finally, white color monodispersed silica NPs have been obtained. Thereafter, the as-prepared silica particles have been directly mixed with paraffin for making nano-SiO_2/paraffin composite PCM. In this manner, a sequence of form shape stable PCM with SiO_2 NP contents of 10, 15, and 20 wt% have been synthesized. These samples have been referred to as SPCM1, SPCM2, and SPCM, respectively.

2 Characterization

The crystalline study of prepared SiO_2 NPs and SiO_2 (20%)/Paraffin composite (SPCM) has been done with a XRD (X-ray diffractometer) and morphological study has been done by FESEM (Field emission scanning electron microscope). Further, the thermal stability of composite has been assessed by TGA (Thermo-Gravimetric Analysis) performed under an N_2 atmosphere from temperature around 30 to 600 °C with a heating rate of 10 °C/min. The phase change properties of paraffin and SPCM have been characterized by DSC under an N_2 atmosphere with a flow rate of 20 ml/min. Numerical integration of the region under the peaks corresponding to the solid–solid and solid–liquid phase transitions has been used to determine the heat storage capacity of paraffin and SPCM. Also, the form-stability of samples has been determined by the leakage test. In this test, samples have been undergoes 130 thermal cycles in the hot air oven at 75 °C and the leakage rate of paraffin can be calculated by the formula [2]

$$L = \frac{W_o - W_n}{W_o} \times 100\%$$

where W_o and W_n are the weight of the sample before and after thermal treatment.

3 Result and Discussion

3.1 Form-Stability Analysis

A seepage test was used to examine the SiO_2/Paraffin composite's ability to the leaking through 130 heat cycles, each lasting an hour. Sample SPCM1 and SPCM2 show high paraffin leakage and leakage continues on increasing thermal cycles as depicted in Fig. 1. However, the leakage rate for the SPCM sample is below ~ 0.15% after 130 thermal cycles. So, when the mass fraction of SiO_2 NPs into paraffin reached 20%, negligible paraffin leakage has been observed, suggesting the excellent form-stability of composite SPCM. Therefore, findings suggested that SiO_2 NPs have a significant effect on the leakage- bearing properties of PCM. This could be due to the high porosity, good absorbent, large surface to volume ratio and good surface chemistry of SiO_2 NPs. The optical photos of the sample are shown in Fig. 2. Further in this paper, only the SPCM sample has been taken for TGA and DSC characterizations.

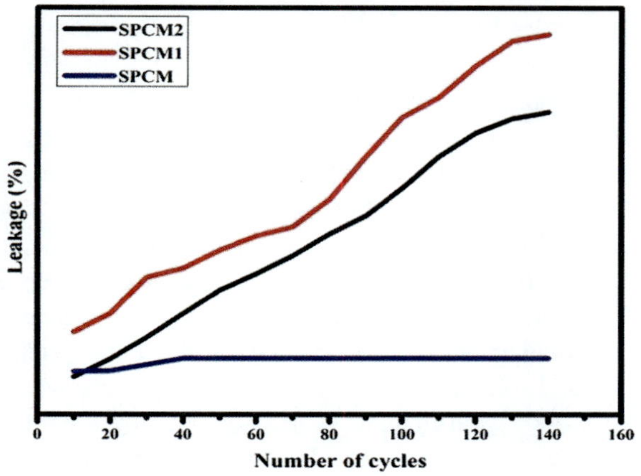

Fig. 1 Leakage rate of SPCM1, SPCM2, and SPCM

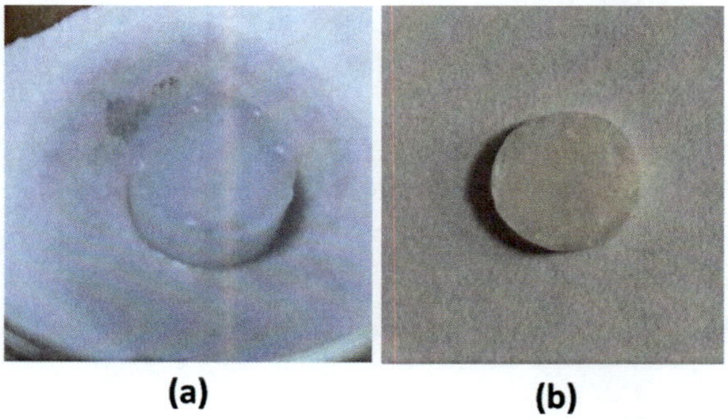

Fig. 2 Optical photographs of **a** paraffin and **b** SPCM sample during the leakage test

3.2 Crystalline and Morphological Analysis

Figure 3a, b has been shown the corresponding XRD diffractogram analysis of the SiO_2 NPs, paraffin, and SPCM composite. For silicon dioxide, just a single wide diffraction peak has been found at diffraction angle 17°–24° (10–11) which is indicative of the amorphous phase of SiO_2 NPs. In Fig. 3b Paraffin shows two narrow, sharp diffraction peaks that appeared at 2θ of 21.37° and 23.72° and some small peaks, respectively. This denotes the crystalline nature of pure paraffin wax, respectively [2]. Some small peaks have been observed due to sonication and overheating during composite formation [5]. The SPCM composite shows all peaks of paraffin which indicate that there is no chemical reaction between the SiO_2 NPs and paraffin PCM.

These samples are scanned using FESEM for the morphological study of SiO_2 NPs and SPCM composite. Figure 4a shows Images of SiO_2 NPs, which evidence that the SiO_2 NPs are homogeneous, uniform, and spherical [9]. While the Fig. 4b shows Dispersion of SiO_2 NPs within paraffin PCM. The structure of paraffin wax as a bundle of stratums and also shows the mixing is homogenous Additionally, the

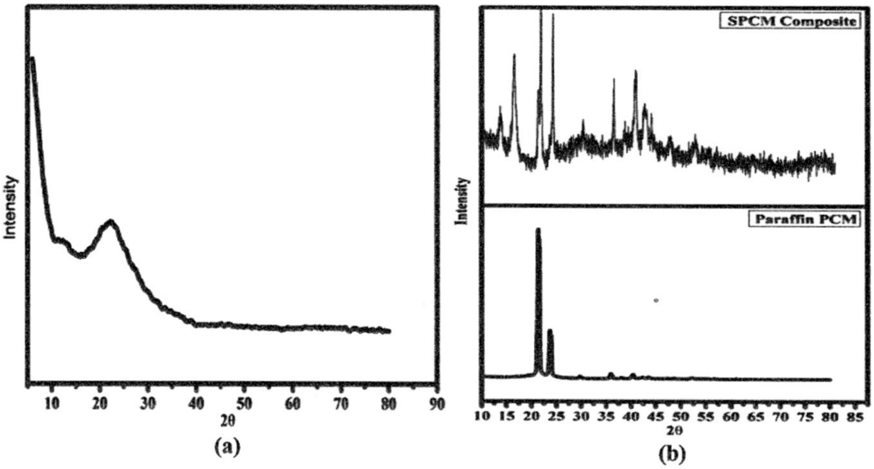

Fig. 3 XRD graph of **a** SiO_2 NPs **b** pure paraffin and SPCM composite

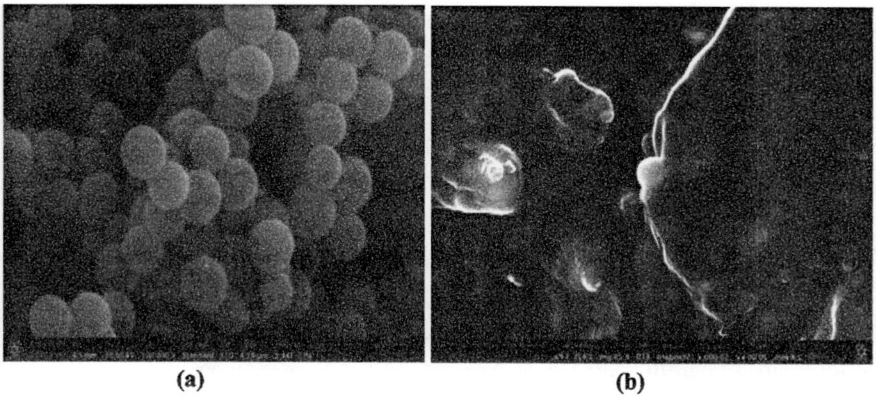

Fig. 4 FESEM image of **a** SiO_2NPs, **b** SPCM composite

photos gathered proved that SiO_2 NP aggregation is negligible, and Fig. 4b ensures proper physical interaction with homogenous mixing of SiO_2 NPs with the base PCM [5].

3.3 Thermogravimetric Analysis

The TGA curves of paraffin and SPCM are displayed in Fig. 5 and show the same pattern. Figure 5 indicates that both pure paraffin and SPCM show one-step decline in the weight loss curve. The degradation begins at around ~ 220 °C and finishes at

Fig. 5 TGA curve of paraffin PCM and SPCM composite

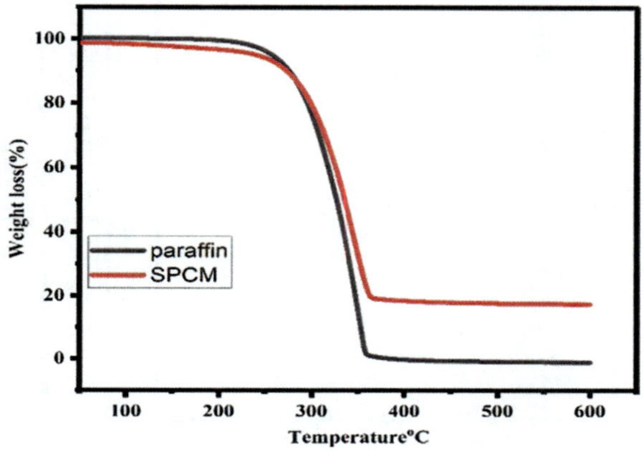

around ~ 350 °C in pure paraffin. While in SPCM composite the degradation started from around 240 °C and ends at 380 °C, respectively. TGA curves reveal that the impregnation of SiO_2 to the paraffin affects the degradation rate. The results revealed that the SPCM maintains good thermal stability up to 240 °C. TGA curve of SPCM composite indicates that the beginning and ending point of degradation temperature enhanced by the homogeneous mixing of SiO_2 NPs into paraffin wax, which means it increases thermal stability and postpones the high-temperature decomposition of paraffin wax.

3.4 Latent Heat Enthalpy and Thermal Reliability

Figure 6 illustrates the DSC thermograms of paraffin and SPCM composite. DSC thermograms of paraffin and SPCM have shown two transition peaks as represented in Fig. 6. The minor peak at ~ 46.64 and 47.09 °C corresponds to a solid-to-solid phase transformation, whereas, the prominent peak at ~ 62.77 and 63.03 °C demonstrates the solid to the liquid phase transition of the paraffin PCM and SPCM composite. The latent heat of paraffin and SPCM composite is found ~ 225.5 J/g and ~ 189 J/g, respectively. There is only 16.18% deterioration in latent heat enthalpy of paraffin has been noticed with the impregnation of SiO_2 NPs (20 wt%). Hence, paraffin PCM and SPCM composite have a mass ratio of about the same value as the latent heat ratio for the composites. It indicates that the SiO_2 NPs in paraffin slightly increase the peak melting point of paraffin.

Additionally, the thermal reliability test has also become significant for checking potential candidature for TES purposes by investigating their long-term stability. For this purpose, the SPCM sample has been selected because in this sample no leakage has been found. The thermal reliability of the sample has been determined via DSC, and the results are displayed in Fig. 7. From the results, the latent heat enthalpy of sample SPCM before/after 130 thermal cycles has been found as 189/141.78 J/g

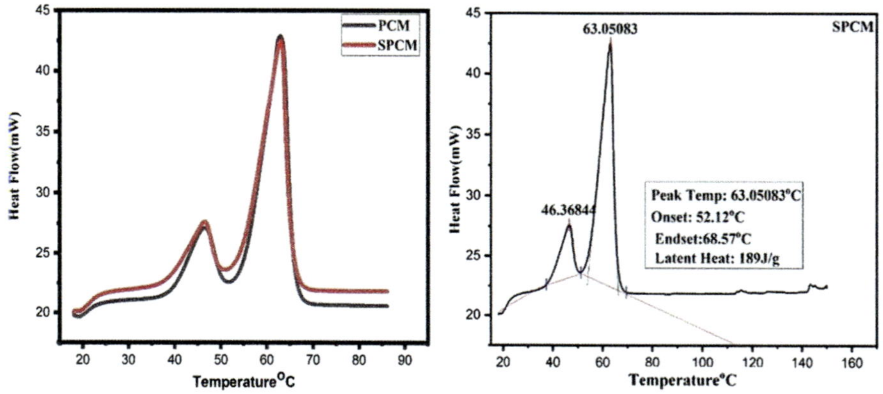

Fig. 6 DSC curve of Paraffin and SPCM composite

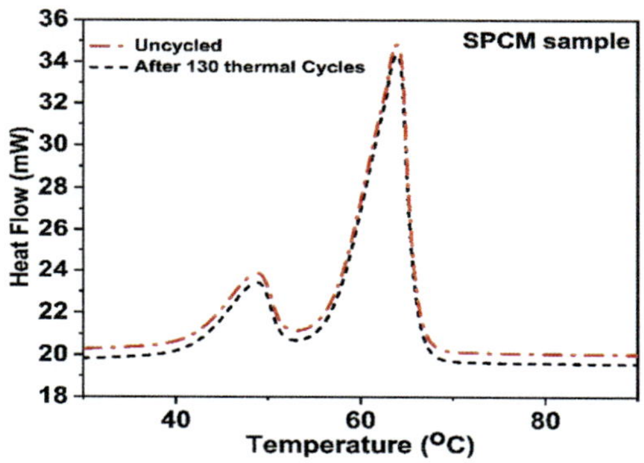

Fig. 7 Thermal reliability of SPCM before and after thermal cycles

respectively. It can be observed that SPCM sample has been shown decrement in their latent heat storage capacity after several thermal treatments which may be due to porous structure of SiO_2 NPs. The developed PCM composite has been shown excellent circulation stability.

4 Conclusion

In the present reported work, the direct mixing of SiO_2 NPs (20% wt.) in paraffin wax has produced an environmentally friendly composite PCM that is both shape and thermally stable. Leakage of paraffin wax is fully stopped by silica nanoparticles. FESEM analysis revealed that the SiO_2 NPs are uniform, monodispersed, and spherical. Silica NPs cease the leakage of paraffin wax approximately 0.15%. DSC and TGA analyses were used to determine the thermal characteristics, and the

DSC curves showed a 16.18% reduction in latent heat compared to pure paraffin. The SPCM sample showed long-term circulation stability and increased thermal stability. According to a TGA analysis of the thermal deterioration of paraffin, the temperature at which paraffin thermally degrades has been raised by the impregnation of SiO_2 NPs, increasing the thermal stability of the paraffin wax. The proposed shape stable and leakage-proof composite material can be the best candidate for TES applications.

References

1. Aulakh JS, Joshi DP (2022) Thermal and morphological study of paraffin/SEBS/expanded graphite composite phase change material for thermal energy storage. Energy Sour, Part A: Recovery, Utilization, Environ Effects 44(1):986–1003
2. Aulakh JS, Joshi DP (2022) Development of paraffin-based shape-stable phase change material for thermal energy storage. Polym Sci, Ser A, 1–10
3. Aulakh JS, Joshi DP (2019) Thermal and heat transfer performance of leakage-proof phase change material. J Emerg Technol Innov Res 6(2):176–179
4. Mahdi MT, Kara IH (2021) Thermophysical characteristic of nano-TiO_2 paraffin wax composite material. J Mech Eng Res Dev 44(6):48–58, ISSN:1024-1752
5. Paneliya S, Khanna S, Utsav, Singh PA, Patel KK, Vanpariya A, Makani HN, Banerjee R (2021) Core-shell paraffin/silica nanocomposite: a promising phase change material for thermal energy storage. Renew Energy 167:591–599
6. Zhou J, Zhao J, Cui Y, Cheng W (2020) Synthesis of bifunctionalnano encapsulated phase change materials with nano-TiO_2 modified polyacrylate shell for thermal energy storage and ultraviolet absorption. Polym Int 69:140–148
7. Quikhalfan M, Sari A, Chehouani H, Brahim B, Bicer B (2019) Preparation and Characterization of nano-enhanced myristic acid using metal oxide nanoparticles for thermal energy storage. Int J Energy Resour 1–16
8. Shah KW (2018) A review on the enhancement of phase change materials from a nanomaterials perspective. Energy Build 175:57–68
9. Kandpal D, Kalele S, Kulkarni SK (2007) Synthesis and characterization of silica-gold core-shell (SiO_2@Au) nanoparticles. Pramana—J Phys 69:277–283
10. Khan Z, Khan ZA, Sewell P (2019) Heat transfer evaluation of metal oxide based nano-PCMs for latent storage system application. Int J Heat Mass Transf 144:118619
11. Li H, Chen H, Li X, Sanjayan GJ (2014) Development of thermal energy storage composite and prevention of PCM leakage. Appl Energy 135:225–233

Performance Enhancement of Pyramid-Shaped Solar Still Using Phase Change Material with Porous Material

Sahil Chauhan, Kunal Gaur, Ajit, and Naveen Sharma

1 Introduction

Today, the clean and hygienic freshwater requirement has accelerated day by day. People are especially present in places where rivers, lakes, groundwater, and other sources of freshwater are inadequate [1–3]. Saltwater makes up around 97% of the total amount of water on the planet, with only 3% of that being drinkable and 2% being frozen and only 1% of water is accessible to human beings [4]. Also, some of the areas on the earth's surface don't have access to fresh water or we can say the drought regions on earth [5]. Furthermore, industrial and human wastes damage rivers, lakes, subterranean water, and wells. There is a greater scope of solar energy in southern India because of the high temperature and therefore developing solar technology to convert saline water to fresh water can impact greater. Rainwater harvesting, water pollution reduction, vacuum distillation, and reverse osmosis are just a few of the ways and strategies for preserving and purifying water, and many more [6, 7].

Solar desalination is one of the approaches used to purify saline water. A solar still is a device that uses solar energy to extract salt from saline water [8, 9]. According to the latest research, it has been discovered that highly salty water or seawater contains 10,000–35,000 ppm of dissolved salt, and in freshwater about 1000 ppm of dissolved salt. The solar desalination technique is used to reduce the amount of salt in water. The major advantage of using the solar desalination process is that regions with a

S. Chauhan (✉) · K. Gaur · Ajit
Department of Mechanical Engineering, Manav Rachna University, Faridabad, Haryana 121004, India
e-mail: sahilchauhan899@gmail.com

Ajit
Department of Mechanical Engineering, GLA University, Mathura, Uttar Pradesh 281406, India

N. Sharma
Department of Mechanical Engineering, Netaji Subhas University of Technology (West Campus), Jaffarpur, Delhi 110073, India

S. Doolla et al. (eds.), *Advances in Clean Energy and Sustainability*,
Green Energy and Technology, https://doi.org/10.1007/978-981-99-2279-6_32

lack of water are mostly high-temperature weather places and temperature is around 40–50 °C and it's an ideal environment for using solar still. But the major problem that comes here is the productivity of a solar still.

Many methods and techniques have been used to improve the production level of freshwater in a short period so they can be applied regularly. Kabeel and Abdelgaied [10] used paraffin wax in solar still with a single slope and the production increased up to 67–68.8% in comparison to traditional solar still. The maximum productivity was recorded as 7.54 L/m^2 when paraffin wax is used and 4.51 L/m^2 without paraffin wax. Sathyamurthya et al. [11] developed a solar still in the shape of a triangular pyramid and used paraffin wax as phase-changing material which helped in increasing the productivity level up to 100%. There are numerous aspects that influence the performance of the square basin pyramid solar still such as the surface area of the basin, the glass cover surface area, solar radiation intensity in the performing weather, latent heat temperature, wind velocity, coating inside the basin metallic box, depth of water, glass cover material. Abu-Hijleh et al. [12] experiment places sponge cubes at the basin of single slope solar still. When compared to solar without sponge cubes, overall water production increased by 273%. Shape-stabilized phase-changing material along with a solar collector in a pyramid solar still can increase the yearly production to 10,000 L/m^2 from 750 L/m^2 for shape-stabilized phase-changing material [13].

Vijayakumar et al. [14] assessed the change in the efficiency of single slope stepped solar still when humidifier-dehumidifier (HDH) is used along with paraffin wax. The production of water increased by 84.4% when solar still HDH and paraffin wax is used when compared to a solar still using only HDH. Kabeel and Abdelgaied [15] employed a thermal conductive graphite absorber plate to construct a pyramid solar still. Pyramid-shaped solar stills with absorber plates and traditional pyramid-shaped stills have efficiency ratings of 69.7–70.98% and 35.34–35.74%, respectively.

Different designs and shapes of solar still (like single slope, double slope, triangular pyramid, square pyramid, stepped having single slope) can also affect the efficiency of solar still. A pyramid-shaped solar still has a higher glass surface area than any other design which increases the production of freshwater. Solar still productivity as per daily usage is still very low and an improved solar still with a higher production level on daily bases is required. Using paraffin wax and clay pots, this experimental work aims to augment the freshwater output of a pyramid-shaped solar still.

2 Experimental Setup and Procedure

Figures 1 and 2 show the schematic and photograph respectively of developed pyramid-shaped solar still with a square basin in the mechanical department, Manav Rachna University, Faridabad (28.3922 °N latitude and 77.3017 °E longitude). The basin's dimensions are 600 mm × 600 mm × 200 mm. The outer basin is made up of a wooden box and the inner basin is made up of a GI sheet. The box is covered by glass inclined at 30° angle. To improve heat absorption, the GI sheet box is painted black. The solar still top is designed in a pyramid shape to increase its surface area for water vapours. Temperatures of the wall, the square basin, the water, and the inner glass were all measured using K-type thermocouples. Wind velocity and sun radiation intensity are measured using an anemometer and a pyranometer, respectively, with an accuracy of ± 0.4 m/s and ± 3 W/m^2 during the working hours of the experiment. Table 1 represents the details of the measuring instrument used in this research work.

The experiments were conducted of three different configurations, as shown in Fig. 3, i.e. Case 1—solar still without paraffin wax and clay pots, Case 2—solar still with paraffin wax and clay pots, and Case 3—solar still with paraffin and black painted clay pots. 500 g of Paraffin wax was kept below the absorber plate and a total of 25 clay pots facing downward were employed over the absorber plate. The thermo-physical properties of PCM and clay pots used here are presented in Table 2.

Initially, saline water is filled inside the basin area through the inlet valve up to 50 mm of water depth. When the solar radiation is diffused to the water, the radiation gets absorbed by the black painted basin box and acts like a solar collector. Vapor formation takes place once the saline water starts heating then the water vapour travels through the condenser area and starts condensing. These vapours slide into the collector channel built around the basin area and get collected into the beaker through the outlet valve. When the saline water evaporates, the amount of salt inside the water vapour gets collected at the bottom of the basin, and hence only pure water vapour is collected inside the collector channel.

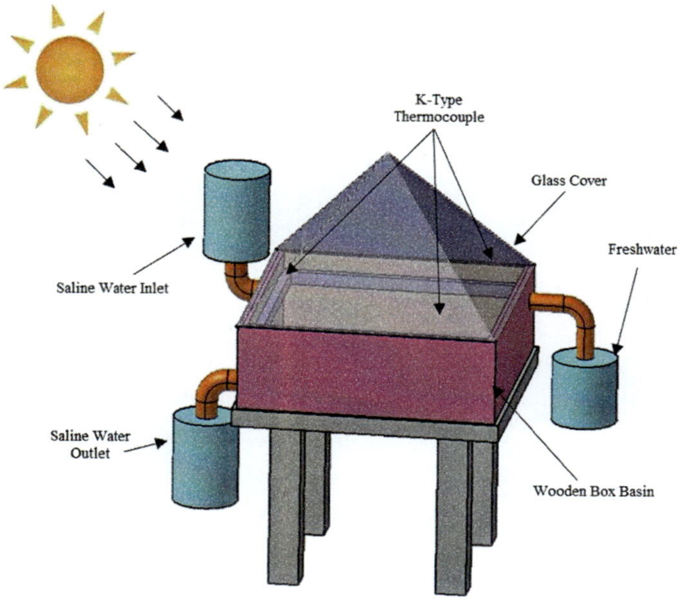

Fig. 1 Schematic of pyramidal solar still

Fig. 2 Photograph of actual solar still

Table 1 Measuring instrument details

Instrument	Range	Accuracy
K-type thermocouple	0–1250 °C	± 3 °C
Solar power meter	0–1999 W/m^2	± 3%
Anemometer	0.4–30 m/s	± 0.4%

Fig. 3 Different configurations

Table 2 Thermo-physical properties of paraffin wax and clay pots [16]

Material	Thermal conductivity (W/m °C)	Specific heat (kJ/kg °C)	Density (g/cm^3)
Paraffin wax (solid-state)	0.24	2.95	0.818
Clay pots	0.5–1.8	0.33	1.41

3 Results and Discussion

Experiments were performed between 8:00 AM to 6:00 PM, from 18 April 2022 to 26 April 2022, measuring the temperature and freshwater production after every hour at Manav Rachna University, Faridabad. Table 3 indicates the weather conditions on experiment days for Case 1, Case 2, and Case 3. The findings were compared for optimum freshwater production on bright days when the intensity of solar radiation variation was around 4% throughout the day.

Pyramid Still without PCM and Porous Material (Case 1)

Figure 4 depicts the surrounding meteorological circumstances of Case 1 solar still, such as sun intensity and temperature in the environment. During the experiment, the intensity of solar radiation grew with the passage of time, reaching a high of 903 W/m^2 at 12:00 noon and then decreasing with the passage of time. This radiation intensity trend was similar on further days also. However, at 2:00 PM the maximum temperature was recorded as 42 °C. It has been observed that the temperature increases as the time progresses attaining its maximum temperature around noontime and decreases as the time further progresses.

Figure 5 depicts the change in temperature and water yield is a function of time for Case 1. It has been noted that the water yield increases in response to a rise in temperature and intensity of solar radiation. The maximum freshwater production was 0.39 L/m^2 during the experiment.

Table 3 Weather conditions on days of experiment

Date	Sunrise (AM)	Sunset (PM)	Temperature (°C)	Wind velocity (km/hr)	Humidity (%)
18/04/2022	5:45	6:45	21–43	19	18–27
22/04/2022	5:45	6:45	22–44	22	17–28
26/04/2022	5:45	6:45	24–44	18	20–29

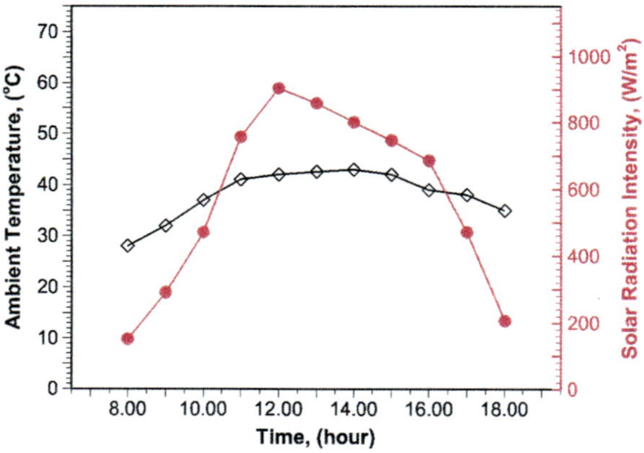

Fig. 4 Change in atmospheric temperature and solar radiation (Case 1)

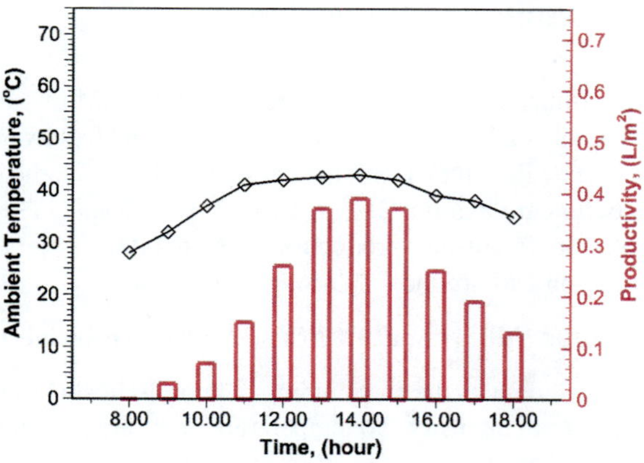

Fig. 5 Change in temperature and water yield with time (Case 1)

Pyramid Still with PCM and Clay Pots (Case 2)

In Case 2, for phase change material, paraffin wax along with 25 clay pots was placed vertically downward at the basin bottom (Fig. 3). Figure 6 depicts the change in temperature of air and intensity of solar radiation. The value of maximum ambient temperature and intensity of solar radiation is 43.5 °C and 905 W/m^2 respectively. As indicated in Fig. 6, the intensity of solar radiation is not uniformly increasing as the time increase because of weather condition that keeps on changing during the experiment. The paraffin wax and clay pots release the absorbed heat when the temperature decreases. When inner glass temperature lower then vapour temperature, the condensation rate will be high, resulting in a higher freshwater production rate. Figure 7 shows the change in freshwater production with respect to ambient temperature. Between 8 and 10 AM in the morning the freshwater production is very low because of the low heat intensity inside the solar still. At 2 PM, the solar still reaches its maximum yield rate of producing 0.55 L/m^2 while in Case 1 it was 0.39 L/m^2, hence pointing out the effectiveness of paraffin wax with clay pots during the experiment.

Pyramid Still with PCM and Black Painted Clay Pots (Case 3)

The change of atmospheric temperature and solar radiation intensity for Case 3 is shown in Fig. 8. Case 3 has a maximum ambient temperature of 44 °C and maximum solar radiation intensity of 902 W/m^2. Climate circumstances such as rain and overcast weather have an impact on solar radiation intensity and atmospheric temperature, resulting in differences. Figure 9 shows the change in freshwater production with respect to ambient temperature. At 2 pm the solar still reaches its maximum yield rate resulting in 0.61 L/m^2 while in Case 2 it was 0.55 L/m^2. This change in freshwater production between Case 2 and Case 3 is mainly because of the black paint on clay pots, indicating the importance of black paint.

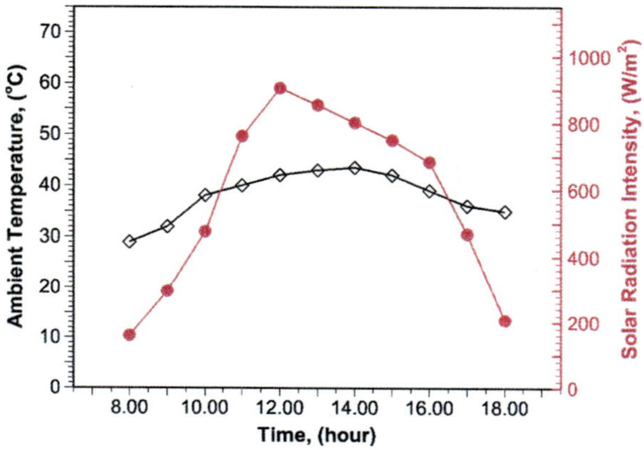

Fig. 6 Change in atmospheric temperature and solar radiation for Case 2

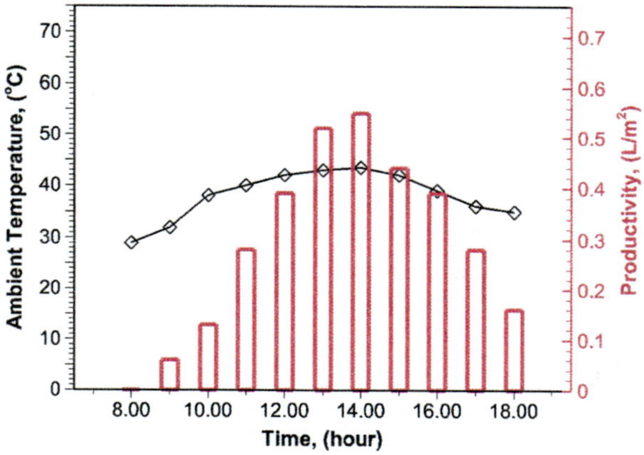

Fig. 7 Change in temperature and water yield with time for Case 2

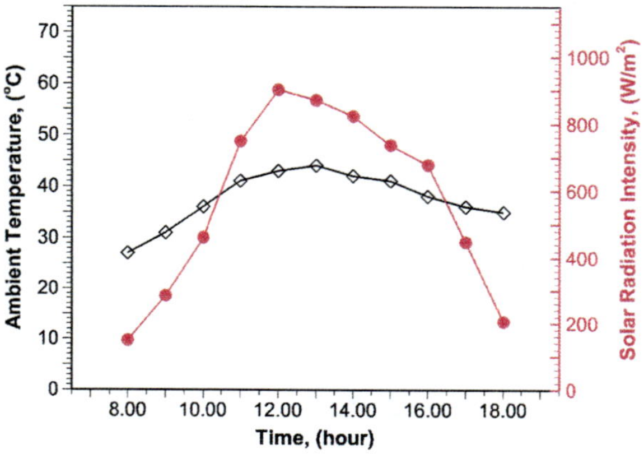

Fig. 8 Change in atmospheric temperature and solar radiation for Case 3

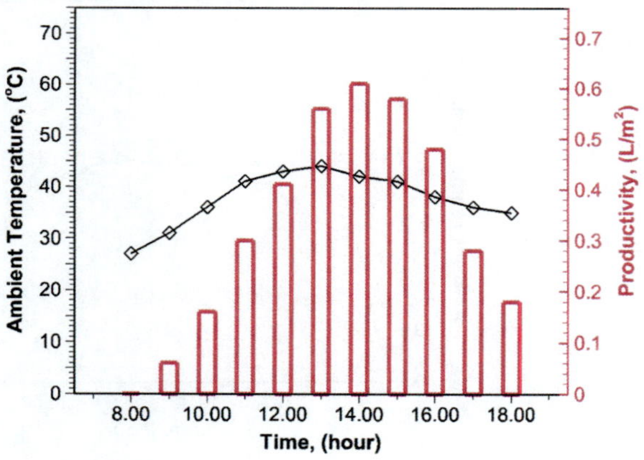

Fig. 9 Change in temperature and water yield with time for Case 3

Productivity Enhancement Comparison

According to Fig. 10, Case 3 produces 3.63 L/m^2 of fresh water, followed by Case 2 (3.20 L/m^2) and Case 1 (2.21 L/m^2). The freshwater production for Case 3 and Case 2, respectively, is 64.25 and 44.79% higher than that of Case 1. Heat is stored in thermal storage materials like paraffin wax and clay pots, which increases the rate of heat transfer, resulting in comparatively high water temperatures at night and, therefore, a higher rate of evaporation that achieves maximum freshwater yield.

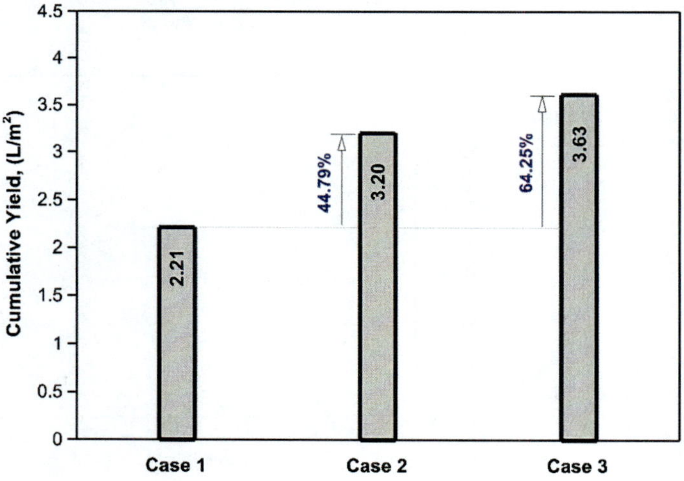

Fig. 10 Cumulative productivity comparison for studied configurations

4 Conclusions

The present work is a performance enhancement experiment to increase the production rate of pyramid-shaped solar still by using paraffin wax with clay pots. The outcome of the experiment represents that paraffin wax, clay pots, and black paint have a prominent role in increasing the daily production of freshwater. Following are the conclusion that is made from the present research work.

- Distillate yield and the rate of evaporation majorly depend upon ambient temperature and solar radiation intensity.
- Use of paraffin wax, clay pots, and black paint in pyramid-shaped solar still leads to enhancement in freshwater production even when the temperature is low after 3 PM and when sunlight is not available.
- The freshwater production rate increases from Case 1 to Case 3, i.e. from 2.21 to 3.63 L/m^2.

References

1. Ang WL, Mohammad AW, Hilal N, Leo CP (2015) A review on the applicability of integrated/hybrid membrane processes in water treatment and desalination plants. Desalination 363:2–18. https://doi.org/10.1016/j.desal.2014.03.008
2. Yousef MS, Hassan H, Ahmed M, Ookawara S (2017) Energy and exergy analysis of single slope passive solar still under Egyptian climate conditions. Energy Procedia 141:18–23. https://doi.org/10.1016/j.egypro.2017.11.005
3. Fathy M, Hassan H, Salem Ahmed M (2018) Experimental study on the effect of coupling parabolic trough collector with double slope solar still on its performance. Solar Energy 163:54–61. https://doi.org/10.1016/j.solener.2018.01.043
4. Al-harahsheh M, Abu-Arabi M, Mousa H, Alzghoul Z (2018) Solar desalination using solar still enhanced by external solar collector and PCM. Appl Therm Eng 128:1030–1040. https://doi.org/10.1016/j.applthermaleng.2017.09.073
5. Khawaji AD, Kutubkhanah IK, Wie JM (2008) Advances in seawater desalination technologies. Desalination 221(1–3):47–69. https://doi.org/10.1016/j.desal.2007.01.067
6. Taghvaei H et al (2014) A thorough investigation of the effects of water depth on the performance of active solar stills. Desalination 347:77–85. https://doi.org/10.1016/j.desal.2014.05.038
7. Kabeel AE, Manokar AM, Sathyamurthy R, Winston DP, El-Agouz SA, Chamkha AJ (2019) A review on different design modifications employed in inclined solar still for enhancing the productivity. J Solar Energy Eng, Trans ASME 141(3). https://doi.org/10.1115/1.4041547
8. Zanganeh P, Goharrizi AS, Ayatollahi S, Feilizadeh M (2019) Productivity enhancement of solar stills by nano-coating of condensing surface. Desalination 454:1–9. https://doi.org/10.1016/j.desal.2018.12.007
9. Yousef MS, Hassan H (2019) An experimental work on the performance of single slope solar still incorporated with latent heat storage system in hot climate conditions. J Clean Prod 209:1396–1410. https://doi.org/10.1016/j.jclepro.2018.11.120
10. Kabeel AE, Abdelgaied M (2016) Improving the performance of solar still by using PCM as a thermal storage medium under Egyptian conditions. Desalination 383:22–28. https://doi.org/10.1016/j.desal.2016.01.006

11. Sathyamurthy R, Nagarajan PK, Subramani J, Vijayakumar D, Mohammed Ashraf Ali K (2014) Effect of water mass on triangular pyramid solar still using phase change material as storage medium. Energy Procedia 61:2224–2228. https://doi.org/10.1016/j.egypro.2014.12.114
12. Abu-Hijleh K, RababaÕh HM Experimental study of a solar still with sponge cubes in basin. [Online]. Available: www.elsevier.com/locate/enconman
13. Le Nian Y, Huo YK, Cheng WL (2021) Study on annual performance of the solar still using shape-stabilized phase change materials with economic analysis. Solar Energy Mater Solar Cells, 230. https://doi.org/10.1016/j.solmat.2021.111263
14. Vijayakumar V, Manu NS, Vasudevan MC, Kiran MV, Rejeesh CR (2021) Phase change materials for improved performance and continuous output in stepped solar stills equipped with HDH. Mater Today: Proc 47:5064–5068. https://doi.org/10.1016/j.matpr.2021.05.089
15. Kabeel AE, Abdelgaied M (2020) Enhancement of pyramid-shaped solar stills performance using a high thermal conductivity absorber plate and cooling the glass cover. Renew Energy 146:769–775. https://doi.org/10.1016/j.renene.2019.07.020
16. Haji-Sheikh A, Eftekhar J, Lou DYS (1983) some thermophysical properties of paraffin wax as a thermal storage medium. Prog Astronaut Aeronaut 86:241–253. https://doi.org/10.2514/5.9781600865596.0241.0253

Performance Analysis of an s-CO_2 Based Solar Flat Plate Collector

Wasim Ashraf, M. Ramgopal, and V. M. Reddy

1 Introduction

Among the renewable energy sources, solar energy is most abundant and is most important. Solar energy has been used extensively hot water generation and power generation over the past decades. Solar water heating systems (SWHS) are clean, sustainable and efficient. Hence they are highly suitable for domestic and commercial hot water generation and space heating [1]. Average water consumption in India is estimated to be 135 L/capita/day [2]. This is mainly required for drinking, cooking, washing of clothes, cleaning of kitchen utensils and flushing of toilets, etc. However, studies show that out of this more than 50% is required as hot water for shower, washing utensils, clothes etc. With the increasing emphasis on the usage of renewable energy, Solar water heating systems are becoming extremely popular all over the world. Depending upon the type of solar thermal collector, circulation of heat transfer fluid (HTF) and method of heat transfer, solar water heaters can be classified as shown in Fig. 1.

According to the mode of heat transfer, solar water heating systems are classified into two main categories: direct and indirect [3]. In a direct solar water heating system, water is heated directly as it flows through the collector. In an indirect system, a heat transfer fluid (HTF) is heated in the collector and then flows through a heat exchanger to heat the water. An indirect water heating system is used where the water is corrosive and/or when the ambient temperatures vary widely leading to freezing of water in the absence of sunlight. The anti-freeze solutions such as ethylene glycol solutions and propylene glycol are commonly used to prevent freezing of water in collectors [4].

W. Ashraf (✉)
School of Energy Science and Engineering, IIT Kharagpur, Kharagpur, India
e-mail: wasimiitkgp92@gmail.com; lncs@springer.com

M. Ramgopal · V. M. Reddy
Department of Mechanical Engineering, IIT Kharagpur, Kharagpur, India

S. Doolla et al. (eds.), *Advances in Clean Energy and Sustainability*,
Green Energy and Technology, https://doi.org/10.1007/978-981-99-2279-6_33

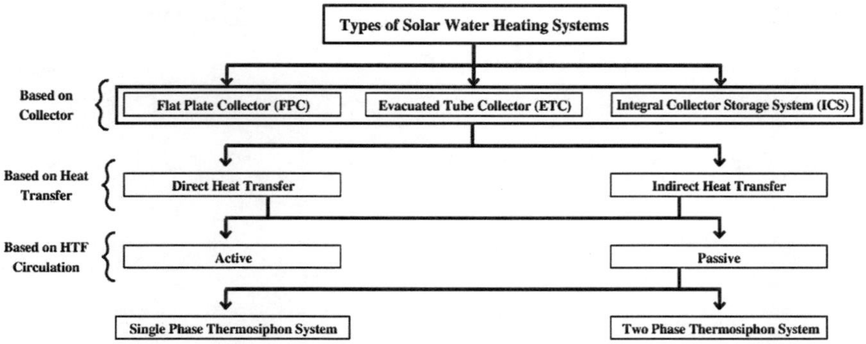

Fig. 1 Classification of solar water heating systems

Generally, flat plate collectors (FPCs) are widely used to produce hot water at temperatures between 20 and 80 °C, while evacuated tube collectors (ETCs) can be used to produce higher temperatures, varying from 50 to 200 °C [5]. FPCs are widely used in solar water heating systems, solar desalination, absorption and adsorption solar refrigeration, industrial process heat and space heating and cooling [6]. The performance of the solar collectors can be improved by using appropriate tube materials, heat transfer fluid and collector design [7]. The selection of a working fluid plays a significant role in the development of an efficient and environmental friendly system that can function even when exposed to low ambient conditions. Water, ethylene glycol/water, propylene glycol/water, silicone oil and air are also used as alternative heat transfer fluids for FPCs [8]. In recent times, studies are carried out on nano-fluids in flat-plate solar collector [8]. There is a growing interest in use of supercritical carbon dioxide (s-CO_2) as HTF in solar collectors as it is a natural working fluid with favourable properties such as non-toxicity, non-flammability, zero ozone depletion potential, zero effective global warming potential, compatibility with common machine construction materials, easy availability and very low price. Furthermore, CO_2 provides superior thermodynamic properties for near critical operation. As the critical point temperature of CO_2 is much lower than other heat transfer fluids (31.1 °C), the fluid will be in supercritical state under collector operating conditions [9, 10].

From the literature review it is found that theoretical and experimental studies on supercritical CO_2 based solar water heaters are relatively scarce, even though it is expected that CO_2 can provide superior performance. Therefore, this study focusses on developing a general theoretical model to assess the performance of a solar FPC for weather data of tropical wet and dry climatic condition of IIT Kharagpur, India with supercritical CO_2 as HTF in a closed loop. The theoretical model is validated with experimental results using water as HTF. A comparative analysis between water and s-CO_2 is also done to get maximum efficiency of the collector. It is observed that the thermal efficiency of the solar water heating system increases with solar insolation, mass flow rate of s-CO_2, ambient temperature. It is also observed that the

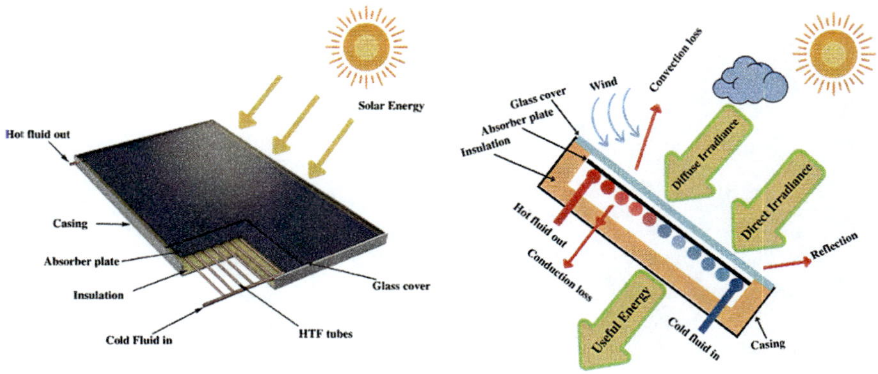

(a) (b)

Fig. 2 **a** Schematic illustration of a flat plate collector and **b** physical phenomena within a flat plate collector

performance of solar flat plate collector can be improved by using s-CO_2 as HTF in comparison to water and other HTF.

2 Experimental Test-Rig

Experiments are performed on a test-rig that uses water as heat transfer fluid. In the test rig, halogen lamps are used to simulate solar radiation. Figure 2a shows the schematic of the flat plate collector, while Fig. 2b shows the mass and energy flows in the collector.

A photograph of the test setup and schematic diagram of the solar water heating (SWH) system using water as a working fluid is shown in the Figs. 3a, b, respectively. The system consists of a flat plate collector (FPC) as a heat collecting source, a hot water storage tank with a cold water reservoir tank, an artificial solar flux generator (halogen lamps), valves, and a data acquisition system. The detailed specifications of the solar flat plate collector are shown in Table 1.

3 Theoretical Analysis and the Model Validation for Forced Circulation Flow

The analysis is carried out assuming fluid flow and heat transfer in steady state, negligible pressure drop inside the riser tubes, uniform solar heat gain throughout the surface of the solar collector, and the loss of heat through front and back are to the same ambient temperature.

The useful heat gain by the working fluid is calculated by the equation:

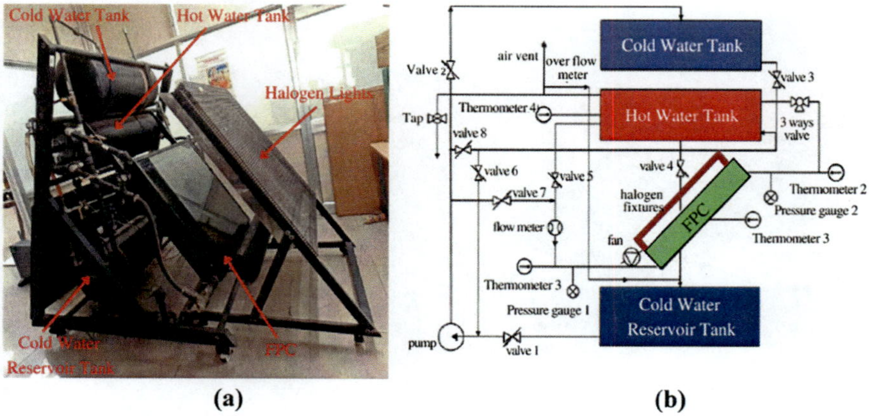

Fig. 3 **a** Photograph of the test setup of SWH system and **b** schematic diagram of the test setup of SWH system

Table 1 Detailed specifications of the solar flat plate collector

S.N.	Parameter	Value
1	Absorber area	0.741 m^2
2	Collector tilt angle	45°
3	Emissivity of the glass cover	0.88
4	Emissivity of the absorber plate	0.12
5	Inner diameter of the tubes	10 mm
6	Inner tubes center to center distance	11 mm
7	Optical efficiency of glass cover	0.95
8	Thermal conductivity of the absorber plate	237 W/mK
9	Thermal conductivity of insulation material	0.04 W/mK
10	Thickness of the back insulation	0.05 m
11	Thickness of the absorber plate	0.12 mm
12	Working fluid in the flow ducts	Water

$$Q_u = \dot{m} C_p (T_{f,out} - T_{f,in}) \tag{1}$$

where, $\dot{m}$, C_p, $T_{f,in}$ and $T_{f,out}$ are the mass flow rate, heat capacity, inlet temperature and outlet temperature of the working fluid, respectively. This equation indicates the useful heat transferred from the collector absorption plate to the working fluid, which depends on heat loss coefficient and optical efficiency.

In terms of heat loss and optical loss the useful heat gain can be written as shown in the following equations. [10]

$$Q_u = A_p\{S - U_l(T_p - T_a)\} \tag{2}$$

$$Q_u = F_r A_p \{S - U_l (T_{f,in} - T_a)\} \tag{3}$$

$$S = \eta_o I_t \tag{4}$$

where, S is solar radiation absorbed by the collector per unit area, η_o is the optical efficiency of the collector and I_t is the solar radiation falling at the top of the solar collector.

The heat removable factor is calculated as:

$$F_r = \frac{\dot{m}}{A_p U_l}\left[1 - \exp\left(-\frac{F' U_l A_p}{\dot{m} C_p}\right)\right] \tag{5}$$

$$F' = \frac{{}^{1}/_{U_l}}{W/[\pi d \alpha_f] + W/[(d + (W - d)F)U_l]} \tag{6}$$

In Eq. (6), F is efficiency factor of the collector defined as,

$$F = \frac{\tan h\left[\frac{m(W-d)}{2}\right]}{\left[\frac{m(W-d)}{2}\right]} \tag{7}$$

where, m is a design parameter and can be calculated as,

$$m = \sqrt{U_l / K_p \delta_p} \tag{8}$$

The overall heat transfer coefficient for the collector is calculated as follows:

$$U_l = U_t + U_b + U_e \tag{9}$$

where, U_t, U_b & U_e are top, bottom and edge loss coefficients respectively for the collector.

$$U_t = \left[\frac{1}{\frac{C}{T_p}\left[\frac{T_p - T_a}{1+f}\right]^e} + \frac{1}{h_w}\right]^{-1} + \left[\frac{\sigma(T_p + T_a)(T_p^2 + T_a^2)}{\frac{1}{\varepsilon_p + 0.00591 h_w} + \frac{1+f+0.133\varepsilon_p}{\varepsilon_g} - 1}\right] \tag{10}$$

In Eq. (10) the following parameters have been used:

$$f = 1.07866(1 + 0.089 h_w - 0.1166 h_w \varepsilon_g) \tag{11}$$

$$C = 520\left(1 - 0.00005\beta^2\right) \tag{12}$$

$$e = 0.430\left(1 - \frac{100}{T_p}\right) \tag{13}$$

where, β is the collector tilt angle, h_w is the wind heat transfer coefficient, ε_g and ε_p are the emissivity of glass cover and absorber plate, respectively.

Heat loss coefficient for the back and edges of the collector can be calculated as:

$$U_b = \frac{k_i}{\delta_b} \tag{14}$$

$$U_e = (L_1 + L_2)L_3 k_i/(L_1 L_2 t_s) \tag{15}$$

where, δ_b is the back insulation thickness (m), k_i is the thermal conductivity of the back insulation material and t_s is the side insulation thickness. L_1, L_2 & L_3 are length, width and height of the collector respectively.

The heat loss rate from the top of the collector is given by:

$$\dot{Q}_{t,l} = h_w(T_c - T_a) + \sigma\varepsilon_g\left(T_c^4 - T_{sky}^4\right) \tag{16}$$

$$U_t = \frac{\dot{Q}_{t,l}}{\left(T_p - T_a\right)} \tag{17}$$

where, T_a, T_c, T_p, and T_{sky} are the ambient, collector surface, absorber plate and sky temperature respectively.

By using meteorological data, collector geometry and Eqs. (2), (3), (9), (10), (15) and (17), the heat loss rate and overall heat transfer coefficient can be calculated for any specific flow rate and fluid inlet temperature. The thermal efficiency of the solar flat plate collector is obtained from:

$$\eta_{th} = \frac{\dot{Q}_u}{A_p I_T} \tag{18}$$

Considering temperature distribution in the collector, the following equation can be derived. [10]

$$(T_{f,out} - T_a - \frac{S}{U_l})/(T_{f,in} - T_a - \frac{S}{U_l}) = (\exp\left(-\frac{F' U_l A_p}{\dot{m} C_p}\right) \tag{19}$$

From the above equation, the water outlet temperature component can be excluded from Eq. (1) and therefore, the equation for energy efficiency of collector can be written as:

$$\eta_{th} = \frac{\dot{m}C_p\left[(T_{f,in} - T_a - \frac{S}{U_l})\left(\exp\left(-\frac{F'U_lA_p}{\dot{m}C_p}\right) - 1\right]}{A_pI_T} \quad (20)$$

4 Results and Discussion

Validation of developed MATLAB code with experimental data

The experimental results for solar FPC are used to validate the developed MATLAB code, considering water as working fluid. The observed variation of thermal efficiency of the solar collector with collector performance coefficient $(T_{fi} - T_a)/I$, mass flow rate and fluid inlet temperature are shown in Figs. 4, 5 and 6a, respectively. As shown in Fig. 4a, the thermal efficiency of the collector reduces with increase in the collector performance coefficient. Due to increase in the temperature difference $(T_{fi} - T_a)$, both the collector performance coefficient as well as the heat losses from the collector increase, which reduces the thermal efficiency of the collector.

As shown in Fig. 4b, increasing the mass flow rate of the working fluid leads to an increase in the thermal efficiency of the collector. Increase in mass flow rate results in a considerable decrease in the absorber plate temperature, which decreases the temperature gradient between the absorber plate and the environment. This results in reduced heat losses and hence higher efficiency.

Further, as it is observed in Fig. 5a, any increase in water inlet temperature leads to a decrease in thermal efficiency of collector. Because of increase in fluid inlet temperature there is an increase in the temperature of absorber plate which in turn leads to an increase in temperature gradient between the absorber plate and the environment and consequently to an increase in heat losses. So, an increase in water inlet temperature leads to a decrease in thermal efficiency of collector.

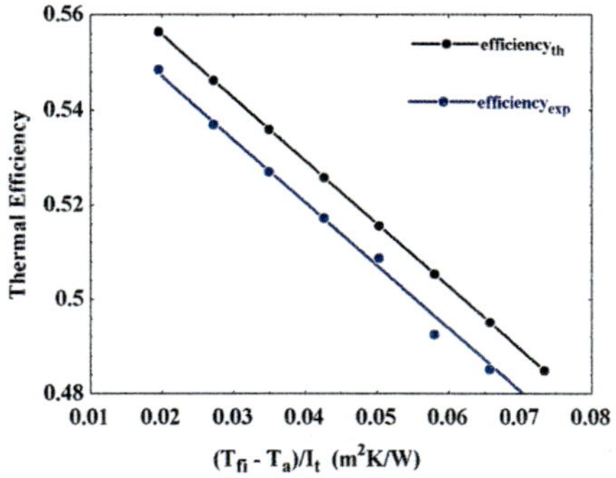

(a)

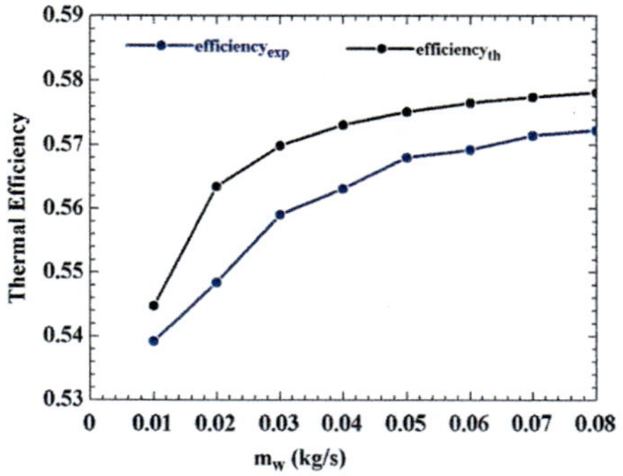

(b)

Fig. 4 Variation of efficiency with **a** collector performance coefficient, **b** mass flow rate

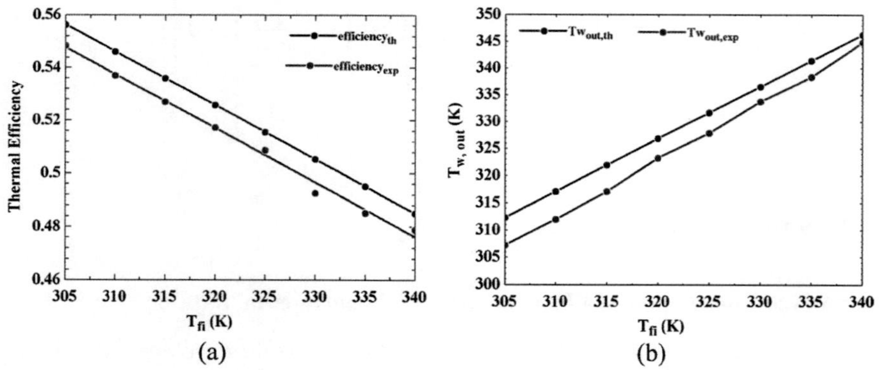

Fig. 5 Variations of **a** thermal efficiency and **b** water outlet temperature with inlet temperature

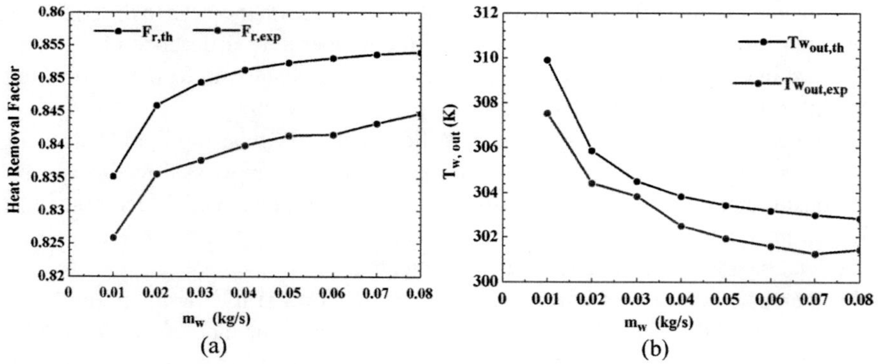

Fig. 6 Variations of **a** heat removal factor and **b** water outlet temperature with mass flow rate

The effect of fluid inlet temperature on fluid outlet temperature is shown in Fig. 5b. As expected, the fluid outlet temperature increases with increase in inlet temperature, even though the efficiency decreases.

From Figs. 4, 5 and 6, it is seen that there is a good qualitative agreement between the theoretical and experimental results. Quantitatively also the difference is well within the experimental uncertainties, which are estimated to be 5.81% for useful heat output and 5.73% for thermal efficiency. This indicates that the mathematical model correctly predicts the performance of the flat plate solar collector. Hence the model is used to predict the performance if FPC with s-CO_2 as working fluid and the results are compared with that of water.

Theoretical results with s-CO_2 as HTF

The s-CO_2 based collector is operated at a pressure of 90 bar. observed variation of thermal efficiency of the collector with collector performance coefficient and fluid inlet temperature are shown in Fig. 7a, b, respectively. The variation in thermal efficiency and fluid outlet temperature with mass flow rate are shown in Fig. 8a, b, respectively.

From Figs. 7 and 8 it is clear that use of s-CO_2 as HTF in place of water results in much superior performance of the flat plate solar collector. This can be attributed to the favorable thermophysical and transport properties of CO_2 as compared to water as shown in Table 2. Due to these favorable properties, the heat transfer coefficient for s-CO_2 is much higher than that of water as shown in Fig. 9. Hence, s-CO_2 can be an excellent HTF in solar collectors, provided the high operating pressures of s-CO_2 are taken care of through proper collector design.

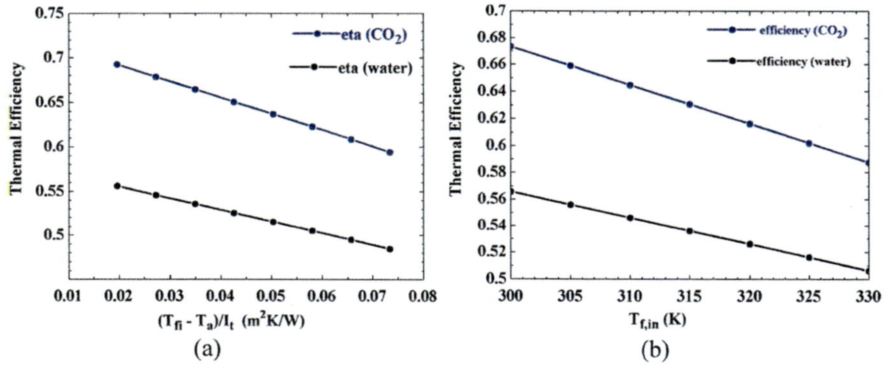

Fig. 7 Variations of thermal efficiency with **a** collector performance coefficient and **b** fluid inlet temperature

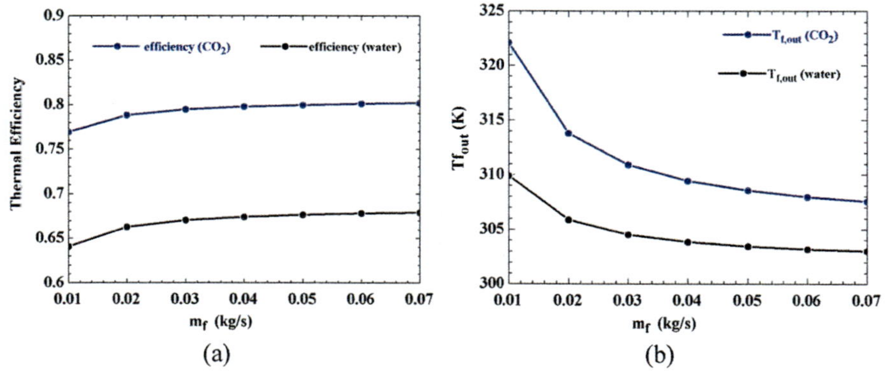

Fig. 8 Variations of **a** thermal efficiency and **b** fluid outlet temperature with mass flow rate

Table 2 Comparison of thermophysical properties of CO_2 and H_2O

Heat transfer fluid	CO_2 (45 °C and 90 bar)	H_2O (45 °C and 1 bar)
Density (kg/m^3)	339.843	990.272
Specific heat (J/kg K)	6132.12	4180
Dynamic viscosity (Pa s)	2.56×10^{-5}	5.97×10^{-4}
Thermal conductivity (W/mK)	0.0497	0.6372

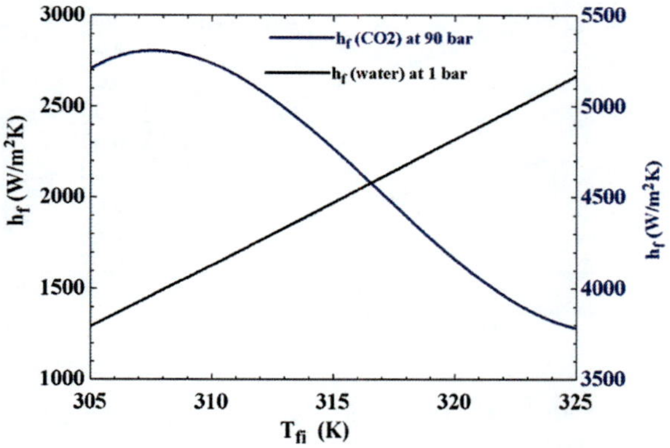

Fig. 9 Variation of fluid side heat transfer coefficient with fluid inlet temperature

5 Conclusions

A simple mathematical model is developed to predict the performance of a flat plate solar collector. The model is validated with experimental results using water as heat transfer fluid. A good qualitative and quantitative agreement is seen between the theoretical and experimental results. Next, theoretical results are obtained using supercritical CO_2 as heat transfer fluid in the collector. The performance of both the fluids are compared at different mass flow rates, collector performance coefficient and fluid inlet temperatures. The theoretical results show that s-CO_2 offers much better thermal performance compared to water under all conditions due to its superior thermophysical and transport properties as compared to the water. However, use of s-CO_2 in solar collectors also poses several challenges due to the high operating pressure of CO_2, possibility of phase change with ambient temperature variation and sharp property variations near critical point. Hence there is a need for more detailed studies on these aspects. In addition, there will be economic implications due to the need for designing the system to withstand the high operating pressures with s-CO_2. However, through proper optimization of tube spacing, tube diameters and fluid flow path it should be possible to employ small diameter tubes which can withstand the high pressures much better with small tube wall thickness. Due to much lower viscosity of s-CO_2 compared to water, small tube diameter tubes can be employed without increasing the pressure drop significantly. However, this needs a detailed thermo-economic analysis of the complete system which will be carried out in future.

References

1. Zhang R, Geoffrey QP, Ni M, Wong JKW (2018) Techno-economic feasibility of solar water heating system: overview and meta-analysis. Sustain Energy Technol Assess 30:164–173
2. The Central Public Health and Environmental Engineering Organisation. http://cpheeo.gov.in
3. Sarwar J, Khan MR, Rehan M, Asim M, Kazim AH (2020) Performance analysis of a flat plate collector to achieve a fixed outlet temperature under semi-arid climatic conditions 2020. Sol Energy 207:503–516
4. Pandey KM, Chaurasiya R (2017) A review on analysis and development of solar flat plate collector. Renew Sustain Energy Rev 67:641–650
5. Kalogirou SA (2004) Solar thermal collectors and applications. Prog Energy Combust Sci 30:231–295
6. Sarkar J (2013) Performance of a flat plate solar thermal collector using supercritical carbon dioxide as heat transfer fluid. Int J Sustain Energy 32 6:531–543
7. Yousefi T, Veysi F, Shojaeizadeh E, Zinadini S (2012) An experimental investigation on the effect of Al_2O_3-H_2O nanofluid on the efficiency of flat plate solar collectors. Renew Energy 39:293–298
8. Yamaguchi H, Sawada N, Suzuki H, Ueda H, Zhang XR (2010) Preliminary study on a solar water heater using supercritical carbon dioxide as working fluid. J SolEnergy Eng 132:011010
9. Sarkar J, Bhattacharyya S, Ramgopal M (2010) A transcritical CO_2 heat pump for simultaneous water cooling and heating. Int J Appl Sci, Eng Technol 6:1
10. Jafarkazemi F, Ahmadifard E (2013) Energetic and exergetic evaluation of flat plate solar collectors. Renew Energy 56:55–63
11. Duffie JA, Beckman WA (2006) Solar engineering of thermal processes. Wiley third ed., New York etc.

Thermal Stratification Characteristics in a Reduced Scale Toroidal Suppression Pool

Sampath Bharadwaj Kota, Seik Mansoor Ali, and Sreenivas Jayanti

1 Introduction

The containment is the last physical barrier that prevents the release of radioactive fission products into the environment during the postulated accident. Several engineered safety features are provided in the reactor to safeguard the containment of the nuclear power plants, and the suppression pool (SP) is one such system provided for removing decay heat and scrubbing the fission products released during the accident. During the station blackout condition encountered in the Fukushima Daiichi reactors (units 2&3), the steam was discharged into the suppression pool via Reactor Core Isolation and Cooling (RCIC) system via blowdown pipes or spargers. The steam vigorously condensed in the suppression pool, thereby losing its latent heat to the enormous pool of water surrounding the spargers. During the accident progression, each unit behaved differently, i.e., the containment pressure rise in unit-3 was steeper than the unit-2 during the first 12 h of the RCIC system [1].

The variation in containment pressure prompted the researchers to dwell on the possible explanations. The reasons were partially attributed to the flooding of the torus room in unit-2 [2], the difference in the RCIC pipe inlet in the suppression pool and the operating duration of safety relief valves [3]. Though unit-2 and unit-3 were considered twin units, unit-2 had a blowdown pipe of a vertically downward single injection, whereas unit-3 had a multi-hole sparger type of horizontal injection. The researchers opined that the horizontal multi-injection sparger could have caused thermal stratification in unit-3 [4].

Further, during stable thermal stratification, the pool's free surface temperature is higher than the bulk temperature, and the partial pressure of steam in the air

S. B. Kota (✉) · S. M. Ali
SRI-Atomic Energy Regulatory Board, Kalpakkam 603102, India
e-mail: ksb@aerb.gov.in

S. B. Kota · S. Jayanti
Department of Chemical Engineering, IIT-Madras, Chennai 600036, India

S. Doolla et al. (eds.), *Advances in Clean Energy and Sustainability*,
Green Energy and Technology, https://doi.org/10.1007/978-981-99-2279-6_34

space is equal to the vapour pressure corresponding to the surface temperature in thermal equilibrium conditions. The higher surface temperature translates to higher containment pressure than if one were to estimate the pressure using the bulk temperature. Furthermore, during the stratification transient, the pool surface temperature can increase to nearly saturation temperature, and the lower portion of the pool can remain at a high subcooled state. It results in a dead volume as the entire pool is not effectively participating in the heat evacuation process. Therefore, considering the dead volume in the suppression pool and higher pool surface temperature, the containment pressure can increase dangerously, risking the containment's integrity.

2 Brief Literature Review and Motivation

Several research groups have pursued experimental and numerical studies on the thermal stratification phenomenon encountered in the suppression pools of the nuclear power plant. The experimental investigations of steam condensation in the subcooled water have been carried out in various geometrical and operating configurations such as pool shapes and vent pipe configurations. The pool configurations include rectangular [5, 6], cylindrical pools of PANDA and POOLEX facilities [7, 8], toroidal [4, 9] and trapezoidal [10] shaped pools. The vent pipe configurations include downward injection blowdown pipe [3, 5, 6, 8, 9], horizontal injection [10], vertically upward injection [7] and multi-hole sparger [3, 4]. In addition, experiments have also been carried out at both sub-atmospheric [4–6] and above atmospheric pressure [11, 12].

Several studies have been carried out to study the thermal stratification phenomenon using the system level codes to model the overall reactor response to the initiating event [8, 13, 14]. Though numerical simulations employing system level codes are computationally less expensive, the codes such as MELCOR, TRACE, and RELAP5 (except for GOTHIC), consider a well-mixed/homogenous pool [13]. Further, several researchers have carried out numerical studies using the CFD tools, and the pioneering work was carried out in this domain by employing SCR methodology [15, 16]. In addition, the CFD simulations have been carried out using the effective heat/momentum source (EHS/EMS) methodology [10, 17] or the two-phase Eulerian method [10].

The investigators focused their studies based on the steam condensation regime, such as complete condensation in the pipe [13, 18], chugging [10, 11, 19], jetting [15, 16] and bubbling [4–6] and the impact of the type of condensation regime on formation and disappearance of thermal stratification.

Understanding thermal stratification in the suppression pool is vital from a reactor safety perspective for a few reasons. Firstly, the containment pressure would be underestimated if the pool is considered well mixed. In the BASF project, the researchers identified that ignoring thermal stratification severely under-estimated rate of increase and maximum pressure by about 160 kPa [1]. Hence it is crucial to account for thermal stratification in the modelling studies. Secondly, in the mixed pool assumption, the NPSH available of the RCIC pump (located at the bottom region) would exceed earlier than in a thermally stratified pool [1, 11].

The available literature on the CFD studies of thermal stratification of suppression pool is mainly for cylindrical/rectangular geometries, which is not an accurate representation of the actual torus-shaped suppression pool present in the Fukushima Daiichi reactor. Therefore, in the present study, a reduced scale whole torus was considered for CFD simulations, and thermal stratification characteristics of the suppression pool are discussed.

3 Numerical Modelling

3.1 Geometry Description and Basis

The Fukushima Daiichi Unit-3 is a BWR with a Mark-1 type of containment comprising a drywell like an inverted lightbulb and a suppression pool chamber that is toroidal in shape. The suppression pool has a major diameter of 33.5 m and a minor diameter of 8.9 m with an inventory of 2980 m^3. The torus's scaled-down (1/20) geometry has a major and a minor diameter of 1.5 m and 0.4 m, respectively, with a volume of 0.592 m^3 half-filled with water. The current study adopts the reduced scale suppression pool geometry from the literature [4]. The dimensions of the torus and steam condensation region (SCR) considered in the current study are given in Fig. 1.

In the present study, thermal stratification simulations were carried out using the effective heat source (EHS) methodology [18]. The EHS model was implemented by incorporating the volumetric heat source equivalent to the latent heat of steam condensation in the pool with a 3.7 kg/h flow rate. Further, it was also assumed that the steam immediately condenses in the vicinity of the sparger and the region in which the steam completely condenses is called "Steam Condensation Region (SCR)" [16–18]. For simplicity, the control volume of SCR was assumed to be cylindrical [16], and the SCR extends radially up to 10 mm. It is a reasonable assumption considering the bubble size in the experiments [4] was in the range of $\approx$ 2–10 mm, and an earlier CFD study employed the penetration length as seven times the injection diameter [20]. Further, the steam condensation also imparts momentum to the suppression pool, but in the present study, it was assumed that as the steam is injected circumferentially, it has a negligible effect on thermal stratification expected in the vertical direction. Further, negligible momentum is introduced into the pool at low steam mass flow rates and can be assumed to be the dominant buoyancy source [13].

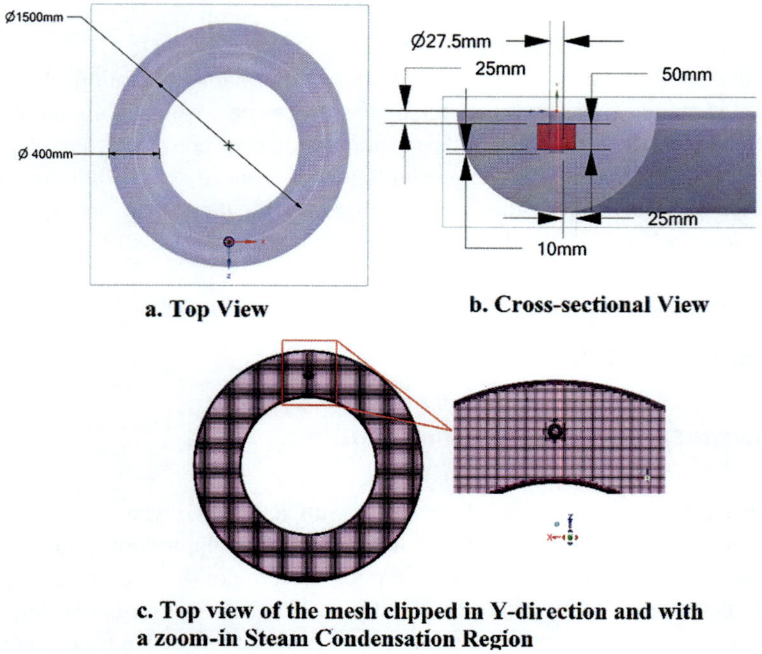

Fig. 1 The top view, cross-sectional view of the geometry depicting the steam condensation region (SCR) for penetration length of 25 mm, **c** mesh

3.2 Grid Sensitivity Study and Case Set-Up

Grid independence study was carried out using meshes with three different cell counts of 196, 512 and 930 k for the case of penetration length of 25 mm. Initially, high penetration length was chosen to ensure that the massive amount of volumetric heat flux provided in the SCR would not increase the temperature in the vicinity to unrealistic values. The mesh was created using the ANSYS mosaic meshing tool, a hexahedral dominant mesh with a lower total face count and higher quality cells. The sizes employed in the current study are given in Table 1.

The SCR region has an average cell size of 5 mm for coarse and 2 mm for fine mesh. Further, the torus region has an average cell size of 10 mm in 1/3rd region

Table 1 Grid sensitivity study details

Case	Cell No. (k)	Grid size	
		Torus	SCR (mm)
A	194	10 mm (1/3 rd); 15 mm (2/3 rd)	5
B	527	10 mm	2.5
C	929	8 mm	2.0

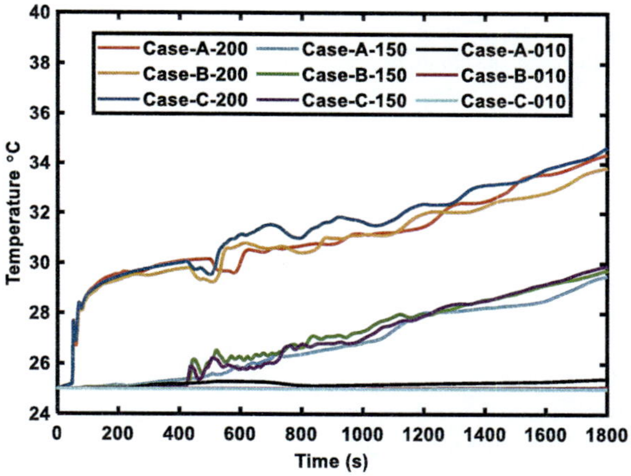

Fig. 2 Grid independence study

surrounding the sparger and 20 mm in farther 2/3rd region from the sparger for coarse mesh and 8 mm cell size for fine mesh. Further, a time-step was selected as 0.01 s for up to 900 s, 0.05 s for 900 < t < 1000 s and 0.1 s for 1000 < t < 1800s. The computational time for the fine mesh simulation was about ten days on the machine using eight cores and 128 GB RAM. In addition, a simulation was also carried out to ascertain the effect of the time step. A constant time-step of 0.1 s was employed, and no impact on the temperature variation was observed. Therefore, a time step of 0.1 s was considered for subsequent simulations. It is observed from Fig. 2 that there is no significant variation in the temporal variation of temperature for the points at different elevations of 200 mm (above the SCR), 150 mm (in the SCR) and 10 mm (below the SCR). These points are located at 45° in an anti-clockwise direction from the sparger location. Therefore, it seemed pragmatic to continue the simulations with 930 k cells in the further simulations with 0.1 s time step. As the volumetric heat flux did not raise the temperature to unrealistic values in the case of 25 mm penetration length, a final value of 10 was chosen to represent the realistic size of the steam condensation region.

The governing equations of continuity, momentum, energy and standard k-ϵ turbulence equations were solved using the SIMPLE algorithm. The SIMPLE algorithm was employed for pressure–velocity coupling available in the Ansys Fluent [21]. This algorithm is a segregated type in which pressure and velocity are updated sequentially. Further, second-order spatial discretization schemes were employed for pressure, momentum, turbulence and energy equations, and a second-order implicit scheme was employed for transient formulation. All the equations (continuity, momentum, turbulence), except for the energy equation, were considered fully converged when the sum of scaled residuals was less than 1e-5 and for energy, it was considered fully converged when the sum of scaled residuals was less than 1e−8. The properties of water employed in the simulation are given in Table 2.

Table 2 Properties of water employed in the simulation

Property	Value
ρ, [kg/m^3]	997.1
C_p, [J/(kg K)]	4182
k, [W/(m K)]	0.6
μ, [kg/(m s)]	0.001003
β, [K^{-1}]	0.0002594

Further, the density variation was modelled using the Boussinesq approximation. The basis of this approximation is that though the temperature-induced density variation is slight, the density variation is significant enough to drive the buoyant motion. Thus, the variation in density is neglected everywhere except in the buoyancy term in the momentum equation. The Boussinesq approximation is valid if $\beta(T\text{-}T_{ref}) \ll 1$.

4 Results and Discussion

4.1 Tunnelling Effect

Due to the presence of volumetric heat flux equivalent to the latent heat of condensation of steam in the vicinity of sparger (SCR), the water in SCR is hotter than the surrounding water. This temperature difference induces buoyancy plumes emanating from the SCR rising to the top of the pool. Once the hot plume reaches the top free surface, it starts extending circumferentially from both clockwise and anticlockwise directions until it reaches the torus's diametrically opposite end. It is observed from Fig. 3 that the fronts collide at around 230 s. This phenomenon is typical of Chimney/Tunnelling effect as reported in the literature [22]. Upon collision of these fronts, they start moving towards the sparger. The circulation pattern observed during the transient simulation can be summarized as: a hot plume rising from the vicinity of the sparger, extending circumferentially, sinking at the opposite ends of the sparger and returning to the sparger at a lower elevation. Though this is a simplified explanation, the established flow pattern seems to be more complex with several recirculation patterns, as observed from the streamlines picture in Fig. 4. These streamlines are coloured with respect to both depth and temperature.

4.2 Average Temperature Distribution

The average temperature distribution in the pool with respect to elevation at various time instances is given in Fig. 5. It can be observed that the bottom-most zone, i.e., from 0–25 mm elevation has only a 0.3 °C increase throughout the 3600 s. It indicates

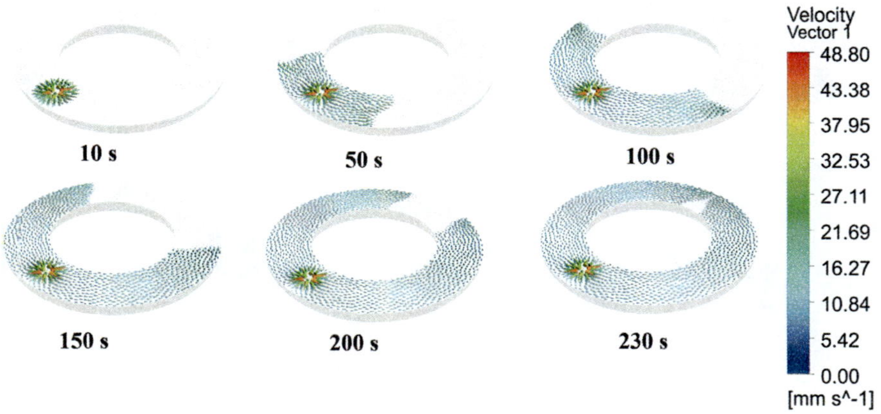

Fig. 3 Tunnelling/Chimney effect observed during early stages of flow development and thermal stratification

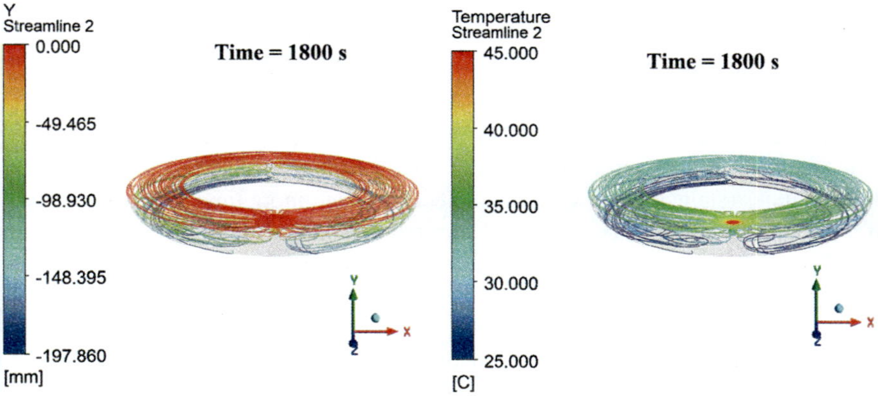

Fig. 4 Streamlines at 1800s coloured with respect to depth and temperature

that the zone is not participating in the heat evacuation process during the natural convection process and can be considered dead volume.

However, in the uppermost zone, i.e., from 175–200 mm, the temperature variation is 11.9 °C, indicating that the zone actively participates in the circulation process. In addition, by considering the initial pool temperature at 25 °C, the surface temperature increased by 13.5 °C within the same period. If one considers the dead volume as the zone with less than a 20% increase ($\Delta T = 13.5 \times 0.2 = 2.7$ °C) in temperature with respect to maximum temperature difference (observed for free surface: $\Delta T =$ 13.5 °C), the dead volume can be observed up to an elevation of 0–75 mm within the given time. This dead volume accounts for 34.54% of the total volume of the pool. Further, it was also observed that the significant temperature variation was in the zones within 75–200 mm, indicating mixing.

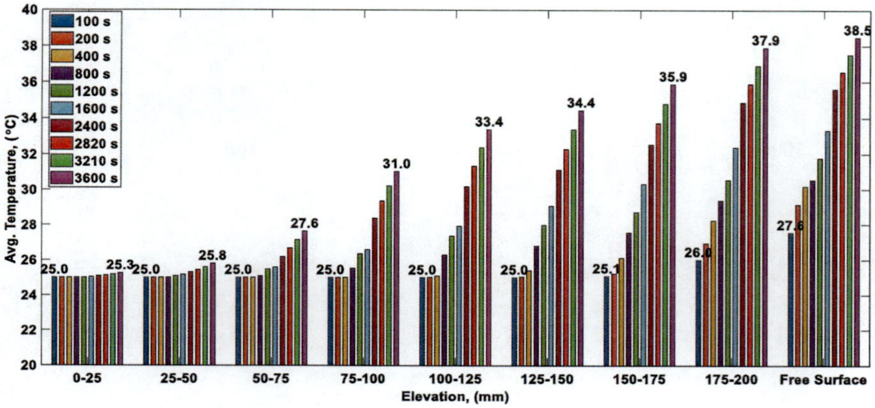

Fig. 5 Average temperature variation with respect to the elevation in the suppression pool

4.3 Volume Fraction Distribution

Drawing an analogy of residence time distribution (RTD) in a non-ideal chemical reactor, temperature distribution graphs are discussed in this section. In an ideal continuous stirred tank reactor (CSTR), the effluent concentration is the same as the concentration in the reactor. However, in a realistic CSTR, due to the presence of dead volumes (stagnant zones), some molecules spend more time than the other set of molecules. Similarly, due to thermal stratification induced mixing, only some part of the pool mixes and the rest of the pool does not participate in the heat removal process.

The volume fraction distribution for a given temperature range is given in Fig. 6, and the following observations can be made from it. Firstly, at the end of 3600 s, 28% of water volume in the pool was with less than a 5 °C temperature rise. Secondly, the majority of the pool, i.e., 39% is in the range of 30–35 °C. Similarly, 30% of water in the pool is 35–40 °C, and 69% is in the range of 30–40 °C by the end of the 3600 s. Further, in the 30–35 °C temperature range, a maximum volume fraction of 0.52 was observed at 2820 s before decreasing to 0.46 by the end of 3210 s. This decrease can be attributed to as the time progresses, the volume fraction of water in the higher temperature zones (35–40 °C) increased. Further, the water above 45 °C constitutes less than 0.1% of the pool. It could be attributed to the higher temperature zones (> 45 °C) being restricted to the SCR in which a volumetric heat source is present.

4.4 Consequence of Thermal Stratification

It can be observed from Fig. 7. that the average pool temperature is less than the pool surface temperature throughout the simulation, and by the end of 3600 s, the

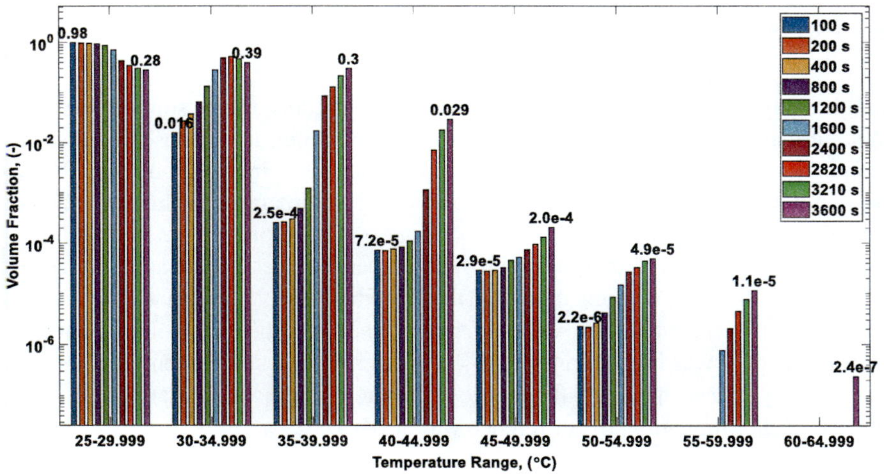

Fig. 6 Volume fraction distribution with respect to the temperature range in the suppression pool

temperature difference increased by ≈6 °C. As mentioned in an earlier section, the pool surface temperature determines the containment pressure rather than the average temperature of the pool. Therefore, it is possible to underpredict the containment pressure if one were to ignore the thermal stratification phenomenon and consider the homogeneous mixing of the pool.

Fig. 7 Temporal variation of pool average temperature and surface average temperature

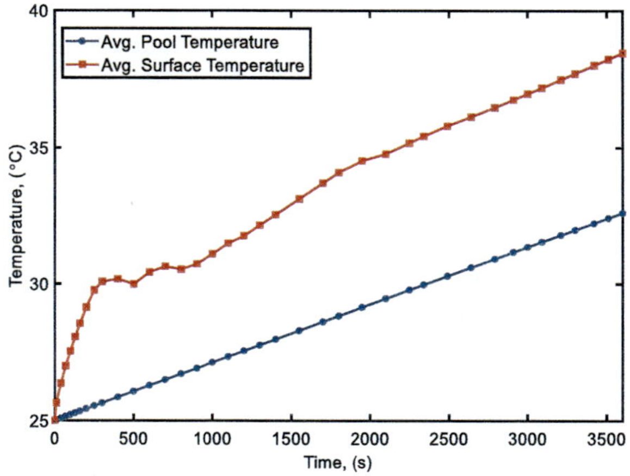

5 Conclusions

In this study, CFD simulations were performed for a reduced scale suppression pool and thermal stratification characteristics were brought out. It was observed that the temperature in the bottom-most zone (0–25 mm elevation) increased by only 0.3 °C throughout 3600 s, indicating that the zone was not participating in the heat evacuation process. This dead volume accounted for about one-third of the total pool volume. Further, it was observed that the significant temperature variation was in the zones within 75–200 mm, indicating mixing in these regions. The volume that had less than a 5 °C increase in temperature was estimated to be 28% of the torus volume. Further, 69% of the pool was in the range of 30–40 °C. Finally, the surface temperature was significantly higher than the bulk pool temperature throughout the transient, and by the end of the simulation, the difference was nearly 6 °C.

References

1. Pellegrini M, Dolganov K, Herranz LE et al (2016) Benchmark study of the accident at the Fukushima Daiichi NPS: best-estimate case comparison. Nucl Technol 196:198–210
2. Mizokami S, Yamada D, Honda T et al (2016) Unsolved issues related to thermal-hydraulics in the suppression chamber during Fukushima Daiichi accident progressions. J Nucl Sci Technol 53:630–638
3. Pellegrini M, Araneo L, Ninokata H et al (2016) Suppression pool testing at the SIET laboratory: experimental investigation of critical phenomena expected in the Fukushima Daiichi suppression chamber. J Nucl Sci Technol 53:614–629
4. Jo B, Erkan N, Takahashi S et al (2016) Thermal stratification in a scaled-down suppression pool of the Fukushima Daiichi nuclear power plants. Nucl Eng Des 305:39–50
5. Song D, Erkan N, Jo B et al (2015) Relationship between thermal stratification and flow patterns in steam-quenching suppression pool. Int J Heat Fluid Flow 56:209–217
6. Song D, Erkan N, Jo B et al (2014) Dimensional analysis of thermal stratification in a suppression pool. Int J Multiph Flow 66:92–100
7. Yang Q, Qiu B, Chen W et al (2021) Experimental study on the influence of buoyancy on steam bubble condensation at low steam mass flux. Exp Therm Fluid Sci. 129:110467
8. Li H, Villanueva W, Puustinen M et al (2018) Thermal stratification and mixing in a suppression pool induced by direct steam injection. Ann Nucl Energy 111:487–498
9. Jo B, Erkan N, Okamoto K (2020) Richardson number criteria for direct-contact-condensation-induced thermal stratification using visualization. Prog Nucl Energy 118:103095
10. Qu X, Revankar ST, Tian M (2018) Numerical investigation on thermal status of a scaled-down suppression pool. Nucl Eng Des 340:183–192
11. Solom M, Vierow KK (2016) Experimental investigation of BWR suppression pool stratification during RCIC system operation. Nucl Eng Des 310:564–569
12. Liu X, Xie X, Meng Z et al (2021) Characteristics of pool thermal stratification induced by steam injected through a vertical blow down pipe under different vessel pressures. Appl Therm Eng 195:117169
13. Cavaluzzi J, Andrs D, Vierow KK (2021) Two-zone stratified wetwell model development and implementation for RELAP-7. Ann Nucl Energy 164:108592
14. Gallego-Marcos I, Villanueva W, Kudinov P (2018) Modelling of pool stratification and mixing induced by steam injection through blowdown pipes. Ann Nucl Energy 112:624–639

15. Kang HS, Song CH (2008) CFD analysis for thermal mixing in a subcooled water tank under a high steam mass flux discharge condition. Nucl Eng Des 238:492–501
16. Moon YT, Lee H Do, Park GC (2009) CFD simulation of steam jet-induced thermal mixing in subcooled water pool. Nucl Eng Des 239:2849–2863
17. Gallego-Marcos I, Kudinov P, Villanueva W et al (2019) Pool stratification and mixing induced by steam injection through spargers: CFD modelling of the PPOOLEX and PANDA experiments. Nucl Eng Des 347:67–85
18. Li H (2014) Approach and development of effective models for simulation of thermal stratification and mixing induced by steam injection into a large pool of water. Royal Institute of Technology
19. Villanueva W, Li H, Puustinen M et al (2015) Generalization of experimental data on amplitude and frequency of oscillations induced by steam injection into a subcooled pool. Nucl Eng Des 295:155–161
20. Gamble RE, Nguyen TT, Shiralkar BS et al (2001) Pressure suppression pool mixing in passive advanced BWR plants. Nucl Eng Des 204:321–336
21. Fluent (2021) Ansys Fluent user's guide. Ansys, Inc.
22. Krepper E, Hicken EF, Jaegers H (2002) Investigation of natural convection in large pools. Int J Heat Fluid Flow 23:359–365

The Impact of Solar Photovoltaic (PV) Rooftop Panels on Temperature Profiles of Surroundings and Urban Thermal Environment

Aishwarya Mandavgane, Sujata Karve, Prajakta Kulkarni, and Namrata Dhamankar

1 Introduction

Incoming solar energy typically is either reflected back to the atmosphere or absorbed, stored, and later re-radiated in the form of latent or sensible heat [1]. Urban heat island (UHI) is a phenomenon that occurs when an urban area has higher temperature compared with its surrounding rural area. Infrastructures, nature of surfaces, vegetation and anthropogenic heat are among the many factors that influence the formation of UHI. Additionally, PV panel surfaces absorb solar insolation due to a decreased albedo. PV panels will re-radiate most of this energy as longwave sensible heat [2] and convert a lesser amount (~ 20%) of this energy into usable electricity. This increased absorption could lead to greater sensible heat efflux that may be trapped under the PV panels [3].

The solar photovoltaic (SPV) sector is booming, with ambitious goals being set all over the world. India is not far behind, with a solar target of 100 gigawatts (GW) by 2022, with solar rooftop accounting for 40 gigawatts of the total [4]. While photovoltaic (PV) renewable energy production has surged, concerns remain about whether PV power adds on to the "heat island" (PVHI) effect, like the increase in ambient temperatures [5]. There is a lack of data as the PVHI effect is real or not. However, very less study has been done in the Indian context, which is crucial considering the country's targets for rooftop installation. When utility-scale PV systems are located near urban centers, increased solar absorption of PV fields compared to surrounding terrain is observed which can warm the ambient air, increase ambient temperatures in the nearby cities, as well as in peri-urban and suburban areas [6]. Similarly, building integrated PV systems are highly absorptive of solar radiation and can heat up to temperatures above those of surrounding structures. The modules thus

A. Mandavgane (✉) · S. Karve · P. Kulkarni · N. Dhamankar
Department of Environmental Architecture and Planning, Dr. B. N. College of Architecture, Pune, India
e-mail: aishwarya.mandavgane@gmail.com

S. Doolla et al. (eds.), *Advances in Clean Energy and Sustainability*,
Green Energy and Technology, https://doi.org/10.1007/978-981-99-2279-6_35

become radiant heat sources for the people and structures beneath them, as well as convective heat sources that can dramatically enhance the temperature of the ambient air in cities, contributing to the urban heat island effect [7].

On contrary, some models have suggested that a cooling effect on the local environment can be caused by the PV system but that depends on the efficiency and placement of the PV panels (Pushpendu Dwivedi, 2020). Therefore, this research is done to understand the relationship between the roof top solar photovoltaic panel installations and their impact on the thermal environment of the surroundings. This study is conducted for a warm and humid climate, case of Pune city, India.

2 Research Question

Considering a future scenario where all the terrace areas in a highly dense as well as sparse neighborhood is covered with solar PV panel to its maximum capacity: Will PVHI be induced due to the installation of rooftop PV in urban neighborhoods? What will be the effect of the PV rooftop installations on the urban thermal environments?

3 Aim

To understand the impact of Solar Photovoltaic (SPV) rooftop installation on the Urban thermal environment in different urban settings.

4 Objective

The objectives of this paper are:

- To study heat island phenomena and its causes
- To study the possibility of roof top solar PV installations in urban neighborhoods with varied characteristics and analyze its effect on surrounding thermal environment.

5 Scope

Different urban settlements would have different impact on the Urban Heat Island effect. Hence a neighborhood area in Pune city is identified with different urban settings (urban density) and analyzed for how the thermal environment of the vicinity is affected by the installations of Rooftop SPV.

6 Limitation

Due to time constraints, it is difficult to take the field measurements for all the seasons. Hence observational approach will not be used.

7 Literature Review

7.1 *Factors Affecting Urban Thermal Environment (UTE)*

At the local, regional, and global scales, human activities have an impact on climate and atmospheric composition. High temperatures, especially in the summer, can have an impact on the environment and quality of life in a community [8]. Physical characteristics or urban forms, the surface properties of the surroundings, as well as anthropogenic emissions, have a large impact on a UTE [9]. The materials used in the surroundings have an impact on the surrounding temperatures. They have unique heat-reflecting, absorbing, and transmitting qualities. Dark surfaces have low solar reflectance values than light surfaces. Thermal emissivity, materials heat carrying capacity are some of the properties that influence the UTE. The urban canyon allows light to reflect back multiple times into the atmosphere, which reduces the cities albedo and can increase temperatures because of its effect on energy absorption, wind flow, and surface capability to emit longwave radiation back into the atmosphere [8].

7.2 *UTE and Solar Photovoltaic*

Current SPV technologies harness solar energy and produce electricity from it, using crystalline technology. However, the efficiency of solar conversion is its major drawback. Crystalline SPV panels account for 85–95% of the sun's absorption. Depending on the conversion efficiency, the SPV panel will convert 13–20% to electricity and the rest will be converted to heat. Through conduction, convetion and radiation this heat get transferred to the surrounding.

Albedo and emissivity assesses the materials reflectivity and absorptivity [10]. SPV panels almost double the peak and average sensible heat flux, which raises the temperature of the urban environment [11]. SPV panels have a low heat capacity and quickly radiate thermal energy into the environment, raising the ambient temperature during the day.

Furthermore, by trapping longwave sunlight, panels obstruct the nocturnal cooling [12]. In winter, when the sun is at a lower altitude, the impact of solar panels on air temperature is rather minimal, according to a study conducted in Paris. SPV installations in densely populated areas of Paris lowered ambient temperature during the day and at night. It was also observed in one of the study that solar panels would

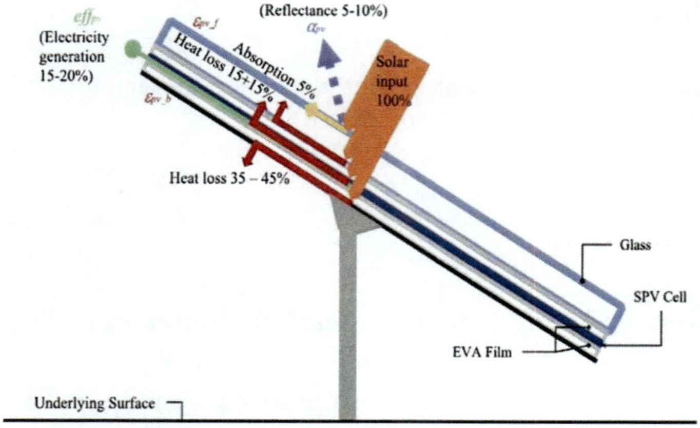

Fig. 1 Heat transfer through SPV module. *Source* Matai [8]

lower the demand for air conditioning in the summer, in addition to reducing UHIs. [5]. Hence some studies concluded that, SPV panel rooftop can be an additional strategy which can help mitigate Urban heat Island [13]. Further study showed that, except for a few sites, a hemispheric temperature decline of 1 °C was seen, whereas warming of up to 1 °C was noted in India, eastern Australia, and the west coast of North America [14] (Fig. 1).

The absence of evaporative cooling owing to a lack of precipitation in both India and eastern Australia could be the cause of the temperature. This means that a study of these three regions at the regional level is required [14]. Henca, an overall analysis of the effect of rooftop PV panels must be done in dense and sparse urban context for India to observe the effect of SPV on UHI.

8 Methodology

Different methods have been used to study the effect of various factors on the thermal environments. These methods can be largely divided into two groups, observational approach and simulation approach [8]. This research focuses on simulation approach done in 3D ENVI-met v4.4.6 model which has the ability to simulate wind flow and MRT [15, 16].

The study commenced with the identification of the study area, mapping it and building a model of the same for further analysis. The study area identification, methods and the tools used are mentioned below.

8.1 Identification of Study Area

The area selection requirements were derived by analyzing the literature reviews studied. A comparative study was done regarding area selection in terms of context, total area under investigation and urban character. The study suggested that.

Umberto Berardi, Jonathan Graham in their paper investigated a neighbourhood in Brampton, Ontario has two small parks, an elementary school, and a shopping plaza. Land cover is mostly impervious, i.e. asphalt or concrete, with grass covering only the parks and a small portion of the parcels associated with each home. The total area under investigation is 0.75 km^2. 837 × 837 m area is simulated in envimet [17]. Another paper by Aiza Cortes, Yuji Murashita, Tomohito Matsuo, Akira Kondo, Hikari Shimadera, Yoshio Inoue evaluated the area of 6–6 Fukushimaku, Osaka City, Japan. This area was chosen because it is in the city centre and exhibits a typical urban setup. The buildings were within the200 × 200 × 69 m analytical area with a horizontal grid interval of 5 m [18].

8.2 Selection of Study Area

The first task was to identify the area for the study. After studying the overall growth pattern of the Pune city, an area which had a combination of dense and sparsely developed residential/mixed use buildings was selected. The area selection was done on the built: unbuilt ratio, vegetation cover, urban sprawl and characteristics etc. The map below shows the area selected. The characteristics of the individual areas are mentioned below (Fig. 2).

Area 1. The size of analysis grid was of 400 × 300 × 75 m with 2 × 2 × 3 m grid cell size. The main buildings were within 300 × 200 m rectangle. This area had a combination of residential, institutional and mixed used development. The area had an institutional campus with 5–6 institutional buildings and an open playground area

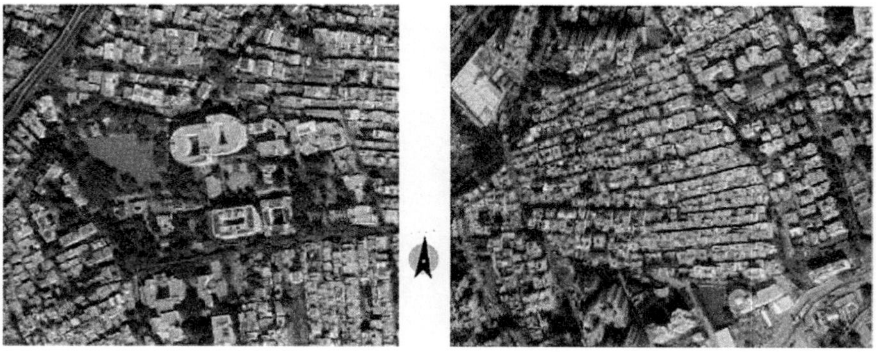

Fig. 2 Study area demarcation Area 1 (left) and Area 2 (right)

of the school and college. Sparse development was observed in this area. This area had high canopy trees of approx. 15–20 m height.

Area 2. The size of analysis grid was of 360 × 260 × 75 m with 2 × 2 × 3 m grid cell size. The main buildings were within 300 × 200 m rectangle. The region had densely cluttered residential and mixed-use development. Very less vegetation was observed in this area.

The trees found were a combination of low height trees and Palm trees (approx. 10–15 m high). The residential buildings had no setback.

8.3 Tools Used for Research

ENVI-met v4.4.6 was used for simulation. QGis software was used to generate high resolution image of the selected area. Majority of the buildings had burned brick wall and cast dense RCC roof. Some of the buildings had G.I. sheet and Mangalore tile roofing as well. Majority of the trees in area1 were highrise i.e., 10–15 m high whereas in Area 2 the vegetation was of low height below 5 m. Simulation was done for the hottest summer month i.e., April. Climate data was entered as shown in Table 1 [19]. Shadow analysis was done in SketchUp. Net roof area was calculated after reducing the core, circulation area and area under shadow. It is observed that in area with sparse development, there is no shadow area on the rooftops. Whereas in dense area, all G and G + 1 structures have more than 50% roof area under shadow. Hence in dense area, all the structures above 7 m ht. are considered for SPV installation.

Simulation for 6 h was done for Daytime and for Night-time. The timings were selected considering the critical hrs. Base case and design case was simulated for 21st April from 9 am to 3 pm for daytime and from 11 pm to 5 am for night-time. PV panel roof assembly was created in ENVI-met consisting of 150 mm RCC cast dense slab with 500 mm airgap with Solar PV panel as top layer. This material was applied to PV available roof area for design case simulation.

Table 1 Climate data for ENVI-met

Climate data	Min	Max
Temperature (°C)	17	42
Humidity (%)	9	82
Wind speed	0.77 m/s	
Wind direction	270°	

8.4 *ENVI-Met Parameters*

The air temperatures and MRT are important for assessing heat islands from ground level to the tops of trees and buildings as they indicate the outdoor thermal comfort assessment [20]. The mean radiant temperature (T_{mrt}), which sums up all global short and long wave radiation fluxes, to which the human body is exposed. It is one of the meteorological parameters governing human energy balance and the thermal comfort [21].

9 Result and Findings

9.1 *Area 1 Results*

Majority building were between G + 2 to G + 5 structures and hence parameters were analyzed at eye level i.e., 1.5 m and average roof level i.e., 13.5 m height. Potential air temperature and MRT were analyzed to understand the impact of PV panels. Simulation results for daytime as well as nighttime were analyzed as the heat gets dissipated at night and may result in higher temperatures. From the results obtained, it was observed that 2 pm and 12am are the critical timings having the highest temperatures. The 2D graphs were generated for base case and design case at both the heights, and the comparison showed the following results: The Tables 2 and 3 shows the results for MRT and potential air temperature for daytime and night-time. Figures 3 to 4 shows the graphs of MRT at the height of roof level (13.5) for base case and design case respectively. The potential air temperature remains constant at eye level as well as roof level during daytime and nighttime. The MRT at both the levels are observed to decrease by an avg. 0.3 °C in design case during day as well as nighttime. Similarly, during night time the graph shows the reduction in MRT for design case when compared with base case. Whereas the PAT experiences negligible change and remains almost constant. Hence the installation of rooftop PV in area 1 reduces the MRT slightly.

Table 2 MRT and PAT readings for Area 1

MRT at 13.5 m	2 pm		12am	
	Min	Max	Min	Max
BASECASE	41.76	77.35	15.15	28.86
DESIGNCASE	40.03	77.16	14.92	28.49
PAT at 13.5 m	2 pm		12am	
	Min	Max	Min	Max
BASECASE	33.37	35.8	28.61	29.92
DESIGNCASE	33.37	35.82	23.36	29.66

Table 3 MRT and PAT result of Area 2

HEIGHT 13.5 m	2 pm		12am	
	Min	Max	Min	Max
MRT BASECASE	39.4	76.51	13.98	25.76
MRT DESIGNCASE	40.42	79.61	13.84	25.34
PAT BASECASE	21.32	35.93	19.85	29.54
PAT DESIGNCASE	33.91	36.33	26.44	29.49

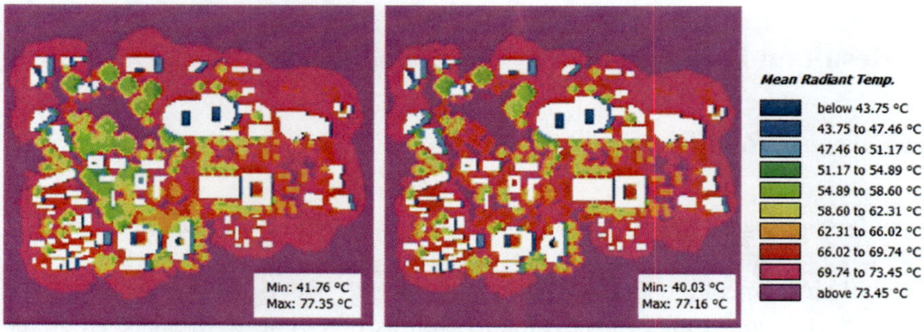

Fig. 3 Base case (left) and design case (right) of MRT at 13.5 m at 2 pm

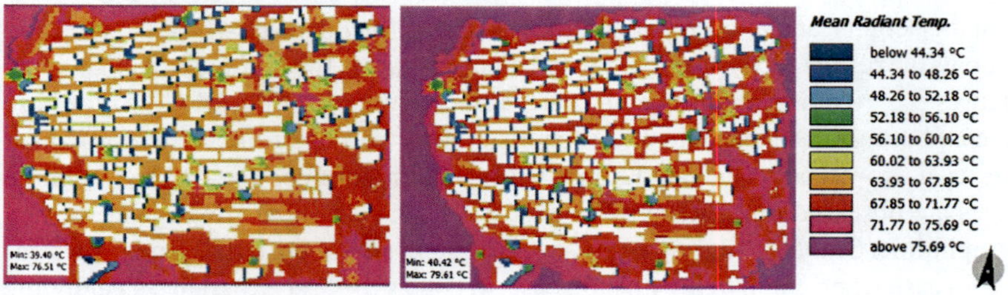

Fig. 4 Base case (left) and design case (right) of MRT at 13.5 m at 2 pm

9.2 Area 2 Results

Majority building were G + 2 and G + 3 structures and hence parameters were analyzed at eye level i.e., 1.5 m and roof level i.e., 13.5 m height. This area has dense cluttered development with very less tress/vegetation. It was observed that the residential development had no setback, and all the buildings were adjoining each other with no or very less spacing in-between the buildings. Potential air temperature and MRT were analyzed to understand the impact of PV panels. The 2D graphs were generated for base case and design case at both the heights, and the comparison showed the following results.

It can be observed that the MRT at daytime increases in the design case where the PV material is applied for the roof. The temperatures rise by 2–3 °C. It can be observed from the maps above, that PV application increases the MRT at roof level and eye level. The potential air temperature during daytime in April at 1.5 m height is observed to increase from 36.85 °C (base case) to 39.16 °C (design case). And at 13.5 m height it is observed to increase by 0.4 °C from base case. During nighttime the MRT decreases by 0.4 °C. The base case maximum MRT is observed to be 25.76 °C and design case to be 25.34 °C. Whereas the PAT at night, an avg increase of 3 °C is observed.

10 Analysis

It is observed that as the urban characteristics vary, the impact of SPV changes. The research was done for two different urban areas one with sparsely spaced buildings with large open spaces and dense vegetation and other with dense urban development with no open spaces and very less vegetation.

Area 1: It is observed that in area with sparsely spaced building the temperatures (MRT and Potential air temperatures) at eyelevel as well as roof level decreases after the application of SPV panels on rooftop. Similar results were observed for nighttime simulations as well. The potential air temperature in this area remains somewhat similar for daytime and decreases for nighttime. This suggests that the PV installation decreases the temperature for sparse urban development.

Area 2: For area with dense settlement, the MRT and Potential air temperature increases due to the installation of rooftop SPV at roof as well as eye level during daytime.

MRT increases at daytime and at night- time the MRT is observed to decrease. Whereas the Potential air temperatures increases during daytime as well as nighttime. This suggests that the urban characteristics, PV installation, building materials and less vegetation cover adds to the urban heat island effect.

11 Discussion

Study regarding 'The Photovoltaic Heat Island Effect: Larger Solar Power Plants Increase Local Temperatures' describes research on the local temperatures of a neighboring desert ecosystem compared to those of PV installations. Depending on the hour of the day and the season, there were substantial temperature changes between various locations, but the PV installation was found to have a temperature that was higher than or on par with other locations. Due to the PVHI effect, the cooling of ambient temperatures in the evening was delayed. Few studies as discussed earlier also indicate decrease in temperatures of the surroundings. Multiple factors are responsible for the changes in UTE.

12 Conclusion

The research shows that the temperatures and the MRT seem to increase in dense settlement which may contribute to the Urban heat island. Whereas the temperatures decrease and cause a cooling effect for sparsely spaced and high vegetation area. It can be concluded that PVHI will be induced in area where all the terrace area in a highly dense urban area has PV installed to its maximum capacity and the vegetation is very less. This shows that PV installation adds on to the "heat island" (PVHI) effect in dense urban areas by increasing the ambient temperatures and mitigation strategies should be discussed before installing PV panels as per the targets set by the government of India.

References

1. Norman J Campbell GS (1998) An introduction to environmental biophysics, 2nd edn. Springer, New York, NY
2. Oke T (1982) The energetic basis of the urban heat island. Q J R Meteorol Soc 108:1–24
3. Burg BR, Ruch PW, Paredes S, Michel B, Burg MB (2015) Placement and efficiency effects on radiative forcing of solar installations. In: AIP conference proceedings 1679, 090001, Switzerland
4. Phillips J (2013) Determining the sustainability of large-scale photovoltaic solar power plants. Renew Sustain Energy Rev 27:435–444
5. Minor RL, Allen NA, Cronin AD, Brooks AE, Pavao-Zuckerman MA, Barron-Gafford GA (2016) The photovoltaic heat Island effect: larger solar power plants increase local temperatures. Sci Rep
6. Broadbent AM, Scott Krayenhoff E, Georgescu M, Sailor DJ (2019) The observed effects of utility-scale photovoltaics on near-surface air temperature and energy balance. J Appl Meterol Climatol 58(5):989–1006
7. Brown KE, Baniassadi A, Pham JV, Sailor DJ, Phelan PE, Brown PP (2020) Effects of rooftop photovoltaics on building cooling demand and sensible heat flux Into the environment for an installation on a white roof. ASME J Eng Sustain Build Cities 1(2):12:22
8. Matai K (2020) Influence of SPV installations on the thermal. Multidisc Sci J, pp 343–357
9. Oke T, Mills G, Christen A, Voogt J (2017) Urban climates. UK: Cambridge University Press, Cambridge
10. Santamouris ASTKM (2011) Using advanced cool materials in the urban built environment to mitigate heat islands and improve thermal comfort conditions. Sol Energy 85(12):3085–3102
11. Pham J, Baniassadi A, Brown K, Heusinger J, Sailor D (2019) Comparing photovoltaic and reflective shade surfaces in the urban environment: effect on surface temperature heat flux and pedestrian thermal comfort. Urban Clim
12. Masson V, Bonhomme M, Salagnac J-L, Briottet X, Lemonsu (2014) Solar panels reduce both global warming and urban heat island. Front Environ Sci pp 1–10
13. Armstrong A, Ostle N, Whitaker J (2016) Solar park microclimate and vegetation management effects on grassland carbon cycling. Environ Res Lett
14. Hu A, Levis S, Meehl G, Han W, Washington W, Oleson K, van Ruijven B, He, M (2016) Impact of solar panels on global climate. Nat Clim Chang, pp 290–294
15. Meloni M, Coccolo S, Kaempf J, Scartezzini J-L, Naboni (2017) An overview of simulation tools for predicting the mean radiant temperature in an outdoor space. In: CISBAT 2017 international conference-future buildings & districts—energy efficiency from, Lausanne, Switzerland

16. Blocken B, Maiheu B, van Heijst GJF Toparlar Y (2017) A review on the CFD analysis of urban microclimate. Renew Sustain Energy Rev J 80:1613–1640
17. Umberto Berardi JG (2020) Investigation of the impacts of microclimate on PV energy efficiency and outdoor thermal comfort. Sustain Cities Soc 62
18. Murashita Y, Matsuo T, Kondo A, Shimadera H, Inoue Y, Cortes A (2015) Numerical evaluation of the effect of photovoltaic cell installation on urban thermal environment. Sustain Cities Soc 19:250–258
19. "Time and Date," January 2018. [Online]. Available: https://www.timeanddate.com/weather/india/pune/historic?month=1&year=2018
20. Mahdavi A, Maleki A (2016) Evaluation of Urban Heat Islands mitigation strategies using 3dimentional urban micro-climate model envi-met. Asian J Civ Eng (BHRC) 17:357–371
21. Lindberg F, Eliasson I, Holmer B, Thorsson S (2007) Different methods for estimating the mean radiant temperature in an outdoor urban setting. Int J Climatol 27:1983–1993
22. Broadbent AM, Scott Krayenhoff E, Georgescu M, Sailor DJ (2019) The observed effects of utility-scale photovoltaics on near-surface air temperature and energy balance. J Appl Meterol Climatol, pp 989–1006
23. Nemet GF (2009) Net radiative forcing from widespread deployment of photovoltaics. Environ Sci Technol

Productivity Improvement of Solar Still Using Cemented Blocks

Naveen Sharma, Shaik Noushad, G. Siva Ram Kumar Reddy, and Ajit

1 Introduction

Energy sources derived from renewable resources can address many pressing issues facing humanity, including climate change, freshwater shortage, environmental destruction, and energy scarcity, and efficient usage of these sources can result in a more sustainable world. Solar energy is a clean, low-cost and abundant source of energy, and effectively applied to numerous applications including electricity generation, desalination/distillation, air and water heating, cooking and drying [1–4].

Globally, fresh water scarcity is becoming increasingly common due to industrialization and population growth. In South India, the scarcity of drinking water presents a huge challenge to society, since the available water cannot accommodate household needs. Industrial wastes have also contaminated potable water resources, including rivers, lakes, wells, and underground water, adding to the concern. The solar distillation method, which uses solar stills (SSs) to produce fresh water from salty water, has been reported to be one of the most promising water purification methods in recent years [4]. Using solar energy, clean water can be extracted from salty using heat transfer mechanisms, including evaporation and condensation, thereby making it ideal for remote communities lacking conventional water treatment systems. Although conventional solar stills often have low water productivity,

N. Sharma (✉)
Department of Mechanical Engineering, Netaji Subhas University of Technology (West Campus), Jaffarpur, Delhi 110073, India
e-mail: sharma.naveen28@yahoo.com

S. Noushad · G. Siva Ram Kumar Reddy
Department of Mechanical Engineering, DVR and Dr. HS MIC College of Technology, Kanchikacherla, AP 521180, India

Ajit
Department of Mechanical Engineering, Manav Rachna University, Faridabad, Haryana 121004, India

S. Doolla et al. (eds.), *Advances in Clean Energy and Sustainability*,
Green Energy and Technology, https://doi.org/10.1007/978-981-99-2279-6_36

practitioners have conducted various studies to improve solar still performance by design modification, or adding fins and SHSMs over basin, or by integrating SSs with solar collectors [5–8].

Many researchers concentrated on using different type of heat storage material for enhancing the yield of the SS [9–18]. In addition to enhancing distillate output during the daytime, energy absorbing material also increased distillate output at night [9]. A study of two slope solar stills with single and double basins with heat storage was presented by Rajaseenivasan et al. [10]. With mild steel pieces, a double basin SS yields 169.17% more than a single basin SS with 80 mm of water depth. Rashidi et al. [11] used a reticular porous material to augment the yield of conventional SS. The SS with porous material was found to produce 17.35% more yield compared to its conventional counterpart. Sakthivel and Arjunan [12] experimented with cotton cloth, as an energy storage material, over the absorber plate during the climatological conditions of Chennai, India. Solar still having cotton cloth of 6 mm thickness has found to be 21.4% more productive than conventional SS. Modi and Modi [13] studied the productivity of double-basin SS with jute and cotton cloth as wick materials. At a depth of 0.02 m, jute cloth was found to have a productivity of 21.46% greater than black cotton cloth. Yarramsetty et al. [14] studied the productivity of pyramid SSs with SHSM, specifically clay pots facing both up and down. Clay pots facing down produced 8.6% more distilled water than those facing up. Moreover, 25 clay pots with facing down demonstrated a maximum productivity of fresh water (2.0 L/m^2) that was 60% higher than that of smooth basin. A recent study by Darbari and Rashidi [15] improved the distillate output of SS by using jute cloth as a porous wick absorber floating in a thermocol basin. In comparison with conventional stills, modified SSs with semicircular-shaped tooth absorbers produced 65% more daily yield.

With its high specific heat capacity and pore holes, sandstone is a good medium for storing thermal energy to improve yield of SS both during the day and at night [16]. A 24-h distillation experiment will yield, respectively, 3.9, 3.41, and 3.0 kg/m^2 of distillate output for sand stones containing SSs, marble pieces containing solar stills, and solar stills alone. Further, Panchal et al. [17] observed that the feed water yield of SS can be increased to 104.68% using SS combining evacuated tubes and calcium stones. Sharshir et al. [18] reported that Salt, as a SHSM, increased the efficiency of solar stills by about 1.5 L/m^2.

Above studies indicate that thermal energy storage materials used in SSs have a remarkable effect on the yield of fresh water. There has not been adequate research done on cemented blocks, used as SHSM inside a pyramid-shaped solar still. With this in mind, the present study examine the consequences of mounting cemented blocks on the absorber plate and black coating over them on the production of Pyramid SS.

2 Experimental Facility and Instrumentation

Figure 1 illustrates the experimental facility installed on the terrace of an autonomous college building located at Vijayawada (16.6834 °N, 80.3904 °E), India. Galvanized iron (GI) sheet is used for the still basin, which has internal dimensions of 0.7 m × 0.7 m and a depth of 0.225 m. GI sheet is painted black so that it can absorb the most solar radiation possible. Additionally, proper insulation is provided on the bottom and sides of the basin to minimize heat loss. Condensing glass is shaped like a pyramid to provide more surface area for vapour to condense. Condensing cover is made of 3.5 mm thick plain glass with a transmission index of 0.9, tilted 16.68° in relation to the horizontal surface. Base of the SS and sidewalls are made from wood with a 22 mm thickness. Solar still is made with meticulous attention to all dimensions and smooth finishing. To improve PSS performance, cemented blocks (length 40 mm, width 40 mm, and height 20 mm) without (CASE-2) and with black coating (CASE-3) are installed in both longitudinal and transverse directions at 50 mm spacing over the absorber (see Fig. 2). As cement block has a high thermal mass, therefore, during the day it absorbs and retains heat, but at night it radiates it out. Cemented blocks were judiciously spaced so that fresh water yield was improved in a noticeable way.

At 30-min intervals, a digital temperature indicator attached to calibrated thermocouples, K-type, is used to accurately record the temperature of the basin, the wall, the water, the vapour, and the inner surface of the glass. An anemometer and a solar power meter are used to acquire wind speed and solar irradiance. To test the standard of briny and cleaned water, pH meters, TDS meters, and conductivity meters are employed. Table 1 shows the measuring instruments' accuracy, range, and standard uncertainty.

Initially, 3 cm of tap water is poured into the SS situated on the roof. When the sun rises, solar radiation enters into the SS via transparent glass and heat is absorbed by the water. The absorption of enough energy causes water vapour to be released during the heating process. Water vapour generated by the generator is condensed

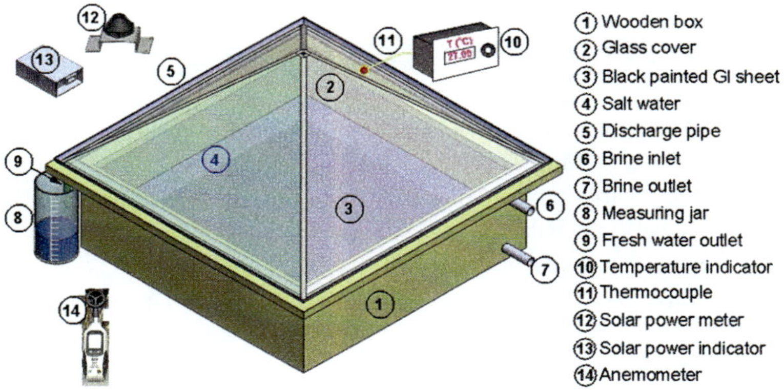

Fig. 1 Three-dimensional model of pyramidal solar still

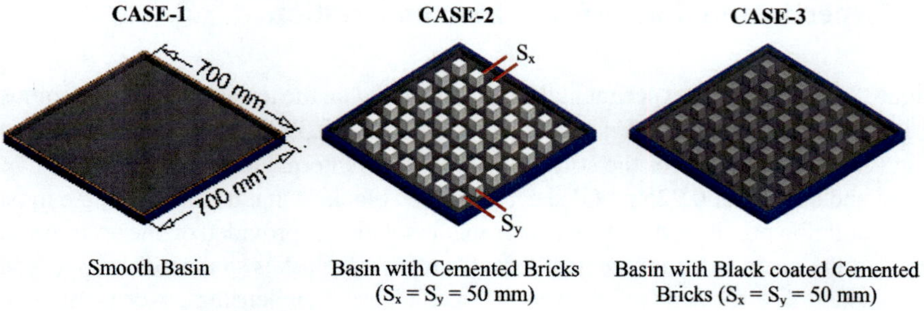

Fig. 2 Studied configurations

Table 1 Measuring instruments' accuracy, range, and standard error

Instruments	Range ± accuracy	Standard error
Solar power meter	$0 - 1800 \pm 5$ W/m^2	$\pm$ 2.88
Anemometer	$1 - 25 \pm 0.2$ m/s	$\pm$ 0.12
Thermocouple	$0 - 275 \pm 1$ °C	$\pm$ 0.58
Temperature indicator	$0 - 1000 \pm 1$ °C	$\pm$ 0.58
Measuring jar	$0 - 1000 \pm 50$ ml	$\pm$ 28.87
pH meter	$0 - 14 \pm 0.01$	$\pm$ 0.006
TDS meter	$0 - 999 \pm 2$ PPM	$\pm$ 1.15
Conductivity meter	$200 - 1000 \pm 0.1$ mS/cm	$\pm$ 0.058

in the condenser unit owing to temperature differences at the glass surface. This condensate is collected in a measuring jar as it slides over inclined glass. In solar stills, water is cleaned by evaporation and condensation, resulting in a condensate that can be used in the home.

3 Results and Discussion

From 7:00 AM to 7:00 PM, experiments are conducted to measure temperatures after every half hour at suitably located sites and hourly distillate yield. Figure 3 shows the hourly distribution of solar irradiance for considered cases during the experimentation period (January 14–17, 2020). As expected, the intensity of solar radiation progressively rises from 7:00 AM, reaching the peak value (792 W/m^2) at 12:00 PM, and then decreasing with passage of time and becoming minimum at 6:00 PM because of sunset. As can be seen from Fig. 3, the variation in solar irradiance for the days of experiments is within a range of 5%, so the experimental data obtained for the investigated cases could be compared to determine how effective heat storage material would be in improving the performance of solar still.

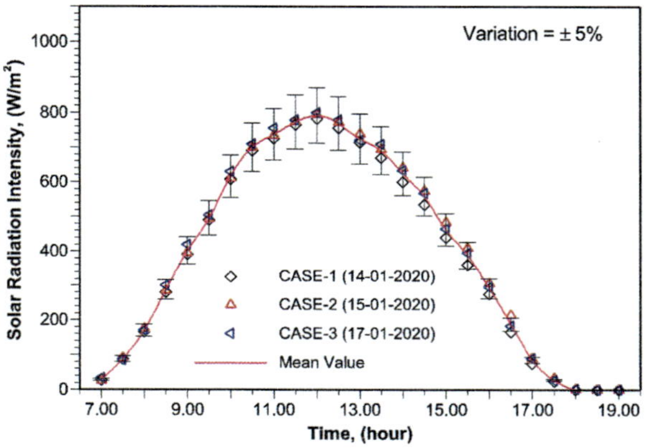

Fig. 3 Solar irradiance distribution for the days of experimentation

Figure 4 shows the temperatures at different locations inside the SS, mainly the basin, inner glass, water and heat storage material temperature that are of prime importance in understanding the distillate yield for investigated cases. All of the temperatures rise first, reaching their maximum around 1:30 PM and then declining thereafter. Even so, a time lag is observed among the highest temperature and solar irradiance values, which can be attributed to the time required to warm the solar still parts.

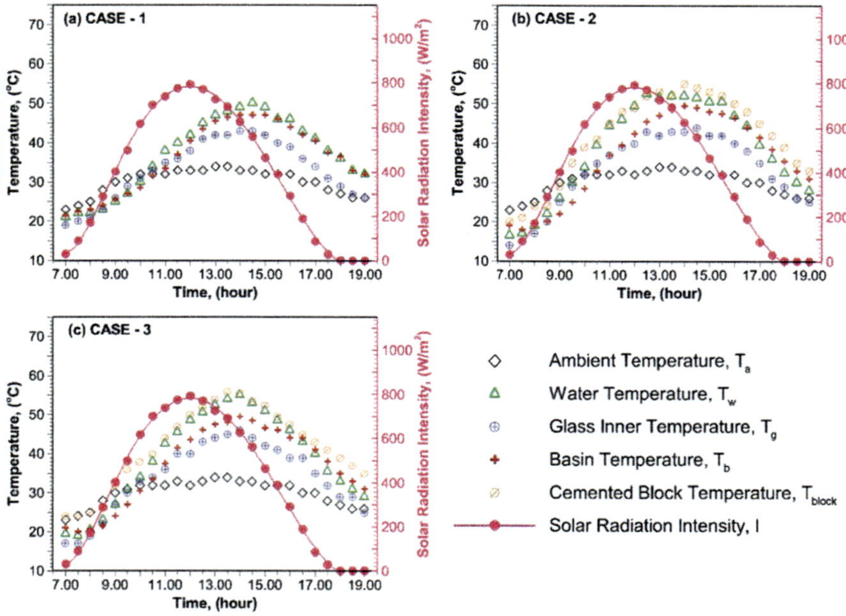

Fig. 4 For different cases, temperature variation inside a solar still

The temperature of salty water is found to be the highest, followed closely by the basin temperature and the lowest temperature for inner glass between 9:30 AM and 5:00 PM for all the cases. As seen in Fig. 4, the temperature of the cemented block (T_{block}) is the highest among all other temperatures for CASE-2 and CASE-3, thereby indicating that the addition of thermal energy storage material over the basin significantly improves water temperature inside the still. CASE-3, CASE-2, and CASE-1 have maximum water temperatures of about 56 °C, 52.5 °C, and 50 °C, respectively.

Owing to the high thermal mass of the cemented blocks, the heat is charged rapidly and discharged slowly, causing higher difference in evaporator and condenser temperatures (Fig. 4). CASE-3 has higher evaporation and condensation rates than CASE-2 and CASE-1, which can be seen as the reason of higher freshwater yield for CASE-3 (see Fig. 5). Figure 5 illustrates how the production flux varies across all cases. The yields of smooth basin SS and cemented blocks SS are close to each other at first, but as time passes, PSS with cemented blocks produces more fresh water. However, after 3:00 PM, the output distillate yield begins to decline. The solar still with black coated cemented blocks produces the most distilled water out of the three. Cemented blocks store a lot of thermal energy, as revealed by the higher values of cemented block temperature in Fig. 4, which transfers this absorbed heat to the salty water, causing a higher rate of evaporation and condensation, resulting in a higher distillate yield for CASE-3 than for CASE-2 and CASE-1.

As can be seen in Fig. 6, CASE-3 produces the highest amount of fresh water, i.e. 1.36 L/m^2, followed by CASE–2 (1.20 L/m^2) and CASE–1 (0.95 L/m^2). For CASE-3 and CASE-2, the distilled water output is 43.16% and 26.32% higher than that for CASE-1. With thermal energy stored in the cemented blocks, the heat transfer rate is increased, leading to comparatively high water temperatures in the evening and, in turn, a rise in evaporation rate that results in maximum freshwater yield.

Compared to just cement blocks, the black coating over the cemented blocks generates a 13.3% increase in distillate yield. Comparing the freshwater productivity enhancement in the present study with the literature, as shown in Table 2, shows that

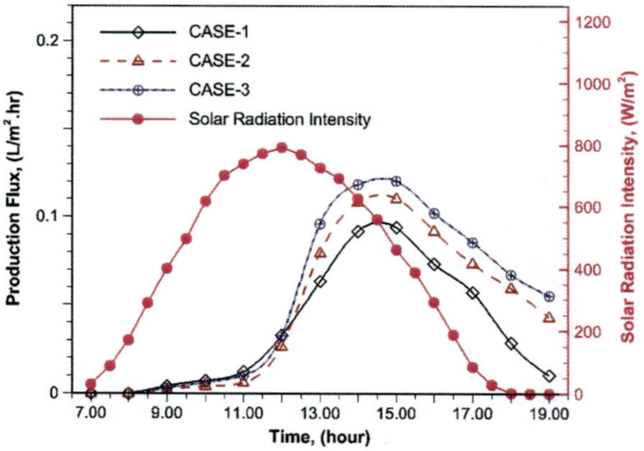

Fig. 5 Hourly variation of distillate output for studied cases

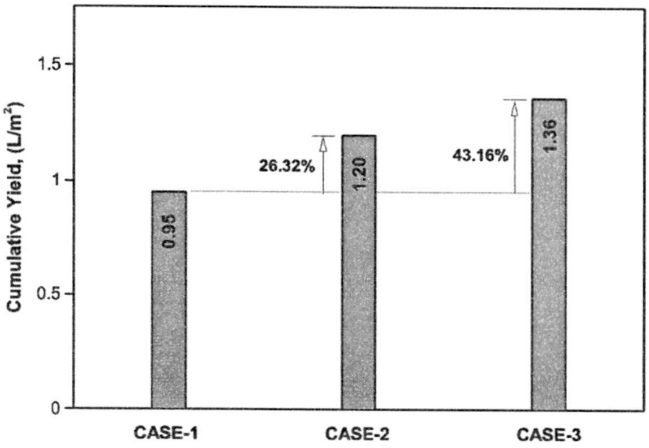

Fig. 6 Productivity variation with and without cemented blocks

Table 2 Comparison of augmentation in fresh water yield with different SHSMs

Study	Solar still	SHSM	Productivity improvement (%)
Sakthivel and Arjunan [12]	Single slope	• 6 mm thick cotton cloth	24.1
Panchal et al. [16]	Single slope	• Sand stones • Marble pieces	30.0 13.7
Sharshir et al. [18]	Single slope	• Graphite flakes with wick • Carbon foam with wick	34.5 28.6
Present study	Square pyramid	• Cemented blocks • Black coated cemented blocks	26.3 43.2

cement blocks produce more fresh water than other SHSMs. Fresh water produced from PSS contains 25 mg/L of TDS, pH of 6.9, and conductivity of 0.095 dSm^{-1}, which are all within the recommended range by WHO. Therefore, the fresh water obtained is fit for domestic applications.

4 Conclusions

This study investigated experimentally the effect of the placement of cement blocks with and without black coating over the absorber of a PSS on the fresh water productivity. The following conclusions are rendered from the results:

- Mounting cemented blocks on the absorbing plate of solar still lead to an increase in daily yield, and the increase is even greater with a black coating applied over the cemented blocks.
- CASE-3 produces the most fresh water daily (1.36 L/m^2), after that CASE-2 (1.20 L/m^2) and CASE-1 (0.95 L/m^2).
- An increase of 43.16% by CASE-3 and 26.32% by CASE-2 in distillate output is achieved in comparison to CASE-1.
- The analysis of energy and economics is in progress and can be seen as the continuation of the present work.

References

1. Sharma N, Varun, Siddhartha (2012) Stochastic techniques used for optimization in solar systems: a review. Renew Sustain Energy Rev 16(3):1399–1411
2. Sunil, Varun, Sharma N (2014) Experimental investigation of the performance of an indirect-mode natural convection solar dryer for drying fenugreek leaves. J Therm Anal Calorim 118(1):523–531
3. Siddhartha, Sharma N, Varun (2012) A particle swarm optimization algorithm for optimization of thermal performance of a smooth flat plate solar air heater. Energy 38:406–413
4. Khawaji AD, Kutubkhanah IK, Wie JM (2008) Advances in seawater desalination technologies. Desalination 221(1–3):47–69
5. Mohiuddin SA, Kaviti AK, Rao TS, Sikarwar VS (2022) Historic review and recent progress in internal design modification in solar stills. Environ Sci Poll Res 1–54. https://doi.org/10.1007/s11356-022-19527-x
6. Bhargva M, Yadav A (2021) Factors affecting the performance of a solar still and productivity enhancement methods: a review. Environ Sci Pollut Res 28(39):54383–54402
7. Sharma N, Noushad S, Reddy GSR, Kumar (2022) Effect of copper fins on fresh water productivity of pyramid solar still. Springer, Singapore, pp 83–91
8. Nehar L, Rahman T, Tuly SS, Rahman S, Sarker MRI, Beg MRA (2022) Thermal performance analysis of a solar still with different absorber plates and external copper condenser. Groundw Sustain Dev 17:100763
9. Kabeel AE, Omara ZM, Essa FA, Abdullah AS, Arunkumar T, Sathyamurthy R (2017) Augmentation of a solar still distillate yield via absorber plate coated with black nanoparticles. Alex Eng J 56(4):433–438
10. Rajaseenivasan T, Elango T, Murugavel KK (2013) Comparative study of double basin and single basin solar stills. Desalination 309:27–31
11. Rashidi S, Rahbar N, Valipour MS, Esfahani JA (2018) Enhancement of solar still by reticular porous media: experimental investigation with exergy and economic analysis. Appl Therm Eng 130:1341–1348
12. Sakthivel TG, Arjunan TV (2019) Thermodynamic performance comparison of single slope solar stills with and without cotton cloth energy storage medium. J Therm Anal Calorim 137(1):351–360
13. Modi KV, Modi JG (2019) Performance of single-slope double-basin solar stills with small pile of wick materials. Appl Therm Eng 149:723–730
14. Yarramsetty N, Sharma N, Narayana ML (2021) Experimental investigation of a pyramid type solar still with porous material: productivity assessment. World J Eng. https://doi.org/10.1108/WJE-02-2021-0096
15. Darbari B, Rashidi S (2022) Performance analysis for single slope solar still enhanced with multi-shaped floating porous absorber. Sustain Energy Technol Assess 50:101854

16. Panchal H, Patel DK, Patel P (2018) Theoretical and experimental performance analysis of sandstones and marble pieces as thermal energy storage materials inside solar stills. Int J Ambient Energy 39(3):221–229
17. Panchal H, Hishan SS, Rahim R, Sadasivuni KK (2020) Solar still with evacuated tubes and calcium stones to enhance the yield: an experimental investigation. Process Saf Environ Prot 142:150–155
18. Sharshir SW, Elsheikh AH, Ellakany YM, Kandeal AW, Edreis EMA, Sathyamurthy R, Thakur AK, Eltawil MA, Hamed MH, Kabeel AE (2020) Improving the performance of solar still using different heat localization materials. Environ Sci Pollut Res 27(11):12332–12344

Analysis of Organic Rankine Cycle Using Various Working Fluids for Low-Grade Waste Heat Recovery

Ayona Biswas and Bijan Kumar Mandal

1 Introduction

Over the last few decades, the excessive usage of fossil fuels and the release of waste of heat energy from exhaust gases from engines, turbines and industrial power plants into the atmosphere have become threats to the environment. Thus, government bodies have paid more attention to low-grade energy sources such as solar, wind, tidal, and geothermal energy sources as an alternative to conventional fossil fuels to the utilization of low-grade energy sources. Only 30% combustion energy of the engine's fuel can be converted into useful work and used to drive vehicles. Approximately 40% of the heat energy is wasted due to exhaust gas emissions [1]. Organic Rankine Cycle (ORC) has potential in power plant sectors over conventional steam power cycles for recovering waste heat energy. Some advantages of ORC are the reduction of emission CO, CO_2, NO_x and other pollutants, economical utilization of energy resources, and low-temperature waste heat recovery.

The selection of organic working fluid for the ORC system is the most important, and several researches have already been carried out on this. Hettiarachchi et al. [2] investigated how to get optimum cycle performance by varying evaporation and condensation temperature. Furthermore, they compared the results for ammonia, n-pentane, R-123, and PF 5050. Saleh et al. [3] investigated the thermal performance of 31 different working fluids for subcritical and supercritical organic rankine cycles. Yamamoto et al. [4] evaluated the optimum operating conditions of ORC using R-123 and water as working fluids. Tchanche et al. [5] made a comparative study between 20 organic working fluids using low-temperature solar ORC systems and considered low-grade waste heat recovery for their further investigation. Hung et al. [6] worked to determine the thermal performance of benzene, ammonia, R-11, R-12, R-134a and R-113. Maizza and Maizza [7] estimated the physical and thermodynamic properties

A. Biswas (✉) · B. K. Mandal
Indian Institute of Engineering Science and Technology Shibpur, Howrah 711103, India
e-mail: ayonabiswas123@gmail.com

S. Doolla et al. (eds.), *Advances in Clean Energy and Sustainability*,
Green Energy and Technology, https://doi.org/10.1007/978-981-99-2279-6_37

of 20 organic working fluids for waste heat recovery applications. Liu et al. [8] evaluated the heat-recovery ratio and thermal efficiency of ORC for ten different working fluids. Wei et al. [9, 10] investigated the system performance of an ORC system at various operating conditions and used R-245fa as the working fluid.

The selection of organic working fluids and defining the optimized operating condition of the ORC system are required to analyse ORC performance to achieve the maximum utilization of waste heat. In this paper, we estimated the operating conditions of an ORC system for recovering the waste heat of PEM fuel cells and industrial waste using a single screw expander. Thermodynamic models of seven different organic working fluids are designed and calculated using DWSIM and REFPFOF software.

2 ORC System Description

The schematic diagram of the ORC is shown in Fig. 1. The various components of a traditional ORC are the pump, screw expander, evaporator, condenser, generator and reservoir. The working fluid is pressurized from the reservoir to the high-pressure pipeline. The waste heat from industrial wastage is added to the evaporator at constant pressure. The working fluid reaches the saturated vapour state. The saturated vapour expands in the single screw expander. In this stage, mechanical energy is converted into electrical power. After that, in the condenser, low-pressure superheated vapour rejects heat and is converted into saturated liquid. In this work, water is considered as the condensing medium at 25 °C.

In the Fig. 1

State point 1: Liquid line (saturated)
State point 3: Vapour line (saturated)
Process 1 → 2: Process through the pump which is an isentropic compression process
Process 2 → 3: Process through the evaporator
Process 3 → 4: Process through the single screw expander
Process 4 → 1: Process through the condenser.

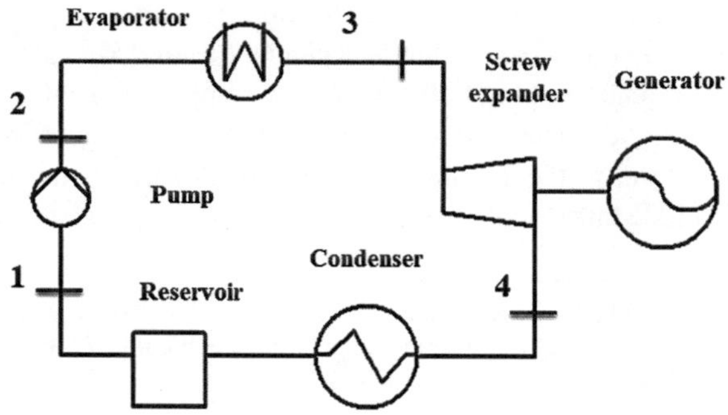

Fig. 1 Schematic diagram of organic rankine cycle

3 Methodology

3.1 Selection of Organic Working Fluid

The points which are mentioned below should be taken into consideration to select the optimized organic working fluid.

- **High efficiency and power output**: First and second law efficiency, net power output need to be as high as possible.
- **Low viscosity**: The viscosity of the working fluid should be low in both liquid and gaseous states to get high heat transfer co-efficient and less friction loss in the heat exchanger.
- **High thermal conductivity**: High thermal conductivity of the working fluids helps to obtain the high heat transfer co-efficient.
- **High vapor density**: Density is inversely proportional to the specific volume and volumetric flow rate. So, if we reduce the density of the density of the fluid then specific volume will become higher and bigger components like: turbine, heat changer will be needed which is very costly. If we increase the specific volume, the pressure decreases and needs high pump work.
- **High condensing pressure**: High condensing pressure of the liquid helps to avoid leakage losses.
- **Positive condensing gauge pressure**: The lowest pressure of the working fluids should be greater the atmospheric pressure to avoid the air infiltration of the fluids.
- **Low melting point**: To avoid the freezing of the working fluid, the melting point should be lower than the lowest ambient temperature.
- **Acceptable evaporating pressure**: Evaporating pressure should be in acceptable range because high pressures usually lead to higher investment costs and increased complexity.
- **High enthalpy**: Enthalpy of the working fluid should be high which leads to high network output.
- **High convective heat coefficient**: High convective heat coefficient increases the heat transfer process between the heat source and the sink.
- **High heat capacity**: If heat capacity of the working fluid increases then mass flow rate decreases and it helps to get better heat recovery from the heat source.
- **Ozone depleting potential (ODP)**: The ozone depleting potential should be as less as possible. ODP should be very low or zero.
- **Global warming potential (GWP)**: GWP need to be low and it is measured with comparison to the CO_2, set to the unity. But some refrigerants can reach GWP value as high as 1000.
- **Higher safety level**: Low toxicity and no flammability of the working fluids are desirable.

3.2 Thermodynamic Analysis

The thermal analysis of the overall system and subsystem is based on some basic assumptions, those are as follows:

- Steady-state operating condition
- No heat loss and pressure drop
- Negligible change in potential energy and kinetic energy
- Saturated working fluid at the outlet of the evaporator and the condenser
- Isentropic pump and expander efficiency.

The work done by the pump is denoted by expression below:

$$W_p = m_f(h_2 - h_1) = \frac{m_f(h_{2s} - h_1)}{\eta_p} \tag{1}$$

The heat addition rate from the waste gas to the evaporator can be expressed as:

$$Q_e = m_f(h_3 - h_1) \tag{2}$$

The work generated by the single screw type expander is written as:

$$W_E = m_f(h_3 - h_{4s})\eta_E\eta_g = m_f(h_3 - h_4)\eta_g \tag{3}$$

The heat rejected by the condenser is equal to:

$$Q_c = m_f(h_4 - h_1) \tag{4}$$

The net power output of the system can be expressed as:

$$W_N = W_E - W_p \tag{5}$$

The thermal efficiency of the system can be expressed as the ratio of net power output and heat input in the evaporator.

$$\eta = \frac{W_N}{Q_e} \tag{6}$$

4 Result and Discussion

4.1 Shortlisting of Working Fluids

Seven working fluids are shortlisted according to their thermodynamic, environmental and physical properties and are enlisted in the Table 1.

The entropy versus temperature diagram of the seven organic working fluids is shown in Fig. 2. R-11 is isentropic fluids, and the remaining are wet fluids. Depending on the slope of the T-S diagram, working fluids are classified into three categories. For the wet fluids, the slope is negative, whereas for dry and isentropic fluids, the slope is positive and infinitely large respectively. The advantage of choosing dry and isentropic fluids is they can be translated into superheated gas states after expansion.

Table 1 Shortlisting of the working fluids

Working fluids	Critical temp.(°C)	Critical pr. (bar)	NBP (°C)	ODP	GWP (100 yrs)	ASHRAE safety group
R 245fa	153	36.1	15	0	858	B1
R 113	213	33.8	48	1	5820	A1
n-pentane	196	33.6	36.2	0	4 ± 2	A3
R 123	183	36.6	27.6	0.02	79	B1
R 11	197	43.7	23.8	1	4660	A1
iso-pentane	187	33.7	27.7	0	4 ± 2	A3
R 1233zd (E)	165.6	34.7	18.3	0	1	A1

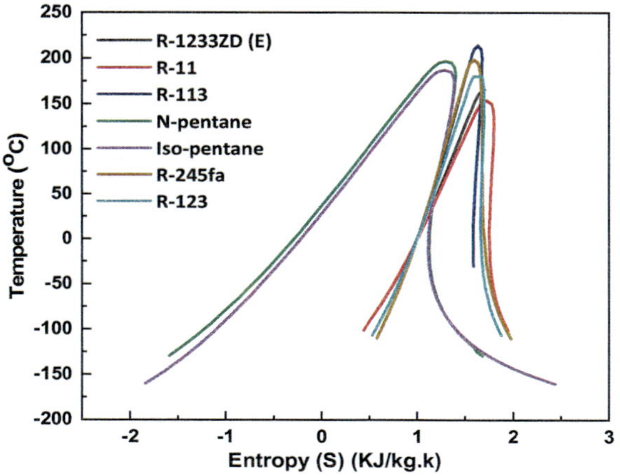

Fig. 2 Entropy-temperature diagram

4.2 Scenario Based Approach

Simulation assumption:

- Operating temperature of scenario 1 = 333 K
- Operating temperature of scenario 2 = 358 K
- Sink Temperature = 298 K with 5 °C pinch point
- Outlet temperature of the condenser = 303 K
- Turbine inlet temperature with 10 °C superheat
- Source of waste heat for scenario 1: PEM fuel cell
- Source of waste heat for scenario 2: Industrial waste
- Turbine efficiency (isentropic) = 70%
- Pump efficiency (isentropic) = 85%
- Generator efficiency = 100%
- Targeted power output for case 1 of both scenarios = 10 KW
- Targeted power output for case 2 of both scenarios = 100 KW (Tables 2, 3, 4 and 5).

Table 2 Efficiency analysis of working fluids at 343 K turbine inlet temperature for 10 KW

Working fluids	m_f(kg/s)	P_e (bar)	P_c(bar)	T_4 (K)	W_{in} (KW)	W_N (KW)	η (%)
R 245fa	0.655	6.5	1.7684	317.469	142.272	10.023	7.04
R 113	1	1.5	0.54061	326.011	175.491	10.074	5.74
n-pentane	0.36	2.5	0.81555	323.707	153.759	10.05	6.54
R 123	0.75	3.5	1.0899	318.057	145.024	9.996	6.89
R 11	0.76	3.5	1.2532	315.449	151.881	9.96	6.56
iso-pentane	0.35	3.5	1.0861	321.947	141.263	9.97	7.05
R 1233zd (E)	0.7	4.54	1.54	319.133	151.537	9.95	6.56

Table 3 Efficiency analysis of working fluids at 343 K turbine inlet temperature for 100 KW

Working fluids	m_f(kg/s)	P_e (bar)	P_c(bar)	T_4 (K)	W_{in} (KW)	W_N (KW)	η (%)
R 245fa	6.55	6.5	1.7684	317.469	1422.72	100.23	7.04
R 113	10	1.5	0.54061	326.011	1754.91	100.74	5.74
n-pentane	3.6	2.5	0.81555	323.707	1537.59	100.5	6.54
R 123	7.5	3.5	1.0899	318.057	1450.24	99.96	6.89
R 11	7.6	3.5	1.2532	315.449	1518.81	99.6	6.56
iso-pentane	3.5	3.5	1.0861	321.947	1412.63	99.7	7.05
R 1233zd (E)	7	4.54	1.54	319.133	1515.37	99.5	6.56

Table 4 Efficiency analysis of working fluids at 368 K turbine inlet temperature for 10 KW

Working fluids	m_f(kg/s)	P_e (bar)	P_c(bar)	T_4 (K)	W_{in} (KW)	W_N (KW)	η (%)
R 245fa	0.435	11	1.7684	326.511	101.542	10.01	9.86
R 113	0.535	3.5	0.54061	334.797	101.787	10.05	9.87
n-pentane	0.216	5	0.81555	335.053	100.693	9.98	9.915
R 123	0.48	6.5	1.0899	327.318	99.5912	9.97	10.01
R 11	0.47	6.5	1.2532	321.11	99.3819	9.98	10.3
iso-pentane	0.235	6	1.0861	336.084	104.348	10.11	9.68
R 1233zd (E)	0.47	7.54	1.54	331.141	99.9464	10.1031	10.1

Table 5 Efficiency analysis of working fluids at 368 K turbine inlet temperature for 100 KW

Working fluids	m_f(kg/s)	P_e (bar)	P_c(bar)	T_4 (K)	W_{in} (KW)	W_N (KW)	η (%)
R 245fa	4.35	11	1.7684	326.511	1015.42	100.1	9.86
R 113	5.35	3.5	0.54061	334.797	1017.87	100.5	9.87
n-pentane	2.16	5	0.81555	335.053	1006.93	99.8	9.915
R 123	4.8	6.5	1.0899	327.318	995.912	99.7	10.01
R 11	4.7	6.5	1.2532	321.11	993.819	99.8	10.3
iso-pentane	2.35	6	1.0861	336.084	1043.48	101.1	9.68
R 1233zd (E)	4.7	7.54	1.54	331.141	999.464	101.031	10.1

4.3 *Effect of Evaporation Pressure*

The variation of net power output and thermal efficiency with evaporation pressure is shown in Figs. 3 and 4. It shows that the net power output as well as thermal efficiency increases with the increase of the evaporation pressure for all working fluids. If evaporation pressure increases then the enthalpy of the expander increases but the working fluid mass flow rate decreases so that the net power output increases. Iso-pentane produces the highest net output power and thermal efficiency at 3.5 bar evaporator pressure due to its better thermodynamic properties.

4.4 *Effect of Heat Addition Rate and Heat Rejection Rate*

The variations of heat addition to evaporator and heat rejection by condenser with turbine outlet temperature are shown in Figs. 5 and 6 respectively. Iso-pentane has highest heat addition rate and heat rejection rate. R-113 has lowest heat addition rate and heat rejection rate.

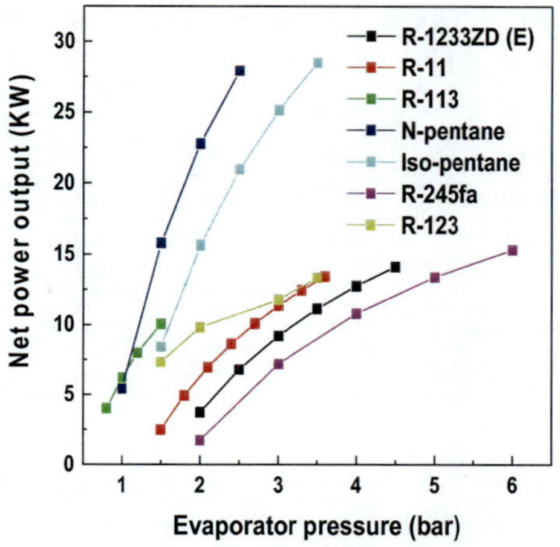

Fig. 3 Net power output versus evaporator pressure

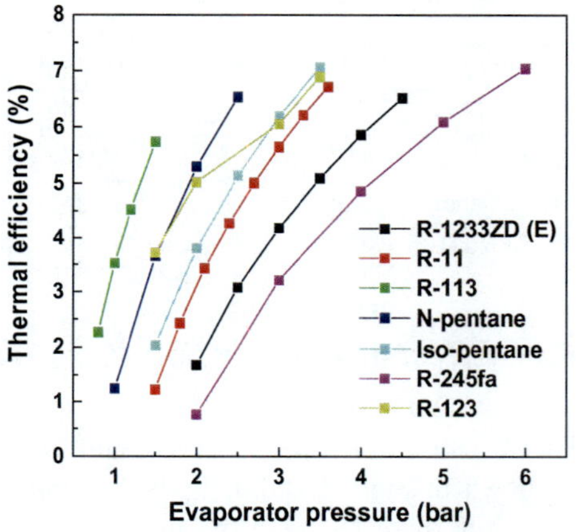

Fig. 4 Thermal efficiency versus evaporator pressure

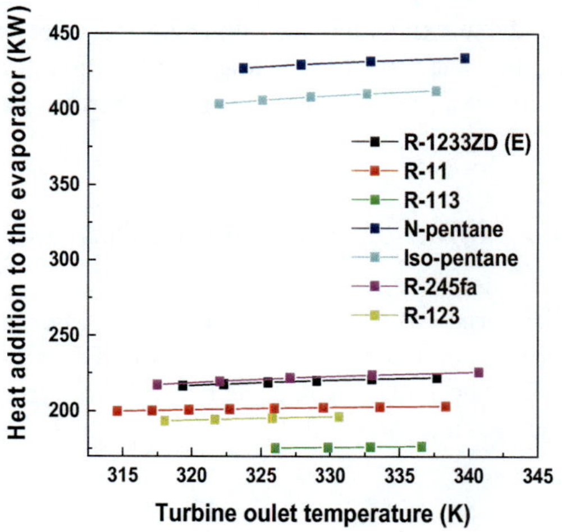

Fig. 5 Heat addition rate versus turbine outlet temperature

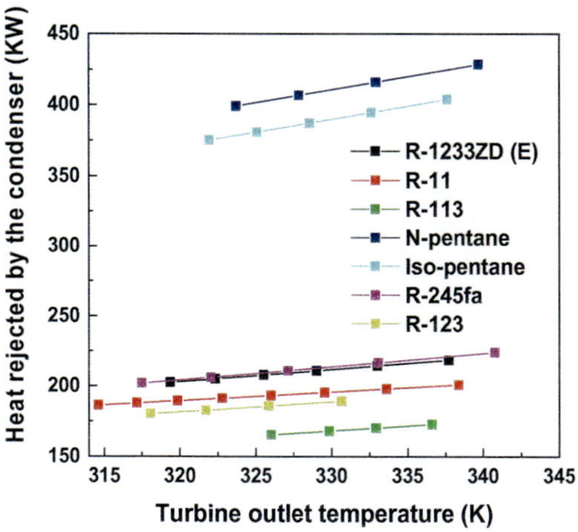

Fig. 6 Heat rejection rate versus turbine outlet temperature

5 Conclusion

From this study on organic Rankine cycle, the following consequences are concluded:

- Selection of organic working fluids plays a vital role for performance analyse of the ORC system.
- Thermal efficiency can be increased by increasing the operating temperature. Thermal efficiency as well as net power output of scenario 2 is greater than scenario 1.
- Iso-pentane has highest heat addition rate and heat rejection rate. R-113 has lowest heat addition rate and heat rejection rate.
- For scenario 1, iso-pentane gives the highest thermal efficiency for both cases. And, for scenario 2, R-1233zd (E) gives the highest thermal efficiency for both cases.
- R 1233zd (E) can be considered as the optimized working fluid due to highest thermal efficiency, low ODP and zero GWP.

The present paper is mainly focused on thermodynamic the performance analysis of ORC using various organic working fluids. We envisaged that the performance can be improved by using internal heat exchanger and increasing the operating temperature. Estimating the cost of ORC system and comparing cost with the other existing system like solar ORC, solar photovoltaic could be pursued for further research.

Abbreviation

GWP Global Warming Potential
IHE Internal Heat Exchanger

NBP	Normal Boiling Point
ODP	Ozone Depleting Potential
ORC	Organic Rankine Cycle
W	Work done (KW)
Q	Heat addition/rejection (KW)
P	Pressure (bar)
T	Temperature (K)
M	Mass flow rate (Kg/s)
S	Specific entropy (KJ/kg.K)
H	Specific enthalpy (KJ/kg)
I	Destruction rate

Greek

η	Efficiency

Subscript

c	Condenser
e	Evaporator
E	Expander
g	Generator
p	Pump
N	Net/total
f	Working fluid
T	Total
in	Input
1	Pump inlet
2	Pump outlet
3	Expander inlet
4	Expander outlet (actual condition)
4 s	Expander outlet (ideal condition)

References

1. Wang EH, Zhang HG, Fan BY, Ouyang MG, Zhao Y, Mu QH (2011) Study of working fluid selection of organic Rankine cycle (ORC) for engine waste heat recovery. Energy 36(5):3406–3418
2. Hettiarachchi HDM, Golubovic M, Worek WM, Ikegami Y (2007) Optimum design criteria for an organic Rankine cycle using low-temperature geothermal heat sources. Energy 32:1698–1706
3. Saleh B, Koglbauer G, Wedland M, Fischer J (2007) Working fluids for low-temperature organic Rankine cycles. Energy 32:1210–1221
4. Yamamoto T, Furuhata T, Arai N, Mori K (2001) Design and testing of the organic Rankine cycle. Energy 26:239–251
5. Tchanche BF, Papadakis G, Lambrinos G, Frangoudakis A (2009) Fluid selection for a low-temperature solar organic Rankine cycle. Appl Therm Eng 29:2468–2476
6. Hung TC, Shai TY, Wang SK (1997) A review of organic Rankine cycles (ORCs) for the recovery of low-grade waste heat. Energy 22:661–667
7. Maizza V, Maizza A (2001) Unconventional working fluids in organic Rankine-cycles for waste energy recovery systems. Applied Thermal Eng 21:381–390
8. Liu BT, Chien KH, Wang CC (2004) Effect of working fluids on organic Rankine cycle for waste heat recovery. Energy 29:1207e17
9. Wei D, Lu X, Lu Z, Gu J (2007) Performance analysis and optimization of organic Rankine cycle (ORC) for waste heat recovery. Energy Convers Manage 48:1113–1119
10. Wei D, Lu X, Lu Z, Gu J (2008) Dynamic modeling and simulation of an organic Rankine cycle (ORC) for waste heat recovery. Appl Therm Eng 28:1216–1224

Parameters Extraction of PEMFC Model Using Evolutionary Based Optimization Algorithms

Rahul Khajuria, Ravita Lamba, and Rajesh Kumar

Abbreviations

N	Number of fuel cells connected in series
E_{nernst}	Nernst voltage [V]
$V_{activation}$	Activation potential [V]
V_{ohmic}	Ohmic potential [V]
$V_{concentration}$	Concentration potential [V]
V_{stack}	Stack output voltage [V]
T	Operating temperature [K]
PH_2	Pressure of hydrogen at anode [bar]
PO_2	Pressure of hydrogen at cathode [bar]
PH_2O	Pressure of water [Bar]
RH_a	Relative humidity at anode
RH_c	Relative humidity at cathode
i_{cell}	Cell current [A]
A	Area of fuel cell [cm^2]
b	Adjusting parametric coefficient [V]
$J_{\max}$	Maximum current density [mA/cm^2]
J	Actual current density [mA/cm^2]
R_m	Membrane resistance [Ω]
R_c	Resistance to the transfer of protons through membrane [Ω]
ζ_i	Parametric coefficients [i = 1,2,3 and 4]
φ	Adjustable parameter of water content in the membrane material
ρm	Specific resistivity [Ω. cm]
SSE	Sum of squared error

R. Khajuria (✉) · R. Lamba · R. Kumar
Malaviya National Institute of Technology, Jaipur 302017, India
e-mail: 2021ree9530@mnit.ac.in

S. Doolla et al. (eds.), *Advances in Clean Energy and Sustainability*,
Green Energy and Technology, https://doi.org/10.1007/978-981-99-2279-6_38

1 Introduction

Energy is one of the main driving force in the current scenario of rapid development. The progress of any country can be evaluated by the amount of energy utilization [1]. Conventional sources of energy are on the verge of being exhausted and their excessive use can cause more threats in parallel to the development. Therefore, alternative sources of energy which are environment friendly, freely available and most importantly nearly zero hazardous need to be adopted in the development of any country [2]. Fuel cell is one of the energy generators which generate electrical power with the help of electrochemical process [3]. Hydrogen energy is used as input in different types of fuel cell which has many advantages over other energy sources. Some other chemical forms of fuels such as hydrocarbons are also used in other types of fuel cell. The working of fuel cell can be related to battery working with a fundamental difference in that the battery operation depends on the materials used, but the fuel cell operates as long as the availability of fuel supply is continuous. Moreover, high efficiency, robust design, environment friendly and low operating temperature are the main advantages of fuel cell over conventional energy sources [4]. Proton Exchange Membrane fuel cell (PEMFC) is one of the most popular design with fast start up, high efficiency, low noise and vibrations and operates at low temperature range, which makes it useful for domestic and transportation applications [5].

A detailed mathematical model should be demonstrated to know about various un- known factors of the fuel cell model and understanding the real processes that occur inside the fuel cell. Therefore, the value of unknown parameter should be evaluated in order to develop optimized fuel cell model with least amount of error. Many scholars have been working on PEMFC modeling recently, and have established several mathematical models of the fuel cell. Amphelett [6] proposed the semi-empirical mathematical model of the PEMFC stack to predict the unknown parameters. The complexity level of the proposed model was high due to presence of multivariable and nonlinearity in the model.

Many researchers used this model to evaluate parameters of fuel cell by using metaheuristic optimization algorithms. Metaheuristic algorithms are inspired from some of the physical laws of nature and behavior of creatures evolving in nature [7]. Metaheuristic methods have the special property of reaching near to the optimal solution for highly complex and non-linear problem. Therefore, different metaheuristic algorithms are found in literature and used for the extraction of fuel cell unknown parameters. Algorithms such as Genetic Algorithm (GA) [8], RNA [9], Hybrid Genetic Algorithm (HGA) [10], Particle Swarm based Optimization (PSO) [11] etc. are used for extracting the model parameters of different fuel cells. To find optimum solution, an objective function is always needed to address the optimization problem. Therefore, in model parameter extraction of fuel cell, voltage based objective function is used [10]. It can be observed from the literature that minimization of sum of square errors (SSE) [12], root mean square error (RMSE) [13], mean square error (MSE) [14] etc. have been used as an objective function to find the optimal values of parameters of fuel cell in different operating conditions. In this paper, minimization of sum of square errors (SSE) of voltage is taken as objective function. The best algorithm suit- able for the optimum modal parameter can be found by the statistical study of all the algorithms used in this work. Various studies in literatures have used mean, standard deviation and plotted box plots [15] to show median and interquartile range (IQR). Algorithm with low median and small IQR is the best algorithm suitable for obtaining solution. In this paper box plot have been plotted for each algorithm and best algorithm among all evolutionary based algorithm is studied.

2 Mathematical Representation of PEMFC

Proton Exchange Membrane fuel cell consists of mainly two electrodes anode and cathode for the supply of hydrogen and oxygen respectively and conducting ionized layer known as polymer electrolyte membrane for movement of ions from anode to cathode and vice-versa. At anode electrons and hydrogen ions are formed due to dissociation of hydrogen and these ions travel through polymer membrane to reach at cathode whereas, electrons travel to cathode via external circuit. At cathode oxygen is combined with electron and hydrogen ions to form water. Therefore, electrical power is generated by the electro chemical action and can be collected via external circuit. Various chemical reactions involved in fuel cell operation are given as below:

$$\text{Anode Reaction}: \ H_2 \rightarrow 2H^+ + 2e^-$$

$$\text{Cathode Reaction}: O_2 + 2H^+ + 2e^- \rightarrow H_2O$$

$$\text{Overall Reaction}: 2H_2 + O_2 \rightarrow H_2O + \text{Heat} + \text{Electricity}$$

The reaction provides water vapor, heat and electricity and electricity in DC form. The polarization I–V curve of a typical fuel cell is governed by three main important losses such as activation losses, ohmic losses and mass transportation losses also known as concentration losses. The starting part of the I–V curve which is known as activation region, behaves non-linear in nature. The central portion which is ohmic region shows the resistance losses like electrical contacts and polymer resistance in a fuel cell and behaves linearly. The lower part of the curve which is concentration region causes a concentration gradient in chemical process and is also non-linear in nature.

A single PEMFC fuel cell has output voltage in DC form and can be defined as follows [15]:

$$V_{cell} = E_{nernst} - V_{activation} - V_{ohmic} - V_{concentration} \tag{1}$$

where, *Enernst* is the irreversible voltage, *Vactivation* are the activation voltage loss,

Vohmic is the ohmic voltage drop and *Vconcentrayion* is the concentration voltage losses.

A single cell is not capable of producing enough voltage as required for applications therefore 'N' number of fuel cells is connected in series to form a fuel cell stack. Fuel cell stack voltage can be written as follows [15]:

$$V_{stack} = N * V_{Cell} \tag{2}$$

Therefore,

$$V_{stack} = N * [E_{nernst} - V_{activation} - V_{ohmic} - V_{concentration}] \tag{3}$$

where, *Enernst*, *Vactivation*, *Vohmic*, *Vconcentration* can be written mathematically as below [15]:

$$\begin{aligned} E_{nernst} =& 1.22 - 8.5 * 10^{-3}(T - 298.15) \\ &+ 4.3085 * 10^{-5} * T[\ln(PH_2 + 0.5PO_2)] \end{aligned} \tag{4}$$

$$PH_2 = 0.5RH_a \times PH_2O\left[\left(\exp\left(\frac{1.635\left(\frac{I_{fc}}{A}\right)}{T_{fc}^{1.334}}\right) \times \frac{PH_a \times PH_2O}{P_a}\right)^{-1} - 1\right] \tag{5}$$

$$PO_2 = 0.5RH_c \times PH_2O\left[\left(\exp\left(\frac{4.192\left(\frac{I_{fc}}{A}\right)}{T_{fc}^{1.334}}\right) \times \frac{PH_a \times PH_2O}{P_c}\right)^{-1} - 1\right] \tag{6}$$

$$V_{activation} = -[\zeta_1 + \zeta_2 * T + \zeta_3 * T * \ln(CO_2) + \zeta_4 * T * \ln(i_{cell})] \tag{7}$$

$$CO_2 = \frac{PO_2}{5.08 \times 10^6 \times \exp\left(\frac{T_{fc}}{498}\right)} \tag{8}$$

$$V_{ohmic} = I[R_m + R_c] \tag{9}$$

where R_m and R_c are electronic and ionic resistance [15].

$$R_m = \rho_m\left(\frac{l}{A}\right) \tag{10}$$

$$\rho_m = \frac{181.6\left[1 + 0.03\left(\frac{I_{fc}}{A}\right) + 0.062\left(\frac{I_{fc}}{A303}\right)^2\left(\frac{T_{fc}}{A}\right)^{2.5}\right]}{\left[\varphi - 0.634 - 3\left(\frac{I_{fc}}{A}\right)\right] \times \exp\left(\frac{4.18(T_{fc}-303)}{T_{fc}}\right)} \tag{11}$$

where, φ is the adjustable parameter of the water content in the material of proton exchange membrane fuel cell.

$$V_{concentration} = -b \ln\left[1 - \frac{J}{J_{\max}}\right] \tag{12}$$

Fuel cell parameter extraction can be done by expressing the optimization problem in minimization of sum of square errors between estimated and experimental voltage of fuel cell stack at N number of data points. Therefore, objective function can be ex- pressed as follows [16]:

$$F_{obj\ \min} = SSE_{\min} = \sum_{i=1}^{N} \left(V_{\exp erimental} - V_{exstimated}\right)^2 \tag{13}$$

Various methods have been implemented to evaluate the model parameters of fuel cell parameter extraction problem. It is tough to solve the problem using earlier developed conventional methods as pertinent equations are non-linear in nature. Therefore, metaheuristic methods are being used now days for evaluation of fuel cell parameters as no need of continuity or differentiability and convexity in given objective functions which is to be optimized. Evolutionary based algorithms are one of the types of metaheuristic method which are stochastic search methods inspired from the natural biological evolution or social behavior of species. These metaheuristic approaches are used to reach at the optimum solution where other traditional approaches are not able to find the solution. Therefore, five evolutionary based algorithms have been used in this article, which are Genetic Algorithm (GA), Mematic Algorithm (MA), Evolutionary Programming (EP), Flower Pollination Algorithm (FPA), and Coral Reefs Optimization (CRO), for optimization of Ned Stack PS6 fuel cell Stack parameters (Table 1).

3 Results and Discussions

In this text the identified parameters are investigated for a well-known commercial PEMFC NedStack PS6 by using five different kinds of evolutionary based algorithms, namely, CRO, GA, MA, EP and FPA respectively. The technical parameters and other operating conditions which govern the working of fuel cell operation are collected from [16], and are listed in given Table 2. Search ranges are given in Table 1 and have not been altered for any algorithm for the fair comparison between different algorithms. Different data points taken for study, voltages corresponding to load currents have been taken from [16]. Each of the algorithms is implemented for 20 times and best obtained results of the estimated parameters for each algorithm are found in this text. In estimation process, the end criterion is taken as 1000 iterations for each algorithm. Optimum values of the seven unknown parameters which have

Table 1 Ranges of values of unknown parameters of NesStack PS6 fuel cell [16]

Parameters	$\zeta 1$	$\zeta 2 \times 10^{-3}$	$\zeta 3 \times 10^{-5}$	$\zeta 4 \times 10^{-5}$	φ	$Rc \times 10^{-4}$	$b \times 10^{-2}$
Lower limit	– 1.997	1.0000	3.6000	– 26.0000	13.0000	1.0000	1.36
Upper limit	– 0.8532	5.0000	9.8000	– 9.5400	23.0000	8.0000	50.00

Table 2 Specifications of NedStack PS6 PEMFC [16]

FC parameters	N	A (cm^2)	l (μm)	J_{max} (mA/cm^2)	T (K)	PH_2	PO_2
NedStack PS6	65	240	178	1200	343	1.0000	1.0000

been found by each algorithm and statistical measurement like best value, worst value, mean and standard deviation is listed in Table 3. Results of each algorithm have been validated by extracting the unknown parameters of NedStack PS6 fuel cell stack. IV curves, Power curves and Efficiency curves are utilized for the validation of obtained optimized unknown parameters of the given fuel cell. Thirty measurements of voltage corresponding to the fuel cell stack current from [16] are used in optimization algorithms. The results of statistical measurements based on each algorithm have been compared with each other and a best evolutionary based algorithm is suggested for the optimization of unknown parameters. Unknown parameters and statistical measure ments have been listed in Table 3. Solution with minimum SSE within the 20 runs for each algorithm are used for obtaining the I–V curves, Power curves and Efficiency curves of the NedStack PS6 as shown in Fig. 1a–c respectivcely. Boxplot for the statistical understanding is also shown in Fig. 2. From Table 3 and in Fig. 1 and it can be seen that MA algorithm gives a less SSE than other adopted evolutionary based algorithm and has a closely matching polarization characteristics to the experimental polarization characteristics. Minimum value of SSE is 2.37, obtained from MA algorithm out of all the evolutionary based algorithms. Further statistical study also reveals that with less value of mean and standard deviation viz. 0.533 and 3.267, the MA algorithm is best algorithm out of given evolutionary based algorithms for the extraction of PEMFC unknown parameters. A close look at the box plot shows that with low median and small IQR range Mematic Algorithm (MA) is best suitable evolutionary based optimization algorithm for fuel cell parameter extraction.

4 Conclusion

In this paper, estimation of unknown parameters of PEMFC is carried out and compared using different evolutionary algorithms such as Genetic Algorithm (GA), Flower Pollination algorithm (FPA), Evolutionary Programming Algorithm (EP), Coral Reefs Optimization algorithm (CRO) and Memetic Algorithm (MA). The main objective of this study is to minimize the sum of squared error between the

Table 3 Best values of PEMFC NedStack PS6 parameters plus SSE values

Parameters	EP	CRO	GA	FPA	MA
$\zeta 1$	– 0.8532	– 0.93325	– 0.998271	– 0.964382	– 0.9225
$\zeta 2 \times 10^{-3}$	3.25438	2.7457	3.133	3.5686	2.8666
$\zeta 3 \times 10^{-5}$	9.8000	4.201618	5.5625	9.75	5.2533
$\zeta 4 \times 10^{-5}$	– 9.5400	– 9.54	– 9.558	9.54	– 9.5400
φ	15.6400	13.9612	15.47	13.000	13.000
$Rc \times 10^{-4}$ (Ω)	3.1296	2.3294	3.38457	1.0000	1.9334
$b \times 10^{-2}$	30.520	45.994	49.98	13.61	24.058
SSE	2.72	2.57	2.48	2.47	2.37
Mean	5.35079	3.6929	3.5756	4.045	3.267
Standard deviation	1.906	1.4638	1.2771	1.156	0.533
SSE best value	2.72	2.57	2.48	2.47	2.37
SSE worst value	10.44	9.29	7.53	5.76	4.41

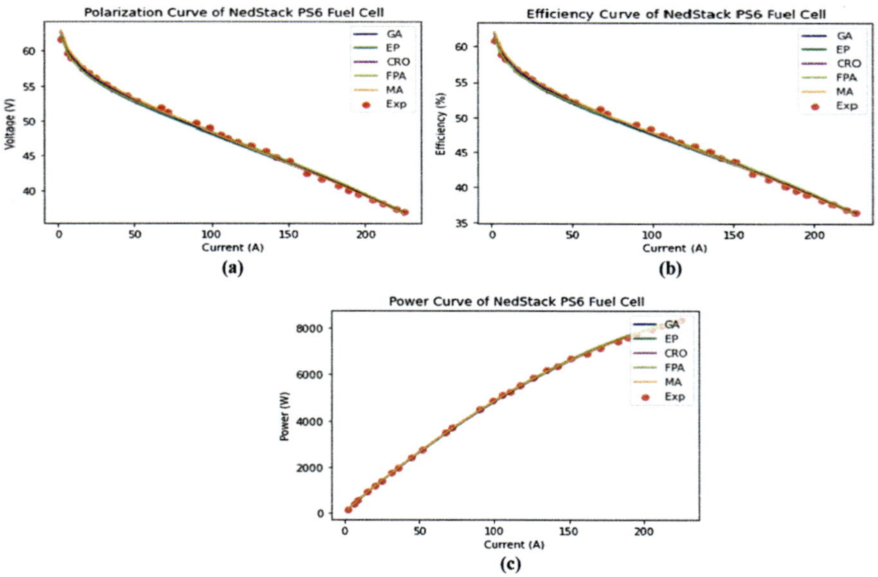

Fig.1 **a** Current–voltage, **b** current-efficiency and **c** current-power curves of NedStack PS6 Fuel cell stack for experimental and different evolutionary based algorithms

experimental and estimated value of voltage points using different evolutionary algorithms. The comparative analysis has been carried out among all the algorithms using statistical analysis and best algorithm is determined. It has been found that the Memetic Algorithm (MA) produced best values of the unknown parameters, out of all evolutionary based algorithm and matched the estimated voltage with the

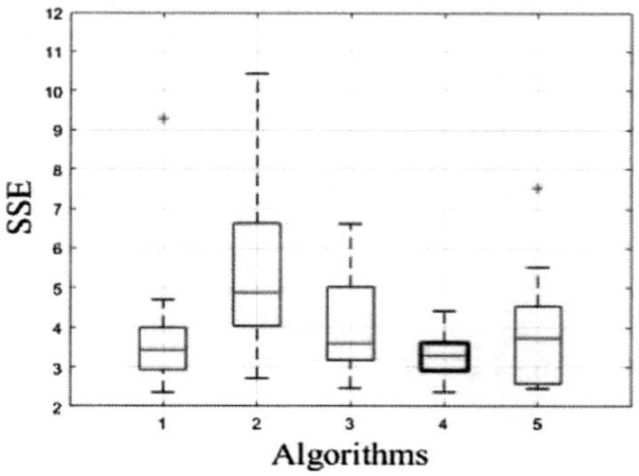

Fig. 2 Box plot for evolutionary based algorithms for NedStack PS6 fuel cell stack

experimental voltage values and different polarization curves such as current voltage curves, current efficiency curve and current power curves of NedStack PS6 (6 kW) fuel cell stack. Statistical studies like mean, standard deviation, best values and worst values have also been determined to confirm the accuracy of the given optimization algorithms. With smallest value of mean and standard deviation, Memetic Algorithm (MA) is the best suitable optimization technique for parameter extraction of proton exchange membrane fuel cell. Further, with the smallest difference between the best and worst values, MA outperformed among all other optimization algorithms. Furthermore, the box plot is shown in which Memetic Algorithm (MA) is having lowest median and low interquartile range (IQR) out of all evolutionary based algorithms. The extension of the current work will be in developing a model to study the impact of aging.

References

1. Sultan HM et al. (2020) Parameter identification of proton exchange membrane fuel cells using an improved salp swarm algorithm. Energy Conver Manage 224
2. Abdin Z, Webb CJ, MacA Gray E (2016) PEM fuel cell model and simulation in Matlab–Simulink based on physical parameters. Energy 116:1131–1144
3. Abdel-Basset M, Mohamed R, Chang V (2021) An efficient parameter estimation algorithm for proton exchange membrane fuel cells. Energies 14.21
4. Amirinejad M, Rowshanzamir S, Eikani MH (2006) Effects of operating parameters on performance of a proton exchange membrane fuel cell. J Power Sour 161.2:872–875
5. Chavan SL, Dhananjay BT (2017) Modeling and performance evaluation of PEM fuel cell by controlling its input parameters. Energy 138:437–445
6. Amphlett JC et al (1994) Parametric modelling of the performance of a 5-kW proton- exchange membrane fuel cell stack. J Power Sources 49(1–3):349–356
7. Sharma A et al. (2021) A novel opposition-based arithmetic optimization algorithm for parameter extraction of PEM fuel cell. Electronics 10.22:2834
8. Mohamed I, Jenkins N (2004) Proton exchange membrane (PEM) fuel cell stack con- figuration using genetic algorithms. J Power Sour 131.1–2:142–146
9. Zhang L, Wang N (2013) An adaptive RNA genetic algorithm for modeling of proton exchange membrane fuel cells. Int J Hydrogen Energy 38.1:219–228

10. Mo ZJ et al. (2006) Parameter optimization for a PEMFC model with a hybrid genetic algorithm. Int J Energy Res 30.8:585–597
11. Askarzadeh A, Rezazadeh A (2011) Optimization of PEMFC model parameters with a modified particle swarm optimization. Int J Energy Res 35.14:1258–1265
12. Menesy AS et al. ((2020)) Effective parameter extraction of different polymer electrolyte membrane fuel cell stack models using a modified artificial ecosystem optimization algorithm. IEEE Access 8:31892–31909
13. Khan SS et al. (2018) Parameter optimization of PEMFC model using back-tracking search algorithm. In: 2018 5th International conference on renewable energy: generation and applications (ICREGA). IEEE
14. Zhang G, Xiao C, Razmjooy N (2020) Optimal parameter extraction of PEM fuel cells by meta-heuristics. Int J Amb Energy 1–10
15. Yang B et al. (2021) Parameter extraction of PEMFC via Bayesian regularization neural network based meta-heuristic algorithms. Energy 228:120592
16. Selem SI, Hasanien HM, El-Fergany AA (2020) Parameters extraction of PEMFC's model using manta rays foraging optimizer. Int J Energy Res 44.6:4629–4640
17. Özdemir MT (2021) Optimal parameter estimation of polymer electrolyte membrane fuel cells model with chaos embedded particle swarm optimization. Int J Hydrogen Energy 46.30:16465–16480
18. Rao Y et al. (2019) Shark smell optimizer applied to identify the optimal param eters of the proton exchange membrane fuel cell model. Energy Conver Manage 182:1–8

A Parametric Optimization for Decision Making of Building Envelope Design: A Case Study of High-Rise Residential Building in Jaipur (India)

Pushpendra Kr. Chaturvedi, Nand Kumar, Ravita Lamba, and Vishakha Nirwal

1 Introduction

Building sector is responsible for 40% of world energy demand and 33% of GHG emissions [1]. According to the IEA net-zero report-2021, Urbanization is growing rapidly, and 75% world population is moving to cities therefore the building floor area is expected to increase by 75% between 2020 and 2050 [2]. The faster rate of expansion in the construction industry puts tremendous strain on electricity supply and demand [3].

India is the world's fastest-growing country in real estate. The total household was estimated at 272 million in 2017 and it is predicted to increase by 328 million by 2027 [4]. Mean this, requires huge demand for electricity in the coming year [5]. According to the Bureau of energy efficiency report-2018–19, building sector (commercial and residential) is responsible for 33% of total electricity and it may fold to two–three times in 2031 [6].

India is 4th largest country in economic growth rate. Although, people want to adopt a higher living standard and well-being. According to the LBNL report, Air conditioning demand becomes doubled incoming 15 years. It is expected that India become the world's 4th largest contributor to electricity demand in 2040 [7, 8].

Currently, India's per capita space cooling demand is 69 kWh/person, and it is expected to increase by 11 times by 2037–38. Ultimately, it increases GHG emissions and contributes leading global temperatures [9]. It is a severe problem for the country and it required transformation in the building sector with available techniques. There are two methods to reduce energy consumption which include Energy Efficiency Management System (EEMS) and Renewable Energy Source (RES). EEMS focuses

P. Kr. Chaturvedi (✉) · N. Kumar
Malaviya National Institute of Technology, Jaipur 302017, India
e-mail: 2020rar9576@mnit.ac.in

R. Lamba · V. Nirwal
Central University, Ajmer, Rajasthan, India

S. Doolla et al. (eds.), *Advances in Clean Energy and Sustainability*,
Green Energy and Technology, https://doi.org/10.1007/978-981-99-2279-6_39

on building passive designs, lighting, and HVAC systems and RES includes solar PV, geothermal, and biomass to fulfill the energy demand [10, 11].

This paper illustrates the live case study of a residential high-rise apartment building situated in Jaipur city (Rajasthan), India. Parametric multi-objective optimization technique is applied for decision making of building envelope design with different wall and roof assembly configurations. The objective of this paper is to identify the best-performing wall and roof scenario while taking into account minimum cooling demand, entire building demand, annual energy performance index (EPI), and hourly peak cooling demand. The novelty of this paper is to access the amount of heat (kW) transmitted inside the building space through different wall and roof configurations by conduction and convection.

Currently, optimization of envelope parameters has an extensive research gap for the development of energy-efficient building designs [12]. It consists of building wall, roof, and glass configurations. It highly affects the occupants' comfort, building energy model, and life cycle behaviour [13]. Commonly three passive technique is used to reduce the building demand that focuses on heat protection, heat modulation, and heat extraction. Heat comes inside the building due to the wall, roof, and glass through conduction, convection, and radiation [14]. The government of India has taken a step toward reducing the energy demand in the residential sector and launched the Energy Conservation Building Code-Residential (ECBC-R) in 2018 [15].

ECBC-R claims that using alternating materials for walls and roofs can reduce total operational energy demand by 30%. To identify the suitability and usability for decision-making of building materials different tools and techniques are available [16, 17]. Parametric and algorithm-based multi-objective optimization approach is highly recommended due to its higher computational speed and flexibility [18]. Parametric optimization is a computer-based programming system in which optimal solutions are generated through automatic feedback of alternative input parameters. Energy-efficient building design focuses on 3 aspects of sustainability, such as social, environmental, and economic. Social aspects concern with thermal, visual, acoustic comfort, and indoor air quality. Environmental parameters are associated with energy efficiency and the lowest CO_2 emissions. Economical parameters influence the life cycle cost of building [19].

Roshmi et al. provide the assessment of different wall and roofing materials in the Indian context. It identifies the lowest embodied and most cost-effective building material for low-income mass housing [20]. Adraino et al. identify the optimal shape, WWR, and orientation for residential apartment buildings and illustrate the impact of these parameters on cooling and lighting demand. It reveals optimal set of mono solutions reduces the energy demand by 40% and enhances the occupants' comfort by 52% compared to the existing scenario [21]. In the same context, some multi-objective optimization-based studies demonstrate the different steps of the optimization methodology for residential building typology. This methodology is summarized into four steps, which include the base model, optimization setup, multicriteria decision making (MCDM), and identification of the robustness of Pareto solutions. These studies assess the different building attributes by exploring algorithmic and parametric optimization approaches and generate a set of mono solutions [22, 23].

Recently, Richard applied a parametric optimization approach and compared the building performance between sequential versus holistic design approaches. Overall, it found that parametric can solve the complex problems associated with building design [24]. It is a more appropriate technique to provide decision-making in the early design phase.

According to five-year plan 2012–2017, India has around 18.78 million housing shortage. Currently, Pradhan Mantri Awas Yojana (PMAY) is the biggest scheme of the Indian government to provide affordable housing for EWS, LIG, and MIG categories households. As per PMAY-2019 report, 8.3 million households are sanctioned of which 2.6 million are completed and remaining need to be constructed in the coming 3 years. So, to achieve this target Indian construction sector is transitioning toward fast-paced and automated construction technologies. This transitioning is not involving the choice of material, layout and architectural design. Therefore, it becomes a threat to household occupants to increase the cost of energy due to increased mechanical cooling load. The objectives of this paper are to identify the effect of different building envelope materials (wall and roof) on building total annual operational demand, maximum cooling demand, and amount of heat passed through building envelope to inside the building.

2 Methodology

Authors identified the type of material used in Indian construction practices in affordable housing through site visits and primary surveys. Currently for wall construction material used such as Redbrick, AAC block, FlyAsh brick, Cast Concrete brick, Resource Efficient Hollow brick, Exposed Red brick + AAC block and for roofing Cement Screed + XPS Insulation + Mother Slab, Mother Slab + Bitumen + Tiles, Mud Phuska + Brick Tile, Mud Phuska + PCC, Foam Concrete + PCC material being used. Figure 1 illustrates the methodological hierarchy adopted for the decision-making of envelope design for a defined case study of the residential building. It is divided into four sections, which include base case simulation process, optimization setup, decision parameters, and identification of the optimal solution. eQUEST version 3.65 is used as optimization tool. It was developed by the US Department of Energy. It is written in the C++ language and DOE-2.2 is used as a supporting tool to simulate the input variable. eQUEST provides the capability to simulate multiple, and alternative simulation cases, where each new case is a parametric variation of the base case. Maximum five alternative design parameters can be optimize with the base case in a parametric run of eQUEST. In this study two design paraeters such as wall and roof are considered for optimization. Total of five roofs and six wall materials are considered and make 30 combinations for optimization, of which one combination comes under the base model. Based on 29 output results, identify the optimal set of wall and roof configurations for energy-efficient building design.

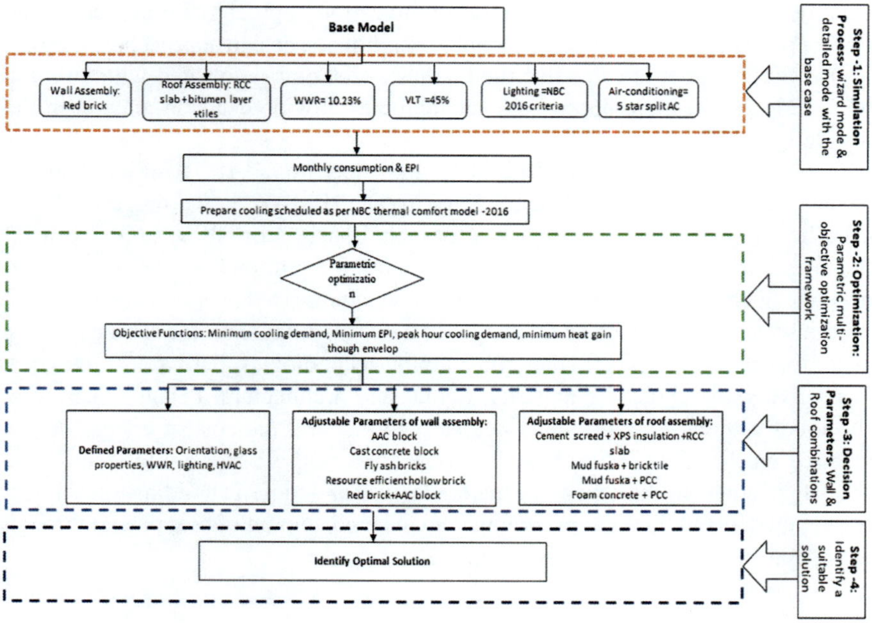

Fig. 1 Methodological structure adopted for parametric simulation

3 Case Study

This case study is a high-rise residential building situated in Jaipur, Rajasthan (India). According to ECBC, Jaipur city comes under a composite climate. The average mean temperature lies between 15 and 34 °C. The case study details are listed in Table 1 and the floor plan of the building and typical apartment plan is shown in Fig. 2a and b respectively.

Table1 Area details of case study

Total site area	10,722.92 m^2
Build-up area	16,390 m^2
Footprint area	1693 m^2
Number of floors	G + 9
Single apartment floor area	Type 1—64.82 m^2
	Type 2—52.31 m^2
Number of towers	6
Projection factor	North (0.75), South (0.58), East (0.71), West (1.11)
WWR	10.31

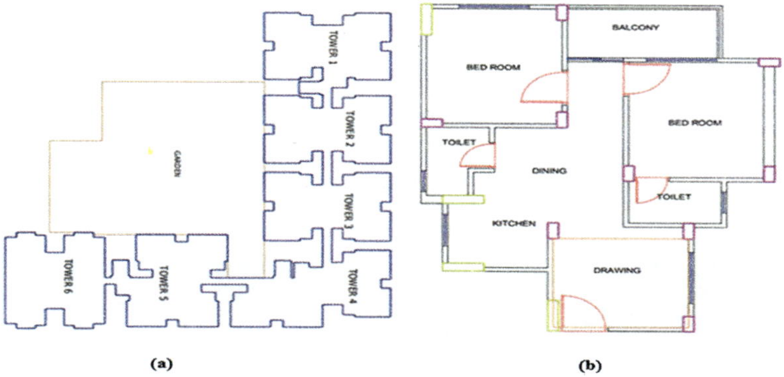

Fig. 2 Floor plan of the building (**a**) and typical apartment plan (**b**)

3.1 Base Case Simulation Process

The base case model is simulated with eQUEST software. Existing parameters collected through survey and other unavailable parameters are considered according to ECBC-R criteria for simulation input. In the detailed simulation mode of the building wizard, different building operational schedules such as occupancy, lighting, cooling, heating, and infiltration are attached. Jaipur city (India) weather file is used as a reference for weather conditions. Cooling and heating schedules attached based on the NBC -2016 thermal comfort model of the air-conditioning building that is shown in Eq. (1) with 90% acceptability rate as predicted mean vote (PMV).

$$T_{indoor\ operative\ temperature} = 0.078\ T_{Mean\ outdoor\ temperature} + 23.25 \pm 1.25 \qquad (1)$$

Table 2 address the base case building model wall, roof assembly, and glass material and their specification. Table 3 address the parameters of the constraint in term of lighting, HVAC, infiltration (air exchange) and occupancy data that are considered as a defined variable in the entire optimization process. Figure 3 shows the eQUEST model of case study.

Table 2 List of base case wall and roof design configuration

Base case model			
Envelop parameters	Physical composition	Number of layers	U value (W/m^2.K)
Wall assembly	Cement plaster + red brick 230 mm + cement plaster	3	2.1
Roof assembly	Mother slab 125 mm + 2 coats of bitumen + tiles 20 mm	3	2.78
Glass	Single-layer glass	White colour	5.67

Table 3 List of constraint parameters for case study

Lighting	0.69 W/ft^2 (NBC-2016) criteria
Occupancy	10.4 m^2/person
Air exchange rate	0.5ACH
HVAC	Split air-conditioning (EER = 15.65) as per BEE 5 star compliance

Fig. 3 eQUEST model of building design

3.2 Objectives Functions

Parametric multi-objective optimization run with different wall and roofing materials. The objective function criteria are defined based on the Jaipur city weather profile. Generally, the maximum operational energy required for cooling due to the city lies in the cooling dominant condition. The objective is to identify the potential of building wall and roofing materials to reduce the building energy demand, cooling demand, and peak cooling demand on an hourly and monthly basis. Furthermore, assess the amount of heat transmitted through the building's walls and roof to inside the building space.

3.3 Decision Parameters

Table 4 addresses the different wall and roof materials with their specification. Table 5 shows the Physical properties of wall and roof material considered during parametric optimization.

Table 4 List of alternating roof and wall materials for optimization

Optimized parameters	Physical composition	U value ($W/m^2.K$)
Roof assembly	Cement screed 20 mm + XPS insulation 50 mm + mother slab 100 mm	0.5
	Mother slab 125 mm + 2 coats of bitumen + tiles 20 mm	2.78
	Mud phuska 50 mm + brick tile 50 mm	2.15
	Mud phuska 115 mm + PCC 25 mm	1.64
	Foam concrete 100 mm + PCC 25 mm	1.16
Wall assembly	Red brick	2.1
	AAC block 250 mm	0.65
	Cast concrete 150 mm	3.4
	Flyash brick 150 mm	1.5
	Resource-efficient hollow brick 200 mm	2.25
	Exposed Red brick 115 mm + air gap 20 mm + AAC block 75 mm + cement Plaster	1.135

Table 5 Physical properties of wall and roof material considered during optimization

Material	Thickness	Density	Specific heat (kJ/kg.K)	Conductivity (W/m.K)
Red brick	230	1760	0.8	0.85
AAC block	250	642	1.24	0.184
Cast concrete	200	2200	1050	1.403
Flyash brick	150	1620	0.93	0.856
Resource-efficient hollow brick	200	1520	0.65	0.631
Cement plaster	15	1762	0.84	0.721
Mud Phuska	115	1622	0.88	0.519
Brick tiles	20	1892	0.88	0.798
RCC mother slab	125			1.411

4 Results and Discussion

Table 6 shows the parametric optimization result of all wall and roof combinations. Authors analyze these results and compare them with a base case scenario and identify optimal set of solutions. Option 2 (wall-AAC block and roof-cement screed + XPS insulation + mother slab) is highly recommended. It is the best performing design scenario in terms of energy-saving and reducing cooling demand. Option 8 (wall-cast concrete and roof- mother slab + bitumen + tiles) is the worst design scenario.

Table 6 Parametric optimization results of 29 combinations, listed in chronological order

Options	Wall U value ($W/m^2.K$)	Roof U value ($W/m^2.K$)	Monthly maximum cooling demand ($kWh \times 10^3$)	Total cooling demand ($kWh \times 10^3$)	Total operational demand ($kWh \times 10^3$)	EPI ($W/m^2/yr.$)	Total cooling load (W/Sq.mtr)	Max. cooling demand in hour (kw)
Base case simulation result								
0	2.1	2.78	121.9	912.1	2178.7	133.33	58.91	17.30
Parametric optimization result								
1	2.1	0.5	116.7	887.80	2154.4	131.83	76.32	17.30
2	0.65		100.4	837.3	2103.9	128.74	48.85	13.36
3	3.4		126.3	923.7	2190.3	134.04	95.68	20.07
4	1.5		111.2	869.3	2136	130.71	65.82	15.74
5	2.25		112.1	911.3	2178	133.27	83.48	17.65
6	1.135		107.2	857.1	2123.7	129.96	58.91	14.70
7	0.65	2.78	106.4	861.9	2128.5	130.25	55.96	13.36
8	3.4		131.3	948.5	2215.1	135.55	101.5	20.07
9	1.5		116.5	893.3	2159.9	132.18	72.06	15.74
10	2.25		123.1	916.5	2183.1	133.59	84.8	17.65
11	1.135		112.7	880.9	2147.5	131.42	65.39	14.70
12	2.1	2.15	102.7	906	2172.6	132.95	80.84	17.33
13	0.65		105	855.8	2122.4	129.88	54.18	13.37
14	3.4		130.1	942.4	2209.0	135.18	100.07	20.07
15	1.5		115.2	887.2	2153.8	131.80	70.46	15.74
16	2.25		121.9	910	2176.6	133.22	83.23	17.76

(continued)

Table 6 (continued)

Options	Wall U value ($W/m^2.K$)	Roof U value ($W/m^2.K$)	Monthly maximum cooling demand (kWh × 10^3)	Total cooling demand (kWh × 10^3)	Total operational demand (kWh × 10^3)	EPI (W/m^2/yr.)	Total cooling load (W/Sq.mtr)	Max. cooling demand in hour (kw)
17	1.135		112.4	874.9	2141.5	131.05	63.5	14.70
18	2.1	1.64	119.6	900.7	2167.3	132.63	79.49	17.29
19	0.65		103.7	850.6	2117.2	129.56	52.63	13.36
20	3.4		129.1	937	2203.6	134.84	98.75	20.07
21	1.5		114.1	882.1	2148.7	131.49	69.05	15.74
22	2.25		120.8	905.1	2171.7	132.90	81.68	17.65
23	1.135		110	869.8	2136.4	130.73	62.20	14.70
24	2.1	1.16	118.4	895.5	2162.1	132.30	78.20	17.30
25	0.65		102.4	845.5	2112.1	129.25	51.06	13.36
26	3.4		128	931.7	2198.3	134.52	97.48	20.07
27	1.5		112.9	876.9	2143.5	131.17	67.65	15.73
28	2.25		117.9	892.1	2158.7	132.10	78.73	17.65
29	1.135		107.2	857.1	2123.7	129.96	58.91	14.70

Figure 4 shows the graphical representation of all optimized combinations accounting for monthly peak cooling demand, total cooling demand and total operational energy demand in (kWh × 10^3). Figure 5 illustrates the eQUEST optimized result of recommended options. It shows that daily cooling demand, monthly cooling demand, and annual operational demand reduce by 21.5, 8.96, and 3.44% compared to base case design.

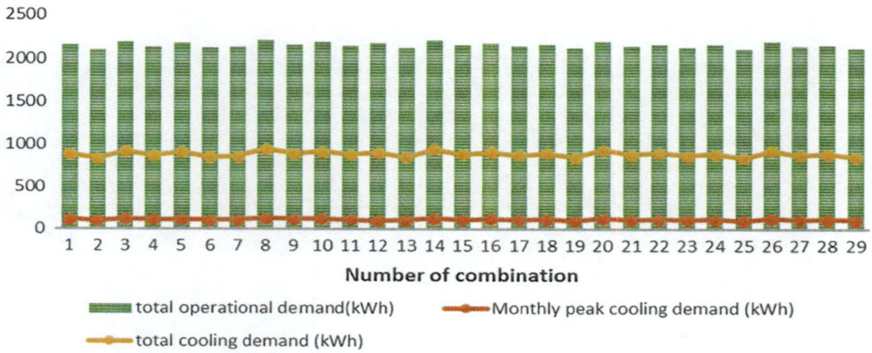

Fig. 4 Cooling demand, peak demand, and total demand for 29 optimized options

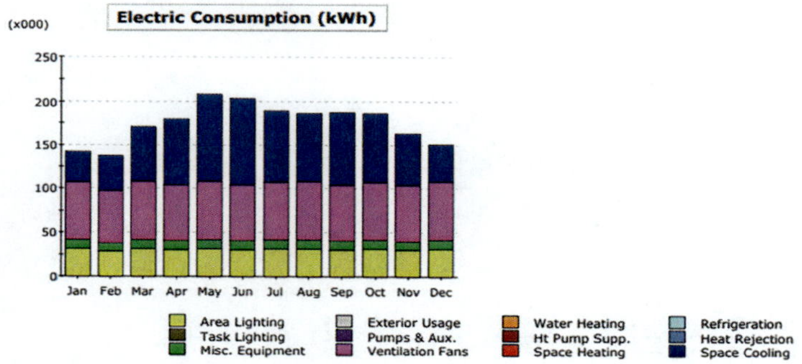

Electric Consumption (kWh x000)

	Jan	Feb	Mar	Apr	May	Jun	Jul	Aug	Sep	Oct	Nov	Dec	Total
Space Cool	34.3	40.1	62.9	75.5	100.4	99.7	82.2	78.5	83.2	78.7	59.1	42.7	837.3
Heat Reject.	-	-	-	-	-	-	-	-	-	-	-	-	-
Refrigeration	-	-	-	-	-	-	-	-	-	-	-	-	-
Space Heat	-	-	-	-	-	-	-	-	-	-	-	-	-
HP Supp.	-	-	-	-	-	-	-	-	-	-	-	-	-
Hot Water	-	-	-	-	-	-	-	-	-	-	-	-	-
Vent. Fans	66.0	59.6	66.0	63.8	66.0	63.8	66.0	66.0	63.8	66.0	63.8	66.0	776.8
Pumps & Aux.	-	-	-	-	-	-	-	-	-	-	-	-	-
Ext. Usage	-	-	-	-	-	-	-	-	-	-	-	-	-
Misc. Equip.	10.1	9.1	10.1	9.8	10.1	9.8	10.1	10.1	9.8	10.1	9.8	10.1	118.9
Task Lights	-	-	-	-	-	-	-	-	-	-	-	-	-
Area Lights	31.4	28.4	31.6	30.5	31.5	30.6	31.4	31.6	30.5	31.4	30.5	31.5	370.9
Total	141.8	137.2	170.6	179.7	207.9	203.9	189.6	186.3	187.4	186.2	163.1	150.2	21,03.9

Fig. 5 eQUEST optimized results for optimal solution combination of wall and roof assembly

Figure 6 shows EPI (W/m^{2}K), hourly peak cooling demand (kW), and space cooling demand (W/Sq/mtr.) of all optimized combinations. It reveals that recommended wall and roof material reduce the EPI, hourly peak cooling demand, and space cooling demand by 3.56, 23.52, and 18% from the base case scenario.

Figures 7 and 8 demonstrate the outdoor heat transmitted through the different roof and wall materials respectively. It shows that roof composition (cement screed + XPS insulation + mother slab) reduces 80% heat loss compared to the base case roof material. In the same context AAC blocks, performance is high compares to other and it reduces the 64% outdoor heat from the base case model.

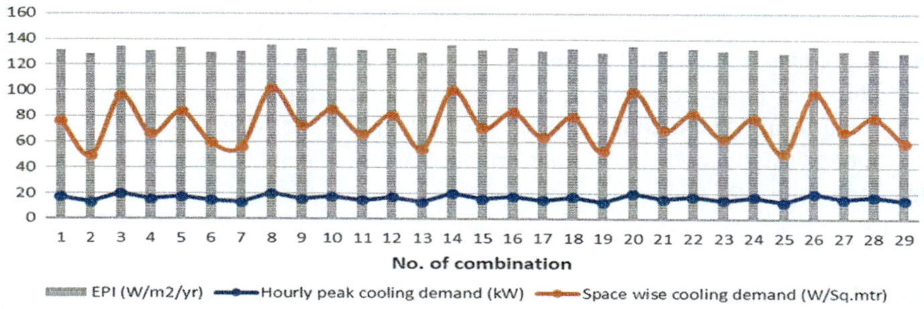

Fig. 6 EPI and hourly peak cooling demand, space-wise cooling demand for 29 combinations

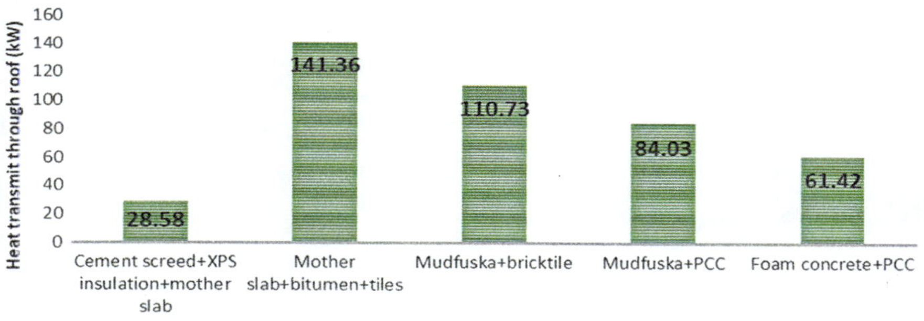

Fig. 7 Amount of heat transmitted through different roofs typology

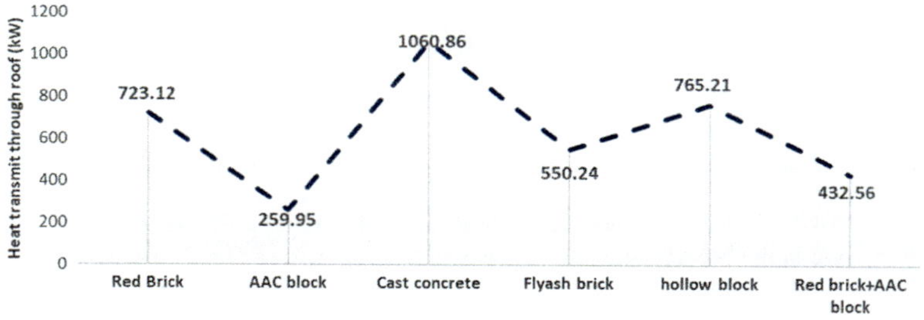

Fig. 8 Amount of heat transmitted through different wall's typology

5 Conclusion

This paper illustrates the application of the parametric multi-objective optimization approach for decision-making of wall and roof design for energy-efficient buildings. Considering the case study of high-rise residential building in Jaipur city (composite climate) different wall and roof assembly material is investigated with parametric optimization. Base case building design scenario compared with optimized wall and roof building design scenario. It reveals that AAC block for wall design and Cement screed + XPS insulation + Mother slab for roofing system have greatest potential to reduce the cooling energy demand, annual energy demand, and peak load cooling demand. This composition of both wall and roof material reduces the conduction of outdoor heat to inside the building space by 64% and 80% respectively. 100 mm AAC block is more stringent in terms of heat conduction compare to other walls material assembly. Cast concrete brick for wall and Mother slab + bitumen + tiles for roofing design have the worst performance in terms of annual energy-saving and this composition of wall and roofing system increases the cooling energy demand and total annual energy demand by 13% and 5.62% respectively from the optimal set of solution. It transmits maximum heat inside the building space which enhances the cooling demand to maintain occupant comfort. This strategies provide the guideline for stockholders, architects, engineers and occupant to selection of wall and roofing materials. This actual building case study in composite climate illustrates the impact of envelope materials in energy consumption and cooling demand. This process can be applied for any building typology and climatic conditions. This research can be explored with different building layout, orientation, glass material, lighting and HVAC system.

References

1. Feng W et al. (Oct 2019) A review of net zero energy buildings in hot and humid climates: experience learned from 34 case study buildings (in English). Renew Sustain Energy Rev 114:24. Art no. 109303. https://doi.org/10.1016/j.rser.2019.109303
2. Energy I, Agency (2021) A roadmap for the global energy sector
3. Urge-Vorsatz D et al. (2020) Advances toward a net-zero global building sector. In: Gadgil A. Tomich TP (eds) Annual review of environment and resources, vol 45. Annual Review of Environment and Resources. Palo Alto: Annual Reviews, pp 227–269
4. Saini L, Meena CS, Raj BP, Agarwal N, Kumar A (2022) Net zero energy consumption building in India: an overview and initiative toward sustainable future (in English). Int J Green Energy, Article 19(5):544–561. https://doi.org/10.1080/15435075.2021.1948417
5. Janda RKKB (2019) India's building stock: towards energy and climate change solutions. Building Res Inf 47
6. Efficiency BOE, GOI (2018) Ministry of power. Impact Energy Efficiency Meas 2018–19. [Online]. Available: beeindia.gov.in
7. Sudhakar MWK, Shanmuga S (2019) Net-zero building designs in hot and humid climates: a state-of-art. Case Studies Thermal Eng
8. Cell O (2019) F. a. C. C. Ministry of Environment, and G. o. India. In: India cooling action plan. [Online]. Available: INDIA COOLING ACTION PLAN.pdf (indiaenvironmentportal.org.in)

9. Economy EE (2017) Thermal comfort for all, sustainable and smart space cooling. [Online]. Available: Thermal-Comfort-for-All.pdf (shaktifoundation.in)
10. Raj BP et al. (Aug 2021) A review on numerical approach to achieve building energy efficiency for energy, economy and environment (3E) benefit (in English). Energies Review 14(15):26. Art no. 4487. https://doi.org/10.3390/en14154487
11. Rey-Hernandez JM, Yousif C, Gatt D, Velasco-Gomez E, San Jose-Alonso J, Rey-Martinez FJ (Sep 2018) Modelling the long-term effect of climate change on a zero energy and carbon dioxide building through energy efficiency and renewables. Energy Buildings 174:85–96. https://doi.org/10.1016/j.enbuild.2018.06.006
12. Bano F, Sehgal V (2019) Finding the gaps and methodology of passive features of building envelope optimization and its requirement for office buildings in India. Thermal Sci Eng Progress 9:66–93, 2019-03-01. https://doi.org/10.1016/j.tsep.2018.11.004
13. Fatima Harkouss FF (2018) Pascal henry. Optimization approaches and climates investigations in NZEB-A review
14. Cabeza LF, Chàfer M (2020) Technological options and strategies towards zero energy buildings contributing to climate change mitigation: a systematic review. Energy Buildings 219:110009. 2020-07-01. https://doi.org/10.1016/j.enbuild.2020.110009
15. m. o. p. Bureau of Energy Efficiency (BEE) (2018) Eco-Niwas Samhita 2018, Energy conservation building code for residential buildings. New delhi
16. Energy conservation building code, I. BEE (2017). [Online]. Available: BEE_ECBC 2017.pdf (beeindia.gov.in)
17. Costa-Carrapiço I, Raslan R, González JN (2020) A systematic review of genetic algorithm-based multi-objective optimisation for building retrofitting strategies towards energy efficiency. Energy Buildings 210:109690, 2020-03-01. https://doi.org/10.1016/j.enbuild.2019.109690
18. Harkouss F, Fardoun F, Biwole PH (2018) "Multi-objective optimization methodology for net zero energy buildings," (in English). J Building Eng Article 16:57–71. https://doi.org/10.1016/j.jobe.2017.12.003
19. Lan L, Wood KL, Yuen C (2019) A holistic design approach for residential net-zero energy buildings: a case study in Singapore (in English). Sustain Cities Soc Article 50:16. Oct 2019, Art no. 101672. https://doi.org/10.1016/j.scs.2019.101672
20. Sen R, Bhattacharya SP, Chattopadhyay S (2021) Are low-income mass housing envelops energy efficient and comfortable? a multi-objective evaluation in warm-humid climate. Energy Buildings. 15(245):111055
21. Ciardiello A, Rosso F, Dell'Olmo J, Ciancio V, Ferrero M, Salata F (2020) Multi-objective approach to the optimization of shape and envelope in building energy design. Appl Energy 15(280):115984
22. Shiel P, Tarantino S, Fischer M (2018) Parametric analysis of design stage building energy performance simulation models. Energy Buildings. 1(172):78–93
23. Zhang J, Liu N, Wang S (2021) Generative design and performance optimization of residential buildings based on parametric algorithm. Energy Buildings. 1(244):111033
24. Gagnon R, Gosselin L, Decker SA (2019) Performance of a sequential versus holistic building design approach using multi-objective optimization. J Building Eng 1(26):100883

Shrinkage Behaviour Studies for the Integration of Low-Temperature Solid Oxide Fuel Cell into Low-Temperature Co-fired Ceramic (LTCC) Technology

C. Prabukumar, K. R. Adithya, Anil Sutar, Khushal Sirsat, Punam Kulkarni, Janardhan Rao Gadde, Vijaya Giramkar, Sriman Tadka, Ranjit Hawaldar, Ranjit V. Kashid, and Shany Joseph

1 Introduction

Solid oxide fuel cells (SOFC) are energy conversion systems that use hydrogen or hydrocarbons and oxygen to generate electricity. The SOFC uses ceramics as the electrolyte and electrodes to operate at high temperatures (500–1000 °C). Unlike other fuel cell systems like proton exchange membrane (PEM) fuel cell, SOFC offers more fuel flexibility, higher fuel and overall efficiency [1]. Other factors limiting the potential of SOFC for commercialization include very high electrolyte sintering temperature, material degradation due to its high operating temperature, and multi-step sintering. The operating temperature of the SOFC can be decreased by using appropriate electrolyte material. The operating temperature is reduced to 550–600 from 1000 °C when Gd doped cerium oxide (GDC) is used instead of yttria-stabilized zirconia (YSZ) [2]. Progress has been achieved in lowering the sintering temperature of the electrolyte by using materials on the nanoscale and the addition of a sintering aid. The Bi_2O_3, Fe_2O_3 and CuO are reported as the effective sintering aid to enhance the densification of the electrolyte [3, 4]. The fuel cell packing is done by using different materials like metals and ceramics. This leads to another major concern of gas leakage associated with improper sealing [5]. It poses a risk of fire explosion when the leaked fuel mixes with oxygen at a high temperature. Measures have been taken to solve the sealing problem using self-healing glass to seal the SOFC [6].

C. Prabukumar (✉) · K. R. Adithya · A. Sutar · K. Sirsat · P. Kulkarni · J. R. Gadde · V. Giramkar · S. Tadka · R. Hawaldar · R. V. Kashid · S. Joseph
Centre for Materials for Electronics Technology (C-MET), Panchawati, Pune 411008, India
e-mail: cprabumech@gmail.com

S. Joseph
e-mail: shany@cmet.gov.in

S. Doolla et al. (eds.), *Advances in Clean Energy and Sustainability*,
Green Energy and Technology, https://doi.org/10.1007/978-981-99-2279-6_40

Low-Temperature Co-fired Ceramics (LTCC) technology has been proved to be an effective technique in fabricating electronic devices. The LTCC tape consists of ceramics and low sintering glass mixture. The sintering of the ceramic particles in the LTCC is possible at as low as 850–900 °C due to the presence of glass. The glass phase melts at low temperature and aids in realizing the densification of the ceramic phase at < 900 °C [7]. So, the hermetic sealing of the fuel cell can be achieved by packing the fuel cell in the LTCC to ensure no leakage of fuel and oxygen gases [8]. Additionally, flow channels for the passage of gases and byproducts, the heaters and thermal sensors can be integrated into LTCC along with the fuel cells. The present study focuses on designing materials to use as electrodes and electrolytes of solid oxide fuel cells that can be packed in the LTCC. The work emphasized the shrinkage behaviour of the fuel cell components to avoid the development of undesired thermal stresses during the co-sintering of the fuel cell and LTCC together. Synthesised of anode CuO–ZnO (C_2Z_8), doped strontium cobaltite (SANC) cathode and Gd doped cerium oxide (GDC) electrolyte are reported in this work. Also, a low-temperature Bi_2O_3 based glass (Bismuth potassium oxide, BKO) is synthesized and used as the sintering aid. The synthesized materials are characterized by the XRD, SEM and particle size analysis.

2 Experimental Details

2.1 Synthesis of Gd Doped Cerium Oxide

Gadolinium doped cerium oxide (GDC) nanoparticles were synthesized following co-precipitation technique. Initially, 0.16 M of $Ce(NO_3)_3.6H_2O$ and 0.04 M $Gd(NO_3)_3.6H_2O$ were dissolved in demineralized (DI) water. The precipitation reaction was started with the addition of 0.425 M oxalic acid dissolved in water into the above-prepared solution. The 165 ml oxalic acid was added drop by drop into the prepared solution using a peristaltic pump. The synthesized product was washed with DI water and dried under an IR lamp. Then, the dried powder was calcinated at 700 °C. Finally, the calcinated powder was ball milled for 48 h by using zirconia balls with acetone as the medium. The ball-milled GDC powder was dried under the IR lamp for further characteristics.

2.2 Synthesis of C_2Z_8

Copper Oxide–zinc oxide (CuO–ZnO or C_2Z_8) composite was prepared by co-precipitation method. In brief, 0.2 M copper nitrate trihydrate and 0.8 M zinc nitrate hexahydrate were dissolved in DI water. Then, 1 M KOH aqueous solution was added to the above solution via the peristaltic pump. The addition of KOH solution was continued till the pH value became 12. The reaction solution was under mechanical stirring throughout this process. At the end of the synthesis process, the obtained precipitate was washed with DI water and IPA. Then the washed powder was dried and calcinated at 800 °C temperature. Finally, the calcinated powder was subjected to ball milling for 24 h using zirconia balls with acetone as the milling medium.

The composite of C_2Z_8-20% GDC was prepared using synthesised C_2Z_8 and GDC. The homogeneous mix of C_2Z_8 and GDC was obtained by ball milling corresponding powder for 12 h in an acetone medium. Then, after drying, this mixture underwent calcination at 600 °C for 3 h. Subsequently, the calcinated C_2Z_8-GDC was subjected to ball milling for 12 h to reduce the particle size.

2.3 *Synthesis of SANC*

$Sr_{0.95}Ag_{0.05}Nb_{0.1}Co_{0.9}O_{3-\delta}$ (SANC) perovskite was synthesized by following the solid-state reaction route. The $AgNbO_3$ perovskite needed for the synthesis of SANC was synthesized by the same method as SANC. $AgNO_3$ and Nb_2O_5 were mixed in a pot mill (wet mixing) for 1 h and calcined at 880 °C for 5 h. Then, the obtained calcine was mixed with $SrCO_3$, Co_3O_4 and Nb_2O_5 in the pot mill for 12 h to get SANC. The synthesized SANC was calcined at 1150 °C for 10 h and milled for 24 h in a planetary mill. Acetone was used as the milling medium for wet mixing and planetary milling.

2.4 *Synthesis of BKO Glass*

The $Bi_{0.9}K_{0.1}O_{1+\delta}$ (BKO) glass was prepared by the wet mixing process using Bi2O3 and K_2CO_3 as the precursors.

2.5 *Characterization*

The structural analysis of the synthesised materials was performed by using the X-ray diffractometer (XRD, Rigaku MiniFlex 600) with Cu-Kα incident radiation (λ = 0.154 nm). The morphological and elemental analysis was carried out by scanning electron microscope (SEM, Hitachi S-4800) attached to the energy dispersive X-ray spectrometer (EDS). The particle size analyzer (Horiba LA-960) was employed to determine the mean particle size of the synthesised materials.

The shrinkage behaviour of the electrolyte, anode and cathode was studied by using the in-house built dilatometer. The materials were used in the pellets form (dia = 6 mm) for this study. The change in linear dimension in response to the temperature (room temperature to 1000 °C) was recorded by dilatometer 3.1.79 software. Electrochemical impedance spectroscopy (EIS) was studied on the electrolyte material using Autolab PGSTAT100 (Metrohm) potentiostat at an open circuit in the frequency of 10 MHz to 1 MHz with 0.2 V AC amplitude. Both the surfaces of pellet (dia = 9 mm, t = 1.34 mm) samples were coated with platinum paste and sintered at 950 °C for this study.

3 Results and Discussion

The obtained XRD pattern of GDC (Gd doped CeO_2) synthesized by the co-precipitate method is shown in Fig. 1a. All the peaks are attributed to the GDC phase ($Ce_{0.7}Gd_{0.3}O_{1.85}$, cubic fluorite, ICDD #00-046-0507). Figure 1b shows the XRD pattern of synthesized C_2Z_8 as an anode material. It is observed that the material is a composite of ZnO (wurtzite, ICDD #00-036-1451) and CuO (monoclinic, ICDD #00-005-0661). The ZnO and CuO show the preferred grain orientation along (101) and (111) planes, respectively. XRD pattern of SANC prepared by the solid-state reaction is shown in Fig. 1c. The material exhibits $SrCoO_{2.29}$ (cubic perovskite, ICDD # 00-039-1083) as the primary phase with $SrCoO_{2.52}$ and $SrCoO_5$ as minor phases. Figure 1d shows the XRD pattern of the synthesized BKO glass. All the peaks belong to the Bi_2O_3 phase (monoclinic, ICDD #00-041-1449).

Figure 2a shows the SEM micrograph of Gd doped CeO_2 (GDC) synthesized by the co-precipitation method followed by ball milling. It is observed that the GDC particles have spherical-like morphology. The particle size analysis of the GDC reveals that the particles are in size range of 50–100 nm with a mean size of 75 nm. Figure 2b and c show the SEM micrographs of C_2Z_8 and SANC after the calcination process. The C_2Z_8 and SANC particles were subjected to ball milling and planetary ball milling, respectively, to reduce their particle size. The reduced particle size of the ball-milled material was measured by particle size analysis. The mean particle size of C_2Z_8 and SANC after milling was 1 μm and 0.85 μm, respectively.

The Nyquist plot of GDC + 30% BKO obtained from electrochemical impedance spectroscopy at different temperatures is shown in Fig. 3a. The Nyquist plot of GDC + 30% BKO at 600 °C is shown separately in Fig. 3b. The impedance spectra of GDC + 30% BKO samples show a curve of an incomplete semi-circle at high-frequency and a semi-circle at low-frequency. The reason for this incomplete semi-circle is due to the upper frequency of the potentiostat (1 MHz). The capacitance value of the high-frequency curve and low-frequency semi-circle was determined by fitting these curves with R and constant phase element (CPE) in parallel. The sample measured at 600 °C shows a capacitance value of 1.02×10^{-9} and 3.8×10^{-5} F for higher and lower frequency curves which are attributed to grain boundary and electrode–electrolyte interface resistance, respectively [9, 10]. The impedance spectra show that the grain boundary resistance decreases with an increased operating temperature owing to enhanced oxygen ionic conductivity [11]. The grain boundary resistance of the electrolyte was measured as 384, 155 and 47 Ω for the temperatures 550, 600

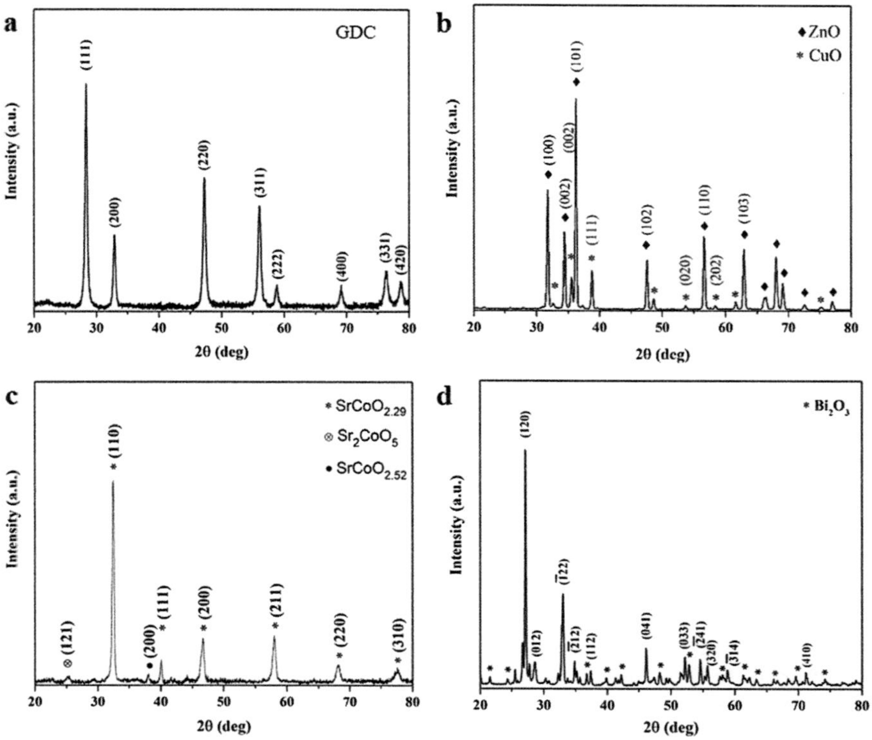

Fig. 1 XRD patterns of **a** Gd doped CeO_2, **b** C_2Z_8, **c** SANC and **d** BKO glass

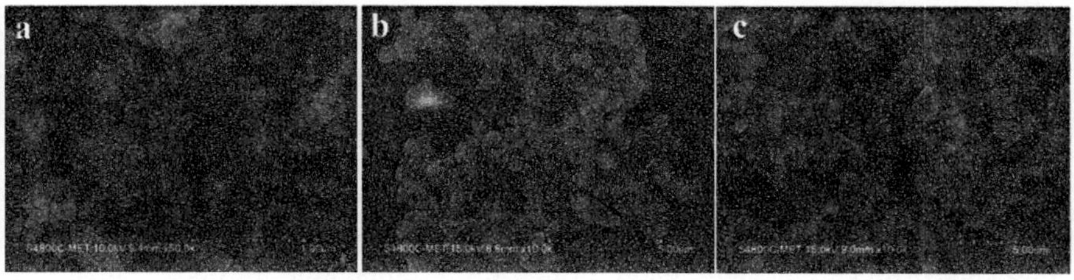

Fig. 2 SEM micrographs of **a** Gd doped CeO_2, **b** C_2Z_8 and **c** SANC

and 650 °C, respectively. The total conductivity of GDC + 30% BKO is calculated by considering the grain-boundary resistance, and the calculated total conductivity of the electrolyte is 1.52×10^{-3} S/cm at the operating temperature of 600 °C.

Figure 4 shows the shrinkage profile of LTCC (Dupont 951), electrolyte, anode and cathode obtained by measuring the change in linear dimension with respect to increasing temperature. The shrinkage of LTCC (Fig. 4a) was started at 675 °C and sintering was completed at around 850–875 °C. This low-temperature sintering of LTCC was due to its composition. The LTCC is made of a glass–ceramic mixture.

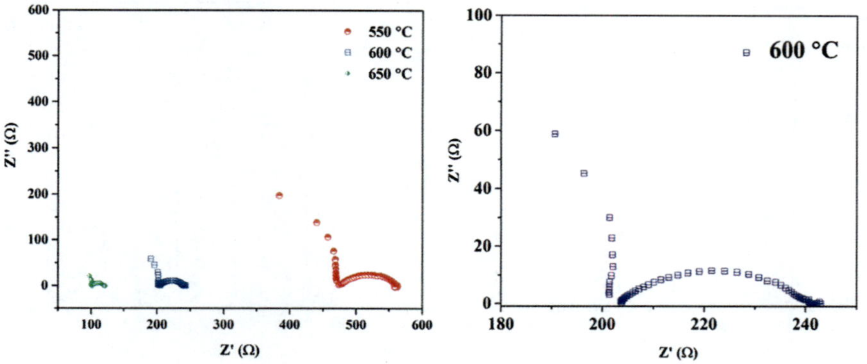

Fig. 3 Nyquist plot of GDC-30% BKO obtained at different temperatures (550, 600 and 650 °C)

The glass has a low melting point temperature that acts as the sintering aid to lower the sintering temperature of LTCC. Figure 4b shows the shrinkage profile of GDC and GDC + 30% BKO glass. It is observed that the shrinkage of GDC begins at the temperature of 765 °C. In order to lower the shrinkage temperature of GDC close to LTCC, the BKO glass was added to GDC (GDC: BKO = 70:30). As the melting temperature of the BKO glass, with Bi_2O_3 as primary content, is relatively low compared to GDC, the shrinkage started early in glass-GDC at 699 °C compared to pure GDC (765 °C) [3]. The glass addition leads to liquid phase sintering of glass-GDC composite where the pores are filled up with the liquid glass. This process may result in the densification of GDC at a lower temperature. A similar observation has been reported on the densification of GDC at lower temperature by adding 0.5% CuO [10].

The shrinkage profiles of C_2Z_8 based anode are shown in Fig. 4c. The shrinkage temperature of pure C_2Z_8 and C_2Z_8 + 20% GDC composite is 861 °C and 806 °C, respectively. In order to decrease the sintering temperature of the anode composite, 10% BKO glass was added to it. It is observed that the shrinkage temperature significantly reduced to 736 °C. The carbon in the form of activated charcoal (12%) was added to the anode-BKO to introduce porosity in the structure. The shrinkage of the anode-glass with carbon started at a further lower temperature of 688 °C than that without carbon. When the temperature was raised above the combustion temperature of carbon (~500 °C for charcoal), the carbon in the anode-glass structure burnt off, leaving the pores behind [11]. The pores produced in the structure have driven the shrinkage at this low temperature.

The $Sr_{0.95}Ag_{0.05}Nb_{0.1}Co_{0.9}O_{3-\delta}$ (SANC) perovskite is an excellent cathode material for solid oxide fuel cell application due to its high oxygen reduction activity [12]. Similar to the GDC electrolyte and C_2Z_8 anode, the linear shrinkage of the SANC cathode against the temperature was obtained by the dilatometry study. The curves were offset for better clarity. The shrinkage temperature of pure SANC was measured as 855 °C from its shrinkage profile (Fig. 4d). The shrinkage temperature of SANC needs to be brought down closer to the electrolyte, anode and LTCC to have thermal

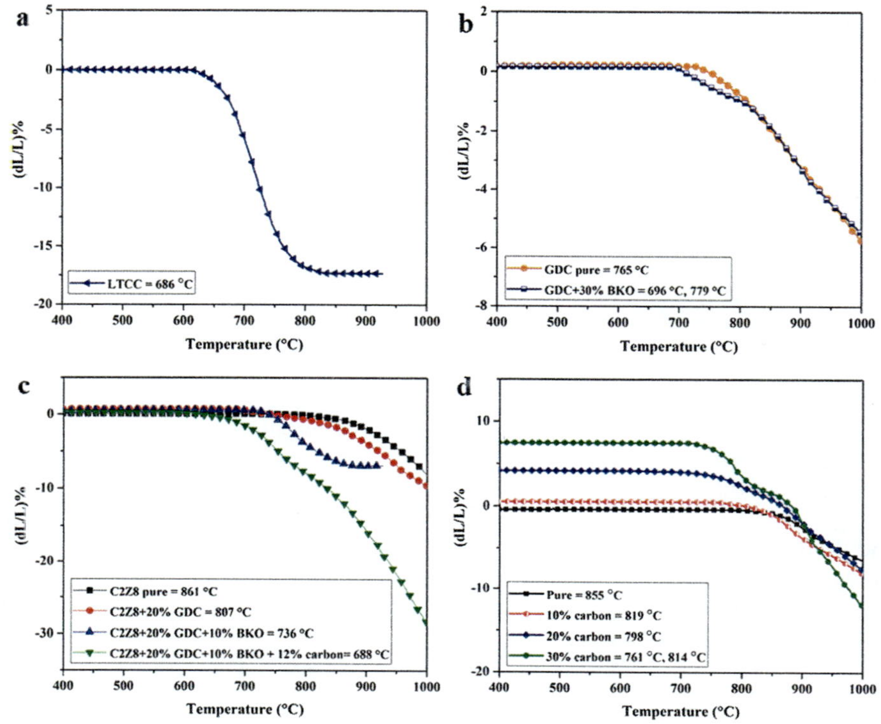

Fig. 4 Shrinkage profile indicating shrinkage temperature of **a** LTCC, **b** electrolyte, **c** anode and **d** cathode

compatibility during the co-firing. The carbon (activated charcoal) was added to SANC in 10, 20 and 30% ratios to study its effect on reducing the shrinkage temperature of SANC. The addition of carbon into SANC showed to lower the shrinkage temperature of SANC like in the case of the C_2Z_8 anode. The shrinkage temperature was decreased from 855 °C to 819 °C and 798 °C with 10% and 20% carbon, respectively. The high However, the addition of carbon above 20% resulted in a shrinkage profile with two distinct slopes at low and high temperatures. So, the ideal ratio of carbon in SANC can be fixed as 20%.

From Fig. 5, it is noted that the shrinkage temperature of C_2Z_8-GDC anode composite (688 °C) optimized with BKO glass and carbon matches that of LTCC (686 °C). With the addition of 30% BKO glass, the shrinkage temperature of GDC was dropped to 696 °C. Considerable reduction in shrinkage temperature was achieved in the case of SANC to match the electrolyte and anode. The temperature was reduced from 855 to 798 °C by introducing carbon into SANC. However, the temperature window between the cathode and other components is still significant. It indicates the scope for further improvement concerned with cathode material. A comprehensive study shall be carried out on the SANC cathode to lower its shrinkage temperature. The shrinkage at 875 °C of electrolyte, anode and cathode material is 2.7%, 12.92%

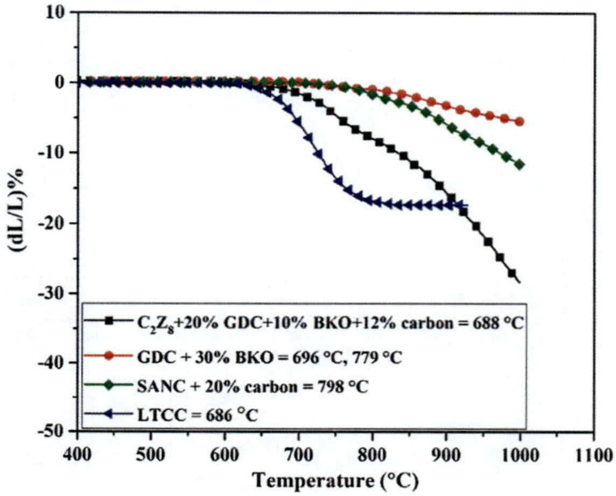

Fig. 5 Shrinkage profile indicating shrinkage temperature of optimized electrolyte, anode, cathode and LTCC

and 4.5%, respectively. The higher shrinkage of anode and cathode over the electrolyte is beneficial as it would drive the densification of the electrolyte at a lower temperature. The electrolyte would experience compressive stress due to the higher shrinkage of the anode and cathode [13].

4 Conclusions

The thermal compatibility between the solid oxide fuel cell components such as GDC electrolyte, CuO–ZnO/GDC composite anode and SANC cathode with LTCC is studied by the dilatometry study. The prepared BKO glass was proved to be good sintering aid to the electrolyte to facilitate the sintering at a lower temperature. The shrinkage temperature of GDC was lowered from 765 to 696 °C after adding 30% BKO glass to it. The shrinkage temperature of the C_2Z_8-GDC composite was reduced from 807 to 688 °C after adding 10% BKO and 12% carbon. The shrinkage temperature of GDC and C_2Z_8-GDC after the additive is closer to the shrinkage temperature of LTCC (686 °C). The shrinkage temperature of pure SANC was measured as 855 °C. However, the addition of carbon showed decreased shrinkage temperature of SANC; the temperature was lowered up to 798 °C with the addition of 20% carbon. There is a need for further improvement on SANC shrinkage to match that of LTCC. The shrinkage studies will be carried out on the optimized materials in the planar form in the future to integrate the planar fuel cell into LTCC.

Acknowledgements The authors are grateful to the Department of Science and Technology (DST), Govt. of India, for funding this project.

References

1. Brandon N (2004) Fuel cells. Encycl Energy 749–758
2. Li Z, Mori T, Yan P, Wu Y, Li Z (2012) Preparation and performance of intermediate-temperature fuel cells based on Gd-doped ceria electrolytes with different compositions. Mater Sci Eng B Solid-State Mater Adv Technol 177(17):1538–1541
3. Takahashi S, Sumi H, Fujishiro Y (2019) Development of co-sintering process for anode-supported solid oxide fuel cells with gadolinia-doped ceria/lanthanum silicate bi-layer electrolyte. Int J Hydrogen Energy 44(41):23377–23383
4. Panthi D, Choi B, Du Y, Tsutsumi A (2017) Lowering the co-sintering temperature of cathode–electrolyte bilayers for micro-tubular solid oxide fuel cells. Ceram Int 43(14):10698–10707
5. Butler MS, Moran CW, Sunderland PB, Axelbaum RL (2009) Limits for hydrogen leaks that can support stable flames. Int J Hydrogen Energy 34(12):5174–5182
6. Liu WN, Sun X, Khaleel MA (2011) Study of geometric stability and structural integrity of self-healing glass seal system used in solid oxide fuel cells. J Power Sources 196(4):1750–1761
7. Induja IJ, Abhilash P, Arun S, Surendran KP, Sebastian MT (2015) LTCC tapes based on Al2O3-BBSZ glass with improved thermal conductivity. Ceram Int 41(10):13572–13581
8. Belavic D, Hrovat M, Dolanc G, Santo Zarnik M, Holc J, Makarovic K ((2012)) Design of LTCC-based ceramic structure for chemical microreactor. Radioengineering 21(1):195–200
9. Bowen MS, Johnson M, McQuade R, Wright B, Kwong KS, Hsieh PY, Cann DP, Woodside C (2020) Electrical properties of gadolinia-doped ceria for electrodes for magnetohydrodynamic energy systems. SN Appl Sci 2(9):1–9
10. Irvine JT, Sinclair DC, West AR (1990) Electroceramics: characterization by impedance spectroscopy. Adv Mater 2(3):132–138
11. Raza R, Wang X, Ma Y, Zhu B (2010) A nanostructure anode ($Cu_{0.2}Zn_{0.8}$) for low-temperature solid oxide fuel cell at 400–600 °C. J Power Sour. 195(24):8067–8070
12. Choolaei M, Bull T, Ramirez Reina T, Amini Horri B (2020) Synthesis and characterisation of nanocrystalline CuO–Fe_2O_3/GDC anode powders for solid oxide fuel cells. Ceram Int 46(10):14776–14786
13. Castro JP, Nobre JRC, Napoli A, Bianchi ML, Moulin JC, Chiou BS, Williams TG, Wood DF, Avena-Bustillos RJ, Orts WJ, Tonoli GHD (2019) Massaranduba sawdust: a potential source of charcoal and activated carbon. Polymers 11(8):1–14
14. Zhu Y, Zhou W, Ran R, Chen Y, Shao Z, Liu M (2016) Promotion of oxygen reduction by exsolved silver nanoparticles on a perovskite scaffold for low-temperature solid oxide fuel cells. Nano Lett 16(1):512–518
15. Lee S, Lee K, Jang YH, Bae J (2017) Fabrication of solid oxide fuel cells (SOFCs) by solvent-controlled co-tape casting technique. Int J Hydrogen Energy 42(3):1648–1660

Effect of Operating Parameters on Biodiesel Yield from Transesterification of Cotton Seed Oil

S. Rupesh, Chris Ben Xavier, and Christy Thomas Sani

1 Introduction

The key challenges driving worldwide interest in biofuels are global warming and price hike of depleting fossil fuel supplies. The environmental and economic concerns associated with conventional fossil fuels can be mitigated with alternative biofuel variants like biodiesel [1]. Biodiesel is a bio-degradable sulphur-free alternative fuel for conventional diesel engines with reduced hydrocarbon and particulate emissions when blended up to 20% by volume with petro-diesel [2, 3]. Transesterification is the process by which fatty acids from feedstock such as oil and animal fat are transmuted to methyl esters, commonly known as biodiesel. Among the various edible and non-edible oil feedstocks, cotton seed oil emerges as a prominent one owing to its lower acid and kinematic viscosity along with enhanced properties when mixed with other oils [4]. The potential of cotton seed oil derived biodiesel as a sustainable alternative fuel was detailed by researchers. A reduction in particulate matter, smoke and carbon emission was identified when cotton seed oil biodiesel blended with octanol and multi-walled carbon nanoparticles is tested on a single cylinder diesel engine [5]. The ASPEN plus process simulator is successfully implemented by several researchers for economic and easy simulation of various energy conversion pathways such as gasification [6, 7], pyrolysis [8], combustion, fermentation [9], anaerobic digestion [10] and transesterification. A kinetic model for transesterification was formulated by Souza et al. [11] for the simulation of biodiesel generation from palm oil. The model was developed by considering only two oil constituents namely, palmitic and oleic acid. Generally, in the simulation of transesterification, the feedstock is represented by one or two of its major constituents. Enzymatic biodiesel production

S. Rupesh (✉)
PES College of Engineering, Mandya, Karnataka, India
e-mail: mailtorupeshs@gmail.com

C. B. Xavier · C. T. Sani
NIT Calicut, Calicut, Kerala, India

S. Doolla et al. (eds.), *Advances in Clean Energy and Sustainability*,
Green Energy and Technology, https://doi.org/10.1007/978-981-99-2279-6_41

from rape seed oil is formulated by modelling triglyceride as trioleate in the kinetic modelling for economic evaluation [12]. In acid catalyst transesterification of acidic oil feedstock is modelled as triolein to study techno-economic feasibility of process alternatives in ASPEN plus [13]. Rape seed oil is represented as triolein and trilinolein in the kinetic modelling of transesterification for the design of supercritical process of biodiesel generation [14].

No work is found to be reported for kinetic modelling of cotton seed oil transesterification by considering its component specific reaction kinetics in ASPEN plus platform. The present simulation deals with the transesterification of cotton seed oil by integrating component specific reaction kinetics.

2 Model Formulation

A kinetic model is developed in ASPEN plus platform to simulate alkaline transesterification of cotton seed oil using methanol in the presence of NaOH catalyst. Cotton seed oil is modelled by considering the triglyceride composition of feedstock reported in the experimental work of Mahdavi and Monajemi [15]. The triglyceride composition of feedstock used to model cottonseed oil is given in Table 1.

Streams 'METHANOL' and 'NAOH', respectively representing alcohol and catalyst, are mixed in the mixer 'MIXER1' and the resulting steam 'MIX2' after methanol recovery is fed to an RCSTR reactor 'R' along with cotton seed oil 'OIL' to simulate transesterification process, as depicted in Fig. 1. The methanol recovery unit 'MRU' is a 5-stage distillation column where the feed enters the 3rd stage, employed for thermal recovery of excess methanol present in the reactor output by using a reboiler duty of 500 cal/s with a reflex ratio of 2.

Vapour-liquid equilibria of the system are represented using Dortmund-UNIFAC (UNIF-DMD) model and the physical properties of feedstock, alcohol, catalyst and products are imported from the ASPEN plus database [16]. Binary interaction parameters for the methanol–water and triglyceride-ester pairings are incorporated from the Non-random two liquid (NRTL) property package database available in the ASPEN plus [17]. The transesterification process is implemented by incorporating composition specific reaction kinetics [18] to cotton seed oil by considering the formation of their corresponding diglyceride, monoglyceride and methyl esters, as given in Table 2. From the output stream 'ROUT' from the reactor, comprised of cotton seed oil methyl ester (COME), unreacted methanol, glycerol and NaOH, the excess

Table 1 Triglyceride composition

Fatty acid	Weight %
Triolein	0.2135
Tripalmitin	0.2465
Trilinolein	0.5215
Tristearate	0.0185

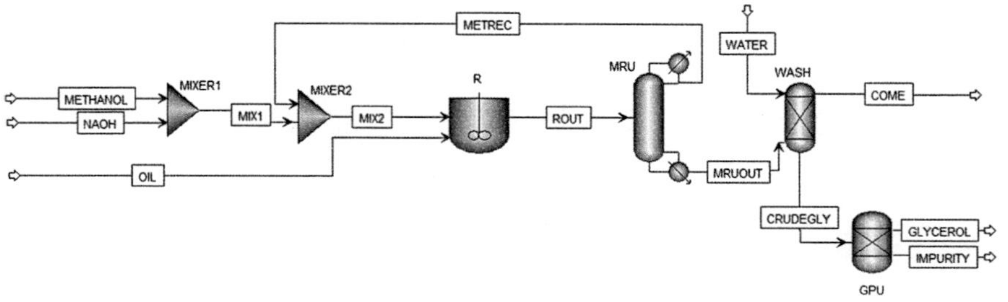

Fig. 1 ASPEN plus flow sheet

methanol is recovered in the methanol recovery unit 'MRU'. The COME is obtained from the resultant stream 'MRUOUT' by glycerol separation and water washing.

The developed model is validated with an average deviation of 3.02% when the model predicted methyl ester conversion is comparedith the corresponding experimental conversion reported by Kuma et al. [17], for linseed oil transesterification

Table 2 Reactions and reaction kinetics

Reactions			
R1	Triglyceride + Methanol $\underset{k_2}{\overset{k_1}{\rightleftarrows}}$ Diglyceride + COME		
R2	Diglyceride + Methanol $\underset{k_4}{\overset{k_3}{\rightleftarrows}}$ Monoglyceride + COME		
R3	Monoglyceride + Methanol $\underset{k_6}{\overset{k_5}{\rightleftarrows}}$ Glycerol + COME		
Reaction rate constants			
k_1	0.049	k_4	133
k_2	0.112	k_5	0.122
k_3	0.226	k_6	0.016

Kinetic rate law

$$\frac{d[TG]}{dt} = -k_1[\mathrm{TG}][\mathrm{A}] + k_2[\mathrm{DG}][\mathrm{COME}]$$

$$\frac{d[DG]}{dt} = k_1[\mathrm{TG}][\mathrm{A}] - k_2[\mathrm{DG}][\mathrm{COME}] - k_3[\mathrm{DG}][\mathrm{A}] + k_4[\mathrm{MG}][\mathrm{COME}]$$

$$\frac{d[MG]}{dt} = k_3[\mathrm{DG}][\mathrm{A}] - k_4[\mathrm{MG}][\mathrm{COME}] - k_5[\mathrm{MG}][\mathrm{A}] + k_6[\mathrm{GLY}][\mathrm{COME}]$$

$$\frac{d[COME]}{dt} = k_1[\mathrm{TG}][\mathrm{A}] + k_3[\mathrm{DG}][\mathrm{A}] + k_5[\mathrm{MG}][\mathrm{A}] - k_2[\mathrm{DG}][\mathrm{COME}] - k_4[\mathrm{MG}][\mathrm{COME}] - k_6[\mathrm{GLY}][\mathrm{COME}]$$

TG—Triglyceride, DG—Diglyceride, MG—Monoglyceride, A—Methanol, GLY—Glycerol, COME—Cottonseed oil methyl ester

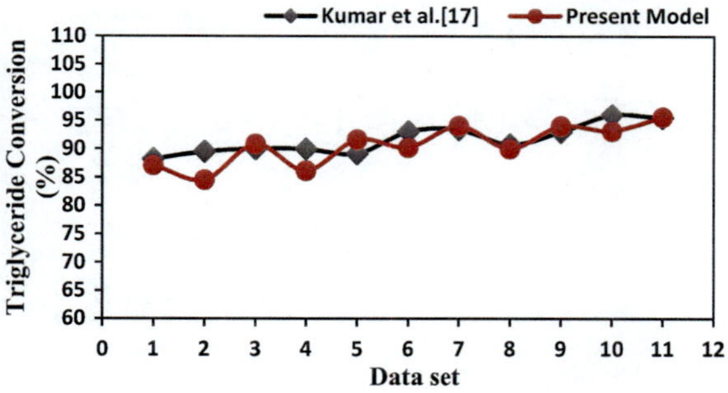

Fig. 2 Model validation

using methanol and NaOH as alcohol and catalyst, respectively. The comparison of experimental and model predicted conversions is depicted in Fig. 2.

3 Results and Discussion

The influence of methanol to oil ratio, reaction time, process temperature and catalyst weight percentage on COME yield is analyzed by using the validated model.

3.1 Effect of Methanol to Oil Ratio on COME Yield

The influence of methanol to oil ratio on biodiesel yield is depicted in Fig. 3. It is found that biodiesel yield increases with increase in M/O ratio. This is attributed to the shifting of the transesterification reaction towards the product side with increase in methanol. The yield is found to be insignificant when the M/O ratio is increased beyond a value of 6:1. However, excess methanol increases the complexity of ester separation and purification along with process cost [19]. Thus, a methanol to oil ratio of 6:1 is identified as a feasible value for transesterification, which is in line with the results reported in experimental works [20–22].

3.2 Effect of Reaction Time on COME Yield

Figure 4 illustrates the effect of reaction time on biodiesel yield. It is observed that biodiesel yield increases with increase in process time and becomes insignificant beyond a duration of 1 h. This is due to the increase in the extent of completion of the transesterification reaction with process time, converting more oil to corresponding methyl esters. Similar results were obtained by Souza et al. [11].

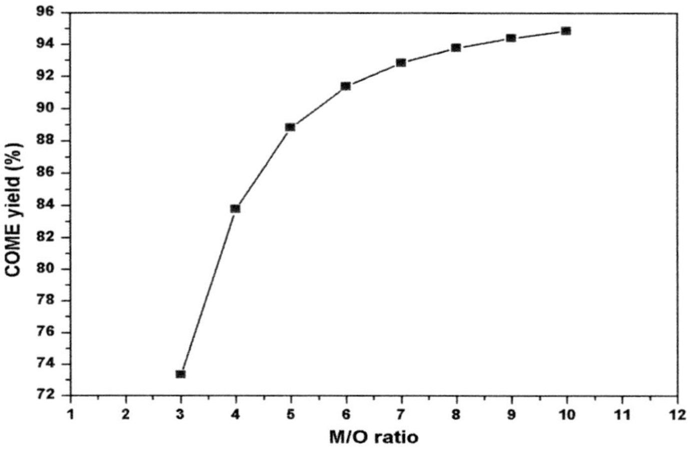

Fig. 3 Effect of methanol to oil ratio on COME yield

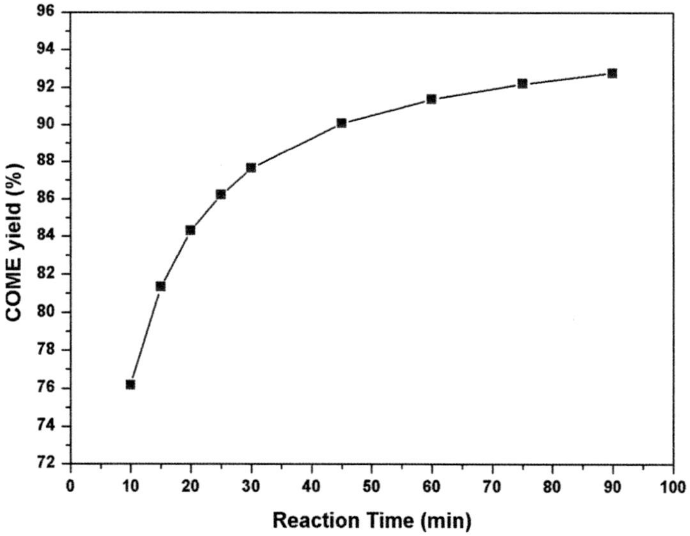

Fig. 4 Effect of reaction time on COME yield

3.3 Effect of Catalyst on COME Yield

The impact of catalyst (NaOH) weight % on COME yield, illustrated in Fig. 5, shows that an increase in catalyst addition leads to an increase in total COME yield by significantly improving the triglyceride conversion. It is observed that an increase in catalyst weight % from 1 to 1.5 leads to an insignificant increase of COME yield by 1.51%,

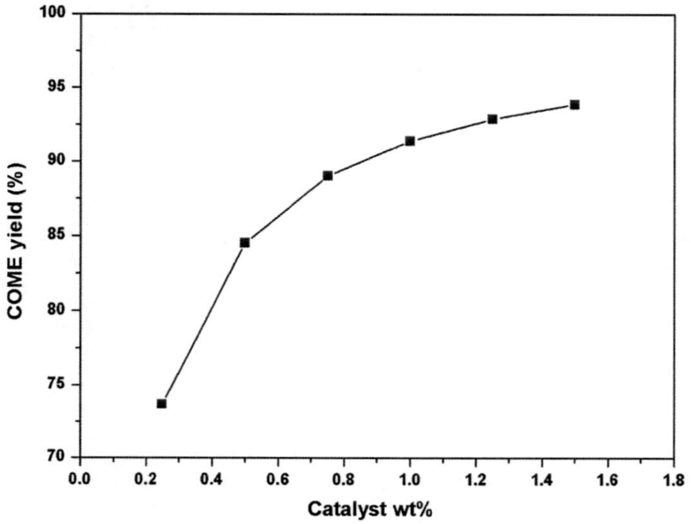

Fig. 5 Effect of catalyst weight percentage on COME yield

thereby restricting the catalyst weight % to 1. Apart from economic benefit, controlling catalyst below 1.5% limits saponification reaction thereby maximizing biodiesel formation by preventing excess soap formation [23, 24]. This is in accordance with the experimental results of Keera et al. [21] in the alkaline transesterification of vegetable oils.

3.4 Effect of Temperature on COME Yield

From Fig. 6 it is identified that COME yield increases with process temperature up to a value of 60 °C and practically invariant thereafter. The viscosity of triglycerides reduces with increase in temperature, thereby enhancing oil –methanol mixing and their reaction. In general, the biodiesel yield is invariant when the reaction temperature approaches the boiling point range of methanol [25]. Here, the temperature corresponding to optimum COME yield is found to be 60 °C, which is in compliance with the observation of Silitonga et al. [26].

4 Conclusions

A constituent specific kinetic model for transesterification of cotton seed oil is developed in ASPEN plus platform by incorporating stepwise conversion of triglycerides to methyl ester and glycerol. The validated model is used to interpret the effect

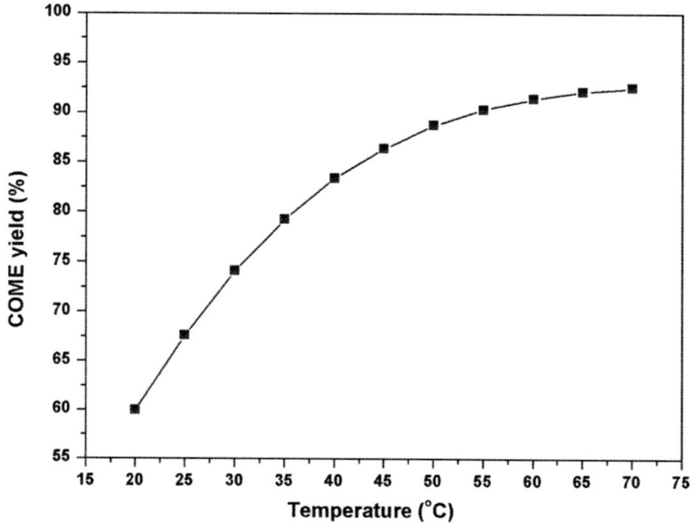

Fig. 6 Effect of process temperature on COME yield

of methanol to oil ratio, reaction time, process temperature and catalyst weight percentage on biodiesel yield. The yield is found to be insignificant when the M/O ratio and reaction time are respectively increased beyond a value of 6:1 and 60 min. Similarly, the practical NaOH catalyst wt.% for transesterification is found to be 1 for COME yield with better economic and saponification paybacks. Moreover, the optimum process temperature, found to be 60 °C, typically impends towards the boiling point of methanol for maximum biodiesel yield. Based on the present model, the optimum parametric values for COME yield are found to be 6:1, 1 h, 1% and 60 °C for M/O ratio, reaction time, catalyst weight percentage and reaction temperature, respectively.

References

1. Komers K, Stloukal R, Machek J, Skopal F (2001) Eur J Lipid Sci Technol 103:363
2. Lee AF, Bennett JA, Manayil JC, Wilson K (2014) Chem Soc Rev 43:7887
3. Islam S, Ahmed AS, Islam A, Aziz SA, Xian LC, Mridha M (2014) J Chem 2014:1
4. Jamshaid M, Masjuki HH, Kalam MA, Zulkifli NWM, Arslan A, Qureshi AA (2022) Energy 239
5. Soudagar MEM, Afzal A, Safaei MR, Manokar AM, EL-Seesy AI, Mujtaba MA, Samuel OD, Badruddin IA, Ahmed W, Shahapurkar K, Goodarzi M (2020) J Therm Anal Calorim 147:525
6. Rupesh S, Gokul Krishnan S (2021) Energy sources. Part A Recover Util Environ Eff 1
7. Rupesh S, Muraleedharan C, Arun P (2016) Resour Technol 2:94
8. Peters JF, Banks SW, Bridgwater AV, Dufour J (2017) Appl Energy 188:595
9. Quintero JA, Cardona CA (2011) Ind Eng Chem Res 50:6205
10. Anaya Menacho W, Mazid AM, Das N (2022) Fuel 309
11. Souza MF, Hirata GF, Batista EAC (2020) Fluid Phase Equilib 525:112792

12. Sotoft LF, Rong BG, Christensen KV, Norddahl B (2010) Bioresour Technol 101:5266
13. Gebremariam SN, Marchetti JM (2018) Energy Convers Manag 174:639
14. Lim Y, Lee HS, Lee YW, Han C (2009) Ind Eng Chem Res 48:5370
15. Mahdavi V, Monajemi A (2014) J Taiwan Inst Chem Eng 45:2286
16. Kuramochi H, Maeda K, Kato S, Osako M, Nakamura K, Ichi Sakai S (2009) Fuel 88:472
17. Kiss AA, Bildea CS (2012) J Chem Technol Biotechnol 87:861
18. Narváez PC, Rincón SM, Sánchez FJ (2007) JAOCS. J Am Oil Chem Soc 84:971
19. Qiu F, Li Y, Yang D, Li X, Sun P (2011) Appl Energy 88:2050
20. Noureddini H, Zhu D (1997) Biocatal Artic 74:1457
21. Keera ST, El Sabagh SM, Taman AR (2011) Fuel 90:42
22. Zhang Y, Dubé MA, McLean DD, Kates M (2003) Bioresour Technol 89:1
23. Abuhabaya A, Fieldhouse J, Brown D (2013) Fuel 103:963
24. Leung DYC, Guo Y (2006) Fuel Process Technol 87:883
25. Agarwal AK (2014) 123
26. Silitonga AS, Mahlia TMI, Ong HC, Riayatsyah TMI, Kusumo F, Ibrahim H, Dharma S, Gumilang D (2017) Energy sources. Part A Recover Util Environ Eff 39:2006

Thermal Performance Analysis of Phase Change Material-Based Plate Finned Heat Sinks for Thermal Management Applications

Pradunmya P. Dutta, Vivek Saxena, and Santosh K. Sahu

1 Introduction

Due to the ever-increasing demand for more and more energy and the requirement of minuscule high energy density systems, passive thermal management systems (PTMS) have found their way into the field of thermal energy storage, such as space systems [1], textile [2, 3], medicine [4], electronics cooling [5, 6] and transportation industries [7, 8] because of having better thermophysical, economic and kinetic properties [9]. PTMS use the solid-liquid transition phase, where heat is absorbed latent form by keeping the temperature at an idle value. The counterpart of PTMS is active thermal management system (ATMS) which is ever-present, but due to its limitation of requiring external power to operate, PTMS has gained popularity. PCM are one of the common materials used in PTMS, as it boasts high latent heat capacity [10–12] which is an attractive property. PCMs have a heat storage capacity that is almost 7–9 times that of the other sensible heat energy systems [13]. The performance of PCM is influenced by the design and material of the heat sinks [14]. There are three distinct categories of PCM, namely, organic, inorganic and eutectic. Organic PCMs have several advantageous properties such as significant value of latent heat and high specific heat, and repeatability or reusability [15]. Despite these advantages, PCM has a downside because of its lower thermal conductivity, which hinders the speed at which heat is transferred. In order to overcome this several approaches are applied to influence the thermal conductivity such as fins, addition of nanoparticles, and metal foams.

Several experimental studies were performed on heat sinks having fins as thermal conductivity enhancer (TCE) and are outlined here. Kumar et al. [16] studied experimentally the thermal performance of PCM based heat sinks incorporating metallic

P. P. Dutta (✉) · V. Saxena · S. K. Sahu
Indian Institute of Technology Indore, Indore 453552, India
e-mail: ms2104103001@iiti.ac.in

S. Doolla et al. (eds.), *Advances in Clean Energy and Sustainability*,
Green Energy and Technology, https://doi.org/10.1007/978-981-99-2279-6_42

foams and internal fins. It was concluded that effective cooling of electronic components is possible by having heat sink with circular pin fin. The same authors [17] also experimentally studied then enhancement in thermal performance of PCM based heat sinks with the addition of nanoparticles. It was reported in this study that heat sinks having square pin fins perform better in thermal management and the addition of nanoparticles greatly enhanced the heat sink performance. Kothari et al. [18] experimentally investigated the thermal performance for heat sinks having plate fins incorporating nanoparticles. It was inferred from the study that heat sink filled with 2% nanoparticles have an enhancement in the operating time by 1.48. Elham et al. [19] studied the effects of proximity of fins on heat sinks thermal performance. It has been reported having wavy and straight fins altogether negatively affected the performance of heat sink rather than heat sinks having all straight or all wavy fins. Gutierrez et al. [20] studied the enhancement in cycling stability of bischofite as thermal energy storage material. It was reported that mixing organic and inorganic PCM has encouraging results in thermal performance.

It is apparent from the literature study that although a few studies have been done by employing PEG-6000 as PCM, more extensive studies into using plate fins have been scarce. The current experimental investigation's goal is to comprehend the melt propagation and thermal performance and optimize the operating time for the heat sinks in use. For this purpose, Heat sink having no plate fin (HSNF), Heat sink having one plate fin (HSOF), Heat sink having two plate fins (HSTF), Heat sink having three plate fins (HSTHF), Heat sink having four plate fins (HSFF), are used by incorporating PCM. Heat sink without fins inherently has less area as compared to heat sinks having fins as TCE, which leads to a lower amount of heat transfer. By incorporating fins, there is a considerable rise in the transfer of heat as the area increases which should lead to a better heat transfer and a clear picture regarding the thermal performance.

2 Experimental Setup

Figure 1 depicts the various modules employed in this investigative study. The heat sink assembly makes use of unfinned and various finned heat sinks, made of aluminium, filled with PCM and is insulated from all sides by ceramic glass wool during the experiments to restrict the loss of heat to the surroundings, so that the maximum amount of heat produced by the plate heater is transferred to the assembly. The heat sinks used have a dimension of $100 \times 100 \times 20$ mm^3. Five different heat sink configuration was used in the experimental study HSNF, HSOF, HSTF, HSTHF and HSFF as depicted in Fig. 2. A plate heater (Sunrise products, India) having a dimension of $100 \times 100 \times 4$ mm^3 is placed in the assembly above which the heat sinks were kept on. The plate heater is connected to a DC power source (Aplab L3260, 0–30 V/0–40 A, India) to provide the required power to the heat sink, mimicking that of an electronic device. The error in measuring the voltage is found to be $\pm$ 0.1 V and that of current is ± 0.2 A. Temperature measurement of the heat sinks is done with the help of k-type thermocouples having an accuracy upto ± 0.3 °C.

A Data Acquisition System (Agilent make 34972 A, 32 channels) is utilised in order to get temperature readings from the thermocouples, which in turn feeds the computer as readable records. A total of 9 thermocouples are used in the study. Four thermocouples (T1, T2, T3, T4) is used in the heat sink base and three thermocouples (T5, T6, T7) are utilised in the heat sink's side wall. Thermocouple (T8) is held inside the insulation to monitor heat loss to the surrounding and one thermocouple (T9) is held to measure ambient temperature. The melt fraction of PCM at a specific interval of time is captured using a Digital Camera (Sony R × 10 MII).

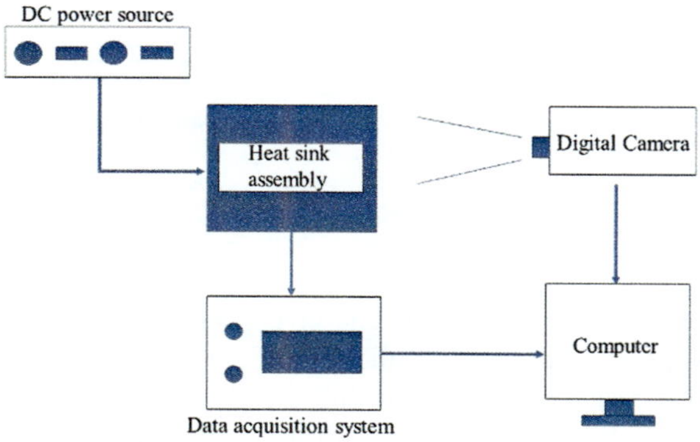

Fig. 1 Depiction of the experimental setup

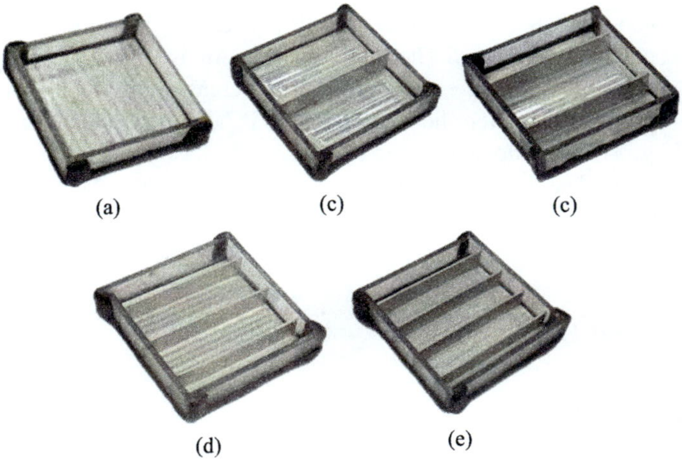

Fig. 2 Different heat sink configurations **a** HSNF, **b** HSOF, **c** HSTF, **d** HSTHF and **e** HSFF

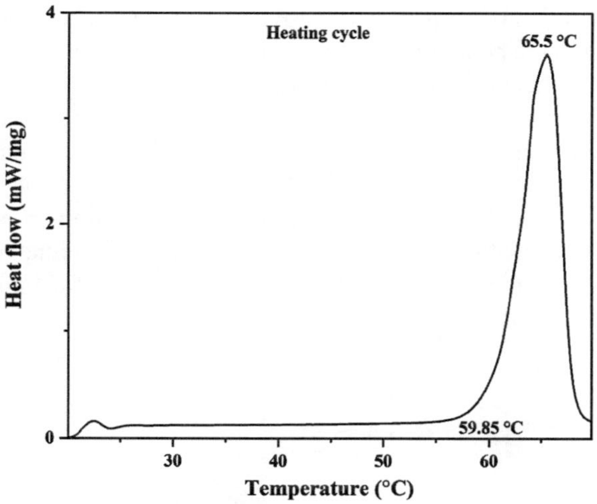

Fig. 3 Differential scanning calorimetry heating of PEG-6000

3 Characterization of PCM

This experimental study uses polyethylene glycol-6000 (PEG-6000) as the PCM. Various thermos-physical property of the sample for instance latent heat, melting range and specific heat are measured using a Differential Scanning Calorimetry (DSC214 Polyma, IIT Indore). Figure 3 illustrates the sample's DSC curve. The DSC testing is carried out at a temperature range of 20–70 °C, with a heating rate of 10 °C/min. The primary peak depicts the transitioning of phase from solid to liquid which starts at a temp of ~59 °C. It is inferred from the graph that the beginning, peak point, and termination of the phase transition occurs at 59.85 °C, 65.50 °C and 69.06 °C, respectively. TEMPOS Thermal Properties Analyzer (METER Group, IIT Indore) is employed to measure the PCM thermal conductivity and is discovered to be 0.201 W/m-K. The latent heat of fusion of PCM in use is determined to be 186.386 kJ/kg. The specific heat at the beginning of phase transition is calculated to be 1.224 J/g °C. The thermophysical properties of PEG-6000, Aluminium (TCE) and plexiglass are given in Table 1.

4 Results and Discussions

4.1 Setup Validation

To validate the experimental setup a comparison is done between the present and previous studies[17]. Figures 4 and 5 compares the current results for the no fin heat sink with and without the presence of paraffin wax with the existing study of Kumar et al. [17]. A heat flux 2.0 kW/m^2 is supplied to the heat sink assemblage

Table 1 Thermophysical properties of PEG-6000, aluminium, and plexiglass [16]

Properties	PEG-6000	Aluminium (TCE)	Plexiglass
Melting temperature (°C)	59.85	660.37	–
Specific heat (kJ/kg-K)	1.224	0.896	1.470
Density (kg/m^3)	1200	2719	–
Thermal conductivity (W/m–K)	0.201	218	0.19
Latent heat (kJ/kg)	186.386	–	–

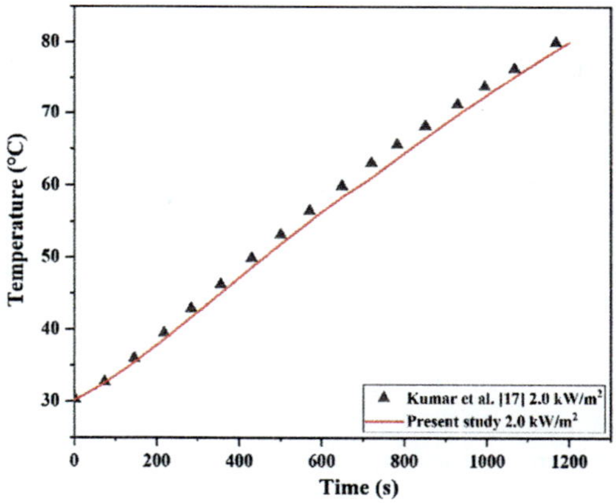

Fig. 4 Temperature–time spread for HSNF without PCM

through a plate heater and time–temperature comparison is done. The current investigation's findings demonstrate a pattern that is comparable to that reported in previous experiments done on dimensionally similar heat sink. A slight difference in the validation maybe accounted for the change in ambient conditions i.e., the initial heating temperature.

4.2 Effect of Heat Sink Arrangement on the Heat Sink's Base Temperature

The present investigation is performed at different heat fluxes of 2.0, 2.5 and 3.0 kW/m^2 and the temperature spread over time for various heat sink layouts is shown in Fig. 6. The average fluctuation of temperature at the bottom of the heat sink is considered for different values of heat flux as mentioned above. To explore

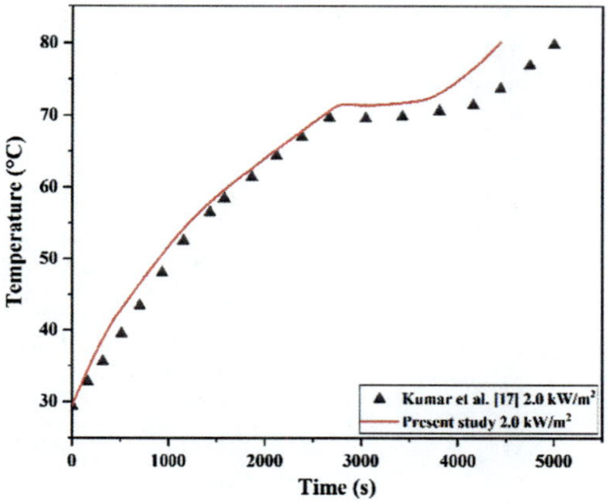

Fig. 5 Temperature–time spread for HSNF with PCM

the impact of different numbers of plate fins on thermal performance, the solid-liquid phase transition of PEG-6000 is compared with different heat flux for five distinct heat sink designs. Figure 6 depicts that there is a distinct variation during the initiation of heating and when phase change occurs. In the initial stage of heating, sensible heating takes places leading to a linear curve. But as soon as melting starts the curve deviates from the original path which shows the transition phase having a better temperature control. The figures also suggests that with the addition of plate fins in the heat sink, keeping the volume of PCM in all the heat sinks at a constant value, there is a larger surface area for heat transmission and as a result, the heat sink's base temperature remains low and helps in reaching a certain temperature level at a later point in time. This delay in reaching a certain set point temperature is also due to more uniform melting in heat sinks with fins which helps in better management of temperature of heat sinks possessing plate fins as compared to heat sinks with no fins.

4.3 *Propagation of Melt Front in the Heat Sinks*

A digital camera is used in the study to record the propagation of melt front of PCM in successive interval of time. To keep it concise, the images are taken for 3.0 kW/m^2 heat flux at regular intervals for various layouts of heat sinks used in the study. The photographs were captured from the time the PCM is solid until it melts completely. The melt propagation in the instance of HSNF is depicted in Fig. 7a and b in the instance of HSOF, Fig. 7c in the instance of HSTF, Fig. 7d in the instance of HSTHF and Fig. 7e in the instance of HSFF. The images shown here are considered in a way that it shows a visible difference between consecutive images of propagation of melt front. As seen from the images, there is a difference in melt propagation in case of various configurations of heat sinks.

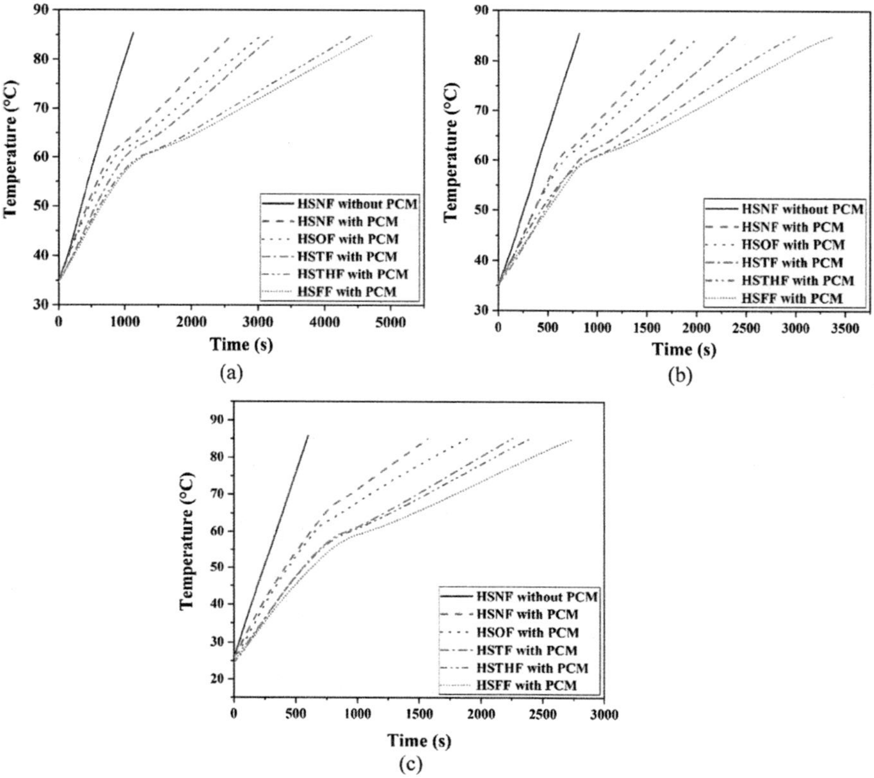

Fig. 6 Temperature spread over time for various heat sink layouts at various heat fluxes **a** 2.0 kW/m^2, **b** 2.5 kW/m^2, **c** 3.0 kW/m^2

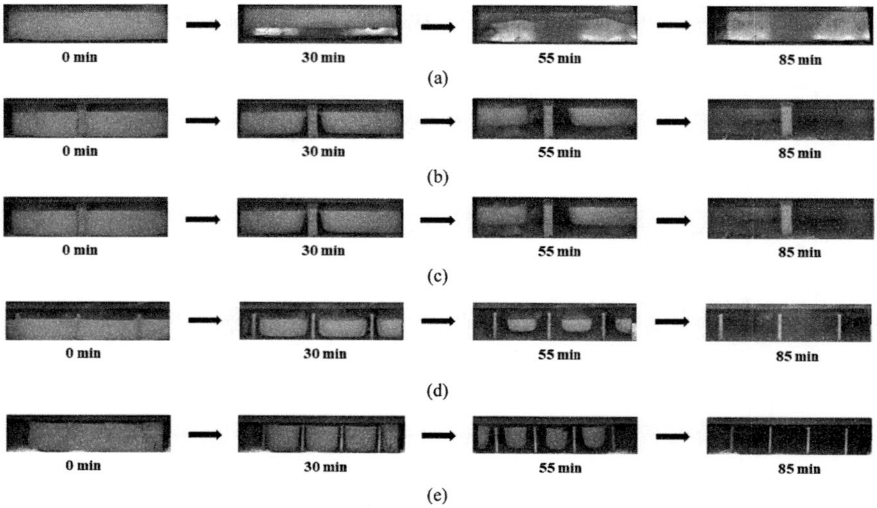

Fig. 7 Propagation of melt front for **a** HSNF, **b** HSOF, **c** HSTF, **d** HSTHF and **e** HSFF

In HSNF, the melt front runs parallel to the bottom plate, whereas in finned heat sinks, the melt front begins near the bottom and runs along the fins before gradually moving towards the centre. This difference between the way melting propagates can be attributed to the number of fins integrated. As more and more fins are added, more area is available for heat transfer and hence melting along with the bottom also occurs near the fins. For the starting few minutes it is observed that melting is slow which can be accounted for the fact that conduction is dominant during the initial phase but as time progresses, convection current sets up and buoyancy forces plays a greater role in the melt propagation of PCM.

4.4 *Enhancement Ratio of Various Heat Sinks for Different Set-Point Temperatures*

In the present experimental study two different set-point temperature, 65°C and 70°C, which are critical for optimum functioning of electronic devices are considered and for different heat flux values of 2.0 , 2.5 and 3.0 kW/m^2. Figure 8 shows that when the magnitude of heat flow increases, the time required to reach a critical SPT decreases. Also, the time decreases for subsequent variation of heat sinks. In case of set-point temperature (SPT) of 65 °C, the decrease in time for HSNF, HSOF, HSTF, HSTHF, HSFF are 420s, 480s, 330s, 690s, 660s respectively. In case of SPT of 70 °C, the decrease in time for HSNF, HSOF, HSTF, HSTHF, HSFF are 590s, 670s, 500s, 1020s, 970s, respectively, from a heat flux of 2.0–3.0 kW/m^2.

The enhancement ratio, which is the ratio of time to reach a critical SPT by PCM based heat sink with TCE to PCM based heat sink without TCE, is used to describe the decrease in time for various heat sink layouts in this study. The enhancement ratio for a 65 and 70 °C critical SPT at different heat flux are represented in Fig. 9. It can

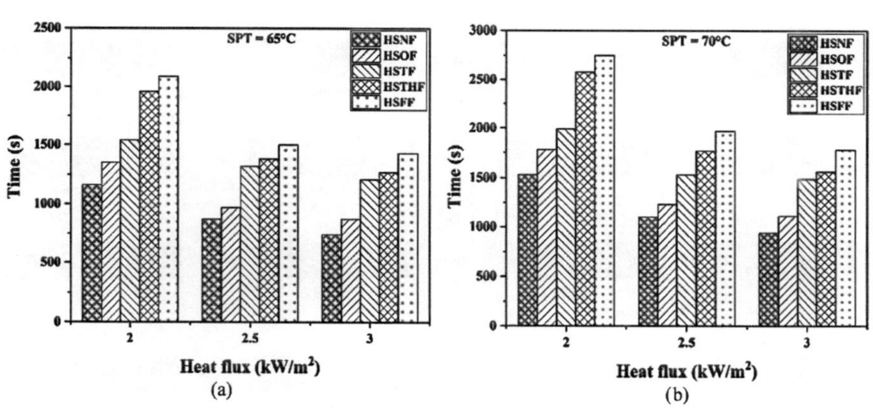

Fig. 8 Time needed to reach critical SPT by various layouts of heat sinks for different values of heat flux **a** 65 °C and **b** 70 °C

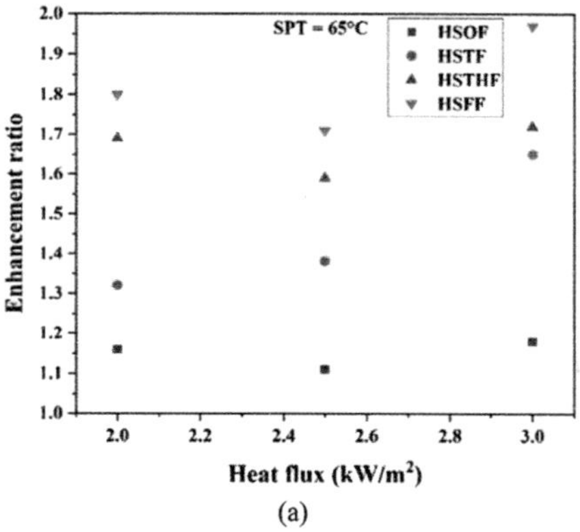

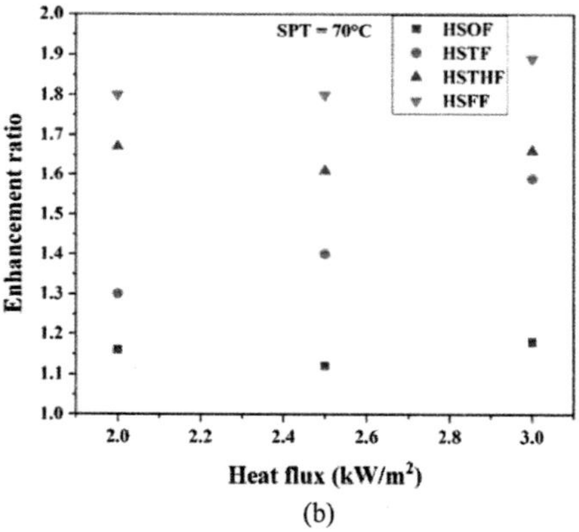

Fig. 9 Enhancement ratio for various layouts of heat sink at different values of heat flux **a** 65 °C and **b** 70 °C

be noted from the graphs that for both critical SPTs, HSFF gives the highest value of enhancement ratio. In HSFF, a maximum enhancement ratio of 1.97 is obtained for a critical SPT of 65 °C and a maximum enhancement ratio of 1.89 is obtained for a critical SPT of 70 °C.

5 Conclusion

The thermal performance of PCM (PEG-6000) based heat sinks was investigated experimentally for five different configurations (HSNF, HSOF, HSTF, HSTHF and HSFF) is performed in this work. The study of the integration of fins into heat sinks is carried out at three different values of heat flux 2.0, 2.5 and 3.0 kW/m^2 and the conclusions from this study are as follows:

(i) For the same time interval, the PCM melts in the order of HSFF-HSTHF-HSTF-HSOF-HSNF, suggesting more heat transfer when there is an addition of plate fins in the heat sinks.
(ii) A heat sink with plate fins lowers the temperature at the bottom for a longer period of time than a heat sink without fins.
(iii) For a SPT of 65 °C, the maximum enhancement ratio obtained is 1.97 for a heat flux of 3.0 kW/m^2 for HSFF.
(iv) For a SPT of 65 °C, the maximum enhancement ratio obtained is 1.89 for a heat flux of 3.0 kW/m^2 for HSFF.

References

1. Kim TY, Hyun BS, Lee JJ, Rhee J (2013) Numerical study of the spacecraft thermal control hardware combining solid-liquid phase change material and a heat pipe. Aerosp Sci Technol 27(1):10–16. https://doi.org/10.1016/J.AST.2012.05.007
2. Sarier N, Onder E (2012) Organic phase change materials and their textile applications: an overview. Thermochim Acta 540:7–60. https://doi.org/10.1016/J.TCA.2012.04.013
3. Demirbağ S, Aksoy SA (2016) Encapsulation of phase change materials by complex coacervation to improve thermal performances and flame retardant properties of the cotton fabrics. Fibers Polymers 17(3):408–417. https://doi.org/10.1007/S12221-016-5113-Z
4. Lv Y, Zou Y, Yang L (2011) Feasibility study for thermal protection by microencapsulated phase change micro/nanoparticles during cryosurgery. Chem Eng Sci 66(17):3941–3953. https://doi.org/10.1016/J.CES.2011.05.031
5. Sebzali MJ, Rubini PA (2007) The impact of using chilled water storage systems on the performance of air cooled chillers in Kuwait. Energy Buildings 39(8):975–984. https://doi.org/10.1016/J.ENBUILD.2006.11.004
6. Kandasamy R, Wang XQ, Mujumdar AS (2007) Application of phase change materials in thermal management of electronics. Appl Therm Eng 27(17–18):2822–2832. https://doi.org/10.1016/J.APPLTHERMALENG.2006.12.013
7. Fang X, Zhang Z, Chen Z (2008) Study on preparation of montmorillonite-based composite phase change materials and their applications in thermal storage building materials. Energy Convers Manage 49(4):718–723. https://doi.org/10.1016/J.ENCONMAN.2007.07.031
8. Li M, Kao H, Wu Z, Tan J (2011) Study on preparation and thermal property of binary fatty acid and the binary fatty acids/diatomite composite phase change materials. Appl Energy 88(5):1606–1612. https://doi.org/10.1016/J.APENERGY.2010.11.001
9. Pavlik Z, A Trnik, Pavlikova M, Keppert M, Černy R (2022) Lime-pozzolan plasters with enhanced thermal capacity. https://publications.waset.org/17270/lime-pozzolan-plasters-with-enhanced-thermal-capacity. Accessed May 05 2022
10. Abhat A (1983) Low temperature latent heat thermal energy storage: heat storage materials. Sol Energy 30(4):313–332. https://doi.org/10.1016/0038-092X(83)90186-X
11. Dincer I (2002) On thermal energy storage systems and applications in buildings. Energy Buildings 34(4):377–388. https://doi.org/10.1016/S0378-7788(01)00126-8
12. Kaygusuz K (1999) The viability of thermal energy storage. Energy Sources 21(8):745–755. https://doi.org/10.1080/00908319950014489
13. Rathore PKS, Shukla SK (2019) Potential of macroencapsulated pcm for thermal energy storage in buildings: a comprehensive review. Constr Build Mater 225:723–744. https://doi.org/10.1016/J.CONBUILDMAT.2019.07.221
14. Raj CR, Suresh S, Bhavsar RR, Singh VK, Govind KA (2020) Influence of fin configurations in the heat transfer effectiveness of solid PCM based thermal control module for satellite avionics: numerical simulations. J Energy Stor 29:101332. https://doi.org/10.1016/J.EST.2020.101332
15. Kothari R, Sahu SK, Kundalwal SI, Mahalkar P (2021) Thermal performance of phase change material–based heat sink for passive cooling of electronic components: an experimental study. Int J Energy Res 45(4):5939–5963. https://doi.org/10.1002/er.6215
16. Kumar A, Kothari R, Sahu SK, Kundalwal SI (Nov. 2021) A comparative study and optimization of phase change material based heat sinks for thermal management of electronic components. J Energy Stor 43. https://doi.org/10.1016/j.est.2021.103224
17. Kumar A, Kothari R, Sahu SK, Kundalwal SI (Jun. 2021) Thermal performance of heat sink using nano-enhanced phase change material (NePCM) for cooling of electronic components. Microelectr Reliab 121. https://doi.org/10.1016/j.microrel.2021.114144
18. Kothari R, Sahu SK, Kundalwal SI (May 2021) Investigation on thermal characteristics of nano enhanced phase change material based finned and unfinned heat sinks for thermal management system. Chem Eng Process - Process Inten 162. https://doi.org/10.1016/j.cep.2021.108328

19. Hosseinirad E, Khoshvaght-Aliabadi M (July, 2021) Proximity effects of straight and wavy fins and their interruptions on performance of heat sinks utilized in battery thermal management. Int J Heat Mass Transfer 173. https://doi.org/10.1016/j.ijheatmasstransfer.2021.121259
20. Gutierrez A, Ushak S, Galleguillos H, Fernandez A, Cabeza LF, Grágeda M (Sep.2015) Use of polyethylene glycol for the improvement of the cycling stability of bischofite as thermal energy storage material. Appl Energy 154:616–621. https://doi.org/10.1016/j.apenergy.2015.05.040

Demystifying the Potential of BIPV in Achieving India's Intended NDC

Aditya Koya, Ajay Shankar, and K. Vijayakumar

1 Introduction

Rapid urbanization and infrastructural development have resulted in rising energy consumption in emerging countries such as India. Buildings are the world's largest energy consumers and are one of the primary sources of carbon emissions as they depend upon fossil fuels for generating electricity [1]. Energy demand in the building industry is increasing at an annual pace of 8% and contributes to 35–40% of total energy consumption [2]. To overcome the environmental problems caused by fossil fuels there exists a need to use renewable energy sources to meet the energy requirements of buildings in smart cities [3]. As a result, both the preservation and protection of our resources for future generations, as well as environmental values, are not jeopardized. Semi-transparent PV kinds are getting prominence for deployment in building walls, roofs, and windows due to their versatile properties, such as supplementing structural materials, offering great insulation, enabling daylight, and generating power [4]. In this viewpoint, Building-integrated photovoltaics (BIPV) can be deployed where the building facade or roof of the building is made with Photovoltaic panels. It is an emerging technology that serves the space constraints buildings have in urban environments.

NDC(s) stand for 'Nationally Determined Contributions', which indicate each country's commitment to cut greenhouse gas emissions. Every five years, countries are expected to review and improve their NDCs, as well as propose more ambitious greenhouse gas reduction actions. India has said that at least 40% of its installed capacity for power generation will be derived from non-fossil fuel-based energy sources and is trying to achieve a projected capacity of 525 GW by 2030 as shown

A. Koya (✉) · A. Shankar · K. Vijayakumar
IIITD&M Kancheepuram, Chennai, India
e-mail: eds20m009@iiitdm.ac.in

K. Vijayakumar
e-mail: vijayakumar@iiitdm.ac.in

S. Doolla et al. (eds.), *Advances in Clean Energy and Sustainability*,
Green Energy and Technology, https://doi.org/10.1007/978-981-99-2279-6_43

in Fig. 2. The principal activity is to execute the energy transition from fossil fuel to renewable energy. Figure 1 manifests that the current status is 106 GW, and by 2022, the target output capacity will be 175 GW. Between April 2000 and March 2021, India's renewable energy sector attracted 10.02 billion dollars in foreign direct investment. India ranks third in the world in terms of renewable energy investments and intentions in 2020, according to the analytics firm British Business Energy. Also, the PM of India has announced some key points at COP 26 summit held in Glasgow. Some of them are:

- To attain a non-fossil energy capacity of 500 GW by 2030 shown in Fig. 2.
- To provide 40% energy requirements through renewable sources.
- Reduce approximately 1 billion carbon emissions by 2030.
- Reduce carbon intensity.
- India aims to achieve the target of Net-Zero by around 2070.

One of the PV's major challenges is an area, with solar panels deployed in broad open spaces to extract the maximum power. Transmission losses arise as a consequence of transmission lines used for delivering power to the load. PV panels can be employed to generate required energy at the load center but often buildings in urban areas have limited rooftop area and other constraints described in Fig. 3 which in turn limits the energy generated from using PV panels alone. These drawbacks are

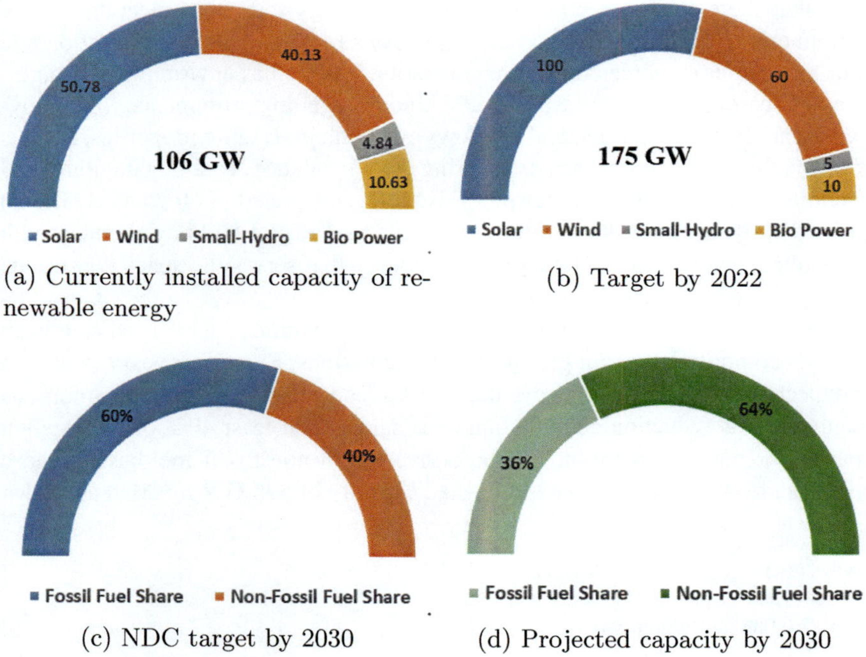

(a) Currently installed capacity of renewable energy

(b) Target by 2022

(c) NDC target by 2030

(d) Projected capacity by 2030

Fig. 1 India's NDC tracking

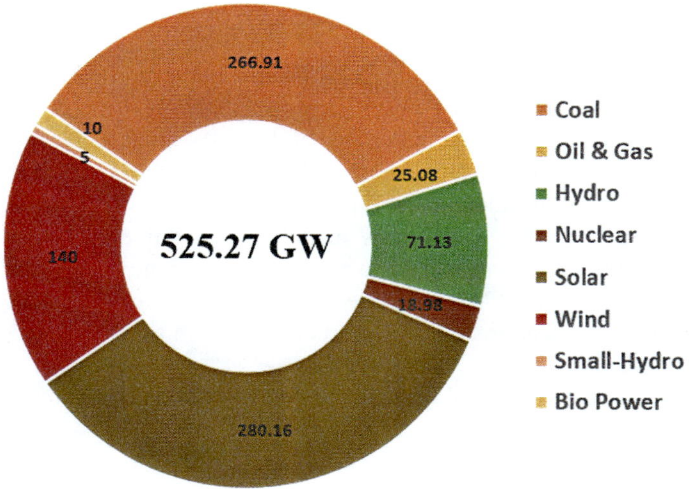

Fig. 2 Projected capacity of total energy generated by 2030

(a) Helipad (b) Roof garden (c) HVAC system

Fig. 3 Challenges of rooftop PV

mitigated by the usage of BIPV. The BIPV system is distinguished by the proximity of energy-generating facilities to the location of energy consumption. Ghosh [5] conducted an economic analysis of BIPV systems as the building envelope in European structures. BIPV is one of the most important characteristics of zero-energy buildings that also improves the built environment's aesthetic appeal. However, it now only accounts for around 1% of the worldwide photovoltaic industry. This necessitates a greater focus on this topic [6]. BIPV modules can switch an energy-hungry building into an on-site sustainable energy generator, minimizing transportation losses and, as a result, the cost of power.

The amount of energy produced by BIPV modules is largely measured by the amount of solar irradiation received from the sun. The energy generated by BIPV modules is affected by the facing of the BIPV module on the building site as well as the local topographical variables [7].

The primary findings of this paper are encapsulated as follows:

1. BIPV module is proposed as an envelope material in the North, East, South, and West directions of the building facade.
2. The solar irradiation falling at the surface of the BIPV module erected vertically at different azimuth angles (North, East, South, and West directions) are estimated using the solar transposition model [3].
3. The energy performance of the BIPV module mounted in each direction of the building facade is presented and compared with the optimally placed PV module.
4. Economic analysis of BIPV mounted vertically at different azimuth angles (North, East, South, and West directions) is presented and compared with the optimally placed PV module.
5. Analysis of the environmental sustainability of BIPV modules mounted vertically as an envelope of the building facade in each direction is presented.

2 System Modeling

The rooftop PV is considered to be optimally placed, whereas the BIPV modules are mounted vertically across the building facade in all directions of the building. The yearly solar irradiation at the geographical location of New Delhi, India was acquired from the European Commission's Photovoltaic Geographical Information System (PVGIS), and the output power was produced from optimally placed rooftop PV and BIPV modules mounted in each direction (North, East, South, and West) is calculated by Eqs. 1 and 2 respectively [3, 8].

The power performance of PV and BIPV modules is formulated as follows:

$$P_{PV,t} = \eta_{PV}\, A_{PV}\, \frac{H_{PV,t}}{H_{stc}} \left[1 + \alpha_p (T_c - T_{stc})\right] \eta_{dr_{PV}} \tag{1}$$

$$P_{BIPV_N,t} = \eta_{BIPV}\, A_{BIPV}\, \frac{H_{BIPV_N,t}}{H_{stc}} \left[1 + \alpha_p (T_c - T_{stc})\right] \eta_{dr_{BIPV}} \tag{2a}$$

$$P_{BIPV_E,t} = \eta_{BIPV}\, A_{BIPV}\, \frac{H_{BIPV_E,t}}{H_{stc}} \left[1 + \alpha_p (T_c - T_{stc})\right] \eta_{dr_{BIPV}} \tag{2b}$$

$$P_{BIPV_S,t} = \eta_{BIPV}\, A_{BIPV}\, \frac{H_{BIPV_S,t}}{H_{stc}} \left[1 + \alpha_p (T_c - T_{stc})\right] \eta_{dr_{BIPV}} \tag{2c}$$

$$P_{BIPV_W,t} = \eta_{BIPV}\, A_{BIPV}\, \frac{H_{BIPV_W,t}}{H_{stc}} \left[1 + \alpha_p (T_c - T_{stc})\right] \eta_{dr_{BIPV}} \tag{2d}$$

where

- η_{PV}, η_{BIPV}: Efficiency of the PV and BIPV module, %
- A_{PV}, A_{BIPV}: Intrinsic area of the PV and BIPV module, m^2
- $H_{PV,t}$: Solar irradiation falling at the surface of optimally placed rooftop PV, kW/m^2
- $H_{BIPV_{N},t}$, $H_{BIPV_{E},t}$, $H_{BIPV_{S},t}$, and $H_{BIPV_{W},t}$: Solar irradiation falling at the surface of BIPV module mounted vertically at north, east, south, and west facade of the building, kW/m^2
- H_{stc}: Solar irradiation under standard test conditions (STC)
- α_p: Temperature coefficient of power, %/°C
- T_c: Historical data of temperature, °C
- T_{stc}: Temperature of the PV module under STC, °C
- $\eta_{dr_{PV}}$, $\eta_{dr_{BIPV}}$: De-rating factor of PV and BIPV module, %.

3 Problem Formulation

The cost of the 100 m^2 BIPV module and the cost of the power converter are included in the proposed PV-BIPV system's economic analysis. Equation 3 is used to compute the present worth (PW) of each factor's cost, which encompasses investment, operation, and maintenance. The total of all the components of PW is the life cycle cost (LCC). Annualized LCC (ALCC) from Eq. 5 utilizing the cost of cumulative present worth PW yields the leveled cost of energy (LCOE). Energy cost calculations are as below:

$$PW = \left(\frac{1+r_f}{1+r_i}\right)\left(1-\left(\frac{1+r_f}{1+r_i}\right)^n\right) \tag{3}$$

$$R = C_0\left(\frac{1+r_f}{1+r_i}\right)^n, \quad M = C_0\, PW \tag{4}$$

$$ALCC = \frac{K+R+M}{PW}, \quad LCOE = \frac{ALCC}{E_T} \tag{5}$$

where

- r_i, r_f: Rate of interest and inflation, %
- C_0, n: Present worth (PW), INR and life of the project, Yr
- K: Initial investment on project, INR
- R, M: PW of replacement and O&M cost, INR
- ALCC: Annualized life cycle cost, INR
- LCOE: Levelized cost of energy, INR/kWh
- E_T: Annual energy generation.

4 Results and Discussions

The output power provided by the BIPV module positioned vertically in all four directions (various azimuth angles for the North, East, West, and South facades) is computed and displayed in Fig. 5. Similarly, the intrinsic area of an ideally located PV module is taken into account, and the PV module's output power is depicted in Fig. 4. Because of their ideal positioning, rooftop PV modules generate more electricity than BIPV modules. Furthermore, as shown in Fig. 5, the BIPV module put on the North façade produces less electricity than the BIPV modules mounted on the building's other sides (Fig. 6).

Energy profiles for PV and BIPV modules have been plotted for all months of the year ranging from January to December. It has been noticed that the PV module generates a relatively higher amount of energy compared with the BIPV module. Energy generated by BIPV modules facing North, East, West, and South directions have been simulated and it's evident out of all directions energy generated by the BIPV module facing North direction is least and energy generated by BIPV module facing South is maximum.

In India, CO_2 emissions per unit of energy are projected to be 0.91–0.95 kg/kWh, SO_2 emissions are 6.94–7.20 g/kWh, and NO emissions are 4.22–4.38 g/kWh [9]. The environmental sustainability study of the PV and BIPV modules presented in Fig. 7 shows that the BIPV module, as an envelope of the building facade, serves to eliminate a significant quantity of GHG from the environment without taking up any functional space. Figure 8 shows the LCOE of a 100 m^2 PV and BIPV module. The BIPV module positioned as an envelope of the building facade delivers energy at a lower rate than an appropriately placed PV module. Furthermore, the BIPV module positioned on the North facade produces the least yearly energy, resulting in energy supply at a higher rate. Similarly, because BIPV on the southern façade creates more energy than BIPV on the other sides of the building, it supplies energy at a reduced amount than other facades (Table 1).

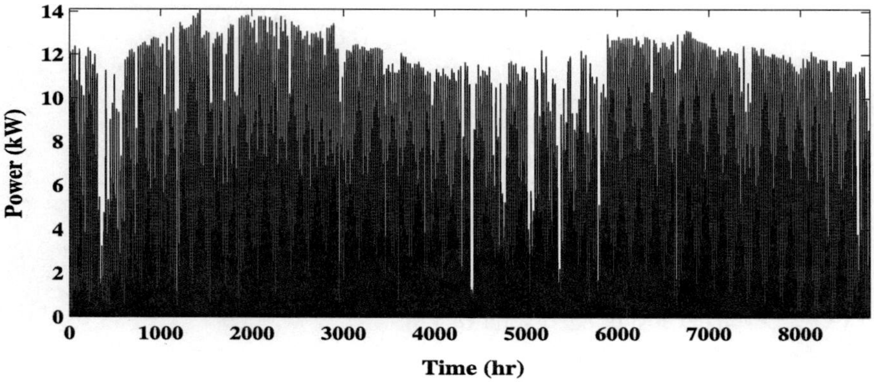

Fig. 4 Power generated by optimally placed PV module

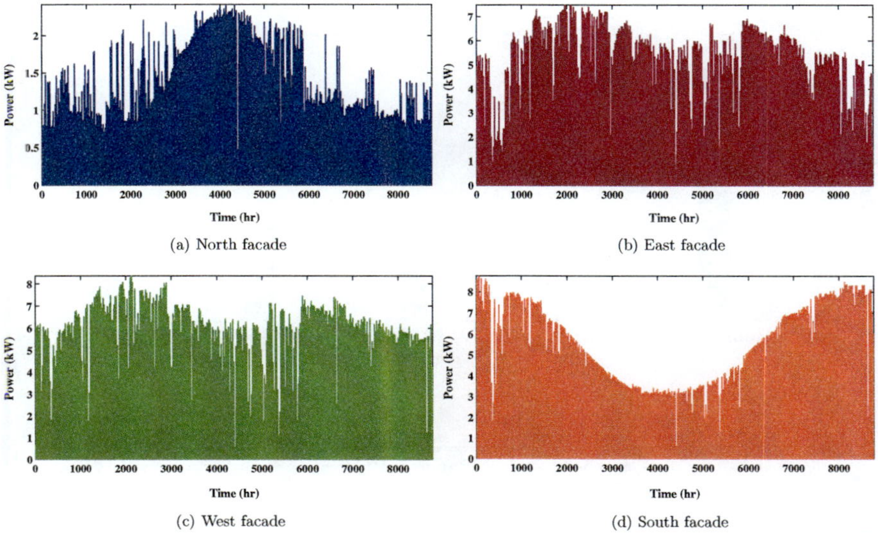

Fig. 5 Output power from 100 m^2 of BIPV module mounted as an envelope of the building facade

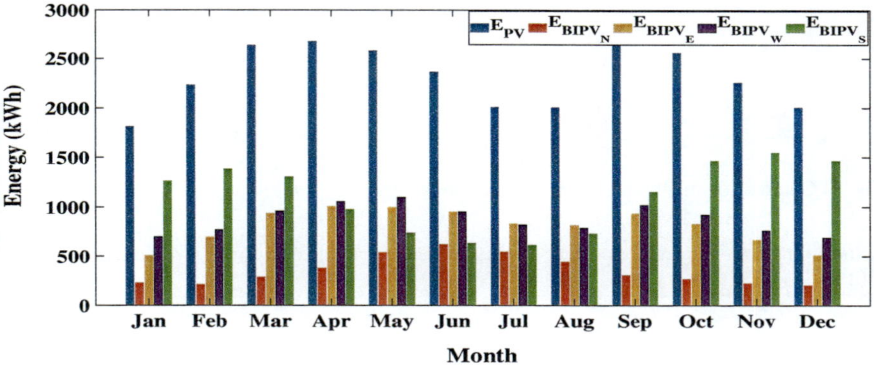

Fig. 6 Energy profile for PV and BIPV modules

Table 2 shows a comparison of the performance of an appropriately positioned PV and a vertically installed BIPV module. PV modules with a 100 m^2 intrinsic area achieve a peak rating of 17.32 kW, whereas BIPV modules with a comparable intrinsic area achieve 12.7 kW. Table 2 further indicates that 100 m^2 of appropriately positioned PV modules produce more energy yearly than BIPV modules.

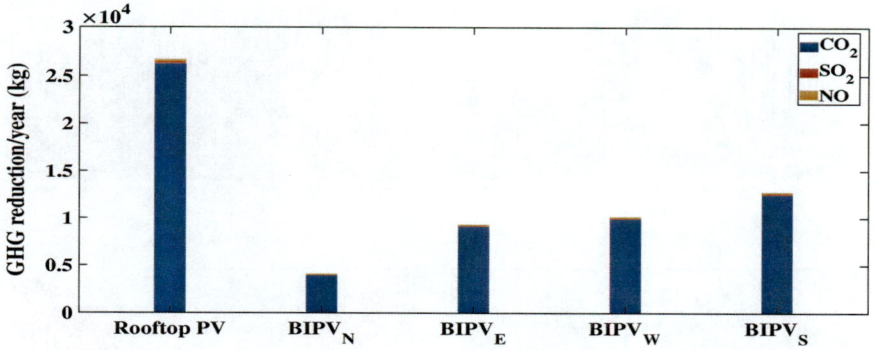

Fig. 7 GHG reduction

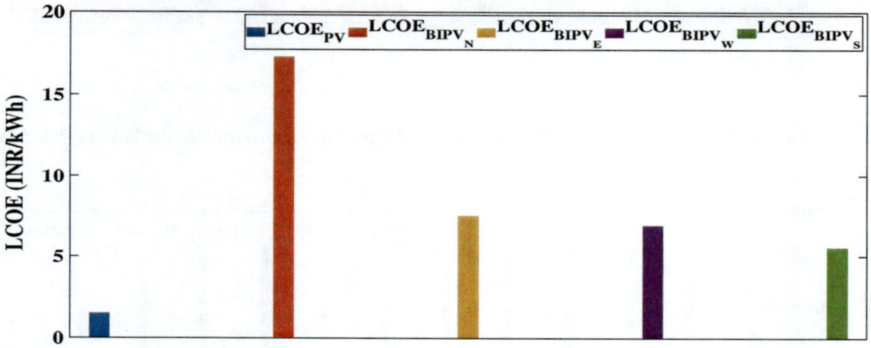

Fig. 8 Comparison of PV and BIPV modules on LCOE

Table 1 Input parameters for power performance and economic analysis of PV and BIPV module [10]

S. No.	Parameters	Value
1	Capital cost of rooftop PV and BIPV module, INR/kW	45,000, 135,000
2	Capital cost of power converter, INR/kW	8400
3	Maintenance cost on PV and BIPV module, INR/Yr	2%
4	Inflation and real interest rate, p.a.	6.5%, 6%
5	Lifespan of rooftop PV, BIPV and converter, Yr	25, 25, 10
6	Efficiency of PV and BIPV module, %	17.32%, 12.7%
7	α_p, H_{stc}	– 0.4%/°C, 1 kW/m^2
8	T_{stc}	25 °C
9	$\eta_{dr_{PV}}$, $\eta_{dr_{BIPV}}$	80%, 85%

Table 2 Comparative analysis on the performance of PV and BIPV module

Solar module		Peak rating (kW)	Annual energy generated (kWh)	ALCC (INR)	LCOE (INR/kWh)
Optimally placed rooftop PV		17.32	27,740	40,688.31	1.47
BIPV placed vertically as an envelope of the building facade	North facade	12.7	4224	72,964.92	17.27
	East facade	12.7	9663	72,964.92	7.55
	West facade	12.7	10,530	72,964.92	6.93
	South facade	12.7	13,227	72,964.92	5.51

5 Conclusion

The techno-economic study of the BIPV module suggested as a building envelope installed on the North, East, South, and West facades are presented in this research. For the energy performance analysis, a 100 m^2 BIPV module is used, which is compared to a similar area of the ideally positioned PV module. The BIPV on the building's south façade is the most energy-efficient, followed by the West, East, and North sides. Although a well-placed PV module delivers more energy than BIPV modules put on the building's façade, PV installation necessitates horizontal space, which is restricted in the urban setting. As a result, BIPV may be used as a secondary source of sustainable energy generation in metropolitan areas. At INR 5.52, 6.94, and 7.56 per kWh, the LCOE from the South, West, and East facades is shown to be cost-effective. BIPV can eliminate a significant amount of GHG emissions from the environment as a building facade envelope, according to an environmental sustainability analysis. India's renewable energy capacity, excluding big hydro, has surpassed 100 GW, and the country is aiming for a 500 GW target by 2030. The infrastructure of high-rise buildings is rising in tandem with urbanization. As a result, BIPV modules have the potential to be a major player in India's ambitious goal of 500 GW of renewable energy installation and carbon neutrality.

References

1. Capuano L (2018) International energy outlook. US Energy Information Administration, Washington, DC, pp 21–25
2. Bureau of Energy Efficiency, Government of India (2021) Energy benchmarks for commercial buildings
3. Shankar A, Vijayakumar K, Chitti Babu B (2021) Techno-economic and energy assessment of building integrated photovoltaic module as an envelope of the building. Int Trans Electr Energy Syst 31(11):e13105
4. Gholami H, Røstvik HN (2020) Economic analysis of BIPV systems as a building envelope material for building skins in Europe. Energy 204:117931

5. Ghosh A (2020) Potential of building integrated and attached/applied photovoltaic (BIPV/BAPV) for adaptive less energy-hungry building's skin: a comprehensive review. J Clean Prod 276:123343
6. Osseweijer FJ, Van Den Hurk LB, Teunissen EJ, van Sark WG (2018) A comparative review of building integrated photovoltaics ecosystems in selected European countries. Renew Sustain Energy Rev 90:1027–1040
7. Shankar A, Vijayakumar K, Babu BC (2021) Energy saving potential through artificial lighting system in PV integrated smart buildings. J Build Eng 43:103080
8. Photovoltaic geographical information system (PVGIS) (2021) https://ec.europa.eu/jrc/en/pvgis
9. Mittal ML, Sharma C, Singh R (2012) Estimates of emissions from coal fired thermal power plants in India. In: 2012 international emission inventory conference, pp 13–16
10. Shankar A, Vijayakumar K, Babu BC, Kaur R (2022) Energy trilemma index-based multiobjective optimal sizing of PV-battery system for a building in tropical savanna climate. IEEE Syst J 1–9. https://doi.org/10.1109/JSYST.2022.3167166

Interface Design and Performance Analysis of Proton Exchange Membrane Fuel Cell using Python

P. V. Kapoor, U. B. Mujumdar, Aniket Lanjewar, and Siddhi Chauhan

1 Introduction

Research and Development of fuel cell (FC) systems are being carried out for various applications such as hybrid power, fuel cell cars, mobile power generation, aerospace, etc. Proton Exchange Membrane fuel cell (PEMFC) proves to be among the reliable systems when it comes to automobile, stationary and portable applications as substitute of internal combustion engine. Due to its higher power density and lower operating temperature, it is highly demandable in automobile applications [1]. The working of typical PEMFC is illustrated in Fig. 1. An FC generates power soon the reaction's chemical energy gets changed to electrical energy through the passage of electrons through the load. Mathematical model with accurate empirical constants is initialized and thermodynamic and electrochemical equations of fuel cell are used for calculating the losses and the equilibrium voltage under varied temperature, and constant inlet pressure of both the reactant gases. Under normal condition, a PEMFC typically produces 0.5–0.9 V. In order to form a stack, multiple cells are connected in series as relatively high power is needed for various applications [2]. PEM fuel cell can be operated for temperatures below 80 °C and its efficiency is in the range of 40–60%.

The cell reaction is as follows in (1) and (2):

$$\text{Anode(oxidation)} : 2H_2 \rightarrow 4H^+ + 4e^- \tag{1}$$

$$\text{Cathode (reduction)} : O_2 + 4H^+ + 4e^- \rightarrow 2H_2O \tag{2}$$

When hydrogen from anode and oxygen from cathode is passed, the electrolyte between them enables the exchange of electrical charges from anode to cathode.

P. V. Kapoor · U. B. Mujumdar · A. Lanjewar (✉) · S. Chauhan
Shri Ramdeobaba College of Engineering and Management, Nagpur, India
e-mail: lanjewarad_1@rknec.edu

S. Doolla et al. (eds.), *Advances in Clean Energy and Sustainability*,
Green Energy and Technology, https://doi.org/10.1007/978-981-99-2279-6_44

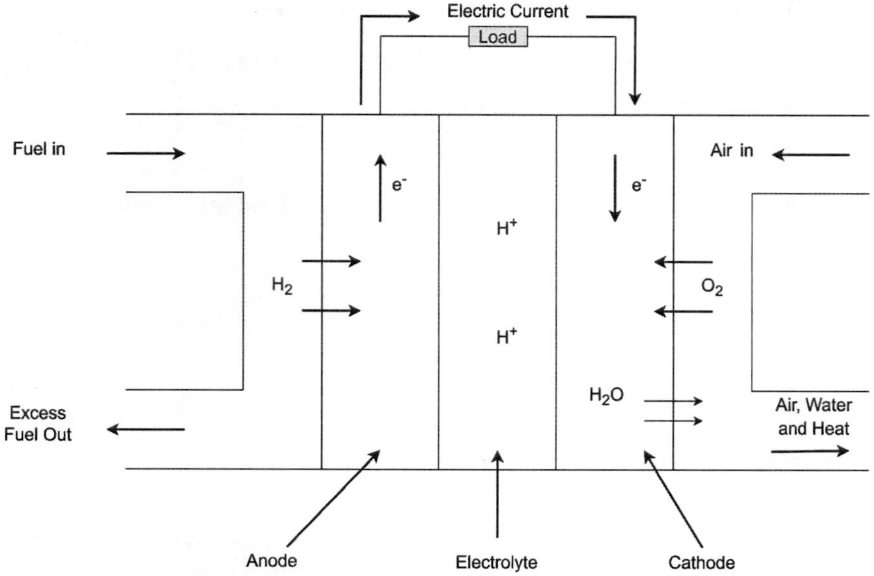

Fig. 1 Working diagram of fuel cell

The fuel reacts to form protons and electrons at the catalytic layer of anode after getting expanded through the electrode [3] which is represented by (1). The transfer of protons occurs at the catalytic layer of the cathode from the electrolyte. At the cathode side, O_2 flows through the plate's channels and gets expanded through the electrode at the cathode's catalytic layer. O_2 expends with H^+(protons) and electrons and the product, liquid water, is obtained with residual heat.

There are different platforms where fuel cell is studied earlier like MATLAB, Pspice software [7]. Python is more appropriate language when it comes to its compatibility with new technologies such as machine learning, which can be used as a tool to predict various parameters like operating voltage, temperature, etc. for different fuel cell stacks. This paper focuses on development of graphical user interface using Qt designer for backend programming which can plot various characteristics, for the specified fuel cell stack, by the user. Taking all the basics aspects into consideration, this paper presents the modelling of fuel cell using Python including the PEMFC's fundamentals. Thus, the proposed GUI helps to plot and analyze steady state as well as dynamic characteristics of various commercial fuel cell stacks like Avista SR-12 and BCS stack at varying operating conditions.

2 Models of Fuel Cell

The PEMFC's actual voltage is lower than the theoretical voltage for several irreversible loss mechanisms [9], generally called over potential or losses. They are namely, Activation loss, concentration loss, and ohmic loss.

2.1 Steady State Model

The total fuel cell voltage output (V) is given by Eq. (3) where, *ENernst* is the open circuit voltage referred as Nernst potential, *Vact* represents activation loss, *Vohm* is ohmic loss and *Vconc* depicts concentration loss [2]. The polarization curve representing all three losses is shown in Fig. 2.

$$V = E_{Nernst} - V_{act} - V_{ohm} - V_{conc} \tag{3}$$

The open circuit voltage can be indicated as Nernst voltage given as [1, 6]:

$$\begin{aligned} E_{Nernst} &= 1.229 - \left(8.5 * 10^{-4}\right) * (T - 298.15) \\ &\quad + \left(4.308 * 10^{-5}\right) * T * [\ln(PH2) + 0.5 \ln(PO2)] \end{aligned} \tag{4}$$

where, T represents cell operating temperature in (K), PH2 and PO2 are the partial pressures of involved reactant gases in (atm). Activation over potential represents the losses caused by kinetics at the surface of the electrodes, which are dominant in low current densities. It is the voltage over potential required to overcome the activation energy. It can be expressed as [5, 6, 8]:

$$V_{act} = \zeta_1 + \zeta_2 T + \zeta_3 T[\ln(CO_2)] + \zeta_4 T[\ln(i)] \tag{5}$$

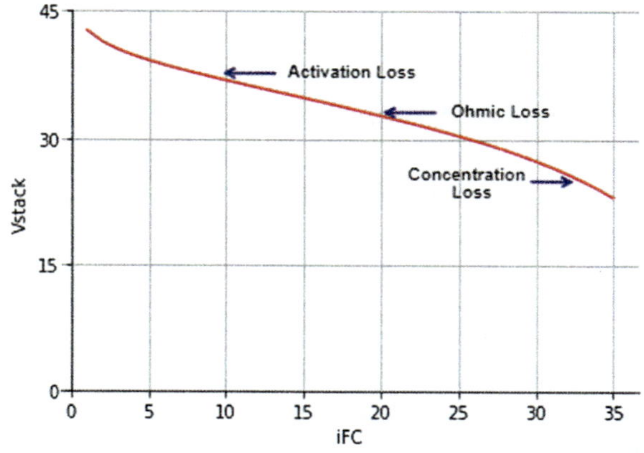

Fig. 2 Polarization curve

where ζ_s are the empirical parametric coefficients, CO_2 is O_2 concentration in (mol/cm^3), i is the fuel cell current in (A) and ζ_2 can be formulated as [6]:

$$\zeta_2 = 0.00286 + 0.0002 * \ln(A) + \left(4.3 * 10^{-5}\right) \ln(CH_2)] \tag{6}$$

where, A, the active cell area in (cm^2) and CH_2 is the reactant fuel's concentration (mol/cm^3). Ohmic loss (V_{ohm}) is due to the natural resistance to the charge flow in conductors caused by resistance to the flow of ions ($Rions$) and electrons along the electrolyte and cell hardware respectively. The voltage drop is essentially proportional to current density hence it is termed as "Ohmic loss". Some important formulae used in calculating this loss are [6, 8]:

$$V_{ohm} = i(R_{ions}) \tag{7}$$

$$R_{ions} = \frac{R_m L}{A} \tag{8}$$

$$R_m = \frac{181.6[1 + 0.03(\frac{i}{A}) + 0.062(\frac{T}{303})^2(\frac{i}{A})^{2.5}]}{[\lambda - 0.634 - 3(\frac{i}{A})]e^{[4.18(\frac{T-303}{T})]}} \tag{9}$$

R_m is the membrane's resistance of the FC and λ, a semi-empirical variable representing the membrane's water content. Concentration losses ($Vconc$) are directly related to the concentration of reactants and products of the reaction. On current consumption from fuel cell, hydrogen and oxygen gets consumed causing reduction in their concentration and pressures. At supposedly higher currents the insufficiency of the reactant flow rates and fuel to match the requisite for reaction rate, causes concentration drop in the cell voltage.

$$Vconc = \frac{RT}{NF} \ln\left(1 - \frac{J_{FC}}{J_{\max}}\right) \tag{10}$$

2.2 Dynamic Model

Whenever two differently charged articles are adjacent, charge build up takes place on their surfaces or load transfers from one to other. The charge layer on the interface electrode/electrolyte functions as a storage of electrical charges and energy and, thus, acts as an electrical capacitor [3]. This effect is of great importance when considering the fuel cell dynamics and is called as charge double layer effect. The results of the Python based dynamic model are presented in Figs. 11 and 12 for BCS and Avista stacks respectively. Thus, PEMFC model has transient activation loss V_{act} which is the solution of the differential equation given below [2]:

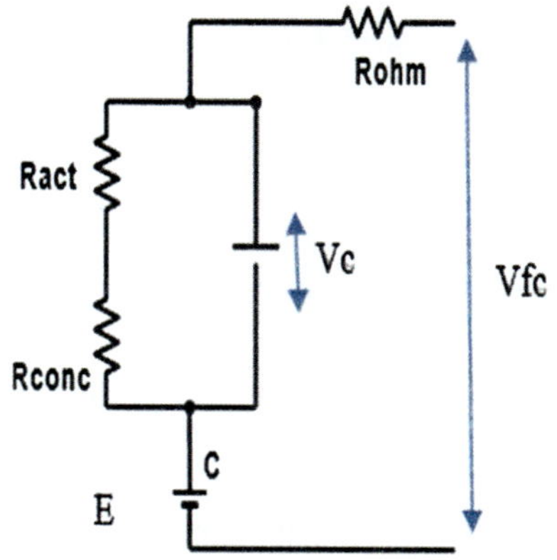

Fig. 3 Equivalent electrical circuit

$$\frac{dVact}{dt} = \frac{i}{C} - \frac{Vact}{RaC} \tag{11}$$

where C, depicts charge double layer capacitance, and R_a, equivalent resistance:

$$R_a = \frac{Vact_0 + Vconc}{i} \tag{12}$$

$V_{act0} = -\eta_{act}$ depicts activation loss at steady state which is applied as the initial condition for differential equation, which can be expressed through electrical equivalent circuit in Fig. 3.

3 Implementation Using Python

3.1 Python as a Modelling Tool

Python is a one of the fastest-growing language in the world. Being an open-source language, number of Python users are increasing exponentially and exploring its utility in various fields thus making it better programming language for practical and research purposes. In this paper the different commercial fuel cell stacks are successfully modelled in Python and a GUI is developed for plotting their various static and dynamic characteristics. Here, NumPy library is used for array creation so that Matplotlib library can input those arrays for plotting electrical characteristics of fuel cell. Finally, PyQt5 library is imported so that different widgets created in the main window of the GUI can be linked with the logic of the code edited.

3.2 Algorithm

This program uses *for* and *while* loop for iterating values of current specifically to inspect the behavior of the fuel cell at different values of temperature. It also plots the dynamic characteristics of fuel cell by importing *odeint* library. The developed graphical user interface is shown in Fig. 4.

Algorithm is as follows:

I. Set a directory of PyQt5 file in command prompt
II. Use "pyuic5 -x GUI.ui -o GUI.py" in command prompt
III. Import Python libraries math, NumPy, array, matplotlib.pyplot, odeint,
IV. Select the type of model using radio button
V. If Avista reference model's radio button is checked, run Avista_SR_12 VI, PI, Efficiency, Dynamic characteristics function blocks
VI. Create arrays to store values of voltage, power and efficiency
VII. Enter the values of temperature t1, t2, t3 in Avista model's function block
VIII. While nt > 0, if (nt = = 1) elseif (nt = = 2) elseif (nt = = 3)
IX. For current in range 0 < i < 42, calculate voltage, power, efficiency
X. Append the arrays for different values of nt (i.e., number of different temperatures) and plot steady state VI, PI characteristics and efficiency curves
XI. Create an array to store stack voltage used for plotting dynamic characteristics
XII. For current in range 0 < i < 36 plot 100 points using linspace function
XIII. Else if BCS stack is checked, run BCS VI, PI, efficiency, dynamic characteristics function blocks
XIV. Repeat step 6–12 with current from 0.1 < i < 30 for steady state characteristics and 0 < i < 26 with step size of 1 for dynamic characteristics of B CS stack.

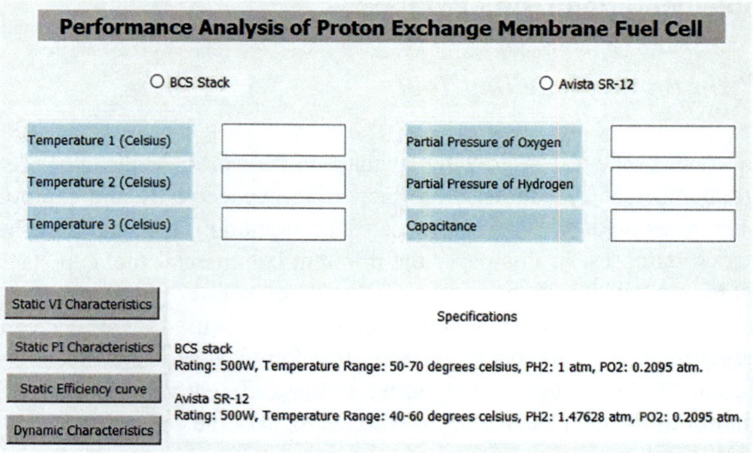

Fig. 4 Developed GUI for PEM fuel cell

4 Reference Models Discussion

The commercial stacks used for validating the electrical characteristics includes Avista SR-12 500 W stack manufactured by Wash based fuel cell company Avista Laboratories (now H2Fuel) and BCS 500 W stack manufactured by American Company BCS Technologies for rated currents of 42 A and 30 A respectively. The parameters involved in these models are stated in Tables 1 and 2 respectively and referred from [3], among which some of them are empirical parameters and others complex to determine.

4.1 Graphical User Interface Development

The graphical user interface is designed to plot static and dynamic electrical characteristics of the fuel cell. It has sets of constants and partial pressures of the reacting fuels defined for both Avista and BCS stacks. The user can select among the given two stacks using the radio button, accordingly the constant's set for the respective stack

Table 1 Avista SR-12 500 W stack parameters

Parameters	Value	Parameters	Value
N	48	ζ_1	− 0.948
T	323 K	ζ_2	$0.00286 + 0.0002 * \ln(A) + (4.3*10^{-5})[\ln(CH_2)]$
A	62.5 cm^2	ζ_3	$7.22 * 10^{-5}$
L	25 μm	ζ_4	$- 1.0615 * 10^{-4}$
PH2	1.4728 atm	λ	23
PO2	0.2095 atm	J_{max}	672 mA/cm^2
B	0.15 V	J_n	22 mA/cm^2
R_C	0.0003 Ω	I_{max}	42 A

Table 2 BCS 500 W stack parameters

Parameters	Value	Parameters	Value
n	32	ζ_1	− 0.948
T	333 K	ζ_2	$0.00286 + 0.0002 * \ln(A) + (4.3*10^{-5})[\ln(CH_2)]$
A	64 cm^2	ζ_3	$7.6*10^{-5}$
l	178 μm	ζ_4	$- 1.93*10^{-4}$
PH2	1 atm	λ	23
PO2	0.2095 atm	J_{max}	469 mA/cm^2
B	0.016 V	J_n	3 mA/cm^2
R_C	0.0003 Ω	I_{max}	30 A

gets selected. For plotting the static characteristics, the GUI takes three values of temperatures as input while for dynamic characteristics it takes charge double layer capacitance as an additional input. Finally, on selecting the type of plot through push buttons that is VI, PI or efficiency curve the respective plots for the given sets of operating parameters is plotted.

5 Result and Analysis

From the electrical characteristics of fuel cell at different temperature it is noticeable that the performance and the efficiency increases with temperature, and same trend is followed by the voltage curves of both the stack models [4]. From the graphs shown in Figs. 5, 6, 7 and 8 generated through GUI at three different operating temperatures, it is inferred that the stack voltage and stack power both increases with increase in temperature for the selected commercial stack models.

Thus, it satisfies the characteristics of the PEMFC. The power and polarization curves are verified by the results in [3] where, curves are plotted for the standard temperatures (50 and 60 °C) of the respective models and their peak powers showed good agreement with the peak powers obtained through the proposed Python model. In this paper results at different operating temperatures (20 and 80 °C) including the standard temperatures are illustrated. The PEMFC's efficiency is about 60% [4] and efficiency of both the Python-based models is within 60% as depicted by the efficiency curves in Figs. 9 and 10. Thus, it validates the proposed Python models. Moreover, Python model for BCS stack gave more appropriate output power while Avista Python model showed all three losses in the polarization curve more suitably. In dynamic characteristics that is Figs. 11 and 12, the activation losses are properly

Fig. 5 BCS VI characteristics

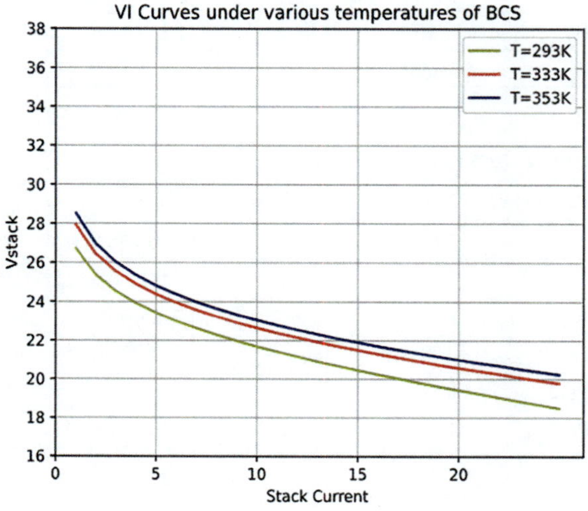

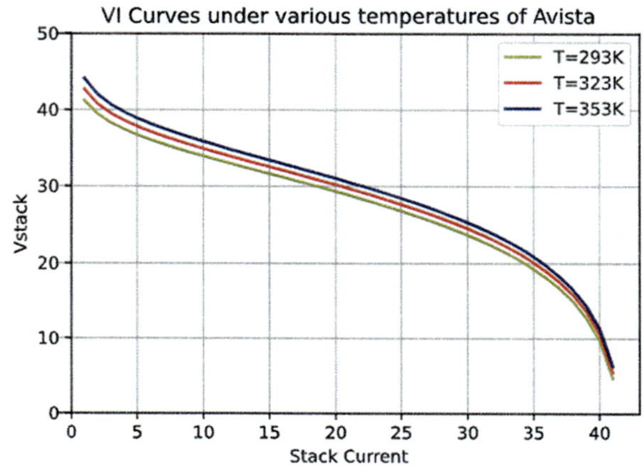

Fig. 6 Avista VI characteristics

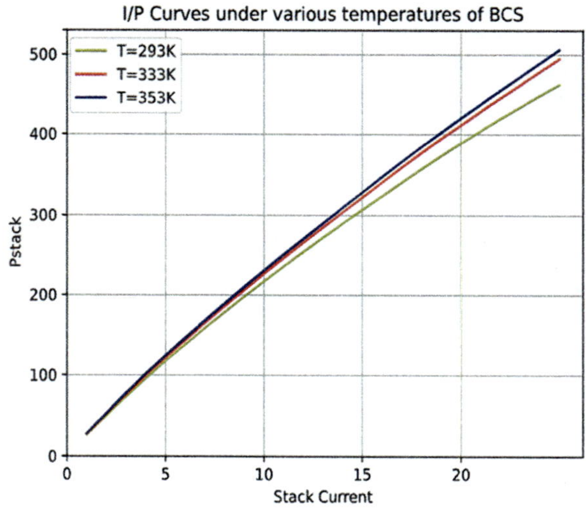

Fig. 7 BCS PI characteristics

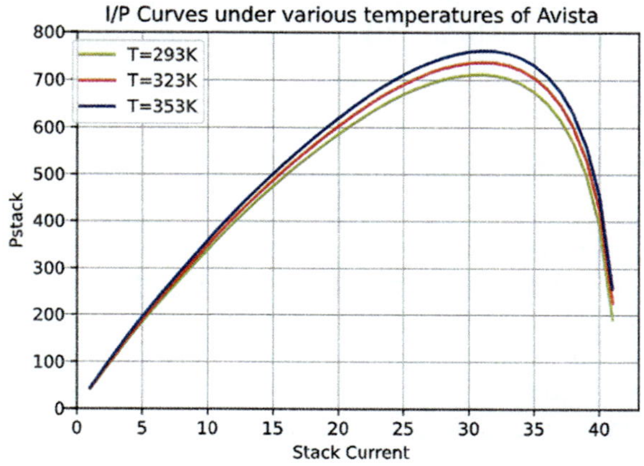

Fig.8 Avista PI characteristics

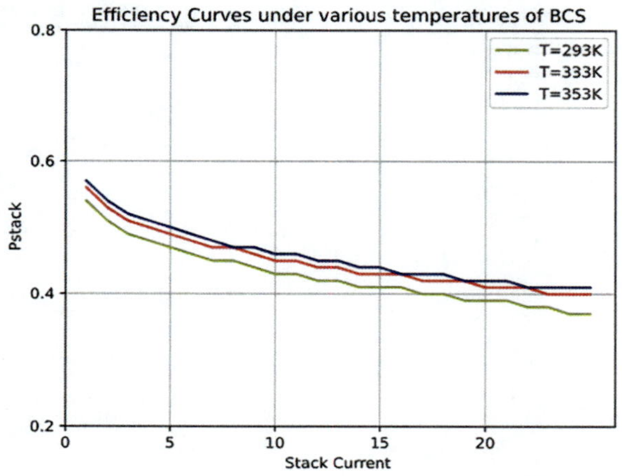

Fig. 9 BCS efficiency curve

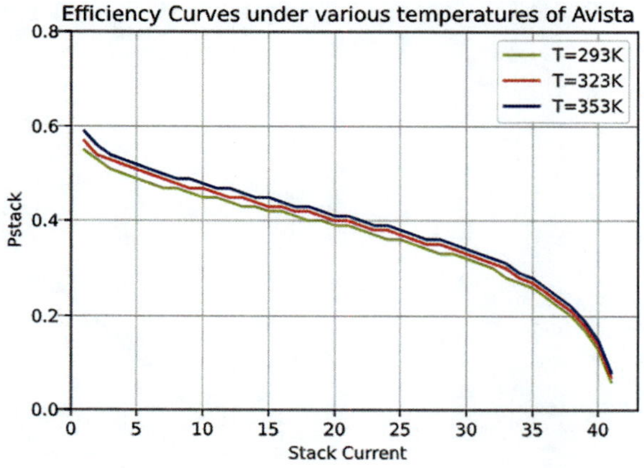

Fig. 10 Avista efficiency curve

visible as compared to the static ones due to the consideration of delay in activation losses caused by the charged double layer effect.

6 Conclusion

In the work presented, Python based FC model and its graphical user interface is successfully designed and implemented. The commercial 500W fuel cell stacks namely, Avista SR-12 stack and BCS stack are referred to analyze steady state along with dynamic characteristics of the model. The GUI designed in Qt designer and imported to a Python environment for backend programming. The activation, ohmic and concentration losses can be evidently observed in the polarization curves. Operating points can be analyzed from the PI curves which can be of benefit for fuel cell's

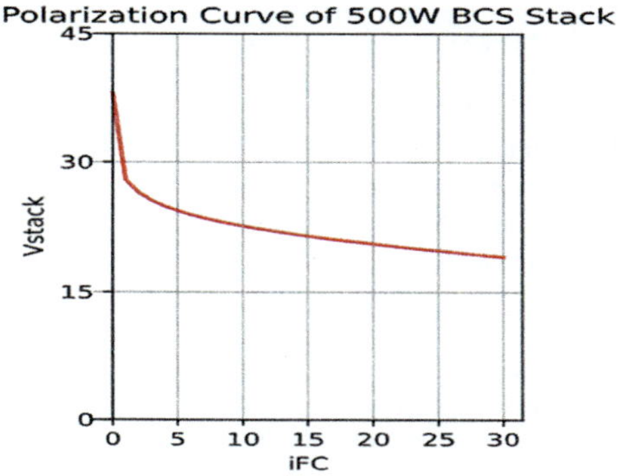

Fig. 11 BCS dynamic characteristics

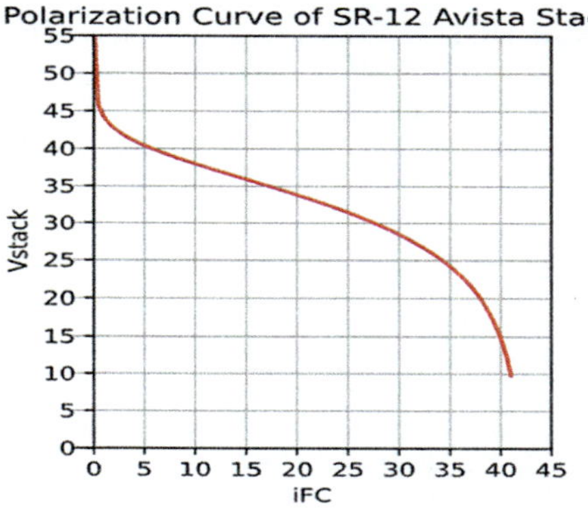

Fig. 12 Avista dynamic characteristics

maximum power point tracking control. The transient behavior's results also show the significance of charge double layer capacitance. Efficiency of both the models is decreasing with increase in current which is validated from the efficiency curves, while it increases along with stack voltage and power as the temperature increases. Graphical user interface makes it convenient for users to input important parameters in particular operating temperatures and constant inlet pressure of both reactant gases for generating datasets for analyzing the fuel cell behavior.

References

1. Amphlett JC, Baumert RFM, Peppley BA, Roberge PR (1995) Performance modeling of the ballard mark IV solid polymer electrolyte fuel cell. Electrochem. Soc. Inc. 142(1):9–15. https://doi.org/10.1149/1.2043959
2. Xiao Y, Agbossou K (2009) Interface design and software development for PEM fuel cell modeling based on matlab/simulink environment. WRI World Congress on Softw Eng 2009:318–322. https://doi.org/10.1109/WCSE.2009.344

3. Correa JM, Farret FA, Canha LN, Simoes MG (2004) An electrochemical- based fuel-cell model suitable for electrical engineering automation approach. IEEE Trans Industr Electron 51(5):1103–1112. https://doi.org/10.1109/TIE.2004.834972
4. Salam MA, Habib MS, Arefin P, Ahmed K, Uddin MS, Hossain T, Papri N Effect of temperature on the performance factors and durability of proton ex-change membrane of hydrogen fuel cell: a narrative review. Mat Sci Res India 17(2)
5. Adzakpa KP, Agbossou K, DubÉ Y, Dostie M, Fournier M, Poulin A (2008) PEM fuel cells modeling and analysis through current and voltage transient behaviors. IEEE Trans Energy Convers 23(2):581–591. https://doi.org/10.1109/TEC.2007.914170
6. Mahmah B, Morsli G, Mounia B, Benmoussa H, Achachera S, Benhamou A et al. (2013) Dynamic performance of fuel cell power module for mobility applications. Engineering 05(02):219e29. http://dx.doi.org/https://doi.org/10.4236/eng.2013.52032
7. Arsov GL (2008) Improved parametric PSpice model of a PEM fuel cell. In: 2008 11th International conference on optimization of electrical and electronic equipment, pp 203–208. https://doi.org/10.1109/OPTIM.2008.4602367
8. Saleh IMM (2015) Modelling, simulation and performance evaluation: PEM fuel cells for high altitude UAS. Sheffield Hallam University (United Kingdom)
9. Karami N, Outbib R, Moubayed N (2012) A low-cost microcontroller based 500-watt PEM fuel cell emulator, IEEE. https://doi.org/10.1109/SysCon.2012.6189439

Comparative Analysis of Performance Parameters of Hydrogen Fuel, Conventional Fuels and Hydrogen Enriched Fuels in an IC Engine

Vinay Prakash Chaudhary, D. B. Lata, Manish Kumar Singh, and Saurav Kumar

1 Introduction

A great fraction about 65% of the global energy demand is fetched from the fossil fuels. This is leading to a major decline in its availability over the world [32]. The transportation sector consumes about 50% of the produced oil globally [2]. In twenty-first century, the major concern is enhancement of energy production with major challenges of climatic conditions [8]. The human health and climate safety demands an alternative source to power the motor vehicles globally [3]. Many researchers' [3, 13] prime motive is on the application field of hydrogen in every sector of the world- industrial, domestic etc. Emphasis on hydrogen usage in petroleum refinery [5, 21], production of ammonia [20, 25], and in metal refinery is also gaining high expectancy. Hydrogen will soon become an alternate fuel for transportation sector seeing consumption, present availability, and environmental effects of conventional fuels [4]. This study is to compare the performance parameters of various fuels compared to hydrogen fuel and hydrogen enriched fuels to find the best possible alternative of conventional fuel. However, emission characteristics and hydrogen storage are major problem. Emission of NO_x, HC, CO, CO_2 etc. from the combustion of fuels will be discussed properly and separately. The different methods of storage of hydrogen have also been discussed in this study.

V. P. Chaudhary (✉) · D. B. Lata · M. K. Singh · S. Kumar
Central University of Jharkhand, Brambe, Ranchi, India
e-mail: vinay.20210201002@cuj.ac.in

S. Doolla et al. (eds.), *Advances in Clean Energy and Sustainability*,
Green Energy and Technology, https://doi.org/10.1007/978-981-99-2279-6_45

Table 1 Properties of hydrogen, diesel, petrol, CNG, LPG, methane

Properties	Hydrogen [6, 14]	Gasoline [14, 26]	Diesel [17]	CNG [30]	LPG [6, 14]
Chemical formula	H_2	C_4–C_{12}	C_8–C_{25}	CH_4 (major), C_2H_6	C_3H_8–C_4H_{10}
Physical state	Gas/Liquid	Liquid	Liquid	Gas	Gas
Density (kg/m^3)	0.084	737	833–881	0.748	1.898
Molecular weight	2.016	110	170	19	50
LHV (MJ/kg)	120	44	42.5	47.14	46.89
HHV (MJ/kg)	142	47.3	46	52.23	49.30
Autoignition temp (K)	858	520	553	873	684
Specific gravity	0.091	0.72–0.78	0.83	0.78	0.53–0.54
A/F ratio	34:1	14.7:1	14.5:1	17.2:1	15.6:1
Thermal conductivity	0.18	0.15	0.13	0.0340	010–0.017
Carbon content (%)	0	87	86	75	81
Hydrogen content (%)	100	13	14	18	19

2 Hydrogen Properties

Hydrogen is lightest and smallest molecule having least density of all the molecules. Hydrogen is about 95% but doesn't exist freely and is mostly available in compound form: H_2, H_2O, NH_3, H_2SO_4 etc. as shown in Table 1.

3 Hydrogen Production

Hydrogen doesn't exist freely in the nature. There is a keen requirement of techniques which can be employed to produce hydrogen. Some renowned methods of hydrogen productions are:

3.1 Steam Methane Reforming

SMR is a process in which a methane reacts with steam in the presence of a catalyst. This reaction produces hydrogen and carbon dioxide. SMR produces about 70–75% hydrogen rich gas with little amount of methane (2–6%), carbon monoxide (7–10%) and carbon dioxide (6–14%). SMR is not a sustainable energy source as it relies on fossil fuels for production. Another limitation of SMR is the production of greenhouse gases effecting environment adversely.

3.2 Partial Oxidation

POX occurs due to the reaction of hydrocarbon fuels with oxygen in less stoichiometric ratio (incomplete combustion). It yields a mixture of H_2 and CO when combusted at 1200–1350 °C. It produces about 48% of total available hydrogen globally. It produces less amount of hydrogen compared to SMR with the limitations of carbon monoxide release in larger amount.

3.3 Hydrogen from Biomass

Hydrogen can be produced from biomass feedstocks mainly by thermochemical and biochemical process. Thermochemical processes are cheap as they are carried out at a very high temperature and easily achieve higher reaction rates favorable for hydrogen production. Gasification and pyrolysis are two main thermochemical process used to produce hydrogen from biomass feedstocks. These processes are operated at higher temperatures and produce a hydrogen rich mixture of gas known as syngas.

3.4 Electrolysis

Electrolysis is the process of splitting the water molecules into H_2 and O_2 molecules by applying electricity in an electrolyzer device. Electrolysis produces pure oxygen with pure hydrogen. Electrolysis is the oldest technique, almost 200 years old [29], still, this process is not very fruitful when to be used on a large scale [7].

Holladay et al. [9] stated that most advanced process for production is reforming of HC fuels. Ni [22] stated that pyrolysis and gasification are most successful and easy methods of hydrogen production. According to Nikolaidis and Poullikkas [23], steam methane reforming was most cost effective.

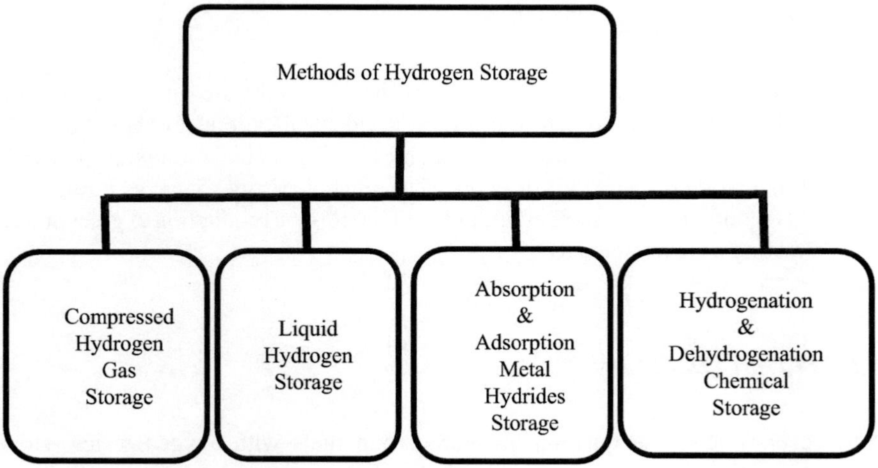

Fig. 1 Block diagram showing methods of hydrogen storage

4 Hydrogen Storage

Hydrogen storage is not an easy task as hydrogen is the most reactive gas and can react with the containers used to store it. However, over the years, many technological advancements did it possible. Figure 1 shows block diagram of ways of storing it.

5 Performance Parameters

Fuel Power—Fuel power is the product of the CV of fuel and the mass flow rate of the fuel. Fuel power is given by,

$$Fuel\ Power = \dot{m}_f \times CV$$

Indicated Power—The amount of power developed in cylinder is the indicated power. The indicated power is represented by the following equation:

$$IP = \frac{K.Pm.L.A.n}{60}$$

Brake Power—The BP is the power available at crankshaft. It is measured using brake mechanism.

$$Brake\ Power = T_{crankshaft} \times \omega_{crankshaft}$$

BSFC—BSFC is a measure of fuel efficiency of engine that burns fuel and produces shaft power. It is the rate of fuel consumption with power produced.

$$BSFC = \frac{\dot{m}_f}{BP}$$

Indicated Thermal Efficiency—It is the ratio of IP generated by the engine to the power generated by the combustion of the fuel. It states the amount of power taken by the piston out of the total fuel power.

$$ITE = \frac{Indicated\ Power}{Fuel\ Power}$$

Brake Thermal Efficiency—It is a measure of overall efficiency of the engine. It is the ratio of available energy in the BP to the fuel power.

$$BTE = \frac{Brake\ Power}{Fuel\ Power}$$

Volumetric Efficiency—It is the ratio of actual volume of air or A/F mixture intake by the piston during suction stroke to the swept volume.

$$\eta_v = \frac{Va}{Vd}$$

BSEC—It is defined as product of bsfc and CV of fuel. It indicates how efficiently fuel energy obtained from given fuel.

$$BSEC = BSFC \times CV$$

Brake mean Effective Pressure—It is the average pressure which if imposed on the pistons uniformly from the top to bottom of each power stroke, would produce desired brake power.

$$bmep = Brake\ Power/\left(Vs\ x\ N'\right)$$

6 Literature Review

Hydrogen Fuel

Homan et al. [10] tested hydrogen as a solo fuel for operating CIes. The autoignition temperature of hydrogen fuel limited the operation range even for a very lean mixture (29:1). Further, he concluded that a diesel engine can be modified with multi spark plugs to provide stability in ignition. The ignition delay became short. Ikegami et al. [12] operated a 4-stroke, water cooled, naturally aspirated, single cylinder with hydrogen fuel. He realized that hydrogen fuel limits the range of operation due to high autoignition temperature. On further modifications, a rough combustion with inappropriate results were seen.

Gasoline

Karthikeyan et al. [16] used Gasoline first and then Hydrogen enriched gasoline. The power output of the engine with hydrogen enriched gasoline was improved from 4.6 to 5.8 HP showing an increment of 18% in generated power. Kumar et al. [19] used gasoline as the fuel for IC engine and added H_2 gas to fuel. BSFC for hydrogen-gasoline showed largest saving in fuel, about 200 gm in 8 kg of fuel for same HP. BTE also improved.

Diesel

Sarvanan et al. [27] enriched diesel fuel with 10–90% H_2 in DI diesel engine at 1500 rpm rated at 3.78 kW power. BTE increases to 29.1% with 90% hydrogen enrichment, but knocking occurred. At 30% hydrogen enrichment, 27.9% of BTE was obtained and that too without knocking. BSEC decreased from 16.7 MJ/kWh for diesel to 12.7 MJ/kWh at 65% of rated load. At full load, 30% hydrogen enrichment gave BSEC of 12.9 MJ/kWh compared to diesel of 14.5 kWh. The BSEC decreases with increase in hydrogen percentage. Sandalci et al. [26] used Diesel fuel and Hydrogen enriched diesel fuel (16% H_2, 36% H_2, 46% H_2) at 1300 rpm speed in a diesel engine. The ITE for diesel fuel was found to be 43.48%. But with the increasing Hydrogen percentage in the diesel, ITE kept on decreasing 3.4, 6.4 and 10.5% at 16, 36 and 46% of hydrogen mixed diesel fuel compared to neat diesel.

CNG

Hora and Agarwal [11] tested HCNG in a single cylinder engine at an engine speed of 1500 rpm. The effect of HCNG on engine characteristics was analysed for 0–30% enrichment of hydrogen into CNG. The BTE at 5.3 bar bmep increased from 25.7 to 28.5% for 0–30% HCNG. At bmep 6.18 bar, BTE varied from 28 to 29.7%. He concluded that with increasing bmep, the BTE gets on increasing with hydrogen enrichment. The BSFC reduced for increased amount of HCNG. BSEC varied from 14.02 MJ/kWh (CNG) to 12.62% (30% HCNG) which showed that increased hydrogen into CNG reduces energy consumption for the same power output. Verma et al. [31] used HCNG in a spark ignited engine with port fuel injectors. The speed was kept constant at 1500 rpm. Hydrogen percentage in CNG for experiment was kept 6, 12.5, 21.25, 33.25 and 100%. He concluded that the BTE increased by 44% at 12.5% Hydrogen in CNG for high loads but for low loads, BTE increased by 20% at 33.25% hydrogen-CNG mixture. The volumetric efficiency decreased due to hydrogen addition.

LPG

Kacem et al. [14] performed an experiment using gasoline, LPG, LPG- Hydrogen, with 10 and 20% H_2. The BT by LPG-Hydrogen fuel increased by 12 and 20% compared to Gasoline and LPG operations respectively. BSFC reduced by 35.37% using LPG-20% H_2. Choi et al. [6] used a single cylinder diesel engine modified into spark ignition with the insertion of spark plug in place of injection nozzle. He employed LPG into the engine cylinder and gradually added hydrogen in the cylinder at 1400 rpm crankshaft speed. The bmep and thermal efficiency fell down as hydrogen amount was increased.

Biofuels

Khatri et al. [17] tested hydrogen-biogas enriched diesel fuel in a CI engine of dual fuel mode at 1500 rpm. He concluded that BTE improved by 3.09% compared to diesel. BSFC decreased from 71.05% (HBD) to 7.89% (HD). Prasad et al. [24] experimented with hydrogen enriched methanol and concluded that pure methanol shows 10.25% increase in the volumetric efficiency compared to gasoline while with hydrogen enrichment, the volumetric efficiency keeps on decreasing. MH5, MH20 shows 1% and 5% reduction in volumetric efficiency respectively compared to gasoline. The BSEC for M100 reduced by 6.3% and at 10% and 20% hydrogen with methanol, increased by 1.4–2.5% compared to methanol. The BP increased with hydrogen addition while BTE tended to increase only for low fraction of hydrogen in methanol. Abed et al. [1] on transesterification of WCO obtained from sunflower to biodiesel, tested it in DI diesel engine at a rated speed of 1500 rpm. He concluded that SFC increased and BTE lowered for all engine speeds compared to diesel fuel. BTE gets on decreasing from B10 to B30. Kulkarni [18] on reviewing papers related to testing of WCO from yellow grease and soyabean oil (20% of each), palm oil, virgin oil etc. concluded that the performance improves and are better than diesel and petrol. Kanth et al. [15] used honge biodiesel blend and diesel enriched with hydrogen in a CI engine. BTE of the engine using HB20 fuel increased by 2.2%, SFC decreased by 6%. Xia et al. [33] prepared bio-diesel from castor oil using transesterification and used potassium hydroxide as a catalyst. BTE was maximum at C50H which was 2% higher than the diesel fuel. Rao et al. ([30] reviewed numerous re-search papers on the influence of biodiesel extracted out of different sources on the performance parameters of the IC engine. He concluded that the thermal efficiency of biodiesel was more enhanced compared to gasoline and diesel.

7 Conclusion and Future Aspects

This study investigated number of fuels which can prove to be an alternative future fuel for transportation. On reviewing these papers, following conclusions are drawn:

- Hydrogen cannot be used as a solo fuel in ICes. The reason behind it is the high autoignition temperature of hydrogen fuel along with high in cylinder temperature. These problems resulted in back firing, pre-ignition and knocking of the engine.
- Hydrogen enriched gasoline enhanced BTE, reduced SFC but the operation of the engine was limited below 5% of hydrogen mixing in gasoline. After this, there was no improvement in thermal efficiency of the engine.
- Hydrogen enriched diesel enhanced the BTE up to 30% of hydrogen enrichment but after that a fall down in ITE and BTE is observed. The BSEC is also lesser compared to diesel at 30% hydrogen in diesel. At 90% hydrogen in diesel gave highest BTE of 29.1% more than diesel but knocking appears to be a persistent problem.
- Hydrogen enriched CNG showed a little boost in the BTE but is observed that with increasing bmep, BTE also increases. The BTE for 12.5 HCNG was highest for high loads but for low loads, 33.25 HCNG was better but BTE decreased compared to high load operation.
- Hydrogen enriched LPG doesn't prove to be a promising dual fuel as the thermal efficiency and bmep of the engine decreased. However, BSFC was optimum compared to any other fuel.
- Hydrogen enriched biofuels are still under research but as far the researches have gone shows that biofuels improve the BTE by a little margin. Still, it is more dependable one because of its availability, renewability and less emission of pollutants in the atmosphere. Biogas, biodiesels from different sources, technological advancements in its production will also improve the performance parameters.

Dual fuel enhanced the performance parameters of the IC engines. Still, the hydrogen enriched fuel is limited to some extent for satisfactory results. Technological advancements or mixing of anti-knocking agents in the dual fuel might resolve these challenges. The major challenge in its adoption as a fuel worldwide is its emission characteristics. The hydrogen enriched biofuels can be looked upon a reliable fuel resolving this issue. Hydrogen low density doesn't allow to store more of compressed hydrogen on board for long distance coverings. Hydrogen storage and production are major areas which require much attention.

References

1. Abed KA, El Morsi AK, Sayed MM, Shaib AAE, Gad MS (2018) Effect of waste cooking-oil biodiesel on performance and exhaust emissions of a diesel engine. Egypt J Pet 27(4):985–989. https://doi.org/10.1016/J.EJPE.2018.02.008
2. Balat M (2008) Potential importance of hydrogen as a future solution to environmental and transportation problems. Int J Hydrogen Energy 33(15):4013–4029. https://doi.org/10.1016/J.IJHYDENE.2008.05.047
3. Balat M, Balat M (2009) Political, economic and environmental impacts of biomass-based hydrogen. Int J Hydrogen Energy 34(9):3589–3603. https://doi.org/10.1016/J.IJHYDENE.2009.02.067
4. Barbir F (2009) Transition to renewable energy systems with hydrogen as an energy carrier. Energy 34(3):308–312. https://doi.org/10.1016/J.ENERGY.2008.07.007
5. Barreto L, Makihira A, Riahi K (2003) The hydrogen economy in the 21st century: a sustainable development scenario. Int J Hydrogen Energy 28(3):267–284. https://doi.org/10.1016/S0360-3199(02)00074-5
6. Choi GH, Chung YJ, Han SB (2005) Performance and emissions characteristics of a hydrogen enriched LPG internal combustion engine at 1400 rpm. Int J Hydrogen Energy 30(1):77–82. https://doi.org/10.1016/j.ijhydene.2004.06.009
7. de Souza RF, Padilha JC, Gonçalves RS, de Souza MO, Rault-Berthelot J (2007) Electrochemical hydrogen production from water electrolysis using ionic liquid as electrolytes: towards the best device. J Power Sour 164(2):792–798. https://doi.org/10.1016/J.JPOWSOUR.2006.11.049
8. Hassmann K, Kühne HM (1993) Primary energy sources for hydrogen production. Int J Hydrogen Energy 18(8):635–640. https://doi.org/10.1016/0360-3199(93)90115-Q
9. Holladay JD, Hu J, King DL, Wang Y (2009) An overview of hydrogen production technologies. Catal Today 139(4):244–260. https://doi.org/10.1016/j.cattod.2008.08.039
10. Homan HS, Reynolds RK, De Boer PCT, McLean WJ (1979) Hydrogen-fueled diesel engine without timed ignition. Int J Hydrogen Energy 4(4):315–325. https://doi.org/10.1016/0360-3199(79)90006-5
11. Hora TS, Agarwal AK (2015) Experimental study of the composition of hydrogen enriched compressed natural gas on engine performance, combustion and emission characteristics. Fuel 160:470–478. https://doi.org/10.1016/j.fuel.2015.07.078
12. Ikegami M, Miwa K, Shioji M (1982) A study of hydrogen fuelled compression ignition engines. Int J Hydrogen Energy 7(4):341–353. https://doi.org/10.1016/0360-3199(82)90127-6
13. Jain IP (2009) Hydrogen the fuel for 21st century. Int J Hydrogen Energy 34(17):7368–7378. https://doi.org/10.1016/J.IJHYDENE.2009.05.093
14. Kacem SH, Jemni MA, Driss Z, Abid MS (2016) The effect of H2 enrichment on in-cylinder flow behavior, engine performances and exhaust emissions: case of LPG-hydrogen engine. Appl Energy 179:961–971. https://doi.org/10.1016/j.apenergy.2016.07.075
15. Kanth S, Debbarma S, Das B (2020) Effect of hydrogen enrichment in the intake air of diesel engine fuelled with honge biodiesel blend and diesel. Int J Hydrogen Energy 45(56):32521–32533. https://doi.org/10.1016/j.ijhydene.2020.08.152
16. Karthikeyan S, Periyasamy M (2021) Impact on the power and performance of an internal combustion engine using hydrogen. Mater Today: Proc. https://doi.org/10.1016/j.matpr.2021.02.356
17. Khatri N, Khatri KK (2020) Hydrogen enrichment on diesel engine with biogas in dual fuel mode. Int J Hydrogen Energy 45(11):7128–7140. https://doi.org/10.1016/j.ijhydene.2019.12.167
18. Kulkarni MG, Dalai AK (2006) Waste cooking oil - an economical source for biodiesel: a review. Indust Eng Chem Res 45(9):2901–2913. https://doi.org/10.1021/ie0510526
19. Kumar GA, Rao GV (2013) Performance characteristics of oxy hydrogen gas on two stroke petrol engine. Int J Eng Trends Technol 6(7). http://www.ijettjournal.org

20. Lattin WC, Utgikar VP (2007) Transition to hydrogen economy in the United States: a 2006 status report. Int J Hydrogen Energy 32(15):3230–3237. https://doi.org/10.1016/J.IJHYDENE.2007.02.004
21. Mueller-Langer F, Tzimas E, Kaltschmitt M, Peteves S (2007) Techno-economic assessment of hydrogen production processes for the hydrogen economy for the short and medium term. Int J Hydrogen Energy 32(16):3797–3810. https://doi.org/10.1016/J.IJHYDENE.2007.05.027
22. Ni M, Leung DYC, Leung MKH, Sumathy K (2006) An overview of hydrogen production from biomass. Fuel Process Technol 87(5):461–472. https://doi.org/10.1016/j.fuproc.2005.11.003
23. Nikolaidis P, Poullikkas A (2017) A comparative overview of hydrogen production processes. In: Renewable and sustainable energy reviews, vol 67. Elsevier Ltd., pp 597–611. https://doi.org/10.1016/j.rser.2016.09.044
24. Nuthan Prasad BS, Pandey JK, Kumar GN (2021) Effect of hydrogen enrichment on performance, combustion, and emission of a methanol fuelled SI engine. Int J Hydrogen Energy 46(49):25294–25307. https://doi.org/10.1016/j.ijhydene.2021.05.039
25. Ramachandran R, Menon RK (1998) An overview of industrial uses of hydrogen. Int J Hydrogen Energy 23(7):593–598. https://doi.org/10.1016/S0360-3199(97)00112-2
26. Sandalci T, Karagöz Y (2014) Experimental investigation of the combustion characteristics, emissions and performance of hydrogen port fuel injection in a diesel engine. Int J Hydrogen Energy 39(32):18480–18489. https://doi.org/10.1016/j.ijhydene.2014.09.044
27. Saravanan N, Nagarajan G (2008) An experimental investigation of hydrogen-enriched air induction in a diesel engine system. Int J Hydrogen Energy 33(6):1769–1775. https://doi.org/10.1016/j.ijhydene.2007.12.065
28. Sharma S, Ghoshal SK (2015) Hydrogen the future transportation fuel: From production to applications. In: Renewable and sustainable energy reviews, vol. 43. Elsevier Ltd., pp 1151–1158. https://doi.org/10.1016/j.rser.2014.11.093
29. Stojić DL, Marčeta MP, Sovilj SP, Miljanić ŠS (2003) Hydrogen generation from water electrolysis—possibilities of energy saving. J Power Sour 118(1–2):315–319. https://doi.org/10.1016/S0378-7753(03)00077-6
30. Rao KVS, Kurbet SN, Kuppast VV (2018) A review on performance of the IC engine using alternative fuels. Mater. Today: Proc. 5(1):1989–1996. https://doi.org/10.1016/j.matpr.2017.11.303
31. Verma G, Prasad RK, Agarwal RA, Jain S, Agarwal AK (2016) Experimental investigations of combustion, performance and emission characteristics of a hydrogen enriched natural gas fuelled prototype spark ignition engine. Fuel 178:209–217. https://doi.org/10.1016/j.fuel.2016.03.022
32. Veziroğlu TN, Şahin S (2008) 21st Century's energy: Hydrogen energy system. Energy Convers Manage 49(7):1820–1831. https://doi.org/10.1016/J.ENCONMAN.2007.08.015
33. Xia C, Brindhadevi K, Elfasakhany A, Alsehli M, Tola S (2021) Performance, combustion and emission analysis of castor oil biodiesel blends enriched with nanoadditives and hydrogen fuel using CI engine. Fuel 306:121541. https://doi.org/10.1016/J.FUEL.2021.121541

CFD Simulation of Portable Thermal Storage Device for Solar Cooking System

Milan Sojitra, Sachin Gupta, Arunendra Kumar Tiwari, Asim Kumar Joshi, and Ramkishore Singh

1 Introduction

Nowadays, energy is essential for human survival. The importance of energy in economic development is crucial since energy and economic activity are inextricably linked. Traditionally, solid fossil fuels like wood and coal have been the major source of energy. The rise in world energy consumption as a result of population growth and the twentieth-century industrial revolution ushered in a transitional period for fossil fuels. It is commonly recognized that to achieve sustainable development, current energy sources such as fossil fuels and nuclear power must be rapidly replaced with renewable energy sources. Which are environmentally friendly and capable of meeting current and future global energy demands without harming the environment. Solar, wind, and biogas are examples of renewable energy sources that might contribute to meeting global energy demands in a sustainable way [1]. TES stores energy at high or low-temperature for later use. It fills the gap between energy demands and supply. There are many types of storage like sensible heat storage (SHS), latent heat storage (LHS), chemical heat storage, and thermochemical heat storage.

SHS is not a good option because the stored energy density is very low compared to the LHS. The efficiency of SHS is lower than the other two storage [1]. The chemical storage device is under development and it is complicated and costly. It also poses environmental issues like toxicity and fires dangers hazards [2]. Hence, LHS is the best option to store energy. The LHS using the PCM has its benefits such as high energy storage density so that small size can be good for required energy storage. Latent heat storage has higher efficiency and almost constant operating temperature [3]. Its per-unit storage capacity is 5 to 14 times higher than SHS materials like water

M. Sojitra · S. Gupta (✉) · A. K. Tiwari · A. K. Joshi · R. Singh
Sardar Patel Renewable Energy Research Institute, Vidyanagar, Gujarat 388120, India
e-mail: sachin.energy123@gmail.com

S. Doolla et al. (eds.), *Advances in Clean Energy and Sustainability*,
Green Energy and Technology, https://doi.org/10.1007/978-981-99-2279-6_46

and rock. The most common geometric configuration of the LHS device is rectangular, cylindrical [4], and spherical. The cylindrical enclosure has high efficiency and also heat loss is low compared to the rectangular enclosure [4].

Sharma et al. [5] experimented to evaluate the thermal performance of an evacuated tube-based solar cooker with the PCM unit. They choose erythritol as a PCM material. They found that evening and night time cooking using the PCM is possible and it does not affect the noon cooking. Saxena et al. [6] experimented with solar cooking using PCM. They studied a different type of PCM material that was used for the solar cooker, and among them, stearic acid (commercial grade) is found to be good latent heat storage which was experimentally tested in a simple box solar cooker.

The complicated nature of PCM (solidification and melting) necessitates numerical analysis to anticipate the PCM behaviors. To anticipate the behavior of PCM many models has been used. ANSYS Fluent is a popular software for computational fluid dynamics (CFD) study for predicting the behavior of PCM [4, 7–14]. Muhammad et al. [4] validated the CFD model for PCM in vertical cylinders, with a maximum discrepancy of 7.5% that model was able to predict melt fraction.

Yadav et al. [15] worked on simulating the melting of a PCM using Ansys (fluent). The melting setup was rectangular and exposed to the constant heat flux of 2500 W/m^2 on the left-hand side of the system. They observed that heat transfers first through conduction and subsequently through natural convection.

Hariharan et al. [16] have explored the paraffin phase-change behavior in a spherical capsule for the TES device. They observed that the solidification is quicker than the melting. Sattari et al. [17] studied the melting behavior of PCM in a spherical capsule using the CFD simulation and validated it with experiments. They also studied the effect of many characteristics such as the capsule's surface temperature, initial temperature, and capsule size. Finally, they concluded that while the initial temperature of the capsule does not affect the melting rate, increasing the surface temperature and decreasing the diameter of the spherical capsule increase the melting rate.

Here as per the literature review limited investigation has been done on solar cooking with high-temperature PCM. So, the primary goal of this research is to determine the charging and discharging characteristics of high-temperature PCM for solar cooking using numeric modeling. Most of the cooking requirements fall in the temperature range of 300 °C. So, sodium nitrate (PCM) has been selected in this study. A model simulation model has been developed to analyse the charging and discharging characteristics of sodium nitrate. To optimize the mesh size and time step, time independence and grid independence tests have also been performed.

2 Methodology

The TES device is cylindrical and made of steel. The sidewall and base of the TES device are assumed to be perfectly insulated. Table 2 shows the properties of the

Table 1 The energy required for various food [19]

Item	Thermal energy required	Cooking temperature (K)
Rice, dal, veg	85–90 kcal/meal	396
Hot water, milk	50 kcal/day	373
Chapattis	50 kcal/chapatti 100 kcal/meal	553

Table 2 Properties and dimensions of the steel and PCM

Properties	Steel	PCM	
		Solid	Liquid
Density (kg/m^3)	8750	2261	2164
Specific heat (J/kg)	542	1110	
Thermal conductivity (W/mK)	45	0.525	0.514
Inner Dimension (D × H) (mm)	150 × 80	–	
Thickness (mm)	1	–	
Thermal expansion coefficient, β (1/K)		2.85×10^{-5} at 580 K	
Mass of PCM (@ 30 °C, solid) (kg)		3.25	
Melting point, T_m (K)		577	
Latent heat of fusion, L (kJ/kg)		172	

TES device (Steel). From Table 1, the temperature required for making the chapattis is approx. 553 K. So, the melting temperature of PCM should be more than 553 K. hence sodium nitrate ($NaNo_3$) has been selected as its melting temperature is 577 K.

2.1 *Problem Formulation*

Here, PCM has been considered an incompressible Newtonian fluid. To reduce the computational time, the problem has been considered as a transient 2-dimensional flow due to the axisymmetric of the system and structured quadratic meshing has been used for simulation as shown in Fig. 1.

Initially, the TES device is considered at ambient temperature i.e. 303 K. Boundary conditions are shown in Table 2. To solve the pressure–velocity gradient a pressure correction, momentum, and energy model PISO, Least square cell-based, PRESTO, and second-order upwind model have been used. Simulation has been run with 50 iterations per time step and 0.1 s time step and validated with Muhammad et al. [4] work. For solidification and melting four different equation has been solved i.e. energy, momentum, turbulence, and wall contact resistance equation (Table 3).

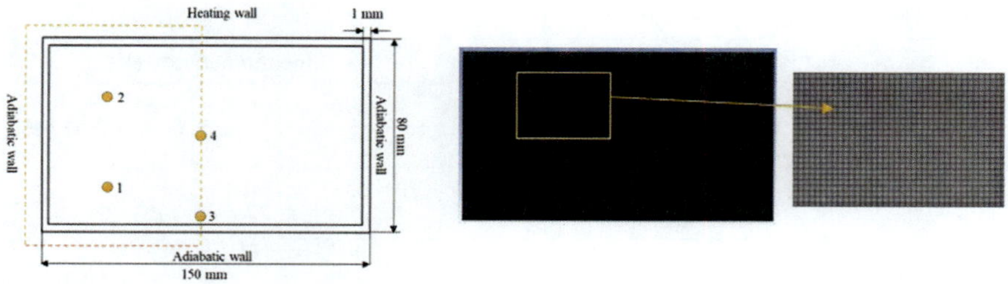

Fig. 1 Schematic diagram of TES device

Table 3 Boundary conditions for the charging process for all surfaces of the TES device

Surface	Boundary condition type	Value
Top wall	Temperature	773 K
Sidewall	Adiabatic	0 W/m^2
Bottom wall	Adiabatic	**0 W/m^2**

3 Result and Discussion

3.1 Validation of the Developed Model

Muhammad et al. [4] did the experiment to find out the pattern of melting of the PCM and they also simulated the experiment and validated it. Their experimental setup with dimensions has been shown in Fig. 2. In this experimental setup side walls, top and bottom walls have been made up of polycarbonate and acrylic respectively. The side and bottom wall have been kept at constant temperatures of 343 K and 305 K respectively and the inside PCM (n-eicosane) has been filled up with the initial temperature of 296 K.

The buoyancy effect was studied using the Boussinesq approximation at 273 K, and it was assumed that the density of the PCM is constant. Thermophysical properties can be specified as functions of temperature using Fluent.

In this case, there is a discontinuity in these characteristics from solid to liquid near the melting point. Numerical instability can be caused by these discontinuities. To define distinct attributes, it was assumed that the change in a property in the mushy zone is linear. The thermophysical characteristics of PCM (n-eicosane) shown in Table 4 were employed.

$$\mu = (9 \times 10^{(-4)} \times T^2 - 0.6529T + 119.94) \times 10^{(-3)} \tag{1}$$

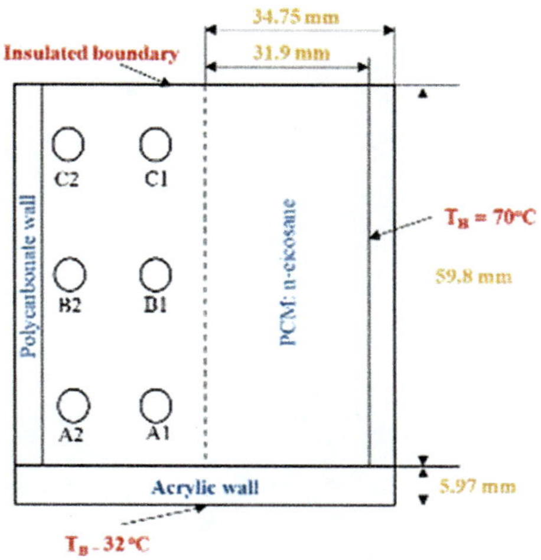

Fig. 2 Experimental setup of Muhammad et al. [4]

Table 4 Properties of n-eicosane

Parameter	Solid (25 °C)	Liquid (50 °C)
Density, ρ (kg/m^3)	910	769
Thermal conductivity, k (W/mK)	0.423	0.146
Specific heat capacity, c_p (J/kg K)	1926	2400
Thermal expansion coefficient, β (1/K)	–	8.161×10^{-4}
Melting point, T_m (°C)	36.4	–
Latent heat of fusion, L (kJ/kg)	**248**	–

3.2 Validation of the Melting Process

Figure 3a is the experimental result of Muhammad et al. [4] and the Fig. 3b is a simulation result of his work and the Fig. 3c is the simulation result of the developed model. In Muhammad et al. [4], the result at 1800 and 2400 s was flatter than the experimental melting profile while in the developed model simulation results are very close to the experimental profile because in this simulation for pressure–velocity coupling PISO method has been used instead of the SIMPLE method which was used by them and geometry was planer instead of symmetry which was used by them. This developed model shows a very close result to experimental work. So for the study of our TES, this developed model has been used.

(a) Experimental result of Muhammad et al [4]

(b) Simulation result of Muhammad et al [4]

(c) Simulation result of this research work for validation

Fig. 3 Validation of the current simulation with the experiment of Muhammad et al. [4] in 600s, 1800s, 2400s

3.3 Grid Independence Test and Time Independence Test

To test the grid independence, a grid independence test has been done. Four different mesh sizes i.e., 0.5 mm, 0.3 mm, 0.1 mm, and 0.05 mm have been selected for the grid independence test. All mesh is quadratic mesh and the number of the node points taken has been shown in Table 5.

Table 5 Nos. of nodes for grid independence test with a various mesh size

Mesh size	0.5 mm	0.3 mm	0.1 mm	0.05 mm
No of nodes	50,215	1,39,502	3,12,261	12,47,770

Table 6 Errors in grid independence and time independence test

Grid independent test		Time independent test	
Mesh size (mm)	Error in liquid fraction concerning 0.05 mm mesh (%)	Timestep for 0.1 mesh size (s)	Error in liquid fraction concerning 0.1 s time step (%)
0.1	1.20		
0.3	17.99	0.5	0.12
0.5	15.06	1	0.23

A time-independent test has also been carried out at different time steps for an optimized element size of 0.1 mm. This experiment was conducted using time increments of 1, 0.5, and 0.1 s. Table 6 shows the error in the grid independence test and the temporal independence test.

From Table 6, it has been observed that the error for 0.1 mm mesh size at 0.5 and 1 s time step is less than 5%, so to save computational time 0.1 mm mesh size and 1 s time step has been selected for the simulation study.

3.4 *Result of the Simulation Study—Charging Process*

The numerical investigation of the charging and discharging of the sodium nitrate in a cylindrical enclosure has been done. The results have been recorded at a regular interval of 1 s. Those results are represented as a contour and graph of liquid fraction, and temperature of the PCM.

Figure 4 represents the liquid fraction contour during the charging of the TES device. The boundary condition at a top wall is a constant temperature condition i.e., 773 K, and other walls consider a perfect insulators. The charging time of the TES device was observed at 22680 s (6.3 h), due to the PCM's poor heat conductivity, this value is quite high. To reduce the charging time high thermal conductivity PCM is desirable. The shape of the liquid fraction while charging looked like an inverted parabola because on both sides PCM is in direct contact with TESD. So conductive heat transfer was taken place.

3.5 *Result of the Simulation Study—Discharging Process*

Discharging of the TES device has been done at 553 K on the top wall because it's suitable for most cooking applications and the rest walls consider perfect insulators. Figure 5 shows the volume average contours for the mass fraction (melt) of the PCM at various time intervals for the discharging of the device. The total discharging time is 9.53 h which is enough high to cook the food. In Fig. 5, blue and red color indicates

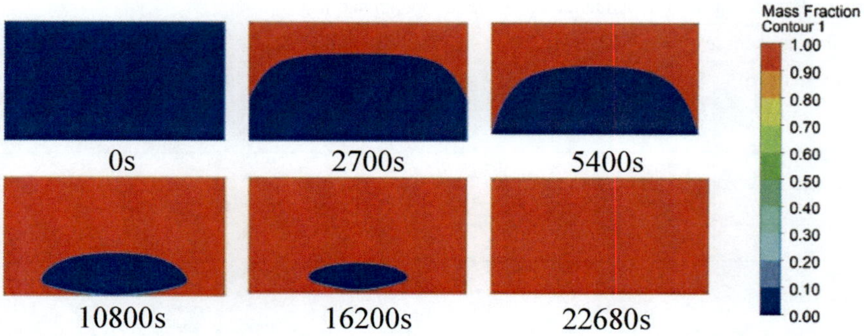

Fig. 4 Liquid fraction contour of the sodium nitrate during the charging process

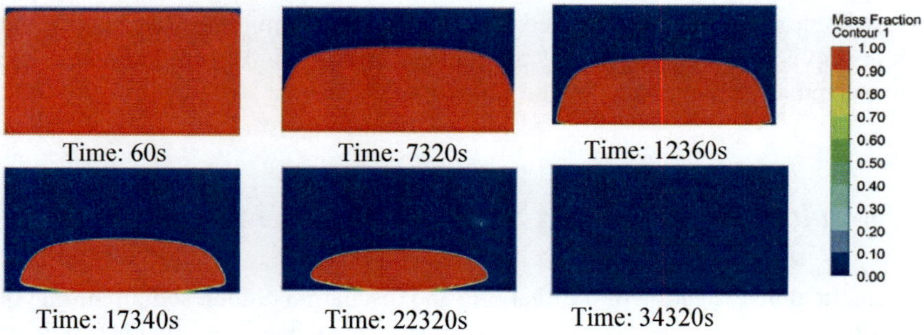

Fig. 5 Liquid fraction contour of the sodium nitrate for discharging process

the solid and liquid states of the PCM respectively. The same shape was found while the TES device was in discharging mode.

Figure 6 depicts the variation of PCM's liquid fraction or mass fraction over time. PCM is initially in the solid form, with no liquid portion. Time required to melt the different solid fractions i.e., 20%, 40%, 60%, and 80% is 900 s, 3885 s, 6945 s, and 10185 s respectively. To charge the device 100% (i.e., full melting of PCM) it took 22,680 s (6.3 h). Initially, the device charge rapidly because conduction plays a major role in heat transfer but after 80% charging, the rate of melting of PCM decreases because convection heat transfer plays dominating role.

The blue line in Fig. 6 indicates the liquid fraction of the PCM during the discharging process. Initially, PCM is in a liquid state i.e., melt fraction is one. When the TES device is exposed to 553 K at the top wall as per earlier discussion, PCM gets solidified. The rate of solidification of the PCM is nearly linear.

Figure 7 depicts the temperature profile over time at various points. It takes 15.83 h to complete one charging and discharging cycle on the TES device. The PCM curve becomes flattered near melting temperature (577 K) and begins to increase again after melting.

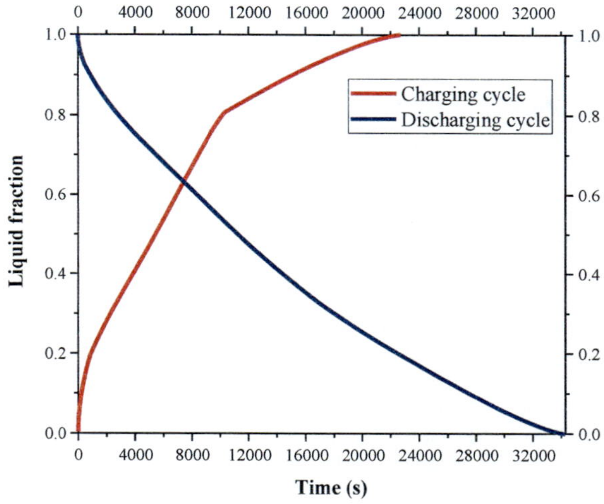

Fig. 6 Liquid fraction versus time for the charging and discharging process

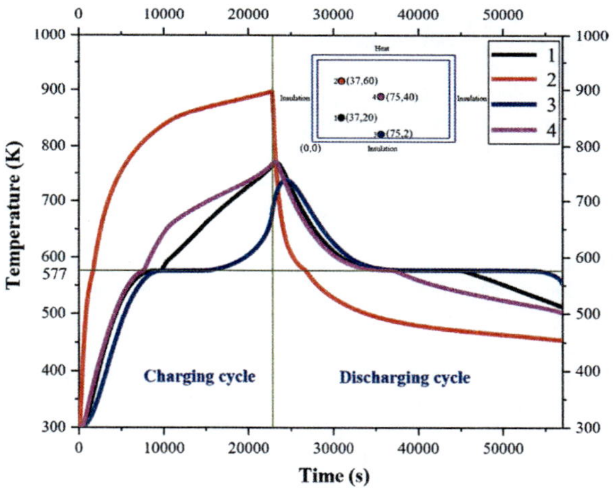

Fig. 7 Temperature versus time graph for the charging and discharging process

4 Conclusion

From the above computational study of the charging and discharging cycle of the PCM i.e., sodium nitrate. it is found that this type of TES device can be charged with solar energy, and it can be utilized for cooking purposes while discharging. The charging and discharging time of the TES device is 6.3 h and 9.53 h respectively. Discharging time is high due to its discharging temperature which is relatively high i.e., 553 K. Charging time is also very high it is due to the PCM's poor conductivity. To overcome this issue, PCM with high thermal conductivity can be used or fin-type arrangement on vessel wall's inner surface for high heat transfer can be done. Various nanomaterials also can be used to increase the thermal conductivity of the PCM.

References

1. Medrano M, Gil A, Martorell I et al (2010) State of the art on high-temperature thermal energy storage for power generation. Part 2-case studies. Renew Sustain Energy Rev 14:56–72. https://doi.org/10.1016/j.rser.2009.07.036
2. ilkington Solar International Gmbh (2000) Survey of thermal storage for parabolic trough power plants period of performance : survey of thermal storage for parabolic trough power plants period of performance. Contract 61. https://doi.org/10.2172/765081
3. Sharma A, Tyagi VV, Chen CR, Buddhi D (2009) Review on thermal energy storage with phase change materials and applications. Renew Sustain Energy Rev 13:318–345. https://doi.org/10.1016/j.rser.2007.10.005
4. Muhammad MD, Badr O, Yeung H (2015) Validation of a CFD melting and solidification model for phase change in vertical cylinders. Numer Heat Transf Part A Appl 68:501–511. https://doi.org/10.1080/10407782.2014.994432
5. Sharma SD, Iwata T, Kitano H, Sagara K (2005) Thermal performance of a solar cooker based on an evacuated tube solar collector with a PCM storage unit. Sol Energy 78:416–426. https://doi.org/10.1016/j.solener.2004.08.001
6. Saxena A, Lath S, Tirth V (2013) Solar cooking by using PCM as a thermal heat storage. Mit Int J Mech Eng 3–2:91–95
7. Al-Abidi AA, Bin Mat S, Sopian K et al (2013) CFD applications for latent heat thermal energy storage: a review. Renew Sustain Energy Rev 20:353–363. https://doi.org/10.1016/j.rser.2012.11.079
8. Al-waeli ATCHhAK, Sopian KhAK (2017) Energy storage: CFD modeling of thermal energy storage for a phase change materials (PCM) added to a PV/T using nanofluid as a coolant. J Sci Eng Res 4:193–202
9. Assis E, Katsman L, Ziskind G, Letan R (2007) Numerical and experimental study of melting in a spherical shell. Int J Heat Mass Transf 50:1790–1804. https://doi.org/10.1016/j.ijheatmasstransfer.2006.10.007
10. Foong CW, Hustad JE, Løvseth J, Nydal OJ (2010) Numerical study of a high temperature latent heat storage (200–300°C) using eutectic nitrate salt of sodium nitrate and potassium nitrate. COMSOL Users Conf. https://doi.org/10.1109/61.329522
11. Koller M, Walter H, Hameter M (2016) Transient numerical simulation of the melting and solidification behavior of NaNo3using awire matrix for enhancing the heat transfer. Energies 9. https://doi.org/10.3390/en9030205
12. Nayak AO, Ramkumar G, Manoj T, Vinod R (2011) Comparative study between experimental analysis and CFD software analysis of PCM material in thermal energy storage system 2
13. Redzuan MCN, Saw CL, Lew WC et al (2017) Numerical simulation of pcm intergrated solar collector storage water heater. ARPN J Eng Appl Sci 12:3363–3367
14. Santoshchavan MRN (2013) CFD analysis on thermal energy storage in phase Change materials using high temperature solution. Int J Eng Res Technol 2:483–490
15. Yadav A, Soni S (2017) Simulation of melting process of a phase change material (PCM) using ANSYS (Fluent). Int Res J Eng Technol 04:2395–2456
16. Hariharan K, Kumar GSS, Kumaresan G, Velraj R (2018) Investigation on phase change behavior of paraffin phase change material in a spherical capsule for solar thermal storage units. Heat Transf Eng 39:775–783. https://doi.org/10.1080/01457632.2017.1341227
17. Sattari H, Mohebbi A, Afsahi MM, Azimi Yancheshme A (2017) Simulation par MFN du processus de fusion des matériaux à changement de phase dans une capsule sphérique. Int J Refrig 73:209–218. https://doi.org/10.1016/j.ijrefrig.2016.09.007
18. Tan FL, Hosseinizadeh SF, Khodadadi JM, Fan L (2009) Experimental and computational study of constrained melting of phase change materials (PCM) inside a spherical capsule. Int J Heat Mass Transf 52:3464–3472. https://doi.org/10.1016/j.ijheatmasstransfer.2009.02.043

Activated Carbon-Graphene Composite/Ethanol-Based Adsorption Refrigeration System: Minimum Regeneration Temperature, Uptake Efficiency, and Cooling Performance

P. R. Chauhan and S. K. Tyagi

1 Introduction

In the past few decades, the mechanical compression-based refrigeration cycle has dominated in HVAC sector (Heating Ventilation and Air Conditioning) due to its reliability and compactness. However, it has mainly two drawbacks: (i) high grade energy consumption, (ii) the usage of traditional refrigerants with the large potential for environmental harm. Although, zero ODP (ozone depletion potential) refrigerants like R-134a and R-22 are now in use, but they have a considerable global warming potential (GWP) [1]. The above-mentioned issues are addressed by heat operated refrigeration and airconditioning cycle such as adsorption cooling cycle. It has several advantages such as the ability to work on ultra-low-temperature low grade energy source [2], the usage of ozoenvironmental friendly working pairs, no moving parts, and peak power demand reduction [3]. Despite these advantages, the traditional adsorption-based cooling system performs poorly due to its intermittent and complex operation, and poor adsorption capacity of adsorbent material as well as the design of sorption reactor [4].

The several highly porous materials such as silica gel, zeolites, activated carbon, and their composites are used as an adsorbent with water, ammonia, ethanol, methanol, CO_2, and R507a refrigerant. An experimental as well numerical work on adsorption kinetics of HFC32 on MaxsorbIII based composite adsorbent is reported considering the thermal diffusivity in both parallel and perpendicular direction of the fibres [5]. The numerical finding demonstrated that the volumetric cooling energy for consolidated composite adsorbent is found be improved up to 78% than that of powdered activated carbon. In their next investigation which is based on the second law analysis in terms of entropy generation, the composite consisting 80%

P. R. Chauhan · S. K. Tyagi (✉)
Solar—Biomass Thermal Science Laboratory, Department of Energy Science and Engineering, Indian Institute of Technology Delhi, Hauz Khas, New Delhi 110016, India
e-mail: sudhirtyagi@yahoo.com

S. Doolla et al. (eds.), *Advances in Clean Energy and Sustainability*,
Green Energy and Technology, https://doi.org/10.1007/978-981-99-2279-6_47

MaxsorbIII, 10% ionic liquid, and 10% Polyvinyl alcohol (PVA) is found to have the lowest irreversibility per unit cooling energy production [6]. The adsorption of HFC404A on Maxsorb III adsorbent is studied at the regeneration temperature ranging from 25 to 75 °C, and the optimal specific cooling power is determined as 275 W/kg [7]. In another investigation, the water adsorption on the zeolite and graphene nanoplatelets (GNP) composite adsorbent is noted to be improved up to 42.8%, as a result, the proposed adsorbent-refrigerant pair improves the cooling performance of adsorption-based heat pump [8]. The synthetization and characterization of activated carbon and graphene nanoplatelets based composite is done to improve the thermal performance of ACS [9]. The numerical modelling for evaluating the minimum, maximum, and optimum regeneration temperature for Solar energy powered adsorption cooling system is reported in reference [10].

According to the research findings from the above discussed literatures, the Maxsorb III based composite adsorbents are significantly more effective in terms of adsorption capabilities and surface area, but to the best knowledge of authors, there is almost no numerical investigation reported in terms specific cooling energy (SCE), volumetric cooling energy (VCE), uptake efficiency, and minimum regeneration temperature for Maxsorb III, PVA, and H25 based composite adsorbents with ethanol pairs for single stage ACS. Therefore, the present work aims to study the effects of various operational parameters on cooling performance and determining theoretical as well as realistic minimum regeneration temperature.

2 Numerical Modelling

The steady state numerical modelling for a single stage system adsorption cooling system is discussed in this section. The two composites having different mass fractions of MaxsorbIII as parent material, PVA as binder, and H25 as thermal conductivity enhancer are taken as adsorbents; and herein the ethanol is selected as refrigerant for thermodynamic cooling assessment. The compositions, and fitting parameters for both composite adsorbents are tabulated in Table 1. In order to study the effect of thermal conductivity enhancer (H25) on performance of the system, the proportions of ingredients are chosen such that the composite 2 is having 20% of H25 while composite 1 is free from H25 keeping equal amount of PVA (i.e., 10%) in their composition. The temperature dependent thermophysical properties of the composites and refrigerant are taken from published literature [5]. The physical significance of higher amount of MaxsorbIII is to enhance the total pore volume which further results in increased total surface area. However, the H25 improves the thermal conductivity up to 5.5 times for composite 2 as reported in available literature.

Table 1 Compositions, and fitting parameters for composite adsorbents

Composite	Compositions	Parameters	Ref.
Composite 1	90wt% MaxsorbIII + 10wt% PVA + 0wt% H25	W_o/(kg/kg) = 1.08, E/(kJ/kg) = 134.42, n/(-) = 1.83, ρ_{pack}/ (kg/m^3) = 316	[9]
Composite 2	70wt% MaxsorbIII + 10wt% PVA + 20wt% H25	W_o/(kg/kg) = 0.86, E/(kJ/kg) = 131.28, n/(-) = 1.86, ρ_{pack}/(kg/m^3) = 460	

2.1 *Equilibrium Uptake Model*

The equilibrium adsorption uptake value is calculated by Dubinin-Astakhov (D-A) model along with Antoine equation at specified pressure and temperature conditions.

$$W = W_0 \exp\left[-\left(\frac{RT}{E}\ln\left(\frac{P_s}{P}\right)\right)^n\right] \tag{1}$$

where W, W_o, P, R, T, E, P_s, and n represent the equilibrium adsorption uptake (kg/kg), the maximum adsorption uptake (kg/kg), equilibrium pressure (kPa), gas constant (kJ/kg-K), temperature (K), activation energy (kJ/kg), saturation pressure (kPa), and index parameter, respectively.

2.2 *Isosteric Heat of Adsorption*

The following expression can be used to calculate the isosteric heat of adsorption, which is the amount of thermal energy released during the landing of refrigerant over the porous surface of the adsorbent.

$$-Q_{st} = RT^2\frac{\partial}{\partial T}(P_s) + E\left\{-\ln\left(\frac{W}{W_o}\right)\right\}^{1/n} \tag{2}$$

2.3 *Cooling Energy Assessment*

The thermodynamic performance analysis of adsorption refrigeration cycle is carried out in terms of specific cooling energy (SCE) and volumetric cooling energy (VCE), given in the following equation [11].

$$SCE = h_{fg} \times (W_{max} - W_{min}) \tag{3}$$

$$\mathrm{VCE} = \rho_{\mathrm{pack}} \times \mathrm{h}_{\mathrm{fg}} \times (\mathrm{W}_{\mathrm{max}} - \mathrm{W}_{\mathrm{min}}) \tag{4}$$

where, W_{max} and W_{min} are the equilibrium adsorption uptake at evaporator temperature and low-pressure level, and regeneration temperature and high-pressure level, respectively. The latent heat of refrigerant and packing density of adsorbent material are represented by h_{fg} and ρ_{pack}, respectively.

2.4 *Estimation of Uptake Efficiency*

The uptake efficiency (η_U), the function of temperature and the thermophysical properties of both adsorbent and refrigerant, can be estimated by following equation to quantify the loss of adsorption owing to vacancies in the adsorbent material.

$$\eta_{\mathrm{U}} = \frac{\Delta \mathrm{W}_{\mathrm{act}}}{\Delta \mathrm{W}} \tag{5}$$

where ΔW_{act} and ΔW represent the change in adsorption uptake for real cycle (with loss of uptake) and ideal cycle (without loss of uptake).

2.5 *Minimum Regeneration Temperature: Theoretical Versus Actual*

In the case of W = 0, the Eq. (6) can be used to calculate the theoretical minimum temperature required for adsorbent bed regeneration [12].

$$\mathrm{T}_{\mathrm{reg,\,min}} = \frac{\mathrm{T}_{\mathrm{ads}} \times \mathrm{T}_{\mathrm{cond}}}{\mathrm{T}_{\mathrm{evap}}} \tag{6}$$

where $T_{reg,\,min}$, T_{ads}, T_{cond}, and T_{evap} denote the minimum regeneration, adsorption, condenser, and evaporator temperature in Kelvin, respectively. Whereas, the actual value of minimum regeneration temperature exists when SCE approaches zero (SCE $\rightarrow$ 0).

3 Results and Discussion

The steady state thermodynamic performance of an adsorption cooling system is examined at regeneration temperature of 90 °C, adsorption and condenser temperature of 30 °C for the application of cooling at evaporator temperature of 0 °C and

10 °C. The renewable energy such as Solar energy, geothermal energy can be utilized as heating source for bed regeneration. The numerical validation in terms of isosteric heat of adsorption is carried out and found to be in close agreement with available literature.

3.1 Evaluation of Q_{st}

The effect of regeneration temperature and adsorption uptake on isosteric heat of adsorption is shown in Fig. 1 for both the promising adsorbent-refrigerant pairs. Adsorption of ethanol on composite 1 and composite 2 decreasing trend with higher uptake due to higher pressure. However, the isosteric heat of adsorption is observed to be decreasing towards higher regeneration temperatures at a certain value of adsorption uptake. For both sorption pairs, the predictions using Eq. (2) accord very well with the calorimetrically determined isosteric heat of adsorption in the available literature. The maximum and minimum isosteric heat of adsorption are found as 1123.76 kJ/kg at w = 0.1 and 941 kJ/kg at w = 1 for composite 1, respectively. The composite 1 requires more heat of adsorption at all values of adsorption uptake in comparison to composite 2. The maximum and minimum heat of adsorption for composite 2 are calculated as 1106.11 kJ/kg at w = 0.1 and 939.87 kJ/kg at w = 0.8, respectively. Unlike composite 1, the composite 2 limits the isosteric heat in term of adsorption uptake from 0.1 to 0.8.

Similarly, at regeneration temperature of 70 °C, the maximum and minimum isosteric heat of adsorption are estimated as 1107.85 kJ/kg at w = 0.1 and 925.08 kJ/kg at w = 1 for composite 1, and 1090.2 kJ/kg at w = 0.1 and 923.96 kJ/kg at w = 0.8 for composite 2, respectively. It can be concluded that composite 2 is good for cooling application whereas composite 1 is more suitable for heat pumping application. Moreover, as heat of adsorption reduces at higher regeneration temperature which results in good feasibility of both composites towards cooling application.

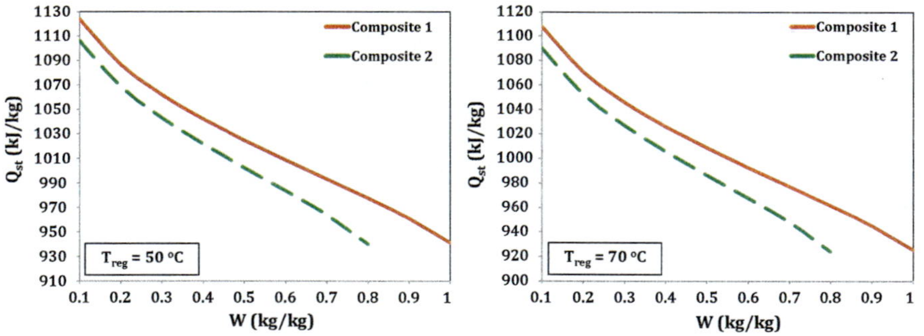

Fig. 1 Q_{st} versus W at T_{reg} of 50 °C and 70 °C for composites 1 and 2

3.2 Estimation of Specific Cooling Energy

The steady state cooling performance study in terms of SCE is also numerically examined for single stage adsorption refrigeration cycle as shown in Fig. 2 for composite 1 and composite 2. The effect of regeneration temperature and evaporation temperature on SCE and VCE is also studied for both the selective working pairs. At a constant evaporation temperature, the SCE value increases almost linearly with increases in regeneration temperature. Also, the numerical findings demonstrated the cooling energy is noted to be increasing towards heat pumping application at a constant value of regeneration temperature whereas it is reverse for cooling or refrigeration application. This happens because of the decrement in amount of uptake with increment in regeneration temperature that results in the increment in the difference between minimum and maximum uptake ($\Delta W = W_{max} - W_{min}$). At regeneration temperature of 90 °C, the SCE values are evaluated as 353.92 kJ/kg at $T_{evap} = 0$ °C and 575.96 kJ/kg at $T_{evap} = 10$ °C for composite 1. However, the maximum values of SCE at 90 °C of regeneration temperature for composite 2 are noted as 285.04 kJ/kg at $T_{evap} = 0$ °C and 526.5 kJ/kg at $T_{evap} = 10$ °C, respectively. This is because the composite 1 requires more heat of adsorption in comparison to composite 2.

The SCE can also be a useful tool to estimate the actual minimum temperature required for regeneration of bed when its value approaches to zero as discussed in modelling section. The actual values of minimum regeneration temperature for composite 1 are obtained as 62.5 °C and 52.2 °C at evaporation temperature of 0 °C and 10 °C, respectively. Whereas, in the case of composite 2, these are found to be 64.7 °C and 52.4 °C at evaporation temperature of 0 °C and 10 °C, respectively.

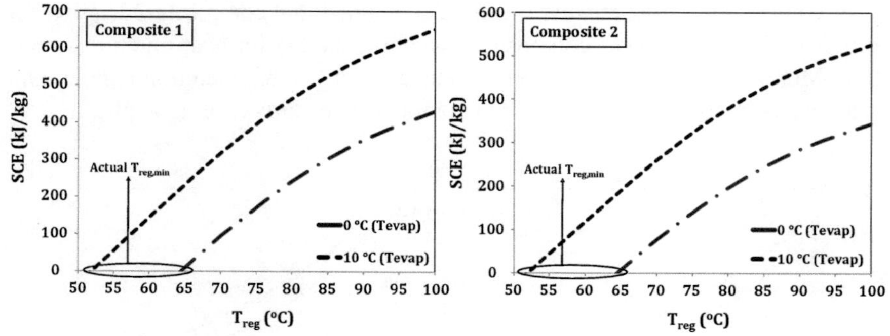

Fig. 2 SCE versus T_{reg} at T_{evap} of 0 °C and 10 °C for composite 1 and 2

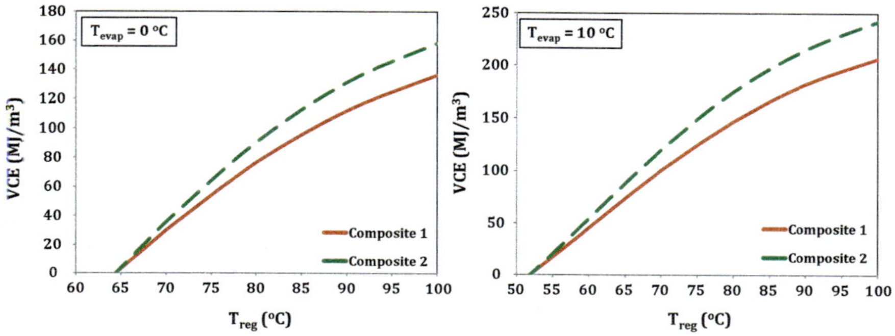

Fig. 3 VCE versus T_{reg} at T_{evap} of 0 °C and 10 oC for composites 1 and 2

3.3 Comparative Analysis Based on Volumetric Cooling Energy

This section presents the performance comparison between both composite adsorbents with ethanol refrigerant in terms of volumetric cooling energy. Figure 3 illustrates the variations in VCE corresponding to regeneration temperature for both the working pairs at evaporator temperature of 0 and 10 °C. The numerical findings demonstrated that VCE increase as regeneration temperature increase and found to be comparatively more for adsorbent 2 due to higher value of its packing density. The maximum values of VCE at 0 °C of evaporation temperature and 90 °C of regeneration temperature are evaluated as 111.84 MJ/m^3 and 131.12 MJ/m^3 for adsorbent 1 and adsorbent 2, respectively. However, at 10 °C of evaporation temperature and 90 °C of regeneration temperature, the maximum values of VCE for adsorbent 1 and adsorbent 2 are estimated as 182 MJ/m^3 and 215.16 MJ/m^3, respectively. The VCE value is obtained greater for composite 2 for all values of regeneration temperature in comparison to composite 1. This is due to the presence of thermal conductivity enhancer in the composition of composite 2. The intersection point of VCE curve at abscissa can also be considered as a criterion for determining the realistic minimum regeneration temperature.

3.4 Comparative Analysis Based on Uptake Efficiency

The uptake efficiency is considered another comparative performance parameter for promising composite 1 and 2 as adsorbent with ethanol refrigerant. Also, it can be applied as a criterion to find the optimal range of regeneration temperature for an adsorption-based cooling system to function efficiently and effectively. Figure 4 shows the variation in uptake efficiency along the regeneration temperature at evaporation temperature of 0 and 10 °C. The results revealed that the uptake efficiency for

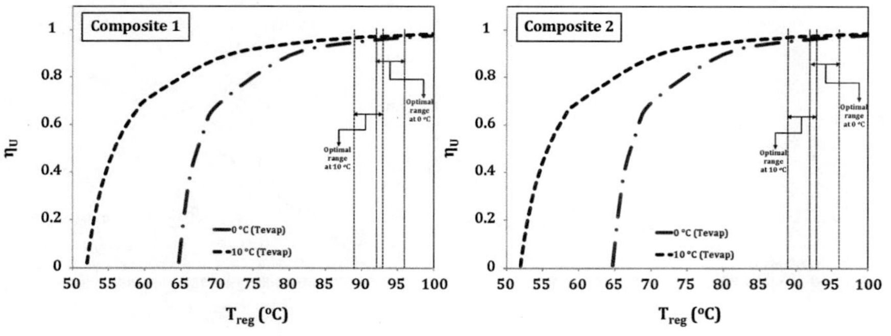

Fig. 4 η_U versus T_{reg} at T_{evap} of 0 and 10 °C for composite 1 and 2

composite 1/ethanol pair is found to be increasing drastically up to 82 °C at evaporation temperature of 0 °C and up to 74 °C at evaporation temperature of 10 °C; later on, it increases gradually. Moreover, the optimal range of regeneration temperature is identified when the percentage difference between two consecutive measurements of uptake efficiency is around 2%.

The uptake efficiency is shown to be higher at high evaporator temperatures for both adsorbents. At evaporator temperature of 0 °C and regeneration temperature of 90 °C, the uptake efficiency is greater for adsorbent 1 than that of adsorbent 2. At evaporator temperature of 10 °C and regeneration temperature of 90 °C, the values of uptake efficiency are determined as 0.96 and 0.97 for composite 1 and composite 2, respectively. The uptake efficiency zero value can also be used to determine a realistic minimum regeneration temperature. As the evaporator temperature reaches 10 °C, the optimal working temperature range shifts slightly to the left when compared to evaporator temperature of 0 °C.

3.5 *Minimum Regeneration Temperature: Theoretical Versus Actual*

In order to do a comparative investigation, another comparison is carried out on the basis of actual and theoretical minimal regeneration temperature. The theoretical minimum regeneration temperature is evaluated using Eq. (6); however, the actual value of minimum regeneration temperature can be estimated by from Fig. 2 at zero value of SCE. It is then observed that for both the working pairs the actual value of minimum regeneration temperature is more than that of the theoretical value at evaporator temperature 0, and 10 °C. Table 2 shows the exact values of actual and theoretical minimal regeneration temperature and the percentage deviation among them at evaporator temperature of 0, and 10 °C.

Table 2 Comparative study for minimum regeneration temperature

Adsorbent—refrigerant	$T_{reg,min}$ in °C (actual)		$T_{reg,min}$ in °C (theoretical)		% deviation w. r. to theoretical	
	T_{evap} = 0 °C	T_{evap} = 10 °C	T_{evap} = 0 °C	T_{evap} = 10 °C	T_{evap} = 0 °C	T_{evap} = 10 °C
Composite 1—ethanol	64.5	52.3	63.3	51.4	1.9	1.8
Composite 2—ethanol	64.7	52.8	63.3	51.4	2.2	2.7

4 Conclusions

The steady state thermodynamic performance in terms of isosteric heat of adsorption, SCE, VCE, uptake efficiency, and minimum regeneration temperature is carried out considering the temperature dependent latent heat of vaporization for a single stage adsorption cooling system using MATLAB R2021b software. The optimal working range of regeneration temperature is defined at evaporation temperature of 0 and 10 °C, and adsorption temperature of 30 °C. The following are concluding remarks from the present numerical work:

- The maximum and minimum isosteric heats of adsorption for composite 1 and composite 2 are predicted to be 1107.85 kJ/kg at w = 0.1 and 925.08 kJ/kg at w = 1 and 1090.2 kJ/kg at w = 0.1 and 923.96 kJ/kg at w = 0.8, respectively. As a result, composite 2 is more suited for cooling applications, while composite 1 is better for heat pumping applications
- The SCE, VCE, and uptake efficiency are found to be increasing with increase in regeneration temperature; however, at a finite value of regeneration temperature, SCE decreases whereas VCE and uptake efficiency improve towards higher evaporator temperature
- At regeneration temperature of 90 °C, the SCE value is evaluated as 353.92 kJ/kg at T_{evap} = 0 °C and 575.96 kJ/kg at T_{evap} = 10 °C for composite 1; however, it is 285.04 kJ/kg at T_{evap} = 0 °C and 526.5 kJ/kg at T_{evap} = 10 °C, respectively for composite 2
- In contrast to SCE and uptake efficiency, VCE is shown to be higher for composite 2 than composite 1 due to a higher value of packing density
- As far as optimal range of uptake efficiency is concerned, the optimum working temperature range for composite 2 shifts slightly leftward than that of composite 1
- The actual value of minimum regeneration temperature is found to be higher than that of the theoretical one. Furthermore, the composite 1 requires lower source temperature than that of composite 2 at all evaporation temperatures.

Acknowledgements One of the authors (PRC) thankfully acknowledges the financial assistance in the form of fellowship due to Department of Energy Science and Engineering, Indian Institute of Technology Delhi.

References

1. Chauhan PR, Kaushik SC, Tyagi SK (2022) Current status and technological advancements in adsorption refrigeration systems: a review. Renew Sustain Energy Rev 154(1):111808
2. Pan QW, Xu J, Ge TS, Wang RZ (2022) Multi-mode integrated system of adsorption refrigeration using desiccant coated heat exchangers for ultra-low grade heat utilization. Energy 238(1):121813
3. Chauhan PR, Verma A, Bhatti SS, Tyagi SK (2019) An overview on mathematical models of adsorption refrigeration system. J Mater Sci Mech Eng 6(1):275–278
4. Chauhan PR, Kaushik SC, Tyagi SK (2022) A review on thermal performance enhancement of green cooling system using different adsorbent/refrigerant pairs. Energy Convers Manage X 14:100225
5. Yagnamurthy S, Rakshit D, Jain S, Rocky KA, Islam MA, Saha BB (2021) Adsorption of difluoromethane onto activated carbon based composites: Thermophysical properties and adsorption characterization. Int J Heat Mass Transf 171(1):121112
6. Yagnamurthy S, Rakshit D, Jain S, Rocky KA, Islam MA, Saha BB (2022) Adsorption of difluoromethane onto activated carbon based composites: Adsorption kinetics, heat of adsorption, cooling performance and irreversibility evaluation. Appl Therm Eng 210:118359
7. Ghazy M, Askalany AA, Saha BB (2020) Maxsorb III/HFC404A as an adsorption pair for renewable energy driven systems. Int J Refrig 120:12–21
8. Rocky KA, Islam MA, Pal A, Ghosh S, Thu K, Saha BB (2020) Experimental investigation of the specific heat capacity of parent materials and composite adsorbents for adsorption heat pumps. Appl Therm Eng 164:114431
9. Pal A, Uddin K, Thu K, Saha BB (2019) Activated carbon and graphene nanoplatelets based novel composite for performance enhancement of adsorption cooling cycle. Energy Convers Manage 180:134–148
10. Banker ND, Dandotiya D, Morthala SV, Gaddam M, Kakileti S (2020) Evaluation of minimum, maximum and optimum source temperature for solar-powered adsorption refrigeration system. Arab J Sci Eng 45(11):9735–9745
11. Yagnamurthy S, Rakshit D, Jain S, Saha BB (2021) Operational envelope and performance enhancement of a two-bed adsorption cooling system. Appl Therm Eng 195:117181
12. Saha BB, El-Sharkawy II, Chakraborty A, Koyama S, Banker ND, Dutta P, Prasad M, Srinivasan K (2006) Evaluation of minimum desorption temperatures of thermal compressors in adsorption refrigeration cycles. Int J Refrig 29(7):1175–1181

Highly Carbonized, Porous Activated Carbon Derived from *Ziziphus Jujuba* for Energy Storage

Senthil Kumar Kandasamy, R. Ramyea, Chandrasekaran Arumugam, V. Sruthi, M. Sudharsan, R. Sugan Raj, and Monika Michalska

1 Introduction

The use of fossil fuels by industries has improved in the level of people in the world, although, it created huge harmful environmental problems. With the development of renewable energy, the use of fossil fuels may be omitted in the future. To balance the need of energy and energy production, energy storage is essential. It involves the conversion and storage, economically. Based on the required ecological conditions, new, economically and environmentally friendly energy conversion and storage systems are essential [1]. Many researchers are working towards this. With the current feature of high capacitance and low voltage limit the supercapacitor has attracted more attention among researchers. By comparing Electrochemical Double Layer Capacitance (EDLC), the need for pseudocapacitor is growing substantially [2].

Tamilselvi1 et al. [3] developed reduced Graphene Oxide (rGO) through catalytic oxidation using low-cost catalyst obtained from coconut shell and coir with a specific capacitance of 111.1 F g^{-1} in 2 M KOH. By calcinations, Sing et al. [4] introduced graphene sheets into carbon paper with 1122 F g^{-1} at 5 mV s^{-1}. Hendriansyah et al. [5] produced activated carbon and graphene from palm oil biomass and carbon nanotube is from palm oil mill effluent. But the composite obtained from palm oil biomass and effluent exhibited a very low specific capacitance of 4.3908 F g^{-1}. Liu et al. [6] developed 3-dimensional porous carbon with interconnected pores

S. K. Kandasamy (✉) · R. Ramyea · C. Arumugam · V. Sruthi · M. Sudharsan · R. S. Raj
Department of Electronics and Communication Engineering, Kongu Engineering College, Perundurai, Erode 638060, India
e-mail: senthilkumar6k@gmail.com

M. Michalska
Department of Chemistry and Physico-Chemical Processes, Faculty of Materials Science and Technology, VŠB-Technical University of Ostrava, 17. Listopadu 2172/15, 708 00 Ostrava-Poruba, Czech Republic

S. Doolla et al. (eds.), *Advances in Clean Energy and Sustainability*,
Green Energy and Technology, https://doi.org/10.1007/978-981-99-2279-6_48

was obtained via carbonization and KOH activation from longan shell. Khan et al. [7] produced hierarchically porous carbon from the natural withered rose through carbonization and chemical activation with a mixture of KOH/KNO_3. Vijayakumar et al. [8] converted wheat flour into high surface area carbon nanosheets with a potential window of 3.2 V. Similarly, several bio-wastes such as banana peel [9], orange peel [10], potato [11, 12] were considered for this application.

Cheng et al. [13] prepared porous carbon from shiitake mushroom and activated it with H_3PO_4 and KOH. This activated carbon showed a large number of micro-, meso pores and interconnected macro-pores with specific surface area (2988 m^2 g^{-1}) [14]. Veerakumar et al. [15] developed porous carbon nano-sheets by carbonization, chemical and physical activation of a paper flower with a specific surface area of 1801 m^2 g^{-1}. Xiaoxia Bai et al. [16] developed porous carbon with interconnected meso-pores from Lignosulphonates wastes. Similarly, bio-char [17] and black liquor [18] were used to make activated carbon with a high surface area. In general, the meso-porous carbon exhibits high specific surface area, pore volume and electrical conductivity, which are highly reliant on the carbonization temperature. Bhat et al. [19] developed porous nano-carbon from agricultural waste peels.

Hong et al. [20] developed 3D porous carbon consisting of micro-, meso- and macropores through $KHCO_3$ activation of shells of chestnut with high specific pore surface area of 2298 m^2 g^{-1}. YanLei Zhang et al. [21] developed porous carbon from waste of *salvia miltiorrhiza* flowers by dry distillation and $NaHCO_3$ activation with a specific surface area of 1715.3 m^2 g^{-1}. Yang et al. [22] developed high heteroatom porous carbons derived from lotus leaves with high capacitance retention of 96.2% even after 10,000 cycles. These porous and heteroatom groups acquire ion transport channels. Srinivasan et al. [23] derived AC from *polyalthia longifolia* seeds by pyrolytic chemical activation. Rajasekaran et al. [24] prepared AC from *eucalyptus globulus* seed by hydrothermal method followed by KOH activation with a specific surface area of 2388.38 m^2 g^{-1}.

Hor et al. [25] derived Activated carbon (AC), from a bio-waste pollen cones and found that KOH showed better performance over $ZnCl_2$. Lee et al. [26] developed carbon aerogel electrodes from jackfruit and durian. Lan et al. [27] prepared AC from walnut shell with KOH activation. Taer et al. [28] prepared pineapple crown AC using one step carbonization and physical activation. Jiang et al. [29] developed conductive carbon by thermal carbonisation using seaweed fiber. And interestingly, higher carbonization temperature led to higher conductivity, as well as minimize other elements such as sulphur and calcium [30]. Ahirrao et al. [31] synthesized AC from *citrus limetta* using facile chemical approach. Yuwei Chen et al. [32] prepared waste newspaper-derived porous AC using simple carbonization and activation techniques. Palisoc et al. [33] developed multi-walled carbon nano-tubes and AC from *Moringa Oleifera* fruit shells.

Martínez-Casillas et al. [34] prepared AC from *pecan* nutshell waste via one-step carbonization/chemical activation using phosphoric acid. Taer et al. [35] developed AC from *acacia* leaves using chemical and physical activation. Le et al. [36] used carbonization followed by in-situ chemical activation to create porous carbon from *areca* palm leaves. Ciftyurek et al. [37] synthesized AC from two different fruit dehydration wastes through hydrothermal carbonization and chemical activation. Liu et al. [38] synthesized N-doped microporous carbon via a facile one-step method by employing natural wood fibers. Qu et al. [39] used residues of corncob to prepare porous carbon by steam activation without pre-carbonization.

Our main aim is to develop activated carbon from *ziziphus jujuba.* The electrochemical measurements were performed with a three-electrode system using pt, Ag/AgCl and working electrode. The working electrode was made by mixing as prepared samples and solution in a ratio similar to our previous work [40–42]. The obtained slurry is coated over a graphite rod and used as an electrode. Instead of carbon cloth, graphite rod is used as a current collector. This graphite rod won't affect the electrochemical nature of the samples. Because, the slurry covered the surface of the graphite rod. The active mass of a working electrode is 25 $\mu g\ cm^{-2}$ after being treated at room temperature. When compared to organic electrolytes, aqueous electrolytes have a higher specific capacitance for activated carbon. Because the prepared material is activated carbon, 6 M KOH is used as an electrolyte. Using an OrigaLys electrochemical workstation, galvanostatic charge discharge (GCD), cyclic voltammetry (CV), and electrochemical impedance spectroscopy (EIS) measurements were done.

2 Experimental Section

The *ziziphus jujuba* was purchased from the local market in Erode, India. Then after consuming the juicy part, the seeds were collected and dried in day light for 3 or 4 days. It was washed with distilled water and dried for one day at room temperature. After drying, it was treated in a muffle furnace at 450° C for 6 h.

Ziziphus jujaba seeds and carbonized seeds are shown in Fig. 1a, b. Over the 300 °C different temperatures were used to optimize the temperature level based on the ash content. At 450° C and also for the period of 6 h duration, the samples exhibited more carbon content, which is needed for an EDLC. It forms the carbonized form of *ziziphus jujuba* seeds and has been activated with KOH by mixing them in a 2:1 mass ratio. Before the activation process, the CJZS sample was stirred in a magnetic stirrer for 2 h at room temperature. Finally, the carbonized sample was washed with ethanol and distilled water, and it was defined as active carbonized *ziziphus jujuba.* The sample was kept in 50% HCl to remove excessive KOH, as well as to wash using distilled water. The obtained samples were dried, ground, and sieved. Finally, the slurry is made using the obtained activated carbon and rubber solution. This slurry is coated over a graphite rod of 2 mm thickness.

Structural analyses were done using X-Ray Diffraction (XRD) and Scanning Electron Microscope-Energy Dispersive X-Ray Analysis (SEM–EDX) measurements. The electrochemical analysis of the electrode was performed using an OrigaLys electrochemical workstation. The electrodes are, Platinum, Ag/AgCl and CJZS.

Fig. 1 **a** *Ziziphus jujuba* seeds **b** optical image of carbonized *ziziphus jujuba* seeds

3 Results and Discussion

XRD patterns were recorded between 5° and 45° with CuKα radiation source. XRD patterns of *ziziphus jujuba* seeds derived KOH activated carbon is shown in Fig. 2. The samples exhibited an amorphous pattern. The appearance of a broad peak between 22° to 24° in the XRD indicates the presence of carbon. No other major peak is observed due to the absence of other elements [42]. A small peak is observed at 43°, suggesting that the activated carbon is layered in nature, which is further confirmed with the SEM morphology. The morphology of the *ziziphus jujuba* seeds derived KOH activated carbon was examined using a JEOL –JSM-6390 scanning electron microscope. The surface morphology of activated carbon with different magnification levels is shown in Fig. 3. At the magnification level of 10,000 the samples exhibited layered morphology with porous structure.

The presence of pores contributes to the improved specific capacitance. The activated carbon was subjected to elemental analysis and the presence of $C = 79.9\%$, $O = 19.12\%$ and $K = 0.8\%$ was observed. Extremely higher carbon content is observed. Thus a higher specific surface area and higher conductivity may be seen. The sample is purely exhibit electric double layer capacitance. At the same time, the presence of oxygen may induce pseudocapacitve properties. A higher carbon content of carbon and a small amount of oxygen may influence the electrode.

The electrode for the electrochemical analysis was fabricated by mixing the samples with the rubber solution. The obtained slurry was coated at the tip of the graphite rod and dried in a room environment to remove the excess solvent. CV analysis of the prepared sample was done for different scan rates such as 5, 10, 25, 50 and 100 mV s^{-1} with the potential window (working potential) of 0 to 1 V. The slightly rectangular shape of the CV curve is observed in Fig. 4. The sample exhibited electrochemical double layer capacitance as a result of a higher carbon content and a lower oxygen content. From the curves, to understand the specific capacitance value of the sample, the absolute area is need to be measured using origin software for the oxidation and reduction process. The specific capacitance is directly proportional to the absolute area obtained from the redox process.

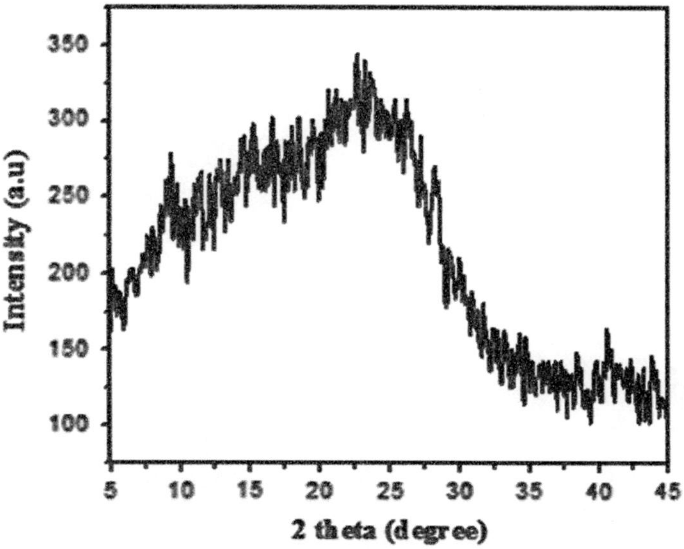

Fig. 2 XRD pattern of KOH activated carbon derived from *ziziphus jujuba* seeds

The curve between potential window and current is plotted from potential CV 4 limits method. From the CV spectra, there is a slight increase in current for the potential from 0 to 4 V. But, once the potential reached 4 V, there is a remarkable increase in current because of the oxidation. When compared to 5 mV s^{-1}, for the scan rate of 5 mV s^{-1} there is an improvement in current. But at the same time for the remaining scan rate the current is reduced because of the instability nature of the material.

Galvanostatic charge and discharge analysis were performed at electrochemical workstation using working electrode, reference electrode, counter electrode and an electrolyte. By maintaining the current, the corresponding potential is calculated and graph is plotted between potential versus time. GCD analysis provides the specific capacitance of prepared samples at different current densities such as 1, 2, 4 and 10 A g^{-1}. The GCD spectra of activated carbon for different current densities are shown in Fig. 5. The CZJS electrode exhibited a specific capacitance of 55.56, 47.62, 16.67, 5.56 F g^{-1} at 1, 2, 4, and 10 A g^{-1}, respectively. This value is compared with the existing work and was identified that the seeds give more specific capacitance. Furthermore, using this value, energy and power density can be calculated. Symmetric nature of CZJS is observed from the GCD analysis at current density over − 0.2 to 0.2 V signifying an ideal capacitive performance. The CZJS electrode possesses excellent capacitive properties. The performance of the CZJS electrodes was significantly improved due to the porous nature (Table 1).

EIS was recorded at the potential of 20 mV. EIS analysis of the CJZS electrode is shown in Fig. 6. By correlating real and imaginary resistance, series resistance and charge transfer resistance can be measured.

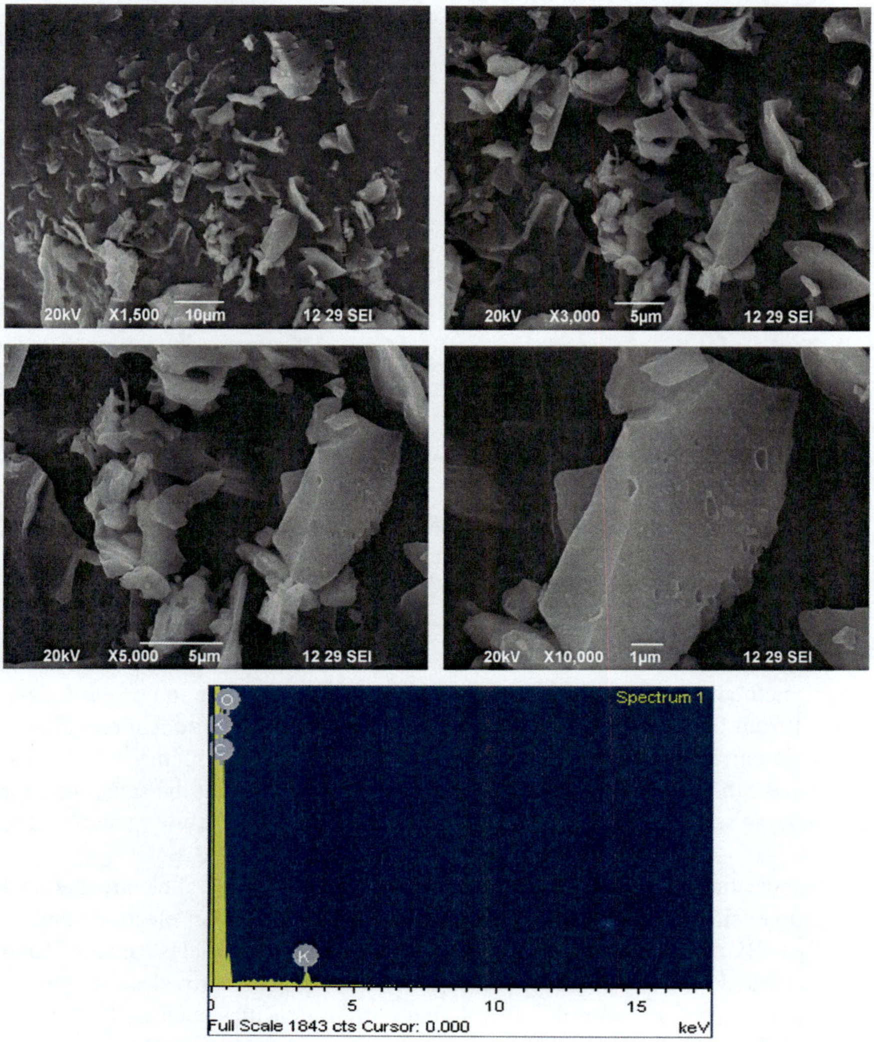

Element	App Conc.	Intensity Corrn.	Weight%	Weight% Sigma	Atomic%
C K	102.61	1.6738	79.90	1.18	84.50
O K	5.02	0.3419	19.12	1.18	15.18
K K	0.78	1.0440	0.98	0.10	0.32
Totals			100.00		

Fig. 3 SEM and EDAX spectra of KOH activated carbon derived from *ziziphus jujuba* seeds

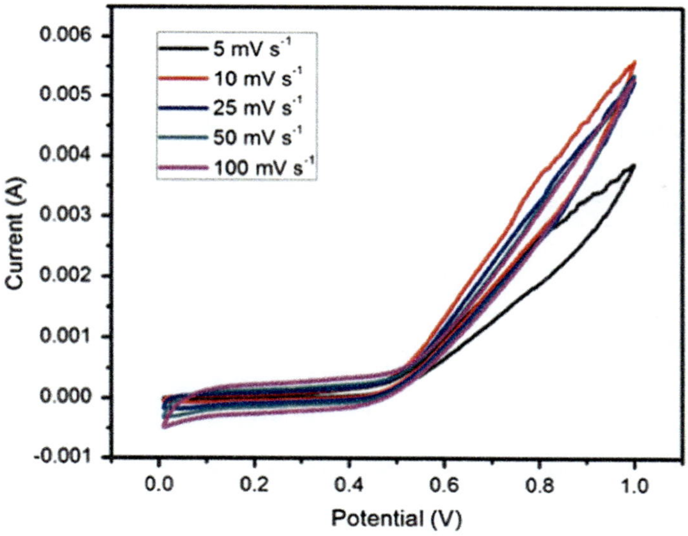

Fig. 4 CV spectra of activated carbon derived from the *ziziphus jujuba* seed

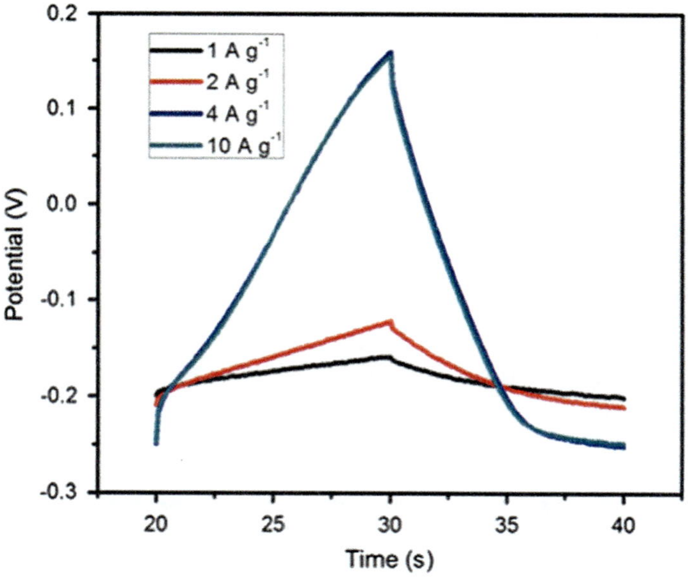

Fig. 5 GCD spectra of activated carbon for different current densities

Table 1 Specific capacitance of various bio-resources

Bio-source	Materials Obtained	Methods	Specific capacitance	R_s and R_{ct}	Refs.
Palm oil biomass	Activated carbon	Hydrothermal carbonization and pyrolysis	1.7554 F g^{-1}	-	[5]
Red cedar wood	Biochar	Pyrolysis and carbonization	12.5 F g^{-1} at 0.5 A g^{-1}	-,-	[43]
Banana peel	Activated carbon	Carbonization and activation	55 F g^{-1} at 1 mV s^{-1}	-,-	[44]
Ficus religiosa leaves	Carbon	Carbonization	3.4 F g^{-1}	-,-	[45]
Lemon waste	Carbon	Hydrothermal method	17.5 F g^{-1}	-,-	[46]
Cotton yarn	Carbon	Carbonization and chemical activation	10.06 F g^{-1} at 150 mV s^{-1}	-,-	[84]
Ziziphus jujuba	Activated carbon	Carbonization and activation	55.56 F g^{-1} at 1 A g^{-1}	25,20	This work

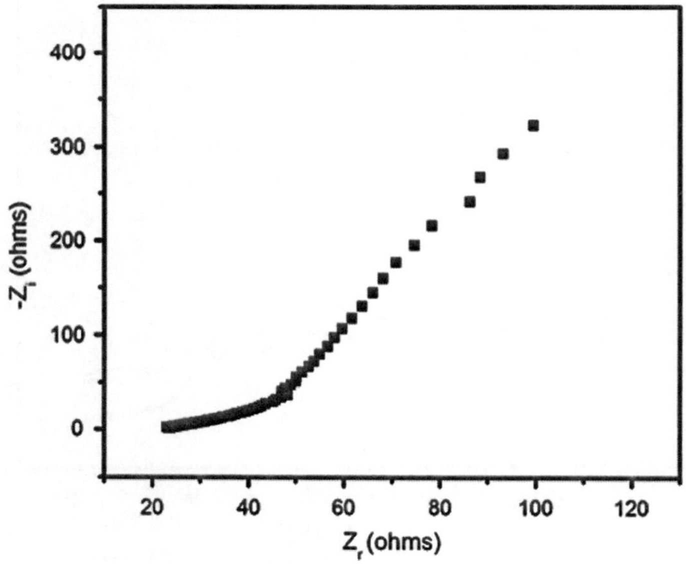

Fig. 6 The EIS spectra of activated carbon depict the Warburg resistance

Charge transfer resistance (R_{ct}) is calculated by subtracting the series resistance from the end point of the semicircle. In general, the Warburg resistance is observed after the end of the semicircle. So, Warburg resistance (45°) is always associated with the charge transfer resistance, and EDLC, is observed from the graph at low frequencies. Linear improvement of resistance is observed in the low-frequency region, indicating an excellent fast ion diffusion behaviour and electrical conductivity in the CJZS electrode. R_s and R_{ct} of the sample are observed to be 25 Ω and 20 Ω, respectively. Hypothetically, the charge transfer resistance may occur because the presence of a low-cost rubber solution in the activated sample adversely affects the electrochemical performance.

4 Conclusion

In this work, uses of *ziziphus jujuba* bio-waste in the energy storage were discussed along with the carbonization and activation process. For the 6 h duration, at the carbonization temperature of 450 °C, a higher carbon content is observed. Then it was activated with KOH by mixing them in a 2:1 mass ratio. A slurry of activated carbon is coated on the tip of the graphite rod for electrochemical measurements. On the basis of XRD measurements, the presence of carbon is justified. Also, an amorphous nature of the sample is found. Through the elemental analysis, a good amount of carbon (79.9%) is observed which is mainly needed condition for a good electrode material along with 19.12% oxygen. Furthermore, it can be easily activated using the activating agent. By analyzing the SEM images, a layered morphology is observed. From these structural analyses, it was concluded that the samples exhibited an amorphous, layered morphology with a higher carbon content that supported the influence of activated carbon electrodes in the supercapacitor. By means of CV, GCD and EIS analysis, this work estimates the electrochemical performance of the CZJS electrode supercapacitor with measured specific capacitance of 55.6 F g^{-1}, series resistance of 25 Ω and charge transfer resistance of 20 Ω. From the experiments, it was found that the CZJS electrode showed a better specific capacitance. Thus, a cheap and eco-friendly material has been developed for use in electrochemical supercapacitor. The scope of this supercapacitor lies in wearable electronics that is very compact portable and also provide a longer shelf life.

Acknowledgements The authors thank the FIST, Department of Science and Technology (SR/FST/COLLEGE-096/2017), India for financial support.

References

1. Kandasamy SK, Singaram KN, Krishnamoorthy H, Arumugam C, Palanisamy S, Kandasamy K, Boddula R, Khan A, Asiri AM, Kolosov AE (2021) Microwave assisted graphene based conducting polymer materials for supercapacitors. Handbook of supercapacitor materials: synthesis, characterization, and applications, Wiley-VCH, New Jersey, United States, 299–321
2. Kandasamy SK, Kandasamy K (2018) J Inorg Organomet Polym 28:559 https://doi.org/10.1007/s10904-018-0779-x
3. Tamilselvi R, Ramesh M, Lekshmi GS, Bazaka O, Levchenko I, Bazaka K, Mandhakini M (2019) Renew Energy 151:731. https://doi.org/10.1016/j.renene.2019.11.072
4. Sing DC, Joseph B, Velmurugan V, Ravuri S, Grace AN (2018) Int J of Nanosci 17:1760023. https://doi.org/10.1142/S0219581X17600237
5. Hendriansyah R, Devianto H, Prakoso T, Widiatmoko P, Nurdin I, Srimurti S, Kusuma KR (2017) In: 4th international conference on electric vehicular technology (ICEVT). https://doi.org/10.1109/ICEVT.2017.8323526
6. Liu Y, Qu X, Huang G, Xing B, Zhang F, Li B, Zhang C, Cao Y (2020) Nanomaterials 10:1. https://doi.org/10.3390/nano10040808
7. Khan A, Senthil RA, Pan J, Sun Y, Liu X (2020) Batteries Supercap 3:1. https://doi.org/10.1002/batt.202000046
8. Manavalan V, Sankar AB, Rohita DS, Nanaji K, Rao TN, Karthik M (2020) Chemistry Select 5:8759. https://doi.org/10.1002/slct.202001877
9. Kandasamy SK, Arumugam C, Vadivel L, Ganapathi M, Nattudurai N, Kandasamy K (2020) Int J Emerg Technol 11:565
10. Devendran M, Kandasamy SK, Palanisamy S, Selvaraj S, Vetrivel R, Selvarajan R, Govindasamy M, Kandasamy K (2020) Int J Electrochem Sci 15:4379. https://doi.org/10.20964/2020.05.08
11. Kandasamy SK, Subramanian B, Krishnamoorthy H, Arumugam C, Suganthi V, Yuvasri M, Shreelogesh D (2021) J New Mate Electrochem Sys 24:78. https://doi.org/10.14447/jnmes.v24i2.a04
12. Kandasamy SK, Arumugam C, Sajitha AS, Rao SP, Selvaraj S, Vetrivel R, Selvarajan R, Alosaimi AM, Khan A, Hussein MA, Asiri AM (2021) J New Mate Electrochem Sys 24:21. https://doi.org/10.14447/jnmes.v24i1.a04
13. Cheng P, Gao S, Zang P, Yang X, Bai Y, Xu H, Liu Z, Lei Z (2015) Carbon 93:315. https://doi.org/10.1016/j.carbon.2015.05.056
14. He Y, Xiang K, Wang Y, Zhou W, Zhu Y, Xiao L, Chen W, Chen X, Chen H, Cheng H, Lu Z (2019) Carbon 153:330. https://doi.org/10.1016/j.carbon.2019.08.022
15. Veerakumar P, Maiyalagan T, Raj BGS, Guruprasad K, Jiang Z, Lin KC (2018) Arab J Chem 13:2995. https://doi.org/10.1016/j.arabjc.2018.08.009
16. Bai X, Wang Z, Luo J, Wu W, Liang Y, Tong X, Zhao Z (2020) Nano Express 15:88. https://doi.org/10.1186/s11671-020-03305-0
17. Nirmaladevi S, Boopathiraja R, Kandasamy SK, Sathishkumar S, Parthibavarman M (2021) Surf Interfaces 27:101548
18. Palanisamy S, Kandasamy SK, Thangmuthu S, Dhinesh KK, Marimuthu P, Prasanna VR, Borje SG (2021)
19. Bhat VS, Kanagavalli P, Sriram G, Prabhu R, John NS, Veerapandian M, Kurkuri M, Hegde G (2020) J Energy Storage 32:101829. https://doi.org/10.1016/j.est.2020.101829
20. Hong P, Liu X, Zhang X, Peng S, Wang Z, Yang Y, Zhao R, Wang Y (2019) Wil Energy Res 44:988. https://doi.org/10.1002/er.4970
21. Zhang YL, Tang Z (2020) Waste Manage 106:250. https://doi.org/10.1016/j.wasman.2020.03.032
22. Huafang Y, Sun X, Zhu H, Yu Y, Zhu Q, Fu Z, Ta S, Wang L, Zhu H, Zhang Q (2020) Ceram Int 46:5811. https://doi.org/10.1016/j.ceramint.2019.11.031
23. Srinivasan R, Elaiyappillai E, Pandian HP, Vengudusamy R, Johnson PM, Chen SM, Karvembu R (2019) J Electro Anal Chem 849:113382. https://doi.org/10.1016/j.jelechem.2019.113382

24. Rajasekaran SJ, Raghavan V (2020) Diamond & Related Materials. 109, 108038. https://doi.org/10.1016/j.diamond.2020.108038
25. Hor AA, Hashmi SA (2020) Electrochimica Acta 356:136826. https://doi.org/10.1016/j.electacta.2020.136826
26. Lee K, Shabnam L, Faisal SN, Hoang VC,Gomes VG (2020) J Energy Storage 27:101152. https://doi.org/10.1016/j.est.2019.101152
27. Lan D, Chen M, Liu Y, Liang Q, Tu W, Chen Y, Liang J, Qiu F (2020) Mater Electron 31:18541. https://doi.org/10.1007/s10854-020-04398-0
28. Taer E, Apriwandi A, Ningsih YS, Taslim RA (2019) Int J Electrochem Sci 14:2462. https://doi.org/10.20964/2019.03.17
29. Jiang L, Han SO, Pirie M, Kim HH, Seong YH, Kim H, Foord JS (2019) Energy Environ 1. https://doi.org/10.1177/0958305X19882398
30. Meng Q, Zhang J, Wang W, Wang H, Jin Z, Zhao K (2020) Ionics 26:3565. https://doi.org/10.1007/s11581-020-03456-1
31. Ahirrao J, Dinesh S, Tambat AB, Pandit NJ (2018) Chem Sel 4:2610. https://doi.org/10.1002/slct.201803417
32. Chen Y, Hu R, Qi J, Sui Y, He Y, Meng Q, Wei F, Ren Y (2019) Material research express 6:95605. https://doi.org/10.1088/2053-1591/ab2d97TI
33. Palisoc S, Dungo JM, Natividad M (2020) Heliyon 6:3202. https://doi.org/10.1016/j.heliyon.2020.e03202
34. Casillas DCM, Gutierrez IM, Ramos CEA, Vidales HIV, Bulnes CAA, Sanchez VHR, Gallegos AKC (2019) Carbon 148:403. https://doi.org/10.1016/j.carbon.2019.04.017
35. Taer E, Natalia K, Apriwandi A, Taslim R, Agustino A, Farma R (2019) Adv Natural Sci Nanosci Nanotech 11:25007. https://doi.org/10.1088/2043-6254/ab8b60
36. Le PE, Nguyen VY, Sahoo SK, Tseng TY, We KH (2020) Energy materials. J Mater Sci 55:10751. https://doi.org/10.1007/s10853-020-04693-5
37. Ciftyurek E, Bragg D, Oginni O, Levelle R, Singh K, Sivanandan L, Sabolsky EM (2018) Environ Pro Sustain Energy 38:13030. https://doi.org/10.1002/ep.13030
38. Liu F, Gao Y, Huang H, Yan C, Chu X, Xu Z, Wang Z, Zhang H, Xiao X, Yang W (2019) J Colloid Interface Sci 548:322. https://doi.org/10.1016/j.jcis.2019.04.005
39. Qu WH, Xu YY, Lu A, Zhang Q, Li WC (2015) Bioresource Technol 189:285. https://doi.org/10.1016/j.biortech.2015.04.005
40. Kumar KS, Chandrasekaran A, Kannan K, Murugesan G (2018) In: International conference on intelligent computing and communication for smart world (I2C2SW), 241, Erode, India (2018). https://doi.org/10.1109/I2C2SW45816.2018.8997149
41. Kandasamy SK, Kandasamy K (2019) Int J Electrochem Sci 14:4718. https://doi.org/10.20964/2019.05.57
42. Kandasamy SK, Kandasamy K (2019) J New Mate Electrochem Sys 22:125. https://doi.org/10.14447/jnmes.v22i3.a02
43. Jiang J, Lei Z, Xinying W, Nancy H, Kishore R, Fanglin C, Shuguo M (2013) Electrochim Acta 113:481–489. https://doi.org/10.1016/j.electacta.2013.09.121
44. Taer E, Taslim R, Aini Z, Hartati SD, Mustika WS (2017) AIP Conf Proc 1801(1):040004. https://doi.org/10.1063/1.4973093
45. Senthilkumar ST, Selvan RK (2015) Chem Electro Chem 2(8):1111–1116. https://doi.org/10.1002/celc.201500090
46. Senthilkumar ST, Fu N, Liu Y, Wang Y, Zhou L, Huang H (2016) Electrochim Acta 211:411–419. https://doi.org/10.1016/j.electacta.2016.06.059

A Novel Wind Speed Forecasting Framework Using Data Preprocessing Based Adversarial Approach

Bala Saibabu Bommidi, Vishalteja Kosana, Kiran Teeparthi, and Santhosh Madasthu

1 Introduction

The major goal of the 26th United Nations Conference on Climate Change (COP26) is achieving worldwide zero emissions. Protracted objectives established by governments are being raised in order to achieve a net zero by 2050. A swift switch to energy sources such as solar and wind is necessary to meet the world's net zero objectives. Global new installations in the wind energy sector are predicted to reach 93 GW in 2020, increasing installed capacity to 743 GW [1]. As a result, wind energy is crucial to reaching net zero objectives. Since wind speed forecasting (WSF) offers the required skills to reduce potential losses, it is essential for better wind energy integration, the development of smart grids, and optimal dispatch scheduling. Because of the unpredictable and intermittent character of wind speed time-series, WSF has proven to be a difficult undertaking that needs extreme attention and prudence.

As a result, wind speed forecasting has long piqued the interest of both academics and business people. Physical techniques, statistical methods, and artificial intelligence (AI) methods are the three basic types of WSF approaches. Physical approaches, in particular, include numeric weather prediction (NWP) approaches, which use geographical and meteorological information for WSF using mathematical formulae [2]. Most case studies make it difficult to gather NWP data such as temperature, humidity, pressure, and topographical structure, among other things.

B. S. Bommidi · V. Kosana · K. Teeparthi (✉)
National Institute of Technology Andhra Pradesh, Tadepalligudem, India
e-mail: kiran.t39@nitandhra.ac.in

B. S. Bommidi
e-mail: pp052008@student.nitandhra.ac.in

S. Madasthu
Energy Production and Infrastructure Center (EPIC), University of North Carolina, Charlotte, NC, USA
e-mail: madasthusanthosh@ieee.org

S. Doolla et al. (eds.), *Advances in Clean Energy and Sustainability*,
Green Energy and Technology, https://doi.org/10.1007/978-981-99-2279-6_49

For example, [3] implements a one-day-ahead WSF model utilizing improved NWP information. However, NWP techniques often have greater long-term prediction capacity, they provide poor short-term prediction outcomes. Statistical approaches like autoregressive (AR) approach [4], AR integrated moving average (ARIMA) approach, and its variations, employ historical data and run quickly, but they often cannot manage high uncertainties in the data. These are easy to implement due to the simple principle of WSF. For example, fractional ARIMA [5] is used to build 24-h and 48-h WSF methods. Another example study develops an offshore WSF model based on the seasonal ARIMA [6]. These statistical models are best suited for making short-term predictions.

The AI approaches are strong, and they can perform nonlinear mapping between past and future data. Individual approaches, hybrid methods, and ensemble methods are available in AI approaches. Individual approaches, such as support vector regression (SVR), have high generalization capabilities for obtaining global solutions, although SVR is restricted in its scalability for big datasets [7]. To solve these drawbacks, artificial neural networks (ANNs) are constructed as computational systems composed of a number of linked units to express connections between variables in a different way than previous techniques. They are differentiated by their balanced processing of data and their dynamic processing capabilities. The back-propagation algorithm has emerged as one of the key components that has contributed to contemporary neural network growth [8]. Hybrid approaches and ensemble methods are used to improve accuracy. Hybrid approaches combine the best characteristics of separate methods [9]. For WSF, a hybrid approach based on fuzzy logic and ANNs is proposed [10]. In recent years, hybrid approaches based on the integration of individual methods and decomposition techniques [11] have been developed to improve prediction accuracy. However, selecting an appropriate decomposition method for feature extraction remains challenging. Optimization techniques such as particle swarm optimization (PSO), genetic algorithm (GA), grey wolf optimizer (GWO), and others are also used to develop hybrid approaches. The authors of [12] proposed a WSF approach using hybrid approach based on feature extraction, and had favourable results. Most models obtain their characteristics implicitly from wind variability. As a consequence, the results generated by WSF techniques have uncertainty associated [13]. In resources development, budgeting, and management, removing uncertainty is a fundamental concern. Most algorithms fail to deliver reliable results for more than 30 min ahead of WSF. The performance declines as the ahead value rises because the models fail to adequately interpret the wind speed to execute WSF. Despite the fact that different methodologies exist in the literature, a WSF method that offers the lowest error while being extremely scalable is still required.

Addressing these, this paper aims mainly to propose a framework that can accurately forecast 1 h ahead (1HA) wind speed. The contributions of this study can be summarized as follows: The residual noise present in the data is eliminated using robust complete ensemble empirical mode decomposition with adaptive noise (CEEMDAN) algorithm. The problem of WSF is articulated as a min-max game based on the generative adversarial network (GAN) framework. This formulation is used to extract the dynamic hidden features in the wind speed time series data distribution. To gain the merits of two robust WSF models, this paper adopted 1 dimensional convolutional neural network (1D-CNN) and long short term memory (LSTM) models and designed this framework for WSF accurately through adversarial training. The proposed framework showed dominating performance for 1HA WSF. The proposed framework performance outperforms eleven reference methodologies by 25%.

The remainder is designed as follows: proposed framework methodology, and working mechanism is presented in Sect. 2. Section 3 provides the experimental results, and discussions for the 1HA WSF. Finally, the conclusions and references follow the results and discussions.

2 Methodology

In this section, the proposed adversarial approach for WSF is demonstrated. The proposed system is classified into data decomposition using the CEEMDAN algorithm and forecasting using the Wasserstein GAN with gradient penalty (WGAN-GP) architecture. Each stage is explained in detail in the subsections that follow.

2.1 Data Preprocessing Using CEEMDAN Algorithm

The wind is largely unpredictable. Because of the unpredictability and non-linearity of the wind, the wind farm data includes a large amount of noise, which impacts the WSF efficiency. The model's efficiency is enhanced by removing noise. As a consequence, this study implemented the robust CEEMDAN for de-noising the input wind speed. Data is divided into eight different intrinsic mode functions (IMFs) using this method. Since this IMF1 signal is seen as noise, it is discarded. The remaining IMFs are concatenated to make a noise-free copy of the original signal. As a consequence, a new noiseless signal is produced. Figure 1 shows the decomposition results using CEEMDAN algorithm.

2.2 Adversarial Approach for WSF

The LSTM and CNN are trained in an adversarial approach for the 1HA WSF. The flowchart of this approach is shown in Fig. 2. The WSF problem is formulated as a min-max game using the GAN framework. This formulation is used to extract dynamically concealed characteristics from the time series data distribution of wind speed. The description of the generator and critic is provided below:

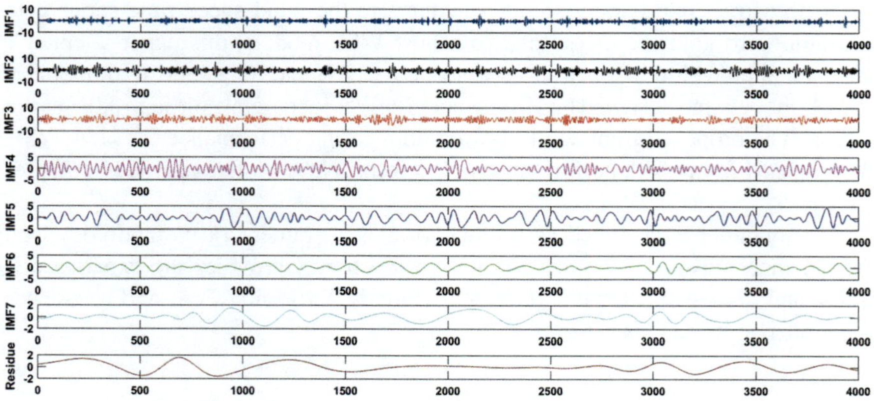

Fig. 1 Result of decomposition using CEEMDAN algorithm

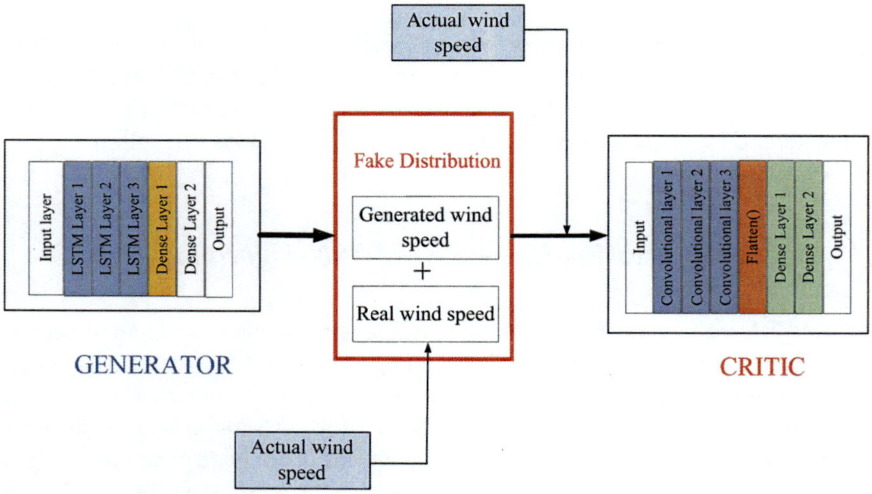

Fig. 2 Flowchart of the proposed adversarial approach for WSF

Generator

Keeping the improved performance of the LSTM in time-series applications, this paper adopted LSTM for the construction of the generator. The input and the outputs of the generator are to be formulated. The input of the generator is constructed by $X = \{x_1, \ldots, x_t\}$, which consists of the wind speed of window size of t hours and the output of the generator is given by Y. To attain the stability of the generator, the architecture consists of three LSTM layers with 128, 64, and 32 hidden neurons and concluded by two dense layers. The number of neurons present at the final layer is equal to 1, i.e. wind speed at $t + 1$. The generator's output is given by $G(X)$, where X denotes the wind speed in the window size.

Critic

The critic's main objective is to discriminate the actual distribution and the generated distribution of the data. To optimize the approach as a robust network, 1D CNN is adopted for the critic. The input of the critic D is fed either from the original distribution of data P_r or from the generated samples P_g from the generator.

There are three 1D convolutional layers present with each layer of filters 128, 64, and 32. The convoluted output is flattened and fed to a dense layer with neurons 16, and the final layer concludes with one output neuron. For all the layers in the critic model, the Activation layer is chosen to be LeakyReLU as it successfully fixes the dying ReLU, i.e. the prevention of vanishing gradient. In addition, LeakyReLU also speeds up the training process. The discriminator used is named as a critic as it is not giving the output as in the case of the normal discriminator. The discriminator in basic GAN produces an output with values ranging from 0 to 1 as a sigmoid activation is used in the concluding layer. Whereas in the case of the proposed framework, the critic is not trained to expel the probability output. Instead to produce a scalar output. Thus, the concluding layer is activated using a linear activation function. As a result, the discriminator is named a critic, which outputs a scalar value that demonstrates how close the generated distribution P_g and real distribution P_r are. The overall loss functions for the generator and the critic used in the proposed framework are formulated in 5 and 6. Overcoming the difficulties faced by the basic GAN, Wasserstein GAN, the proposed framework employed the gradient policy to enforce the Lipschitz constraint to speed up the training process. Because a lower batch size speeds up training, and improves the performance of the WSF, the batch size is chosen as 64, and the weights are optimized using the Adam optimizer at a learning rate of 0.001.

$$Critic's\ Loss : \frac{1}{N}\sum_{i=1}^{N}\left[D(y^i) - D(G(x^i)) + \lambda E\left(||\nabla D_{y^i \sim x^i}||_2 - 1\right)^2\right] \quad (1)$$

$$Generator's\ Loss : -\frac{1}{N}\sum_{i=1}^{N}(D(G(x^i)) \quad (2)$$

where $G(x^i)$ indicates the output of the generated samples. The negative sign $(-)$ is used to maximize the loss function as discussed in the theoretical study. The historical wind speed data is concatenated with the generated output and given as the input to the critic. This is a crucial step that enhances the training of the critic and performance. The hyper parameter λ is chosen as 10 to ensure better training. The generator is trained twice in this research, but the critic is trained just once. The purpose for this is to increase the generator's predicting performance (Table 1).

Table 1 The statistics at the Oregon wind farm

Mean	Median	Skewness	SD	Range	Minimum	Maximum
7.75	7.67	0.15	2.80	14.29	0.09	15.28

SD standard deviation

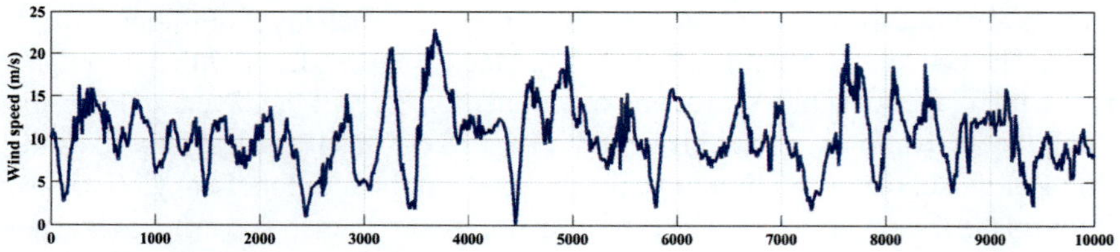

Fig. 3 The original wind speed variation at Oregon wind farm

3 Analytical Study

A wind speed dataset from a wind farm in Oregon, USA, is used in this study [14]. Wind suffers from repeated highs and lows due to the wind farm's geographical location as shown in Fig. 3. Thus, it makes the WSF a difficult process. So, this dataset is selected for evaluating the proposed framework for 1HA WSF.

3.1 Evaluation Criteria

In this work, primary statistical evaluation metrics such as mean absolute error (MAE), mean squared error (MSE), and root mean squared error (RMSE) are used to compare the hybrid approach with other benchmark approaches. The formulas are given in Eqs. 3–6. The better the forecasting method, the lesser the prediction error. The R^2 values indicate how well the forecasting method fit the given data. The greater the R^2 value, the better the prediction approach.

$$RMSE = \sqrt{\frac{1}{N}\sum_{i=1}^{M} E(i)^2} \tag{3}$$

$$MAE = \frac{1}{N}\sum_{n=1}^{M} |E(i)| \tag{4}$$

$$MSE = \frac{1}{N}\sum_{i=1}^{M} E(i)^2 \tag{5}$$

$$R^2 = 1 - \sum_{i=1}^{N} \frac{E(i)^2}{(t_i - \overline{y})^2} \tag{6}$$

where, $E(i)$: t(i) – y(i); N: total samples; $t(i)$: predicted value; $y(i)$: actual value; $\overline{y}$: Average value of actual wind speed.

3.2 Results and Discussions

To evaluate the performance of the proposed framework, in this paper, two categories of approaches are compared. Methods, such as CNN, MLP, GRU, RNN, and LSTM, fall into the first group of models. All these models from the first category are hybridized using the CEEMDAN algorithm for denoising. As a result CEEMDAN-LSTM, CEEMDAN-CNN, CEEMDAN-MLP, CEEMDAN-GRU, and CEEMDAN-RNN fall into the second hybrid category of models. The two category models are trained over a fixed number of epochs of 100. The parameters of the models are determined using the trial and error procedure. The hyperparameters of the CEEMDAN algorithm are preset. The proposed framework and all the reference models are tested for 1HA WSF, and the performance indices obtained are shown in Table 2. The metrics are illustrated through bar charts through Figs. 4 and 5. The illustration of the predicted output of the proposed framework with the original speed is shown in Fig. 6. Figure 7 shows the 1HA WSF results of the models with CEEMDAN algorithm and proposed approach.

Based on Table 2 and Figs. 4, 5, 6, 7, the proposed framework achieved the best results compared to the first category models and second category models with CEEMDAN decomposition. Compared to the MSE, CNN has a value of 2.0841, which is the least among the first category models. However, CNN's dominant per-

Table 2 Comparison of evaluation criteria on eleven models for 1HA WSF

Model	MSE	RMSE	MAE	R^2 score
CNN	2.0841	1.2761	1.0067	0.8307
MLP	2.4022	1.5499	1.2344	0.7862
GRU	2.1675	1.4722	1.0706	0.8069
LSTM	2.1576	1.4689	1.054	0.8178
RNN	2.3031	1.5176	1.0889	0.7948
CEEMDAN-LSTM	1.2978	1.1392	0.8164	0.8797
CEEMDAN-CNN	1.0361	1.0179	0.7569	0.9017
CEEMDAN-MLP	1.8241	1.3506	0.993	0.8312
CEEMDAN-RNN	1.6195	1.2726	0.9471	0.8499
CEEMDAN-GRU	1.3589	1.1657	0.8583	0.8741
Proposed framework	0.772	0.8786	0.6459	0.9317

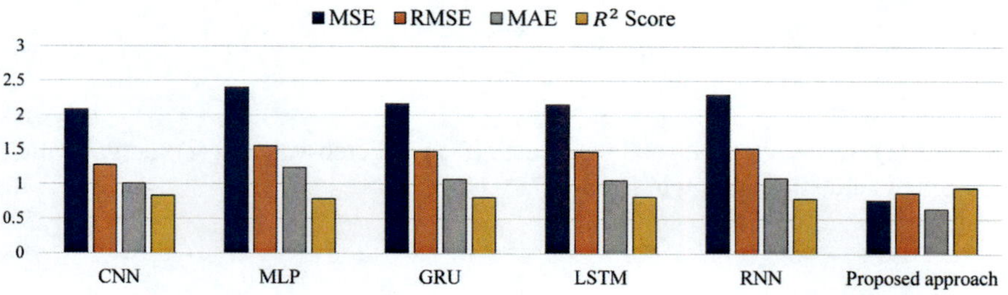

Fig. 4 Comparison of evaluation metrics of proposed approach and models without CEEMDAN for 1HA WSF

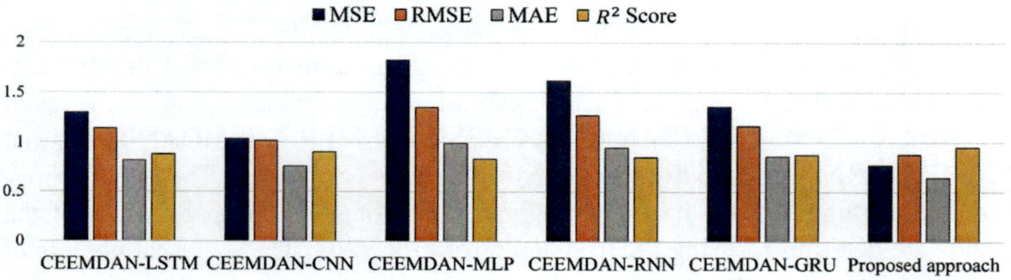

Fig. 5 Comparison of evaluation metrics of proposed approach and models with CEEMDAN for 1HA WSF

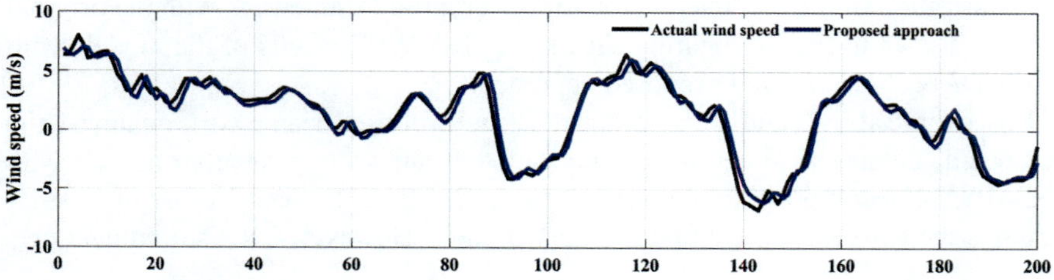

Fig. 6 Predicted 1HA wind speed by the proposed adversarial approach and original speed

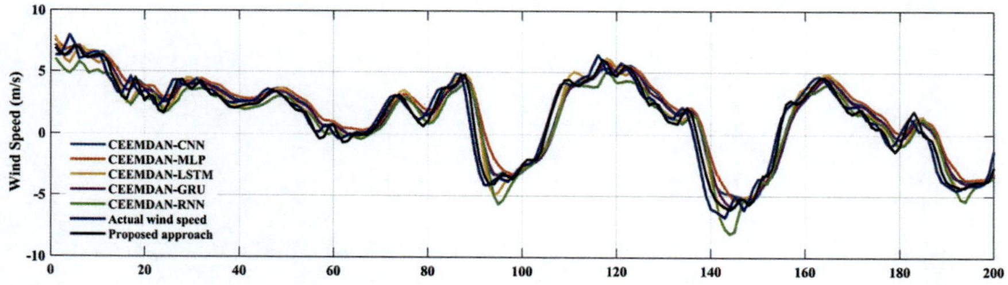

Fig. 7 1HA WSF result of the WSF approaches with CEEMDAN algorithm proposed adversarial approach

formance is confined to only the first category. When comparing MSE, RMSE, MAE, and R^2 scores, category 2 models with data denoising technique outperformed category 1 models without denoising technique. In the second category of models, the proposed framework obtained even better metrics than all the hybrid WSF approaches. For example, when MSE values are considered, the proposed framework achieved 0.772, which is less than the CEEMDAN-CNN approach. The CEEMDAN-CNN and CEEMDAN-LSTM approaches are ranked second and third in the MSE, respectively. The last two spots are occupied by the first category RNN and MLP models due to the poor interpretation of the characteristics of wind speed of 1HA. Similarly, comparing the RMSE values, the proposed framework obtained the best value, followed by the CEEMDAN-CNN, and CEEMDAN-LSTM approaches. The improvement in the performance of the proposed framework over the second best approach is around 25%, and 13% in the MSE, and RMSE, respectively. Considering the MAE values, the last two spots are occupied by the RNN and MLP approaches. The proposed framework outperformed the best reference model by 14%. The proposed framework achieved a 0.9317 in the R^2 score, which is higher than the other compared models, demonstrating the efficiency in learning the hidden properties from the 1HA wind speed data. From Fig. 6, the predicted speed of the proposed framework is inline with the actual wind speed, whereas in Fig. 7, it is evident that all the comparative hybrid approaches are highly deviating from the actual wind speed. Based on the evaluation indices, the proposed framework predicts the accurate wind speed than the eleven comparative models for 1HA WSF due to excellent feature extraction, denoising, and feature interpretation.

4 Conclusions

In this paper, a hybrid and novel approach for accurate wind speed forecasting is proposed. The WSF problem is formulated as a min-max game using the GAN framework. This formulation is used to extract the dynamic hidden features in the wind speed time series data distribution. The proposed framework is classified into denoising input data using the CEEMDAN algorithm and wind speed prediction using the adversarial framework. This paper adopted two robust forecasting models, CNN, and LSTM, to develop the proposed adversarial approach. A case study is carried out using wind data from a wind farm located in Oregon to assess the performance of the proposed framework. Various performance measures are used to conduct a comparative analysis of the proposed framework with other relevant models. The performance is improved by around 25% over the eleven comparative approaches.

However, the proposed methodology does have a limitation. For instance, a time-consuming process of trial and error is used to choose the hyperparameters for the proposed framework. Therefore, the future study will examine heuristic optimization techniques for the automated parameter selection for choosing and optimizing the model parameters. Future research will also examine into how the proposed framework can be used to perform the probabilistic wind speed forecasting.

References

1. Global Wind Energy Council (2021) Global wind report 2021, 25 Mar 2021
2. Cassola F, Burlando M (2012) Wind speed and wind energy forecast through Kalman filtering of numerical weather prediction model output. Appl Energy 99:154–166
3. Zhao X, Liu J, Yu D, Chang J (2018) One-day-ahead probabilistic wind speed forecast based on optimized numerical weather prediction data. Energy Convers Manage 164:560–569
4. Koivisto M, Seppänen J, Mellin I, Ekström J, Millar J, Mammarella I, Komppula M, Lehtonen M (2016) Wind speed modeling using a vector autoregressive process with a time-dependent intercept term. Int J Electr Power Energy Syst 77:91–99
5. Kavasseri RG, Seetharaman K (2009) Day-ahead wind speed forecasting using f-ARIMA models. Renew Energy 34(5):1388–1393
6. Liu X, Lin Z, Feng Z (2021) Short-term offshore wind speed forecast by seasonal ARIMA—a comparison against GRU and LSTM. Energy 227:120492
7. Liu M, Cao Z, Zhang J, Wang L, Huang C, Luo X (2020) Short-term wind speed forecasting based on the Jaya-SVM model. Int J Electr Power Energy Syst 121:106056
8. Tang Z, Zhao G, Ouyang T (2021) Two-phase deep learning model for short-term wind direction forecasting. Renew Energy 173:1005–1016
9. Nair KR, Vanitha V, Jisma M (2017) Forecasting of wind speed using ANN, ARIMA and hybrid models. In: 2017 international conference on intelligent computing, instrumentation and control technologies (ICICICT). IEEE, pp 170–175
10. Monfared M, Rastegar H, Kojabadi HM (2009) A new strategy for wind speed forecasting using artificial intelligent methods. Renew Energy 34(3):845–848
11. Emeksiz C, Tan M (2022) Multi-step wind speed forecasting and Hurst analysis using novel hybrid secondary decomposition approach. Energy 238:121764
12. Kosana V, Teeparthi K, Madasthu S, Kumar S (2022) A novel reinforced online model selection using Q-learning technique for wind speed prediction. Sustain Energy Technol Assess 49:101780
13. Liu D, Niu D, Wang H, Fan L (2014) Short-term wind speed forecasting using wavelet transform and support vector machines optimized by genetic algorithm. Renew Energy 62:592–597
14. NREL. www.nrel.gov. Accessed 6 Jan 2021

Comparison of Different Types of Coal Powerplants in India Retrofitted with Calcium Looping Based CCS System

Srinath Haran, Anand B. Rao, and Rangan Banerjee

Nomenclature

Acronyms

ASU	Air separation unit
BFPT	Boiler feed pump turbine
Cyc	Cyclone
FWH	Feed water heater
G	Generator
HPT	High pressure turbine
RH	Reheater
SH	Superheater
TRL	Technology readiness level

S. Haran (✉) · A. B. Rao · R. Banerjee
Interdisciplinary Programme (IDP) in Climate Studies, Indian Institute of Technology Bombay, Mumbai 400076, India
e-mail: 174400002@iitb.ac.in

A. B. Rao
Centre for Technology Alternatives for Rural Areas (CTARA), Indian Institute of Technology Bombay, Mumbai 400076, India

R. Banerjee
Department of Energy Science and Engineering, Indian Institute of Technology Bombay, Mumbai 400076, India

Department of Energy Science and Engineering, Indian Institute of Technology Delhi, Hauz Khas, New Delhi 110016, India

S. Doolla et al. (eds.), *Advances in Clean Energy and Sustainability*,
Green Energy and Technology, https://doi.org/10.1007/978-981-99-2279-6_50

1 Introduction

CO_2 emissions from coal powerplants are primarily responsible for the problem of global warming and subsequent climate change. The use of carbon capture and storage (CCS) technologies can enable effective CO_2 capture from coal powerplants [1]. Although various CCS technologies are available for CO_2 capture at coal powerplants, they are at different stages of technology development. MEA (monoethanolamine), a liquid solvent based system is the only commercial technology that is available as of today [2]. However, many new and advanced CCS technologies (having the capability to show improved performance) are coming up. The current paper deals with one such relatively new CCS technology, the calcium looping based post combustion CCS system [3].

1.1 Calcium Looping Based CCS System

The calcium looping (CaL) based CCS system is a post combustion CO_2 capture system and consists of the carbonator and the calciner reactors where the carbonation (exothermic) and calcination (endothermic) reactions take place, as can be seen from Eqs. (1) and (2) respectively [3].

$$CaO(s) + CO_2(g) \rightarrow CaCO_3(s) + Heat,\ \ \Delta H_{298K} = -178\ kJ/mol\ of\ CaCO_3 \quad (1)$$

$$CaCO_3(s) \rightarrow CaO(s) + CO_2(g) - Heat,\ \ \Delta H_{298K} = +178\ kJ/mol\ of\ CaCO_3 \quad (2)$$

The schematic diagram of the calcium looping system is as shown in Fig. 1 [4]. In the carbonator, the CO_2 from the coal powerplant flue gas reacts with calcium oxide (CaO) at around 650 °C and gets converted to calcium carbonate ($CaCO_3$). The $CaCO_3$ is then sent to the calciner where, heat supplied via coal oxycombustion helps in regeneration of CaO (at around 900 °C) and a concentrated stream of CO_2 is generated, which is compressed and sent for storage. As the calcium looping process takes place at high temperatures, various high temperature heat sources can be used for additional power generation through a secondary steam cycle [3]. The calcium looping based CCS technology has developed rapidly to TRL 6 (pilot plant) [2].

1.2 Research Gaps and Objective of the Study

Many process modeling studies dealing with thermodynamic and/or technoeconomic assessment of different types of coal powerplants (subcritical/supercritical/ultrasupercritical) retrofitted with calcium looping based CCS system, using different types of coal varieties and using different steam conditions

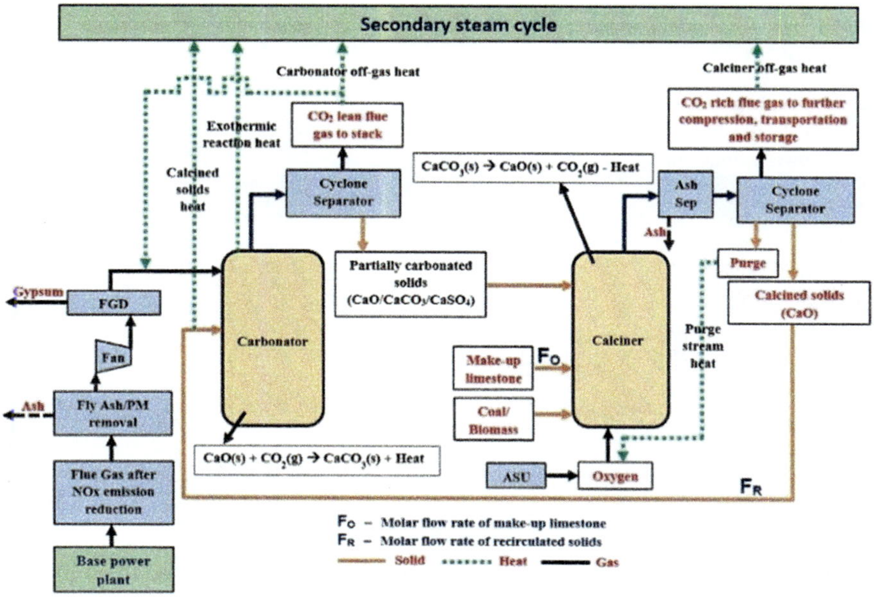

Fig. 1 Schematic diagram of the calcium looping based CCS system

in the secondary steam cycle can be found in literature [4–6]. However, a comparative thermodynamic assessment of different types coal powerplants retrofitted with calcium looping based CCS system has not yet been performed.

The objective of the current paper is to perform a comparative thermodynamic assessment of different types of coal powerplants (such as subcritical, supercritical and ultrasupercritical) retrofitted with calcium looping based post-combustion CCS technology. The results of this comparative thermodynamic assessment would help to assess the overall feasibility the calcium looping based CCS system by measuring various performance parameters such as net power output, overall (electric) efficiency, CO_2 emissions avoided and primary energy consumed.

2 Methodology and Assumptions

The current study involves modeling, simulation and thermodynamic assessment of (a) different types of coal powerplants (subcritical, supercritical and ultrasupercritical and (b) different types of coal powerplants retrofitted with calcium looping based CCS system (referred to as "integrated systems"). The data and assumptions for different types of coal powerplants have been taken from [7]. It is to be noted that two (subcritical and supercritical) of these three plants are coal powerplants currently operating in India. Some of the important details about different types of coal powerplants and the CCS system used in this study can be seen from Table 1.

Table 1 Details about different types of coal powerplants analyzed in this study

Parameter	Case 1	Case 2	Case 3
Type of steam cycle in base powerplant	Subcritical	Supercritical	Ultrasupercritical
Gross power output (MWe)	500	660	430
Steam conditions (HPT inlet pressure/HPT inlet temperature/Reheat temperature)	166.7 bar/537 °C/537 °C	242.2 bar/537 °C/565 °C	350 bar/700 °C/720 °C
CCS technology employed	Calcium looping		
Secondary steam cycle conditions	Supercritical		
Coal used	Indian high ash coal		

The modeling and simulation has been performed using Aspen Plus [8] software. For thermodynamic assessment of different coal powerplants, the property methods used are "PR-BM" (Peng-Robinson equation of state with Boston-Mathias alpha function) for air, coal and flue gas streams and "STEAMNBS" steam tables for steam and water streams respectively. For thermodynamic assessment of calcium looping based CCS system, the property methods used are "RK-SOAVE" for air, coal, flue gas and other solid streams and "STEAMNBS" steam tables for steam and water streams of secondary steam cycle. The model parameters used for simulation of calcium looping based CCS system are calculated as per equations given in [4]. The performance parameters evaluated and compared for different types of coal powerplants and the integrated systems include (a) Net power output, (b) Overall (electric) efficiency, and (c) Net specific CO_2 emissions. For the integrated system, the efficiency penalty and "SPECCA" (Specific Primary Energy Consumption for CO_2 Avoided) are also evaluated. SPECCA is defined as the ratio of difference in net heat rate of the powerplant before and after CO_2 capture to the difference in net specific CO_2 emissions [4, 9]. The unit of SPECCA is MJ_{HHV}/kg_{CO2}.

2.1 *Working of the Integrated System (Base Powerplant + Calcium Looping Based CCS System + Secondary Steam Cycle)*

The integrated system schematic diagram is as shown in Fig. 2 [4]. In the base powerplant, flue gas undergoes NO_x, PM and SO_x removal. Further, preheating of the flue gas (stream no. 4) takes place by exchanging heat with the CO_2-lean flue gas (stream no. 7) in a heat exchanger (HX-8) before it enters the carbonator. The

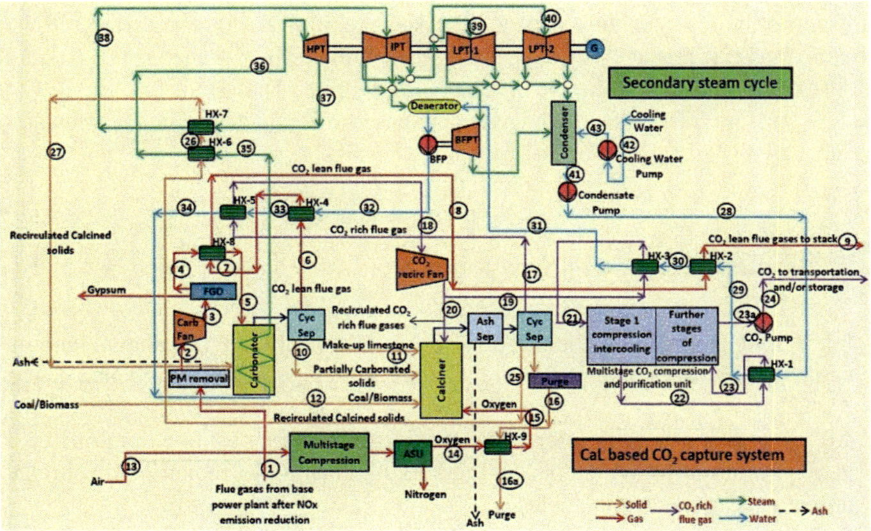

Fig. 2 Schematic diagram of the integrated system (base powerplant + calcium looping system + secondary steam cycle)

preheated flue gas (stream no. 5) enters the carbonator, where CO_2 from the flue gas reacts with recirculated calcined solids rich in CaO (stream no. 27). The solid–gas mixture then passes through a cyclone separator. The hot CO_2 lean flue gas which is separated, undergoes multiple heat exchange processes before being emitted to the atmosphere whereas the partially carbonated solids (stream no. 10) are transported to the calciner where the calcination reaction (separation of $CaCO_3$ into CaO and CO_2) takes place.

The high temperature heat required for the calcination reaction is achieved via oxycombustion of Indian high ash coal. The solid–gas mixture from the calciner exit is sent first to an ash handling unit and then to a cyclone separator. The flue gas rich in CO_2 generated from the calcination reaction undergoes multiple heat exchange processes and a part it is recirculated back to the calciner, whereas the remaining part undergoes compression in a multistage compressor and is assumed to be transported through a pipeline to a suitable geological site (where it is eventually stored). In each cycle, part of the deactivated solids is removed as purge (Stream 16a). The reduced sorbent activity and removal of purge is compensated by adding make-up limestone (stream no. 11) in the calciner. The remaining part of calcined solids (after exchanging heat with steam from the secondary steam cycle) are then recirculated back to the carbonator to participate in the subsequent CO_2 capture cycle.

The high temperature CO_2-lean, CO_2-rich flue gases and the calcined solids exchange heat with the low and high pressure feedwater streams and steam in the secondary steam cycle, as can be seen from Fig. 2. The carbonator generates steam (stream no. 35) utilizing the exothermic reaction heat. The steam exiting the carbonator is superheated (by exchanging heat with the hot calcined solids (stream no. 25)) and sent to the high pressure turbine. The steam exiting the high pressure turbine (stream no. 37) is reheated (by exchanging heat with the hot calcined solids (stream no. 26)) and sent to the intermediate pressure turbine. The steam further goes to the low pressure turbine and then to the condenser to complete the steam/feedwater cycle.

2.2 *Assumptions of the Study*

- It has been assumed that the location of all powerplants and the calcium looping system is India, where the average ambient temperature is assumed to be 33 °C [7].
- All powerplants and the calcium looping system operate using a domestic coal variety, referred to as "Indian high ash (HA) coal". The Indian high ash coal properties can be seen in Table 2 [7].
- In all the powerplants, NO_x removal takes place with the help of selective catalytic reduction (SCR) unit, SO_x removal by wet flue gas desulphurization (FGD) unit and PM removal by electrostatic precipitator (ESP). Details regarding these pollution control devices can be referred from [10]. Each coal powerplant retrofitted with these pollution control devices is known as the "base powerplant (PP)".
- The CO_2 capture efficiency in the carbonator is assumed to be 90% for all the three cases [4]. Assumptions for carbonator, calciner, solid handling devices, ASU, CO_2 compression unit, secondary steam cycle can be found in [4] .
- The gross power output (rated capacity) of the subcritical, supercritical and ultra-supercritical coal powerplants are different, as can be seen from Table 1. Hence, in order to ensure a fair comparison, the assessment has been performed by assuming the base powerplant gross power output as 660 MW_e for all the three cases.

Table 2 Properties of Indian high ash (HA) coal

	As received (wt %)	Dry basis (wt %)
Carbon	34.46	39.16
Hydrogen	2.43	2.76
Oxygen	6.97	7.92
Nitrogen	0.69	0.78
Sulfur	0.45	0.51
Ash	43	48.87
Moisture	12	–
Higher heating value (HHV) (MJ/kg)	13.96	15.86

3 Results and Discussion

Table 3 shows some of the important results for the different types of coal powerplants and their corresponding integrated system. As compared to Case 1 and Case 2, the coal consumption (kg/MWh$_{net}$) is the lowest for Case 3 for both base powerplant as well as the integrated system. The coal consumption in the base powerplant is low for Case 3, as it operates at higher steam conditions (temperature and pressure), which also results in lower CO_2 emissions. Due to low CO_2 emission, lesser amounts of gas and solids circulate in the calcium looping system for Case 3, thereby leading to less amount of coal required for regeneration of sorbent in the calciner. Hence, coal consumption for the integrated system is the lowest for Case 3.

The secondary steam cycle gross power output relative to base powerplant gross power output is the highest for Case 1 (74.9%) as compared to Case 2 (71.8%) and Case 3 (62.6%). This is because of the high coal consumption for Case 1, which leads to high CO_2 emissions in the base powerplant, thereby making available a larger volume of gas and solids (in the calcium looping system) for heat exchange with water/steam in the secondary steam cycle.

In the following sub-sections, the various performance parameters evaluated are shown as a comparison for the three different cases.

3.1 *Net Power Output*

Figure 3 shows comparison of power output for the three different cases. The net power output of integrated system for Case 1 (subcritical plant) increases by 47% in comparison to the base powerplant net power output, whereas the corresponding net power output increase for Case 2 (supercritical plant) and Case 3 (ultrasupercritical plant) are 45% and 39.9% respectively. The higher net power output increase for Case 1 (in comparison to base powerplant net power output) is mainly due to higher amount of power generated in the secondary steam cycle.

Thus, retrofitting the existing coal powerplants with calcium looping based CCS system would transform them into near zero emission plants and also increase their capacity of electricity generation. The existing coal powerplant fleet in India is dominated by subcritical plants, many of them nearing their end of life. Thus, retrofitting these old subcritical units with calcium looping based CCS system could give them a

Table 3 Some of the important results for the three different cases

	Case 1	Case 2	Case 3
Coal consumed in base powerplant (kg/MWh$_{net}$)	642.6	616.2	545.7
Coal consumed in integrated system (kg/MWh$_{net}$)	825.8	802.6	737.0
Gross power output from secondary steam cycle (MWe)	494.6	473.9	412.9

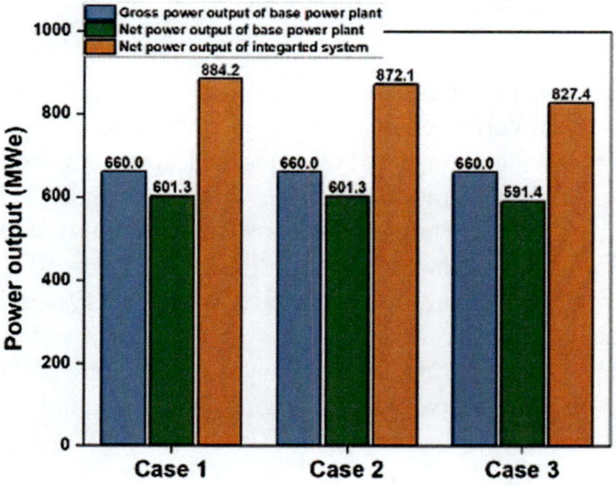

Fig. 3 Comparison of net power output for different cases

new lease of life. The calcium looping based CCS system has the potential to address the twin issues of environmental protection and energy security.

3.2 *Overall (Electric) Efficiency*

Figure 4 shows comparison of overall (electric) efficiency for base powerplant and the integrated system for the three different cases. The highest overall (electric) efficiency for base powerplant (41.6%) as well as the integrated system (30.8%) is obtained for the Case 3 (ultrasupercritical plant), whereas the lowest overall (electric) efficiency is obtained for Case 1 (subcritical plant). However, Case 1 has the lowest efficiency penalty (7.8% points) as compared to Case 2 (8.6% points) and Case 1 (10.8% points). The efficiency penalty is the lowest for Case 1 because it has the highest increase in gross power output of the secondary steam cycle relative to the base powerplant gross power output and at the same time, it has the lowest increase in coal consumption of integrated system relative to the coal consumption of base powerplant.

3.3 *Net Specific CO_2 Emissions*

Figure 5 shows the net specific CO_2 emissions of base powerplant and the integrated system for the three different cases. The CO_2 emissions avoided for the three cases are almost the same (i.e. around 87–88%).

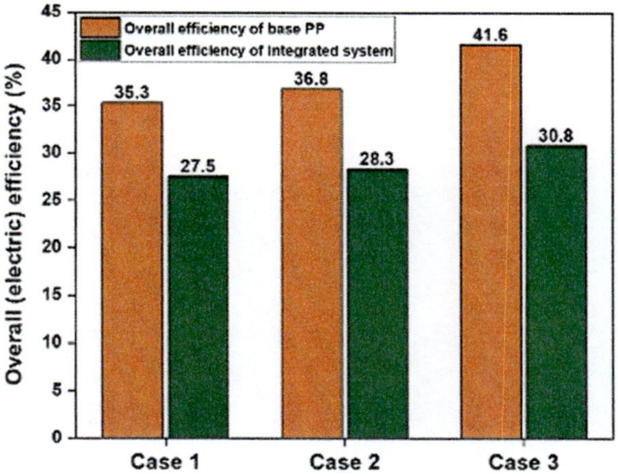

Fig. 4 Comparison of overall (electric) efficiency for different cases

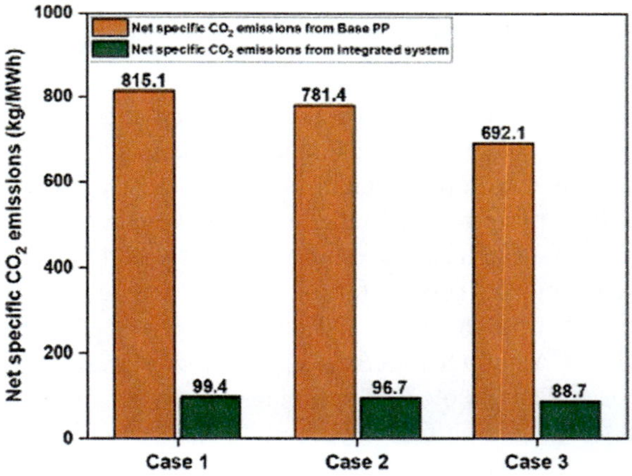

Fig. 5 Net specific CO_2 emissions for different cases

3.4 SPECCA

Figure 6 shows SPECCA value for all the three cases. It is clear that Case 1 has the lowest value of SPECCA (4.06 MJ_{HHV}/kg_{CO2}). This is because Case 1 (28.5%) has the lowest increase in net plant heat rate of the integrated system relative to the net plant heat rate of the base powerplant as compared to Case 2 (30.3%) and Case 3 (35.1%), even as the CO_2 emissions avoided is almost the same for all the three cases (i.e. around 87%).

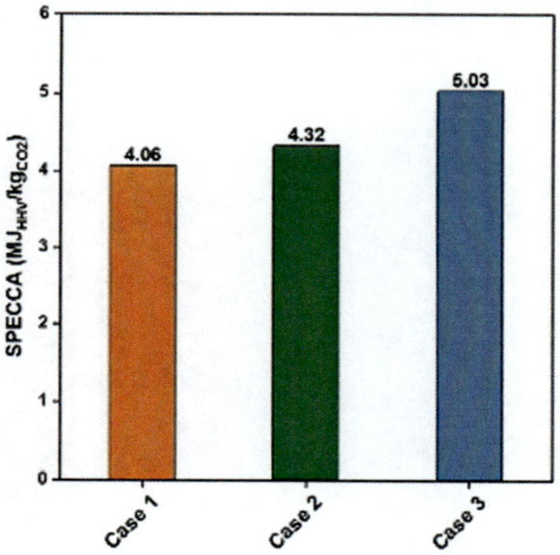

Fig. 6 Comparison of SPECCA for different cases

4 Conclusion

The current paper deals with thermodynamic assessment and comparison of different types of coal powerplants (subcritical, supercritical and ultrasupercritical) in India retrofitted with the calcium looping based CCS system. Modeling and simulation of base powerplant and the integrated system (for all three cases) were performed using Aspen Plus. The results show that Case 1 (subcritical plant) has the highest net power output increase for the integrated system (47%) relative to the base powerplant net power output and lowest values of efficiency penalty (7.8% points) and SPECCA (4.06 MJ_{HHV}/kg_{CO2}). The highest overall (electric) efficiency for the base powerplant (41.6%) and the integrated system (30.8%) is obtained for Case 3, but it has the highest efficiency penalty (10.8% points). Thus, the use of calcium looping based CCS system at existing coal powerplants in India, especially at the old subcritical units would transform them into near zero emission plants and also increase their capacity of electricity generation, thereby having the potential to address the twin issues of environmental protection and energy security. This is the first study for comparative thermodynamic assessment of different types of coal powerplants retrofitted with calcium looping based CCS system. Future studies could look at the comparative economic assessment of the different types of coal powerplants retrofitted with calcium looping based CCS system and also at techno-economic assessment of each type of powerplant retrofitted with different CCS technologies.

References

1. GCCSI (2022) Understanding CCS. https://www.globalccsinstitute.com/why-ccs/what-is-ccs/. Last Accessed 06 May 2022
2. Bui M et al (2018) Carbon capture and storage (CCS): the way forward. Energy Environ Sci 11:1062–1176
3. Shimizu T et al (1999) A twin fluid-bed reactor for removal of CO_2 from combustion processes. Chem Eng Res Des 77:62–68
4. Haran S et al (2021) Techno-economic analysis of a 660 MWe supercritical coal power plant in India retrofitted with calcium looping (CaL) based CO_2 capture system. Int J Greenhouse Gas Control 112:103522
5. Astolfi M et al (2019) Improved flexibility and economics of calcium looping power plants by thermochemical energy storage. Int J Greenhouse Gas Control 83:140–155
6. Martínez I et al (2018) CO_2 capture in existing power plants using second generation Ca-looping systems firing biomass in the calciner. J Clean Prod 187:638–649
7. Suresh MVJJ et al (2010) 3-E analysis of advanced power plants based on high ash coal. Int J Energy Res 34:716–735
8. Aspen Technology Inc (2022) Aspen Plus V8.4. https://www.aspentech.com/products/engineering/aspen-plus. Last accessed 6 May 2022
9. Campanari S et al (2011) Application of MCFCs for active CO_2 capture within natural gas combined cycles. Energy Procedia 4:1235–1242
10. Srinivasan S et al (2018) Benefit cost analysis of emission standards for coal-based thermal power plants in India (CSTEP-Report-2018-06)

Performance Analysis of Bifacial PV Module in Different Climatic Zones of India

Deepak Yadav, Birinchi Bora, Arup Dhar, Mugala Naveen Kumar, Jai Prakash, and Chandan Banerjee

1 Introduction

India's use of renewable energy is steadily increasing as it strives to become a carbon- neutral nation. India has a target to install 100 GW of solar energy by 31st December, 2022. Solar PV community is putting efforts to increase the energy output of PV module in the limited land size. Highly efficient PV module such as HIT and Sunpower PV module technologies are used in the field for this purpose [1]. The costs of these highly efficient module technologies are higher as compared to other module technologies. One of the solutions to reduce the cost of energy generation is to use the bifacial module technology. For India, Bifacial is a new technology and it is not well understood that how this technology will behave in Indian climate in terms of performance and reliability. So, it is very important to estimate the energy generation of the bifacial modules at different climatic zones of India. Very limited literature is available for this technology in the Indian context so it is crucial to study the energy generation and potential of Bifacial Module in Indian context.

The performance of bifacial module depends on the albedo, tilt angle, height from the ground and, latitude of the installation site along with other parameters as compared to mono facial module. Sreenath et al. [2] evaluated the performance of a bifacial solar photovoltaic system under various albedo settings in 2021. He compared bifacial and mono-facial PV modules for four sites on the Pekan campus of the University Malaysia Pahang (UMP), which is located in the Malaysian state of Pahang. The specific yield for bifacial module is 10.58% higher than that estimated for monofacial. Varun and Manikandan et al. [3] evaluated the performance of Bifacial and Monofacial modules in vertical and latitude mounting in South India in 2020 using PVsyst. They claimed that the Latitude tilted Bifacial Panel produced 1000 kW more than the North-South panel's vertical orientation. Urdaneta et al. [4] employed

D. Yadav (✉) · B. Bora · A. Dhar · M. N. Kumar · J. Prakash · C. Banerjee
National Institute of Solar Energy, Gurugram, India
e-mail: deepak.yadav@nise.res.in

S. Doolla et al. (eds.), *Advances in Clean Energy and Sustainability*,
Green Energy and Technology, https://doi.org/10.1007/978-981-99-2279-6_51

a fixed-tilt system and a horizontal single-axis tracking system in the year 2021 to run 321 simulations of three sites with various latitudes, resulting in six multiple linear regression equations. In 2021, Christopher Pike and colleagues released the first comprehensive evaluation of south-facing bifacial and vertical east-west facing bifacial PV modules, as well as, south-facing monofacial and bifacial photovoltaic (PV) modules, in Alaska [5]. With the help of comprehensive opto-electro-thermal simulation framework, Sun et al. [6] performed a worldwide study and optimization for a range of module combinations. Their research found that for an albedo of 0.25, which is common for flora and soil groundcover, the bifacial gain with respect to ground-mounted bifacial module is less than 10% over the world. By increasing albedo to 0.5 with artificial reflectors (e.g., white concrete), the bifacial gain can be doubled to 20%. Additionally, elevating the module 1 m above from the ground can increase the bifacial gain to 30%.

People are doing performance analysis of bifacial module for different applications: Agrivoltaics, fencing, sound barrier. In 2021, Robledo et al. [7] has done the simulation of the energy production of a vertical bifacial photovoltaic module used for agrivoltaic project in France. A new simulation tool was presented that offers a fair balance of accuracy and speed of computation. The use of the simulation tool was demonstrated in a case study of a proposed PV plant in the south of France. It was proposed that the PR for bifacial PV modules should be computed using effective bifa- cial irradiation as a baseline. In 2021, Martnez et al. [8] presents a power rating procedure for hybrid bifacial CPV/flat-plate PV modules based on the CPV, IEC 62670, flat-plate PV, IEC 60904 and IEC 60891 and bifacial PV, IEC TS 60904-1-2 standards. This includes the proposed standard test and operating conditions (STC and SOC) for the terrestrial conversion of the reference AM1.5g spectrum and the filtering criteria required to do so. Using System Advisor Model (SAM) software from the National Renewable Energy Laboratory (NREL), Okere and Iqbal conducted a Comparison of Emerging Solar PV Modules for Utility-Scale PV Installation from a Techno-Economic Perspective. The simulation results showed that the bifacial PV technology is very suitable choice for the same climatic conditions and places with a high likelihood of high reflection. In 2021, Kopecek and Libal [9] summaries the current state of bifacial photovoltaics (PV) and reported that bifacial module technology is the suitable option for agriphotovoltaic with lowest electricity generation cost. Rokonuzzaman et al. [10] undertook a performance analysis of a bifacial module-based time variable multilevel solar panel system (MSPS) in 2021. They reported that a bifacial multilevel solar panel produces 29.37% more energy than a single mono-facial multilevel solar panel. In its 2021 report [11], the International Energy Agency (IEA) reported a bifacial PV modelling comparison to assess the state of the art of bifacial PV performance models. It includes a synopsis of eleven bifacial field test sites from around the world, as well as field results examples.

Based on the above literature review, it has been observed that there are very few studies available related to performance analysis of bifacial PV module in Indian context. In this study, performance analysis of bifacial PV module for 100 cities of different climatic zones has been done in terms of optimal orientation and bifacial gain with respect to monofacial module. The results are likely to help the Customer, Policy Maker, Designer, Researcher in decision making, policy planning and experimentation.

2 Methodology

The simulated performance analysis of 100 different cities of India has been done using SAM software (version SAM 2021.12.02) from the National Renewable Energy Laboratory (NREL). The cities were selected across different climatic zones of India. Before estimating the energy yield, optimal tilt angle of the site has been estimated by using the equation reported by Sun et al. [6]. The equations are:

$$\beta_o = a \times \text{Lat} + b \tag{1}$$

$$a = 0.86 - 0.57 \times \text{A} \times \exp\,(-\text{H/L}) \tag{2}$$

$$b = 4.5 + 62 \times \text{A} \times \exp(-\text{H/L}) \tag{3}$$

$$If\ \beta_o \geq 90^{\circ},\ \beta_{opt} = 90^{\circ} \text{ and } If\ \beta_o < 90^{\circ},\ \beta_{opt} = \beta_o$$

where β_{Opt} is Optimal Tilt Angle, L at is Latitude, H is Ground Clearance Height, L is Module Length and A is Albedo. For Gurugram, $\beta_{Opt} = 37°$ as calculated from the above equations.

Using Eqs (1), (2) and (3), the Optimal tilt of 100 cities of India has been estimated and used to simulate the annual energy output using SAM.

For the simulation the following assumptions are made:

The bifacial gain (BG) for each site has been calculated as per the following formula:

$$BG = \frac{(Y_b - Y_m) \times 100\%}{Y_m} \tag{4}$$

where BG is bifacial gain, *Yb* is annual energy yield of bifacial module, *Ym* is annual energy yield of monofacial module for the site.

3 Results and discussions

The equation used for estimating the optimal tilt angle of installation has been verified by simulating the performance of bifacial module at National Institute of Solar Energy (NISE), Gurugram for different south facing tilt angle using SAM. The latitude and

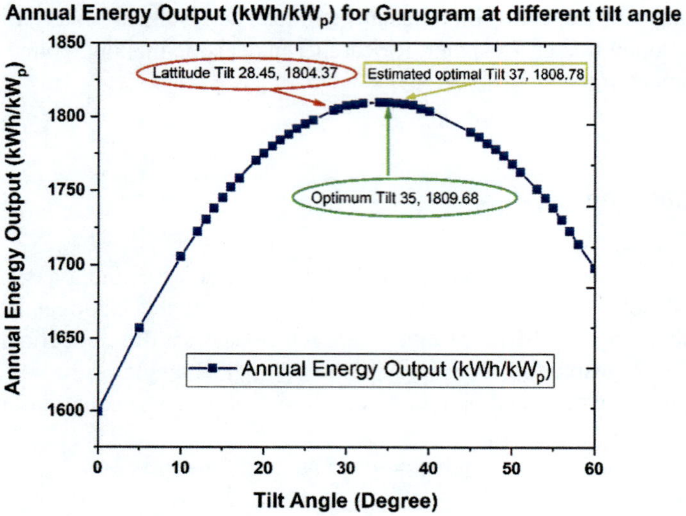

Fig. 1 Annual energy output (kWh/kWp) for NISE, Gurugram at different tilt angle

longitude of the site is 28.42 °N, 77.14 °E respectively. Figure 1 shows the output of bifacial PV plant (kWh/kWp) at different tilt angle for Gurugram. It has been observed that for tilt angle of 35° the energy output is maximum with generation of 1809.68 kWh/kWp, which is in agreement with the optimal tilt angle estimated by using the Eqs (1), (2) and (3). The difference in annual energy output for tilt angle 35° and 37° for Gurugram is only 0.9 kWh/kWp. Therefore, optimal tilt of a particular location can be estimated from the three equations.

The optimum tilt angle for bifacial module also depends on latitude, Elevation and albedo of the site. This has been studied for NISE, Gurugram and shown in Fig. 2. It has been observed that with the increasing albedo the optimum tilt is also increasing for a fixed elevation and latitude. The optimum tilt angle decreases with increasing elevation. At very high elevation the effect of tilt angle and albedo is very less.

In this simulation study the elevation and albedo has been kept fixed. However, in actual installation these parameters will be different as per different installation site. Using the parameters in Table 1, the performance of monofacial and bifacial module were estimated. Figure 3 shows the specific yield (kWh/kWp) for 100 different cities of India for the bifacial module at optimal tilt. The annexure contains more details of every location. Maximum energy output of 1985 to 2325 kWh/yr was observed for bifacial module for nine different cities of India. Minimum energy output of 1427 to 1730 kWh/yr was observed for bifacial module for nine different cities of India. Pasighat has minimum annual energy output while Aksai Chin has maximum Annual Energy output.

The percentage of bifacial gain was estimated from annual energy output data of mono facial and bifacial module using the equation mentioned above. Figure 4 shows the percentage bifacial gain for 100 different cities of India. Maximum bifacial gain of

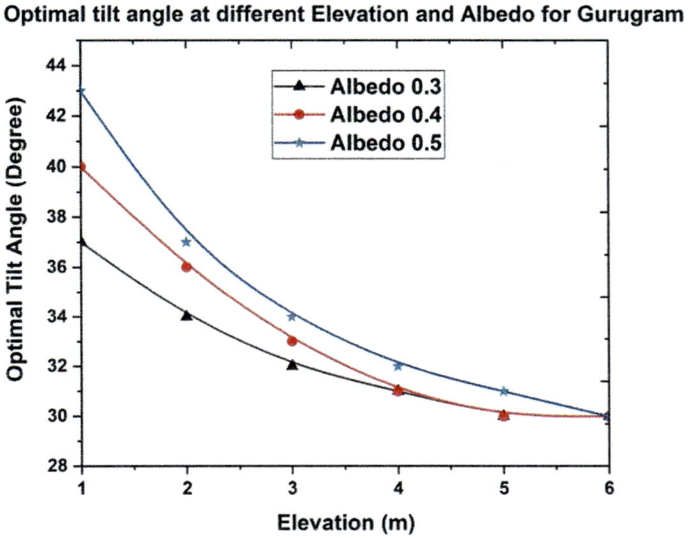

Fig. 2 Optimal tilt angle at different elevation and Albedo for Gurugram

Table 1 Parameters used for simulation

S. No	Parameter	Value
1	Albedo (A)	0.3
2	Bifaciality	0.8
3	Ground clearance height (H)	1 m
4	Module length (L)	1.98 m
5	Module width	1 m
6	Module wattage	365.36 W
7	Module technology	Mono-crystalline

11.35–11.73% was observed for five different cities of India. Minimum bifacial gain of 7.89–9.16% was observed for six different cities of India. Itanagar has maximum bifacial gain while Gwalior has minimum bifacial gain. The excess land area used for bifacial power plant per kW is only 0.56 m^2. Figures 3 and 4 were made through google my map website [12].

The bifacial gain is maximum when the irradiance falling in bifacial module is maximum with respect to the monofacial module. The irradiation falling over the bifacial module at optimal tilt and at latitude tilt for mono facial module against the bifacial gain is plotted for 7 randomly selected different cities are presented in Fig. 5. It can be seen that the ratio between the irradiation absorbed by the bifacial module and the monofacial module is coherent with the bifacial gain. This concludes that the installation of the bifacial PV modules at the optimum tilt will enhances the light falling over the module and therefore the output energy. However, the optimum tilt

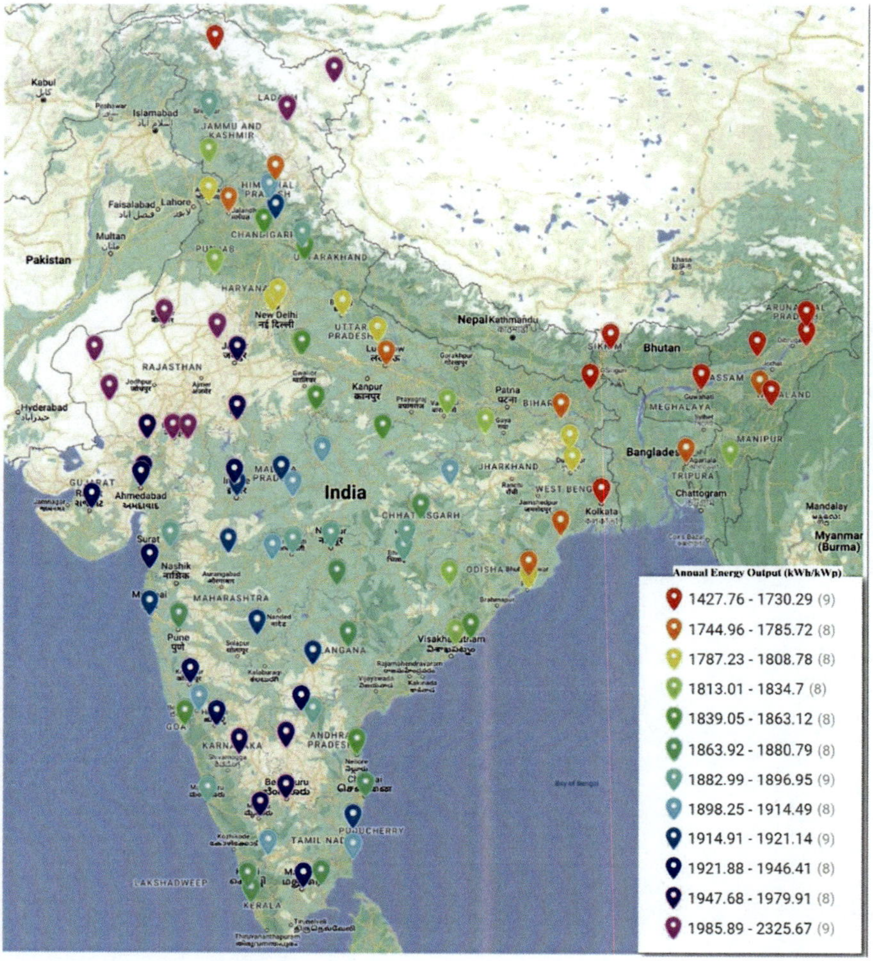

Fig. 3 Annual energy output (kWh/kWp) of bifacial PV power plant for 100 cities in India

will be different for different ground clearance height and based on that the total energy falling over the bifacial module will increase as compared to monofacial module.

The Performance of bifacial module is subject to change with varying albedo, module height from the ground and module bifaciality. For Sikar (Rajasthan), the bifaciality gain has been plotted against the diffuse to global irradiance ratio and shown in Fig. 6. It has been observed that with increase in diffuse to global irradiance ratio, the bifacial gain increases.

In this simulation study some parameters are fixed. Optimum design parameters as per site conditions are required for better output. The effect of temperature difference in the bifacial module with respect to the monofacial counterpart is also planned in

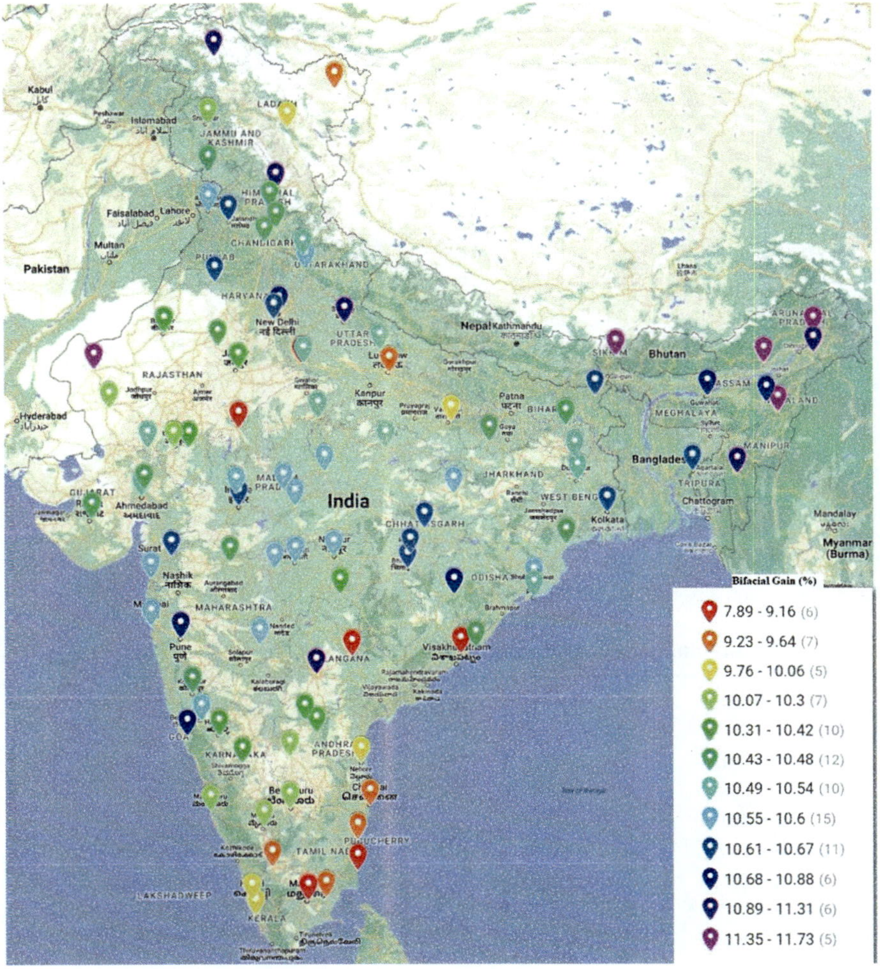

Fig. 4 Percentage bifacial gain for 100 cities of India

near future. Additionally, the future research may include Optimum tilt at different height, study of different albedo and economic viability etc.

4 Conclusion

For India, bifacial module is a new technology and PV community little know-how about its behavior in India in terms of performance and reliability. So, it is very important to estimate the energy generation of the bifacial modules at different locations of India. Through this work authors have tried to put some light on the

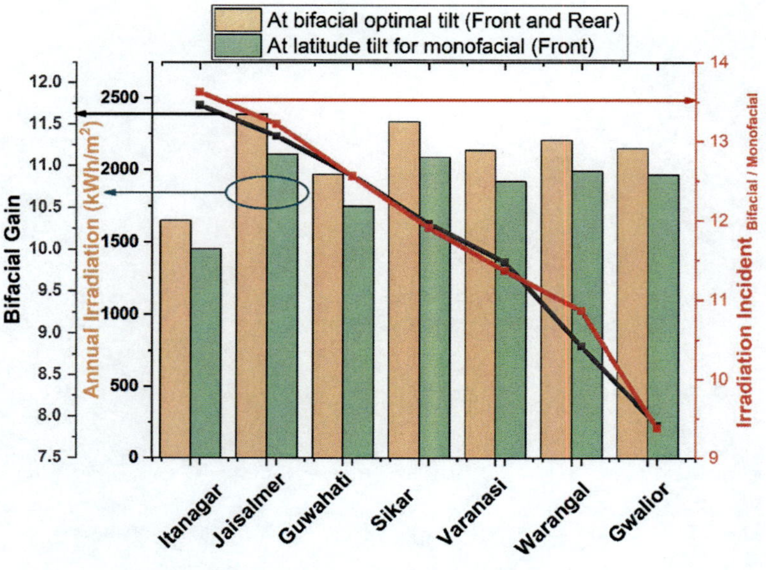

Fig. 5 Irradiation absorbed by the bifacial and monofacial module versus bifacial gain

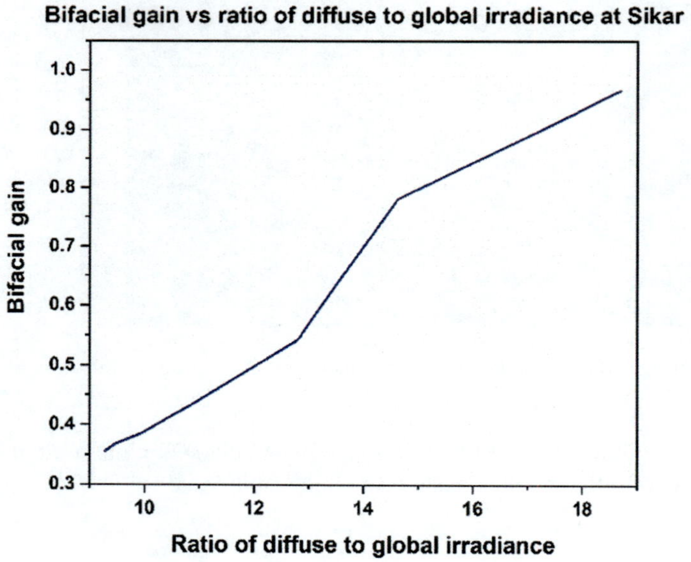

Fig. 6 Bifacial gain versus ratio of diffuse to global irradiance at Sikar for a particular day

performance of bifacial PV modules as compared to that of monofacial counterpart. The simulations are done on SAM using optimum tilt angle for every particular location. The simulated annual energy output of bifacial module at optimal tilt is studied for 100 Cities of India using SAM software. Bifacial Gain varies from 7.89 to 11.83% for 100 Cities of different climatic zones of India. The excess land area used for bifacial power plant per KW is only 0.56 m^2. It has been observed that Pasighat has minimum annual energy output while Aksai Chin has maximum Annual Energy output for bifacial module at the optimal tilt position. Itanagar has maximum bifacial gain while Gwalior has minimum gain as compared to mono facial module. The future research may include optimal installation parameters for bifacial module in terms of tilt angle, ground clearance height, albedo effect to draw maximum output for a particular site and prepare a potential map for India.

References

1. Most efficient solar panels 2022—Clean Energy Reviews. (2022) https://www.cleanenergyreviews.info/blog/most-efficient-solar-panels. Accessed May 21 2022
2. Sreenath S, Sudhakar K, Yusop AF (2021) Performance assessment of conceptual bifacial solar PV system in varying albedo conditions. IOP Conf Ser Mater Sci Eng 1078(1):012033. https://doi.org/10.1088/1757-899x/1078/1/012033
3. Manikandan S, Varun M, Manikandan S (2020) Performance evaluation of bifacial and mono-facial modules in vertical and latitude mounting at South India using PVsyst. IOP Conf Ser Mater Sci Eng. 912(4). https://doi.org/10.1088/1757-899X/912/4/042066
4. Urdaneta LG, Suárez LP, Aguilera JV (2021) Marginal contribution of factors to energy gains of bifacial modules. Ing Energética 42(1): 1–12. Available https://dialnet.unirioja.es/descarga/articulo/7791580.pdf%0A. https://dialnet.unirioja.es/servlet/extart?codigo=7791580
5. Pike C, Whitney E, Wilber M, Stein JS (2021) Field performance of south-facing and east-west facing bifacial modules in the arctic. Energies 14(4):1–15. https://doi.org/10.3390/en14041210
6. Sun X, Khan MR, Deline C, Alam MA (2018) Optimization and performance of bifacial solar modules: a global perspective. Appl Energy 212:1601–1610. https://doi.org/10.1016/j.apenergy.2017.12.041
7. Robledo J et al (2021) European photovoltaic solar energy conference and exhibition (EU PVSEC), September 2021, pp 1–8
8. Martínez JF, Steiner M, Wiesenfarth M, Siefer G, Glunz SW, Dimroth F (2021) Power rating procedure of hybrid concentrator/flat-plate photovoltaic bifacial modules. Prog Photovoltaics Res Appl 29(6):614–629. https://doi.org/10.1002/pip.3410
9. Kopecek R, Libal J (2021) Bifacial photovoltaics 2021: status, opportunities and challenges. Energies 14(8). https://doi.org/10.3390/en14082076
10. Das A, Ali M, Shuvo M, Zahan D (2021) Performance study of bifacial module based time varying Multilevel Solar Panel System (MSPS). Available: http://dspace.bracu.ac.bd/xmlui/handle/10361/14797%0A. http://dspace.bracu.ac.bd/xmlui/bitstream/handle/10361/14797/16121153,16121147,16121134,16321052_EEE.pdf?isAllowed=y&sequence=1
11. Stein JS et al (2021) Bifacial PV modules and systems experience and results from international research and pilot applications. Available: https://iea-pvps.org/wp-content/uploads/2021/04/IEA-PVPS-T13-14_2021-Bifacial-Photovoltaic-Modules-and-Systems-report.pdf
12. "My Maps – About – Google Maps." https://www.google.com/maps/about/mymaps/. (Accessed May 21 2022)

Comparative Thermal Performance Analysis of the RCC Envelope with a Low Thermal Transmittance (U-Value) Envelope

Brijesh Pandey, Shatakshi Suman, Prabhat Sharma, and Suneet Singh

1 Introduction

In formwork construction technologies (e.g., tunnel and MIVAN), steel or aluminum forms are used to construct walls and slabs in one continuous concrete pour [1, 2]. In this system, traditional beams and columns constructions are eliminated. Instead, walls and slabs are cast in one operation at the site using specially designed, easy-to-handle lightweight pre-engineered aluminum forms [3]. Continuous concrete pouring into the formwork and reinforcing the concrete with steel makes the envelope structure a monolithic reinforced cast concrete (RCC) structure.

Formwork construction technologies are proven cost-effective compared to the conventional construction methods in the literature. Because the formwork can be used repetitively, the project duration can be reduced to half, and the number of required workers is less than the conventional construction methods [4, 5]. Based on the economic costs and project duration, formwork construction technologies have been recommended for India's mass housing project [2].

This study aims to compare the thermal performance of two construction technologies; first, aluminum formwork, and second, mandated by the Eco Niwas Samhita (ENS) residential building code. The aluminum formwork technology, on the one hand, facilitates the speedy and economical construction of the built environment but performs poorly in terms of thermal performance, whereas the ENS mandated construction technology costs more in terms of money and construction time of the built environment but does well in terms of thermal performance. The mandated wall thickness for these two construction technologies is different. Therefore, two different

B. Pandey (✉) · P. Sharma · S. Singh
Department of Energy Science and Engineering, Indian Institute of Technology Bombay, Mumbai 400076, India
e-mail: brijeshpandeyby715@gmail.com

S. Suman
Alliance for an Energy Efficient Economy, New Delhi 110024, India

S. Doolla et al. (eds.), *Advances in Clean Energy and Sustainability*,
Green Energy and Technology, https://doi.org/10.1007/978-981-99-2279-6_52

wall thicknesses have been considered. Considering the heat stress report of the Indian subcontinent [6], it has become essential to bring into focus the re-evaluation of the policy of promoting speedy and economical but poor thermal performance construction technologies over thermally well-performed construction technologies.

Here, a monolithic RCC structure that can be constructed through different formwork technologies (tunnel and MIVAN) has been analyzed from the perspective of thermal performance. Thermal performance has been characterized by the operative temperature, the average temperature of interior and exterior walls and roof surfaces, and residential envelope transmittance value (RETV). Operative temperature can be defined as the average of the mean radiant temperature and mean air temperature of the built environment. RETV has been defined in Sect. 3.2. It is recommended that the maximum value of RETV should be 15 W/m^2.

A dwelling unit from the open domain has been selected to analyze the thermal performance [7]. The thermal performance of the RCC structured dwelling unit has been compared with the dwelling unit constructed with low U-value construction materials (recommended in Eco Niwas Samhita (ENS)—an energy conservation and building code for residential buildings in India [8]). The low U-value construction materials have been selected in such a way that it maintains the RETV within 15 W/m^2. The low U-value construction materials have been referred to as ENS complaint materials throughout this study. The comparative analysis for the thermal performance has been performed by considering the dwelling unit located in the composite climatic condition of India. From the analysis, it has been concluded that an envelope constructed through the ENS complaint materials reduces the peak indoor operative temperature by ~ 5 °C compared to the monolithic RCC structure.

In addition, this study also assesses the impact of different thermal properties of the materials and shading measures on the operative temperature of the dwelling unit and its envelope thermal transmittance (RETV) by exploring different construction materials and overhang (as a shading measure) depths in different climatic conditions.

2 System Description

A dwelling unit (shown in Fig. 1b) with a floor area of 28 m^2 is considered to analyze the thermal performance. The dwelling unit is assumed to be a stand-alone, ground-connected structure with no surrounding buildings. All the doors and windows of the dwelling unit are assumed to be closed, and no electrical appliances in the dwelling unit have been considered. Also, there is no occupancy in the building. The ground surface temperature has been assumed to be 2 °C below the monthly mean outdoor air temperature. The dwelling unit has been assumed to be constructed with RCC and ENS compliant materials for the comparative thermal performance analysis. The schematic of the RCC and ENS complaint envelope is shown in Fig. 1c, d. The envelope details and thermal properties of the RCC and ENS complaint materials have been given in Tables 1 and 2.

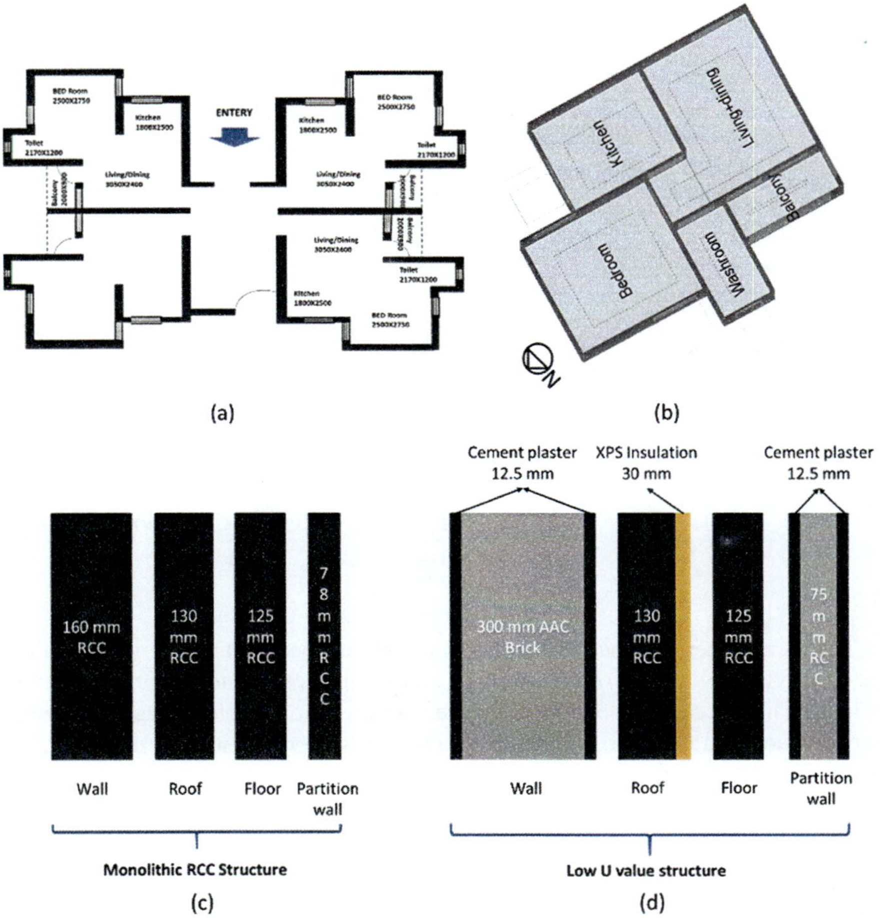

Fig. 1 **a** floor plan of the dwelling unit, **b** 3D model of the dwelling unit, **c** construction details of monolithic RCC structure, and **d** construction details of ENS compliant structure

Table 1 Material's details of the RCC structure

RCC (reinforced cast concrete) details			
Envelope details		Thermal properties	
Components	Description	Type	Description
Concrete wall thickness	160 mm	Thermal conductivity	1.5 (W/m–K)
Roof and floor thickness	130 mm, 125 mm	Specific heat	880 (J/kg-K)
Partition thickness	78 mm	Density	2288 (kg/m^3)

Table 2 Material's details of the ENS compliant materials

ENS compliant material's details			
Envelope details		Thermal properties	
Components	Description	Type	Description
AAC brick wall thickness	300 mm	Thermal conductivity (AAC brick)	0.184 (W/m–K)
Partition thickness	160 mm	Specific heat (AAC brick)	1240 (J/kg-K)
Cement plaster thickness (inside and outside)	12.5 mm	Density (AAC brick)	642 (kg/m^3)
Floor and roof thickness (RCC)	125 mm, 130 mm	Thermal conductivity (cement)	0.72 (W/m–K)
Roof insulation layer (XPS)	30 mm	Specific heat (cement)	920 (J/kg-K)
		Density (cement)	1650 (kg/m^3)
		Thermal conductivity (XPS)	0.046 (W/m–K)
		Specific heat (XPS)	1400 (J/kg-K)
		Density (XPS)	10 (kg/m^3)

The thermal performance of the dwelling unit has been analyzed by simulating the EnergyPlus [9] model of the dwelling unit for the composite climatic condition of India. The place corresponding to the weather condition is selected as New Delhi. The weather data corresponding to New Delhi has been taken from the ISHRAE weather data [10].

3 Methodology

3.1 *EnergyPlus Model*

To analyze the thermal properties of the dwelling unit, it has been modelled in EnergyPlus. EnergyPlus is a building energy simulation tool that works on the fundamental principle of energy balance. Each zone (volumetric space) is assumed as one node containing uniform properties of state variables, e.g., temperature, pressure, and density. For the one node temperature T_z of the air inside the zone, an energy balance can be given as

$$m_z c_z \frac{dT_z}{dt} = \sum_{i=1}^{N_s} Q_i + \sum_{i=1}^{N_{surface}} h_i A_i (T_{si} - T_z) + \sum_{i=1}^{N_{zones}} m_i c_p (T_{zi} - T_z) + m_{inf} c_p (T_\infty - T_z) + Q_{sys} \quad (1)$$

where the left hand side represents the energy stored in zone air. In the right hand side, the first, second, third, fourth, and fifth term represents the sum of the internal convective loads, the sum of convective heat transfer from the envelope, heat transfer due to interzone air mixing, heat transfer due to infiltration of outside air and air system output, respectively. The above equation results in the temperature T_z of the air inside each zone.

3.2 Residential Envelope Transmittance Value (RETV)

RETV characterizes the thermal performance of the building envelope (except the roof) [8]. Low RETV value helps in reducing the heat gain from the building envelope and therefore improves thermal comfort. RETV can be represented as

$$RETV = \frac{1}{A_{envelope}}\left[\left\{a \times \sum_{i=1}^{n}\left(A_{opaque_i} \times U_{opaque_i} \times \omega_i\right)\right\} + \left\{b \times \sum_{i=1}^{n}\left(A_{non-opaque_i} \times U_{non-opaque_i} \times \omega_i\right)\right\} + \left\{c \times \sum_{i=1}^{n}\left(A_{non-opaque_i} \times SHGC_{eq_i} \times \omega_i\right)\right\}\right] \quad (2)$$

where $A_{envelope}$ is the envelope area (excluding the roof) of dwelling units, A_{opaque_i} areas of different opaque building envelope components, U_{opaque_i} is thermal transmittance values of different opaque building envelope components, $A_{non-opaque_i}$ is the areas of different non-opaque building envelope components, $U_{non-opaque_i}$ is the thermal transmittance values of different non-opaque building envelope components, $SHGC_{eq_i}$ is equivalent solar heat gain coefficient values of different non-opaque building envelope components, ω_i is the orientation factor of respective opaque and non-opaque building envelope components, and a, b, and c are the coefficients for the different climate zones. The data regarding the orientation factor (ω_i) and the coefficients ($a, b$and$c$) are provided in ref. [8].

4 Results and Discussion

4.1 Thermal Performance Analysis

This section compares the zone averaged operative temperature, average interior and exterior surface temperature of the wall, and roof surfaces' temperature of the dwelling unit for the RCC and ENS compliant envelope. A simulation has been performed for the composite climatic condition (New Delhi) between 7th June and 10th June. The simulation period is based on the peak summer period identified by the outdoor dry bulb temperature.

Figure 2 compares the zone-averaged operative temperature of the dwelling unit constructed using RCC and ENS compliant materials. The figure shows that the peak indoor operative temperature for the RCC structure follows the peak of the dry-bulb temperature. There is a lag of ~ 3 h between the peaks of dry bulb temperature and indoor operative temperature. Whereas, the figure shows that the peak of indoor operative temperature for the low U-value envelope is ~ 5 °C below the dry-bulb temperature and ~ 6 °C below the indoor operative temperature of the RCC envelope. The figure also shows that the lag between indoor operative temperature and dry-bulb temperature peaks for the ENS compliant envelope is ~ 1.5 h. The reason for the reduced peak in the case of a ENS compliant envelope is the low thermal transmissivity of the material, which limits the heat flow through the envelope from the outside towards the inside of the built environment. The less lagging hour in the case of an ENS compliant envelope is the material's low density compared to the RCC structure, which limits the heat storage in the envelope.

Figure 3a compares the average temperature of the interior and exterior walls for the case of RCC and ENS compliant envelopes. The figure shows that the exterior walls' average temperature of the ENS compliant materials is higher than the RCC structured walls. The reason for this is that the low thermal conductivity of the material in the case of an ENS compliant envelope limits the heat flow from the outer surface to the inner surface, and heat accumulates at the outer surfaces, whereas in the case of the RCC structured walls, the high thermal conductivity of the materials allows the heat flow from the exterior surface to the interior surface continuously and that leads to less exterior surface temperature. Similarly, during night-time, the trough of the exterior walls' temperature for the ENS compliant materials is less

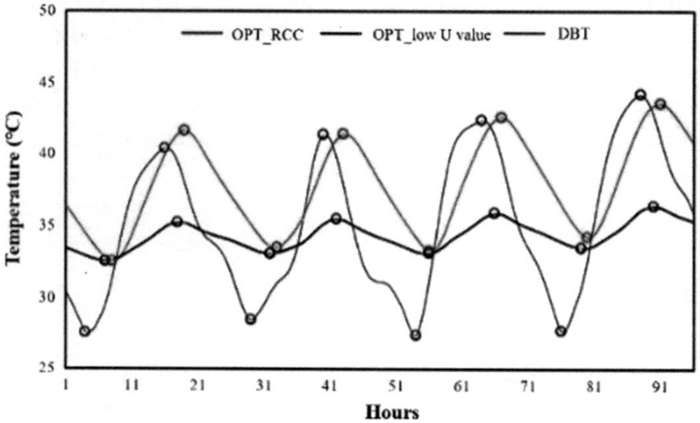

Fig. 2 Comparison of zone averaged operative temperature of the dwelling unit constructed through RCC structure and ENS compliant materials. OPT: operative temperature. DBT: dry bulb temperature

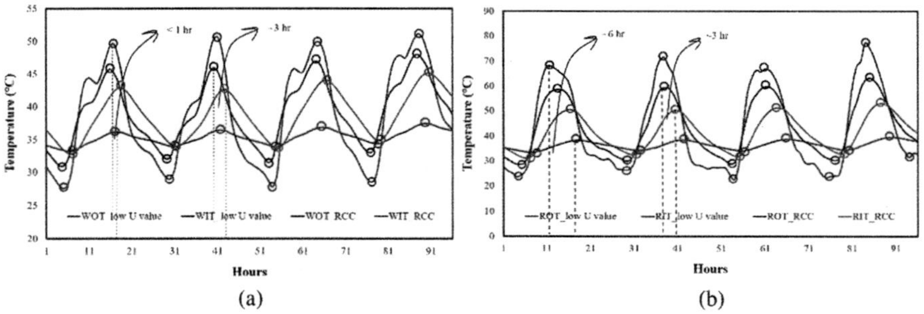

Fig. 3 Comparison of average walls' and roof surfaces' temperature of the dwelling unit constructed through the RCC structure and ENS compliant materials. WOT: wall outside temperature. WIT: wall inside temperature. ROT: roof outside temperature. RIT: roof inside temperature

compared to the RCC structured walls. The reason is similar that the low thermal conductivity of the material in the case of ENS compliant materials does not allow the inside heat of the built environment to come outside through walls, whereas, in the case of the RCC structured walls, the high thermal conductivity allows the inside heat of the built environment to come outside through envelope which results in the increase of the exterior walls' temperature during the night. The figure also shows that the peak of the interior walls' temperature is ~ 5 °C higher in the case of the RCC structured wall compared to the ENS compliant materials. The reason is similar to as explained above.

Figure 3b shows similar phenomena (as in the case of the wall) of surfaces' temperature for the roof. The higher time lag between the interior and exterior roof temperature peaks in the case of an ENS compliant compared to the RCC structured roof is due to an insulation layer.

4.2 Thermal Comfort Analysis

This section calculates thermal discomfort hours for the dwelling unit constructed through RCC and ENS compliant materials. To estimate the thermal discomfort hours, simulations have been performed for both the case scenarios for the summer months (1st April–30th June) of composite climate (New Delhi). A CBE thermal comfort calculator for adaptive comfort has been used to calculate the upper-temperature limits of the adaptive thermal comfort model with 0.9 m/s air velocity.

Figure 4 shows the comparison of zone averaged operative temperature for the case of RCC structured and ENS compliant materials constructed dwelling unit. The figure also shows the 90% acceptability limit (solid black line) of the adaptive thermal comfort model with 0.9 m/s air velocity. Quantitatively the discomfort hours are 1805 and 1427 h for the case of RCC structured and ENS compliant materials

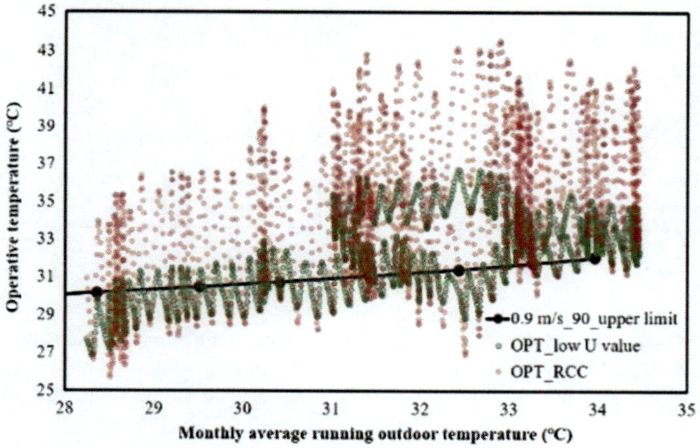

Fig. 4 Comparison of the zone averaged operative temperature of the dwelling units constructed through RCC and ENS compliant materials. OPT: operative temperature

constructed dwelling units, respectively. This indicates that when the construction material is changed from RCC to ENS compliant materials, a 20.9% reduction in discomfort hours is obtained.

4.3 Effect of the Materials' Thermal Properties on Operative Temperature

The effect of density, specific heat, and thermal conductivity on operative temperature has been analyzed through sensitivity analysis using the Morris method [11]. The material's thermal properties (density, specific heat, and thermal conductivity) have been considered uniformly distributed. The range of the thermal properties of the materials used for the Monte Carlo simulation to determine the sensitivity for operative temperature is given in Table 3.

Figure 5 shows the sensitivity of the different thermal properties of the material for the operative temperature of the built environment in different climatic conditions (composite-New Delhi, warm and humid-Mumbai, temperate-Bengaluru, hot & dry-Jodhpur and cold-Shillong). The X-axis of the figure shows the main effect of the

Table 3 Specification of construction materials for sensitivity analysis

Thermal properties	Range	Distribution
Specific heat	600–3500 (J/kg-K)	Uniform
Thermal conductivity	0.1–3 (W/m–K)	Uniform
Density	600–3500 (kg/m^3)	Uniform

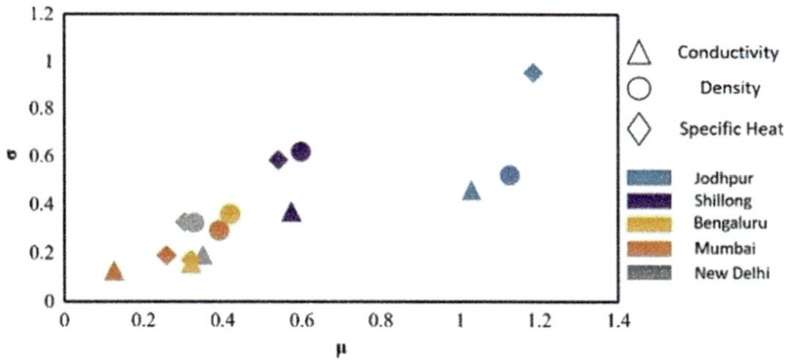

Fig. 5 Scatter diagram of Morris sensitivity analysis for operative temperature

sensitivity of the single parameter. The Y-axis shows the interactive effect among parameters for the operative temperature.

The sensitivity analysis for the Jodhpur shows that specific heat is most sensitive to the operative temperature, and also it affects the operative temperature very non-linearly. The second and third most sensitive parameter for the operative temperature is the density and thermal conductivity of the materials. In the physical sense, the operative temperature can be controlled more effectively by varying the specific heat of the materials than density and thermal conductivity.

Similarly, it can be inferred from the figure that the different thermal properties of the construction materials have different levels of sensitivity to the operative temperature in different climatic conditions. Hence, a detailed thermal analysis by varying the thermal properties of the construction materials should be performed before selecting the construction material for each climate.

4.4 Effect of the Materials and Shading Measures on RETV

This section analyzes the effect of different thermal properties of the construction materials and shading measures on RETV through sensitivity analysis. RETV is calculated using Eq. (2). RETV has been calculated for all the climatic conditions except for the cold climate, as it is not required for the cold climate. The material's thermal properties (density, specific heat, and thermal conductivity) and overhang depth have been considered uniformly distributed. The range of the materials' thermal properties is the same as in Table 3, and the overhang's depth is varied between 0.2 and 1.5 m.

The sensitivity analysis for the RETV in different climatic conditions is shown in Fig. 6. The figure shows that the shading measure (overhang) is very non-linearly affecting the RETV in all climatic conditions. The figure shows that in the temperate climate (Bengaluru), the specific heat is most sensitive to RETV, followed by the density and thermal conductivity of the materials. For the hot and dry climate (Jodhpur), the thermal conductivity of the materials is most sensitive to RETV,

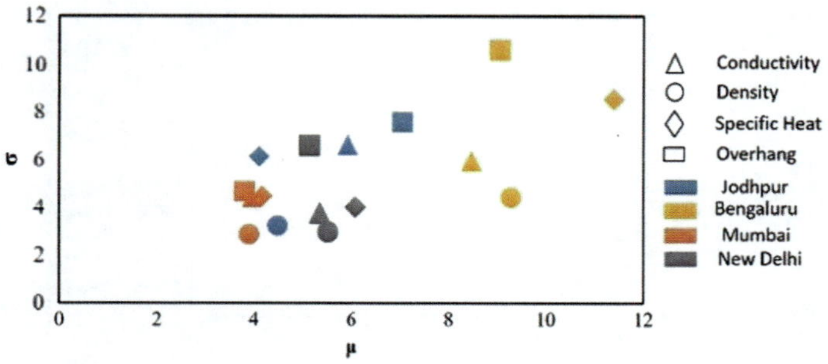

Fig. 6 Scatter diagram of Morris sensitivity analysis for RETV

followed by the density and specific heat. The figure shows that the relative importance of the density is more than the specific heat. However, the specific heat affects the RETV more non-linearly compared to the density of the material. In the composite (Delhi) and warm & humid climate (Mumbai), there is a slight difference between the relative importance of the density, specific heat, and thermal conductivity in terms of sensitivity to RETV.

In a physical sense, if the RETV has to be maintained within the prescribed limit (i.e., 15 W/m^2) as recommended by the ENS, then the construction materials have to be selected by analyzing first the specific heat followed by the density and thermal conductivity of the materials.

5 Conclusions

The thermal performance analysis concludes that the ENS compliant building envelope is better than the RCC structured envelope due to its ability to maintain thermal comfort and low-temperature fluctuations in the built environment.

This study shows through thermal performance analysis of the RCC structured dwelling unit that formwork construction technologies, e.g., tunnel and MIVAN, that result in the monolithic RCC structure, should be considered cautiously before promoting it as a solution for the mass housing projects in India. On the one hand, formwork technologies have benefits in terms of speed of construction and modular designs; research and design interventions are needed to negate the impact of poor thermal performance during the operational phase.

From the sensitivity analysis, it has been concluded that different thermal properties of the construction material affect the indoor operative temperature and RETV differently in different climatic conditions. It has also been concluded that shading measures significantly affect the RETV values and should be appropriately designed.

References

1. Kushal P, Ajitkumar J, Nikhil S (2015) MIVAN technology. Int J Eng Tech Res ISSN: 2321-0869, 3(3)
2. https://ghtc-india.gov.in/Content/LHP-Rajkot.html. Last accessed 21 May 2022
3. Patil P, Mundhada P (2022) Comparative analysis of MIVAN formwork and conventional formwork. In: Kolhe ML, Jaju SB, Diagavane PM (eds) Smart technologies for energy, environment and sustainable development 2. Springer Proceedings in Energy
4. Akshay G (2018) Time and cost optimization of construction project using MIVAN technology. Int J Eng Res Appl (IJERA) 8:70–74
5. Ray P, Bera DK, Rath AK (2021) Comparison between the tunnel form system formwork and the MIVAN formwork system in a multi-unit building project. In: Das B, Barbhuiya S, Gupta R, Saha P (eds) Recent developments in sustainable infrastructure. Lecture notes in civil engineering, 75
6. https://www.nrdc.org/sites/default/files/india-heat-resilience-20220406.pdf. Last accessed 21 May 2022
7. https://wbhousingboard.in/home/project_location_map/download/eastern_noop_grove.pdf, Last accessed 1 April 2022
8. https://www.econiwas.com/pdf/publication/ECBC_BOOK_Web.pdf. Last accessed 1 May 2022
9. Crawely D (2001) EnergyPlus: creating a new-generation building energy simulation program. Energy Build 33:319–331
10. https://energyplus.net/weather-region/asia_wmo_region_2/IND. Last accessed 1 April 2022
11. Morris MD (1991) Factorial sampling plans for preliminary computational experiments. Technometrics 33:161–174

Application of Metaheuristic Techniques in Optimal Parameter Estimation of Solid Oxide Fuel Cell

Rahul Khajuria, Ravita Lamba, Rajesh Kumar, and Srinivas Yelisetti

Abbreviation

N	Number of fuel cells connected in series
$V_{out,stack}$	Output voltage of stack [V]
$V_{out,cell}$	Output voltage of cell [V]
P_{H2}	Pressure of hydrogen at anode [atm]
P_{H2o}	Pressure of oxygen at cathode [atm]
E	Open circuit voltage of fuel cell [V]
E^0	Reversible voltage of fuel cell [V]
V_{conc}	Concentration over-potential voltage [V]
V_{ohm}	Ohmic voltage of fuel cell [V]
V_{act}	Activation potential of fuel cell [V]
R	Universal gas constant [J mol^{-1} K^{-1}]
F	Faraday's constant [C mol^{-1}]
T	Temperature of fuel cell [K]
I_l	Limiting current [mA]
I_{ae}	Anode exchange current density [mA]
I_{ce}	Cathode current exchange density [mA]
I_{load}	Load current [mA]
A	Parametric constant [V]
B	Parametric constant [V]
R_{ohm}	Resistance [Ω]
SSE	Sum of square errors
$V_{Experimental}$	Experimental value of voltage [V]

R. Khajuria (✉) · R. Lamba · R. Kumar · S. Yelisetti
Malaviya National Institute of Technology, Jaipur 302017, India
e-mail: Khajuriarahul414@gmail.com

S. Doolla et al. (eds.), *Advances in Clean Energy and Sustainability*,
Green Energy and Technology, https://doi.org/10.1007/978-981-99-2279-6_53

$V_{Estimated}$	Estimated value of voltage [V]
l_i	Lower bound of limits
u_i	Upper bound of limits

1 Introduction

Excessive energy production from fossil fuel based sources causes several alarming issues worldwide like global warming, negative environmental conditions and depletion of resources at a fast rate [1]. Renewable energy sources have the advantage of being environment friendly, less environment pollution, low carbon emission and above all they can be proven as an innovative tool for sustainable development [2]. Solid oxide fuel cell is one of the renewable energy resources which use the chemical energy of fuels such as hydrogen, hydrocarbons etc. at a very high temperature and convert this form of energy into electrical energy [3]. Solid oxide fuel cell has several advantages of being highly efficient even at high temperatures (600–1000 °C), robust, environment friendly due to water and heat as a byproduct, flexible with fuel input and guaranteed high efficiency (typically 60–70%). All these properties of solid oxide fuel cell make them a reliable energy generator with many applications such as distributed generation, auxiliary power units, co-generation, transportation, waste water treatment and space applications [4].

Solid oxide fuel cell works on the principle of electro-chemical process which means they convert the chemical energy of input fuel directly to electrical energy by means chemical reactions inside a fuel cell. Understanding the electrochemical dynamics and behavior of solid oxide fuel cells at different operating conditions needs the accurate modeling [5]. There are several unknown parameters which govern the working of solid oxide fuel cell, but are not mentioned on the datasheet provided by the manufacturer. Therefore identification of these unknown parameters is essential and a model is needed to understand the behavior of these parameters [6]. Since, modeling of SOFC is a complex, non-linear, multivariable and multimodal problem therefore identification of unknown parameters is not possible with classical and traditional mathematical approaches [7]. Metaheuristic optimization approaches are the powerful tools for solving a complex and non-linear problem as they have the properties of reaching to near the optimum solution [8]. They reach to the global maxima and minima depending upon the problem and converge to solution at a very fast rate without sticking into local maxima and minima respectively [9].

Evolutionary based algorithms are one of the metaheuristic optimization methods which can solve the multivariable problems [10]. Recent literatures have used several metaheuristic algorithms such as particle swarm optimization [11], hybrid algorithms

[12], converged grass fibrous root optimization [13], teaching learning based algorithms [14] etc. for identification of unknown parameters. In this paper, seven evolutionary based algorithms have been used to identify the unknown parameters and also they have been compared to each other to evaluate which algorithm is more suitable for problem identification. Problem identification needs an objective function to be defined. Recent literatures have used voltage based objective function [15] expressed as minimization of root mean square error, mean square error, sum of squared error, mean absolute error etc. In this study we have used a voltage based objective function described as minimization of sum of squared errors subjected to constraints. To validate the authenticity of the algorithm optimization for used objective function, statistical studies are needed to perform for calculating mean and standard deviation. For representing the statistical study on a plot to indicate median value ad interquartile range (IQR), box plot is needed to drawn for various algorithm. In this study box plot is also shown to validate the best suitable algorithm for unknown parameter identification.

2 SOFC Model and Objective Function

Output stack voltage of a solid oxide fuel cell (SOFC) can be written as [15]:

$$V_{out,stack} = N * V_{cell} \tag{1}$$

where, N is number of cells connected in series to obtain desired output stack voltage.

Therefore output voltage of a single cell is given as [16]:

$$V_{out,cell} = E_{o.c} - V_{act} - V_{conc} - V_{ohm} \tag{2}$$

where, $E_{o.c}$ is open circuit voltage and V_{act}, V_{ohm} are activation, concentration and ohmic voltage losses respectively and can be written as follows [16]:

$$E_{oc} = E_0 + \frac{RT}{4F} ln\left[\frac{(PH_2)^2 PO_2}{(PH_2O)^2}\right] \tag{3}$$

E_o is reversible voltage *R, F, T* are gas constant, Faraday's constant and temperature respectively. P_{H2}, P_{O2} and P_{H2O} are saturated pressure of hydrogen, oxygen and water respectively.

$$V_{act} = A\left(\text{Sinh}^{-1}\left(\frac{I_{load}}{2I_{ae}}\right) + \text{Sinh}^{-1}\left(\frac{I_{load}}{2I_{ce}}\right)\right) \tag{4}$$

$$V_{conc} = Bln\left(1 - \frac{I_{load}}{I_l}\right) \tag{5}$$

$$V_{ohm} = I_{load} R_{ohm} \tag{6}$$

where, I_l, I_{load}, I_{ae}, I_{ce}, represent the limiting current, load current, anode exchange current and cathode exchange current respectively; R_{ohm} is ionic resistance; A and B are coefficients. Estimation of the unknown parameters of SOFC can be done by expressing the voltage based objective function as a minimization of sum of squared error optimization problem at '*N*' number of measured experimental data points subjected to constraints. Therefore, objective function subjected to constraints is given as follows [16]:

$$F_{obj_{\min}} = SSE_{\min} = \sum_{i=1}^{N} \left(V_{\exp erimental} - V_{exstimated} \right)^2 \tag{7}$$

$$Subjected\, to\, Constraints: \quad l_i \le u_i, \quad I_{load.n} \le I_l\, and\, I_{ce} \le I_{ae}$$
$$Where,\ n = 1, 2, 3, 4, \ldots, N\, and\, i = 1, 2, 3, \ldots, 7$$

where '*N*' denotes number of experimental data points; l_i and u_i denotes the lower and upper bounds of the range respectively.

Various methods have been implemented to extract SOFC parameters. It is difficult to evaluate these parameters with classical approaches as parameter estimation of a solid oxide fuel cell is a non-linear, multimodal and multivariable problem. Therefore, metaheuristic methods having advantages of solving complex and non-linear problem and to reach near to optimum solution are used. In this paper, evolutionary based algorithms have been adopted to reach near to the optimum solution. Evolutionary based algorithms are inspired from the human evolution or species and are easy to implement in solving the complex problems. Seven evolutionary based algorithms namely, Differential Evolutionary algorithm (DE) Genetic Algorithm (GA), Evolutionary Programming (EP), Flower Pollination Algorithm (FPA), Evolutionary Strategies (ES), Memetic Algorithm (MA) and Coral Reefs Optimization (CRO) are used to find the unknown parameters of given SOFC from reference [17, 18].

3 Results and Discussion

In this work, Solid oxide fuel cell developed by Siemens Energy [17] is used to evaluate the unknown parameters and to develop a model with the help of evolutionary algorithms, which exactly matches with the experimental data obtained from [17, 18]. The working parameters of SOFC are given in Table 1. Also, the ranges of unknown parameters are listed in Table 2. In this paper, seven evolutionary based metaheuristic algorithms namely Genetic Algorithm (GA), Evolutionary Programming (EP), Flower Pollination Algorithm (FPA), Evolutionary Strategies (ES), Differential Evolution Algorithm (DE), Memetic Algorithm (MA) and Coral Reefs Optimization (CRO) has been used to evaluate the unknown parameters of SOFC. The aim of this study is to find the best metaheuristic algorithm among given algorithms so as to predict the value of unknown parameters of solid oxide fuel cell which greatly matches the obtained polarization curve with the estimated polarization curve by using these algorithms. Each evolutionary based algorithm is run for 20 individual times with population size of 100 and for 1000 iterations. For each run of particular metaheuristic algorithm, values of unknown parameters and sum of squared errors have been determined. After 20 runs, the best value among all runs for all algorithms has been compared. The algorithm with lesser value of sum of squared error is said to be the best among all algorithms. For each individual algorithm mentioned, statistical measures have been performed such as mean, standard deviation, best value and worst value. The box plot has also been plotted between sum of squared error and metaheuristic algorithms used to validate the authenticity of the algorithm. In box plot, with small value of median and inter-quartile range, best algorithm among all given algorithms can be found. Values of seven unknown parameters measured with evolutionary based algorithms mentioned, along with Sum of Squared Error (SSE), best value, worst value, mean and standard deviation has been listed in Table 3. A great closeness to the experimental and estimated values can be checked from the polarization characteristics curves such as current–voltage (I-V) curves, efficiency curves and power curves of solid oxide fuel cell. Experimental values of solid oxide fuel are taken from [17, 18], and are used to validate the model with applied evolutionary algorithm. Current–Voltage curves, current- power curves and current-efficiency curves have been plotted for each mentioned metaheuristic algorithm and are shown in Figs. 1a–c respectively. With a close look in the Fig. 1, the reader can understand that differential evolutionary based algorithm (DE) is suitable algorithm to predict the values of unknown parameters of solid oxide fuel cell among all other evolutionary based algorithm mentioned as estimated values are greatly matching with the experimental obtained values by using differential evolutionary based algorithms. Further, statistical study reveals that with less value of mean and standard deviation, differential evolutionary based algorithm is best evolutionary based algorithm among all other algorithm. Further, a box plot has also been plotted as shown in Fig. 2, and with the help of box plot it can be seen that with a very small interquartile range (IQR) and median, differential evolutionary algorithm is more suitable among all other evolutionary based algorithm used in this study.

Table 1 Specifications of solid oxide fuel cell [16]

FC parameters	R (J/mol K)	F (C/mol)	T (K)	P_{H2O} (atm)	P_{H2} (atm)	P_{O2} (atm)
SOFC [34]	8.314	96,487	873	1.000	1.0000	1.0000

Table 2 Search range values of unknown parameters of solid oxide fuel cell [16]

Parameters	E_{oc} (V)	A (V)	I_{ae}(mA/cm^2)	I_{ce} (mA/cm^2)	B(V)	I_l (mA/cm^2)	R (kΩ/cm^2)
Lower limit	0	0	0	0	0	0	0
Upper limit	1.2	1	100	100	1	10,000	1

Table 3 Estimated values of unknown parameter of solid oxide fuel cell by using different metaheuristic algorithms

Parameters	EP	CRO	GA	FPA	MA	ES	DE
E_{oc} (V)	1.1229	1.112	1.132	1.179	1.112	1.184	1.20
A (V)	5.04E − 02	1.804E − 01	9.054E − 03	2.618E − 03	6.769E − 02	6.33E − 02	2.618E − 02
I_{ae} (mA/cm^2)	1.2E + 01	9.92E + 01	1.783E + 01	9.07E + 01	2.67E + 01	8.546E + 01	6.488E + 01
I_{ce}. (mA/cm^2)	2.3E + 01	2.03E + 01	5.893E + 01	4.395E + 01	6.675E + 01	1.575E + 01	1.000E + 01
B (V)	4.81E − 01	1.000E + 0	9.995E − 01	2.064E − 01	5.33E − 01	1.000E + 0	7.888E − 01
I_l (mA/cm^2)	6.4E + 03	1.00E + 03	1.000E + 03	7.977E + 03	3.400E + 03	7.068E + 03	6.323E + 03
R (Ω)	1.00E − 03	1.000E − 03	3.224E − 03	1.00E − 03	1.000E − 03	1.000E − 03	1.000E − 03
SSE	2.61E − 05	1.74E − 04	1.39E − 04	1.76E − 05	4.72E − 04	3.03E − 05	8.33E − 06
Mean	7.09E − 03	4.849E − 04	5.485E − 03	1.511E − 04	1.193E − 04	1.223E − 03	1.57E − 05
Standard deviation	4.13E − 03	2.743E − 04	5.341E − 03	1.77E − 034	2.665E − 04	1.069E − 03	6.792E − 06
Best value	2.61E − 05	1.74E − 04	1.39E − 04	1.76E − 05	4.72E − 04	3.03E − 05	8.33E − 06
Worst value	5.14E − 02	1.151E − 03	2.53E − 03	6.75E − 04	1.409E − 03	3.41E − 03	3.37E − 05

4 Conclusion

In this paper, different evolutionary based optimization algorithms have been used to solve a complex and non-linear problem of identification of SOFC parameter based on given model. Seven different evolutionary based metaheuristic algorithms viz. GA, DE, ES, EP, MA, CRO and FPA, were used to estimate unknown parameters of given solid oxide fuel cell. Voltage based objective function expressed as minimization sum of squared error was used to validate the estimated results to the experimental ones. Polarization characteristics such as current- voltage curves, power curves and

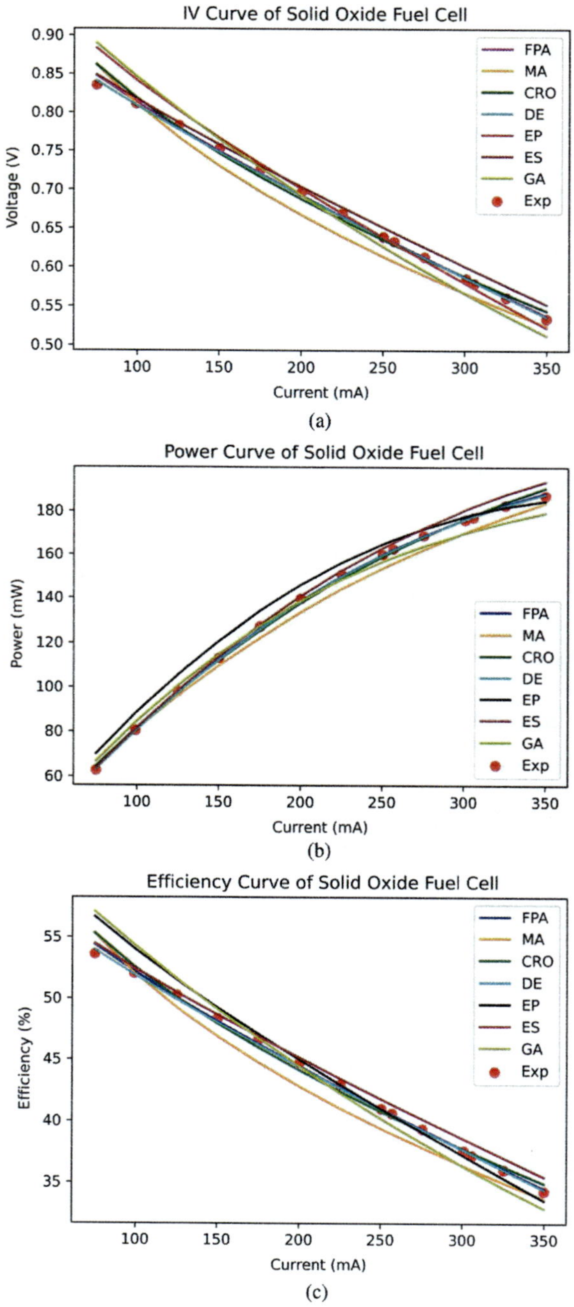

Fig. 1 **a** Current–Voltage, **b** Current-Power and **c** Current-Efficiency curves of solid oxide fuel cell for experimental data and estimated data by different metaheuristic algorithms

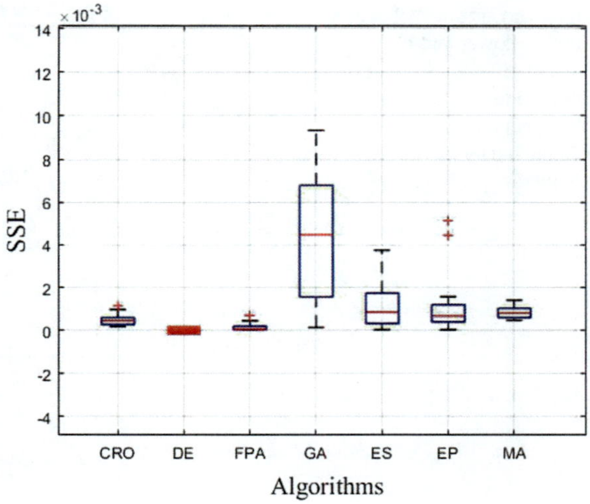

Fig. 2 Box plot for different metaheuristic algorithms

efficiency curves are also plotted to match the experimental characteristics and we found that differential evolutionary based algorithm is best evolutionary based algorithm among all with highest matching characteristics. Also, statistical studies such as mean, standard deviation, best value and worst values confirm the accuracy of DE. Furthermore, in box plot study, with smallest range of IQR and median DE is best among all other evolutionary algorithm used in this study.

References

1. Yakout AH et al (2022) Comparison among different recent metaheuristic algorithms for parameters estimation of solid oxide fuel cell: steady-state and dynamic models. Alexandria Eng J 61(11):8507–8523
2. Abaza A, Ragab AES, Saeed A, Bayoumi A (2020) Optimal parameter estimation of solid oxide fuel cell model using coyote optimization algorithm. Recent advances in engineering mathematics and physics. Springer, Cham, 135–149
3. Peksen M (2015) Numerical thermomechanical modelling of solid oxide fuel cells. Prog Energy Combust Sci 48:1–20
4. Selimovic A (2002) Modelling of solid oxide fuel cells applied to the analysis of integrated systems with gas turbines
5. Bove R, Ubertini S eds (2008) Modeling solid oxide fuel cells: methods, procedures and techniques. Springer Science & Business Media
6. Savioli J, Watson GW (2020) Computational modelling of solid oxide fuel cells. Curr Opin Electrochem 21:14–21
7. Wang K et al (2011) A review on solid oxide fuel cell models. Int J Hydrogen Energy 36(12):7212–7228
8. Luo R, Shafiee M (2021) The application of metaheuristics in optimal parameter identification of solid oxide fuel cell. Energy Rep 7:2563–2573
9. Chitsaz A et al (2018) Exergoenvironmental comparison of internal reforming against external reforming in a cogeneration system based on solid oxide fuel cell using an evolutionary algorithm. Energy 144:420–431

10. Sadeghi M et al (2015) Thermoeconomic optimization using an evolutionary algorithm of a trigeneration system driven by a solid oxide fuel cell. Energy 89:191–204
11. Yousri D, Hasanien HM, Fathy A (2021) Parameters identification of solid oxide fuel cell for static and dynamic simulation using comprehensive learning dynamic multi-swarm marine predators algorithm. Energy Convers Manage 228:113692
12. Bai Q, Li H (2022) The application of hybrid cuckoo search-grey wolf optimization algorithm in optimal parameters identification of solid oxide fuel cell. Int J Hydrogen Energy 47(9):6200–6216
13. Shi H, Li J, Zafetti N (2020) New optimized technique for unknown parameters selection of SOFC using converged grass fibrous root optimization algorithm. Energy Rep 6:1428–1437
14. Xiong G et al (2018) Parameter identification of solid oxide fuel cells with ranking teaching-learning based algorithm. Energy Convers Manage 174:126–137
15. Jiang B, Wang N, Wang L (2014) Parameter identification for solid oxide fuel cells using cooperative barebone particle swarm optimization with hybrid learning. Int J Hydrogen Energy 39(1):532–542
16. Xiong G et al (2021) Optimal identification of solid oxide fuel cell parameters using a competitive hybrid differential evolution and Jaya algorithm. Int J Hydrogen Energy 46(9):6720–6733
17. Pierre J. Siemens energy. In: 11th annual SECA workshop 2010 Jul 27
18. Xiong G et al (2020) A simplified competitive swarm optimizer for parameter identification of solid oxide fuel cells. Energy Convers Manage 203:112204

Hydrodynamics Study of Electrode Intrusion Effects in Hierarchical Interdigitated Flow Field Design for Vanadium Redox Flow Batteries

Sanjay Kumar, Vivek Kumar Barnwal, and Ila Jogesh Ramala Sarkar

1 Introduction

Currently, the world is focusing more towards renewable energy sources as compared to fossil fuels for electricity generation to reduce toxic gases emission. Vanadium redox flow batteries (VRFB) have been demonstrated up to several MW scale capacity storage devices as compared to other energy storage devices [1, 2]. The performance of VRFB depends on many factors. Electrode thickness and flow field designs are important factors for determining the performance of a redox flow battery (RFB). A well-designed flow field ensures the uniform distribution of reactants throughout the cell area, provides a convective feed of reactants to the reaction zone, and minimizes its pressure drop [1]. Flow field pattern is one of the key parameters for better circulation of reactants into the reaction zone, thus reducing mass transfer polarisation of the cell. Optimization of electrode intrusion will give higher battery efficiency and it will help to reduce diffusional length for ions transport [3]. Uniform distribution of electrolytes on the entire surface of electrode is essential for optimal performance of the cell. Thus, minimum pressure and consistent circulation of electrolytes over electrochemically active areas are the main flow-related issues associated with flow field design. Several computational fluid dynamics (CFD) studies have been done on those aspects of flow fields in the area of fuel cells, and to a lesser extent, for RFB applications. Interdigitated flow field (IFF) design showed improved electrochemical performance with compared to serpentine flow field design [4]. Interdigitated flow field is identical to parallel channels, other than the inlet and outlet legends are not linked to each other. The solution is required to flow throughout a permeable medium beneath the rib. Prasad et al. [4] and Prasad and Jayanti [5] evaluated fluid dynamics

S. Kumar (✉) · I. J. R. Sarkar
Department of Chemical Engineering, Faculty of Technology, Marwadi University, Rajkot, India
e-mail: sksanjay.chemical@gmail.com

V. K. Barnwal
Department of Chemical Engineering, BIT Sindri, Dhanbad, India

S. Doolla et al. (eds.), *Advances in Clean Energy and Sustainability*,
Green Energy and Technology, https://doi.org/10.1007/978-981-99-2279-6_54

of the interdigitated flow field with CFD simulations and concluded that the pressure drop characteristics and circulation through flow channels are such that the uneven flow was likely over a certain situation. Prasad and Jayanti [6] calculated the consequence of flow chemistry aspects on flow circulation and channel outline on water vapor supply in the cathode region by a two-module conceptualization. The Hierarchical interdigitated flow field is a modified version of interdigitated flow field design to reduce the pressure drop and to improve further electrochemical performance [7]. In my previous studies [8, 9], the effects of channel dimension on hydrodynamics effects in serpentine flow field have been completed.

In this work, we report a detailed CFD simulation study of intrusion effects on pressure drop and electrolyte circulation in a hierarchical interdigitated flow field of size 105 mm × 105 mm. Three different electrodes intrusions were employed (0.5, 1, and 1.5 mm thickness) and kept a constant active electrode thickness of 2.5 mm for each case. The performances have been compared with normal IFF for the same electrode intrusion thickness. The effects of flow rate on electrode intrusion have been investigated and compared to normal interdigitated flow field design.

2 Simulation Study

Computational fluid dynamics simulation has been used to study the pressure drop and velocity circulation in the active electrode for VRFB applications. ANSYS 2021 R2 version is used for our simulation studies. The geometry of hierarchical interdigitated flow field design were created using CFD software as shown in Fig. 1. The overall size of the computational domain was 105 mm × 105 mm and a compressed porous electrode of 2.5 mm thickness was created on the back side of electrode for this study. Three different thicknesses of electrode intrusions of 0.5, 1, and 1.5 mm were used into flow channels to check the velocity gradient and overall pressure drop within the cell. In this simulation, water is used as a working medium for all the modeling cases. Inlet and outlet pipes were created to support the electrolyte flow in the flow channel. At first, electrolyte flow through the inlet pipe to flow channel, then due to diffusion and convection it passes through the entire electrode area and then finally exit through flow channel outlet.

A number of governing equations have been used for the simulation studies. The Continuity Eq. (1) and Navier–Stokes equation Eq. (2) were used for the flow channels as well as for electrodes and Darcy's law has employed in the porous region [8].

$$\nabla.(\rho v) = 0 \tag{1}$$

$$\rho(v.\nabla)v = -\nabla p + \nabla.\left[\mu\left(\nabla v + (\nabla v)^T\right) - \frac{2}{3}\mu(\nabla.v)I\right] \tag{2}$$

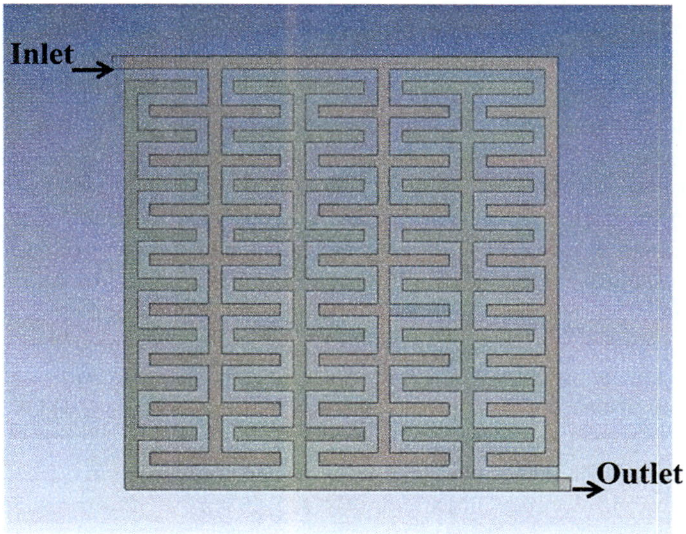

Fig. 1 Hierarchical interdigitated flow field design (105 mm × 105 mm)

In Eqs. (1) and (2), υ and ρ represent the velocity vector, density, and pressure, viscosity are denoted by p and μ, and I represent the identity matrix. In Eq. (2), Darcy's law is incorporated and simplified onto the Brinkman Eq. (3) and represented as:

$$\frac{\mu}{K}v = -\nabla p + \frac{1}{\varepsilon}\nabla.\left[\mu\left(\nabla v + (\nabla v)^T\right) - \frac{2}{3}\mu(\nabla.v)I\right] \quad (3)$$

The permeability values were taken from Kumar et al. [2], as 4.27×10^{-11} used for electrode under the rib and 5.58×10^{-11} used for under the flow channel respectively. For solving the governing equations, no-slip boundary situations were used at the wall, zero gauge pressure at the outlet, and velocity is used in the inlet.

3 Results and Discussion

3.1 Effect of Intrusion on Pressure Drop

Predicted CFD simulated results for the pressure drop for three different electrode intrusions for an active cell area of 105 mm × 105 mm were presented on behalf of both the flow field design. In earlier work [2], normal type of interdigitated flow field has showed maldistribution of electrolyte in the larger cell, which leads to poor electrochemical performance. Lower pressure drop from the cell will reduces the pumping loss and therefore an interdigitated flow field design fitted into this as compared to other flow field design [1]. A new design of hierarchical interdigitated flow field has showed in Fig. 1. The predicted pressure drop at different inlet flow rates and three different electrode intrusions are compared with normal interdigitated flow field as shown in Fig. 2. The predicted pressure drop at 0.5 mm electrode intrusion has shown similar value of pressure drop in Fig. 2a. Here the flow area of 2.5 and 0.5 mm were covered with an electrode, so one can expect that the pressure drop will be lower because flow resistance is less here. If we increase the electrode intrusion from 0.5 to 1 mm, then electrolyte flow area reduces from 2.5 to 2 mm. The flow resistance increase with the increase of electrode intrusion and so, more amount to electrolyte penetrate in the electrode region, thus higher circulation will be predicted in this case. The predicted pressure drop value for 1 mm electrode intrusion is shown in Fig. 2b. In this case also the predicted pressure drop is attained, which achieved in both the flow field designs. In case of 1.5 mm electrode intrusion, the predicted pressure drop is 10% higher in case of HIFF as compared to normal IFF. The flow area is 1.5 mm and electrode intrusion area is also 1.5 mm, so, flow resistance will be highest as shown in Fig. 2c. In case of HIFF, electrolyte travel more area as compared to normal IFF. So, one can expect that pressure drop will be highest in case of HIFF.

As we increase the electrode intrusion into the flow channel, flow area reduces in flow channel, so, more resistance is created in the flow path, thus pressure drop increases within the cell.

Electrolyte flow pattern has been investigated for both the designs of HIFF and IFF. In case of HIFF, flow area is more due to increased number of channels than IFF. Velocity contour at mid-plane of the porous electrode for all the three-electrode compressions are plotted and it is shown in Fig. 3. In the case of a normal interdigitated flow field, the electrolyte circulation is higher but at the end some maldistribution of electrolyte occurs as shown in Fig. 3a, b, but in case of modified HIFF, there is no maldistribution of electrolytes observe throughout the active electrode. As we increase the electrode compression from 0.5 to 1 mm, the electrolyte circulation rate is higher as shown in Fig. 3c, d.

In case of IFF, last few channels showed higher maldistribution, thus, it is estimated that most of the electrolytes will not participate in the electrochemical reaction, but in case of HIFF, the maldistribution of electrolytes did not occur at all. A similar pattern was also observed in the case of 1.5 mm of intrusion as shown in Fig. 3e, f.

For better knowledge of how much electrolyte circulation rate happens in electrode region, we have considered a section line in the middle of a electrode intrusion (1 mm case) to plot velocity distribution along with the channel length. Figure 4 showed the comparison of predicted velocity distribution in HIFF and normal IFF at mid-height of the flow channel. In case of normal IFF, the velocity distribution is almost uniform but at the end of channel, a higher velocity rate is observed. HIFF shows a higher velocity circulation rate due to more number of L-shaped channels. A large velocity variations is noticed throughout the porous electrode because of mid channels high circulation and less circulation in L-shape channels. Further work will be carried out for the investigation of electrochemical performance using experiments as well as numerically.

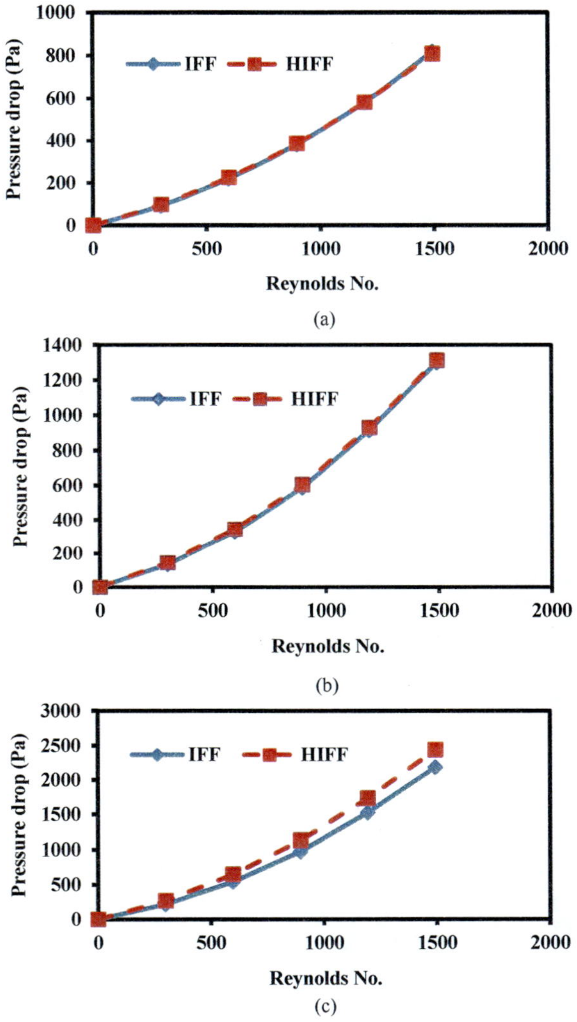

Fig. 2 Predicted pressure drop versus Reynolds number for 0.5, 1, 1.5 mm electrode intrusion

4 Conclusions

Hydrodynamic behavior through CFD simulations has been examined in case of Hierarchical interdigitated flow field with an active region of 110 cm^2 and compared with normal interdigitated flow field of similar active area. In hydrodynamic studies, three different electrode intrusions of 0.5, 1, and 1.5 mm thickness were used respectively. Different inlet flow rates have been used to check the effect of pressure drop under laminar conditions. The predicted pressure drop showed similar values at 0.5 and 1 mm electrode intrusion cases, but 1.5 mm electrode intrusion had a 2% higher pressure drop due to more resistance created in the flow path. Further investigations

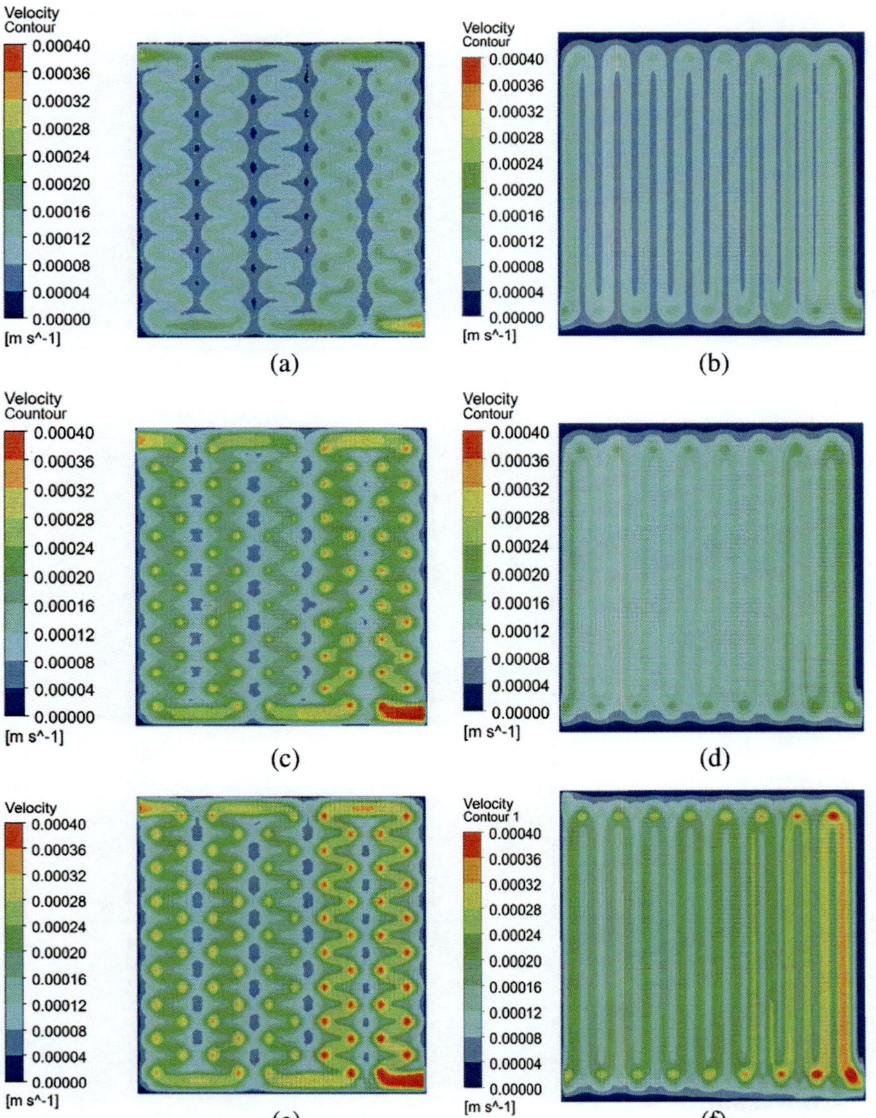

Fig. 3 Predicted velocity contour at mid plane of electrode **a** 0.5 mm **c** 1 mm and **e** 1.5 mm electrode intrusion for HFF and **b** 0.5 mm **d** 1 mm and **f** 1.5 mm electrode intrusion for IFF

were carried out to check the electrolyte circulation rate within the active porous zone for velocity of 0.1 m/s in all three electrode intrusions. Normal IFF design showed almost uniform electrolyte circulation rate through flow channel but in case of HIFF showed high circulation effects in all the three electrode intrusions cases. Hence, it

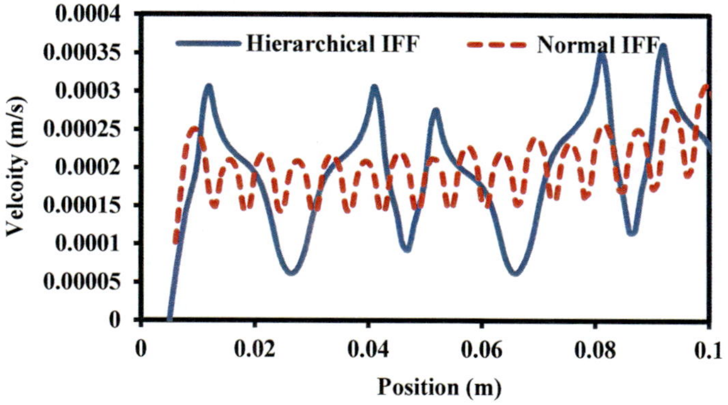

Fig. 4 Predicted velocity distribution at mid-height in HIFF and normal IFF

can be concluded that HIFF design can be employed for an improved electrochemical performance for stack-level studies.

References

1. Kumar S, Jayanti S (2016) Effect of flow field on the performance of an all-Vanadium redox flow battery. J Power Sources 307:782–787
2. Kumar S, Jayanti S (2017) Effect of electrode intrusion on pressure drop and electrochemical performance of an all-Vanadium redox flow battery. J Power Sources 360:548–558
3. Latha TJ, Jayanti S (2014) Hydrodynamic analysis of flow fields for redox flow battery applications. J Appl Electrochem 44:995–1006
4. Prasad KBS, Maharudrayya S, Jayanti S (2006) Flow maldistribution in interdigitated channels used in PEM fuel cells. J Power Sources 159:595–604
5. Prasad KBS, Jayanti S (2008) Effect of channel-to-channel cross-flow on local flooding in serpentine flow-fields. J Power Sources 180, 227–231
6. Prasad KBS, Jayanti S (2007) Multi-component formulation on water evacuation in interdigitated flow field. In: ASME conference on Fuel Cell Sci. Eng. Tec., New York, USA
7. Zenga Y, Lia F, Lua F, Zhoub X, Yuana Y, Caoa X, Xianga B (2019) A hierarchical interdigitated flow field design for scale-up of high performance redox flow batteries. Appl Energy 238:435–444
8. Kumar S, Agarwal V, Barnwal VK, Sahu S, Singh A (2022) Optimization of channel and rib width dimensions in serpentine flow field using CFD simulation for Vanadium redox flow battery applications. Energy Storage e2.349
9. Kumar S, Jayanti S, Singh A (2022) Electrolyte circulation effects in electrochemical performance for different flow fields of effect of all-Vanadium redox flow battery. Energy Storage e336

Investigation of Structural and Electrochemical Properties for Orange Peel Derived Carbon

Simple, K. K. Kushwaha, Shweta Tanwar, and A. L. Sharma

1 Introduction

In today's century demands for energy are rising day by day because of the increasing population. Most people are using non-renewable energy sources like coal, fossil fuels, etc. But they are limited and about to diminish [1]. For that reason, there is a need for renewable energy sources like batteries or supercapacitors (SCs). Batteries are not useful when there is a need for a high quantity of energy delivered in a short stretch of time, so SCs are better in this case [2]. Researchers are working on SCs nowadays. Supercapacitors have high cycle life and high power density as compared to batteries. They are just a modified version of capacitors which is similar in working but differ in the way they are made. They have high specific capacitance as compared to conventional capacitors. SCs are of two types, one is pseudocapacitors and another is electrostatic double-layer capacitors (EDLC) [3]. They both differ in their mechanism and the electrode material. Pseudocapacitors store the charge electrochemically by transferring ions in between effective electrode surface and electrolytes [4]. Materials utilized for the preparation of electrodes are conducting polymers, transition metal oxides, hydrides, sulphides and hydroxides. EDLCs stores the charge electrostatically as the name suggests. There is a formation of the double layer of charges around the electrode that's why they are called electrostatic double-layer capacitors [5]. Their electrodes are made up of carbon or carbon-derivatives like activated carbon, carbon nanotubes, graphene, etc. Among all of them activated carbon (AC) is the great promising candidate due to its easy availability, cost-effectiveness, and high porosity. AC can be obtained from any source but biomass wastes are best for it because they will be eco-friendly and less expensive. In Subramanian et al. the AC was derived from banana fibres with zinc chloride ($ZnCl_2$) and potassium hydroxide (KOH) [6], Lokesh et al. used peanut shell as carbon precursor and activated it with

Simple · K. K. Kushwaha · S. Tanwar · A. L. Sharma (✉)
Department of Physics, Central University of Punjab, Bathinda, Punjab 151401, India
e-mail: alsharma@cup.edu.in

S. Doolla et al. (eds.), *Advances in Clean Energy and Sustainability*,
Green Energy and Technology, https://doi.org/10.1007/978-981-99-2279-6_55

sodium hydroxide (NaOH) [7], Chen et al. activated the carbon derived from rotten potatoes [8], Arthi et al. obtained the activated carbon from ginger [9], Ajay et al. used bamboo as a precursor for carbon [10]. From all of these works, we are inspired to use orange peels as our precursor because it is highly fibrous, porous, and also easily available. So, in this present report, we have prepared carbon by using peels of orange which we have bought from the local market of Bathinda, Punjab. Activated it with different molarity of phosphoric acid (H_3PO_4) and then characterized it by utilizing powder XRD, FTIR, and electrochemical techniques.

2 Synthesis

2.1 Preparation of Activated Carbon

Some oranges were bought from the local fruit market of Bathinda, Punjab. The hydrofluoric acid (HF, 40%) was purchased from Sigma Aldrich. The orange peels were taken and washed thoroughly with deionized water for removal of dust particles and impurities which are soluble in water. Those peels are then dried under sunlight for 2–3 days before placing them in vaccum oven at 70 °C for 24 h. Those dried peels were then ground and sieved to get a fine powder form. After that carbonization is done at a temperature of 300 °C for 1 h in a muffle furnace, the carbonized sample is then washed with hydrofluoric acid (HF, 40%) and distilled water to make its pH neutral. For activation different molarities of phosphoric acid (H_3PO_4, 85%) were used and they are 0.5, 1 and 1.5 M. After impregnation we kept it for 1 day (24 h) at ambient temperature followed by 3 h of sintering at 300 °C in furnace (Fig. 1).

2.2 Fabrication of Electrodes

The electrode of prepared samples for their electrochemical properties measurements are prepared following the steps as follows. The electrochemical properties were measured in a two-electrode assembly for the prepared sample. We have taken prepared AC, carbon black, and polyvinylidene fluoride (PVDF) in the ratio of 70:20:10. Made a mixture of them by adding solvent N-Methyl-2-pyrrolidone (NMP) and ground it using a mortar pestle. Now coat this slurry on nickel foam (diameter 1 cm) and let them dry in the oven at 60 °C for 7–8 h. Take Whatman paper and put 6 M potassium hydroxide (KOH) on it dropwise, sandwiched this separator in between those electrodes by using a cell stand. Our sample is now ready for their electrochemical measurements.

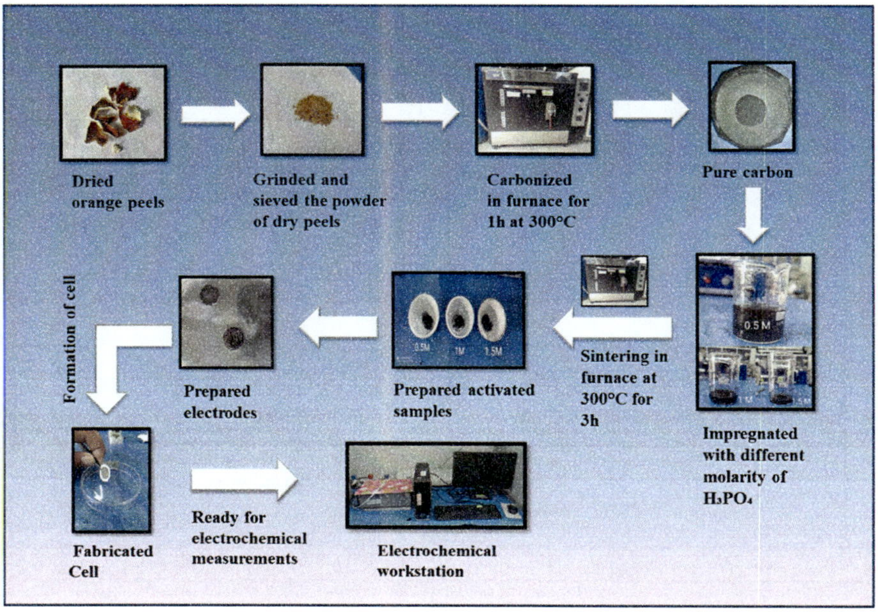

Fig. 1 Flowchart for the preparation of activated carbon (AC) using orange peels

3 Characterization of Materials

Powder x-ray diffraction (XRD) on model PANalytical Empyrean in Bragg reflection geometry with radiations of Cu K_α ($\lambda = 0.154$ nm) confirmed the samples' crystalline nature. For the determination of various chemical groups and bonds present in the pure and activated samples fourier transform infrared (FTIR) spectroscopy (Bruker Tensor 27 Model: NEXUS-870) was conducted in transmittance spectrum with wavenumbers ranging from 400 to 4000 cm^{-1}. Electrochemical techniques such CV, GCD, and EIS were also performed by using Model: CHI-760 electrochemical work station equipment.

4 Discussion of Findings

4.1 XRD Analysis

Crystalline nature is determined by using the XRD characterization technique. Cu-Kα radiation ($\lambda = 0.154$ nm) was used for powder XRD. In Fig. 2a we can see that the two peaks were found at 25° and 43.6° corresponding to (002) and (100) planes. These broad peaks indicate and confirm the amorphous nature of carbon [11]. Also,

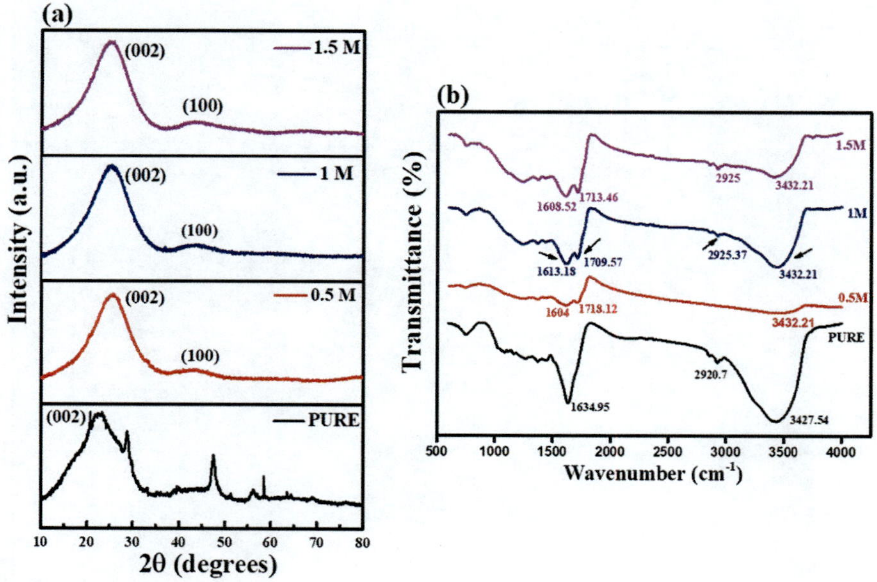

Fig. 2 **a** XRD and **b** FTIR plot for each sample

there are some extra peaks in pure carbon XRD showing impurities (source is not known) that are absent in activated carbon.

4.2 FTIR Analysis

We carried out FTIR to find the different chemical bonds, organic compounds, and to determine the structural of the prepared material. Each sample is mixed with KBr powder individually and then subjected to an FTIR instrument by making pellets. In Fig. 2b we could observe the graph plotted between wavenumber and transmittance. The range taken for the spectrum is 400 to 4000 cm^{-1} [1]. The broad peaks located at 3432.2 cm^{-1} show the OH stretching vibrations [12]. The peak at value 2925.3 cm^{-1} has a C-H bond like appearance [13]. The peaks arising at 1613.18 and 1709.57 cm^{-1} symbolize C = C bond associated with the aromatic ring and carbonyl group respectively [12]. The former peak is absent in pure carbon showing that there is no such carbonyl group before the activation process (Table 1).

Table 1 Chemical bonds observed in FTIR spectrum

Wavenumber (cm^{-1})	Chemical bond	Reference
3432.21	O–H stretching vibrations	[12]
2925.37	C-H bond	[14]
1709.57	C=O group	[12]
1613.18	C =C bond of an aromatic ring	[12]

4.3 *Electrochemical Analysis*

Electrochemical techniques such as CV, GCD, and EIS were carried out by utilizing CHI-760 electrochemical workstation instrument at room temperature. The electrolyte between the electrodes is 6 M KOH and the system is a two-electrode assembly. The CV and GCD were done within 1 V (from − 0.5 V to + 0.5 V) by varying scan rate from 20 to 100 mV s^{-1} and current density changing from 0.3 to 1 A g $^{-1}$. The frequency band for EIS was 1 Hz to 0.1 MHz.

Cyclic Voltammetry (CV)

CV is done for both pure and activated samples at scan rates varying from 20 to 100 mV s^{-1}. The graph is plotted by using origin software (originPro 2016 64-bit). The below-mentioned Eq. (1) is used to compute the value of specific capacitance from the CV graph.

$$C_S = \frac{A}{(V_2 - V_1) * m * k} \quad \left(F\,g^{-1}\right) \tag{1}$$

where A denotes the area covered by the CV curve, (V_2-V_1) is the 1 V potential window (from − 0.5 to + 0.5), m is the active mass of 3 mg, and k is the scan rate (in mV s^{-1}).

After activating our carbon, it was found that the specific capacitance increases and its highest value is 5.55 F g^{-1} for the 1 M sample at a scan rate of 20 mV s^{-1}. If we increase the molarity of activating agent i.e. H_3PO_4 from 0.5 to 1 M there is a rise in capacitance showing higher specific surface area due to the creation of additional pores. But if we go from 1 to 1.5 M decrease in capacitance value is seen. This is due to the reason that by increasing the molarity from 1 to1.5 M there is an enhancement in the number of micropores which reduces the effective surface area for the accommodation of ions. The 1 M sample has wider pores as compared to the 1.5 M sample which can accommodate more ions. Figure 3 given below shows the CV graph for pure carbon and AC.

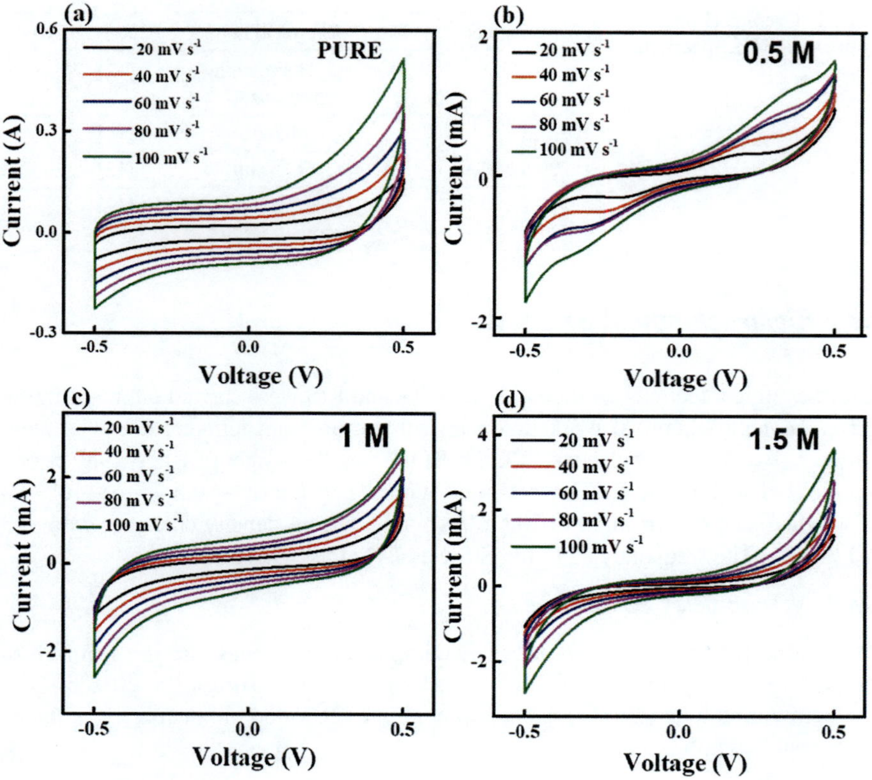

Fig. 3 **a** CV graph of pure carbon, **b** CV graph for 0.5 M sample, **c** CV graph for 1 M sample, **d** CV graph for 1.5 M sample at scan rates varying from 20 to 100 mV s^{-1}

Galvanostatic charge–discharge (GCD) and Electrochemical Impedance spectroscopy (EIS)

GCD is carried out for pure and activated carbon at unlike current densities. The measurements of specific capacity, specific energy, and specific power are done by using the Eqs. (2), (3) and (4) respectively,

$$C_s = \frac{I * \Delta t}{m * \Delta V} \quad \left(F\,g^{-1}\right) \tag{2}$$

$$E = \frac{C * \Delta V^2}{7.2} \quad \left(W\,h\,kg^{-1}\right) \tag{3}$$

$$P = \frac{E * 3600}{\Delta t} \quad \left(W\,kg^{-1}\right) \tag{4}$$

where I represents the current, Δt represents the discharging time and ΔV is the corresponding voltage difference. The maximum capacitance is found at a current density of 0.3 A g^{-1} for the 1 M sample which is 3.84 F g^{-1}. At a current density of 1 A g^{-1} specific energy and specific power are found to be 0.189 and 486.8 W Kg^{-1} respectively. Figure 4 given shows the graphs of GCD for all samples.

The comparison of electrochemical behavior between pure carbon and all the activated samples is shown in Fig. 5. It is noticed from the Fig. 5a, b that the value of specific capacitance is decreasing with the rise in scan rate and current density respectively. By plotting the bar graph Fig. 5c for activated samples we have analyzed that for sample 1 M the specific capacitance is maximum as compared to others in CV as well as GCD. Figure 5d repesents the Nyquist plot for all the prepared samples. Table 2 shows the computed values of series resistance (R_b) and charge transfer resistance (R_{ct}).

Ragone plot for 1 M sample [Fig. 5e] shows that there is a rise in power density with the decrease in specific energy value. The maximum specific power obtained is 486.85 W Kg^{-1} with energy density of 0.189 Wh Kg^{-1}. Figure 5f shows that among

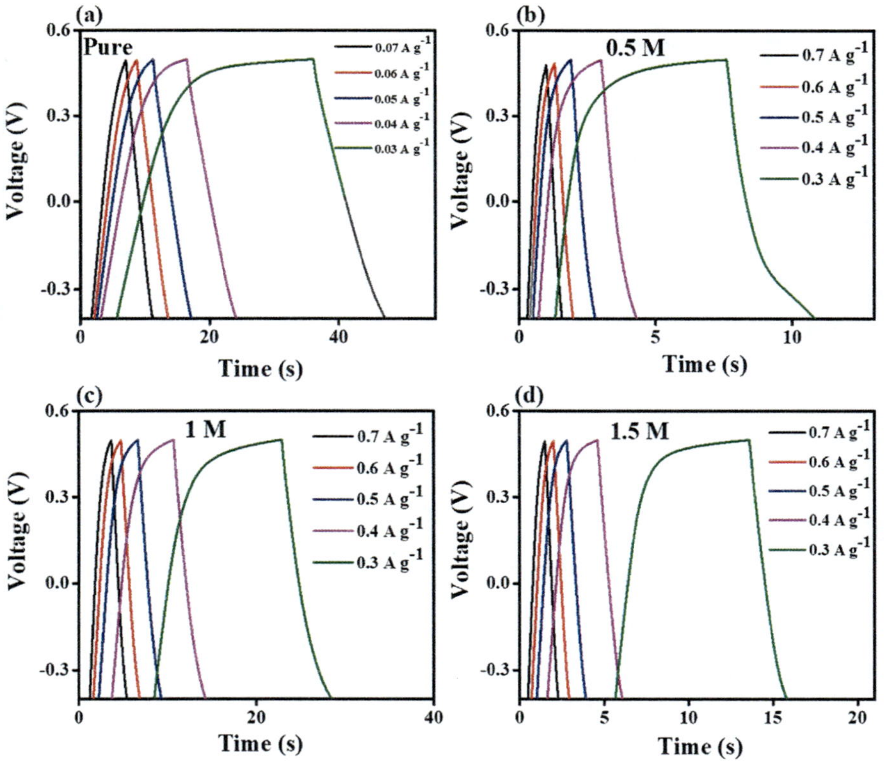

Fig. 4 **a** GCD graph for pure sample at current densities ranging from 0.07 to 0.03 A g^{-1}, **b** GCD graph for 0.5 M sample, **c** GCD graph for 1 M sample, **d** GCD graph for 1.5 M sample at unlike current densities i.e. from 0.3 to 1 A g^{-1}

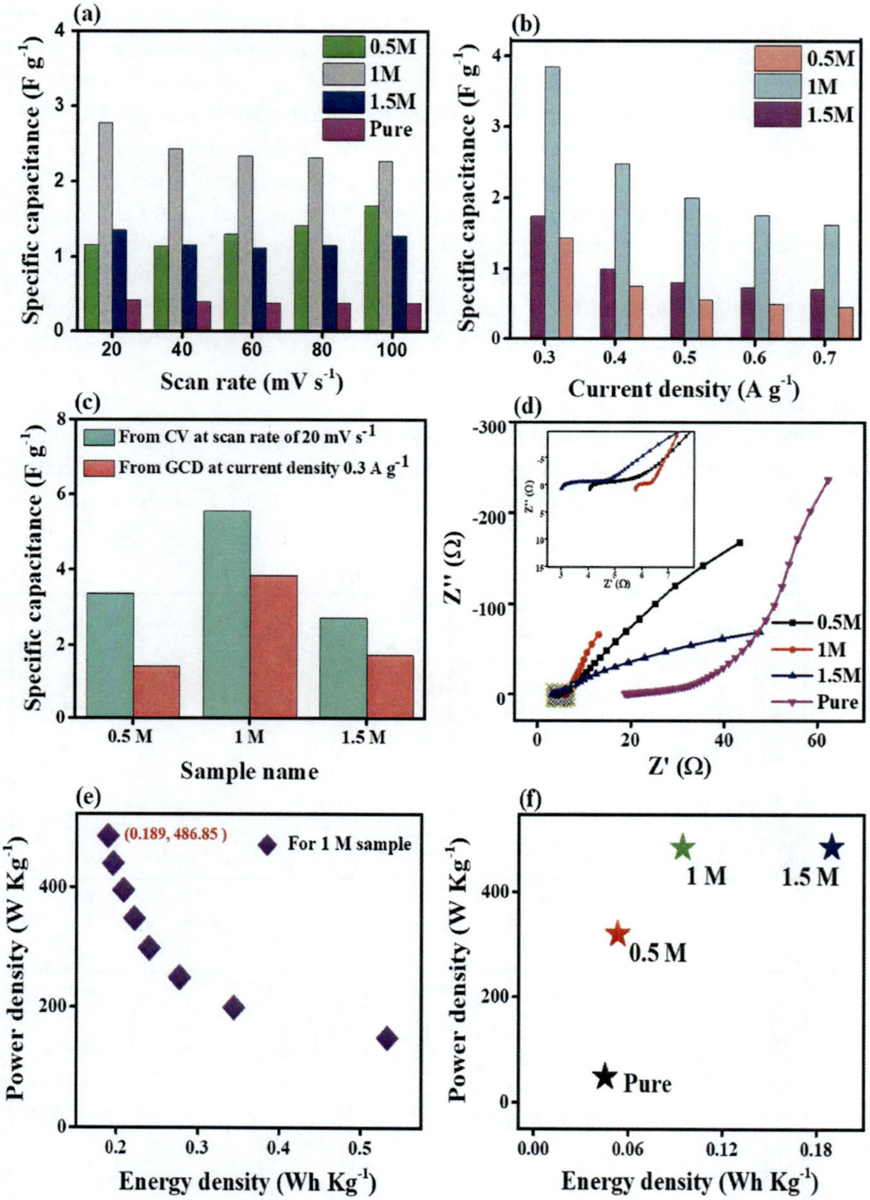

Fig. 5 **a, b, c** Specific capacitance versus scan rate, current density, and sample name respectively for all the prepared samples **d** Nyquist plot **e** Ragone plot of 1 M sample **f** power density versus energy density graph for each sample

Table 2 Resistance calculations from EIS graph

Sample name	R_s (Ω)	R_{ct} (Ω)
Pure	18.72	0.63
0.5 M	4.05	0.086
1 M	5.77	0.04
1.5 M	2.9	0.815

Table 3 Measured data of specific capacity, specific energy and specific power for all samples

Sample name	C_s (F g^{-1}) (at 20 mV s^{-1})	C_s (F g^{-1}) (at 0.3 A g^{-1})	Energy density (Wh Kg^{-1})	Power density (W Kg^{-1})
Pure	0.85	0.43	0.059	14.97
0.5 M	3.36	1.44	0.053	320.07
1 M	5.55	3.84	0.189	486.85
1.5 M	2.72	1.74	0.094	485.69

all the samples the sample of 1 M shows a high power density. All the measurements of specific capacity, energy density, and power density for each sample are shown in Table 3. From this table we have found that 1 M sample is giving maximum value of capacitance i.e. 5.55 F g^{-1} at a scan rate of 20 mV s^{-1}. So from all the results and calculations, we have found that H_3PO_4 of 1 M will be a promising candidate for activating the pure carbon which is derived from orange peels. It shows better electrochemical behavior as relative to other samples.

5 Conclusion

The conclusion is that we have prepared activated carbon (AC) from orange peels by activating it with H_3PO_4. It has been observed that increasing the concentration of activating agents there is an impact on the electrochemical behavior of the prepared carbon material. Initially from 0.5 to 1 M capacitance value rises but from 1 to 1.5 M its value decreases. So activation with 1 M solution of H_3PO_4 will be the best process giving maximum capacitance value i.e. 5.55 F g^{-1} at a scan rate of 20 mV s^{-1}. At a current density of 1 A g^{-1} the specific power is 486.8 W Kg^{-1}. The maximum capacitance value from commercial activated carbon is 25–30 F g^{-1} [21]. Our obtained value is quite less than the values obtained from commercial materials. But it can be improved by using inert atmosphere in furnace.

Acknowledgements One of the authors Simple is thankful for all the Ph.D. Research Scholar of Solid State Ionics Laboratory, Department of Physics, Central University of Punjab (CUPB), Bathinda. Authors also acknowledge Central instrumentation lab (CIL) and CUPB for the material characterization such as XRD and FTIR.

References

1. Tanwar S, Singh N, Sharma AL (2022) Structural and electrochemical performance of carbon coated molybdenum selenide nanocomposite for supercapacitor applications. J Energy Storage 45:103797
2. Tanwar S et al (2021) High efficient carbon coated TiO2 electrode for ultra-capacitor applications. J Phys D Appl Phys 55(5):055501
3. Tanwar S, Singh N, Sharma AL (2022) Structural, microstructural and electrochemical properties of carbonaceous nanocomposite for supercapacitor applications. Adv Funct Mater Devices. Springer, Singapore, pp 123–130
4. Kour S, Tanwar S, Sharma AL (2022) MnO2 nanorod loaded activated carbon for high-performance supercapacitors. J Alloys Compd 910:164834
5. Singh N et al (2022) High efficient activated carbon-based asymmetric electrode for energy storage devices. Mater Today Proc 57:5–10
6. Subramanian V et al (2007) Supercapacitors from activated carbon derived from banana fibers. J Phys Chem C 111(20):7527–7531
7. Pandey L et al (2021) Fabrication of activated carbon electrodes derived from peanut shell for high-performance supercapacitors. Biomass Convers Biorefinery, pp 1–10
8. Chen X et al (2016) Three-dimensional activated carbon recycled from rotten potatoes for high-performance supercapacitors. Waste Biomass Valorization 7(3):551–557
9. Gopalakrishnan A, Raju TD, Badhulika S (2020) Green synthesis of nitrogen, sulfur-co-doped worm-like hierarchical porous carbon derived from ginger for outstanding supercapacitor performance. Carbon 168:209–219
10. Ajay KM, Dinesh MN (2021) Performance studies of bamboo based nano activated carbon electrode material for supercapacitor applications. Mater Today Proc 46:4510–4514
11. Ranaweera CK et al (2017) Orange-peel-derived carbon: designing sustainable and high-performance supercapacitor electrodes. C 3(3):25
12. Devendran M et al (2020) Preparation of chemically modified porous carbon networks derived from Citrus sinensis flavedos as electrode material for supercapacitor. Int J Electrochem Sci 15(4):4379–4387
13. Nandiyanto AB, Dani RO, Ragadhita R (2019) How to read and interpret FTIR spectroscope of organic material. Indonesian J Sci Technol 4(1):97–118
14. Zapata B et al (2009) Thermo-kinetics study of orange peel in air. J Therm Anal Calorim 98(1):309–315
15. Wei Q et al (2019) Preparation and electrochemical performance of orange peel based-activated carbons activated by different activators. Colloids Surf A Physicochem Eng Aspects 574:221–227
16. Wan L et al (2020) Facile preparation of porous carbons derived from orange peel via basic copper carbonate activation for supercapacitors. J Alloys Compd 823:153747
17. Ajay KM et al (2021) Electrochemical investigations on low cost KOH activated carbon derived from orange-peel and polyaniline for hybrid supercapacitors. Inorg Chem Commun 127:108523
18. Xu J et al (2014) Preparation and characterization of activated carbon from reedy grass leaves by chemical activation with H3PO4. Appl Surf Sci 320:674–680
19. Saini S, Chand P, Joshi A (2021) Biomass derived carbon for supercapacitor applications. J Energy Storage 39:102646
20. Tanwar S, Singh N, Sharma AL (2022) Aging impact of Se powder on the electrochemical properties of molybdenum selenide: supercapacitor application. Mater Today Proc 57:94–99
21. Obreja, Vasile VN, Dinescu A, Obreja AC (2010) Activated carbon based electrodes in commercial supercapacitors and their performance. Int Rev Electr Eng 5(1)

To Be or not to Be a Prosumer: Understanding the Economics of Rooftop Solar PV in India

Afsal Najeeb, Aishwarya S. Sherla, and Anand B. Rao

1 Introduction

With the aim of accelerating the uptake of Rooftop Solar PV (RSPV) systems, the Central and State governments have provided a wide variety of incentives including capital subsidies, feed in tariffs, special metering arrangements etc. A rooftop solar PV (RSPV) system is an arrangement where the electricity consumer installs a solar photovoltaic system on the rooftop (or premises). RSPV systems connected to the utility grid can serve the connected load of the consumer and feed in the excess electricity generation back to the utility grid. The energy flow would be measured by a bi-directional energy meter. Net metering and gross metering are the two prominent paradigms of settlement of energy use in RSPV arrangements. In a net metering paradigm, the electricity generated by the RSPV system is consumed by the consumer in real time. Any excess generation beyond the demand of the consumer is injected into the utility grid. The consumer is free to draw power from the grid at times when demand is greater than RSPV generation. The utility defines a settlement period (mostly one or two months) at the end of which the consumer is billed for the net amount of electricity used from the utility grid. In case there has been a net injection of electricity, the consumer will be entitled to benefits as described by the tariff policy of the particular utility.

A. Najeeb (✉) · A. B. Rao
Centre for Technology Alternatives for Rural Areas (CTARA), Indian Institute of Technology Bombay, Mumbai, Maharashtra 400076, India
e-mail: 204354005@iitb.ac.in

A. S. Sherla
Department of Energy Science and Engineering, Indian Institute of Technology Bombay, Mumbai, Maharashtra 400076, India

S. Doolla et al. (eds.), *Advances in Clean Energy and Sustainability*,
Green Energy and Technology, https://doi.org/10.1007/978-981-99-2279-6_56

1.1 Economic Viability of RSPV Systems for Residential Consumers

The Ministry of New and Renewable Energy (MNRE), Government of India provides a capital subsidy of 40% to RSPV systems to encourage intake. Guidelines for application, installation, testing and commissioning of RSPV systems have been well defined to avoid regulatory roadblocks through a dedicated online portal [1]. Despite a policy environment that has been considered to be conducive, the uptake of RSPV systems have been limited in the residential sector. The profitability of an RSPV system for an individual consumer depends on its ability to offset payments towards electricity bill, generate a new income stream or assurance of suture profits. Consumer tariffs and incentives for excess generation are also significant determinants of profitability. Thus, the bottom-line economic viability of an RSPV system for an individual consumer becomes an interplay of multiple factors.

While there are many studies considering the potential and performance of RSPV systems, most studies focus on the non-residential sector and on the technical performance. This paper aims to evaluate the economic viability of RSPV systems for residential consumers considering state level policies, central level policies, geographic location and self-consumption. The results would aid in understanding the right policy mix of incentives and limitations that would incentivize adoption of RSPV systems by residential consumers. Using real time values of costs, prices and incentives from the consumer perspective, the study tries to provide a transparent analysis of economic feasibility that would aid decisions making on adoption of RSPV systems.

2 Objectives and Methodology

2.1 Objectives

The objective of the stud is to understand the economic viability of RSPV systems for residential consumers for different locations, different levels of consumption, different incentives for excess injection and different tariff rates. Net Present Benefits (NPB) and Levelized Cost of Electricity (LCOE) would be used as indicators of economic viability.

2.2 Methodology

The analysis of economic viability and performance of the RSPV system was done by simulating the system on HOMER Pro. HOMER Pro uses location, demand load pattern, technical and financial parameters of the system as inputs and simulates

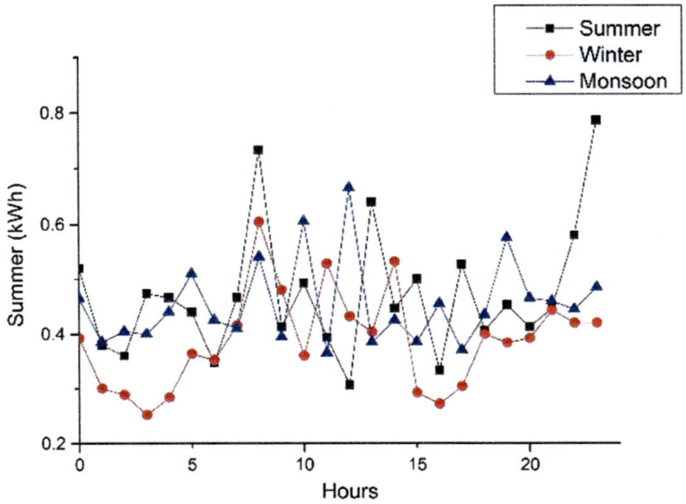

Fig. 1 Load pattern of household for Mumbai

the performance of the system for various sensitivity cases [2]. It also provides optimization considering minimization of net present costs. Five locations viz. a viz. Mumbai, Kolkata, Chennai, Lucknow and Shimla were chosen for the analysis with Mumbai as the base case for detailed analysis. Insolation values for each location was obtained from Global Radiation Database [3]. Household load profiles for all locations were obtained from NEEM dashboard. The seasonal average load pattern for Mumbai is shown in Fig. 1. A standard discount rate of 10% was used in all cases [4].

The size of the system, consumption of the household, grid rate of electricity, capital subsidy available and the feed in tariff were assumed as input parameters for sensitivity analysis. Levelized cost of electricity (LCOE) and Net Present Benefits (NPB) were chosen as the metrics to understand economic viability. For the purpose of this study, both parameters have been defined as shown in Eqs. (1) and (2). For the purpose of this analysis, LCOE can be defined as

$$LCOE = C_{anntot}/E_{served} \tag{1}$$

where, LCOE is the Levelized Cost of Electricity (Rs. /kWh), C_{anntot} is the total annualized cost of the system (in Rs. /year) and E_{served} is the total electrical load served (in kWh/year)

$$NPB = NPC_{grid} - NPC_{RSPV} \tag{2}$$

where NPB refers to the Net Present Benefits that would be delivered by a proposed system (Rs.), NPC_{grid} is the Net Present cost for the consumer with grid supply alone (in Rs.) and NPCRSPV is the Net Present Cost for the consumer with a specific

RSPV system installed (in Rs.) [5]. LCOE indicates the average rate of electricity (in Rs. per kWh) the consumer will have to pay for consumption over a period of 25 years. Net Present Cost of the grid and RSPV systems indicate the present value of all the costs each system throughout the lifetime considered minus the present values of the revenues generated [6]. The metric is an indicator of the net benefits a consumer can derive from making an upfront investment in installing an RSPV system.

3 Results and Discussion

3.1 Analysis of Base Case

In the case of Mumbai, for a household that consumes 10 kWh, the optimal size of the system has been calculated to be 3 kW. Considering the tariff of electricity, a household consuming 10 kWh/day pays an effective tariff od Rs. 6.19 /kWh considering the fixed charge, the energy charge and other duties and surcharges. The feed in tariff would be Rs. 3 and 40% capital subsidy would be available. For such a household, the optimal system size as calculated by HOMER is 3 kW. Over a period of 25 years, the system would provide a net cost savings of Rs. 97,517. The levelized cost of electricity will effectively be reduced to Rs. 1.98/kWh from Rs. 6.19/kWh the household currently pays. The RSPV system will require an investment of Rs. 117,000 and can be considered economically viable with an IRR of 15%, an ROI of 11% and a simple payback period of 6.5 years.

Figure 2 shows the variation of Net Present benefits (NPB) and Levelized Cost of Electricity (LCOE) with system size. The NPB increases till the optimal size of 3 kW and then decreases. Whereas the LCOE goes on decreasing with increase in size as there is more renewable generation. Beyond the optimal size of 3 kW, the system generates useful electricity but the economic value of the generated electricity is not sufficient to recover the additional capital cost. The decrease in NPB for higher capacities is due to the fact that the incentives for excess electricity supply from the RSPV is much lower than the tariff for grid electricity paid by the consumer. Thus, it can be seen that with the price and incentive structure of Mumbai, oversized systems do not make economic sense. In effect, the net metering arrangement in such locations encourages offsetting of grid consumption with RSPV consumption; but does not encourage a consumer to be a net supplier of energy. The consumer can, however be a net power importer but with sub optimal economic incentives.

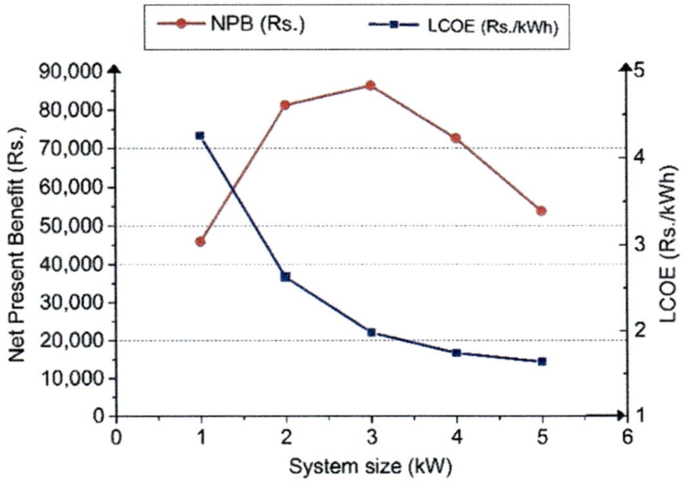

Fig. 2 Variation of LCOE and NPB with system size

3.2 Variation with Consumption

Considering households with higher consumption, the optimal sizing of the RSPV system increases as shown in the Fig. 2. Households with consumption of about 5 units per day would require only an RSPV system of size 1 kW. The non linearity in the system size with increasing consumption is due to step wise increase in PV capacity assumed (Fig. 3).

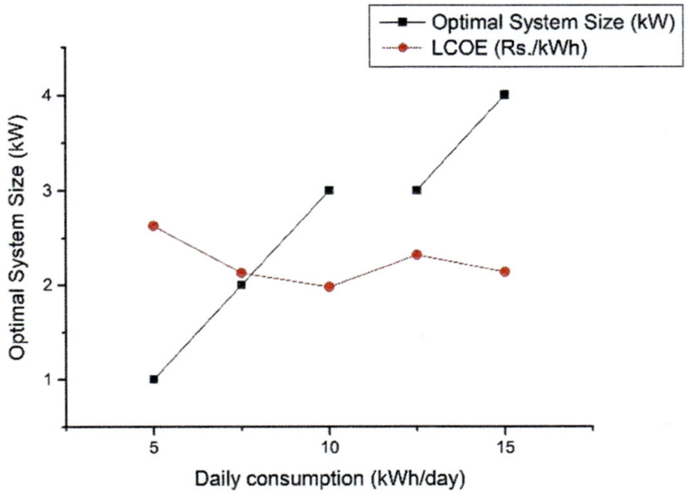

Fig. 3 Variation of optimal system size and LCEO with consumption

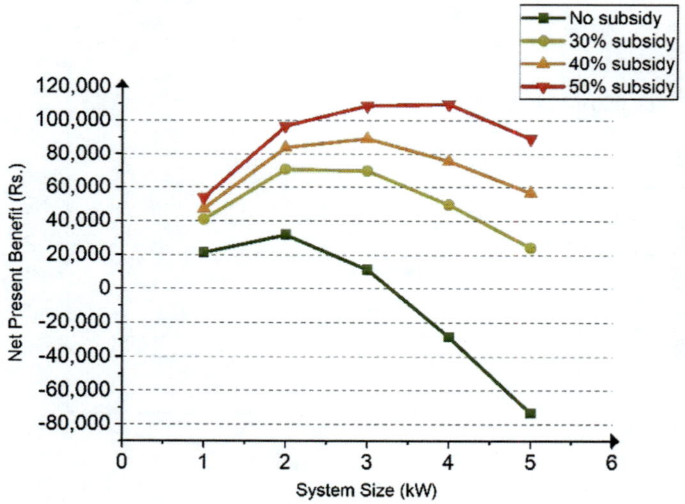

Fig. 4 Variation of net present benefits (NPB) with subsidy for 10 kWh/day consumption

Figures A.1 and A.2 (Annexure[1]) shows the optimal system configuration for the cases of 10 and 15 kWh per day. We find that using RSPV makes economic sense as long as grid prices are more than Rs. 4.50 per unit and there is a non zero assured buy back. For households with consumption of 15 kWh per day, the feasible region increases as shown in Fig. 4

Similarly, considering the optimal sizing of the system, we find that RSPV systems are not necessary in cases where the cost of grid electricity is only close to Rs. 3 per unit. Thus, RSPV systems cannot be considered as a profitable economic investment for a household with low electricity tariffs, irrespective of the level of consumption. High subsidies on residential consumption may thus reduce the effective tariff borne by the consumer and discourage adoption of RSPV. Thus, it would be oxymorons for any level of government to have policies that subsidies electricity consumption while encouraging adoption of RSPV systems by households.

3.3 *Effect of Subsidy and Feed in Tariff*

Capital subsidy and feed in tariff are the two significant and direct policy levers available to modify the pattern of adoption of RSPV systems by residential consumers. While the benefits of RSPV adoption could be significant and easily discernable for industrial and commercial consumers, residential consumers need strong signals of significant cost savings to encourage adoption. The capital subsidy provided makes reduces the initial investment by the consumer making adoption easier. It also has an

[1] Available at shorturl.at/cjzHJ.

important effect on the economic viability of the system. Presently, a capital subsidy of 40% is provided throughout the country in addition to state level subsidies in some states [7]. Figures 5 and 6 shows the difference in feasibility of installing an RSPV system with and without capital subsidy at a constant consumption level of 10 kWh/day.

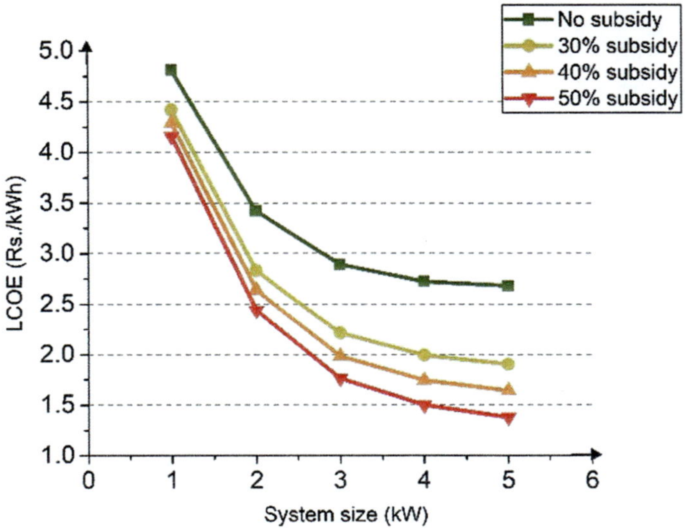

Fig. 5 Variation of LCOE with subsidy

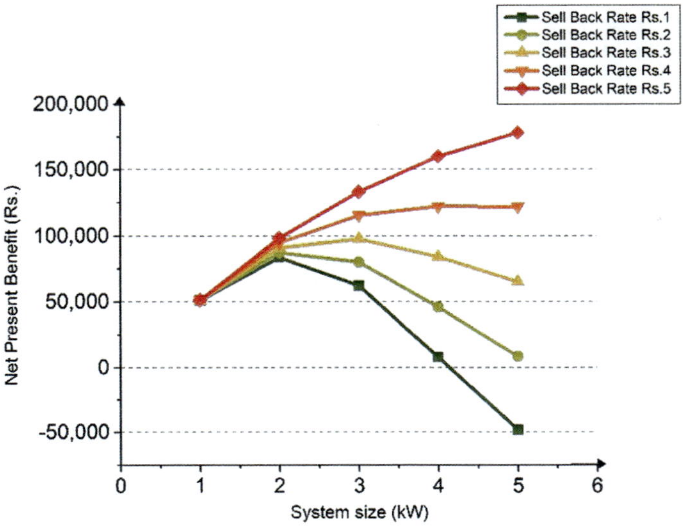

Fig. 6 Variation of NPB with FIT

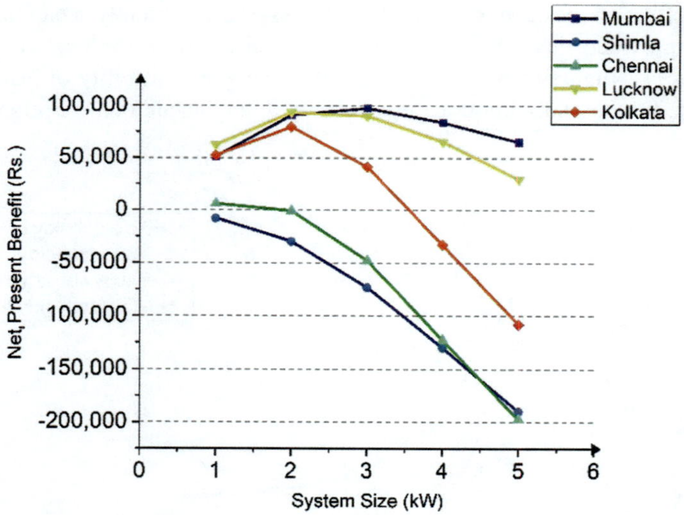

Fig. 7 Variation of NPB across locations

Without the capital subsidy and at low grid costs, there is no economic sense to installing an RSPV system as the optimal size of the system calculated is 0 kW (Figs. A5 and A6, Annexure). In this case, the cost of energy also rises considerably to Rs. 3.42 for the base case considered. Without subsidy, systems above 3 kW incur a net cost as indicated in Fig. 7 and does not make economic sense. Even in cases with subsidy, the net benefits decrease after attaining an optimum high. The subsidy has a similar effect on the levelized cost of electricity which varies by about Rs. 0.60 to Rs. 1.22 based on the availability of subsidy. In some sense, subsidy is incentivizing oversized systems.

Figure 8 shows the variation of LCOE with capital subsidy. We find that the trend of decreasing LCOE with increasing system size for different levels of capital subsidy. There is marginal different of less than 5% in LCOE between the levels of subsidy assumed. There is a steep decrease in LCOE till the optimal system size. Beyond the optimal system size, the variation in LCOE is marginal as excess injection is not incentivized considerably.

In the case of consumers with higher consumption, systems of up to 4 kW makes economic sense, even without the subsidy as shown in Fig. A3 and A4 (Annexure). There is scope for net benefits of more than Rs. 20,000 even without subsidy. Thus, for consumers with higher consumption, there is enough incentive to move towards an RSPV system even in the absence of upfront capital subsidy.

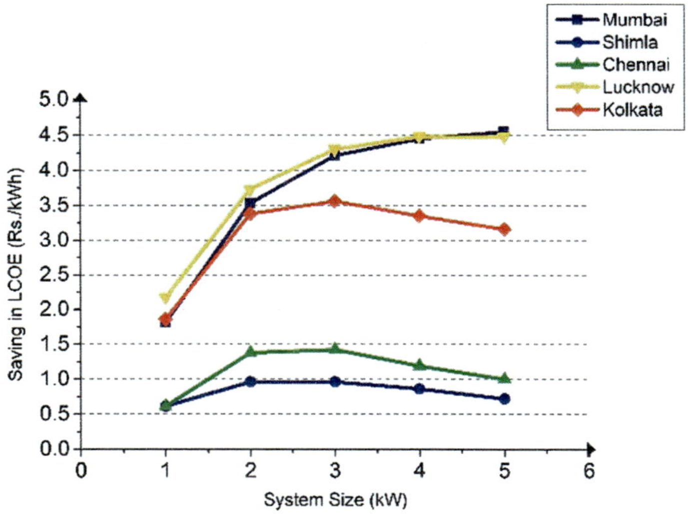

Fig. 8 Saving in LCOE across locations

3.4 Variation with Feed in Tariff

Feed in tariff is the second policy lever available to encourage or discourage the adoption of RSPV systems. For a household with consumption level of 10 kWh/day, the LCOE varies from Rs. 1.5 to Rs. 3 for different sell back rates. The variation of NPB is shown in Fig. 5. Feed In tariff is the only direct revenue stream available to the consumer as self-consumption leads to cost saving rather than revenue generation. We find that there is considerable difference in the effective price of electricity for the consumer even at low levels of feed in tariff. Similar parameters for a 15 kW system are shown in Fig. A7 and A8 (Annexure). We find that the trend observed in the case of 10 kWh/day consumption continues with a scaling effect. The difference in LCOE and NPB is marginal at lower capacities and is discernable only at higher system sizes.

4 Comparison between States

To understand the economic viability of RSPV under the present policy environment, four cities viz. a viz. Shimla (31.1 °N 77.2 °E), Mumbai (19.1 °N 72.9 °E), Chennai (13.1 °N 80.3 °E), Kolkata (22.6 °N, 88.4 °E) and Lucknow (26.8 °N, 80.9 °E) were considered for a comparative analysis. The variation of solar insolation across the selected locations asobtained from NREL (2018) are visualised in the Fig. A7 in Annexure [3]. RSPV systmes for all the locations were modelled using HOMER considering the site specific values for insolation, solar degradation, effect of dust

Table 1 Details of locations

Location	State	Climatic zone	Equivalent price of electricity (Rs. /kWh)	Fee in tariff/incentive
Shimla	Himachal Pradesh	Montane	3.07	30% APPC
Lucknow	Uttar Pradesh	Humid subtropical	6.42	Rs. 2 per unit
Mumbai	Maharashtra	Tropical wet	6.19	APPC
Kolkata	West Bengal	Tropical wet and dry	6.12	None
Chennai	Tamil Nadu	Tropical wet and dry	4.06	None

and derating factors [8]. From the household side, consumption patterns of each of the locations were obtained from the NEEM dashboard [9]. Mumbai has a higher insolation compared to other locations. Locations at higher latitudes have lower insolation.

As the locations fell under different states, the residential sector tariffs and feed in incentives for these locations were also different. The details are presented in Table 1. As the tariff structures varied across consumption, the case of a household consuming 10 kWh per day or 300 units per month was considered as the base case. The equivalent price for electricity per unit takes into account the fixed charges and other costs additional to energy charges.

We find that the states vary widely in both the cost of electricity and the incentives for excess generation. In Kolkata and Chennai, a consumer can utilize a maximum of his consumption from the RSPV system as the excess injection is treated as inadvertent injection with no benefits conferred. In Mumbai, the incentive is presently tied to the Average Pooled Power Purchase Cost (APPC) which is the cost incurred by the DISCOM in procuring and delivering electricity. In Uttar Pradesh and Himachal Pradesh, the incentives are lower than typical APPC levels. The cost of electricity is lowest in Tamil Nadu and Himachal Pradesh which has high government subsidies on the domestic sector. The other locations have comparable tariffs with variation less than 8% between themselves.

4.1 Comparison of Economic Viability of the RSPV Systems

The economic viability of the systems were analysed using the Net Present Benefits and the Levelised Cost of Electricity (LCOE) defined as in Sect. 3. The following Fig. 16 shows the variation of Net Present Benefits for different system sizes for a household with 10 kWh per day consumption. We find that for Shimla and Chennai which have very high incentives on gird electricity, RSPV systems are not profitable even at small sizes. At higher capacities, the sunk in capital cost does not generate any return (as the FIT is absent or too low) and it does not make economic sense for a consumer to invest in an RSPV system. Kolkata has a high tariff for electricity, but there are no incentives for feeding excess electricity into the grid. Here, optimally sized systmes become economically viable while over sized systems lead to considerable losses for the consumer. In locations Lucknow and Mumbai shows the highest economic viability of RSPV systems over a larger range of system sizes. In these locations, optimally sized systems deliver the maximum economic benefits but oversized systems (upto about twice the size of optimal size) does not lead to net economic loss. Households might tend to oversize the systems in expectation of rise in future demand, rise in electricity prices or considering convinience or improper estimation. In the absence of compensation mechanisms, customers with oversied systmes will be heavily penalised inspite of contributing to renevwable penetration on the grid.

Considering the LCOE for optimal capacity across different locations (Fig. A.10, Annexure), we find that for states with incentive mechanism, the LCOE decreases continuously as the system size increases. In states with no incentive mechanisms, the LCOE increases after the optimal system size. But as the tariffs of the present cost of electricity borne by the consumers in these locations are different, the saving in LCOE (Fig. 18) would be a better indicator of the profitability of the system from a consumer's perspective. We find that the savings in LCOE vary from Rs. 0.50/kWh to Rs. 1/kWh for states with subsidized tariffs to about Rs. 4.50/kWh in states with regular tariffs and incentive mechanisms. The saving in LCOE is the least for locations Shimla and Chennai with high subsidies on consumer tariff. In Kolkata, where there is no incentive for excess injection, the saving in LCOE decreases after the optimal system size. In the case of other locations, the consumer can derive savings in LCOE across the range of capacities even though the benefits for system sizes beyond the optimal capacity are minimal.

5 Conclusion

The paper analyses the economic viability of Rooftop Solar PV systems from a consumer standpoint using Levelized Cost of Electricity (LCOE) and Net Present Benefits (NPB) as the metrics for profitability. Different locations throughout the country with different levels of solar insolation, tariffs and incentive mechanisms have been considered. The analysis shows that capital subsidy is critical to economic viability of RSPV systems in the residential sector even for high consumption households. In the case of Mumbai which can be considered the base case, an optimally sized RSPV system would generate an ROI of 15% and lead to reduction in cost of electricity by up to Rs. 4/kWh. The lack of proportionate incentives for feed in electricity over and above self-consumption indicate that the present policy environment encourages offsetting of self-consumption through RSPV systems but does not encourage injection of excess electricity to the grid. This limits the ability or interest of consumers to install RSPV systems beyond the level of self-consumption to become net exporters of electricity.

References

1. NREL (2010) India solar map. Nrel
2. NREL (2022) HOMER documentation
3. NREL (2022) NSRDB: National solar radiation database. Website https://nsrdb.nrel.gov/
4. Yadav SK, Bajpai U (2020) Energy, economic and environmental performance of a solar rooftop photovoltaic system in India. Int J Sustain Energy 39:51–66
5. HOMER Energy LLC (2022) HOMER pro documentation—LCOE. Website
6. HOMER Energy LLC (2022) HOMER pro documentation—NPC. Webiste https://www.homerenergy.com/products/pro/docs/latest/total_net_present_cost.html
7. MNRE (2015) Grid connected rooftop systems. 10
8. Golive YR, Zacharaiah S, Bhaduri S, Dubey R (2018) All-India survey of photovoltaic module reliability. Executive Summary 217
9. Bureau of Energy Efficiency (2022) National energy end use monitoring dashboard. https://needashboard.in/

Thermal Analysis of Ceramic Coated Piston Crown Used in a Diesel Engine

Sameer Murlidhar Telote, A. K. Aadhithiyan, R. Sreeraj, and S. Anbarasu

Abbreviation

SI	Spark ignition
C_p	Specific heat capacity (kJ/kg K)
TBC	Thermal barrier coatings
TEC	Thermal expansion coefficient
TDC	Top dead center
k	Thermal conductivity (W/m°C)
HC	Hydrocarbon
CO	Carbon monoxide
FEA	Finite element analysis
FEA	Finite element analysis
HP	Horsepower
IC	Internal combustion
NO	Nitrous oxide
BDC	Bottom dead center
FEM	Finite element modelling

1 Introduction

The energy supplied to an engine is lost for cooling the engine and in the exhaust; only 33.33% of the energy is converted into some useful work. So, if this loss is somehow prevented, then an IC engine's efficiency and performance can be boosted. Insulating

S. M. Telote · A. K. Aadhithiyan · R. Sreeraj · S. Anbarasu (✉)
Hydrogen Energy Systems Research Facility, Department of Mechanical Engineering, NIT Rourkela, Odisha 769008, India
e-mail: anbarasus@nitrkl.ac.in

S. Doolla et al. (eds.), *Advances in Clean Energy and Sustainability*,
Green Energy and Technology, https://doi.org/10.1007/978-981-99-2279-6_57

the component of an IC engine, such as the piston crown in a cylinder lining valves etc., by TBC helps to lower the heat transfer. The performance and efficiency can be increased since the piston is one of the main components of an IC engine; this new technology of insulating the surface by a TBC can be used on the piston head. Much work has been done on improving the engine's efficiency by combining thermal barrier coverings with different TBC materials and sizes. Yaochen et al. [1] tested the piston with ANSYS software to detect decay, heat distribution, and pressure on the pistons. Reddy et al. [2] tested a thermal analysis on a standard diesel piston. Rajam et al. [3] did the piston upgrade work. They process the piston in a working model considered in this current work on the basic model. Bhattacharya et al. [4] operated on an internal combustion engine piston having a power output of 6500 W at 5500 rotations per minute. Aluminium 4032 alloy is used as an engine piston material in the design and analysis. Wankhede et al. [5] calculate the pressure and temperature values on the upper piston area. The piston structure model was created using the CATIA software. After it is imported into Hyper Mesh for integration purposes. Solanki et al. [6] have done a detailed analysis and design of efficiency of the hybrid piston having four strokes (7.35 kW) single-cylinder 10HP (IC) Diesel Engine. In this work, they have used lightweight alloy such as aluminium alloy and a better-strength steel piston head and on the piston wall. Using finite element modelling, they studied the pressure variation over the piston and assessed the engine's actual condition during the combustion process. Stress due to the combustion of fuel is considered to avoid failure.

Vibhandik et al. [7] investigated the behaviour of pistons in an IC engine with different loading conditions. The Geometric Piston is created with CAD software. The TATA MOTORS four diesel engine piston is used in this work. Prabhala et al. [8] replaced the steel components with aluminium for weight reduction, but the strength of the aluminium is meagre, so they considered aluminium alloy 1060 instead. Rao et al. [9] built a piston of a diesel engine. They use Brass, Cast Aluminum, and Aluminum as piston materials. Structural analysis was performed on the piston using pressure to identify the piston strength using three materials. Temperature analysis was performed to determine the temperature distribution. Aadhithiyan and Anbarasu [10] experimentally studied the changes in volumetric efficiency, isothermal efficiency and specific power consumption of a reciprocating air compressor with and without hard chromium plating on the cylinder liner. They observed that the volumetric and isothermal efficiency increased by 76% and 97%, respectively, whereas specific power consumption decreased by 49%. Vishal et al. [11] worked-on piston design and analysis. This is where production processes, analysis of piston, and piston design were studied. The investigation was done to measure the temporal piston temperature in different parts of the piston, from cold starting to solid-state, and compare the results of the standard feature analysis. Cerit et al. [12] investigated the stress and temperature distribution of a spark-ignition engine piston partially ceramic coated. They performed a heat analysis of standard pistons using the software ANSYS. A water-cooled SI engine has a single cylinder in standard and coated conditions. The results showed that the temperature at the surface of the bounded area of the piston raised to 100 °C, resulting in an increase in the mixed temperature of the gasoline at the crack and wall surface. Therefore, the cold initial HC emissions are significantly lower linked to the conventional engine without disrupting the engine's performance. A 43.2% maximum reduction was found in the hydrocarbon emission compared to a conventional engine. The materials used in the manufacturing of the piston are aluminium alloy. But it also has the disadvantage of notwithstanding extreme temperatures and is not resistant to rough wear. When the piston crown is exposed to very high temperatures, cyclic fatigue causes piston failure.

From the literature review, it is evident that using thermal barrier coating on the IC engine piston reduces the heat losses exposed to high temperatures and increases the cooling rate of the parts like piston crown, liner, plate, and piston rings. The combustion chamber with insulated protection from the thermal barrier coating affects the whole process. Due to this, the engine's exhaust emission characteristics and performance improve. The main objective of the current investigation is to find out the thermal behaviour of the three different TBC ceramic coatings (Yttria-stabilized zirconia (YSZ), Titanium dioxide (TiO_2) and Hard chromium), which are applied on the piston top surface of the aluminium A360 piston using ANSYS software.

2 Methodology

The following steps are considered for analyses of the piston.

- First, the engineering data section describes piston materials' thermal and mechanical properties and ceramic coatings.
- The thickness of the TBC coating, piston bore and piston width and 3D piston model was created in ANSYS Workbench.
- Afterwards, the types of material with their properties were distributed in the ANSYS model.
- This geometry generated default meshing with a 0.8 mm element size.

2.1 Thermal Barrier Coating (TBC)

The steady-state analysis of both coated and uncoated piston is done, and the temperature distribution of the top surface of the piston crown and the substrate are investigated. Owing to the lower thermal conductivity of the thermal barrier coating material and the temperature of the top surface of the piston increases. This TBC material also helps reduce heat loss and dangerous emissions such as nitrous oxide and hydrocarbons. Since the thermal durability of the TBC material is high, no cooling is required for the components. Three thermal barrier coatings (Yttria-stabilized zirconia (YSZ), Titanium dioxide (TiO_2) and Hard chromium (HC)) with three different thicknesses (0.04 cm, 0.08 cm, and 0.12 cm) are analyzed using the finite element technique. The properties of all the three TBCs are given in Table 1.

Table 1 Material properties of aluminium A360 alloy, YSZ and Ti O_2 and hard chromium

Piston material	Specific weight (g/cm^3)	Thermal conductivity (W/mK)	Youngs modulus (GPa)	Melting temperature (°C)	Poisson's ratio	Compressive strength (MPa)
Aluminium A360 alloy	2650	113	71	760	0.33	180
Yttria-stabilized zirconia (YSZ)	6	2	205	2800	0.3	2200
Titanium Dioxide (TiO_2)	4	11	230	1870	0.27	680
Hard chromium	7	69	144	1850	0.19	390

2.2 Analytical Analysis

The partial differential equation (PDE) given below is also called the equation of heat. In the coordinate system, when used as a function and time T(x, y, z, t), it satisfies this equation as follows [15].

$$\boldsymbol{K_x \frac{\partial^2 T}{\partial x^2} + K_y \frac{\partial^2 T}{\partial y^2} + K_z \frac{\partial^2 T}{\partial z^2} + Q = \rho C_p \frac{\partial T}{\partial t}} \tag{1}$$

where k is thermal conductivity, Q is the heat source rate in an area in W\\$m^3$ and specific heat Cp at constant pressure in J\\$m^3$ °C. Following are the natural as well as normal boundary conditions that are used [12].

$$\boldsymbol{T(x, y, z, t) = T_1(x, y, z, t)} \tag{2}$$

$$\boldsymbol{k_n \frac{\partial T}{\partial n} + q_p + h(T - T_\infty) + \sigma\varepsilon\left(T^4 - T_\infty^4\right) = 0} \tag{3}$$

where $\boldsymbol{k_n}$ is normal to surface thermal conductivity, h is the convection coefficient in W/m^2 °C, q_p (x, y, z, t) is the heat flux in W/m^2, ε is emissivity, σ is Steffen Boltzmann constant in w/m^2C^4, and T_∞ is the temperature of the ambient for radiation or convection. For the heat transfer analysis, initial boundary conditions need to be specified before the rest of the boundary conditions.

$$\boldsymbol{T(x, y, z, 0) = T_{in}(x, y, z)} \tag{4}$$

Using various techniques, Eq. (1) is reduced into the following form [15].

$$\boldsymbol{CT + kT = F} \tag{5}$$

where F is load and for steady-state analysis, the value of F is zero, and k is (effective) thermal conductivity. Hence temperature distribution is calculated using the above equation in the domain.

3 Finite Element Modelling of the Piston

The 3D piston model, including the metal substrate and ceramic coating, contains approximately 428,943 nodes and 252,229 elements. The rings and the piston material are cast iron and Aluminum alloy, respectively. These substances are considered elastic isotropic, and linear. Selected TBCs stabilize the zirconium usually used in thermal barrier coatings due to its excellent thermal protection properties, high-temperature resistance, and durability. These structures were helpful in various engineering programs. Building materials sometimes have different thermos-mechanical properties in a different directions. The TBC contains a metal-ceramic suspension and might not be isotropic. Heat components TBC with ceramic mud have flawed structures arising from continuous penetration of mass of completely dissolved particles (Figs. 3 and 4 and Table 2).

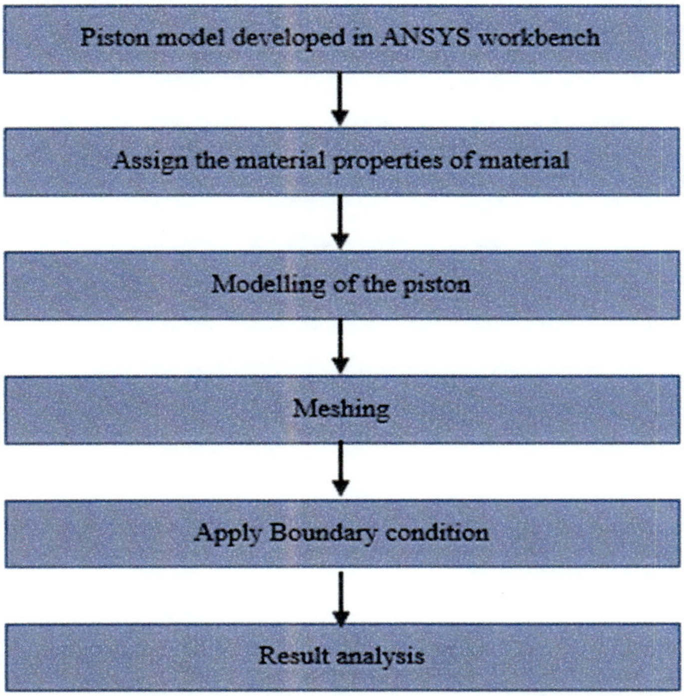

Fig. 1 Methodology

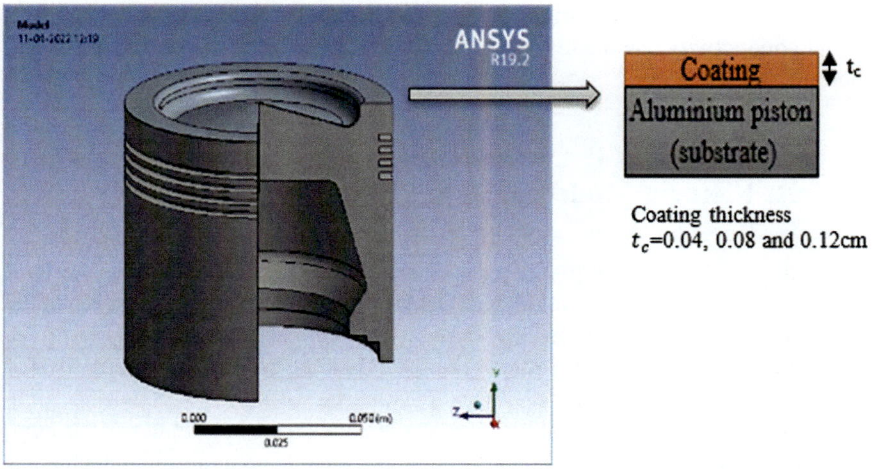

Fig. 2 Piston model in ANSYS

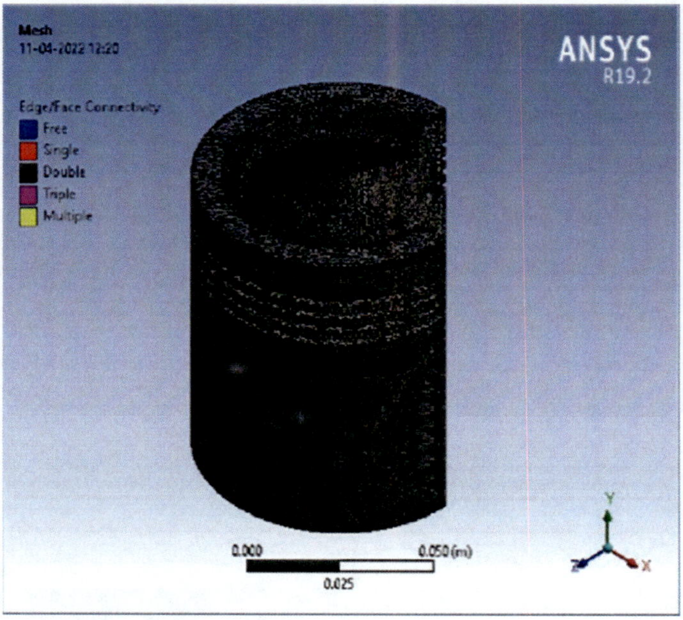

Fig. 3 Meshing of the piston model

Few assumptions are considered before applying all boundary conditions:

- The conductive rate of heat transfer to the oil film is ignored
- No cavitation of piston rings and skirt
- The heat transfer effect due to the motion of the piston is neglected

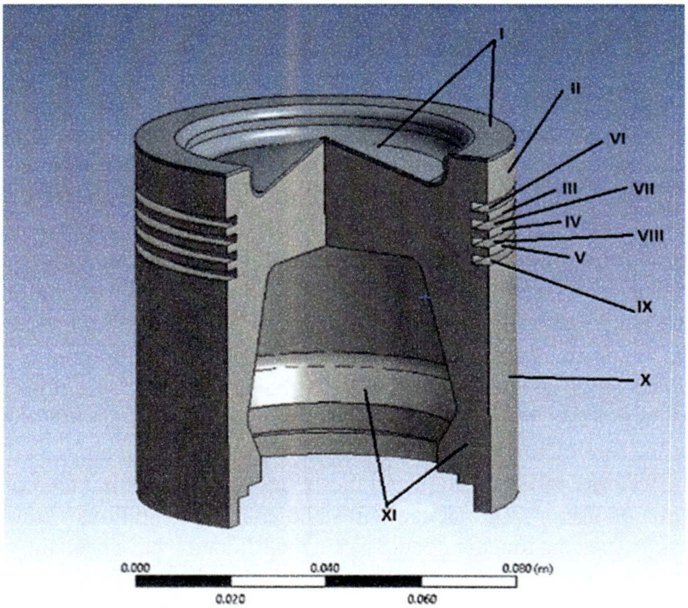

Fig. 4 Boundary conditions used in FEA

Table 2 Convection boundary conditions

Region	Convective coefficient (kW/m^{2}K)	Temperature (K)	Region	Convective coefficient (kW/m^{2}K)	Temperature (K)
I	0.7	973	VII	0.4	453
II	0.5	498	VIII	0.4	433
III	0.4	453	IX	0.4	413
IV	0.4	443	X	1.5	383
V	0.4	433	XI	1.5	973
VI	0.4	473			

- Twisting of the piston rings is neglected.

Thermal properties of the piston material are constant with respect to time in the analysis; therefore, the FEM tool with steady-state analysis can be applied to the three TBC-coated engine pistons to determine the temperature distribution over the surface. After using the boundary condition, all sets of Piston designs were analyzed with a steady-state-thermal work tool.

3.1 A Steady-State Analysis of Piston

The current work includes the steady-state (numerical) analysis of the piston's internal combustion engine. Aluminium A360 alloy is used as the piston material, and the three ceramic coatings are Yttria-stabilized zirconia (YSZ), Titanium dioxide (TiO_2) and Hard chromium (HC) with three different thicknesses of 0.04, 0.08, and 0.12 cm each. Firstly, the materials are assigned in the ANSYS geometry workbench. The 3D piston model was created with fine meshing having default type settings with an element size of 0.8 mm in the ANSYS workbench. The investigation of three TBCs on the aluminium alloy piston was analyzed to identify the thickness of the coating, which can reduce the surface pressure and intermetallic pressure and the possibility of coating disintegration. The FEM calculation was based on the same 3D model shown in Fig. 3. The piston is modelled such that it will reduce the number of elements and the calculation time. The element in equilibrium plays a crucial role in obtaining accurate results. The piston mesh and ceramic coating are constructed of 10-node material in a way that can withstand extreme conditions without losing much accuracy. The heat transfer conditions in the IC engine have been the subject of study because of some complications. The selected boundary condition is convection boundary condition as the main heat transfer method to analyze the condition.

4 Results and Discussion

The IC engine piston has three thermal barrier coatings (Yttria-stabilized zirconia (YSZ), Titanium dioxide (TiO_2) and Hard chromium (HC)) with three different thicknesses (0.04, 0.08, and 0.12 cm) is analyzed using the finite element technique. The stress variation and temperature distribution for all the cases, i.e., for uncoated and coated pistons, were discussed. The uncoated piston temperature contour is shown in Fig. 5. This figure shows that the peak temperature was obtained at the piston crown centre and areas near the edge of the piston bowl. It is because of the circumferential heat flux subjected to the piston crown. Also, the centre of the piston crown has attained the maximum temperature, and it is lowest at the bottom of the piston. More accurately, the maximum temperature found on the pistol crown was 264.06 °C, and the minimum was 121.47 °C. The temperature increases towards the piston's edge and from the centre of the crown and top surface to the bottom of the piston bowl (Table 3).

After processing the solution, the temperature at the piston head and inside substrate thermal analysis was discussed. As part of the thermal analysis, these results were obtained for all pistons. The steady-state thermal analysis of the three ceramic thermal barriers (HC, YSZ, and TiO_2) coated pistons of IC engines were studied, and their results were compared. Using identical boundary conditions, the temperature contours are shown in Fig. 6a–c for the thicknesses of 0.04 cm for YSZ, TiO_2 and HC each. The peak temperature occurred on mid of the top of the piston on the YSZ,

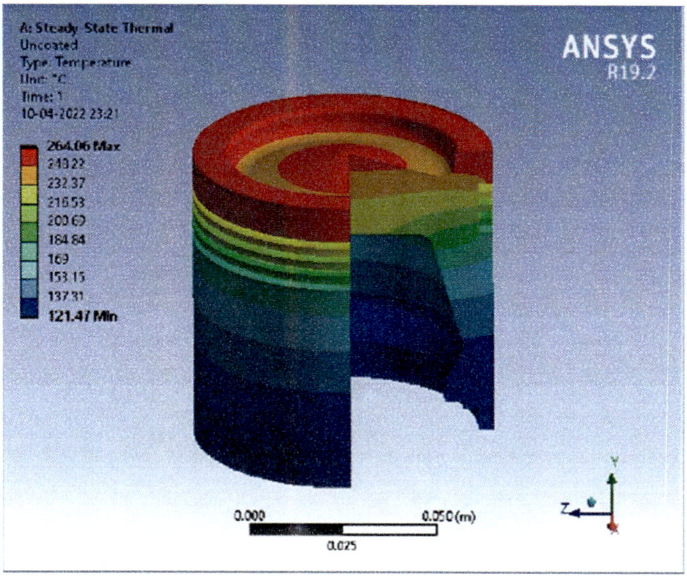

Fig. 5 Uncoated piston temperature contour

Table 3 Temperature values at the top of the piston

Ceramic materials	Temperature at 0.04 cm coating thickness (°C)		Temperature at 0.08 cm coating thickness (°C)		Temperature at 0.12 cm coating thickness (°C)	
	Max	Min	Max	Min	Max	Min
YSZ	323.04	120.93	365.06	120.1	401.32	119.41
TiO_2	283.33	121.7	293.98	121.48	304.52	121.27
HC	274.69	121.86	276.17	121.8	277.56	121.74

TiO_2 and HC coating surface were 323.04 °C, 283.33 °C and 274.69 °C, respectively, and the minimum temperature occurred were 120.93 °C, 121.7 °C and 121.86 °C near the bottom of the piston.

Figure 7a–c show the temperature contours of 0.08 cm ceramic coatings for YSZ, TiO_2 and Hard chromium each. The maximum temperature observed is 365.06 °C, 293.9 °C and 276.17 °C, respectively, and the minimum temperature observed is 120.1 °C, 121.47 °C, and 121.8 °C, respectively. Figure 8a–c show the temperature contours of 0.12 cm thickness ceramic coatings for YSZ, TiO_2 and Hard chromium each. The maximum temperature observed is 401.32 °C, 304.5 °C, and 277.56 °C, respectively, and the minimum temperature observed is 119.41 °C, 121.2 °C, and 121.74 °C, respectively. Even though the thickness of the ceramic coating increases, the rate of the increase in the temperature isn't the same. The temperature at the centre of the piston crown is found to be maximum and is 22.23%, 38.24%, and 51.98% for the thickness of 0.04 cm, 0.08 cm, and 0.12 cm, respectively for YSZ.

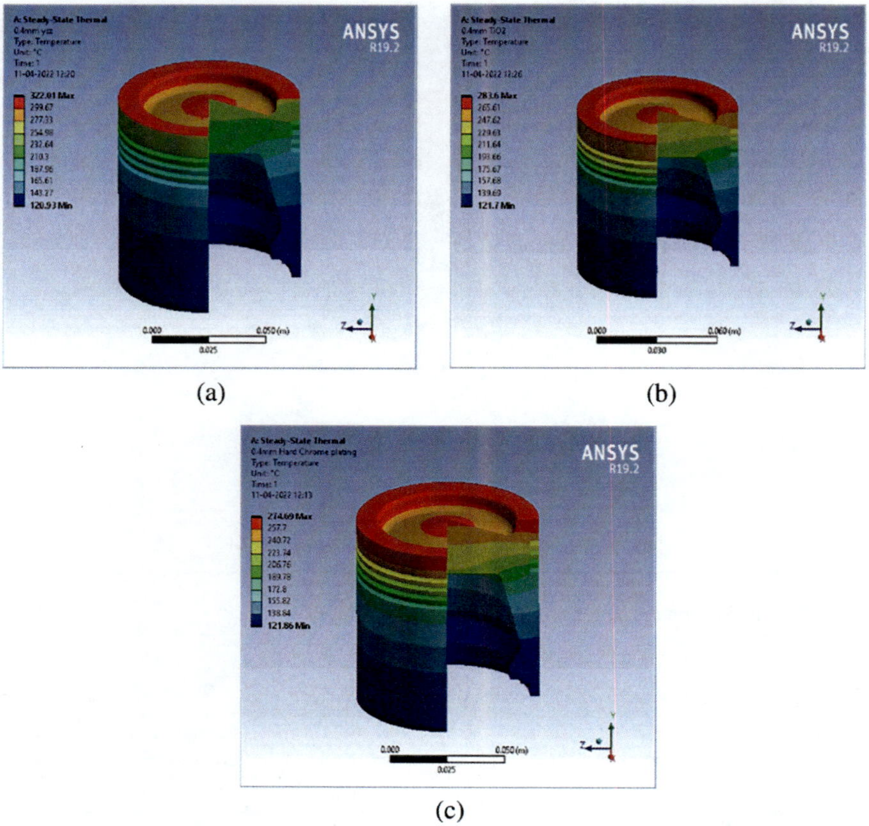

Fig. 6 The temperature contour of 0.04 cm coating thickness of **a** Yttria-stabilized zirconia (YSZ), **b** Titanium dioxide (TiO_2) and **c** Hard chromium

Similarly, it is 7.2%, 11.23%, and 15.32% for the thickness of 0.04 cm, 0.08 cm, and 0.12 cm respectively for TiO_2 and it is 4.02%, 4.58%, and 5.11% for the thickness of 0.04 cm, 0.08 cm, and 0.12 cm respectively for Hard chromium.

The temperature contours of aluminium A360 alloy substrate for pistons are shown. It is noted that the maximum temperature values in the centre of the piston substrate are 260.44 °C, 271.4 °C, and 272.77 °C, respectively, corresponding to the 0.04 cm coating of YSZ, TiO_2 and Hard chromium. Since material strength depends on the peak temperature attained inside the metal. The piston strength is improved by reducing the operating temperature, which leads to piston growth. Therefore, it is necessary to keep the temperature of the metal substrate at the lowest (Table 4).

A line OA has been created on the geometry of the piston model, as shown in Fig. 9. It starts from the piston centre O and ends at the piston's outer age, including the bowl's length at A. the line is on the half of the piston head surface. The graph of temperature on the head of the piston versus the radial distance of this line for all the thermal barrier coating of YSZ with an uncoated piston is shown below (Fig. 10).

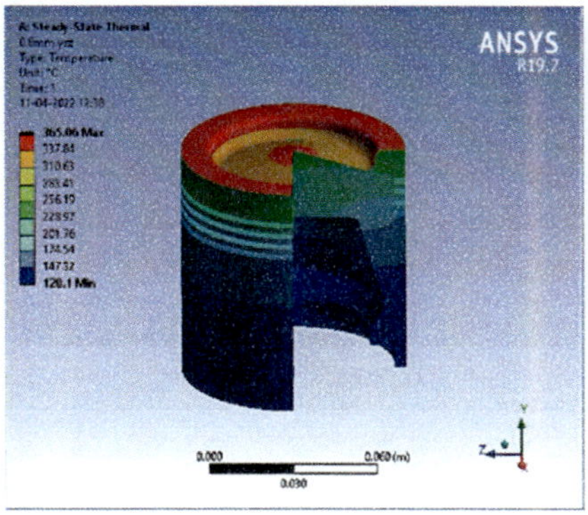

(a)

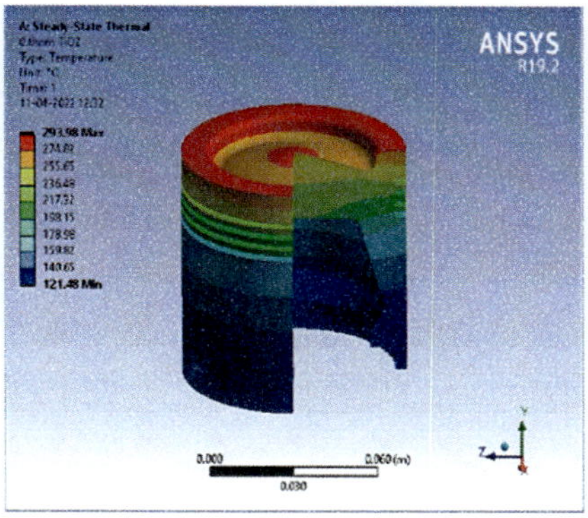

(b)

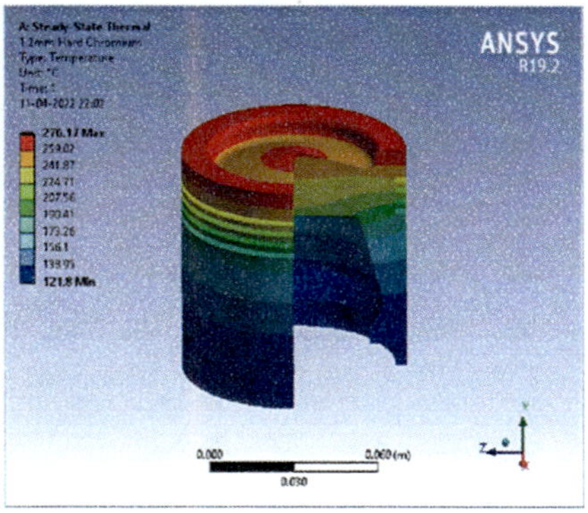

(c)

Fig. 7 The temperature contour of 0.08 cm coating thickness of **a** Yttria-stabilized zirconia (YSZ), **b** titanium dioxide (TiO_2) and **c** Hard chromium

The distribution of the surface temperature of the aluminium A360 alloy substrate of a ceramic-bound piston with a different thickness of cover against the radial distance in the OA line is set out in Fig. 11b. Notably, the temperature found on the aluminium metal substrate is much smaller than that of the piston top surface. It is because of the excellent properties provided by the TBC materials. The highest temperatures occur in the centre of the piston crown at 1.37%, 5.43%, and 8.94% of thickness with dimensions of 0.04 cm, 0.08 cm, and 0.12 cm, respectively in YSZ. The reduction of the temperature over the surface of the piston provides a positive contribution to the strength of the piston metal. It thus extends the life of the IC engine piston.

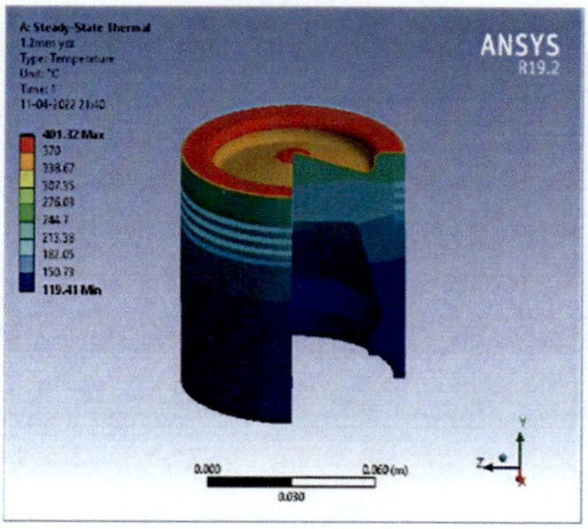

(a)

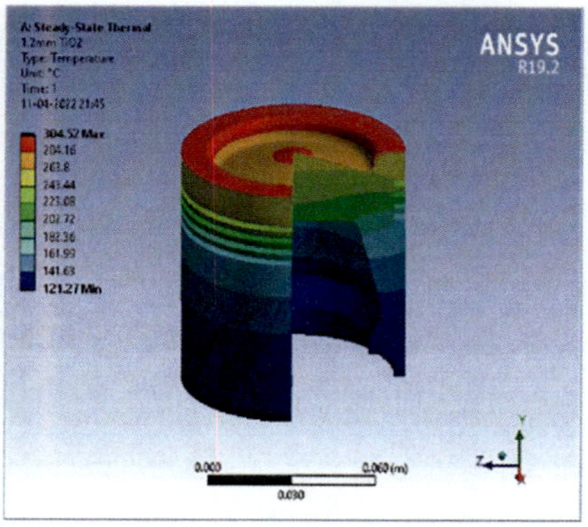

(b)

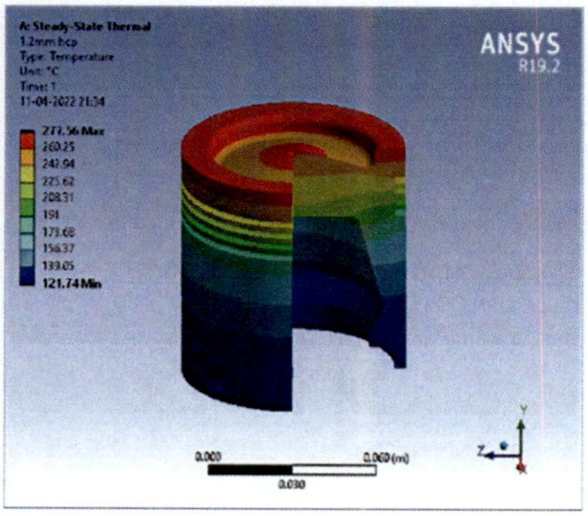

(c)

Fig. 8 The temperature contour of 0.12 cm coating thickness of **a** Yttria-stabilized zirconia (YSZ), **b** titanium dioxide (TiO_2) and **c** hard chromium

Table 4 Maximum substrate temperature

Ceramic materials	Max substrate temperature $t_c = 0.04$ cm (°C)	Max substrate temperature $t_c = 0.08$ cm (°C)	Max substrate temperature $t_c = 0.12$ cm (°C)
YSZ	260.44	249.7	240.44
TiO_2	271.4	269.44	267.39
HC	272.77	272.28	271.65

5 Conclusion

The numerical analysis of the three thermal barrier coating (TBC) materials, i.e., YSZ, TiO_2 and Hard chromium, has been studied in this work. From the finite element method (FEM) of all the coating thicknesses, it is found that the thickness of the thermal barrier coating material used on the piston directly affects the temperature

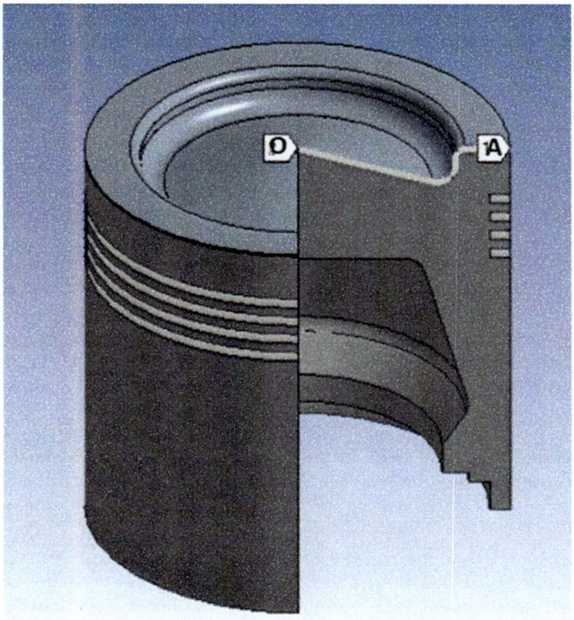

Fig. 9 Piston model with path OA

on the top of the piston surface. The YSZ coated piston has attained the maximum temperature among the piston cases considered in the current work. Among all the coatings considered, the peak temperature is obtained at the piston crown centre and areas near the edge of the piston bowl. The temperature found on the top of the piston surface was higher than that of the uncoated surface. The numerical analysis reveals the peak temperature achieved on the YSZ piston head is 323.04 °C, 365.06 °C, and 401.32 °C for 0.04 cm, 0.08 cm, and 0.12 cm thickness, respectively, compared to the piston without TBC coating. This rise in the temperature on the top of the piston crown increases the engine's thermal efficiency.

Since the strength of the material depends on the operating temperature, it is necessary to keep the substrate temperature at the lowest. The substrate peak temperature found on the YSZ piston are 260.44 °C, 249.7 °C, and 240.44 °C for the thickness of 0.04 cm, 0.08 cm, and 0.12 cm, respectively. Compared to the uncoated piston, the maximum temperature reduction percentage is found to be 1.37%, 5.43% and 8.94% for the thickness of 0.04 cm, 0.08 cm, and 0.12 cm, respectively. This reduction in substrate temperature increases the strength of the piston, and the life of the piston increases. The YSZ thermal barrier coating performs better than that titanium dioxide (TiO_2) and hard chromium coatings.

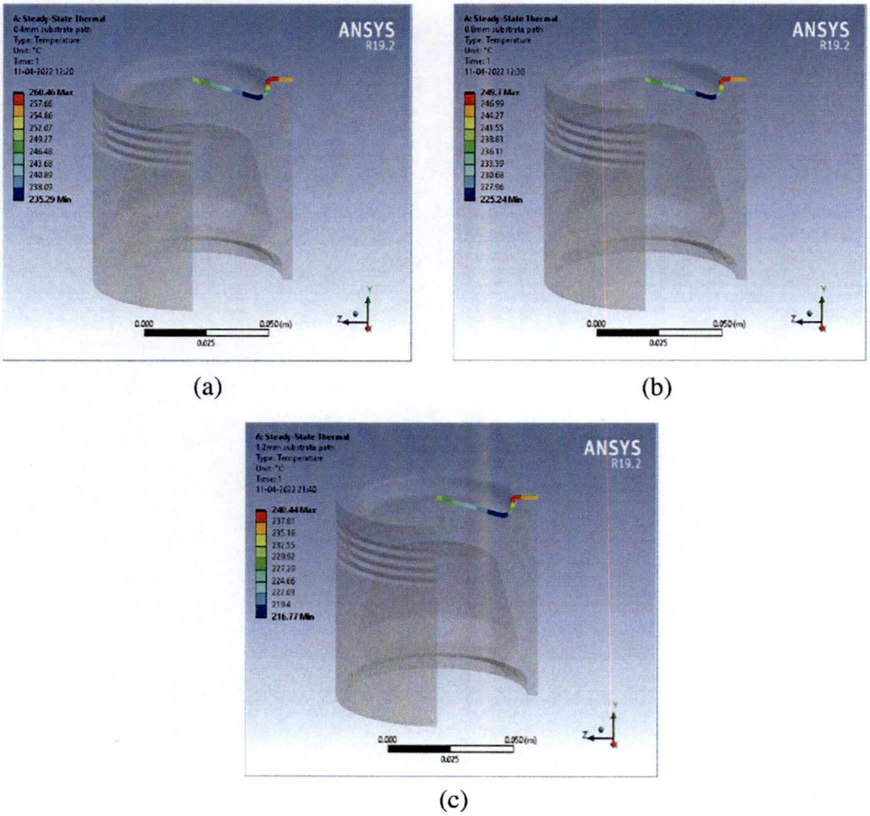

Fig. 10 Temperature contour of the substrate surface at 0.04 cm, 0.08 cm and 0.12 cm coating thickness of **a** Yttria-stabilized zirconia (YSZ), **b** titanium dioxide (TiO_2) and **c** hard chromium

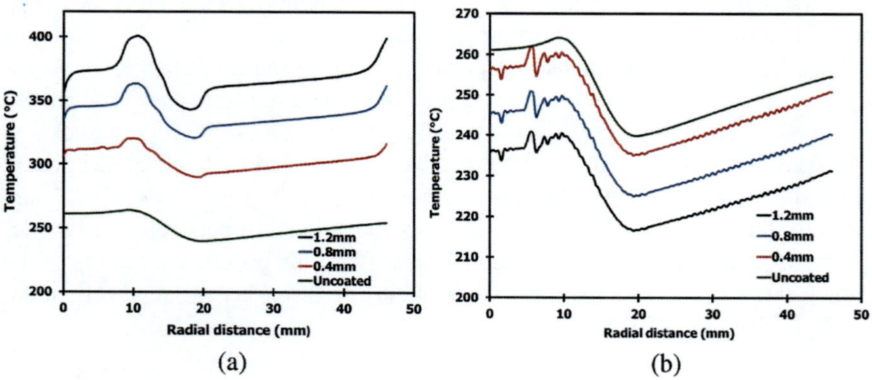

Fig. 11 **a** Temperature at the top of the piston versus radial distance plot of Yttria-stabilized zirconia (YSZ), **b** substrate temperature of the piston versus radial distance plot of Yttria-stabilized zirconia (YSZ)

References

1. Yang M, Xu Y (2013) Structure design and simulation of titanium engine piston based on thermal-mechanical coupling model
2. Reddy SS, Kumar DBSP (2013) Thermal analysis and optimization of ic engine piston using finite element method. Int J Innov Res Sci Eng Technol 2(12):7834–7843
3. Rajam CV, Murthy PVK, Krishna MM (2014) Non-linear static structural analysis of optimized piston for bio-fuel using ANSYS. Int J Manage IT Eng 4(1):148
4. Bhattacharya S, Basu A, Chowdhury S, Upadhyaya YS (2014) Analysis of piston of two stroke engine. Int J Res Eng Technol 3(06):642–648
5. Zolekar V, Wankhade DL (2013) Finite element analysis and optimization of ic engine piston using RADIOSS and OptiStruct. In: altair technology conference
6. Solanki AB, Patel MCH, Makawana AY (2014) Design analysis and optimization of hybrid piston for 4 stroke single cylinder 10 HP (7.35 Kw) diesel engine–a review
7. Vibhandik D, Pradhan A, Mhaskar S, Sukthankar N, Dhale A (2014) Design analysis and optimization of piston and determination of its thermal stresses using CAE tools. Int J Eng Sci Res Technol, pp 375–381
8. Prabhala SK, Kumar KSR (2012) Design and weight optimization of IC engine. Int J Adv Eng Res Stud (IJAERS) 2(1):56–58
9. Rao KV, Hasu B (2014) Modelling, analysis and optimization of diesel engine piston. IJREAT Int J Res Eng Adv Technol 2(1), ISSN: 2320–8791. www.ijreat.org
10. Aadhithiyan AK, Anbarasu S (2018) Studies on the hard chrome plating in reciprocating air compressors. In: Proceedings of the ASME 2018 international mechanical engineering congress and exposition. Volume 8A: heat transfer and thermal engineering 52118, V08AT10A043
11. Vaishali RN, Khamankar SD (2015) Stress analysis of piston using pressure load and thermal load. Int J Mech Eng 3(8):1–8
12. Cerit M, Ayhan V, Parlak A, Yasar H (2011) Thermal analysis of a partially ceramic coated piston: effect on cold start HC emission in a spark ignition engine. Appl Therm Eng 31(2–3):336–341
13. Sharma JK, Raj R, Kumar S, Jain RK, Pandey M (2021) Finite element modelling of Lanthanum Cerate (La2Ce2O7) coated piston used in a diesel engine. Case Stud Thermal Eng 25:100865
14. Sivakumar G, Kumar SS (2014) Investigation on effect of Yttria stabilized Zirconia coated piston crown on performance and emission characteristics of a diesel engine. Alex Eng J 53(4):787–794
15. Thermodynamics and Fluid Mechanics Group and Annand WJD (1963) Heat transfer in the cylinders of reciprocating internal combustion engines. Proc Inst Mech Eng 177(1):973–996

Experimental Study on Catalytic Pyrolysis of Waste Polypropylene at Different Temperatures

Ravindra Kumar, Payal Das, Anup Kumar Sadhukhan, Rohit Kumar Singh, Biswajit Ruj, and P. Gupta

1 Introduction

Plastics are essential component of modern civilization. In a day to day life water bottle, chairs, carry bags, car parts, decorative items all are made of plastics. They are mainly produced from numerous petro-chemical hydrocarbons blended with many essences i.e. flame-retardants, lubricants and flow promoters, these make it difficult to bio-degrade [1]. Worldwide plastics production reached about 368 MMT in 2019 [2]. China is the world's largest manufacturer as well as producer of plastic waste, monthly production ranges between 6 and 8 MMT, Whereas USA is the 2nd biggest manufacturer of plastic waste in the ecosphere. In India plastic waste generation during the year 2019–20 is approximately 34.7 MMT per year. In India maximum quantity of plastic waste is generated in the state of Maharashtra followed by Tamil Nadu and Punjab [3]. Plastic waste recycling into new plastic mould is very limited, so plastic waste management is one of the major challenges for most of the countries. The most developing countries follow the landfill disposal for effective plastic waste management [4]. The landfills disposal is also associated with various problems to the nearby habitats such as excess insects, can damage the crops of nearby field and also causes different types of diseases [5]. Besides, the rapid urbanization waste has scope for land filling due to shortage of available deserted lands.

Plastic has very long chain of carbon so it has very high calorific value but direct combustion of plastic is very hazards due emissions of many toxic chemicals and Poly aromatic hydrocarbons leading to huge environmental pollution. In recent years, the energy transformation from waste plastic has been a good option to fully utilization of waste to meet the increased energy demands. Pyrolysis is one of the method

R. Kumar · P. Das · A. K. Sadhukhan (✉) · P. Gupta
Department of Chemical Engineering, NIT Durgapur, Durgapur 713209, India
e-mail: anupkumar.sadhukhan@che.nitdgp.ac.in

R. K. Singh · B. Ruj
Environmental Engineering Group, CSIR-CMERI, Durgapur 713209, India

S. Doolla et al. (eds.), *Advances in Clean Energy and Sustainability*,
Green Energy and Technology, https://doi.org/10.1007/978-981-99-2279-6_58

to recover the energy from the plastics. Pyrolysis is the thermochemical process, in which long molecular chain carbon breaks into less complex molecules by the application of proper heat and pressure in the absence of oxygen or presence of inert gases. During pyrolysis of plastic 3 components are generally obtained—liquid (oil, commonly known as pyrolytic oil), gas (fuel gases) and solid (char). pyrolysis is modern technique to recover energies from wastes rather than incineration and landfilling. The pyrolytic oil can be used for many applications after examining the fuel properties [6]. The char could be utilized in treatment of waste water plants, it can also be an active component for production of carbonaceous substances (like MWCNT etc.) and gases for process heating or for other industries. The major advantages of pyrolysis are transformation of low grade energy density materials into higher grade energy density products [7]. In recent research the researchers move to catalytic pyrolysis due to less temperature (400–6500 °C) requirement compare with thermal pyrolysis (300–900 °C) [8]. In addition, the catalytic pyrolysis enhances the product quality (enhanced quality oil).

Mastral et al. [9] uses HDPE in a fluidized bed reactor at operating temperature of 650 °C and they observed, production of liquid yield around 68.5 and 31.5wt.% gaseous and remaining is solid products. Onwudili et al. [10] used LDPE in pressurized batch reactor (0.8–4.3 MPa) at a temperature of 425 °C and acquired 89.5% liquid yield, 10% gaseous and 0.5% solid (char). His results show that pressure also influences the configuration of pyrolysis product. Bagri and Williams [11] uses LDPE in fixed-bed reactor at a temperature of 500 °C with a heating rate of 10 °C/min. The experiment was performed for a period of 20 min in an inert medium by the application of nitrogen gas. They observed 95 wt.% of liquid yield with low gaseous yield and little amount of char. Panda et al. [12] observed highest liquid yield of 82.85% by weight at 500 ºC with PP in a semi batch reactor. Silica alumina catalyst shows better performance as compared to kaolin towards pyrolytic oil and reduces the end reaction temperature. The maximum pyrolytic oil obtained by silica alumina and kaolin catalyst are 91% and 89.5% respectively at a temperature of 500 °C with plastic to catalyst ratio is 3:1 for both the catalyst. Pyrolysis of waste plastic is the 2nd best practices after the implementation into road building and concrete structure according to Energy, Economy and Environmental (EEE) index [13]. Pyrolytic oil obtained from pyrolysis has good commercial application in automotive fuel (diesel and petrol) and C_2-C_4 olefins, which have a massive demand in petrochemical industry [14]. The adopted temperature ranges by the different researchers, for obtaining yields fraction is 400–700 °C.

This work aims to investigate experimentally the pyrolysis characteristics of waste PP and assess the effect of temperature and catalyst on product distribution of liquid, gas and solid. The TGA result shows the degradation behaviour of waste PP with catalyst. The pyrolytic oil is characterized by identifying the various functional groups present in it.

2 Materials and Methods

2.1 Materials

Waste PP was collected from the dump yard of Durgapur Municipal Corporation (DMC), West Bengal, India. The wastes were collected and categorized based on identification resin code. The waste PP disposable glasses were mainly considered for raw material. The materials collected cleaned with tap water and dried under sunshine for 2 h. The dried up glasses were cut into flakes by a shredder and the flakes were used in the pyrolysis. The sample mass used in the reactor is 100 gm. for each batch. The pyrolysis experiments were performed with only waste PP and with kaolin as catalyst. During catalytic pyrolysis, the catalyst mixed with the sample uniformly and fed to the Pyrolyzer for pyrolysis process.

2.2 Experimental Setup

The pyrolysis experiments were performed in a semi-batch tube-shaped reactor having a capacity of 500 ml. The main reaction zone is associated with electrical heating coils, the current through the heating coil is controlled by a PID controller, which in turn controls the reactor temperature. The overhead lead of the reactor is tightly placed at the reactor top and ensures no air or gas entry/exit from the reactor. It is connected with an inclined gas outlet tube properly jacked with the cold water cooling system. This ensures any condensable gas during the passage through the gas collecting line. The outlet of the gas collecting line is fitted with an oil collecting container. While one line is attached with a gas collecting chamber which can withstand pressure more than atmospheric pressure. The non-condensable gases are collected in the gas chamber, the temperature is kept as low as possible by cooling it with ice bath and the pressure of the gas chamber is read from the pressure gauge attached with the chamber. During experiment the reactor is heated by external heating source i.e. tube-shaped electrical furnace and the controller at the desired temperature. The schematic diagram of the experimental setup is shows in the Fig. 1. For each run 100 gm of sample was taken. The experimental temperature was maintained in the range 450-600 °C. Plastic was fed into the Pyrolyzer with catalyst and purging was done by nitrogen at a rate of 100 ml min-1 up to a period of 15–20 min to create a reductive/inert atmosphere for the pyrolysis process. Heat is applied externally on Pyrolyzer by heating the wall of reactor. On heating in absence of air/oxidant, the material in the reactor undergoes decomposition to produce gases along with condensable which leaves the reactor due to the high temperature inside it passed through the long gas collecting tubular path, on cooling to 10–15 °C the volatiles condensed in the form of liquid oil. The mass of the liquid products was measured and stored for further analysis. Similarly, the non-condensable gas stream passes through water column to remove solid particles and ashes which may present

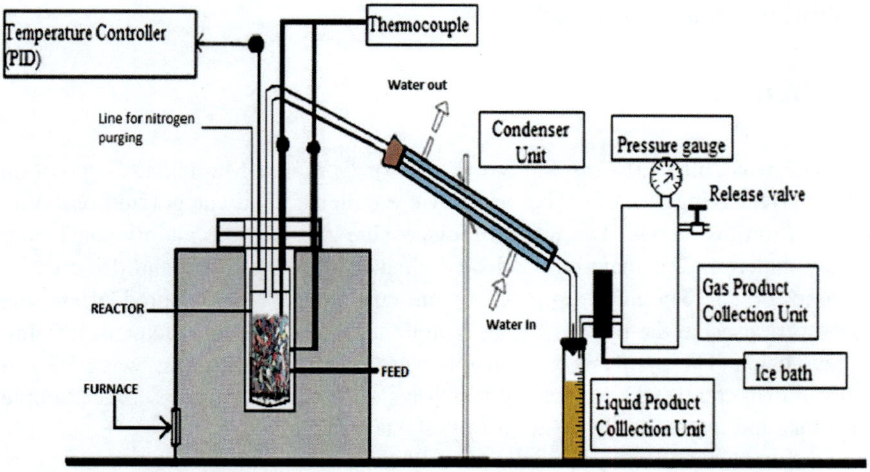

Fig. 1 Experimental setup of pyrolyzer

as contaminants in the waste stream. At the end of experiment, the solid residue (char) deposited at the bottom of the reactor was cooled, collected, and weighed. The weight of the non-condensable gaseous was calculated from its volume, temperature and pressure data using ideal gas equation (Fig. 1).

3 Result and Discussion

3.1 TGA Analysis

The decomposition temperature during pyrolysis, directs the nature and quantity of product acquired during thermal degradation of PP wastes. So, TGA analysis carried out to investigate the degradation behaviour of materials in the reaction conditions, which is used during pyrolysis. It was performed to observe the weight loss of the plastic with temperature and time. The main constituents of a Thermo gravimetric analyzer are a controlled ceramic furnace attached to microbalance and a data recorder. The procedure involved heating of the plastic mass (in nitrogen) at heating rate 10 °C min^{-1} from room temperature to 600 °C; the sample was hold for 2 min at the final temperature to observe it further degradation. During catalyst treatment kaolin was mixed in the proportion of 10:1.

The non-isothermal degradation characteristics are presented by TG and DTG curves for WPP and the same is shown in Fig. 2a, b at a heating rate of 10 °C/min. Both the curves signified the same trend as degradation occurred in two stages for WPP without and with catalyst.

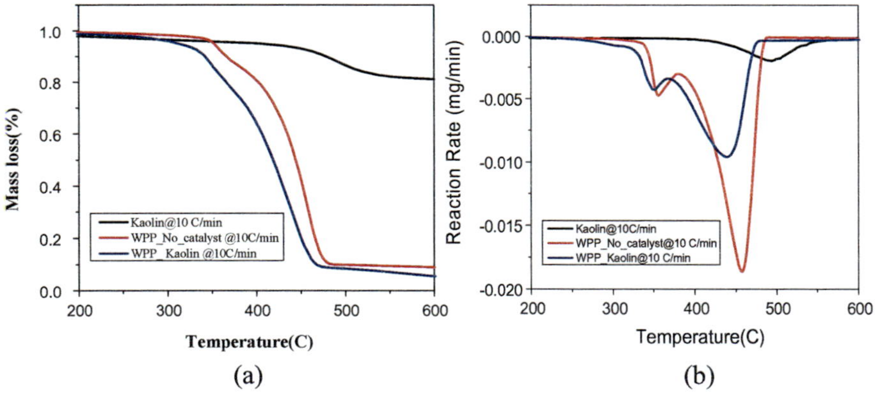

Fig. 2 **a** Mass loss versus temperature **b** reaction rate versus temperature

Generally, degradation of WPP proceeds through two stages of degradation while degradation of kaolin mainly follows only one stage. The degradation of the WPP without catalyst starts at a temperature of 3500 °C and ends at 4700 °C and approx. 10% residue remains in the crucible whereas with catalyst degradation of WPP starts at 3000 °C and ends at 4600 °C and approx. 8% residue remains in the crucible. From the above observation, it is clear that fewer amounts of char are obtained and also decomposition temperature reduces. The degradation of WPP without a catalyst degrades slightly higher temperature and showed increased reaction rate over shorter period compared to the WPP with a catalyst.

3.2 Yield Product Analysis from Semi-batch Reactor

The yield of various pyrolysis products (Liquid, gas and solid) obtained from lab-scale semi-batch paralyzer are summarized in Table 1 and the corresponding graphical representation is furnished in Fig. 3a, b. In all the cases pyrolytic oil (PO) and gaseous products were the major products with negligible amount of char. Oil yield by weight was maximum at temperature of 550 °C for WPP with catalyst while the gaseous product is more at 450 °C after interaction with kaolin. The PO at different temperature varies from light hydrocarbon oil (at low temperature) to moderately heavy hydrocarbon oil (at high temp). In comparison between catalytic and non-catalytic pyrolysis at a different temperature, the oil yield decreases with the addition of kaolin, with the enhanced gaseous yield. As the catalytic pyrolysis lowers the degradation rate (Fig. 2b) the lighter oil is obtained. The lighter oil (low density) is expected to be of better quality of oil and its yield is enhanced by the addition of catalyst. Kaolin is acidic in nature, so it promotes the production of smaller chain hydrocarbon due to its high cracking capability. By the addition of catalyst, the active

Table 1 Product distribution of waste PP with and without catalyst at different temperatures

S. No.	Temperature	Oil (ml)	Oil (gm)	Density (gm/cc)	Gas (gm)	Char (gm)
1	Waste PP@450 °C	113	82.5	0.730	7.1	10.4
2	Waste PP@500 °C	117	86.5	0.739	4.85	8.65
3	Waste PP@550 °C	111	83.5	0.752	5.2	11.3
4	Waste PP@600 °C	109	82.5	0.757	7.88	9.62
5	Waste PP_kaolin@450 °C	102	73.24	0.718	15.11	11.68
6	Waste PP_kaolin@500 °C	98	70.8	0.723	14.7	14.5
7	Waste PP_kaolin@550 °C	106	79.28	0.748	12.8	7.92
8	Waste PP_kaolin@600 °C	106	78.46	0.741	13.04	8.5

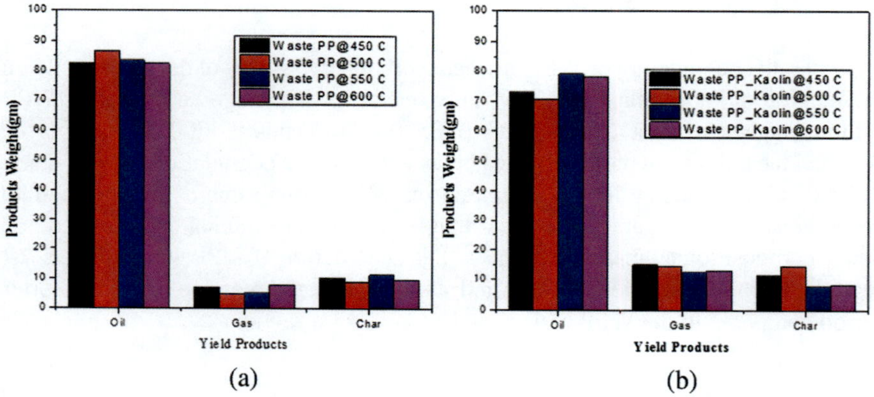

Fig. 3 **a** Waste PP without catalyst at different temperature **b** waste PP with catalyst at different temperature

surface area provides site for decomposition of WPP and it provides the chance of decomposition at lower temperature. So more non-condensable gaseous are obtained by rigorous cracking [12].

4 FTIR Analysis

The chemical composition of PO sample was examined by using FTIR spectroscopy. The FTIR spectra of PO, pyrolysed at three temperatures (450, 500 and 550 °C) in presence of catalyst showed almost identical FTIR spectra. The data revealed that the hydrocarbon components mainly consists of alkanes, alkanes and methyl groups. The peaks of FTIR (Fig. 4) spectra are found to be consistent. The area of the

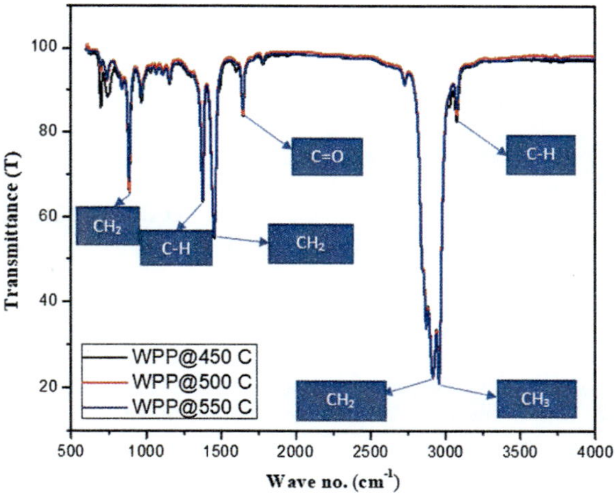

Fig. 4 FTIR spectra for PO at different temperatures in presence of catalyst

spectrum varies from 500 cm^{-1} to 4000 cm^{-1} confirmed the occurrence of -CH_3, -CH_2 and C-H groups of extremely aliphatic components in PO. The PO obtained from pyrolysis at a given temperature has been repeated thrice. Hardly any deviation of the FTIR spectra has been detected for PO from various run under a given temperature. Even the PO obtained from different pyrolysis temperature, the occurrences of the various peaks of PO constituents remained almost unchanged. It means there is hardly any influence of temperature on composition of pyrolytic oil from given raw material. The PO obtained without catalyst also showed the similar FTIR peaks but with different peak intensity. This indicates the alternation of relative proportions of various hydrocarbons. The quantitative compositional analysis by GC–MS of PO with and without catalyst is expected to provide the better insight regarding the composition characteristics of PO. The same is aimed in our future research endeavor.

5 Conclusion

- The pyrolysis investigations were conducted in the semi-batch reactor over the temperature range 450–600 °C with WPP to catalyst ratio of 10:1.
- The optimum temperature for maximum PO yield was 500 °C without catalyst while the same was increased to 550 °C in presence of catalyst, due to slower reaction rate.
- The PO with catalyst is lighter one than that obtained in absence of catalyst.
- FTIR analysis of PO showed the presence of alkanes, alkanes and methyl groups.

References

1. Ma C, Yu J, Wang B, Song Z, Xiang J, Hu S et al (2017) Catalytic pyrolysis of flame retarded high impact polystyrene over various solid acid catalysts. Fuel Process Technol 155:32–41
2. Plastics Europe Market Research Group (PEMRG) 2021
3. Annual Report 2019–20 on Implementation of Plastic Waste Management Rules (2016) CENTRAL POLLUTION CONTROL BOARD DELHI
4. Gandidi IM, Susila MD, Mustofa A, Pambudi NA (2018) Thermal-catalytic cracking of real MSW into bio-crude oil. J Energy Inst 91:304–310
5. Alexandra LC (2012) Municipal solid waste: turning a problem into resource waste: the challenges facing developing countries. Urban Specialist. World Bank. 2–4
6. Sharuddin SDA, Faisal A, Wan MAWD, Mohabid KA (2016) A review on pyrolysis of plastic wastes. Energy Convers Manage 115, 308–326
7. Das P, Tiwari P (2018) Valorization of packaging plastic waste by slow pyrolysis. Resour Conserv Recycl 128:69–77
8. Miandad R, Rehan M, Barakat MA, Khan AAH, Ismail IMI, Dhavamani J, Gardy J, Nizami AS (2019) Catalytic pyrolysis of plastic waste: moving toward pyrolysis base bio refineries. Frontiers Energy Res Bioenergy Biofuels 7(17):1–17
9. Mastral FJ, Esperanza E, Juste M (2001) Pyrolysis of high density polyethylene in a fluidized bed reactor, influence of the temperature and residence time. J Anal Appl Pyrol 63:1–15
10. Onwudili JA, Insura N, Williams PT (2009) Composition of products from the pyrolysis of polyethylene and polystyrene in a closed batch reactor: effects of temperature and residence time. J Anal Appl Pyrol 86:293–303
11. Bagri R, Williams PT (2001) Catalytic pyrolysis of polyethylene. J Anal Appl Pyrol 63:29–41
12. Panda AK, Singh RK (2011) Catalytic performances of kaolin and silica alumina in the thermal degradation of polypropylene. J Fuel Chem Technol 39(3):198–202
13. Gopinath KP, Nagarajan VM, Krishnan A, Malolan R (2020) A critical review on the influence of energy, environmental and economic factors on various processes used to handle and recycle plastic wastes: Development of a comprehensive index. J Cleaner Prod 274:123031
14. Soliman A, Farag HA, Nassef E, Amer A, ElTaweel Y (2020) Pyrolysis of low-density polyethylene waste plastics using mixtures of catalysts. J Mater Cycles Waste Manage

Enhanced Electrochemical Performance of Si by CNF Material for Li-Ion Battery

Yashkumar Patel, Anjali Vanpariya, and Indrajit Mukhopadhyay

1 Introduction

Electronic device updates have accelerated in recent years. With the rapid progress of portable electronic gadgets and electric vehicles, the LiBs in energy storage has played a major role, and the innovation of high energy density and power density energy storage systems is needed. For a long time, graphite materials have been employed as anodes in LIBs. The theoretical lithium intercalation capacity of graphite is just around 370 mAh/g [1, 2]. The graphite anodes, on the other hand can keep good cycle performance due to their minimal swelling during lithium intercalation. The attraction of graphite materials as anodes for lithium-ion batteries is their long cycle life. However, we still need to build innovative anode materials with excellent energy density for future uses in electric vehicles and huge rechargeable batteries.

Because of its large specific capacity of 4200 $mAhg^{-1}$, low insertion voltage, and natural abundance, Si is a more appealing anode material than graphite. With a composition of 4.2 lithium versus one Si, Si electrochemically forms the lithium-alloys corresponding to $Li_{4.2}Si$. The ratio of lithium to graphite is just 1:4. This implies that Si has ten times the capacity of graphite [3–5]. As a result, Si has been proposed as a possible anodic material for LiBs. In comparison to graphite anodes, Si has a working voltage of 0.5 V. It's also rich in reserves, as well as being both safe and harmless [6, 7]. In fact, this is a reversible chemical reaction with modest modifications in the host structure, resulting in greater reversibility. The interaction of Si with Li, on the other hand, forms numerous alloy-type compounds [8, 9] and, as a result, causes large volume changes during the lithiation/delithiation process. The biggest drawback of Si is its massive volume expansion (400%) when lithium ions are inserted. The large volume change can cause Si to disintegrate and have a

Y. Patel · A. Vanpariya · I. Mukhopadhyay (✉)
Department of Solar Energy, Solar Research and Development Centre, Pandit Deendayal Energy University, Gandhinagar, Gujarat 382421, India
e-mail: indrajit.m@sse.pdpu.ac.in

S. Doolla et al. (eds.), *Advances in Clean Energy and Sustainability*,
Green Energy and Technology, https://doi.org/10.1007/978-981-99-2279-6_59

short cycle life [10–15]. To begin, the massive volume expansion will cause some Si to pulverize, reducing electrical conductivity and causing Si to break electrical contact. The electrode structure will then be destroyed, and the active material will split. The size of nano-Si is extremely tiny. This shows that the sum of volume change is small, which reduces Si cracking during charge–discharge cycles. The electrons could not transfer to the current because the pure Si-based anode has a limited electrical conductivity. Researchers have conducted extensive study to improve their ability in response to the problem. Many better approaches exist, including three-dimensional porous Si particles, core–shell/amorphous Si nanowires, Si metal oxide composite, Si carbon core shell structure, graphene as a buffer [12, 16–20] and so on. Currently, Si/carbon composite anodes are being explored as a viable anodic material to substitute graphite materials. LIBs have widely utilised carbon materials such as graphene [5, 21, 22], carbon nano-fibers [23], and carbon nanotubes [24–26]. The general aim of this approach is to enhance electrochemical performance.

Carbon materials act as a conductive framework, permitting electrons to go through carbon as opposed to high-resistivity silicon. Consequently, putting in place a large and well-connected carbon network could have positive results. We recently discovered that CNF could improve the cyclic stability of these elements in lithium cells in a simple and effective way. [27, 28] Cellulose is the most plentiful and sustainable polymer resource on the planet today. It is mostly utilized as a building material in the form of wood or textile fibers like cotton, as well as paper and board. Cellulose is found in a variety of plants, frequently in combination with other biopolymers (typically known to as lignocellulosic materials), although commercialized cellulose is mostly produced from exploited materials like wood or naturally pure sources including such cotton [29, 30]. Micrometric cellulose fibers, nanocelluloses (i.e., nano fibrillated cellulose and cellulose nanocrystals), and water/solvent soluble cellulose derivatives are all examples of cellulose. Because of its high strength and stiffness paired with reduced weight, biodegradability and environmental friendliness, nanocelluloses have received an increasing amount of attention for the manufacture of cellulose-based nanocomposite materials during the last ten years [31, 32] thus far.

We used ultrathin cellulose nanofibers as carbon sources to create extensively connected SiNp via carbon network. Furthermore, cellulose is abundant and inexpensive, making it an excellent option for use in batteries. We hypothesize that a densely coupled carbon network can effectively inhibit SiNp volume expansion while simultaneously providing electron conductivity. The novel kind of anode's capacity and cycle efficiency were studied.

2 Experimental

2.1 Materials

Silica sol (SiO_2) and Magnesium (Mg) powder (99%, 100–200 mesh powder) pur-chased from Astron chemicals, India. Ethanol, Polypropylene, copper foil (10 micron), and cellulose nanofiber (CNF) purchased from Sigma Aldrich. DI water (18.2 m) was used throughout the investigation.

2.2 Synthesis of Completely Reduced Silicon Nanoparticles

The hydrothermal synthesis of Si from SiO_2 was previously described. After hydrothermal obtained material mixed submerged in 1 $molL^{-1}$ of hydrochloric acid for a period of 8 h. This was to eliminate by items like MgO and different oxides. From that point onward, the mixture was completely washed with DI water. After that remaining mixture consist of SiNp was dissolved in a solution of 1 wt.% dilute HF for 15 min so that any unreacted silica and other impurities that were formed during the reduction process could be eliminated [26].

2.3 Electrode Fabrication and Application

SiNp, CNF, Carbon black (CB), and PVDF in weight percent ratios of 80:0:10:10 and 60:20:10:10, respectively, were used to make bare SiNp and SiNp/CNF composite electrodes. In isopropanol, SiNp and CNF were combined and swirled constantly until all the liquid had evaporated. After adding conductive CB and PVDF to the composite, an electrode for electrochemical characterization was created by coating a on a copper foil substrate using the doctor blade approach and then drying it in a hot air oven at 60 °C for half an hour. The electrochemical test was carried out in a split cell that was stored inside an argon-filled glove box. The negative electrode was SiNp/CNF coated on a copper substrate, the counter electrode was Li foil, and the separator between the negative and positive electrodes was polypropylene. The electrolyte was 1 M LiPF6 in EC:DMC (1:1). The electrochemical test was performed at room temperature in the glove box. Using a Bio-logic device (BCS-810), cyclic voltammetry measurements were taken in the indicated potential at scan rates ranging from 0.1 to 10 mVs^{-1}. The amplitude of the electrochemical impedance spectra was 10 mV at open circuit potential, and the frequency range was 10 MHz to 10 kHz (OCP).

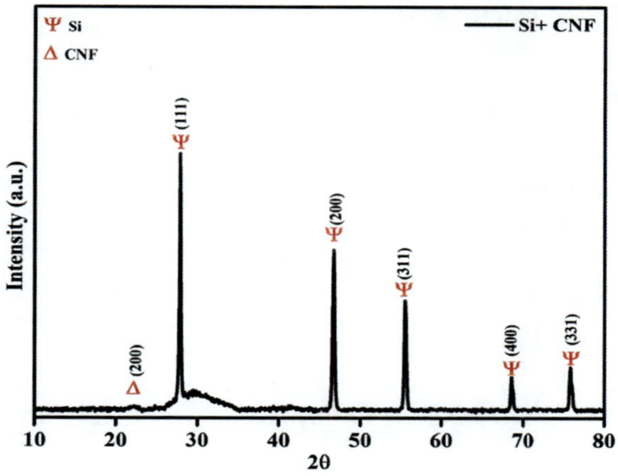

Fig. 1 XRD pattern of the SiNp/CNF electrode

3 Results and Discussion

3.1 *Structural Analysis*

X-ray diffraction analysis was used to investigate the structural properties of the Si/CNF electrode (XRD) in Fig. 1, XRD graphs are clearly visible peaks 28°, 47°, 56°, 69°, and 76°, corresponding to (111), (220), (311), (400), and (331) crystalline Si. (JCPDS card no. 27–1402). This indicates that SiNp crystallized. The broad area at 30°and little peak at 23° relate to the existence of amorphous CNF [32] XRD patterns proves that SiNp/CNF electrode was successfully prepared by physical mixing.

3.2 *Electron Microscopy Analysis*

The morphological properties of the SiNp/CNF network was investigated by field emission scanning electron microscopy (FE-SEM, Zeiss, Ultra 55). Figure 2a depicts SiNp equally integrated into CNF and shares majority of the distribution at lower magnification image. A close look at higher magnification in Fig. 2b reveals that there are local formation of very few segregated nanofiber domains. However, majority of the SiNp/CNF composite is made up of irregularly cross-linked microfibers, which are accumulations of CNF nanofibers with SiNp.

Between CNF and SiNp, a variety of connections are formed, CNF can provide extra linkages between SiNp, and its incorporation enhances the mechanical strength of SiNp, potentially boosting the rate capacity of SiNp as anodes by accumulating the volume expansion during charge–discharge.

Fig. 2 FE-SEM image of SiNp/CNF before charge–discharge performance **a** lower magnification, **b** higher magnification

3.3 Electrochemical Characterization

The cyclic voltametric (CV) curves are given in Fig. 3a. It was done for each of the seven cycles using a half-cell with a voltage window of 0.01 to 2.5 V (versus Li/Li^+) and at various scan rates in the range of 0.1 to 10 mVs^{-1}. At a scan rate of 0.1 mVs^{-1}, broad hump revels the growth of the electrolyte reduction on electrode surface which called solid electrolyte interface (SEI) layer and 2nd broad peak indicates the insertion of Li in Si. After 1st cycle broad hump disappeared due to stabilized SEI formation by addition of CNF material and observed decrease the double layer capacitance. This electrochemical behavior matches the charge/discharge profiles shown in galvanostatic charge/discharge profiles. Figure 3b shows the first cycle rate performance of SiNp/CNF and bare-SiNp, which were both tested at the present rate of 0.1 Ag^{-1}. SiNp/CNF samples had clearly more regulated SEI growth than bare-SiNp samples. This is because CNF increases overall performance by regulating SEI generation on the Si surface. When it comes to Si, the fast drop of voltage caused by SEI production, as well as the large swelling of SiNp, de-stabilize the voltage in the SiNp electrode, implying that CNF provides mechanical strength to control SEI. Cycling tests were used to assess long-term cell function (Fig. 5c). SiNp/CNF demonstrated good cycling regularity as well as capacity retention of 28.52% (517 mAg^{-1}) at 0.1 Ag^{-1} after 100th cycles and a coulombic efficiency of 96.2%. The first cycle discharge capacity was 1813 mAh g^{-1} with a Coulombic efficiency of 68.27%, which was produced by the growth of the SEI layer as well as the reduction of the SiO_x generated on the Si species surface. The electrode fluctuates for 20 cycles after that, exhibiting the electrode's good cycling stability and coulombic efficiency of 90%.

Figure 4 shows the rate performance of each phase of the SiNp/CNF sample and bare silicon for 5 cycles at 0.1, 0.5, 0.8 and 1 Ag^{-1}. Afterwards, the test was again performed at 0.1 Ag^{-1} for test capacity restoration following high-rate charge/discharge. It has been found that, at every first phase, SiNp/CNF offers reversible capacities of 1813, 1094, 820, 569 and 1803 $mAhg^{-1}$ while bare Si offers

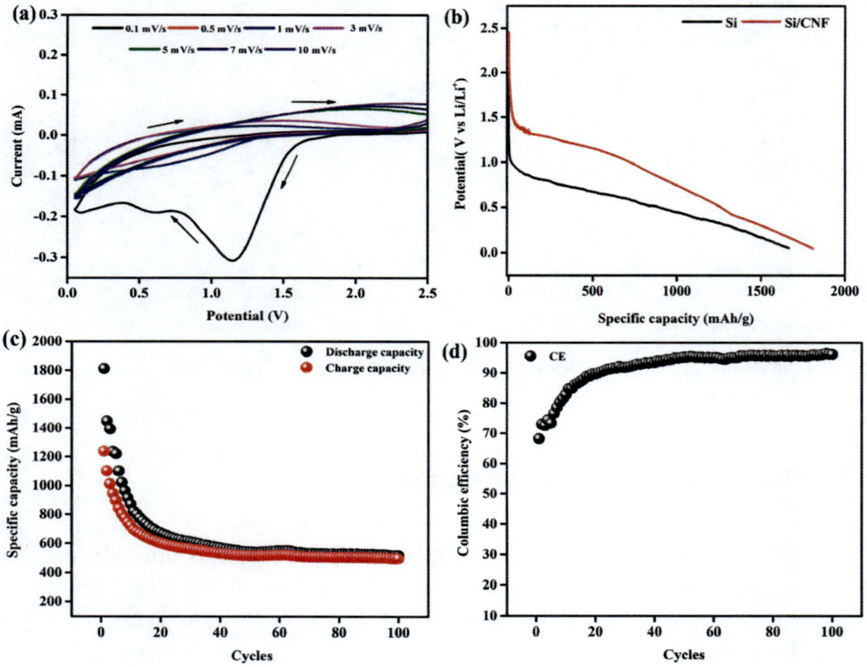

Fig. 3 Electrochemical performance of SiNp/CNF electrode **a** cyclic voltammetry **b** first cycle charge- discharge performance **c** cyclic performance at current density 0.1 A/g and **d** columbic efficiency

first reversible capacities of 1666, 812, 591, 387 and 1552 $mAhg^{-1}$. The loss in capacity at 1 Ag^{-1} is due to a lack of time for lithium ions to interact with active materials. Kim et al. [32] Comparison of rate capabilities for SiNp/CNF electrode with bare Si electrode clearly shows that CNF is preventing the fast degradation of silicon as well as retained high discharge capacity. The significantly coupled SiNp via carbon network that boosts the mechanical properties and cumulative volume expansion of SiNp after charge discharge are responsible for Si-excellent CNFs rate performance. We believe that the high contact area of the CNF stops each silicon nanoparticle from breaking surface contact, preventing fast degradation while charging and discharging. Furthermore, CNF wrapping SiNp prevent silicon from expanding during lithiation. SiNp are protected from pulverization via volume expansion control via CNF.

Figure 5 depicts SEM images of the SiNp/CNF electrode after 100th cycles at 0.1 Ag^{-1}. The morphology of spherical SiNp was retained after 100th cycles, as seen in Fig. 5a. Figure 5a also discloses the formation of SEI layer throughout the composite net-work. A close look image of SiNp/CNF reveals that SiNp are linked and wrapped by a CNF (Fig. 5b). It is revealed that CNF control the volume expansion and SEI formation after 100th cycle. As a result of this understanding, the well-connected CNF network are operate to suppress a volume variation of silicon, extending the life of electrochemical cells.

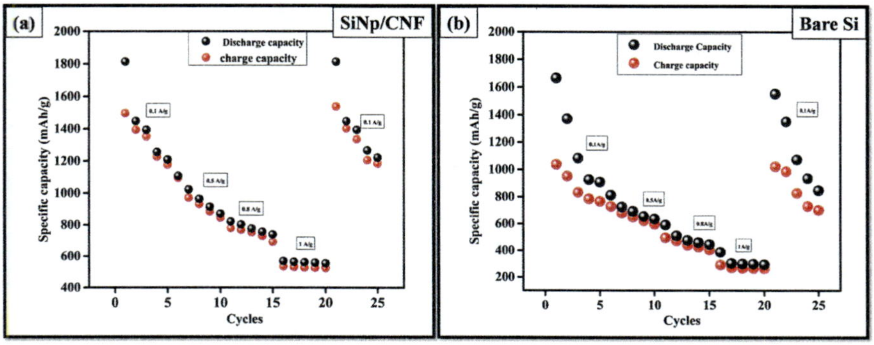

Fig. 4 Rate capability test at different current rate of **a** SiNp/CNF electrode **b** bare Si

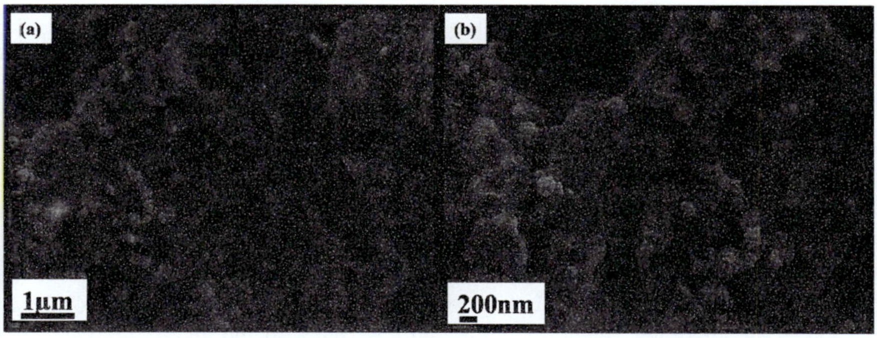

Fig. 5 Morphology of SiNp/CNF electrode after cycling performance **a** lower magnification **b** higher magnification

3.4 Electrochemical Impedance Analysis

As seen in the inset graph in Fig. 6, the EIS of SiNp/CNF electrodes was also fitted. Before cycling, a semicircle and a line can be identified in the EIS pattern in the high and low frequency regions, respectively (Fig. 6). In the high-frequency range, the charge-transfer resistance (R_{ct}) is represented by a semicircle, whereas in the low-frequency range, the diffusion resistance (R_s) is represented by a sloping line. Yu et al. [33] After 100th cycles, the Nyquist plot displays two semicircles and a 45 linear diffusion drift (Fig. 6). The semicircles depict Li^+ ion transport through the R_{ct}. The fitted values of R_s and R_{ct} for before cycling 76 Ω and 288 Ω and after cycling 77 Ω and 709 Ω respectively. This indicates the fastest charge transfer kinetics, which are facilitated by the fibrous structure of SiNp/superior CNF.

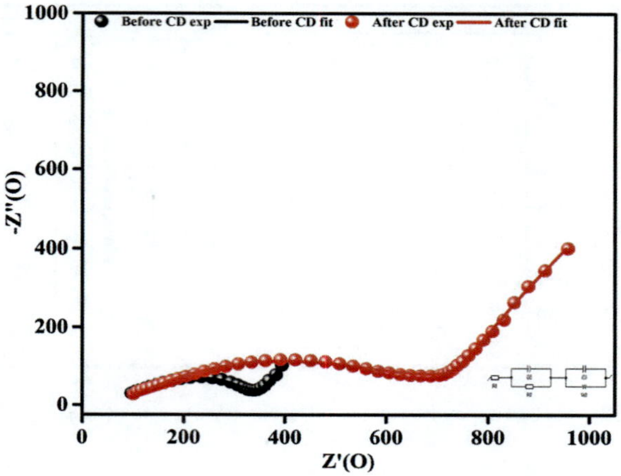

Fig. 6 Electrochemical impedance spectra of SiNp/CNF anode

4 Conclusion

We synthesized an extensively connected Si using the CNF carbon network. The synthetic approach is quite easy, and it uses cellulose, which is inexpensive and abundant, making it ideal for battery use. This lithium-ion battery anode material has a high reversible discharge capacity and provides outstanding performance (517 $mAhg^{-1}$ at 0.1 Ag^{-1}). We believe that the carbon network produced from ultrathin CNFs, which imparts mechanical properties to SiNp, is largely responsible for these extraordinary outcomes. By aggregating volume expansion and maintaining electric contact among SiNp, a densely connected CNF—carbon network can reduce the probability of losing contact during lithiation. Furthermore, when contrasted to a bare Si electrode, this system ensured SEI generation control. In comparison to electrodes constructed from pure Nano silicon, using Nano sized Si and CNF alone is a simple, appropriate method for generating composite electrodes with very superior electrochemical behavior in lithium batteries.

5 Credit Authorship Contribution Statement

Yashkumar Patel: Conceptualization, Visualization, Methodology, Study, Validation, Writing—review and editing. Anjali Vanpariya: Review and editing, Data curation. Indrajit Mukhopadhyay: Visualization, Investigation, Supervision, Writing—review and editing.

6 Declaration of Competing Interest

It is hereby declared that the authors have no known competing financial interests or relationships that may have appeared to affect their work in this paper.

Acknowledgements The authors would like to thank the SERB, Department of Science and Technology, Government of India for providing grants through project no. (SERB/2018/002067) and (DST/TMD/MES/2517/32(G)) to carry out the present work. And, thanks Department of Science and Technology, Government of India for Financial support under project no. DST/INSPIRE Fellowship/2019/IF190880. We also acknowledge the Solar Research and Development Center (SRDC) at Pandit Deendayal Energy University, which provided all the facilities needed for conducting this research.

References

1. Jieyi Y, Song G, Xinglong D (2017) Electrochemical performance of Si nanoribbons as anode material for Li-ion battery synthesized by Arc-discharge plasma. Chinese J Mater Res 31(3):161–167
2. De las Casas C, Li W (2012) A review of application of carbon nanotubes for lithium ion battery anode material. J Power Sources, pp 208, 74–85
3. Lv Q, Liu Y, Ma T, Zhu W, Qiu X (2015) Hollow structured silicon anodes with stabilized solid electrolyte interphase film for lithium-ion batteries. ACS Appl Mater Interfaces 7(42):23501–23506
4. Liang B, Liu Y, Xu Y (2014) Silicon-based materials as high capacity anodes for next generation lithium ion batteries. J Power Sources 267:469–490
5. Chang C-B, Tsai C-Y, Chen K-T, Tuan H-Y (2021) Solution-grown phosphorus-hyperdoped silicon nanowires/carbon nanotube bilayer fabric as a high-performance lithium-ion battery anode. ACS Appl Energy Mater 4(4):3160–3168
6. Winter M, Besenhard JO, Spahr ME, Novak P (1998) Insertion electrode materials for rechargeable lithium batteries. Adv Mater 10(10):725–763
7. Vanpariya A, Lellala K, Bhagat D, Mukhopadhyay I (2022) Electrochemical deposition of Si nano-spheres from water contaminated ionic liquid at room temperature: Structural evolution and growth mechanism. J Electroanal Chem 910:116175
8. Hatchard T, Dahn J (2004) In situ XRD and electrochemical study of the reaction of lithium with amorphous silicon. J Electrochem Soc 151(6):A838
9. Obrovac M, Krause L (2006) Reversible cycling of crystalline silicon powder. J Electrochem Soc 154(2):A103
10. Zhang Y, Zhang X, Zhang H, Zhao Z, Li F, Liu C, Cheng H (2006) Composite anode material of silicon/graphite/carbon nanotubes for Li-ion batteries. Electrochim Acta 51(23):4994–5000
11. Cao X, Chuan X, Li S, Huang D, Cao G (2016) Hollow silica spheres embedded in a porous carbon matrix and its superior performance as the anode for lithium-ion batteries. Part Part Syst Charact 33(2):110–117
12. Gao H, Hou F, Zheng X, Liu J, Guo A, Yang D, Gong Y (2015) Electrochemical property studies of carbon nanotube films fabricated by CVD method as anode materials for lithium-ion battery applications. Vacuum 112:1–4
13. Lee SW, Gallant BM, Lee Y, Yoshida N, Kim DY, Yamada Y, Noda S, Yamada A, Shao-Horn Y (2012) Self-standing positive electrodes of oxidized few-walled carbon nanotubes for light-weight and high-power lithium batteries. Energy Environ Sci 5(1):5437–5444

14. Scrosati B, Garche J (2010) Lithium batteries: status, prospects and future. J Power Sources 195(9):2419–2430
15. Zhou Z, Xu Y, Liu W, Niu L (2010) High capacity Si/DC/MWCNTs nanocomposite anode materials for lithium ion batteries. J Alloys Compd 493(1–2):636–639
16. Zheng D, Wu C, Li J, Guan L (2013) Chemically shortened multi-walled carbon nanotubes used as anode materials for lithium-ion batteries. Physica E: Low-Dimensional Syst Nanostruct 53:155–160
17. Zuo P, Yin G, Yang Z, Wang Z, Cheng X, Jia D, Du C (2009) Improvement of cycle performance for silicon/carbon composite used as anode for lithium ion batteries. Mater Chem Phys 115(2–3):757–760
18. Luo F, Chu G, Xia X, Liu B, Zheng J, Li J, Li H, Gu C, Chen L (2015) Thick solid electrolyte interphases grown on silicon nanocone anodes during slow cycling and their negative effects on the performance of Li-ion batteries. Nanoscale 7(17):7651–7658
19. Dimov N, Xia Y, Yoshio M (2007) Practical silicon-based composite anodes for lithium-ion batteries: fundamental and technological features. J Power Sour 171(2):886–893
20. Chan CK, Peng H, Liu G, McIlwrath K, Zhang XF, Huggins RA, Cui Y (2008) High-performance lithium battery anodes using silicon nanowires. Nature Nanotechnol 3(1):31–35
21. Zhou X, Yin YX, Wan LJ, Guo YG (2012) Self-assembled nanocomposite of silicon nanoparticles encapsulated in graphene through electrostatic attraction for lithium-ion batteries. Adv Energy Mater 2(9):1086–1090
22. Chang J, Huang X, Zhou G, Cui S, Hallac PB, Jiang J, Hurley PT, Chen J (2014) Multilayered Si nanoparticle/reduced graphene oxide hybrid as a high-performance lithium-ion battery anode. Adv Mater 26(5):758–764
23. Si Q, Hanai K, Ichikawa T, Hirano A, Imanishi N, Takeda Y, Yamamoto O (2010) A high performance silicon/carbon composite anode with carbon nanofiber for lithium-ion batteries. J Power Sour 195(6):1720–1725
24. Wang W, Kumta PN (2010) Nanostructured hybrid silicon/carbon nanotube heterostructures: reversible high-capacity lithium-ion anodes. ACS Nano 4(4):2233–2241
25. Fu K, Yildiz O, Bhanushali H, Wang Y, Stano K, Xue L, Zhang X, Bradford PD (2013) Aligned carbon nanotube-silicon sheets: a novel nano-architecture for flexible lithium ion battery electrodes. Adv Mater 25(36):5109–5114
26. Vanpariya A, Marathey P, Khanna S, Patel R, Mukhopadhyay I (2021) Hydrothermal synthesis of silicon nanosphere embedded on carbon nanotubes for high-performance lithium-ion batteries. Int J Nanotechnol 18(5–8):483–493
27. Saint J, Morcrette M, Larcher D, Laffont L, Beattie S, Pérès JP, Talaga D, Couzi M, Tarascon JM (2007) Towards a fundamental understanding of the improved electrochemical performance of silicon–carbon composites. Adv Funct Mater 17(11):1765–1774
28. Caballero Á, Morales J, Sánchez L (2005) Tin nanoparticles formed in the presence of cellulose fibers exhibit excellent electrochemical performance as anode materials in lithium-ion batteries. Electrochem Solid-State Lett 8(9):A464
29. Klemm D, Kramer F, Moritz S, Lindström T, Ankerfors M, Gray D, Dorris A (2011) Nanocelluloses: a new family of nature-based materials. Angewandte Chemie Int Edition 50(24):5438–5466
30. Jabbour L, Bongiovanni R, Chaussy D, Gerbaldi C, Beneventi D (2013) Cellulose-based Li-ion batteries: a review. Cellulose 20(4):1523–1545
31. Saito T, Hirota M, Tamura N, Kimura S, Fukuzumi H, Heux L, Isogai A (2009) Individualization of nano-sized plant cellulose fibrils by direct surface carboxylation using TEMPO catalyst under neutral conditions. Biomacromol 10(7):1992–1996
32. Kim JM, Guccini V, Seong K-D, Oh J, Salazar-Alvarez G, Piao Y (2017) Extensively interconnected silicon nanoparticles via carbon network derived from ultrathin cellulose nanofibers as high performance lithium ion battery anodes. Carbon 118:8–17
33. Yu J, Zhang C, Wu W, Cai Y, Zhang Y (2021) Nodes-connected silicon-carbon nanofibrous hybrids anodes for lithium-ion batteries. Appl Surf Sci 548:148944

Optimization of Circulation Power in First Wall of Breeding Blanket Using He-CO_2 Gas Mixture as a Replacement of Helium

Ankit Gandhi, Deepak Sharma, Nimesh Gajjar, and Paritosh Chaudhri

1 Background

1.1 Nuclear Fusion

Nuclear fusion is a process in which two or more atoms fuse to form a heavier atom accompanied by the release of energy. Two isotopes of hydrogen, namely deuterium (D) and tritium (T) fuse together to produce 17.6 meV of energy (14.1 meV neutron + 3.5 meV alpha particle) [1].

$$D + T \rightarrow^{4} He + N \tag{1}$$

This release of energy in form of plasma shall be slowed down to produce heat and breed tritium. The first component to see this energy is known as the First Wall (FW) of the breeding blanket. The schematic of a typical breeding blanket module is shown in Fig. 1 [2]. It has a U-shape structure known as FW which faces the plasma energy and this energy is extracted by high-pressure, high-temperature Helium gas. Each channel has square dimensions of 15 mm × 15 mm. Helium enters FW at 300 °C and 8 MPa pressure.

A. Gandhi (✉) · D. Sharma · P. Chaudhri
Institute for Plasma Research, Bhat Gandhinagar Gujarat, Gandhinagar, India
e-mail: ankit2020@gmail.com

N. Gajjar
Gandhinagar Institute of Technology, Moti Bhoyan Gandhinagar Gujarat, Gandhinagar, India

P. Chaudhri
Homi Bhabha National Institute, Anushaktinagar, Mumbai, India

S. Doolla et al. (eds.), *Advances in Clean Energy and Sustainability*,
Green Energy and Technology, https://doi.org/10.1007/978-981-99-2279-6_60

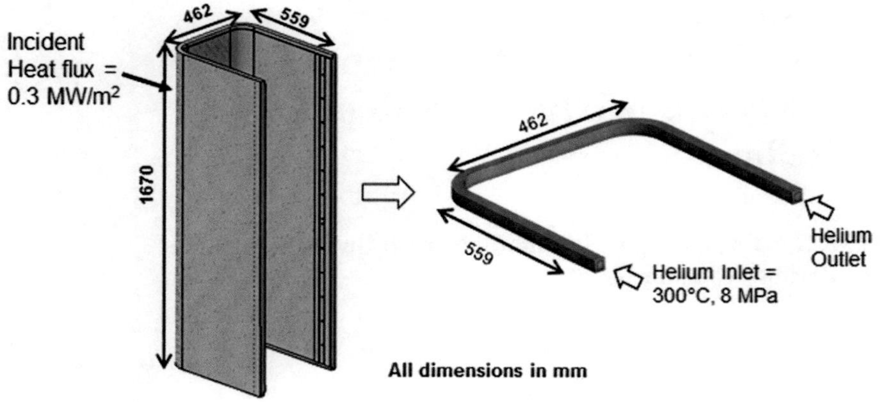

Fig. 1 Schematic layout of first wall and details of single channel

1.2 Coolants of High-Temperature Reactors

The available choices of gas coolants in High-Temperature Reactors are Argon, Krypton, Xenon, Carbon dioxide (CO_2) and Helium.

Argon, Krypton and Xenon are noble gases with molecular weights of 40, 83.8 and 131.3 g/mole respectively. They are colorless, odorless gas and chemically inert. The chemical inertness makes them a good choice as a coolant for high-temperature reactors. El-Genk et al. [3] found that among the noble gases available Argon has the lowest heat capacity except Helium. Krypton has better heat capacity than Argon due to its higher molecular weight. Xenon has the highest heat carrying capacity among noble gases. But on the other hand, Xenon has an issue of activation in irradiated conditions due to its large neutron cross-section. Also, Xenon gas shows a higher pressure drop due to its high molecular weight.

Nonbol [4] mentions that CO_2 is chosen as a coolant for Advanced Gas cooled Reactor (AGR). CO_2 gas has a proven track record of operation at high-pressure, high-temperature conditions with good thermal stability and compatibility with structural materials. Due to high density of CO_2 gas, the circulation power is found to be reduced while maintaining reactor cooling.

Melese [5] mentions that Helium gas is proven as coolant in high-temperature applications as early as 1950s. Helium has low neutron cross section, chemically inert and good thermal performance. The principal disadvantage of Helium gas is low density.

Thermodynamic Comparison of Gas Coolants Properties

Gas coolants in high temperature reactors operate at high temperatures to have maximum heat extraction and to attain high efficiency. The operating pressure and temperature of a typical FW structure in a breeding blanket are approximately 700 K

Table 1 Thermo-physical properties of gas coolants at 700 K and 8 MPa [7]

S. No	Parameters	He	CO_2	Kr	Ar	Xe
1	Molecular weight (g $mole^{-1}$)	4.003	44	83.80	40	131.29
2	Density (kg m^{-3})	5.4	60.4	113.5	53.7	181.7
3	Thermal conductivity(Wm^{-1} K^{-1})	0.285	0.051	0.019	0.035	0.012
4	Dynamic viscosity (Pa-s)	3.6×10^{-5}	3.2×10^{-5}	5.1×10^{-5}	4.4×10^{-5}	5.0×10^{-5}
5	Heat capacity ($kJkg^{-1}$ K^{-1})	5.19	1.16	0.25	0.53	0.17

and 8 MPa. Table 1 presents the thermo-physical properties of gas coolants at 700 K and 8 MPa. The above data is retrieved from NIST [6].

It is obvious from Table 1 that Helium has better thermal and transport properties (high heat capacity, high thermal conductivity and low viscosity) except for low density. Other noble gases (Xe, Kr, and Ar) have high density than Helium but poor heat capacity and thermal conductivity. CO_2 gas has not only a higher density than Helium but also its heat capacity and thermal conductivity is reasonable compared to other noble gases.

1.3 Possibility to Replace Helium Gas with He-CO_2 Binary Gas Mixture

To resolve the issue of Helium low density, a viable solution is to mix Helium gas with a moderately heavy gas at an optimized mole fraction.

Bammert [8] compares He-Ne, He-N_2 and He-CO_2 gas mixtures on closed-cycle gas turbines. He found out that with an increasing faction of CO_2 gas (28% and more) there is an increase in overall efficiency while keeping the maximum inlet turbine temperature to 800 °C. Also, the mixtures result in compact dimensions of turbomachines (turbine and compressor) compared to pure Helium gas.

Lee et al. [9] have proposed He-CO_2 mixture for heat removal of the Korean Blanket Module. Numerical analysis was performed on FW channels using CFD tool and found that at a 0.4 mol fraction of CO_2, the circulation power gets reduced to 13% compared to pure Helium.

2 Thermal–Hydraulic Analysis of Single Channel of FW

2.1 Input Parameters and Methodology

ANSYS CFX tool is used to estimate velocity distribution, pressure drop and temperature distribution in a single channel of FW using Helium and He-CO_2 mixture.

The input parameters and boundary conditions for the single channel of FW are as listed in Table 2 [10].

- Gnielinski equation [11] is used to calculate Nusselt number given below:

$$\mathrm{Nu} = \frac{0.125 \times \varepsilon \times (\mathrm{Re} - 1000)\,\mathrm{Pr}}{1 + 12.7 \times \left(\mathrm{Pr}^{\frac{2}{3}} - 1\right) \times \sqrt{0.125 \times \varepsilon}} \tag{2}$$

$$\varepsilon = \frac{1}{(1.82 \log \mathrm{Re} - 1.64)^2} \tag{3}$$

- Frictional pressure drop due to flow length and smooth/sharp bends of circuit.

$$\Delta P = \frac{\rho f l v^2}{2D} \tag{4}$$

$$\Delta P = \mathrm{K}\frac{\rho v^2}{2} \tag{5}$$

where K is the concentrated loss factor calculated from data and correlation in [12].

For smooth bends:

$$\mathrm{K} = \mathrm{A1} \times \mathrm{B1} \times \mathrm{C1} + 0.00035 \times \mathrm{R0} \times \mathrm{D} \times \delta 0 \tag{6}$$

A1, B1, C1 are angle dependent functions given in [12] R0, the curvature radius, D the section hydraulic diameter, δ° is the angle in degrees.

For sharp bends K = C1 * A * ξ1 C1, A and ξ1 are functions given in [12].

Table 2 Input parameters and boundary conditions for FW design

S. No	Parameter	Value
1	Incident heat flux	0.3 MW/m^2
2	Helium inlet temperature	300 °C
3	Helium pressure	8 MPa
4	Helium flow rate	0.0125 kg/s

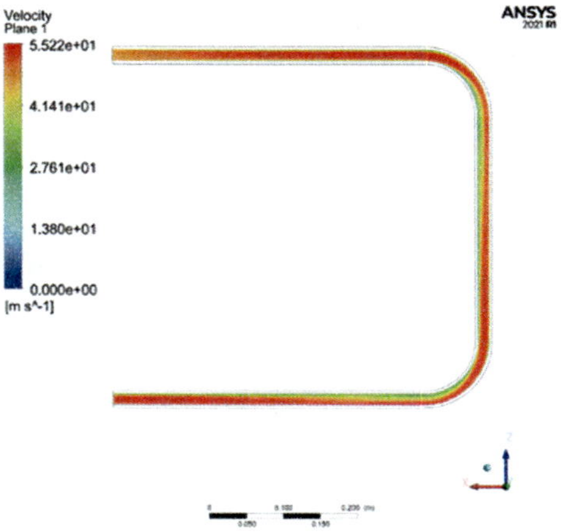

Fig. 2 Velocity profile distribution for a single channel of FW

- Pumping power per channel (P_p) is calculated as [9]:

$$Pp = \Delta P \times u \times A \tag{7}$$

where u = Average flow velocity, m/s and A = Cross-section area of the flow channel, m^2

2.2 *Using Helium Gas*

Results of Thermal–Hydraulic Analysis for Single Channel of Helium Gas

2.2.1 Velocity Profile

The velocity profile in the FW channel is shown in Fig. 3. Average helium is close to ~ 50 m/s in single channel of FW (Fig. 2).

2.2.2 Pressure Drop

The pressure profile in the FW channel is shown in Fig. 4. Pressure drop in channel is estimated to be 0.104 bar (Fig. 3).

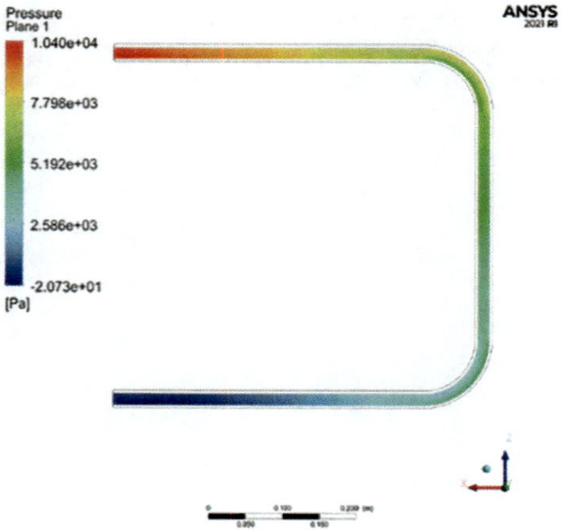

Fig. 3 Pressure drop profile distribution for a single channel of FW

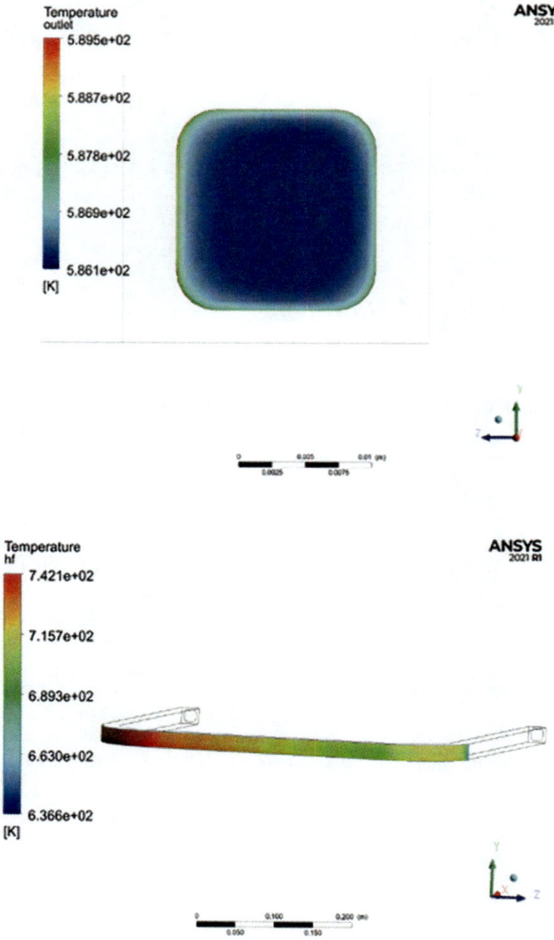

Fig. 4 Outlet temperature distribution and wall temperature for a single channel of FW

2.2.3 Temperature Distribution

The average outlet Helium temperature is 313 °C and the maximum FW channel temperature is ~ 469 °C as obtained from CFX are shown in Fig. 4.

Pumping power and Heat Transfer Coefficient (HTC) are analytically calculated using Eqs. (2) and (7) for Helium were found to be 332 W and 4055 W/m^2-K respectively.

2.3 *Using He-CO_2 Gas Mixture*

A similar analysis is performed for a single channel of FW using He-CO_2 mixture at various mole fractions. The input parameters will be similar to the Helium case. The results are presented in Table 3.

Figure 5 shows the profile of pumping power and mole fraction of CO_2 for a single channel of FW. The minimum circulation power is observed at 0.6 mol fraction of CO_2 gas in the He-CO_2 mixture.

3 Results and Discussions

Literature survey shows that to mitigate the issue of the low density of Helium gas, various gas mixtures like He-CO_2, He–Xe, He-Kr, He–Ne, He-N_2 etc. are potential options as a replacement of Helium gas.

He-CO_2 gas mixture is found as the best solution among other alternatives due to their better thermal and transport properties, proven track record for stable thermal operation in high-temperature operation in nuclear environments, easy availability and good compatibility with structural material.

Numerical performance of a single channel of FW was investigated for He-CO_2 mixture and benchmarked against Helium gas. The following conclusions are drawn:

- The benchmarked Helium case has a pumping power of 332 W and HTC value is 4055 W/m^2-K
- Comparable HTC values (3520–4355 W/m^2-K) is observed in 0.4–0.45 mol fraction of CO_2 gas in the He-CO_2 mixture
- The minimum circulation power (7.45 W) is observed for 0.6 mol fraction of CO_2 gas in the He-CO_2 mixture which is almost 85% less compared to pure Helium.

The savings in circulation power have a potential saving in terms of capital and operational cost. It can be summarized that He-CO_2 mixture in mole fraction range of 0.4–0.6 may be used as replacement of Helium gas for heat extraction in breeding blankets.

Table 3 Thermal-hydraulics results for single channel of FW using He-CO_2 mixture

S. No	He-CO_2 mixture	Parameter	Value
1	60:40	Reynolds number	266,921
2		Prandtl number	0.555
3		Friction factor	0.0147
4		Nusselt number	330.5
5		Average outlet temperature (°C)	317
6		HTC (W/m^2-K)	4590
7		Pressure drop in single channel (bar)	0.11
8		Pumping power required (W)	70.3
9	50:50	Reynolds number	238,570
10		Prandtl number	0.633
11		Friction factor	0.015
12		Nusselt number	331.6
13		Average outlet temperature (°C)	319
14		HTC (W/m^2-K)	3600
15		Pressure drop in single channel (bar)	0.075
16		Pumping power required (W)	36.2
17	40:60	Reynolds number	123,054
18		Prandtl Number	0.482
19		Friction factor	0.017
20		Nusselt number	164
21		Average outlet temperature (°C)	322
22		HTC (W/m^2-K)	2100
23		Pressure drop in single channel (bar)	0.03
24		Pumping power required (W)	9.45
25		Reynolds number	138,078
26		Prandtl number	0.424
27		Friction factor	0.016
28	35:65	Nusselt number	163.8
29		Average outlet temperature (°C)	323
30		HTC (W/m^2-K)	2218
31		Pressure drop in single channel (bar)	0.04
32		Pumping power required (W)	16.3
33		Reynolds number	146,053
34		Prandtl number	0.373
35		Friction factor	0.016
36	30:70	Nusselt number	156

(continued)

Table 3 (continued)

S. No	He-CO_2 mixture	Parameter	Value
37		Average outlet temperature (°C)	325
38		HTC (W/m^2-K)	2234
39		Pressure drop in single channel (bar)	0.05
40		Pumping power required (W)	24.8
41		Reynolds number	159,218
42		Prandtl number	0.335
43		Friction factor	0.016
44	25:75	Nusselt number	154.2
45		Average outlet temperature (°C)	327
46		HTC (W/m^2-K)	2251
47		Pressure drop in single channel (bar)	0.06
48		Pumping power required (W)	36.4

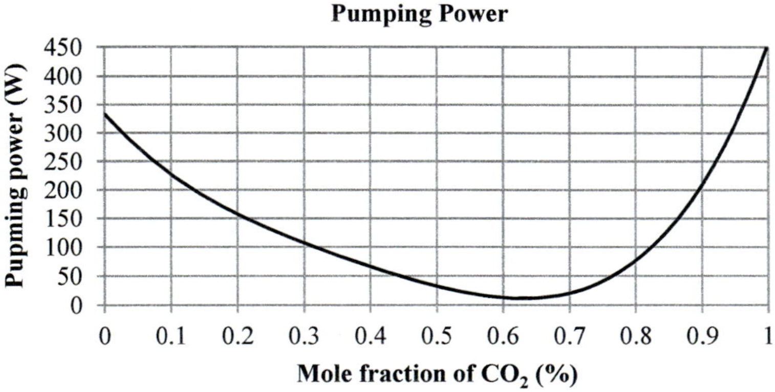

Fig. 5 Pumping power profile in a single channel for He-CO_2 mixture

References

1. Sharma AN (2015) PhD Thesis, studies on quench characteristics of superconducting magnets of SST-1. Homi Bhabha National Institute
2. Kumar SR et al (2016) Engineering design and analysis of Indian LLCB TBM set. Fusion Eng Design (111):1581–1586
3. El-Genk MS et al (2008) On the use of noble gases and binary mixtures as reactor coolants and CBC working fluids. Energy Convers Manage 49:1882–1891
4. Nonbol E Description of the advanced gas cooled type of reactor (AGR). Riso National Laboratory Roskilde, Denmark, www.risoe-dk/nks
5. Melese G, Katz R (1984) Thermal and flow design of helium-cooled reactors. ISBN 0-89448 027 8, American Nuclear Soc, pp 367
6. NIST Chemistry WebBook. https://webbook.nist.gov/chemistry/fluid/

7. Gandhi A, Chaudhuri P, Gajjar N, Joshi N (2022) A survey on application of noble gases and its binary mixtures in high temperature gas cooled reactors. GIT- J Eng Technol 14:129–133
8. Bammert K, Klein R (2015) The influence of He-Ne, He-N_2, and He-CO_2 gas mixtures on closed-cycle gas turbines. In: ASME 1974 international gas turbine conference and products show conference proceedings 74:1–9
9. Lee Y-G et al (2015) Reduction of circulation power for Helium-cooled fusion reactor blanket using additive CO_2 gas. Fusion Eng Des 100:436–442
10. Sharma D et al (2018) Design update and thermal-hydraulics of LLCB TBM first wall. Fusion Sci Technol 134:51–61
11. Gnielinski V (1976) New equations for heat and mass transfer in turbulent pipe and channel flow, Int Chem Eng 16th Edn, pp 359–368
12. Idelchik IE (1980) Handbook of hydraulic resistance, 3rd Edn

Heat Transfer Enhancement of Metal Hydride Based Hydrogen Storage Device Using Nano-fluids

R. Sreeraj, A. K. Aadhithiyan, Prateek Sahoo, and S. Anbarasu

Nomenclature

C	Rate constant
C_p	Specific heat capacity
E	Energy of activation
ΔH	Enthalpy of formation
ID	Inner diameter
OD	Outer diameter
k	Thermal conductivity
M	Molar mass
R_u	Universal gas constant
S	Source term
ΔS	Entropy of formation
P	Pressure

Greek letters

ε	Porosity
k	Permeability
μ	Viscosity
ρ	Density
T	Temperature
t	Time

R. Sreeraj · A. K. Aadhithiyan · P. Sahoo · S. Anbarasu (✉)
Hydrogen Energy Systems Research Facility, Department of Mechanical Engineering, NIT Rourkela, Odisha 769008, India
e-mail: anbarasus@nitrkl.ac.in

S. Doolla et al. (eds.), *Advances in Clean Energy and Sustainability*,
Green Energy and Technology, https://doi.org/10.1007/978-981-99-2279-6_61

$\vec{u}$ Hydrogen gas velocity
wt% Hydrogen weight percentage
X Hydrogen saturation level

Subscripts

a Absorption process
eff Effective
eq Equilibrium
g Hydrogen gas
i Initial state
m Metal hydride
ref Reference
sat Saturation
bf Base fluid
nf Nanofluid
np Nanoparticle

1 Introduction

Hydrogen is an impressive secondary energy carrier which can deliver or store a tremendous amount of energy. The significance of a hydrogen-based economy has been greatly advertised to reduce the need of fossil fuels. In an atmospheric condition, hydrogen is found to be in gaseous form. In pure form, hydrogen is extracted by the process of thermolysis, electrolysis, etc. [1, 2]. Therefore, fossil fuel-based energy resources can be substituted with hydrogen and it leads to a hydrogen-based energy economy. However, the usage of hydrogen is limited owing to production and its' storage.

The metal hydride reactor (MHR) provides a safe medium for hydrogen storage. It has high hydrogen storage capacities and can be handled with ease. Metal hydrides require lesser energy for storing hydrogen than the other forms of storage (i.e., liquid, pressurized gas). It also has a minor weight and less volume than liquid hydrogen storage option. Metal hydrides can release hydrogen at relatively low temperatures and moderate pressures. However, its storage performance is limited due to its inferior thermal conductivity.

During the hydrogen absorption process, the thermal effects are greatly influence the use of MHR for hydrogen storage. Moreover, the slow kinetics of sorption is an essential test in the usage of metal hydrides in the storage of hydrogen. Over the past several years, many investigations have been carried out on the hydride reactor's performance by adjusting the thermo-physical properties and enhancing the performance of the hydride bed. Recently, Sreeraj et al. [3] reported a concise review of the various methods adopted on metal hydride reactors.

Several numerical studies on thermal dynamics and hydrogen transfer of MHR have reported using various geometry configurations, nanofluids, phase change materials, doped metals, and so on. A two-dimensional numerical model of a cylindrical tank with a concentric tube equipped with fins developed by Askri et al. [4] for improving hydrogen storage in metal hydride tanks. The model was used to investigate the effect of tank wall thermal mass and other parametric variables on the hydriding process. The addition of fin structures to the central concentric tube enhanced hydrogen storage time by 80%. The impact of substituting circular heat fins with conical fins, as well as embedded straight heat transfer fluid tubes on MHR have numerically studied by Satyaki et al. [4]. The investigation revealed that the improvement was more pronounced near the water inlet regions, and the optimised design for 5 kg $LaNi_5$ with 6 tubes and 19 fins resulted in a 380 s absorption time. Mellouli et al. [5] presented a two-dimensional numerical model for optimising heat and mass transfer in metal hydride storage tanks for fuel cell vehicles using fin-type tube heat exchangers. The fins increased the rate of heat transfer, resulting in a 66% storage time. Kaplan [6] developed three different cylindrical reactors to test the heat transfer mechanisms and their effects on the charging process of an MHR at different pressures. The charging of the metal hydride reactors was dependent on the effective heat transfer coefficient; a higher heat transfer rate results in faster reaction kinetics, i.e., heat transfer in MHR is much improved in reactors with heat transfer fluid frameworks rather than heat fins. The effect of embedded cooling tubes (ECT) on MHR for 2.75 kg $LmNi_{4.91}Sn_{0.15}$ was studied both numerically and also experimentaly by Anbarasu et al. [7–9]. The absorption time for 60 ECT was 720 and 540 s at 30 °C and 25 bar inlet conditions. Muthukumar et al. [10] evaluated metal hydride hydrogen storage devices by different parametric values such as absorption temperature, hydrogen supply pressure, and heat transfer coefficient. The rate of hydrogen absorption increased with increasing hydrogen supply pressure and decreased with increasing absorption temperature, with the heat transfer coefficient having a negligible effect. Chung et al. [11] experimentally demonstrated that the heat pipes affected the transfer of heat in a hydrogen storage tank using $LaNi_5$ as the media of storage. It was observed that the rate of hydrogen storage was improved by the heat pipes in both absorption and desorption processes. With an initial condition of 10 bar and 293 K, the reactor was able to complete the absorption and desorption at 2600 and 900 s respectively. Rahul et al. [12] gave a mathematical model of a metal hydride reactor filled with $MmNi_{4.6}$ $Al_{0.4}$ and incorporated nano-fluids as the heat transfer fluid for the hydrogen sorption process. It was observed that the CuO/H_2O nano-fluid showed the best improvement in both absorption time and heat transfer rate by 9.5% and 10% respectively.

It is observed from the literatures that most of the reported studies have been focued on improving the heat transfer enhacment within the MHR bed using various heat exchanging configurations [3]. Few studies were reported on the heat transfer medium and their flow parameters [13]. Hence, there is a huge scope for studing the influence of heat transfer medium and their effects on the hydrogen storage performance in MHR. In the present investigation, different nano particles viz., Al_2O_3/H_2O, CuO/H_2O, MgO/H_2O, CeO_2/H_2O, and TiO_2/H_2O were considered and also performance characteristics of 1 kg $LaNi_5$ based MHR is presented.

2 Mathematical model

Based on the assumptions adopted by Anbarasu et al. [9], the governing equations used for the present assessment of hydrogen absorption in the hydride bed as follows.

Equations concerning the conservation of mass:

The conservation of mass of hydrogen is given by Eq. (1) for the alloy with uniform porosity [4].

$$\varepsilon\frac{\partial\rho_g}{\partial t} + \nabla\left(\rho_g\cdot\vec{u}\right) = -S_{mass} \tag{1}$$

where the density of hydrogen gas was calculated from the ideal gas equation.

Similarly, the conservation of mass of metal hydride is written by Eq. (2) [4].

$$(1-\varepsilon)\frac{\partial\rho_m}{\partial t} = S_{mass} \tag{2}$$

The volume of the metal hydride bed assumed to be constant, and only the increase in density was factored in. Therefore, the mass source is given by Eq. (3) [4].

$$S_{mass} = C_a\exp(-\frac{E_a}{R_u.T})\ln(\frac{P}{P_{eq}})(\rho' - \rho'_{sat}) \tag{3}$$

The effective densities of the hydride bed at saturation and at any instant of time are given by Eq. (4) and (5), respectively.

$$\rho'_{sat} = \rho_{sat}(1-\varepsilon) \tag{4}$$

$$\rho'_m = \rho_m(1-\varepsilon) \tag{5}$$

Equation (6) gives the saturated density of metal hydride density [4].

$$\rho_{sat} = \rho_i(1 + wt\%_{sat}) \tag{6}$$

The equilibrium pressure for absorption (P_{eq}), calculated by the van't Hoff equation given as.

Equation (7) [4].

$$P_{eq} = P_{ref} \exp\left(-\frac{\Delta H}{R_u.T} + \frac{\Delta S}{R_u} + (\varphi + \varphi_0)\tan\{\pi(X - 0.5)\} + \frac{\beta}{2}\right) \quad (7)$$

where X represents the saturation level of hydrogen absorption and corresponds to the mass of hydrogen gas within the metal hydride bed.

Equations concerning the conservation of energy:

The conservation of energy equation is given by Eq. (8) [4],

$$(\rho C_P)_{eff}\frac{\partial T}{\partial t} + (\rho C_P)_g(\vec{u} \cdot \nabla T) = \nabla(k_{eff} \cdot \nabla T) + S_{heat} \quad (8)$$

The heat source term S_{heat} is calculated using the relation below,

$$S_{heat} = S_{mass}\frac{\Delta H}{M_g} \quad (9)$$

The effective heat capacity and thermal conductivity of porous metal hydride are given by the Eq. (10–11) [4]

$$(\rho C_P)_{eff} + \varepsilon(\rho C_P)_g = (1 - \varepsilon)(\rho C_P)_m \quad (10)$$

$$k_{eff} = \varepsilon k_g + (1 - \varepsilon)k_m \quad (11)$$

The velocity of hydrogen gas inside the metal hydride bed was calculated using Darcy equation: [4].

$$\vec{u} = -\frac{\kappa}{\mu_g} \cdot \nabla P \quad (12)$$

Nano-fluid equations:

When nanoparticles are added to the base fluid, thermo-physical properties such as density, specific heat capacity, thermal conductivity, and viscosity are improved over the base fluid. Correlation equations were used to determine the properties. The density and specific heat of the nano-fluid are given by Eqs. (13–14) [4].

$$\rho_{nf} = (1 - \theta)\rho_{bf} + \theta\rho_{np} \quad (13)$$

$$(C_p)_{nf} = (1 - \theta)(C_p)_{bf} + \theta(C_p)_{np} \quad (14)$$

Table 1 listed the thermophysical properties of a few selected nanofluids and base fluids which were used to compute the present work.

Table 1 Properties of the nanofluids and base fluid [12]

Fluid	Density (kg/m^3)	Specific heat (J/kg-K)	Thermal conductivity (W/m–K)	Viscosity (N/m-s)	Vol%
H_2O	998	4182	0.6	0.001003	5
Al_2O_3/H_2O	1148.1	3606.79	0.6890	0.001678	5
CuO/H_2O	1264.1	3274.20	0.6895	0.001463	5
MgO/H_2O	1127.1	3657.59	0.69086	0.001859	5
CeO_2/H_2O	1156	3583	0.747	0.00094	3
TiO_2/H_2O	1160.6	4007.21	0.806	0.00155	5

3 Results and discussion

3.1 Grid Independent Test

Meshes with three different grid sizes were developed using the COMSOL Multiphysics 5.6 software's meshing tool. The distinct grid sizes included coarse, extra coarse, and extremely coarse elements. Figure 1 depicts the average bed temperature over 1000 s for hydrogen absorption using the grid sizes specified above. Figure 1 shows the results of the extremely coarse and extra coarse meshes overlapping. As a result, for the current study, the simulation results generated with an extremely coarse mesh were found to be grid-independent. This extremely coarse mesh was used in numerical simulations for further analysis.

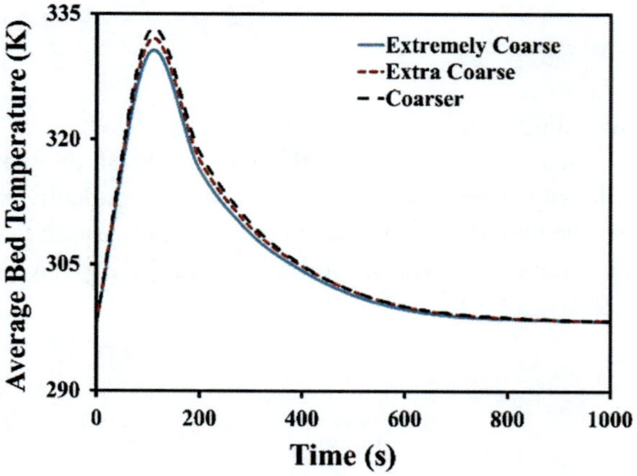

Fig. 1 Grid independency test

Table 2 Calculated properties of the nano-fluids and base fluid for ECT configuration

Fluid	Reynold's no	Prandtl no	Nusselt no	Convective heat transfer coefficient (W/m^2-K)
H_2O	31,592	6.9909	62.029	5861
Al_2O_3/H_2O	23,100	8.2605	59.108	6413
CuO/H_2O	27,433	6.9473	59.084	6415
MgO/H_2O	19,250	9.842	58.968	6415
CeO_2/H_2O	39,046	4.5087	57.557	6770
TiO_2/H_2O	20,136	9.3179	58.78	7285

3.2 *Effect of Nano-Fluids on Average Hydride Bed Temperature and Duration of Hydrogen Absorption*

Tables 1 and 2 briefly listed the calculated properties of the five nano-fluids selected for present numerical simulation. The addition of nanoparticles significantly improved properties such as the Reynolds number, Prandtl number, Nusselt number, and the heat transfer coefficient. Figure 2, 3, 4, 5 and 7 shows the effect of various nano-fluids on the average temperature of the hydride bed and the time required to achieve 90% hydrogen weight concentration at 15 bar supply pressure. The Al_2O_3/H_2O, MgO/H_2O, CuO/H_2O, TiO_2/H_2O, and CeO_2/H_2O nano-fluids with different concentrations of volume were chosen and results were obtained and correlated with base fluid (water). The rate of hydrogen absorption in the hydride bed was observed to be faster for nano-fluids than for the base fluid, water. The use of nano-fluids improved the heat transfer properties, which improved the rate of hydrogen absorption in the hydride bed. The highest rate of absorption was observed for CuO/H_2O nano-fluid at a volume concentration of 5%, and as a result, the CuO/H_2O nano-fluid was selected for sensitivity analysis. In terms of heat transfer rate, the results showed that CuO/H_2O outperformed other nanofluids and base fluids. CuO/H_2O demonstrated a higher heat transfer coefficient than other cooling fluids and a greater improvement in the rate of cooling throughout the absorption process at a supply pressure of 15 bar. The CuO/H_2O nano-fluid increased the rate of heat transfer for metal hydride reactor up to 6.85% for ECT arrangement and up to 7.32% for spiral tube arrangement (Fig. 6 and Table 3).

3.3 *Effects of Hydrogen Supply Pressure for Both ECT and Spiral Configuration*

The simulation results were correlated for 90% hydrogen weight concentration–time with CuO/H_2O nano-fluid of 5% volume concentration for ECT and Spiral configurations at different hydrogen supply pressures. The hydrogen supply pressure was

Fig. 2 Effect of H_2O base-fluid on average bed temperature and time for 90% hydrogen weight concentration

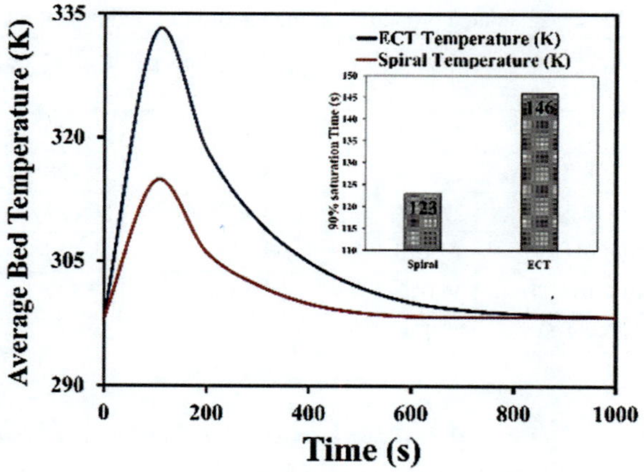

Fig. 3 Effect of Al_2O_3/H_2O nano-fluid on average bed temperature and time for 90% hydrogen weight concentration

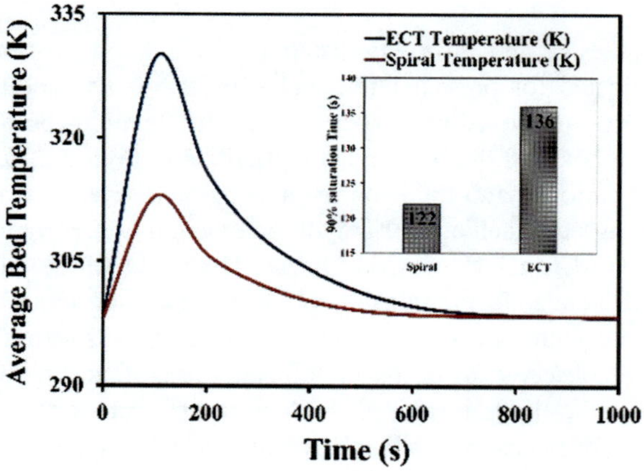

Fig. 4 Effect of CuO/H_2O nano-fluid on average bed temperature and time for 90% hydrogen weight concentration

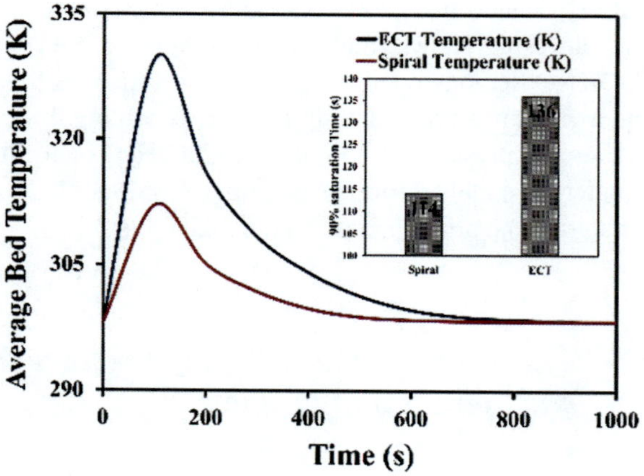

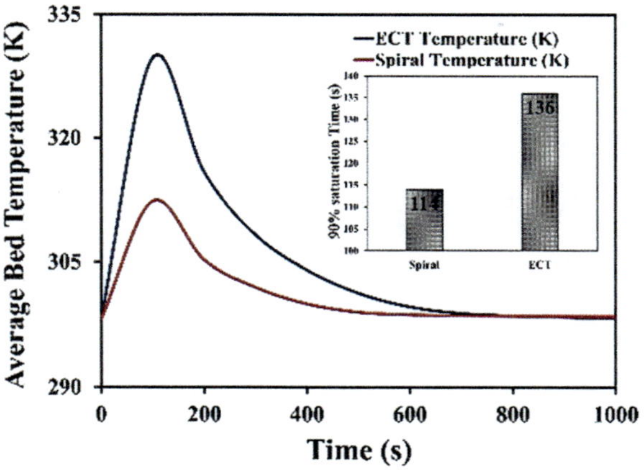

Fig. 5 Effect of MgO/H_2O nano-fluid on average bed temperature and time for 90% hydrogen weight concentration

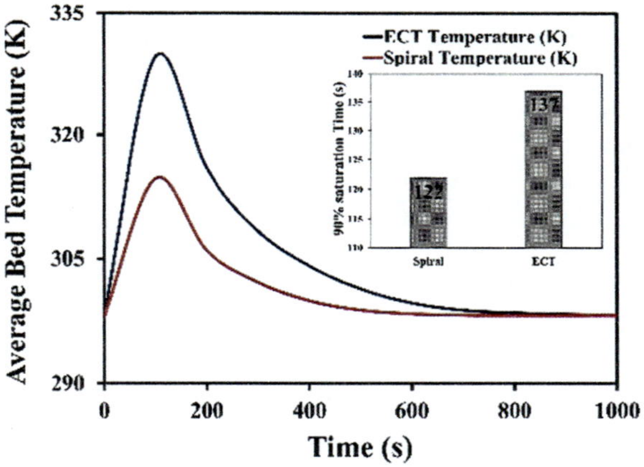

Fig. 6 Effect of CeO_2/H_2O nano-fluid on average bed temperature and time for 90% hydrogen weight concentration

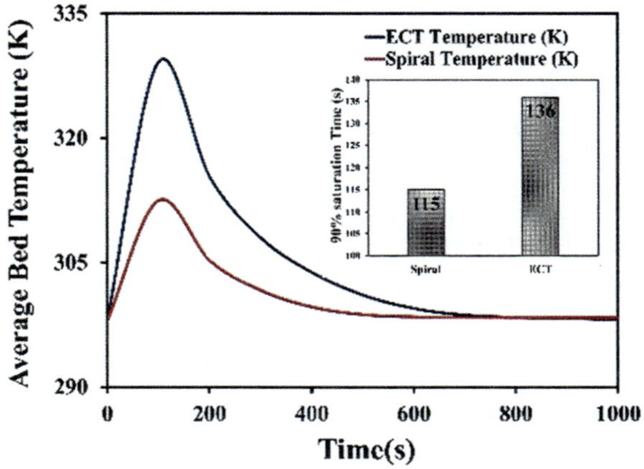

Fig. 7 Effect of TiO_2/H_2O nano-fluid on average bed temperature and time for 90% hydrogen weight concentration

Table 3 Properties of the nano-fluids and base fluid for Spiral configuration

Fluid	Nusselt no			Convective heat transfer coefficient (W/m^2-K)		
	N_1	N_2	N_3	h_1	h_2	h_3
H_2O	318.16	292.37	276.17	30,062	27,625	26,095
Al_2O_3/H_2O	254.21	233.6	220.66	27,583	25,347	23,942
CuO/H_2O	278.33	255.76	241.59	30,221	27,771	26,233
MgO/H_2O	230.13	211.48	199.76	25,038	23,008	21,733
CeO_2/H_2O	325.1	298.74	282.19	38,244	35,143	33,196
TiO_2/H_2O	234.78	215.74	203.79	29,097	26,738	25,257

taken as 5 bar, 10 bar, and 15 bar respectively, and the inlet temperature of the cooling fluid was kept at 298 K. Figure 8 highlighted the effect of three different supply pressure on 90% hydrogen weight concentration–time with CuO/H_2O nano-fluid. The 90% hydrogen weight concentration time was found to decrease with an increase in the supply pressure of hydrogen for CuO/H_2O nano-fluid at ECT and Spiral configurations. For all hydrogen supply pressures, the time of hydrogen absorption with CuO/H_2O nano-fluid for Spiral configuration was found to be shorter than that of ECT configuration. Because the potential difference between the pressure of the hydrogen supply and the average hydride bed temperature was large, the spiral configuration outperformed the ECT configuration in terms of heat extraction. The reaction kinetics of the hydride bed were slower at low hydrogen supply pressures compared to high hydrogen supply pressures for both ECT and Spiral configurations. When compared to water, the CuO/H_2O nano-fluid with ECT configuration improves hydrogen saturation time by 6.85%, while the CuO/H_2O nano-fluid with Spiral configuration enhances hydrogen saturation time by 7.32%.

3.4 *Effects of Inlet Temperature for Both ECT and Spiral Configuration*

The simulation results were correlated for 90% hydrogen weight concentration time with CuO/H_2O nano-fluid of 5% volume concentration for ECT and Spiral configurations at different inlet cooling fluid temperatures. The inlet temperature of CuO/H_2O nano-fluid was taken as 293 K, 298 K, and 303 K, and the hydrogen supply pressure was kept at 15 bar. Figure 9 depicts the effect of the three inlet temperatures on 90% hydrogen weight concentration–time with CuO/H_2O nano-fluid. It was observed that on lowering the inlet temperature for both ECT and Spiral configurations, the reaction kinetics of the hydride bed increased and the removal of heat was enhanced leading to a decrease in absorption time. The results also highlighted that the time of absorption for CuO/H_2O nano-fluid was lower in the Spiral configuration than in the ECT configuration due to an improved heat transfer rate.

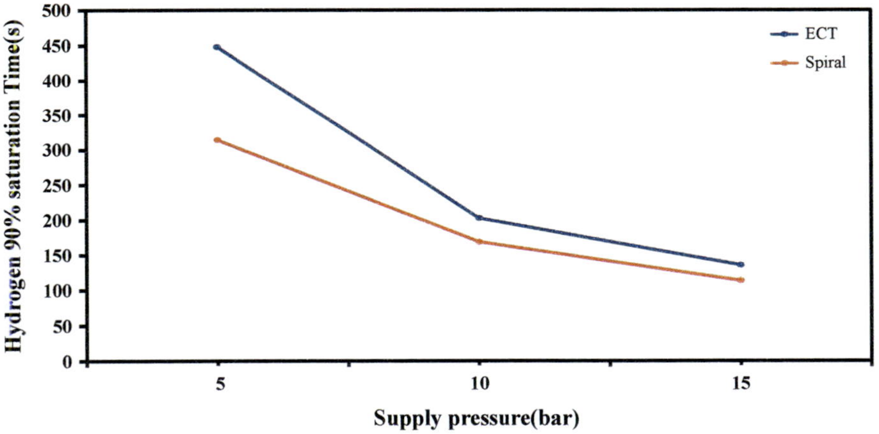

Fig. 8 Effect of supply pressure on 90% hydrogen weight concentration–time with CuO/H_2O nano-fluid at ECT and Spiral configurations

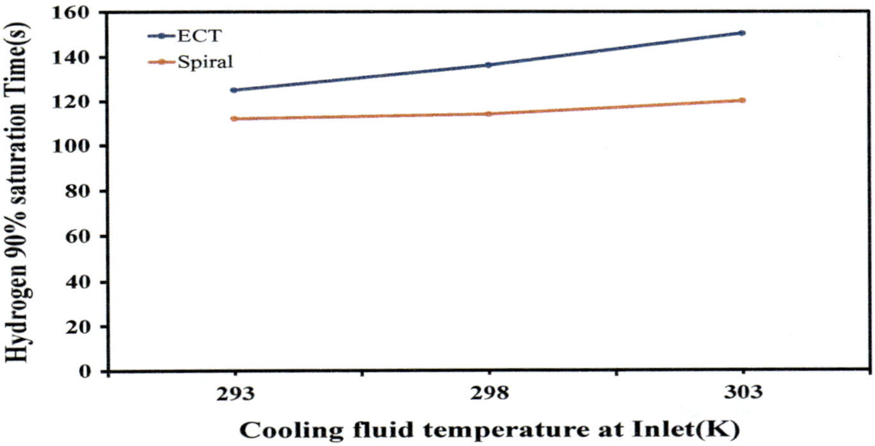

Fig. 9 Effect of inlet temperature on 90% hydrogen weight concentration–time with CuO/H_2O nano-fluid at ECT and spiral configurations

3.5 Comparison of the Temperature Profile of ECT and Spiral Configurations with CuO/H2O Nano-Fluid

The temperature profile of ECT and spiral configuration at different intervals are given in Figs. 10 and 11. From Figs. 10 and 11, it is observed that the rise and fall in temperature of CuO/H_2O nano-fluid are faster in Spiral configuration as compared to ECT configurations.

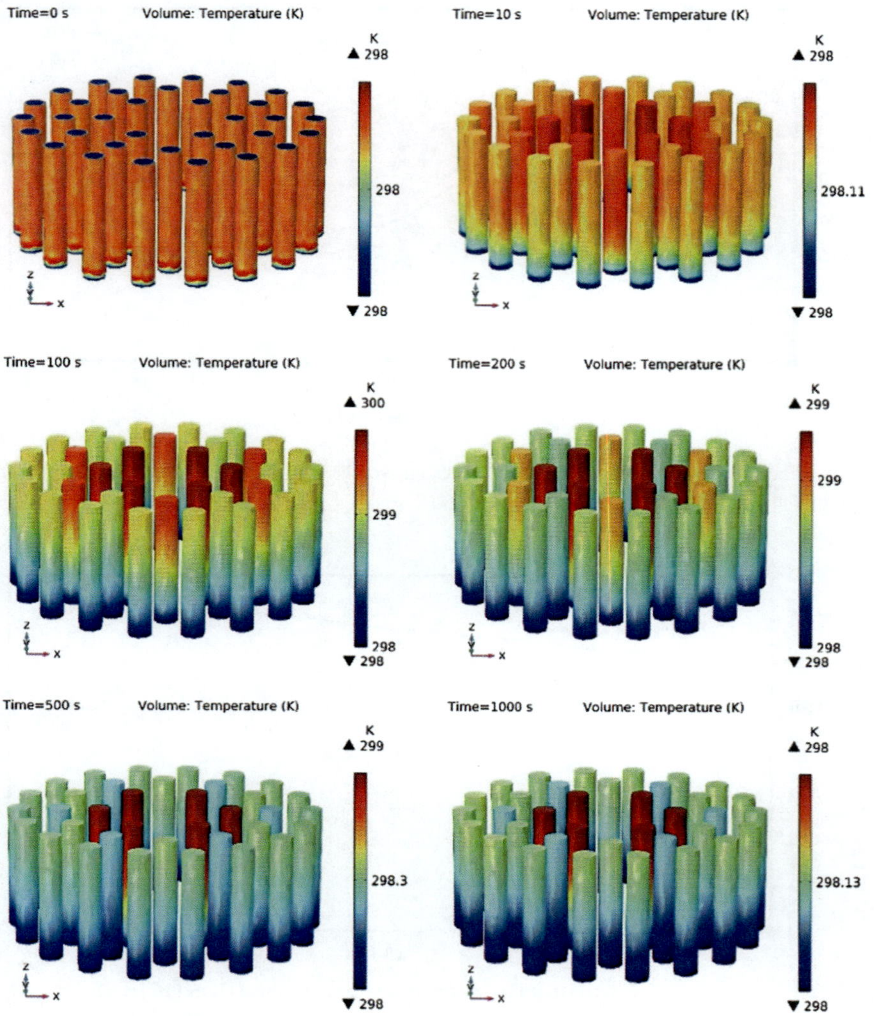

Fig. 10 The temperature profile of ECT configuration with CuO/H_2O at $t = 0$ s, 10 s, 100 s, 200 s, 500 s, 1000 s

As the hydrogenation process begins, the alloy's temperature rises to a maximum in 10 s in both cases. However, the metal hydride's poor thermal conductivity retards the process. The temperature of the alloy decreases over time due to the circulation of heat transfer fluid and the alloy's temperature is lowest near the cooling tubes. The region where the heat transfer fluid enters is lower temperature than the region near the heat transfer fluid outlet. The spiral tube frameworks in these reactors increased the heat transfer rate and reduced the reaction time better than ECT, as shown by the contours.

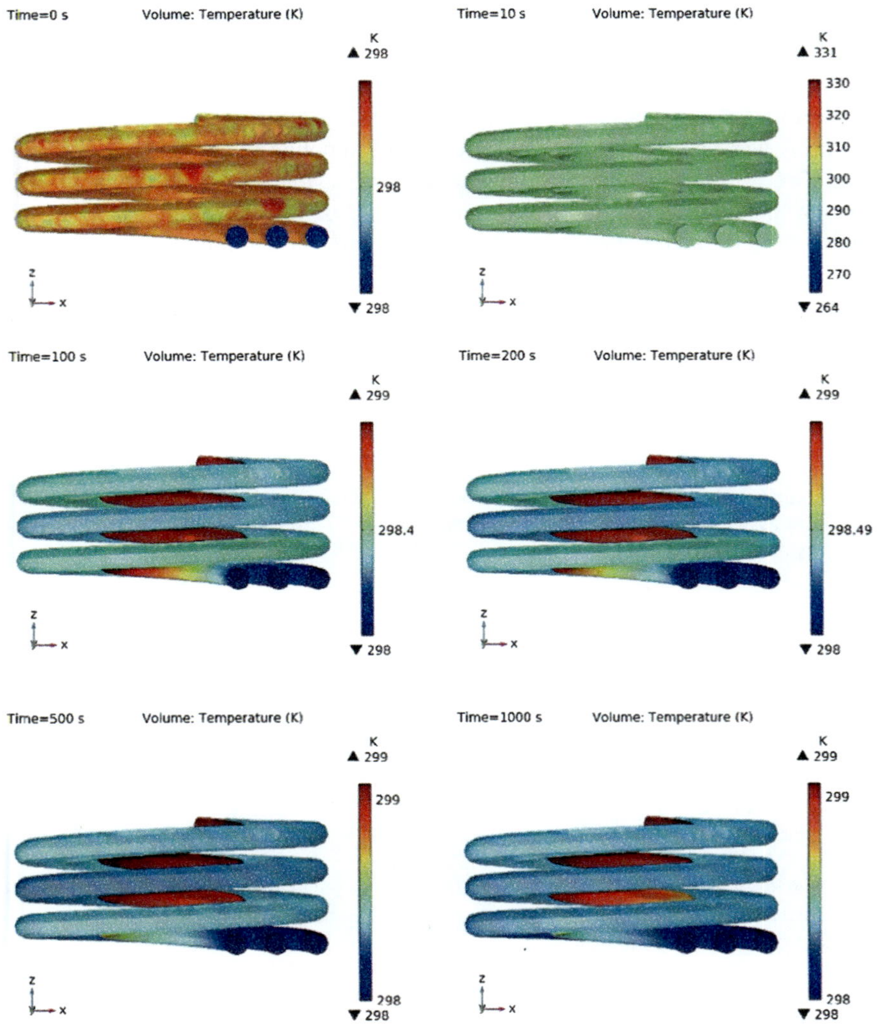

Fig. 11 The temperature profile of Spiral configuration with CuO/H_2O at t = 0 s, 10 s, 100 s, 200 s, 500 s, 1000 s

4 Conclusion

A thermal model of metal hydride reactor with $LaNi_5$ was developed for the absorption process of hydrogen in the current work for embedded cooling tubes and Spiral configurations. Five different nano-fluids were then selected as the cooling fluid to further development of thermal model and the results were showcased. The nano-fluids selected for analysis were Al_2O_3/H_2O, CuO/H_2O, MgO/H_2O, CeO_2/H_2O, and

TiO_2/H_2O. The CuO/H_2O nano-fluid with a volume concentration of 5% demonstrated a superior rate of absorption in comparison to other nano-fluids. For the supply pressure of 15 bar and the CuO/H_2O nano-fluid with a volume concentration of 5% exhibited a higher rate of absorption and it was also noticed that the time of absorption was improved by 6.85% for embedded cooling tubes configuration and by 7.32% for Spiral configuration. It was also observed that the CuO/H_2O nano-fluid with a volume concentration of 5% was thermally preferable to other nano-fluids which were selected for analysis. The rate of heat transfer was found to be enhanced with an increase in the pressure of the hydrogen supply. The time of absorption was found to be improved with a decrease in the inlet temperature of nano-fluid. Moreover, the Spiral configuration presented a faster rate of heat transfer in comparison to the embedded cooling tubes configuration. Hence, 5 vol% CuO/H_2O nano-fluid gave superior efficacy of metal hydride reactor for hydrogen storage in the analysis considered for the present work.

Acknowledgements The authors gratefully acknowledge the financial assistance provided by the Department of Science and Technology (Project title: DST- IIT Bombay Energy Storage Platform on Hydrogen, DST No: DST/TMD/MECSP/2K17/14).

References

1. Dincer I (2012) Green methods for hydrogen production. Int J Hydrogen Energy 37:1954–1971. https://doi.org/10.1016/j.ijhydene.2011.03.173
2. Dincer I, Acar C (2014) Review and evaluation of hydrogen production methods for better sustainability. Int J Hydrogen Energy 40:11094–11111. https://doi.org/10.1016/j.ijhydene.2014.12.035
3. Sreeraj R, Aadhithiyan AK, Anbarasu S (2022) Integration of thermal augmentation methods in hydride beds for metal hydride based hydrogen storage systems : review and recommendation. J Energy Storage 52:105039. https://doi.org/10.1016/j.est.2022.105039
4. Chandra S, Sharma P, Muthukumar P, Tatiparti SSV (2020) Modeling and numerical simulation of a 5 kg LaNi5-based hydrogen storage reactor with internal conical fins. Int J Hydrogen Energy 45:8794–8809. https://doi.org/10.1016/j.ijhydene.2020.01.115
5. Mellouli S, Askri F, Dhaou H, Jemni A, Nasrallah SB (2009) Numerical simulation of heat and mass transfer in metal hydride hydrogen storage tanks for fuel cell vehicles. Int J Hydrogen Energy 35:1693–1705. https://doi.org/10.1016/j.ijhydene.2009.12.052
6. Kaplan Y (2009) Effect of design parameters on enhancement of hydrogen charging in metal hydride reactors. Int J Hydrogen Energy 34:2288–2294. https://doi.org/10.1016/j.ijhydene.2008.12.096
7. Anbarasu S, Muthukumar P, Mishra SC (2014) Tests on LmNi4.91Sn0.15 based solid state hydrogen storage device with embedded cooling tubes—part A: absorption process. Int J Hydrogen Energy 39:3342–3351. https://doi.org/10.1016/j.ijhydene.2013.12.090
8. AnbarasuS, Muthukumar P, Mishra SC (2014) Tests on LmNi4.91Sn0.15 based solid state hydrogen storage device with embedded cooling tubes—part B: desorption process. Int J Hydrogen Energy 39:4966–4972. https://doi.org/10.1016/j.ijhydene.2014.01.039
9. Anbarasu S, Muthukumar P, Mishra SC (2014) Thermal modeling of LmNi4.91Sn0.15 based solid state hydrogen storage device with embedded cooling tubes. Int J Hydrogen Energy 39:15549–15562. https://doi.org/10.1016/j.ijhydene.2014.07.088

10. Muthukumar P, Satheesh A, Madhavakrishna U, Dewan A (2009) Numerical investigation of coupled heat and mass transfer during desorption of hydrogen in metal hydride beds. Energy Convers Manag 50:69–75. https://doi.org/10.1016/j.enconman.2008.08.028
11. Chung CA, Yang SW, Yang CY, Hsu CW, Chiu PY (2013) Experimental study on the hydrogen charge and discharge rates of metal hydride tanks using heat pipes to enhance heat transfer. Appl Energy 103:581–587. https://doi.org/10.1016/j.apenergy.2012.10.024
12. Urunkar RU, Patil SD (2021) Enhancement of heat and mass transfer characteristics of metal hydride reactor for hydrogen storage using various nanofluids. Int J Hydrogen Energy 46:19486–19497. https://doi.org/10.1016/j.ijhydene.2021.03.090
13. Afzal M, Mane R, Sharma P (2017) Heat transfer techniques in metal hydride hydrogen storage: a review. Int J Hydrogen Energy 42:30661–30682. https://doi.org/10.1016/j.ijhydene.2017.10.166

Potential Use of Paddy Stubble as an Energy Source in Indian Cement Industry

Bibekananda Mohapatra, Prateek Sharma, Kapil Kukreja, S. K. Chaturvedi, and Pratik N. Sheth

1 Introduction

India generates about 500 million tonnes of crop residue every year and the major share is from cereals [1, 2]. Paddy Stubble (one of the crop residues) is referred to as the short parts of paddy that are left standing after the grain is harvested. This leftover part in the form of paddy stubble is burnt as such in fields of Punjab and Haryana which leads to the worst air quality in the national capital Delhi and vicinity during the onset of winter and poses a serious environmental pollution issue. Total emission due to paddy and wheat straw burning across India is contributed by the state of Punjab and Haryana to the tune of around 48% [3]. The smoke of stubble constitutes harmful gases such as carbon monoxide, nitrogen oxides, sulfur dioxide, and methane along with particulate matter, and when this is inhaled it causes damage to the respiratory tract. It has been estimated that burning 1 tonne of paddy straw releases 1460 kg of carbon dioxide (CO_2), over 60 kg of carbon monoxide (CO), 2 kg of oxides of sulfur (SO_X), and 3 kg of the particulate matter directly contributing to environmental pollution [4]. Moreover, there are other emissions like CH_4, N_2O, NOx, and aerosols [5]. The stagnant wind situation aggravates the problem. It has been found that people also suffered irritation in the eyes, nose, and throat [6].

The problem of stubble burning is not unique to India and many countries deal with the menace of stubble burning in East Asia. Nations like China, Philippines, Malaysia, Thailand, Japan, Indonesia, and Nepal are producing bio-energy and compost using stubble and crop residues [7]. In the Philippines, Philippine Rice Research Institute

B. Mohapatra · P. Sharma (✉) · K. Kukreja · S. K. Chaturvedi
Centre for Mining, Engineering, Environment and Plant Operation, National Council for Cement and Building Materials, Ballabgarh, Haryana 121004, India
e-mail: ncbprateek@gmail.com

P. Sharma · P. N. Sheth
Chemical Engineering Department, Birla Institute of Technology and Science—Pilani, Pilani Campus, Pilani, Rajasthan 333031, India

S. Doolla et al. (eds.), *Advances in Clean Energy and Sustainability*,
Green Energy and Technology, https://doi.org/10.1007/978-981-99-2279-6_62

(PRRI) organizes the 'Dayami Festival' annually to spread awareness to curb rice straw burning. Paddy straw use as natural fertilizer, bio-energy fuel, and a substrate for mushroom cultivation is promoted on the sidelines of the festival [8]. China's straw yield in 2014 was 700 million tonnes and it too faces the problem of straw burning as a large amount of straw is burnt openly. In China, the production of electricity and methane is promoted instead of straw burning and 5 to 10% conversion to biofuel is in pipeline [9, 10].

1.1 Current Status of Paddy Stubble Utilization in India

Out of the 11.3 million tonnes of total stubble generated, 1.34 million tonnes and 9.96 million tonnes are burnt in Haryana and Punjab respectively [11]. The stubble burning data of last year indicates that only around 15% of the total cases of stubble burning were from the state of Haryana [12]. Several steps have been taken by the government to stop the rampant burning of stubble and to utilize it in the industry for various purposes. Stubble waste-based power plants are being set up and two such plants of 18 MW each have been set up in Ferozpur [13] and Faridkot districts [14]. India and Sweden are jointly working on a pilot project to convert paddy straw to green coal based on torrefaction technology at the National Agricultural Food Biotechnology institute. The target is to process 30,000 tonnes of stubble per annum [15]. Sukhbir Agro Energy Ltd. (SAEL) has collaborated with the Indian Institute of Technology (IIT) Madras, Chennai for a project that aims to establish the optimal operation of power plants with 100 percent paddy stubble or mixing co-firing paddy stubble with other biomass fuels [16]. Paddy stubble conversion to ethanol and bio CNG is also under the feasibility stage to set up biorefineries in India and this will consume around 3 lakh tonnes of paddy stubble. Central electricity authority (CEA), the ministry of new and renewable energy (MNRE), and the ministry of petroleum and natural gas (MoPNG) are exploring the utilization of paddy stubble for co-firing in power plant boilers, biogas/power production, and 20% ethanol blending in gasoline as a transportation fuel respectively [17]. All options discussed in this paragraph are targeting a small fraction of paddy stubble which cannot solve the problem completely. Thus, a workable solution like the co-processing of paddy stubble as an alternative fuel in cement manufacture is proposed.

Coal and petcoke are the major fuels for the Indian cement industry and current fuel consumption is around 56.9 million tonnes per annum in terms of coal equivalent excluding captive power plants [18]. The heat energy that is being wasted by stubble burning can be gainfully utilized in cement rotary kilns, reducing the existing fossil fuel consumption. The share of biomass as an alternative fuel in the Indian cement industry was around 24% in 2017 [19] which is significant. Simultaneously, the global share of biomass in the cement industry has also grown from 0.27% in 1990 to 5.6% in 2016 [20]. Several researchers discussed issues related to stubble burning, government regulations, different ways of stubble management, and its utilization in different forms. But paddy stubble utilization as an alternative

Table 1 Characterization results of stubble

Fuel	Dry ash-free basis					Air-dried basis			
	C	H	O	N	S	VM	Ash	Moisture	HHV (MJ/kg)
S-1	47.91	7.31	43.87	0.80	0.11	67.03	15.21	3.3	14.78
S-2	45.32	6.86	46.93	0.81	0.07	66.14	14.60	3.87	14.81

Table 2 Paddy stubble ash analysis

Wet analysis	Chemical composition (%)										
	LOI	SiO_2	Fe_2O_3	Al_2O_3	CaO	MgO	SO_3	Cl	TiO_2	P_2O_5	Mn_2O_3
S-1	10.51	65.52	0.60	2.03	3.35	4.51	1.29	4.64	Traces	0.10	0.22
S-2	10.92	60.19	1.26	2.29	4.66	2.35	1.11	4.76	Traces	0.10	0.29

fuel in the cement industry has not been explored yet and needs in-depth analysis. Accordingly, a techno-economic model is to be chalked out. The objective of this study is (a) characterization of paddy stubble (b) prepare a techno-economic model for its utilization in the Indian cement industry highlighting the challenges and opportunities.

2 Materials and Methods

Eighteen samples of paddy stubble in total were collected from four districts of Haryana (Palwal, Faridabad, Panipat, Kurukshetra) and two districts of Punjab (Patiala, Rajpura) for characterization. The purpose of the visit was to get first-hand information from farmers on the stubble-burning issues. The results of the proximate and ultimate analysis of S-1 (Haryana) and S-2 (Punjab) are tabulated in Table 1 and 2.

The volatile matter of ~ 68% and an ash content of ~ 15%, making it a high potential alternative fuel and at par with the other agro wastes and biomass presently used in the cement plants.

3 Co-processing of Paddy Stubble in Cement Plants

Co-processing of paddy stubble in cement plants can have several benefits. The higher alkali content in stubble ash provides an additional opportunity for cement plants to maximize the use of petcoke as fuel to avoid process problems. Moreover, cement plants facing coating problems in their kiln system due to high sulphur content can utilize paddy stubble as an alkali-rich source to maintain the alkali to the sulphur ratio in clinker which is a key parameter monitored during operation. However, alkali in cement needs to be controlled as high alkali affects concrete properties. The paddy stubble ash has a very high chloride content of more than 4% which may be due to high alkali chlorides. Hence, effective pre-processing and optimization are required to utilize stubble in cement plants. Plants can only operate at low TSR% or they may have to opt for kiln bypass systems till a technological breakthrough comes to tackle the high chloride problem. One of the reasons for the less applicability of paddy stubble as animal feedstock is high silica content which can also be corroborated with high silica in ash content as per Table 3. However, this stubble ash can act as a source of silica for clinker manufacture with certain adjustments in the raw mix design.

Table 3 Basis for technical model and tentative investment for establishing DLCC and SLCC facilities

S. No	Description	Value	Unit
1	Total stubble generation in Punjab in 1 month time	9,960,000	Tonnes
2	Target Palletisation (50% of total)	4,980,000	Tonnes
3	Estimated Pellets Production (10% moisture loss)	4,482,000	Tonnes
4	Total time duration to produce and dispatch the pellets from DLCC to SLCC	6	Months
5	Production required	24,900	Tonnes/day
6	Machines production capacity per day	30,000	Tonnes/day
7	Proposed production capacity	40,000	Tonnes/day
8	Total pelletization production lines to be installed at DLCC (2 × 10 tph)	100	Nos
9	Cost of 2 × 10 tph pellet production line in EPC (excluding land cost)	~ 10	Crore
10	Total storage capacity at DLCC	3,300,000	Tonnes/day
11	Facility development cost at DLCC for baled storage and pelletization (Total 22 nos. DLCC)	~ 1300 to 1400	Crores
12	Total storage capacity at SLCC	100,000	Tonnes
13	Facility development cost at SLCC	~ 100	Crores
14	Total investment for SLCC and DLCC (11 + 13)	1400 to 1500	Crore
15	Dispatch in the month of stubble generation (besides storage capacity)	1,200,000	Tonnes
16	Total storage cum handling capacity (10 + 12 + 14)	4,600,000	Tonnes

3.1 Potential of Utilizing Biomass in Cement Plants-A Way Forward Towards Atmanirbhar Bharat

As per FY 19–20 data, only 10% of the cement sector fuel requirement is met by domestic coal [18]. The rest of the fuel (imported coal and some portion of petcoke) is imported from different countries [21]. Preliminary work has been done to estimate the potential of stubble consumption in the current scenario. Considering the total clinker production of 235 million tonnes having fuel consumption of 56.9 million tonnes, 9.96 million tonnes of paddy stubble will help in achieving 15% TSR in the cement industry. This excludes the fuel requirement of captive power plants. This will help in the reduction of fuel import quantity with one step forward towards Atmanirbhar Bharat. To make it viable, railway rake shall be more suitable for long-distance transportation, covering cement plants all over India since mechanized railway unloading facility (wagon tippler) is available in most cement plants. Further, there is a requirement to develop pre-processing and co-processing technologies to maximize the utilization of paddy stubble as fuel overcoming its limitations. Several thermal treatment technologies can be applied to improve the quality of stubble as fuel. Torrefaction and gasification are some of them. Torrefaction is a thermochemical process in an inert or oxygen-limited environment where biomass is slowly heated to within a specified temperature range and retained there for a stipulated time such that it results in near-complete degradation of its hemicellulose content while maximizing the mass and energy yield of the solid product. Torrified biomass with an energy content of up to 23 MJ/kg with sulphur and chlorine reduction by 80–90% has been reported in the literature as compared to raw biomass [22]. Gasification of paddy stubble as an alternate fuel can be a source of power generation or heat source and 300 kWh of electrical energy can be generated through one tonne of paddy straw gasification [23]. The major advantage of the gasification technology in the cement industry is the removal of stubble ash before entering the kiln system and consistent heat value in the form of syngas.

3.2 Challenges for Utilizing Paddy Stubble in Indian Cement Plants

Since the source location of stubble burning i.e., Punjab and Haryana don't have an integrated cement plant, logistics is one of the key issues for stubble co-processing in cement plants. The nearby states Rajasthan and Himachal Pradesh have a total integrated cement capacity of ~ 81 MTPA [24]. However, the utilization of stubble in the cement plants of Rajasthan and Himachal Pradesh (HP) only, does not solve the problem due to the huge quantity of stubble generation in comparison to the installed cement capacity in these states. Plants in HP are in hilly areas which also adds to the landed cost of the material. Moreover, stubble shredding and handling facilities are not available at all plant premises which requires investment. A large quantity of stubble shall be generated within 10–15 days which needs a huge storage area due to the low bulk density of the material and considering round-the-year stubble consumption by cement plants. The collection of stubble from farms is itself an uphill task as there is no economic model and incentivization policy at present. Moreover, kiln main burner firing with AF is a limitation in the current scenario [25]. Given the above, it is vital to develop a suitable techno-economic model to process the stubble and make it a lucrative fuel for the cement industry considering system design, storage, logistics, etc.

4 Proposed Model

A model (Fig. 1) has been developed in this paper to utilize paddy stubble in the form of pellets in cement plants in India considering paddy stubble collection from fields to disposal. Punjab is burning 9.96 million tonnes of stubble every year which is much higher as compared to Haryana i.e., 1.34 million tonnes, hence this solution is being proposed with consideration of Punjab state only.

It is proposed to hire one logistic management firm (LMF) and one processing and operation management firm (POM) to deal with the problem. Both the firms will work with effective coordination and carry out the following activities; (1) Identification of a district-level stubble collection centre (DLCC) in each of the 22 districts of Punjab producing stubble (2) LMF shall workout the stubble collection plan along with the farmers in advance before crops are harvested (3) Custom hiring centers shall be set up to coordinate with DLCC and LMF to provide balers on rental basis to prepare compact bales (4) These paddy bales shall be transported from farms to DLCC in tractor/trolley, where POM shall process the stubble into pellets.

Pelletisation of biomass like paddy stubble consists of pretreatment, pelletization, and post-treatment steps. The first step is to remove unwanted materials like stones, pebbles, etc. by filtration followed by proper storage to prevent contamination. Moisture in stubble needs to be in the range of 10–15% to be fed to the palletization system

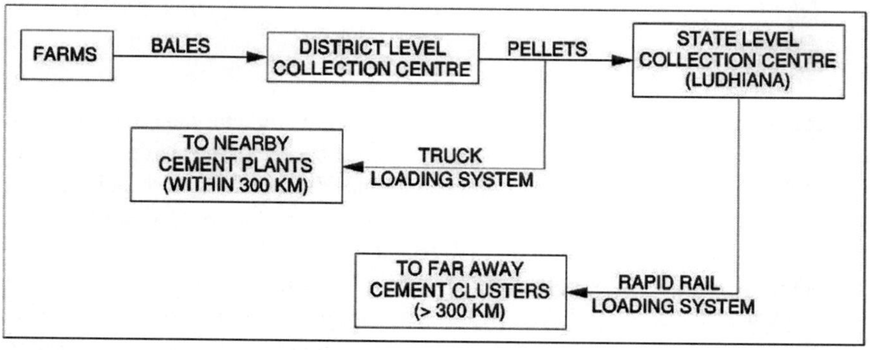

Fig. 1 Flow diagram for proposed technical model

since some moisture content helps in improving binding properties. Then the material is ground to around 3 mm by a hammer mill and sent to a pellet mill where due to high pressure and frictional forces, the material gets fused to form pellets. The electricity consumption cost depends upon various factors like the type of material feed and moisture content.

It is envisaged that after pelletization, paddy stubble heat value shall be around 16.74 MJ/kg. This enables convenient handling and economical transportation, easy to load in railway rake/trucks [26], and fast unloading at cement plants as the majority of cement plants are having wagon tipplers and can safely store for a long time. These pellets shall be transported through trucks to the state-level stubble collection centre (SLCC) or directly to the nearest cement plants located in Himachal Pradesh. It is proposed that SLCC may be located near the sharing boundary of Ludhiana, Jalandhar, Firozpur, and Moga which is the centre of Punjab state.

At the SLCC, trucks would unload stubble pellets into dump hopper to be further transported and stored in silos of RCC construction. Four nos. silos of 25,000 tonnes capacity are envisaged. From the bottom of the silos, the material would be extracted and fed to the rapid wagon loading system for the online loading of pellets in railway wagons. The rapid loading system shall load a complete rake in an hr time and has the potential to load more than 15 rakes per day.

Table 3 indicates the basis for the development of the technical model and tentative investment for establishing DLCC and SLCC facilities.

The total tentative investment for establishing 22 nos. DLCC and 1 no. SLCC has estimated at around 1400 to 1500 crore and the production cost of pellets is 4473 Rs/tonne. Logistics also plays a critical role in the overall cost scenario. The entire Indian cement industry is distributed in the form of clusters depending on limestone deposits. Major cement clusters are—Gulbarga (Karnataka), Satna (Madhya Pradesh), Yerranguntla Nalgonda (Telangana), (Andhra Pradesh), Chandrapur (Maharashtra), Bilaspur (Chhattisgarh), and Chanderia (Rajasthan) [24]. Preliminary logistics have been worked out to get ballpark figures for transportation costs from farms to these cement plant clusters.

The transportation from farms to DLCC is to be within a district of Punjab, hence the average cost of transportation has been assumed to be less than Rs. 500/tonne. The stubble pellets shall be transported by truck from DLCC to SLCC with a distance ranging from 20 to 200 km at Rs. 2.58/tonne/km costing to Rs. 516/tonne.

Further transportation from SLCC to the respective cement clusters shall be through railway rakes for long distances or trucks for short distances. For truck transportation from DLCC directly to nearby cement plants (3 nos.) located in Himachal Pradesh within a 300 km range, transportation cost works out to Rs. 500–776/tonne.

For railway transportation from SLCC to farther cement clusters, the railway freight charges vary from Rs. 515 to Rs. 1420/tonne stubble pellet. Three out of seven clusters are far-sighted locations having freight charges of more than Rs. 1300/tonne.

Thus, total logistics cost from farms to cement plant clusters shall vary from Rs. 1531–2436/tonne considering railway transportation from SLCC to cement clusters and Rs. 1500–1800/tonne considering truck transportation from SLCC to nearby cement plants in Himachal Pradesh Henceforth, landed cost of stubble pellets at cement plants on an average would vary from 6000–7450 Rs/tonne. However, this cost may be further reduced when the pelletization will be done on a larger scale. As the cement plant may get the pellets at the same cost as coal, it will be an attractive solution for the cement industry concerning the reduction in carbon footprint since biomass is carbon neutral and make cement more sustainable. Apart from solving the stubble burning problem, this solution will create huge employment opportunities in rural areas. Facilities developed at DLCC and SLCC may also be used for the transportation of other grains when pellets dispatches are done (for the remaining 6 months of the year). The cost estimated in the table-6 is tentative and a detailed project report needs to be carried out to have the real picture. The above cost does not include the land cost or land lease cost which needs government intervention and may also be the deciding factor for the above proposal. The major advantage of this proposal is transportation through railway rake which includes the landed cost of stubble within the reach of different cement clusters all over the country. One of the studies estimated that the health and economic costs of crop residue burning in northern India lead to an economic loss of over Rs. 2 lakh crores/annum [27]. Comparing this cost with the proposed one-time investment of 1300 to 1400 crores for establishing DLCC and SLCC facilities, it is almost negligible and makes the proposal more attractive for execution.

5 Conclusions

The proposed model seems attractive and beneficial with the consideration of the environmental and public health issue and would be a win–win situation for farmers, cement plants, and government agencies. It is anticipated that once a suitable economic model is developed and other challenges as discussed above are resolved, cement plants can take up the paddy stubble for co-processing. However, achieving high TSR shall be quite challenging due to high alkali and chloride content. Moreover, the cost of pellets at the cement plant shall be the major concern of the industry hence industry may expect some monetary support from the government to co-process the pellets in the cement plant in a lucrative manner. As the majority of pellets shall be received by cement plants in a very short period (as this is a seasonal problem), hence additional investment for storage and co-processing facilities is expected. Further enhancement of TSR beyond a certain range through pellets utilization will demand the investment of some additional system like kiln bypass, new generation burner, etc. Cement plants located far away from north India may expect some subsidy on railway freight. The government can look upon the investment as a viability gap fund (VGF) which can be borne by respective states or central government.

Acknowledgements The authors wish to thank the support of the Centre for Cement Research and Testing of M/s National Council for Cement and Building Materials for paddy stubble characterization.

References

1. Ministry of Agriculture, National policy (2014) for management of crop residues (NPMCR) D.o.A. Cooperation, Editor
2. Bhuvaneshwari S, Hettiarachchi H, Meegoda JN (2019) Crop residue burning in india: policy challenges and potential solutions. Int J Environ Res Public Health 16:832
3. Manjeet M, Kumar JSS (2019) Crop residue burning: issue and management for climate-smart agriculture in NCR region, India. Asian J Agric Extension Econ Sociol, pp 1–11
4. Singh J (2018) Paddy and wheat stubble blazing in Haryana and Punjab states of India: a menace for environmental health. Environ Qual Manage 28:47–53
5. Gupta PK, Sahai S, Singh N, Dixit CK, Singh DP, Sharma C, Tiwari MK, Gupta RK, Garg SC (2004) Residue burning in rice–wheat cropping system: causes and implications. Curr Sci 87:1713–1717
6. Study confirms adverse health impact of stubble burning. https://timesofindia.indiatimes.com/city/chandigarh/study-confirms-adverse-health-impact-of-stubble-burning/articleshow/66000859.cms. Accessed 21 March 2022
7. Subramani MR (2022) Power from biomass, bio-energy, compost: how other countries manage the issue of stubble burning. https://swarajyamag.com/ideas/power-from-biomass-bio-energy-compost-how-other-countries-manage-the-issue-of-stubble-burning#:~:text=In%20countries%20such%20as%20Australia,biomass%2C%20using%20these%20crop%20residues.&text=The%20Indian%20government%20does%20offer,find%20them%20a%20little%20costly. Accessed 4 April 2022
8. Anish T (2022) The answer to paddy-straw burning problem may be found abroad. Down to earth. https://www.downtoearth.org.in/news/agriculture/the-answer-to-paddy-straw-burning-problem-may-be-found-abroad-62635. Accessed 5 May 2022
9. Qu C, Li B, Wu H (2012) Controlling air pollution from straw burning in china calls for efficient recycling. Environ Sci Technol 46:7934–7936
10. Ren J, Yu P, Xu X (2019) Straw utilization in China—status and recommendations 11, 1762
11. Ministry of agriculture and farmers welfare, Review of the scheme "Promotion of agricultural mechanisation for in-situ management of crop residue in states of Punjab, Haryana, Uttar Pradesh and NCT of Delhi", c.f.w. Agriculture, Editor. (2019): New Delhi
12. Kapil S (2019) On the cusp, in Down to earth, pp 32–33
13. SMS (2018) Envocare limited: final environmental impact assessment report and environmental management plan
14. SMS (2018) Envocare Limited: Executive summary of draft environmental impact assessment report for proposed 18 MW biomass based thermal power plant at village sedha singh wala, tehsil-Jaito, district-Faridkot, Punjab
15. India, Sweden Begin Pilot Project To End Stubble Burning And Turn Agri-Waste Into Biocoal. https://swarajyamag.com/insta/india-sweden-begin-pilot-project-to-end-stubble-burning-and-turn-agri-waste-into-biocoal. Accessed 5 Jan 2022
16. IIT Madras is planning to power thermal stations with paddy straw, could reduce stubble burning https://www.indiatoday.in/education-today/news/story/iit-madras-planning-to-power-thermal-stations-with-paddy-straw-reduce-stubble-burning-1628272-2019-12-14. Accessed 21 Feb 2022
17. Datta A, Emmanuel M, Ram NDS (2020) Discussion paper: crop residue management: Solution to achieve better air quality. The Energy Resource Institute: New Delhi
18. Assocham (2020) How to make commercial coal mining successful. in Webinar
19. World business council for sustainable development: low carbon technology Roadmap for the Indian cement sector: status review (2018)
20. Alsop P (2019) The cement plant operations handbook. Warners (Midlands) Plc, UK
21. India cement industry bosses expect to import more high-sulphur petcoke - ET EnergyWorld. https://energy.economictimes.indiatimes.com/news/coal/india-cement-industry-bosses-expect-to-import-more-high-sulphur-petcoke/74218132. Accessed 10 May 2022

22. Moser W (2019) Decarbonisation quick wins. International cement review. (2019). pp 62–64
23. Lohan SK, Jat HS, Yadav AK, Sidhu HS, Jat ML, Choudhary M, Peter JK, Sharma PC (2018) Burning issues of paddy residue management in north-west states of India. Renew Sustain Energy Rev 81:693–706
24. Saxena A, Chaturvedi S, Ojha P, Agarwal S, Mittal A, Sharma P, Ahamad G, Mazumdar R, Arora V, Kalyani K, Anupam BS, Kukreja K, Ahmed R, Kaura BAP (2019) Compendium The Cement Industry India (2019), National council for cement and building materials, India. pp 251–260
25. Pieper C, Liedmann B, Wirtz S, Scherer V, Bodendiek N, Schaefer S (2020) Interaction of the combustion of refuse derived fuel with the clinker bed in rotary cement kilns: a numerical study. Fuel 266:138–144
26. Pradhan P, Mahajani S, Arora A (2018) Production and utilization of fuel pellets from biomass: a review. Fuel Process Technol 181:215–232
27. Business today (2022) Air pollution from stubble burning costs India $30 billion annually: study. https://www.businesstoday.in/latest/economy-politics/story/air-pollution-from-stubble-burning-costs-india-30-billion-annually-study-175502-2019-03-04. Accessed 24 June 2022

Sustainable Biofuel Solution for Industrial and Commercial Sectors: A Stochastic Green Supply Chain Design Approach

Kapil Gumte

1 Introduction

The demand for energy is increasing which conventional fuels alone cannot satisfy, hence there is a need for non-conventional resources like solar, wind, geothermal, tidal and biomass, etc. Out of various non-conventional renewable energy resources, the current paper focuses on biomass and biofuels for developing nations like India which has 70% of the population in rural areas [1] and is mostly dependent on agriculture and forestry. In developing nations like India, Brazil, etc. it is mandatory to blend bioethanol with gasoline to reduce the economic burden and enhance the local economy for farmers. Apart from handling the economical demand and supply ratio of energy and fuels, it is also necessary to have a clean green source of energy that must be environmentally friendly. Hence to achieve these energy targets strategic supply chain network design (SCND) acts as a vital tool. The supply chain (SC) aims to obtain the raw resources to move, process, treat and distribute through different layers in sequence to reach the final demand destination at minimum cost to maximize the profit by optimally managing the supply, utilities, labor, government rules, and it's every element.

To comprehend the current work, it is necessary to understand the earlier researcher's work in this domain. For Sweden, de Jong et al. [2] developed SC with major four cost reduction strategies with distributed unimodal truck transport and found that biofuel production cost is higher than fossil fuel for which stochasticity study was missing. Kesharwai et al. [2] developed SC for the Missouri United States with second-generation biomass for economic viability and environmental sustainability using a centralized and decentralized strategy to reduce the overall cost. But the model lacked stochasticity and volatility. Carvajal et al. [3] developed stochastic SC for sugarcane biomass focusing on agriculture sites for sowing,

K. Gumte (✉)
Indian Institute of Management Jammu, Jammu, India
e-mail: kapil@iimj.ac.in

S. Doolla et al. (eds.), *Advances in Clean Energy and Sustainability*,
Green Energy and Technology, https://doi.org/10.1007/978-981-99-2279-6_63

growing, and harvesting in Columbia. They tried to handle the uncertainty of climate conditions using the two stage stochastic programming (TSSP) method which got limitations of tractability when the number of uncertain parameters was increased. When the uncertainty parameters i.e. biomass demand, and price, are represented in a possibilistic manner [4] used triangular fuzzy uncertainty to implement biogas supply chain design for Iran where site location and connectivity are done. Here getting the threshold value of operations was not possible due fuzzy nature. Further readers can go through the paper by Atashbar et al. [5] for a detailed review of the bio supply chain involving upstream biomass procurement, pre-processing, bio-refineries, harvest planning, midstream inventory control, transport, and downstream demand distribution. Here various modeling approaches and their applications with past, present, and future directions are given.

From the literature overview, it is found that (knowledge gap) the SCND model involving probability-based uncertainty amalgamated with demand forecasting, time value of money, external imports, and carbon credits with government rules of blending 20% bioethanol to gasoline [6] that too specifically for developing like India is missing.

The current paper covers the knowledge gap by providing the following contribution:

1. The stochasticity is incorporated in biomass feed supply due to its variations across the geography using chance-constrained programming (CCP) methodology to perform the techno-economic risk calculation.
2. Technically the model developed optimally locates the sites and connects them with appropriate site capacity using a deterministic equivalent form of uncertainty.
3. External imports are blended to avoid stock-outs situations in case of insufficient local production. The imported bioethanol maintains the quantity as well as octane number quality of demanded bioethanol product.
4. Environmentally, pollution is calculated in terms of greenhouse gas emissions (GHGe). The GHGe for biofuels is compared with fossil fuels GHGe to obtain the GHGe saving and convert the saving's into revenue generation carbon credit.
5. Economically the profits are calculated via net present value or net present worth (NPW) where biofuel demand is forecasted from 2018 to 2026 and inflation is accommodated using the value of money with time.
6. Further the yield point or threshold point of the economic condition is found for biofuel projects to run with minimum profit involving stochasticity in developing nations using India as a case study to attract investors and the commercial sector with calculated risk.

Next, Sect. 2 elaborates on problem formulation and model development followed by the stochastic handling approach in Sect. 3. Section 4 talks about the data and discuss the results. Last but not the least, Sect. 5 concludes the summary of the entire work.

2 Problem Formulation and Developed Model Explanation

The basic concept for the model is taken from [7] which is deterministic in nature. Here the model undergoes changes to make it stochastic.

Four layered supply chain is developed for single product bioethanol in the sequence i.e. supplier, manufacturer, distributor, and retail customer (see Fig. 1). The four raw material at the supplier is taken from 2nd generation lignocellulose biomass source namely bagasse, corn stover, bamboo, and wood chip. At manufacturing sites, the raw biomass material is processed and converted to bioethanol. External imports are connected to distributors, where inventories are also kept. For implementation India territory is chosen, which is divided into 12 zones (see Fig. 2) with each zone having 3 units of supplier sites g_1–g_{36}, manufacturing sites g_{37}–g_{72}, distributor sites g_{73}–g_{108}, 1 unit of each retailer site g_{109}–g_{120}, and 2 external import sites g_{121}–g_{122}. Hence total of 122 potential sites exists among which the developed model (see Sect. 3) will choose the optimal sites and connectivity across the time horizon of 2018–2026. It is assumed that there exists train connectivity at each site for connectivity within and across the zones for raw material and product movement. Further, the material flows from supplier to retailer direction and money flows from retailer to supplier direction.

It is recommended to go through Appendix Nomenclature for acronym comprehension. The model is developed with an optimization approach where objective net present value (NPV) is used to calculate the profit which involves the value of money with time and constant depreciation (Eq. 1). The major source of earning is via retail selling price $SePr_{p,t}$ for the demand $Dem_{p,r,t}$ and carbon credits $GHGeRv_t$.

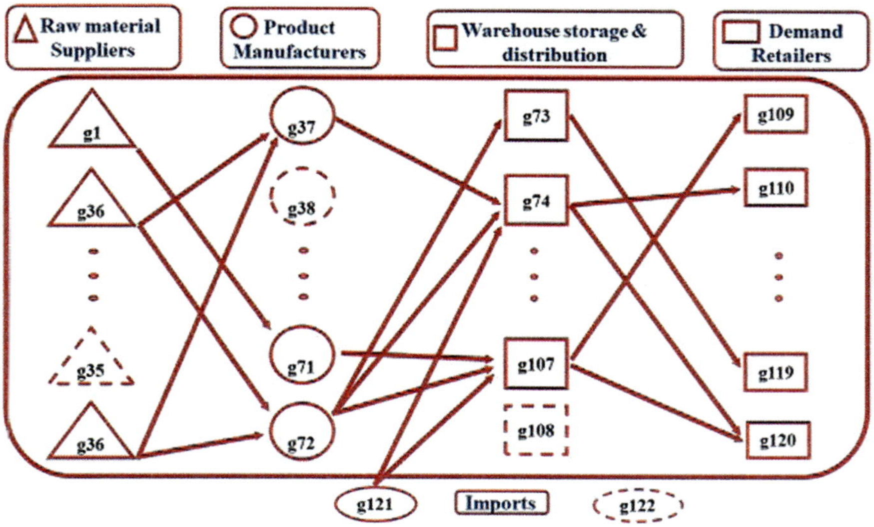

Fig. 1 Schematics of flexible supply chain network design

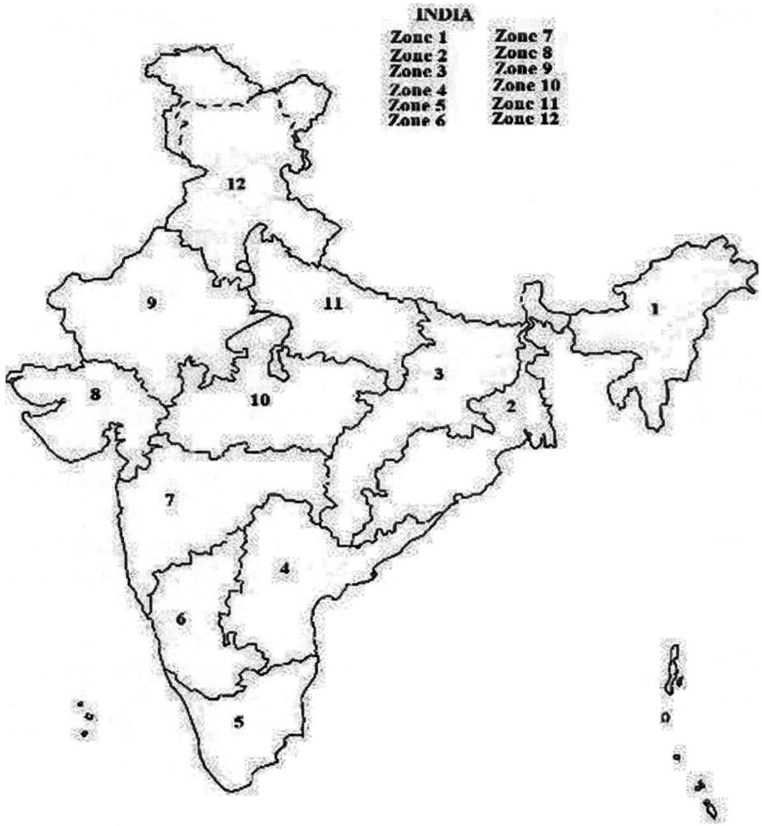

Fig. 2 Indian geography divided into 12 zones

The equation involves operating *Opx* (Eq. 3) composed of transport TrC_t, inventory $InvC_t$, production PrC_t and import cost ImC_t, and *Cpx* (Eq. 4) expenditure involving infrastructure cost. The depreciation *Dpr* used the constant method (Eq. 5). Max *NPW* where

$$NPW = \sum_t \frac{1}{(1+\Omega)^{N-1}}((Ern_t - Opx_t - Dpr) \times (1 - \Psi) + Dpr) - \sum_t \frac{1}{(1+\Omega)^{N-1}}(Cpx_t) \tag{1}$$

$$Ern_t = \sum_{p,ru} Dem_{p,r,t} \times SePr_{p,t} + GHGeRv_t \tag{2}$$

$$Opx_t = TrC_t + InvC_t + PrC_t + ImC_t \tag{3}$$

$$Cpx_t = InfC_t \tag{4}$$

$$Dpr = \sum_t \frac{i}{\left((1+i)^N - 1\right)}(InfC_t - 0.2 \times InfC_t) \tag{5}$$

Equations 6–13 represents binary constraints {0, 1} across supply chain nodes indicating connectivity and site location for e.g. if $Xsm_{s,m,t}$ exits between supplier s and manufacturer m i.e. 1, then site $Ys_{s,t}$ also exits at supplier s and $Ymu_{mu,t}$ manufacturer m. Similarly, other binary constraints follow the logic.

$$Xsm_{s,m,t} \le Ys_{s,t}, \quad \forall t, \forall s, \forall m \tag{6}$$

$$Xsm_{s,m,t} \le Ym_{m,t}, \quad \forall t, \forall s, \forall m \tag{7}$$

$$Xmd_{m,d,t} \le Ym_{m,t}, \quad \forall t, \forall m, \forall d \tag{8}$$

$$Xmd_{m,d,t} \le Yd_{d,t}, \quad \forall t, \forall m, \forall d \tag{9}$$

$$Xduru_{d,r,t} \le Yd_{d,t}, \quad \forall t, \forall d, \forall r \tag{10}$$

$$Xduru_{d,r,t} \le Yr_{r,t}, \quad \forall t, \forall d, \forall r \tag{11}$$

$$Ximdu_{im,d,t} \le Yim_{im,t}, \quad \forall t, \forall im, \forall d \tag{12}$$

$$Ximdu_{im,d,t} \le Yd_{d,t}, \quad \forall t, \forall im, \forall d \tag{13}$$

Equations 14–29 shows the flow balance across the supply chain. Equation 14 shows the feed $Fed_{f,s,t}$ getting distributed to the manufacturer with feed capacity $FedMax_{f,s,t}$ (Eq. 15) and transport quantity $QsmlMax_{f,s,m,l,t}$ limitations (Eq. 16).

$$Fed_{f,s,t} = \sum_m Qsml_{f,s,m,l,t}, \quad \forall f, \forall s, \forall l, \forall t, m \in g \tag{14}$$

$$\begin{aligned} FedMin_{f,s,t} \times Ys_{s,t} &\le Fed_{f,s,t} \\ &\le FedMax_{f,s,t} \times Ys_{s,t}, \quad \forall f, \forall s, \forall t \end{aligned} \tag{15}$$

$$\begin{aligned} QsmMin_{f,s,m,l,t} \times Xsm_{s,m,t} &\le Qsml_{f,s,m,l,t} \\ &\le QsmlMax_{f,s,m,l,t} \times Xsm_{s,m,t}, \quad \forall s\_ f_{s,f}, \forall m, \forall l, \forall t \end{aligned} \tag{16}$$

Once the feed reaches the manufacturing site $Qsml_{f,s,m,l,t}$, the pre-treatment and conversion $con_{f,tec,p}$ happen using the corresponding technology *tec* as per the feed availability for product $Pmtec_{p,m,tec,t}$ bioethanol manufacturing (Eq. 17). The total bioethanol $Pm_{p,m,t}$ (Eq. 18) has the production capacity $PmMax_{p,m,t}$ (Eq. 19) constraint is transported $Qmdl_{p,m,d,l,t}$ to next layer (Eq. 20) with transport capacity limitations $QmdlMax_{p,m,d,l,t}$ (Eq. 21).

$$\sum_{s} Qsml_{f,s,m,l,t} \times con_{f,tec,p} = Pmtec_{p,m,tec,t}, \quad \forall m, \forall tec, \forall p, \forall l, \forall t, \forall s\_f_{s,f}, s \in g \tag{17}$$

$$\sum_{tec} Pmtec_{p,m,tec,t} = Pm_{p,m,t}, \quad \forall p, \forall m, \forall t, tec \in Tec \tag{18}$$

$$PmMin_{p,m,t} \times Ym_{m,t} \le Pm_{p,m,t} \le PmMax_{p,m,t} \times Ym_{m,t}, \quad \forall p, \forall m, \forall t \tag{19}$$

$$Pm_{p,m,t} = \sum_{d} Qmdl_{p,m,d,l,t}, \quad \forall p, \forall m, \forall l, \forall t, d \in g \tag{20}$$

$$QmdlMin_{p,m,d,l,t} \times Xmd_{m,dut} \le Qmdl_{p,m,d,l,t} \le QmdlMax_{p,m,d,l,t} \times Xmd_{m,d,t} \quad \forall p, \forall m, \forall d, \forall l, \forall t \tag{21}$$

To take care of stock out the situation, external bioethanol imports $imt_{p,im,t}$ are distributed $Qimdl_{p,im,du,l,t}$ across inventories of distribution centers (Eq. 22), following the import capacity $imtMax_{p,im,t}$ (Eq. 23) and transport quantity $QimdlMax_{p,im,d,l,t}$ limitation (Eq. 24).

$$imt_{p,im,t} = \sum_{d} Qimdl_{p,im,du,l,t}, \quad \forall p, \forall im, \forall l, \forall t, d \in g \tag{22}$$

$$imtMin_{p,im,t} \times Yim_{im,t} \le imt_{p,im,t} \le imtMax_{p,im,t} \times Yim_{im,t}, \quad \forall p, \forall im, \forall t \tag{23}$$

$$QimdlMin_{p,im,d,l,t} \times Ximd_{im,d,t} \le Qimdl_{p,im,d,l,t} \le QimdlMax_{p,im,d,l,t} \times Ximd_{im,d,t}, \quad \forall p, \forall im, \forall d, \forall l, \forall t \tag{24}$$

At distribution sites, the inventory $invd_{p,d,t}$ receives the input from the previous time period $invd_{p,d,t-1}$, manufacturers $Qmdl_{p,m,d,l,t}$ and importers $\sum_{im} Qimdl_{p,im,d,l,t}$ and gives output to retailer sites $Qdrl_{p,d,r,l,t}$ (Eq. 25) with inventory capacity $invdMax_{p,du,t}$ (Eq. 26) and transport quantity $QdrlMax_{p,d,r,l,t}$ limitations (Eq. 27).

$$invd_{p,d,t} = invd_{p,d,t-1} + \sum mQmdl_{p,m,d,l,t} + \sum imQimdl_{p,im,d,l,t} - \sum_{r} Qdrl_{p,d,r,l,t}, \quad \forall p, \forall d, \forall t, \forall l, \forall r, \forall t, m \in g, r \in g \tag{25}$$

$$invdMin_{p,d,t} \times Yd_{d,t} \leq invd_{p,d,t} \leq invdMax_{p,du,t} \times Yd_{d,t}, \quad \forall p, \forall d, \forall l, \forall t \tag{26}$$

$$QdrlMin_{p,d,r,l,t} \times Xdr_{d,r,t} \leq Qdrl_{p,d,r,l,t} \leq QdrlMax_{p,d,r,l,t} \times Xdr_{d,r,t}, \quad \forall p, \forall d, \forall r, \forall l, \forall t \tag{27}$$

The imports are added using a linear mixing method to blend with local bioethanol to achieve the needed octane number (ON) quality (Eq. 28).

$$\sum_{im} imt_{p,im,t} \times \text{ON of imported product} + \sum_{du} invd_{p,d,t} \times \text{ON of indigenous product} = \text{ON needed} \times \left(\sum_{im} imt_{p,im,t} + \sum_{d} invd_{p,d,t} \right), \quad \forall p, \forall t, im \in g, d \in g \tag{28}$$

Finally, all the product reaches retailer to meet the demand (Eq. 29)

$$\sum_{d} Qdrl_{p,d,r,l,t} \geq Dem_{p,r,t}, \quad \forall p, \forall r, \forall l, \forall t, d \in g \tag{29}$$

Next Eqs. 30–33 shows the greenhouse gas emission calculations. Equation 30 calculates the total greenhouse emission $GHGe_t$ during cultivation at the suppliers, during manufacturing, and during transportation using corresponding emission factors $fc_{f,t}$, $fm_{m,t}$ and $fl_{f,l,t}$. Next in Eq. 31, greenhouse gas emissions due to fossil fuel petrol $GHGeF_t$ are calculated and compared with greenhouse gas emissions for bioethanol $GHGe_t$, where the savings $GHGeSav_t$ due to bioethanol are obtained, which is further converted into revenue $GHGeRv_t$ using unit carbon credit value $Cval_t$.

$$GHGe_t = \sum_{f,s,t} fc_{f,t} \times Fed_{f,s,t} + \sum_{p,m,t} fm_{m,t} \times Pm_{p,m,t} + fl_{p,l,t} \left[\sum_{f,s,m,l,t} \left(dsm_{s,m} \times Qsml_{f,s,m,l,t} \right) + \sum_{p,m,d,l,t} \left(dmd_{m,d} \times Qmdl_{p,m,d,l,t} \right) \right.$$

$$+ \sum_{p,im,d,l,t} \left(fimdl_{p,l,t} \times dimd_{im,d} \times Qimdl_{p,im,d,l,t} \right)$$

$$\left. + \sum_{p,d,r,l,t} \left(fdrl_{p,l,t} \times ddr_{d,r} \times Qdrl_{p,d,r,l,t} \right) \right] \tag{30}$$

$$GHGeF_t = GasCa \times GasE \times \sum_{p,r} Dem_{p,r,t} \tag{31}$$

$$GHGeSav_t = GHGeF_t - GHGe_t \tag{32}$$

$$GHGeRv_t = GHGeSav_t \times Cval_t \tag{33}$$

Next, Eqs. 34–38 represents the detailed costing done (see Eqs. 3 and 4). Transport cost TrC_t is calculated using unit transport cost $Utc_{l,t}$, distance traveled and quantity transferred (Eq. 34). Similar unit concept is used for infrastructure $InfC_t$ (Eq. 35), inventory $InvC_t$ (Eq. 36), production PrC_t (Eq. 37), and import cost $ImpC_t$ (Eq. 38).

$$TrC_t = Utrc_{f,l,t} \left[\sum_{f,s,m,l} dsm_{s,m} \times Qsml_{f,s,m,l,t} \right.$$

$$+ \sum_{p,m,d,l} dmd_{m,d} \times Qmdl_{p,m,d,l,t} + \sum_{p,d,r,l} ddr_{d,r}$$

$$\left. \times Qdrl_{p,d,r,l,t} + \sum_{p,im,d,l} dimd_{im,d} \times Qimdl_{p,im,d,l,t} \right],$$

$$\forall t, f \in F, p \in P, l \in L, s \in g, mu \in g, d \in g, im \in g, r \in g \tag{34}$$

$$InfC_t = \sum_{s} Ys_{s,t} \times Cs_{s,t} + \sum_{mu} Ym_{m,t} \times Cm_{m,t}$$

$$+ \sum_{d} Yd_{d,t} \times Cd_{d,t} + \sum_{r} Yr_{r,t} \times Cr_{r,t}$$

$$+ \sum_{im} Yim_{im,t} \times Cim_{im,t},$$

$$\forall t, s \in g, m \in g, d \in g, im \in g, r \in g \tag{35}$$

$$InvC_t = \sum_{p,d} invd_{p,d,t} \times UinvC_{p,d,t}, \quad \forall t, p \in P, d \in g \tag{36}$$

$$PrC_t = \sum_{p,m} UpC_{p,t} \times Pm_{p,m,t}, \quad \forall t, p \in P, m \in g \tag{37}$$

$$ImpC_t = \sum_{p,im} UimC_{p,t} \times imt_{p,im,t}, \quad \forall t, p \in P, im \in g \tag{38}$$

3 Uncertainty Handling: Chance Constraint Programming

Chance constrained programming (CCP) is an uncertainty handling technique where equations (constraint or objective) having the uncertain parameter are dealt in probabilistic manner i.e. probability of equation satisfaction varying between 0 and 1 and is shown with extent of probability satisfaction (α). Due to stochasticity, the constraints may not always be satisfied as in the case of the deterministic approach. The parameter showing the uncertainty is assumed to follow the law of large numbers leading to normal distribution. Computationally, the optimization model cannot be solved directly due to stochasticity hence the equations are converted to deterministic equivalent form using the probability concept. For e.g. consider:

$$f(x) \text{ as objective} \tag{39}$$

$$h(x, \beta) = 0 \text{ as equivalent constraint having uncertain parameter } \beta \tag{40}$$

$$g(x, \beta) \geq 0 \text{ as unequal constraint having uncertain parameter } \beta \tag{41}$$

Applying probability on uncertain equations.

$$P(h(x, \beta) = 0) > \alpha_v \tag{42}$$

$$P(g(x, \beta) \geq 0) > \alpha_w \tag{43}$$

where P shows the probability of constraint satisfaction, α_v and α_w shows the extent of the probability satisfaction, which is also the area covered under the normal distribution. Next when the deterministic equivalent equation is applied the equations change their form as:

$$h\big(x, mean(\beta) + F^{-1}(\alpha_v) * standard\ deviation(\beta)\big) = 0 \tag{44}$$

$$g\big(x, mean(\beta) + F^{-1}(\alpha_w) * standard\ deviation(\beta)\big) \geq 0 \tag{45}$$

$F^{-1}(\alpha)$ is also termed a reliability factor *rf*. Numerically, if 80% of the constraint is satisfied i.e. $\alpha = 0.8$ (area covered by normal distribution), the corresponding $F^{-1}(\alpha)$ becomes 0.84. Now coming back to model equations, the uncertainty is kept at the biomass feed supply parameter $FedMax_{f,s,t}$ which varies depending on seasonality, water availability, climate, and geographical terrain (Eq. 15). Hence deterministic equivalent form i.e.

$$g\left[x, mean\left(FedMax_{f,s,t}\right) + F^{-1}(\alpha_w) * standard\,deviation\left(FedMax_{f,s,t}\right)\right] \geq 0 \tag{46}$$

The data collection of model is not shown due to space limitations but can be made available if required.

4 Results and Discussion

The developed model is complex, NP-hard, and follows mixed-integer linear programming (MILP) formulation which is solved using general algebraic modeling system (GAMS) programming latest version 38.3.0 on i7 processor, 32 Gb ram system. Further complexity can be understood as the model contains 55,757 single equations, 51,346 single variables, 28,962 discrete variables, and 171,070 non-zero variables.

4.1 Deterministic Versus Stochastic Study—Techno-economical Approach

Table 1 shows the comparison between deterministic case and stochastic case with 70% constraint satisfaction (reliability factor $rf = 0.52$). The value of 70% was chosen after performing the reliability factor range (0–100) effect on NPV to find the NPV maximum at this value (see Fig. 3). It can be observed that the total number of supply chain sites for stochastic handling case is 374 (28.04%) less than deterministic case 520, which logically implies lower infrastructure cost (52.89% less) for sites i.e. ₹ 7.1042E+10 wrt ₹ 1.5079E+11. Because of the lower infrastructure cost and its corresponding lower count, more inventory is stored at the distributor and more transport happens across the existing sites leading to higher inventory costs and transport costs to meet the rising demand. The results show less production cost in the uncertainty case indicating less manufacturing which in turn increases the externally imported product by 3.06 times to satisfy the demand (see Fig. 4). Even though the import cost is higher, the overall operating expenditure of ₹ 1.5875E+11 is still less than the deterministic case of ₹ 1.5908E+11. This happens majorly due to infrastructure and production cost combination having a considerably lower cost effect (see Fig. 4). Overall profit across the time horizon for uncertainty handling case ₹ 2.0520E+12 is found better than (2% more than) deterministic case ₹ 2.0120E+12.

Table 1 Comparing uncertainty versus determinist case in tabular form

	Stochastic ($\alpha = 70\%$)	Deterministic
Total site selected (count)	374	520
Transport cost (₹)	2.4212E+10	2.2724E+10
Inventory cost (₹)	1.6478E+09	5.3888E+08
Infrastructure cost (₹)	7.1042E+10	1.5079E+11
Production cost (₹)	3.9248E+10	1.0518E+11
Import cost (₹)	9.3620E+10	3.0614E+10
Operating expenditure (₹)	1.5875E+11	1.5908E+11
Capital expenditure (₹)	7.1042E+10	1.5079E+11
Earning (₹)	4.8619E+12	4.8536E+12
Depreciation (₹)	4.1852E+09	8.8832E+09
NPV (₹)	2.0520E+12	2.0120E+12
GHGe (tCO_2e)	3.7970E+06	7.0756E+06
GHGe revenue (₹)	7.0431E+11	6.9587E+11

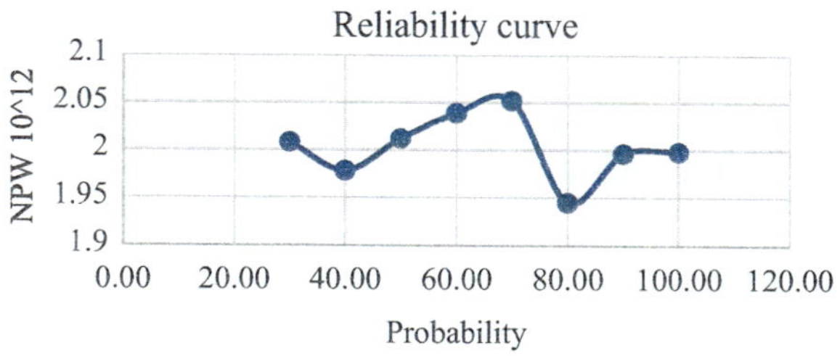

Fig. 3 Variation of profit NPV wrt extent of constraint satisfaction

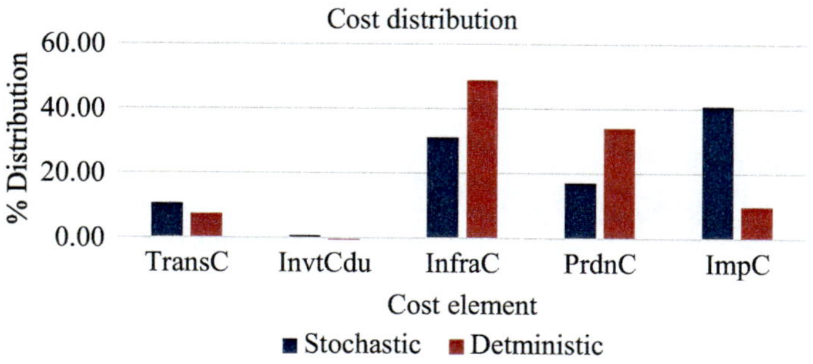

Fig. 4 Cost component comparison and distribution across SC for stochastic and deterministic case in %

4.2 Environmental GHGe Calculation

Coming to the environment component via greenhouse gas emission calculation across the life cycle assessment (LCA) using the well to tank approach, it is found that the stochasticity handling case generates less pollution i.e. 3.7970E+06 (46.34%) tCO_2e wrt deterministic case 7.0756E+06 tCO_2e (see Table 1). The main reason is the decreased value of production $Pm_{p,m,t}$ contribute to fewer emissions (see Eq. 30) with better GHGe saving and higher GHGe revenue ₹ 7.0431E+11 wrt 6.9587E+11.

4.3 Threshold Value

From Fig. 3, it can be observed that the constraint satisfaction level for feed supply was varied for the NPV values. Here it is found that when constraint satisfaction reaches 20% (Eq. 46) i.e. $rf - 0.84$, the MILP model gives an integer infeasible solution. This sets the biomass threshold value of 20% and above for probability to get satisfied for economic feasibility across the SC for the industrial and commercial sectors to invest in the biofuel domain.

5 Conclusion

A stochastic NP-hard time series-based MILP-based multi-feed, multi-echelon, multitime period bio supply chain network has been developed which tries to achieve sustainability via the combined effect of technical, economical, and environmental aspect. The external import of bioethanol is blended with indigenous bioethanol to maintain the quantity and octane number quality simultaneously even though per unit import cost is considerable to avoid loss of higher retail revenue and customer satisfaction level. The biomass uncertainty is handled using CCP via the probabilistic distribution of data and its deterministic equivalent form to be solved using optimization. Detailed cost distribution is studied for both stochastic and deterministic cases where the stochastic case is found to be advantageous (2% more profit NPV). Further, the GHGe study shows a 46.34% reduction in pollution due to CCP usage wrt deterministic case. Overall, it is found that the threshold value of 20% and above biomass constraint satisfaction is necessary for the SC project to run successfully at its minimum profit level.

Appendix: Nomenclature

The nomenclature section along with Sect. 2 having modeling equations explains the acronym used. In Eq. 1, Ω shows the factor of discount for inculcating the money's time value, Ψ shows the goods and service (GST) tax imposed on biofuel, i shows interest rate as a fraction of depreciation per annum, N shows the total time period horizon involved. Next, t show the annual time period across the project time horizon. In Eqs. 6–13, X shows the connectivity binary variable $\{0, 1\}$ where it is followed by supply chain layer name for e.g. s as suppler, m as manufacturer, d as distributor, im as importer, r as the retailer, and g as the set of all locations used across all the equations. With similar logic, Y shows the binary variable $\{0, 1\}$ for site location, and in all equations Q shows the quantity transferred. Min shows the minimum value and Max shows the maximum value. Further, subscript p shows the product; f shows the feed; l shows the transport medium; f shows the feed. Further in Eqs. 34–38, U shows the unit cost value. This way reader can comprehend the acronym meaning by combining Sect. 2 details and Appendix Nomenclature.

References

1. Natarajan K et al (2015) Biomass resource assessment and existing biomass use in the Madhya Pradesh, Maharashtra, and Tamil Nadu states of India. Challenges 6(1):158–172. https://doi.org/10.3390/challe6010158
2. de Jong S, Hoefnagels R, Wetterlund E, Pettersson K, Faaij A, Junginger M (2017) Cost optimization of biofuel production—the impact of scale, integration, transport and supply chain configurations. Appl Energy 195:1055–1070. https://doi.org/10.1016/j.apenergy.2017.03.109
3. Carvajal J, Sarache W, Costa Y (2019) Addressing a robust decision in the sugarcane supply chain: introduction of a new agricultural investment project in Colombia. Comput Electron Agric 157:77–89. https://doi.org/10.1016/j.compag.2018.12.030
4. Khishtandar S (2019) Simulation based evolutionary algorithms for fuzzy chance-constrained biogas supply chain design. Appl Energy 236:183–195. https://doi.org/10.1016/j.apenergy.2018.11.092
5. Atashbar NZ, Labadie N, Prins C (2018) Modelling and optimisation of biomass supply chains: a review. Int J Prod Res 56(10):3482–3506. https://doi.org/10.1080/00207543.2017.1343506
6. Dey B, Roy B, Datta S, Singh KG (2021) Comprehensive overview and proposal of strategies for the ethanol sector in India. Biomass Convers Biorefin
7. Gumte KM, Mitra K (2019) Bio-supply chain network design to tackle ethanol deficiency in India: a mathematical framework. J Clean Prod 234:208–224. https://doi.org/10.1016/j.jclepro.2019.06.160

Technique of Utilization of Coal Waste in an Efficient and Effective Way

Shivanchal Mishra, Subodh Ranjan Vajesnayee, and Nand Kumar Tiwari

1 Introduction

India is vast country and the consumption of energy is also at large scale. This energy requirement is fulfilled by various alternative sources like coal based thermal power plant, solar energy, wind energy, hydraulic energy, nuclear power plant, among all mentioned sources most of the energy consumption is fulfilled by coal based thermal power plant by which large amount of fly ash is produced, and create lots of problem like requirement of dumping area and large environmental hazardous problems like air pollution and leaching. Generally, there are 3 types of waste materials produced from thermal power plants: (I) bottom ash, (II) fly ash, and (III) pond ash. Bottom ash is heavy in weight. Due to its self weight, it cannot blow with air and settle down in bottom of boiler. Fly ash is light weight material which blows with air and these light weighted and small particles are captured by electro-static precipitator. When these two ashes are mixed together and after mixing of both types of ashes, this mixer is carried by water and dumped at a place, called 'Pond Ash'. Generally, volume of pond ash is larger than other type of ashes (Fly ash and bottom ash).

Lots of transportation network are developed in India. In several roads network fly ash is also used as an embankment-fill material. Various researches are already done for utilisation of fly ash: Ram and Mohanty [1] find out the geotechnical properties of fly ash and various methods to utilisation of fly ash. Chandra and Krishnaiah [2] is utilised fly ash in red mud to stabilized red mud. Mahdi et al. [3] finds out tensile strength, flexural strength of fly ash with an additive ash of rice husk ash (RHA) and

S. Mishra (✉) · S. R. Vajesnayee · N. K. Tiwari
Department of Civil Engineering, NIT Kurukshetra, Kurukshetra, Haryana 136119, India
e-mail: shivanchalmishra83@gmail.com

S. R. Vajesnayee
e-mail: sranjan@nitkkr.ac.in

N. K. Tiwari
e-mail: nand@nitkkr.ac.in

S. Doolla et al. (eds.), *Advances in Clean Energy and Sustainability*,
Green Energy and Technology, https://doi.org/10.1007/978-981-99-2279-6_64

stabilization of fly ash. Anand and Sarkar [4] find out bearing ratio (CBR) of fly ash then fly ash can utilised as an embankment fill material. Turan et al. [5] is worked on utilisation of fly ash and stabilization of clay by fly ash. Li et al. [6] and Taveriyanto et al. [7] is worked on utilisation of fly ash. All these studies are emphasised on that with the utilisation of fly ash: cost of embankment construction is reduced and scarcity of availability of conventional material like clay is reduced and requirement of disposal problem of fly ash is also resolved.

There are two types of Fly ash: (I) Class-C, and (II) Class-F. This classification is based on calcium content available in fly ash. Class-C fly ash contains large amount of calcium content (CaO) so it possess highly reactive nature. It can react with water either lime to be added or not added. The percentage of calcium content in Class-F fly ash is low when compared with Class-C fly ash. There are various researches are already done for improvement of strength of fly ash. Some researches are done on stabilization of fly ash with the help of bentonite. Another additive like gypsum can used to improve strength of Class-F fly ash [8].

In this paper, compaction characteristics (optimum moisture content (OMC) and Maximum dry density (MDD)) and bearing ratio (CBR) of fly ash stabilized with an additive (RHA) (3–15%) are find out and also find its suitability for embankment-fill material for road construction. For this purpose, there are two type of tests conducted: (I) standard proctor test (SPT) for compaction characteristics: optimum moisture content (OMC) and maximum dry density (MDD) are find out for various proportion of additive (RHA), (II) CBR test: CBR values are determined for various proportions of additive (RHA) and also analysis the impact of soaking on CBR value of fly ash with different–different proportions of additive (RHA).

After above literature review, the following objectives are set-out:

(I) To find out compaction characteristics (OMC and MDD) of fly ash stabilized with additive (RHA),
(II) Analysis the influence of RHA on bearing ratio (CBR) of the stabilized fly ash,
(III) Analysis the impact of curing on CBR value of the stabilized fly ash,
(IV) To find out suitability of fly ash as an embankment-fill material for pavement construction.

2 Materials

Three types of material are used in this report, are mentioned in this section which is given below:

Table 1 The geotechnical properties of fly ash

Sr. No.	Geotechnical properties	Value
1	D_{10}	0.065 mm
2	D_{30}	0.0875 mm
3	D_{60}	0.1118 mm
4	C_u	1.97
5	C_c	1.08
6	LL and PL	Non-plastic
7	OMC	32.50%
8	$Y_{d, max}$ (gm/cm^3)	1.079
9	G	1.861

2.1 *Fly Ash*

Fly ash, collected from Coal based thermal power plant, Jind road, Assan, Panipat, Haryana. The geotechnical characteristics of fly ash are given below in Table 1:

2.2 *Rice Husk Ash (RHA)*

RHA, used in this paper has been collected from Jain rice mill, Kurukshetra, Haryana. It is greyish-black in colour, porous in nature with very high specific surface area and light in weight. It is obtained from rice mill at burning temperature of 550–800 °C.

3 Test Procedures and Experimental Program

For analysis of the geotechnical characteristics of fly ash, and also determined CBR value of fly ash with an additive (RHA), the experiment work are performed in following steps: (I) Geotechnical properties of fly ash (by grain-size distribution, specific gravity test), (II) standard proctor test (SPT) of fly ash alone and with an additive (RHA), find out OMC as well as MDD, (III) CBR test of fly ash with various proportion of additive (RHA) for both soaked and unsoaked condition.

The various proportions of rice husk ash (RHA) are added into the fly ash, the proportions of RHA are 3.0, 6.0, 9.0, 12.0, 15.0% and to analysis the curing, curing periods are considered 4, 7, 14 days. For standard proctor test (SPT), there are total six different proportion of additive (RHA) are added in fly ash then find out MDD and OMC for each proportion. For CBR test total 6 different unsoaked test and 18 soaked tests (4, 7, and 14 days) are conducted.

3.1 Moisture Density Relationships of Stabilised Fly Ash

In many cases, the geotechnical properties of fly ash are specified in terms of MDD and OMC. As per requirement of field, heavy or light proctor tests are conducted. Subgrade should be compacted on 95% of MDD for road and bridge works. In this paper, light standard proctor test (SPT) (IS2720-part 7) is conducted and to find out the value of MDD and optimum OMC of fly ash with different proportions of additive (RHA).

3.2 California Bearing Ratio (CBR) Test

CBR value is an important parameter for sub-grade or flexible pavements design. In case of flexible pavement design with the help of CBR value of subgrade, thickness of pavement can find out. To analysis bearing ratio of fly ash with an additive (RHA), unsoaked as well as soaked both type of tests are conducted then also find out suitability of fly ash with an additive (RHA) as a pavement embankment-fill material, CBR test is conducted according to IS2720: part-16. The sample is soaked in water for different-different soaking periods to determine the impact of curing periods on CBR value of sample. Two surcharge weights each of them 2.5 kg are put on specimen at the time of soaking and testing. The mould of CBR test has diameter of 15.2 cm and height of 17.8 cm. The specimen for CBR is prepared at OMC and MDD which is obtained from standard proctor test (SPT). After proper compaction and preparation of specimen, mould is sealed for curing process. The prepared specimen is cured in humidity control chamber at relative humidity > 95% and at a temperature of 30 ± 1 °C. The specimen is cured for 4, 7, 14 days to analysis the impact of curing on CBR value. After curing, tests are performed on mechanical loading frame. Penetration value by vertical dial gauge and proving ring values are note down, movable plate is moving with a constant rate of 1.2 mm/min. The diameter of piston which is attach with proving ring is 5.0 cm. Reading of proving ring and dial gauge are note down very carefully.

4 Results of Experiment

The experiment results of standard proctor test (SPT) and California bearing test (CBR) are shown below:

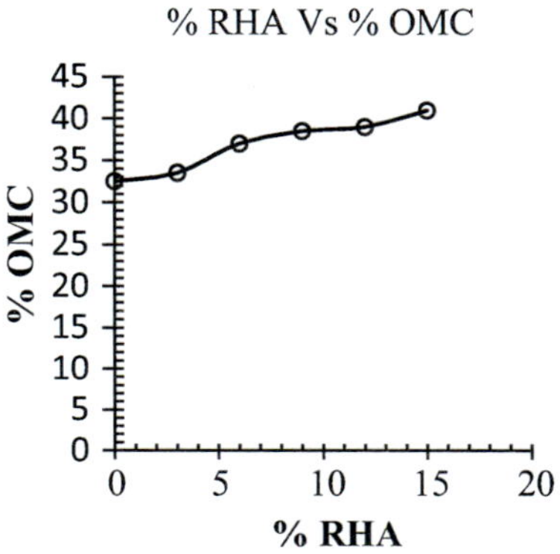

Fig. 1 The effect of proportion of RHA on OMC

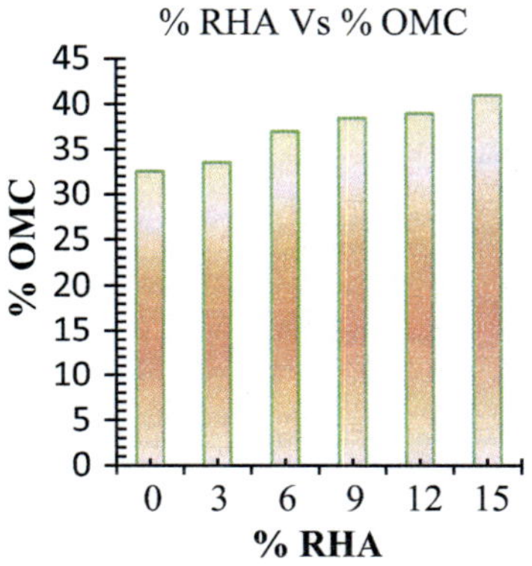

Fig. 2 The effect of proportion of RHA on OMC

4.1 Standard Proctor Test (SPT) Results

OMC and MDD are determined from SPT of the fly ash, stabilized with different proportion of additive (3.0, 6.0, 9.0, 12.0, and 15.0%) is shown in Figs. 1, 2, 3 and 4. The value of OMC of RHA is increased because of RHA has high specific surface area so RHA can absorbed large quantity of water and increase in OMC of fly ash and there is no time given for pozzalonic reaction, MDD value is reduced significantly. The values of OMC are varied from 32.5 to 41% and the values of MDD are varied from 1.079 to 0.93 gm/cm^3.

4.2 CBR Test

The CBR values of fly ash, stabilized with an additive (RHA) are discussed in this section for both unsoaked and soaked conditions. The CBR values are shown in Figs. 5, 6, 7 and 8 at different curing periods (4, 7, 14 days). The increase in CBR

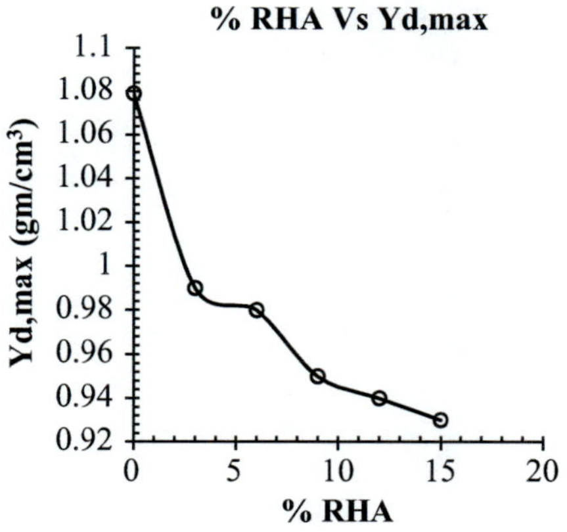

Fig. 3 The impact of various proportion of RHA on MDD of fly ash

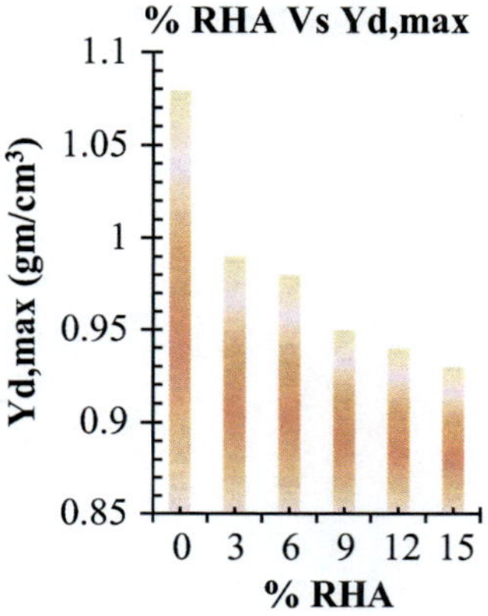

Fig. 4 The impact of various proportion of RHA on MDD of fly ash

values are observed due to cementitious properties of RHA and fly ash take place. For 3–12% RHA, the value of CBR is increased and CBR value is decreased for 12–15% of RHA. The values of CBR are significantly increased as curing periods (4, 7, 14 days) increased.

5 Summery and Conclusions

The SPT and CBR tests are conducted within geotechnical lab. The values of MDD and MDD are found out from SPT and the values of CBR are obtained from CBR test with 0, 4, 7, 14 days of curing. Both type of CBR test are conducted unsoaked and soaked test. The following conclusions are obtained from analysing the experiment data:

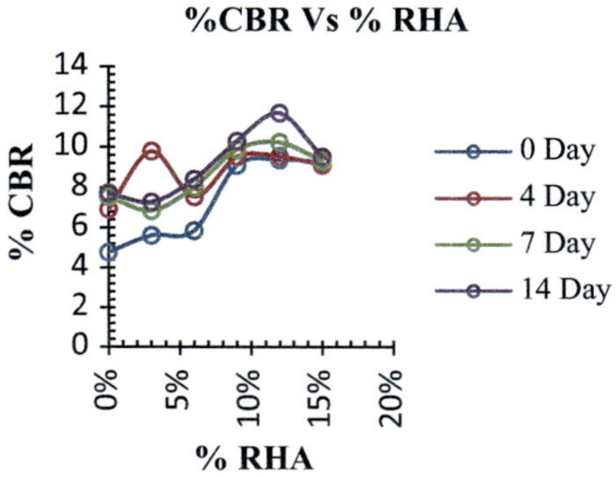

Fig. 5 The effect of RHA on CBR value of fly ash

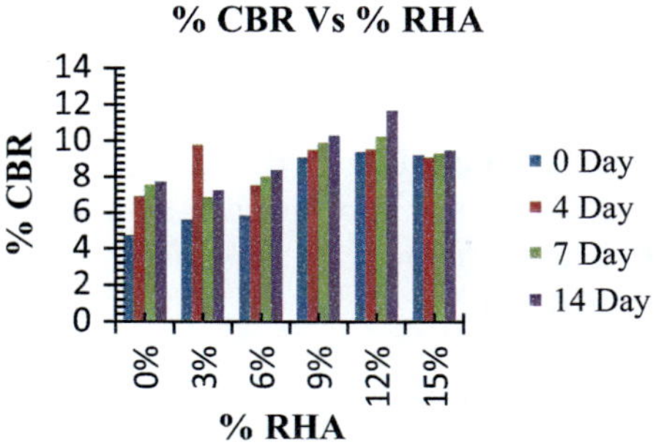

Fig. 6 The effect of RHA on CBR value of fly ash

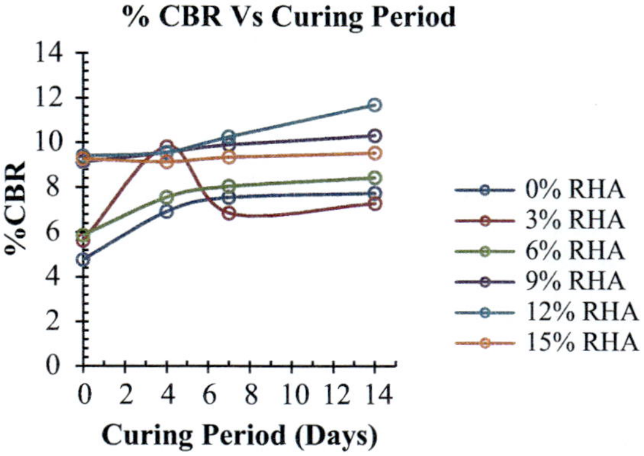

Fig. 7 The impact of curing on CBR value of fly ash

The values of OMC are increased from 32.50 to 41% when proportion of RHA is increased. The maximum value of optimum moisture content is obtained 41% at 15% rice husk ash.

- The values of maximum dry density are decreased from 1.079 to 0.93 gm/cm^3 when proportion of rice husk ash increased. The maximum value of MDD is determined 1.079 at 0% rice husk ash.

Fig. 8 The impact of curing on CBR value of fly ash

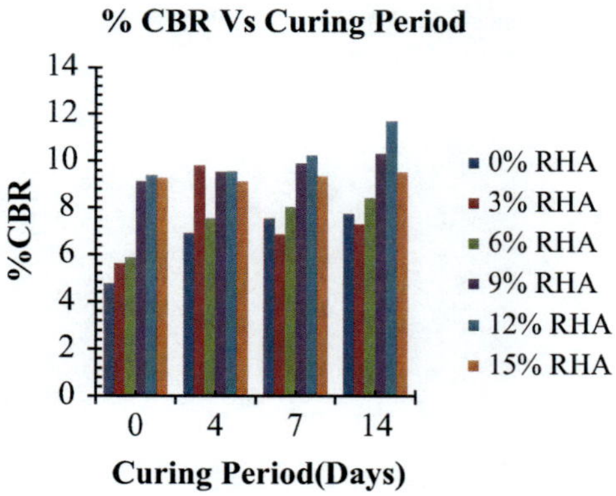

- Bearing ratio of fly ash stabilized with rice husk ash is increased at a certain limit with increased in proportion of rice husk ash (3.0, 6.0, 9.0, and 12%) and beyond 12% of rice husk ash the value of bearing ratio is decreased. The maximum value of CBR value is obtained 11.682% at 12% rice husk ash.
- Bearing ratio of fly ash is increased with increase in curing period (4, 7, and 14 days). The maximum value of CBR is obtained 11.682% at 14 days with 12% RHA.
- For construction of pavement, the value of CBR is atleast 10%, and coal based waste material fly ash with 12% RHA is perfect for overlaying material.

References

1. Ram AK, Mohanty S (2022) State of the art review on physiochemical and engineering characteristics of fly ash and its applications. Int J Coal Sci Technol 9(1):1–25
2. Chandra KS, Krishnaiah S (2022) Strength and leaching characteristics of red mud (bauxite residue) as a geomaterial in synergy with fly ash and gypsum. Transp Res Interdisc Perspect 13:100566
3. Mahdi SN, Hossiney N, Abdullah MMAB (2022) Strength and durability properties of geopolymer paver blocks made with fly ash and brick kiln rice husk ash. Case Stud Constr Mater 16:e00800
4. Anand A, Sarkar R (2022) Bearing capacity of spatially variable unsaturated fly ash deposit using random field theory. Curr Sci 122(5):542
5. Turan C, Javadi AA, Vinai R (2022) Effects of class C and class F fly ash on mechanical and microstructural behavior of clay soil—a comparative study. Materials 15(5):1845
6. Li X, Guo Y, Sharma R, Singh A, Zhang H, Zhang J, Fu Y (2022) Utilization of different grain size of municipal solid waste bottom ash in high-performance mortars. Sustainability 14(7):4263
7. Taveriyanto A, Andiyarto HTC, Saefullah MV, Haliza MP (2022) The utilization of gypsum board and fly ash waste on brick in terms of compressive strength to reduce environmental pollution. IOP Conf Ser: Earth Environ Sci 969(1):012012. IOP Publishing
8. Ghosh A, Subbarao C (2007) Strength characteristics of class F fly ash modified with lime and gypsum. J Geotech Geoenviron Eng 133(7):757–766

Enhancement of the Performance of Dye-Sensitized Solar Cell by Integrating with Ternary Photonic Crystal

J. R. Sofia and K. S. Joseph Wilson

1 Introduction

From the perspective of the increasing energy demand of the world and the continuously depleting conventional energy resources, a renewable and clean energy source such as solar energy is very much needed. Dye-sensitized solar cells (DSSCs) belonging to the third-generation solar cells, has attracted greater attention due to the cost-friendliness and easy manufacturing process. DSSCs are widely researched owing to their light weight, flexibility and ability to work under low-light conditions [1–4].

DSSC comprises of the transparent conducting oxide (TCO), semiconductor, dye sensitizer, electrolyte and counter electrode [1]. When the photons are incident on the solar cell, the dye molecules get excited. The photoelectrons generated, rapidly get transported by diffusion to the conduction band of the semiconductor and then to the TCO, which is collected by the external load. The redox electrolyte compensates the oxidized dye. Thus, photons are absorbed and electricity is generated.

Recent research on DSSC has achieved the device efficiency of 14.2% [5] and has to evolve for commercialization. Various studies reported, have shown the efficiency of DSSCs can be increased by combating the challenges of the larger electron recombination rates and the optical losses such as reflective losses and lesser absorption [2, 6]. The performance of the DSSC can be improved by adopting light trapping strategies such as usage of photonic crystal based Anti-reflective coatings (ARC) and Distributed Back Reflectors (DBR) to help in the localization of light within the solar cell [7, 8].

Photonic crystals are periodic structures of alternative high and low refractive indices, which control the propagation of electromagnetic waves incident on them.

J. R. Sofia (✉) · K. S. Joseph Wilson
PG and Research Department of Physics, Arul Anandar College, Karumathur, Madurai, India
e-mail: jrsofia@aactni.edu.in

S. Doolla et al. (eds.), *Advances in Clean Energy and Sustainability*,
Green Energy and Technology, https://doi.org/10.1007/978-981-99-2279-6_65

The photonic band gap of the photonic crystals depends on the refractive indices of the layers and on the structural differences of the photonic crystal [9].

Depending on the reflection and transmission coefficients of the photonic crystal, they can be employed as ARC or DBR. When photonic crystal is used as back reflector at the rear side of the solar cells, the photons leakage is reduced and enhancement in the light absorption of solar cells is reported. Thus, optical losses suffered by the solar cells can be mitigated by the use of photonic crystals [10–13].

Ternary photonic crystals of structure $(ABC)^N$ are made up of three layers of different refractive indices arranged periodically. Research reports have shown that ternary photonic crystals have wider photonic band gap (PBG) in the visible region of the spectrum [14–21]. Also, ternary photonic crystals show an advantage of generating this PBG with lesser number of periods. In this work, one dimensional ternary photonic crystal (1D-TPC) is used as DBR in DSSC and the improvement in the light absorption is calculated and hence the improvement in J_{sc} of the DSSC is estimated.

2 Theory and Calculations

The integrated system comprises of DSSC with 1D Ternary Photonic crystal (1D-TPC), used as DBR as shown in Fig. 1. The DSSC consists of the ruthenium complex dye adsorbed TiO_2 as photoelectrode, potassium iodide electrolyte, with platinum as the counter electrode. The 1D Ternary Photonic crystal is of the structure Dielectric-Semiconductor-Dielectric, comprising of SiO_2–n-Si–TiO_2 layers. The transfer matrix method (TMM) is employed in the quantifying the absorbance of the DSSC with and without the 1D-TPC.

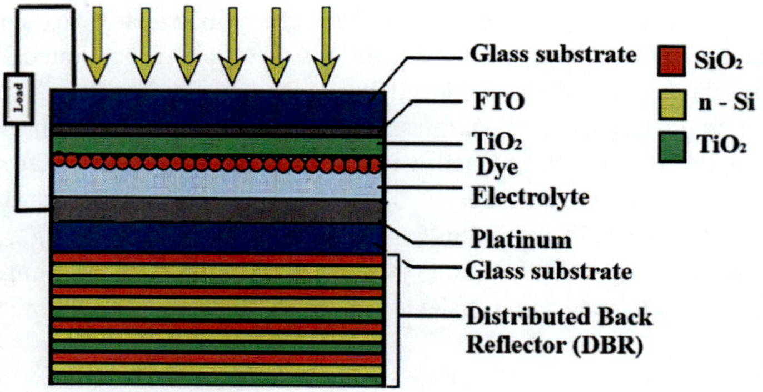

Fig. 1 Schematic diagram of the DSSC integrated with 1D-TPC

2.1 Refractive Indices of the Layers of the DSSC

The following DSSC parameters are employed in this work. The refractive index and thickness of the front and back glass substrates are $n_{sub} = 1.51$ and $d_{sub} = 0.5$ cm. The refractive index and thickness of FTO, TiO_2, electrolyte and platinum are given by $n_{FTO} = 1.80$ and $d_{FTO} = 400$ nm; $n_{TiO2} = 1.95$ and $d_{TiO2} = 15$ μm; $n_{electrolyte} = 1.433$, and $d_{electrolyte} = 50$ μm, $n_{platinum} = 2.33$ and $d_{platinum} = 200$ nm respectively.

The composite refractive index n_{PE} of dye adsorbed photoelectrode of the DSSC can be written as [22]

$$n_{PE} = n_{TiO_2} + \beta(\lambda)\, j \tag{1}$$

The real part of the n_{PE} is determined by the TiO_2, while the imaginary part $\beta(\lambda)$ is determined by the ruthenium dye adsorbed into the TiO_2 matrix. The imaginary part of the refractive index n_{PE}, i.e., $\beta(\lambda)$ is frequency dependent and is given by

$$\beta(\lambda) = \beta_0 \exp\left(1 - z(\lambda) - e^{-z(\lambda)}\right);\ \text{where } z(\lambda) = \frac{\lambda - \lambda_0}{\Delta\lambda} \tag{2}$$

The parameter β_0 represents the degree of dye loaded in the TiO_2 matrix, and the values of β_0 are assumed to be 0.004, 0.0055 and 0.008 which are used to analyze the results for different levels of dye adsorption in the DSSC. The parameter $z(\lambda)$ is dependent on the ruthenium complex and the values of $\lambda_0 = 538$ nm and $\Delta\lambda = 64.16$ nm are used [22].

2.2 Calculation of Refractive Index of n-Si Layer Using Plasma Model

The 1D-TPC, used as DBR, comprises of SiO_2–n-Si–TiO_2 layers, with the refractive indices and thicknesses as $n_{SiO2} = 1.43$ and $d_{SiO2} = 50$ nm and $n_{TiO2} = 1.92$ and $d_{TiO2} = 45$ nm. The refractive index of the n-Si is estimated using the plasma model.

The dielectric function of the semiconductor (n-Si) described by the plasma model is given by [23]

$$\varepsilon_{n-Si}(\omega) = \varepsilon_\infty \left(1 - \frac{\omega_{pe}^2}{\omega^2 - \frac{j\omega}{\tau_e}} - \frac{\omega_{ph}^2}{\omega^2 - \frac{j\omega}{\tau_h}}\right) \tag{3}$$

where, ε_∞ is high-frequency dielectric constant.

The scattering lifetimes of the holes τ_h and electrons τ_e in a doped semiconductor is given by

$$\tau_e = \frac{m_e \mu_e}{e};\ \tau_h = \frac{m_h \mu_h}{e} \tag{4}$$

Table 1 Parameters used to calculate the refractive index of n-Si layer

Parameters	Values
ε_∞	11.7
N_i	$1.5 \times 10^{16} m^{-3}$
m_e	$0.26 m_0$
m_{hh}	$0.49 m_0$
m_{lh}	$0.16 m_0$
m_0	9.1×10^{-31} kg
ε_0	$8.854 \times 10^{-12} N^{-1}C^2m^{-2}$
N_d	$1.0 \times 10^{21} m^{-3}$

where μ_e and μ_h are electron and hole mobilities and 'e' is the electronic charge.

The mobility of the electrons and holes, μ_e and μ_h are written as

$$\mu_e = 92 + \frac{1268}{1 + \left(\frac{N_d}{1.3 \times 10^{17}}\right)^{0.91}};\ \mu_h = 54.3 + \frac{406.9}{1 + \left(\frac{N_d}{2.35 \times 10^{17}}\right)^{0.88}} \tag{5}$$

The electron and hole plasma frequencies, ω_{pe} and ω_{ph} can be written as

$$\omega_{pe} = \left(\frac{n_e e^2}{m_e \varepsilon_\infty \varepsilon_0}\right)^{1/2};\ \omega_{ph} = \left(\frac{n_h e^2}{m_h \varepsilon_\infty \varepsilon_0}\right)^{1/2} \tag{6}$$

The effective mass of the holes m_h is calculated from effective masses of the light holes m_{lh} and heavy holes m_{hh} as

$$m_h = m_{hh}\frac{1 + r^{3/2}}{r + r^{3/2}};\ \text{where, } r = m_{hh}/m_{lh} \tag{7}$$

The electron concentration n_e and hole concentration n_h of the semiconductor matrix after addition of the dopants is given by,

$$n_e = \sqrt{N_i^2 + \frac{N_d^2}{4}} + \frac{N_d}{2};\ n_h = \sqrt{N_i^2 + \frac{N_d^2}{4}} - \frac{N_d}{2} \tag{8}$$

where N_i is the intrinsic carrier density and N_d is the donor density. The refractive index of the n-Si is $n_{n-Si} = \sqrt{\varepsilon_{n-Si}}$ (Table 1).

2.3 Transfer Matrix Method

According to the Transfer Matrix Method (TMM), the transfer matrix of each of the layer is

$$M_i = \begin{bmatrix} \cos\Delta_i & \frac{j}{n_i}\sin\Delta_i \\ jn_i\sin\Delta_i & \cos\Delta_i \end{bmatrix} \tag{9}$$

where the phase difference $\Delta_i = \left(\frac{2\pi}{\lambda}\right)n_i d_i \cos\theta_i$ in which n_i, d_i and θ_i are the refractive index, thickness of ith layer in the system and angle of incidence at the interface.

The transfer matrix of the 1D-TPC, comprising of SiO_2–n-Si–TiO_2 layers is defined as

$$M_{1D\text{-}TPC} = M_{SiO_2} \cdot M_{n\text{-}Si} \cdot M_{TiO_2} \tag{10}$$

The transfer matrix of the DSSC without the 1D-TPC, comprising of the glass substrate, FTO, dye adsorbed TiO_2 as photoelectrode, electrolyte, platinum as counter electrode and glass substrate is defined as

$$M_{DSSC} = M_{sub} \cdot M_{FTO} \cdot M_{TiO_2+dye} \cdot M_{electrolyte} \cdot M_{platinum} \cdot M_{sub} \tag{11}$$

The transfer matrix of the DSSC with the 1D ternary photonic crystal, comprising of the glass substrate, FTO, dye adsorbed TiO_2 as photoelectrode, electrolyte, platinum as counter electrode and glass substrate followed by the N-periods of the ternary photonic crystal is defined as

$$\begin{aligned} M_{DSSC\text{-}1D\text{-}TPC} = {} & M_{sub} \cdot M_{FTO} \cdot M_{TiO_2+dye} \cdot M_{electrolyte} \cdot M_{platinum} \cdot M_{sub} \\ & \cdot \left(M_{SiO_2} \cdot M_{n\text{-}Si} \cdot M_{TiO_2}\right)^N \end{aligned} \tag{12}$$

The system can be represented by the matrix

$$M_{total} = \begin{bmatrix} m_{11} & m_{12} \\ m_{21} & m_{22} \end{bmatrix} \tag{13}$$

Using the total matrix, the Reflectance (R) and Transmittance (T) of the system can be estimated by

$$R = |r|^2,\ r = \frac{n_1 m_{11} + n_1 n_s m_{12} - m_{21} - n_s m_{22}}{n_1 m_{11} + n_1 n_s m_{12} + m_{21} - n_s m_{22}} \tag{14}$$

$$T = \mathrm{Re}\left(\frac{n_s}{n_1}\right)|t|^2,\ t = \frac{2n_1}{n_1 m_{11} + n_1 n_s m_{12} + m_{21} - n_s m_{22}} \tag{15}$$

where n_1 and n_s are the refractive indices of first and last layers of the system.

The Absorbance (A) of the system can be calculated from

$$A = 1 - R - T \tag{16}$$

Initially, the reflectance of the 1D ternary photonic crystal is estimated (by using $M_{total} = M_{1D\text{-}TPC}$). The reflectance and transmittance of the DSSC is calculated, (by using $M_{total} = M_{DSSC}$) and the absorbance of the DSSC is estimated using Eq. (16). Then, the reflectance and transmittance of the DSSC with the 1D-TPC is calculated (using $M_{total} = M_{DSSC\text{-}1D\text{-}TPC}$) and absorbance of the DSSC integrated with the 1D-TPC is calculated.

2.4 Analysis of Short-Circuit Current Density (J_{sc})

As the PBG of the 1D-TPC lies in the visible region of the spectrum, when such photonic crystal is used as DBR in the solar cell, it reflects back the photons which were not absorbed and tend to leak through the DSSC. Thus, the DBR considerably reduces the optical losses by the localization of photons within the solar cell.

The Light Harvesting efficiency (LHE) can be written as

$$LHE = A = 1 - R - T \tag{17}$$

The relation between the LHE of the DSSC and Incident Photon-to Current Efficiency (IPCE) is given by

$$IPCE(\lambda) = \phi(\lambda)\,\xi(\lambda)\,LHE(\lambda) \tag{18}$$

The short-circuit current density (J_{sc}) of the solar cell is given by

$$Jsc = \int q\,IPCE(\lambda)\,F(\lambda)\,d\lambda \tag{19}$$

where $\phi(\lambda)$ is the electron-transfer yield, $\xi(\lambda)$ is the charge collection efficiency and $F(\lambda)$ is the ratio of the solar spectral irradiance and the photon energy.

The improvement in short-circuit current density (J_{sc}) of DSSC attained by coupling of the ID Photonic crystal is given by [19, 22, 24]

$$\Delta J_{sc} = \frac{\int q\phi(\lambda)\xi(\lambda)LHE_{DSSC\text{-}1D\text{-}TPC}(\lambda)F(\lambda)d\lambda - \int q\phi(\lambda)\xi(\lambda)LHE_{DSSC}(\lambda)F(\lambda)d\lambda}{\int q\phi(\lambda)\xi(\lambda)LHE_{DSSC}(\lambda)F(\lambda)d\lambda} \tag{20}$$

LHE_{DSSC} (λ) is the LHE of the DSSC without the 1D-TPC. $LHE_{DSSC\text{-}1D\text{-}TPC}$ (λ) is the LHE of the DSSC integrated with the 1D-TPC. By assuming that $\phi(\lambda)$ and $\xi(\lambda)$ are equal to 1, the improvement in the absorption of photons is analyzed and hence, the increase in J_{sc} of DSSC is studied.

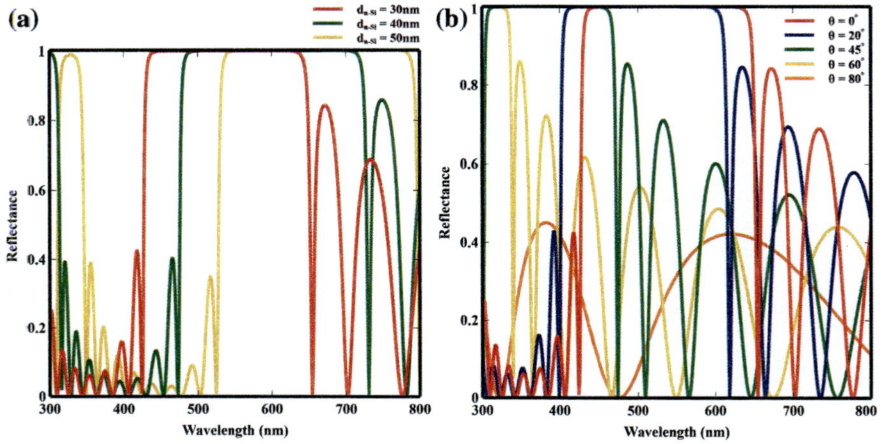

Fig. 2 **a** Reflectance of the 1D-TPC with varying thickness of 'n-Si' layer, **b** reflectance of the 1D-TPC with varying incident angles

3 Results and Discussion

3.1 Reflectance of 1D-Ternary Photonic Crystal

By employing the transfer matrix method as defined by the Eqs. (9–16), the individual matrices of layers SiO_2–n-Si–TiO_2 are defined and the reflectance of the 1D-TPC has been calculated. Since, the photonic band gap of the 1D-TPC lies in the visible region, it facilitates the 1D-TPC be used as DBR in DSSC. When the thickness of the 'n-Si' layer of the photonic crystal is varied, the PBG shifted towards the infrared region of the spectrum as in Fig. 2a.

When the reflectance of the 1D-TPC is plotted for various angles of incidence, the PBG of the photonic crystal shifted towards the ultraviolet region with increasing angle of incidence as shown in Fig. 2b. The reflectance for different density of dopants (N_d) in the n-Si layer of the photonic crystal is studied and is shown in Fig. 3. It is found the PBG slightly becomes narrower when N_d is increased.

3.2 Analysis of Absorbance of the DSSC with and Without DBR

The light absorption of the DSSC with and without the 1D-TPC is estimated by the TMM. The variation of light absorption with wavelength for different degrees of dye loading as represented by $\beta_0 = 0.004$, 0.005 and 0.008 is shown in Fig. 4a–c.

The scattered plots of absorbance resemble the IPCE graphs of theoretical study carried out by Lozano et al. [22].

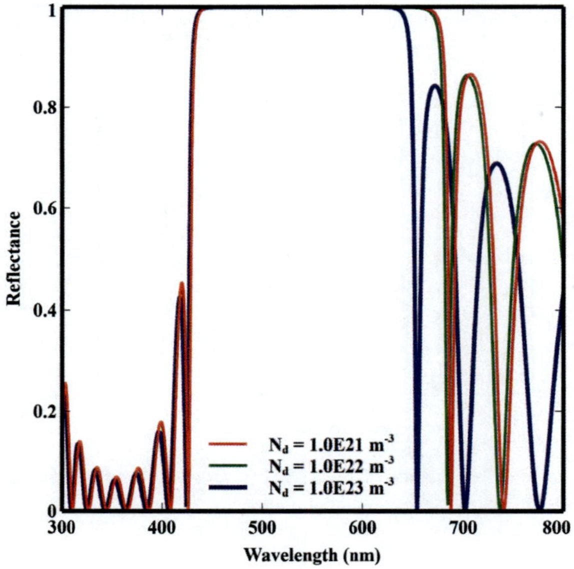

Fig. 3 Reflectance of the 1D-TPC with varying doping density (N_d) of n-Si layer

It is found that the light absorption increases with dye loading in the photoelectrode. The integration of Ag as DBR provides greater absorption of photons in the entire range of the visible spectrum as shown in Fig. 4a. But, the 1D-TPC as DBR provides significant improvement in the absorbance and exhibits a superior performance than Ag as DBR. This can be interpreted as the proposed DBR can play a vital role in improving the performance of DSSCs with lesser dye loading.

3.3 Enhancement in Short-Circuit Current Density (J_{sc}) of DSSC by Incorporating 1D Ternary Photonic Crystal

The improvement in the J_{sc} of the DSSC by the 1D-TPC is calculated by the Eqs. (17–20). With increase in the degree of dye loading (β_0), light absorption in the DSSC increases and hence short-circuit current density (J_{sc}) increases. When the 1D-TPC is integrated with the DSSC, a sudden improvement in the J_{sc} ($\Delta J_{sc}\%$) occurs. The variation of the $\Delta J_{sc}\%$ with number of periods of the 1D-TPC (n) is studied and is given in Fig. 5a. It is found that the $\Delta J_{sc}\%$ increases with number of periods and becomes constant for higher values. It is also found that the $\Delta J_{sc}\%$ depends on dye loading level. It shows with $\beta_0 = 0.004$, $\Delta J_{sc}\%$ gives the maximum value, which is nearly 150%. Hence, it is concluded that $\Delta J_{sc}\%$ becomes more effective for low dye loading level (β_0). By varying the angle of incidence, $\Delta J_{sc}\%$ of DSSC is calculated and plotted in Fig. 5b, which reveals that the $\Delta J_{sc}\%$ decreases abruptly with the angle of incidence.

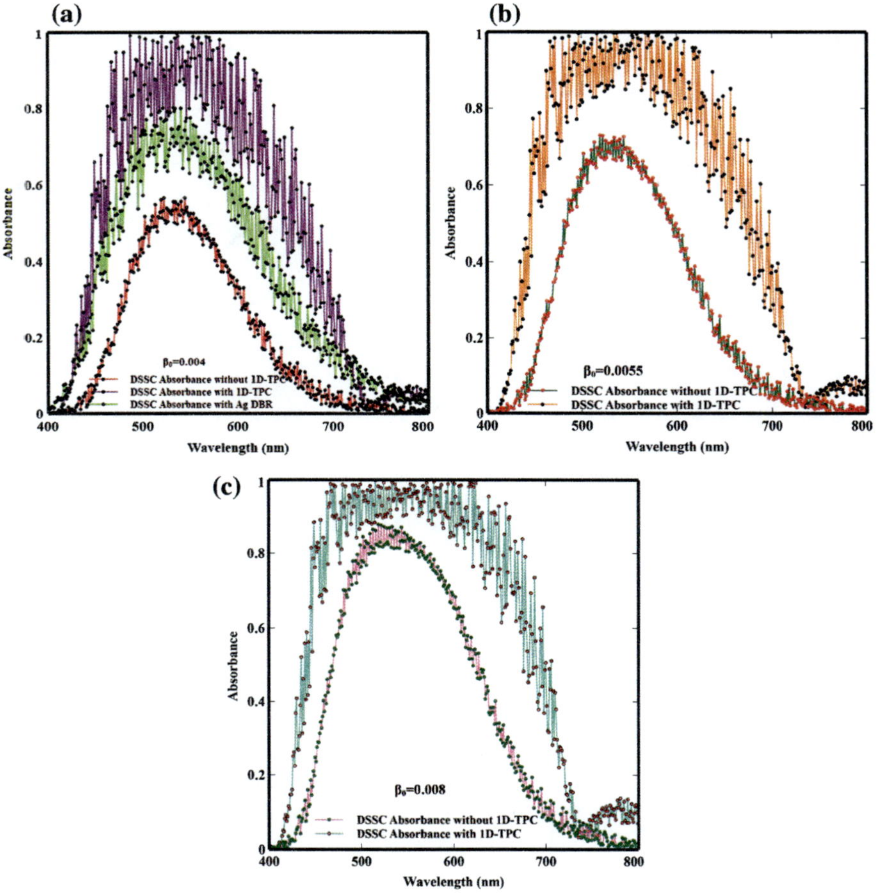

Fig. 4 **a** Absorbance of the DSSC without DBR, with 1D-TPC as DBR and silver as DBR for degree of dye loading $\beta_0 = 0.004$; **b** absorbance of the DSSC with and without 1D-TPC for degree of dye loading $\beta_0 = 0.0055$; **c** absorbance of the DSSC with and without 1D-TPC for degree of dye loading $\beta_0 = 0.008$

4 Conclusion

A theoretical integration of 1D-Ternary Photonic crystal comprising of SiO_2–n-Si–TiO_2 layers with DSSC has been carried out. By using transfer matrix method (TMM), it has been observed that the photonic band gap (PBG) of 1D-TPC lies in the visible region. Hence it has been utilized as DBR. From the absorbance calculations for the DSSC with and without the DBR, the short-circuit current density is estimated. A significant improvement in the short-circuit current density is observed for proposed DSSC with the 1D-TPC as DBR.

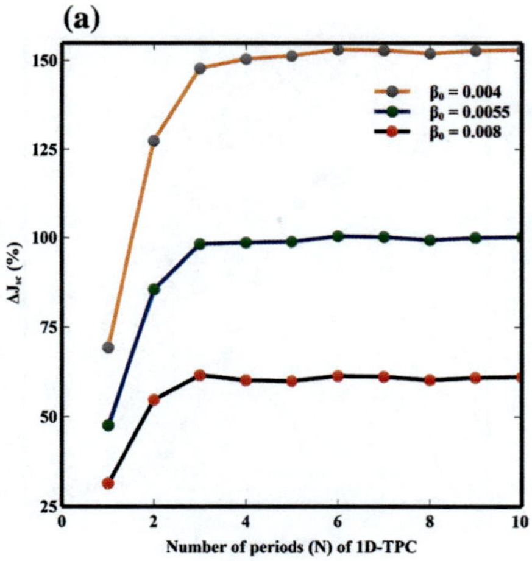

Fig. 5 **a** Improvement of J_{sc} with the number of periods (N) of the 1D-TPC of DSSC, **b** improvement of J_{sc} with different angle of incidence of the integrated system

References

1. O'Regan B, Grätzel M (1991) A low-cost, high-efficiency solar cell based on dye-sensitized colloidal TiO_2 films. Nature 353:737–740
2. Hardin BE, Snaith HJ, McGehee MD (2012) The renaissance of dye-sensitized solar cells. Nat Photonics 6(3):162–169
3. Fakharuddin A, Jose R, Brown TM, Fabregat-Santiago F, Bisquert J (2014) A perspective on the production of dye-sensitized solar modules. Energy Environ Sci 7:3952–3981
4. Bavarian M, Nejati S, Lau KKS, Lee D, Soroush M (2014) Theoretical and experimental study of a dye-sensitized solar cell
5. Ji J-M, Zhou H, Eom YK, Kim CH, Kim HK (2020) 14.2% efficiency dye-sensitized solar cells by co-sensitizing novel Thieno[3,2-b]indole-based organic dyes with a promising porphyrin sensitizer. Adv Energy Mater 10(15):2000124
6. Nishimura S et al (2003) Standing wave enhancement of red absorbance and photocurrent in dye-sensitized titanium dioxide photoelectrodes coupled to photonic crystals. J Am Chem Soc 125:6306–6310
7. Liu W, Ma H, Walsh A (2019) Advance in photonic crystal solar cells. Renew Sustain Energy Rev 116:109436
8. López-López C, Colodrero S, Calvo ME, Míguez H (2013) Angular response of photonic crystal-based dye-sensitized solar cells. Energy Environ Sci 6:1260–1266
9. Doghmosh N, Taya SA, Upadhyay A, Olaimat MM, Colak I (2021) Enhancement of optical visible wavelength region selective reflector for photovoltaic cell applications using a ternary photonic crystal. Optik Int J Light Electron Opt 243:16749
10. Curtin B, Biswas R, Dalal V (2009) Photonic crystal based back reflectors for light management and enhanced absorption in amorphous silicon solar cells. Appl Phys Lett 95:231102
11. Colodrero S, Mihi A, Anta JA, Ocana M, Míguez H (2009) Experimental demonstration of the mechanism of light harvesting enhancement in photonic-crystal-based dye-sensitized solar cells. J Phys Chem C 113(4):1150–1154
12. Florescu M, Lee H, Puscsu I, Pralle M, Florescu L, Ting DZ, Dowling JP (2007) Improving solar cell efficiency using photonic band-gap materials. Sol Energy Mater Sol Cells 91(17):1599–1610

13. Fathima MI, Joseph Wilson KS (2021) Efficiency enhancement in dye sensitized solar cell using 1D photonic crystal. Silicon 5
14. Pandey GN, Kumar N, Thapa KB, Ojha SP (2016) Reflectance properties of one-dimensional metal-dielectric ternary photonic crystal. AIP Conf Proc 1728:020310
15. Abadla MM, Tabaza NA, Tabaza W, Ramanujam NR, Wilson KJ, Vigneswaran D, Taya SA (2019) Properties of ternary photonic crystal consisting of dielectric/plasma/dielectric as a lattice period. Optik 185:784–793
16. Chang TH, Wu PH, Chen SH, Chan CH, Lee CC, Chen CC, Su YK (2009) Opt Express 17:6519
17. Zeng L, Yi Y, Hong C, Liu J, Feng N, Duan X, Kimerling LC, Alamariu BA (2006) Appl Phys Lett 89:111111
18. O'Brien P, Kherani NP, Zukotynski S, Ozin GA, Vekris E, Tetreault N, Chutinan A, John S, Mihi A, Mı´guez H (2007) Adv Mater 19:4177
19. Mihi A, Calvo ME, Anta JA, Mı´guez H (2008) J Phys Chem C 112:13
20. Ko DH, Tumbleston JR, Zhang L, Williams S, DeSimone JM, Lopez R, Samulski ET (2009) Nano Lett 9:2742
21. Prieto I, Galiana B, Postigo PA, Algora C, Martı´nez LJ, Rey-Stolle I (2009) Appl Phys Lett 94:191102
22. Lozano G, Colodrero S, Caulier O, Calvo ME, Míguez H (2010) Theoretical analysis of the performance of one-dimensional photonic crystal-based dye-sensitized solar cells. J Phys Chem C 114:3681–3687
23. Liu C-C, Wu C-J (2013) Near Infrared filtering properties in photonic crystal containing extrinsic and dispersive semiconductor defect. Prog Electromagn Res 137:359–370
24. Mihi A, Míguez H (2005) Origin of light-harvesting enhancement in colloidal-photonic-crystal-based dye-sensitized solar cells. J Phys Chem B 109(33):15968–15976

Study of Compatible Anode for Silicate-Based Cathode Material

Ravi Vikash Pateriya, Shweta Tanwar, and A. L. Sharma

1 Introduction

In recent years because of technological development, the energy demand has increased at a rapid rate. The traditional source of energy fossil fuels is of limited content and may vanish in upcoming years. Hence there is a need for alternative renewable energy sources like solar energy, windmill, etc. But due to the inconsistency in the availability of most renewable energy sources, energy storage is very important. Researchers and scientists have developed various energy storage devices to meet energy needs [1, 2]. Among various energy storage devices available like batteries, supercapacitors, fuel cells, etc. rechargeable Li-ion batteries are very promising candidates for fulfilling energy demands [3]. Li-ion batteries are superior to other devices. Li-ion batteries are of prominent use as they have high power density, high storage capability, better leakage current, constant voltage, cost-effectiveness, etc. making them the most favorable contender for energy storage [4]. Li-ion battery has four components: cathode (positive electrode), anode (negative electrode), electrolyte, and separator [5]. The electrochemical reaction takes place at the electrodes. The cathode is generally an electron acceptor having a high electronegativity and the work of the anode is to give electrons while discharging. While charging the process gets reversed. An electrolyte provides a medium for the migration of ions and separates the electrodes to avoid short-circuiting. Among all the components of Li-ion batteries, the cathode plays a very important role in enhancing energy density and life cycle. Based on their structure, the cathode materials can be divided into various categories:

R. V. Pateriya (✉) · S. Tanwar · A. L. Sharma
Department of Physics, Central University of Punjab, Bathinda 151401, India
e-mail: cyrusravivikashp@gmail.com

A. L. Sharma
e-mail: alsharma@cup.edu.in

S. Doolla et al. (eds.), *Advances in Clean Energy and Sustainability*,
Green Energy and Technology, https://doi.org/10.1007/978-981-99-2279-6_66

A. Layered structure
B. Olivine structure
C. Spinel structure.

$LiCoO_2$ has been used as a high-energy material for long time. Now there is a need to look for other high-energy materials for large-scale production. For this Li_2MSiO_4 type orthosilicate cathode materials were introduced in 2009 by Prof. Goodenough due to their high theoretical capacity (333 mAh g^{-1}), which is calculated as

$$\mathbf{C = n\,F/m} \tag{1}$$

where 'n' is a number of Li^+ ions participating in intercalation, F stands for Faraday constant, and 'M' is the molecular mass of active material. From the formula, the capacity directly depends upon 'n'. In orthosilicate cathode materials, two Li^+ ions are intercalated during reaction in comparison to other cathode materials.

However, poor conductivity and suitable anode for Li_2MSiO_4 type material have always been a point of concern. In this paper, we put forward the synthesis of Li_2MnSiO_4 cathode material by hydrothermal reaction and verified performance of prepared material using different carbon anode material. The structural analysis and bonding of the prepared sample were analyzed using XRD and FTIR. The electrochemical properties of cathode material were analyzed using the electrochemical workstation. The electrochemical results of Li_2MnSiO_4 were compared by using activated carbon and graphite as anode by cell assembly with Li_2MnSiO_4 cathode material. CV, EIS and GCD results were discussed and compared for battery use by cell assembly.

2 Material and Methodology

Li_2MnSiO_4 material was synthesized using lithium hydroxide (LiOH), Tetraethyl orthosilicate (TEOS), and Manganese acetate tetrahydrate as precursors in stoichiometric amount in a molar ratio of 4:1:1 respectively. All the materials were brought from Sigma Aldrich and directly used without further purification, and deionized double-distilled water was used for making all the solutions. First, manganese acetate tetrahydrate was dissolved in 12 ml Double distilled water followed by the addition of tetraethyl orthosilicate under constant stirring for 1 h. Finally, well-grounded LiOH powder was added to the solution and further stirred for half-hour. The resulted solution mixture was transferred to a 25 ml Teflon line stainless steel autoclave. The autoclave was then put into a temperature-controlled furnace at 180 °C for 45 h. The furnace was allowed to cool naturally and the obtained product was first centrifuged and dried at 70 °C for 24 h using a vacuum oven (Fig. 1).

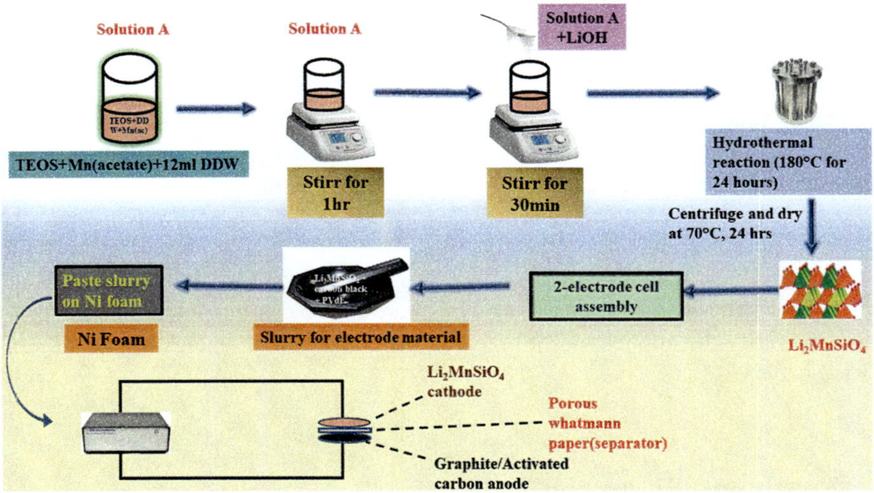

Fig. 1 Flow chart of the stepwise synthesis process of Li_2MnSiO_4

3 Characterization

The crystallinity of synthesized electrode material was verified using X-ray diffraction (XRD) by Model: PANalytical Empyrean X-ray diffractometer using CuK_α radiation of wavelength 1.540598 Å. The personalization of various chemical and functional bond was observed using Fourier transform infrared (FTIR) in transmittance spectra in wavenumber range of 600–4000 cm^{-1} by Bruker Tensor 27 Model: NEXUS-870.

3.1 Electrochemical Properties Measurement

The electrochemical analysis of prepared Li_2MnSiO_4 sample was measured by using CHI 760 instrument at room temperature. For electrochemical measurement the cathode and anode were coated on nickel foam. Li_2MnSiO_4 was used as cathode material while activated carbon and was used as anode. Whatmann paper was used as a separator and 1 M $LiPF_6$ in a mixture of EC-DEC-DMC (1:1:1 volume) as the electrolyte. The cell testing was performed by pasting slurry of composite of active material, binder polyviniylidene fluoride (PVdF) and carbon black in 80:10:10 ratio with N-Methyl-2-pyrrolidone (NMP) on nickel foam. The area of nickel foam was 1 * 1 cm^2. The prepared cathode was then dried for 12 h in vacuum oven at 60 °C. The activated carbon anode was prepared by same procedure as mentioned above. The load mass of material on nickel foam was 2 mg. The electrochemical measurements were performed using electrochemical workstation CHI-760 by sandwiching cathode and anode along with separator and electrolyte between stainless steel electrodes. Specific capacity was calculated using cyclic voltammetry (CV). Electrochemical impedance spectroscopy (EIS) measurements were performed from 1 Hz to 1 MHz frequency range with 1 mV voltage supply. The impedance plot is a semi-circle whose left intercept on Z' axis represents system resistance (R_s) an right intercept represents charge transfer resistance (R_{ct}) according to equation

$$Z' = R_s + R_{ct} + \sigma\omega^{-1/2} \quad (2)$$

Galvanostatic charge–discharge analysis were done by chronopotentiometry technique to verify the application of prepared material for high energy application of Li-ion battery.

4 Result and Discussion

The structural characterization of prepared Li_2MnSiO_4 cathode material was done by XRD and FTIR spectroscopy. XRD was used to confirm crystalline structure of prepared material and FTIR was used to identify the type of material/bonds present in sample.

The crystal structure of prepared Li_2MnSiO_4 material was confirmed by XRD pattern with JCPDS card no. 04-014-1657. The XRD spectrum showed well-defined, sharp and intense peaks as shown in Fig. 2a. The major peaks were observed at Braggs angle (2θ) 25.47, 32.92, 38.09 and 55.24°, corresponding hkl values and d-spacing was calculated by Braggs law

$$n\lambda = 2d\sin\theta \quad (3)$$

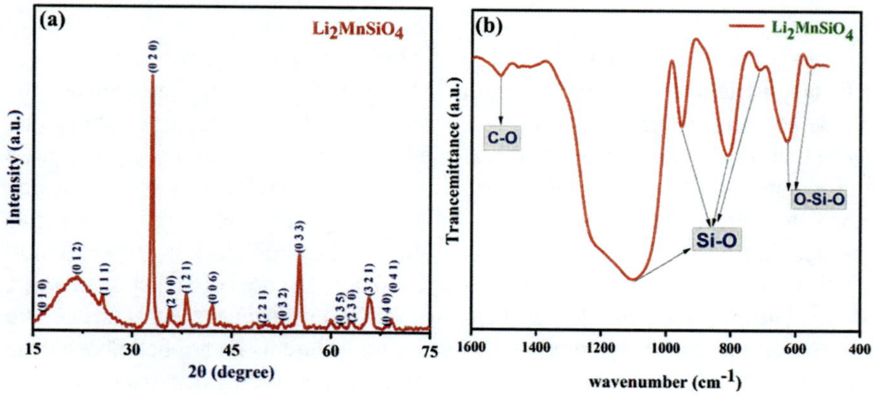

Fig. 2 **a** XRD plot of Li_2MnSiO_4, **b** FTIR plot of Li_2MnSiO_4 sample

Table 1 FTIR peak assignment for Li_2MnSiO_4 cathode material

Wavenumber (cm^{-1})	530	578	890	926	1090	1510
Assignment	O–Si–O	O–Si–O	Si–O	Si–O	Si–O	C–O

In XRD pattern there is no reflection of peaks of starting material. The crystalline peaks are well indexed according to orthorhombic structure with $Pmn2_1$ space group confirming the formation of single phase Li_2MnSiO_4.

The FTIR spectrum of prepared Li_2MnSiO_4 sample in range of 400–1600 cm^{-1} are shown in Fig. 2b. The position of 530, 578, 926 and 890 cm^{-1} are transmittance peaks of [SiO_4]. The 926 and 890 cm^{-1} represent ν3 and ν1 mode of Si–O bonds. The transmittance band around 530 and 578 cm^{-1} are assigned to O–Si–O bond [6]. The peak at 1510 cm^{-1} attribute to C–O bond. From FTIR spectrum we can conclude that there is no major impurity present in to the prepared Li_2MnSiO_4 sample. Table 1 shows the different functional group and corresponding wavenumber obtained from FTIR spectrum.

4.1 Cyclic Voltammetry Analysis

The electrochemical properties of the prepared Li_2MnSiO_4 sample were evaluated in two electrode system using different carbon materials as anode and $LiPF_6$ as electrolyte with help of electrochemical workstation. Figure 3 shows the CV plots of the prepared cell using different anode at a scan rates. Figure 3a represents the CV curve of activated carbon and Fig. 3b represents CV curve of graphite. The CV curve of activated carbon as anode exhibits a good current and cover large area. The assembled cell with activated carbon anode delivered the specific capacity of 53.07 mAh g^{-1} at scan rate of 5 mV s^{-1} which is satisfactory for Li-ion battery. While in comparison to activated carbon, cell with anode as graphite delivered specific capacity of 10.21 mAh g^{-1} at same scan rate representing that activated carbon is better anode material in comparison to graphite for Li-ion battery. The plot of specific capacity of cell v/s scan rate with activated carbon as anode and graphite anode is shown in Fig. 5a (Fig. 4).

4.2 Electrochemical Impedance Spectroscopy (EIS)

The Nyquist plot of EIS measurement is shown in Fig. 3. The assembled cell showed a better. For activated carbon, the diameter of semicircle demonstrates good contact between electrolyte and electrode resulting to faster kinetics and increased ion tolerability. The Nyquist plot for activated carbon and its fitted circuit is shown in Fig. 4a. The R_{ct} value of 6 Ω was calculated for activated carbon. The low value of R_{ct}

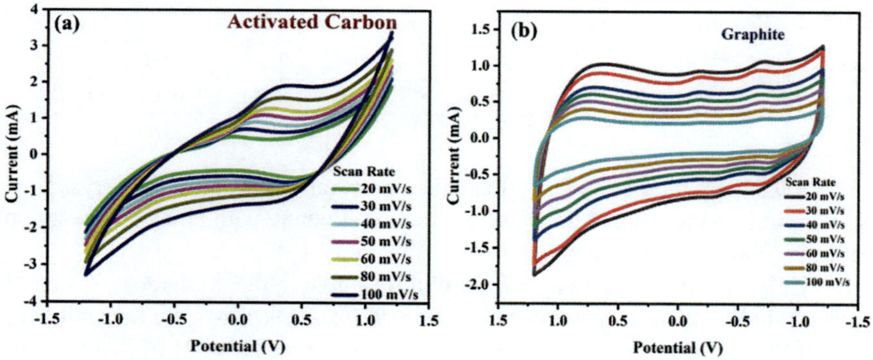

Fig. 3 Cyclic voltammetry curve of Li_2MnSiO_4 cathode and **a** activated carbon as anode **b** graphite as anode at different scan rates

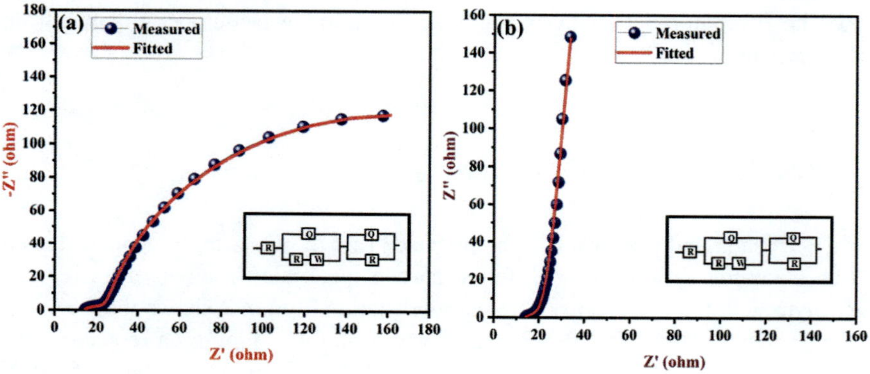

Fig. 4 Nyquist plot of Li_2MnSiO_4 cathode along with fitted circuit and **a** activated carbon anode, **b** graphite anode

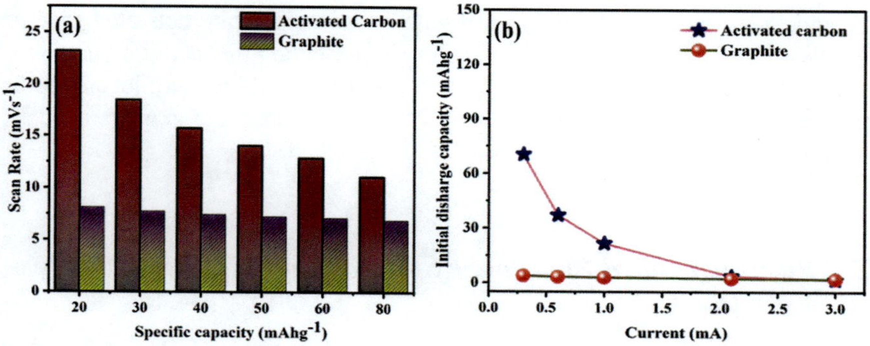

Fig. 5 **a** Specific capacity comparison of prepared cells, **b** initial discharge capacity curve of cell with activated carbon and graphite anodes

represents better electrochemical performances of the prepared sample. The cell with graphite anode showed charge transfer resistance (R_{ct}) value of about 7 Ω. The Nyquist plot for graphite anode indicates that assembled cell is showing some capacitive behaviour as shown in Fig. 4b.

4.3 Charge Discharge Analysis

The charging-discharging is a key factor for battery performance. The galvanostatic charge discharge analysis of different prepared cell was investigated at different currents. Cell with activated carbon anode exhibited an initial specific discharge capacity of 70.54 mAh g^{-1} at current 0.3 mA with coulombic efficiency of 92.64% and capacity retention of 84.31% after 100 cycles. Cell with graphite anode delivered initial specific discharge capacity of 3.97 mAh g^{-1} at 0.3 mA current with coulombic efficiency of 89.02% and capacity retention of 92% after 100 cycles. The initial discharge capacities of different cell are shown in Fig. 5b.

5 Conclusion

In this study, the Li_2MnSiO_4 cathode material was prepared successfully via hydrothermal route. The XRD analysis confirmed the formation of orthorhombic polymorph of Li_2MnSiO_4 in $Pmn2_1$ space group. With the material synthesis testification of cathode with activated carbon and graphite as anode material was confirmed. As anode activated carbon gave better results with Li_2MnSiO_4 in comparison to graphite. The electrochemical results of Li_2MnSiO_4 with activated carbon and graphite as anode suggests that activated carbon showed better results as compared to graphite. Hence, we can conclude that activated carbon can be a good anode option for Li_2MnSiO_4 type materials for Li-ion battery use in future.

Acknowledgements One of the author Ravi vikash pateriya want to thanks to UGC for providing JRF for doing research work. Ravi vikash pateriya also wants to thank Central Instrumentation Lab central university of Punjab, Bathinda to help in XRD and FTIR characterization.

References

1. Arya A, Sharma AL (2020) A glimpse on all-solid-state Li-ion battery (ASSLIB) performance based on novel solid polymer electrolytes: a topical review. J Mater Sci 55:6242–6304. https://doi.org/10.1007/s10853-020-04434-8
2. Tanwar S, Arya A, Gaur A, Sharma AL (2021) Transition metal dichalcogenide (TMDs) electrodes for supercapacitors: a comprehensive review. J Phys Condens Matter 33:303002. https://doi.org/10.1088/1361-648X/abfb3c

3. Arya A, Sharma AL (2017) Polymer electrolytes for lithium ion batteries: a critical study. Ionics. https://doi.org/10.1007/s11581-016-1908-6
4. Arya A, Sharma AL (2019) Electrolyte for energy storage/conversion (Li +, Na +, Mg 2+) devices based on PVC and their associated polymer: a comprehensive review. J Solid State Electrochem 23:997–1059. https://doi.org/10.1007/s10008-019-04203-x
5. Pateriya RV, Tanwar S, Arya A, Sharma AL (2022) Polymer composites for Lithium-Ion batteries. In: Polymer electrolytes and their composites for energy storage/conversion devices. CRC Press, pp 149–176
6. Luo S, Wang M, Zhu X, Geng G (2012) Hydrothermal synthesis of Li2MnSiO4 powders as a cathode material for lithium ion cells. Key Eng Mater 512–515:1588–1591. https://doi.org/10.4028/www.scientific.net/KEM.512-515.1588

Experimental Investigation on Phase Change Material Enhanced Pin Finned Heat Sinks for Thermal Management Applications

Vivek Saxena, Aastha Luthra, Pradunmya P. Dutta, Santosh K. Sahu, and Shailesh I. Kundalwal

1 Introduction

Electronic components that are being made nowadays are designed to provide enhanced speed and increased compactness, because of which the threshold for heat flux dissipation has increased. Higher heat production leads to higher working temperatures and hence, a higher failure probability [1, 2]. Thus, it is required to design a system having higher cooling efficiency to decrease failures and increase the life of these components. Current thermal management systems incorporate active cooling techniques such as the liquid-cooled system, where an active source of energy needs to be provided to run the system [3, 4]. These systems require more sophisticated design, noisy operation, and active maintenance. As a result, the development of a passive thermal management system has become a focus of study. Phase change materials (PCM) have depicted immense potential in the development of passive thermal management systems. PCM is generally classified as organic and inorganic, out of which the former is more popular because of the excellent properties such as high latent heat of fusion, high specific energy, non-toxicity, negligible subcooling, and enhanced thermal stability. Integration of PCM-enhanced thermal systems is limited because of the lower thermal conductivity of the PCM.

By using a variety of methods, including metallic fins, nanoparticles, and metallic foams, significant attempts have been made to improve the PCM's overall thermal conductivity. Various theoretical models have been proposed to model the overall thermal conductivity of composite PCM [5]. Rehman et al. [6] studied PCM embedded with copper and Iron-Nickel foams under various thermal loading conditions. The composite PCM reported 17% reduction in the base temperature compared

V. Saxena · A. Luthra · P. P. Dutta (✉) · S. K. Sahu · S. I. Kundalwal
Indian Institute of Technology Indore, Indore, India
e-mail: ms2104103001@iiti.ac.in

V. Saxena
e-mail: phd2001103003@iiti.ac.in

S. Doolla et al. (eds.), *Advances in Clean Energy and Sustainability*,
Green Energy and Technology, https://doi.org/10.1007/978-981-99-2279-6_67

to pure PCM. In one of the studies by Kothari et al. [7], test data for plate-finned heat sinks embedded filled with PCM was reported. In their experiments, the thermal performance of one, two, and three finned heat sink with metallic foams was compared to the unfinned heat sink when all were saturated with PCM. It was reported that the two finned heat sinks with metal foam outperform all others in terms of the enhancement ratio. The effectiveness of PCM-based cooling systems' heat transmission has been the subject of several research [8–18]. Baby and Balaji conducted an experimental investigation on the thermal performance of heat sinks using plate fins and pin fins [8]. The HS with pin fins outperforms the HS with plate fins, according to the authors. In their analysis of the impact of pin fin thickness on thermal effectiveness, Ali and Arshad [13] discovered that HS with a 3 mm fin thickness performs at its best. According to their investigation on the impact of pin fin thickness, HS with a 2 mm thick fin provides the best thermal performance. As per the report, the inline design improves thermal performance. Various studies reported the performance of NePCM with HS to be influenced by nano-particle characteristics (concentration, size, and type) [19, 20], fin properties, heat load, and the targeted temperature value [13–18]. Baby and Balaji [21] demonstrated that pin fin heat sinks show superior thermal performance, a 24X enhancement factor for a 7W power level. Srikanth et al. [22] employed multi-objective optimization to study the heat transfer performance of PCM based 72 pin fin heat sink using 4 discrete heaters. This study established the accountability of discrete heating for the optimization of heat sinks. Pakrouh et al. [23] numerically analyzed the parameters of PCM based pin fin heat sinks. The base temperature, as well as the heat sink's operating time, are significantly reduced, according to the results, when the number, thickness, and height of fins are increased. Usman et al. [24] studied finned and unfinned heat sinks based on PCM for passive cooling of electronics. The results show that the triangular inline pin-fin heat sink design is the most common and that the RT-44 PCM is the most efficient PCM for passive thermal control of electronic devices. Rukh et al. [25] experimentally investigated to improve heat transfer from round pin heat sinks utilizing PCM based on N-eicosane. The results suggested that the best thermal performance was shown by the 3 mm pin-fin arrangement, which outperformed all other heat sinks in terms of various performance parameters such as enhancement ratio, conductance and heat storage capacity. Yang et al. [26] numerically and experimentally investigated PCM (lower melting point) with internal finned heat sink. The study showed that E-BiInSn demonstrated significantly better thermal performance than octadecanol in each scenario.

For different applications, the heat-load and set target temperature requirements vary. It can be seen from the literature that a host of experimental studies are available, but the test data for different targeted temperature ranges with circular pin finned and square pin finned-based heat sinks embedded with NePCM are not significantly available [6, 7, 9–11]. The aim of the present experimental investigation is to evaluate the thermal performance of heat sinks for different targeted temperature ranges and concentrations of the nano-particles. To compare the test data in the present investigation, heat sinks without fins (HSNF), pin fins of circular shape (HSCPF) and pin fins of square shape (HSSPF) are investigated with metallic nano-particles

(Al_2O_3 and CuO). The input heat flux for the experiments has been taken at a constant value of 2.5 kW/m^2, and two different concentrations of nano-particles (Ø = 0.5 and 1.0) are considered. The targeted temperature range for the present investigation is taken to be 65 and 85 °C, which lies in the operating temperature range of electronic gadgets. The comparison is presented with respect to the base case, i.e., heat sink without fins and PCM without the addition of nano-particles for the same heat flux conditions.

2 Experimental Setup

The test facility modules used in this research project are depicted in Fig. 1. The dimensions of heat sinks are 100 × 100 × 20 mm^3 and the dimensions of the heater plate (maker: Sunrise, India) is 100 × 100 × 4 mm^3. The heater plate is place below the heat sinks and a DC power source was used (Aplab L3260, 0–30 V/0–40 A, India) to maintain the constant heat flux conditions during the experiments. The top portion of the heat sinks were covered with acrylic sheet material and the heat sink assemblies were insulted using glass wool to eliminate the surroundings heat loss. The PCM (paraffin wax) was purchase from Sigma-Aldrich and the nano-particles from Nanoshel India. The heat sinks (HSNF, HSSPF and HSCPF) are depicted in the Fig. 2. K type thermocouples were used to capture the real time temperature data during the experiments and a Data Acquisition System (Agilent 34972A, 32 channels) that fed the computer as readable records. The base of the heat sink was provided four thermocouples (T1–T4) including two (T5–T6) in the side walls and (T7–T8) in the top acrylic sheet. It is discovered that there is a measurement inaccuracy of ± 0.1 V for voltage and ± 0.2 A for current. K-type thermocouples with an accuracy of up to ± 0.3 °C are used to measure the heat sinks' temperature.

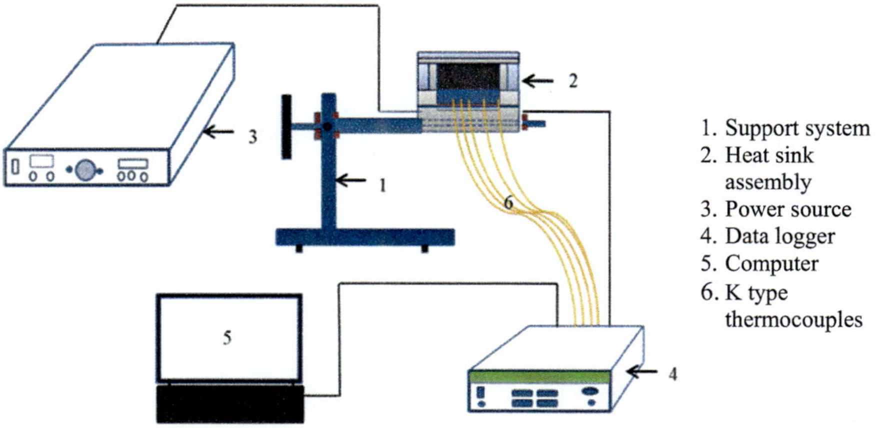

Fig. 1 Schematic for the experimental setup

(a) HSNF (b) HS SPF (c) HS CPF

Fig. 2 Different heat sink configurations

3 Material Characterization

It is important to determine the thermophysical properties of the material used in the heat sink assembly. To investigate the properties of PCM/NePCM the differential scanning calorimetry (DSC 8000, Perkin-Elmer, USA) was performed. A heating rate of 10 °C/min was selected and the heating curve is presented in Fig. 3. It is worth noting that adding CuO/Al_2O_3 nano-particles into pure PCM has no major effect on the melting temperature of the material. The addition of CuO/Al_2O_3 nano-particles to pure PCM, on the other hand, reduces the latent heat of fusion because of this the energy storage capacity decreases. Similar observations have been reported by earlier researchers as well [27, 28].

The thermophysical properties for PCM, heat sink material (Aluminium) and acrylic material is listed in Table 1.

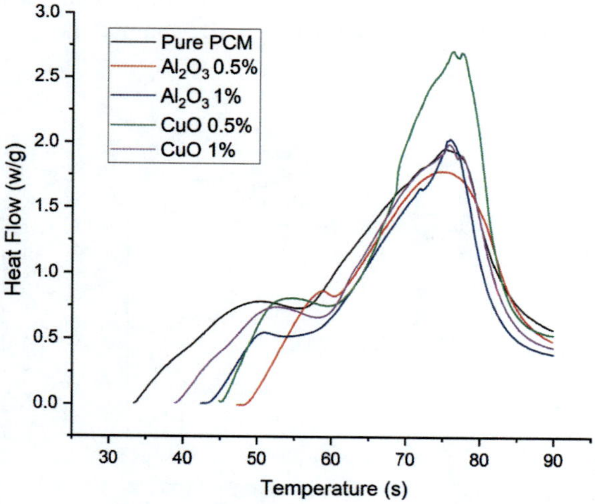

Fig. 3 DSC heating test

Table 1 Thermophysical properties [27, 28]

Properties	Paraffin wax	CuO	Al_2O_3	Aluminium	Plexiglass
Melting temperature (°C)	53.8	–	–	660.37	–
Specific heat (kJ/kg − K)	2.4	0.551	0.765	0.896	1.470
Density (kg/m^3)	770 (l) 839 (s)	790	3600	2719	–
Thermal conductivity (W/m − K)	0.22	33	36	218	0.19
Latent heat (kJ/kg)	237.5	–	–	–	–

4 Results and Discussions

4.1 Setup Validation

A comparison between the current and prior studies [17] was used to validate the experimental setting. The current results for the unfinned heat sink with and without the inclusion of Paraffin wax are compared to those of Kothari et al. [17] in Figs. 4 and 5. A plate heater provided a heat flow of 2.0 and 2.7 kW/m^2 to the heat sink assembly, and time-temperature measurements were taken. The results of the current analysis show a pattern that is similar to that shown in prior tests on heat sinks with similar dimensions. The variation in ambient conditions, i.e., the initial heating temperature, could be the reason for the minor variance in validation.

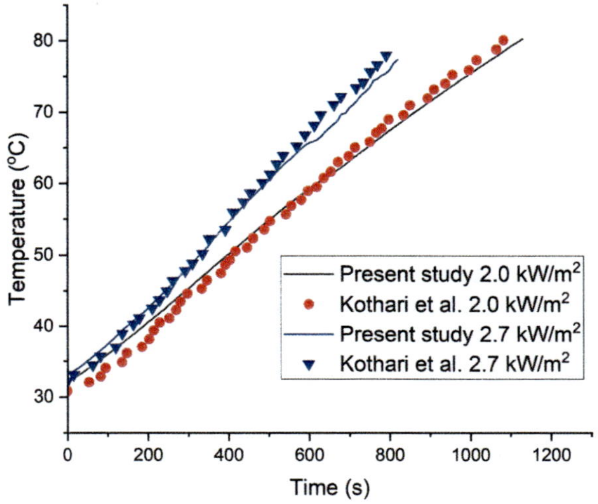

Fig. 4 Setup validation (without PCM)

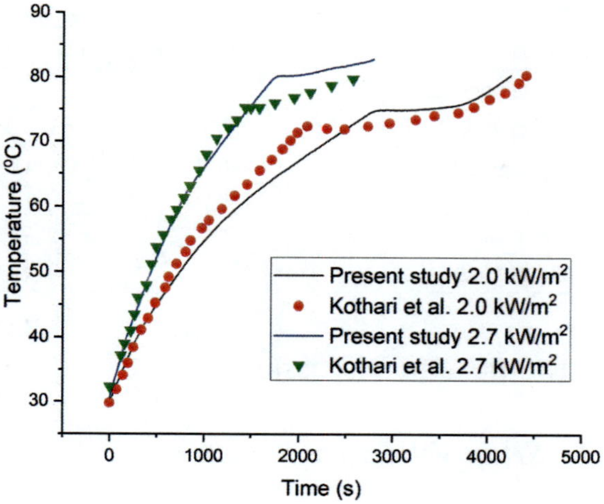

Fig. 5 Setup validation (with PCM)

4.2 Effect of Heat Sink Configuration on the Base Temperature

The results depicting the effect of heat sink type for various combination of phase change material/composites are borated below. The results presents the thermal performance comparison of pure PCM with the NePCMs. The average change in the base temperature of the heat sinks are taken into account and the data reported is based on the time to reach the target temperature of 90 °C. In the case of pure PCM, it can be seen that the heat sink with square pin fin and heat sink without fins has the same time to achieve the desired temperature. However, the average base temperature is more in the case of heat sink without fins.

It can be seen from the Fig. 6b–e, the addition of the nano-particles has effect on the time to achieve the desired temperature. In any event, the heat sink with square pin fins has the highest heat transfer effectiveness in terms of maintaining a lower base temperature, followed by the heat sink with circular pin fins and the heat sink without fins. The variation in base temperature between square and circular pin finned is negligible after the latent heating phase; this could be due to the prevalence of convection currents.

4.3 Effect of NePCM Type and Concentration

The temperature-time response of the individual heat sinks for all the infiltration cases of the PCM/NePCM are discussed in this section. Figure 7a depicts the test data for heat sink without any fins (HSNF) and it can be seen that the maximum time to reach the target temperature of 90 °C is reported in the case of pure PCM and

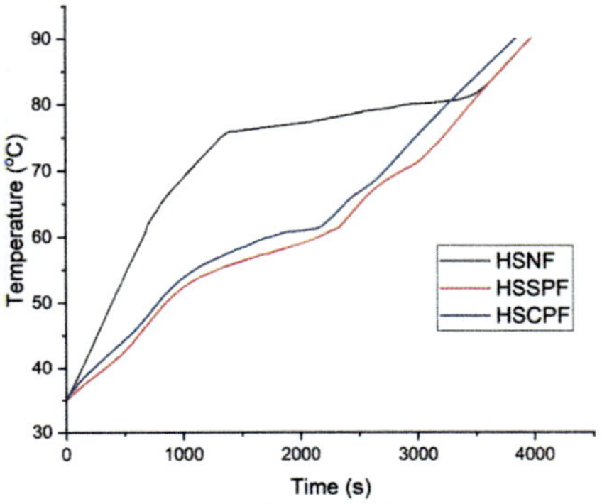

(a) Pure PCM

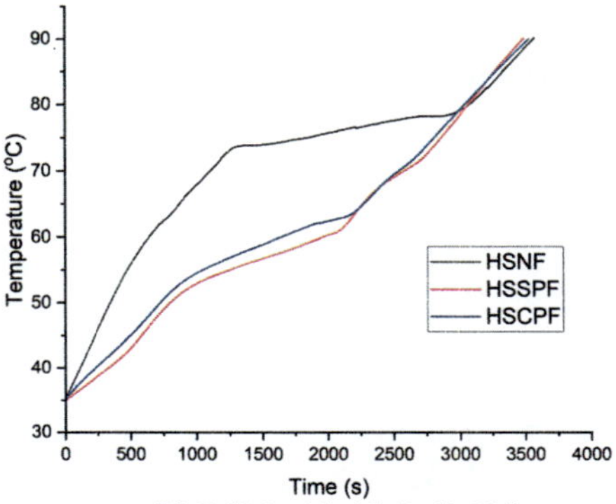

(b) PCM with Al_2O_3 (0.5%)

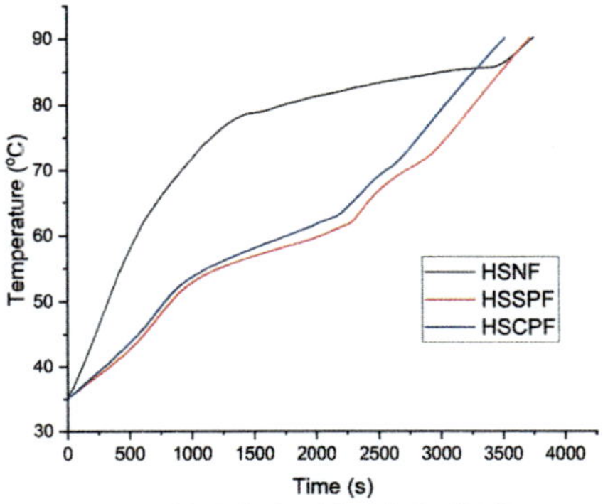

(c) PCM with Al_2O_3 (1%)

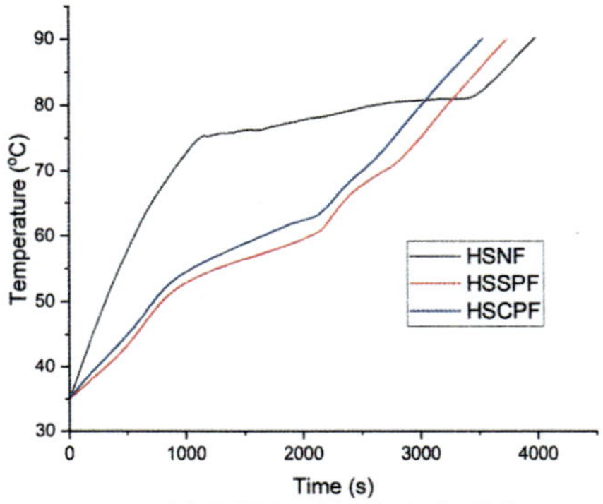

(d) PCM with CuO (0.5%)

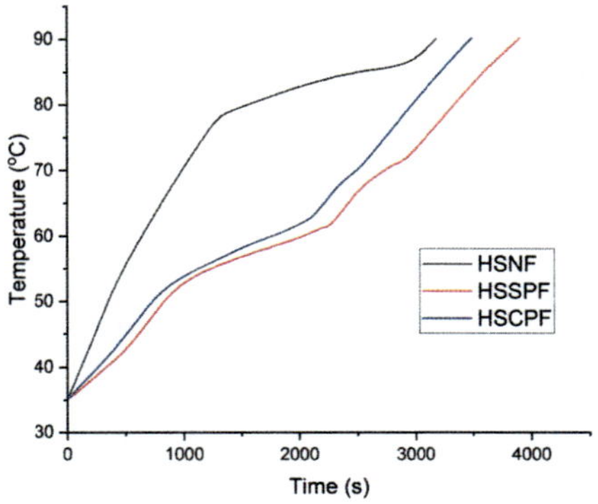

(e) PCM with CuO (1%)

Fig. 6 Time–temperature distribution for different materials

CuO (0.5%) followed by Al_2O_3 (1%), Al_2O_3 (0.5%) and CuO (1%). In case of heat sink with square pin fins (HSSPF) the maximum time is reported by pure PCM, CuO (1%), CuO (0.5%), Al_2O_3 (1%) and Al_2O_3 (0.5%), respectively. For heat sink with circular pin fins (HSCPF), the thermal performance is seen maximum in the case of pure PCM and then no significant changes are seen in the case of NePCM. It is to be noted that the longer the time to reach a specified targeted temperature better is the performance of the thermal management system. NePCM embedded heat sinks are unable to absorb more heat than pure PCM embedded heat sinks, despite the fact that the thermal conductivity generally enhances as the concentration of nanoparticles increases. Possible reasons for such observations could be the increased thermal resistance linking the surface of heat sink and NePCM; furthermore, the addition of nano-particles alters the viscosity of the PCM, which decreases the convection dominant heat transfer, and finally, the decrement in the latent heat of fusion [29].

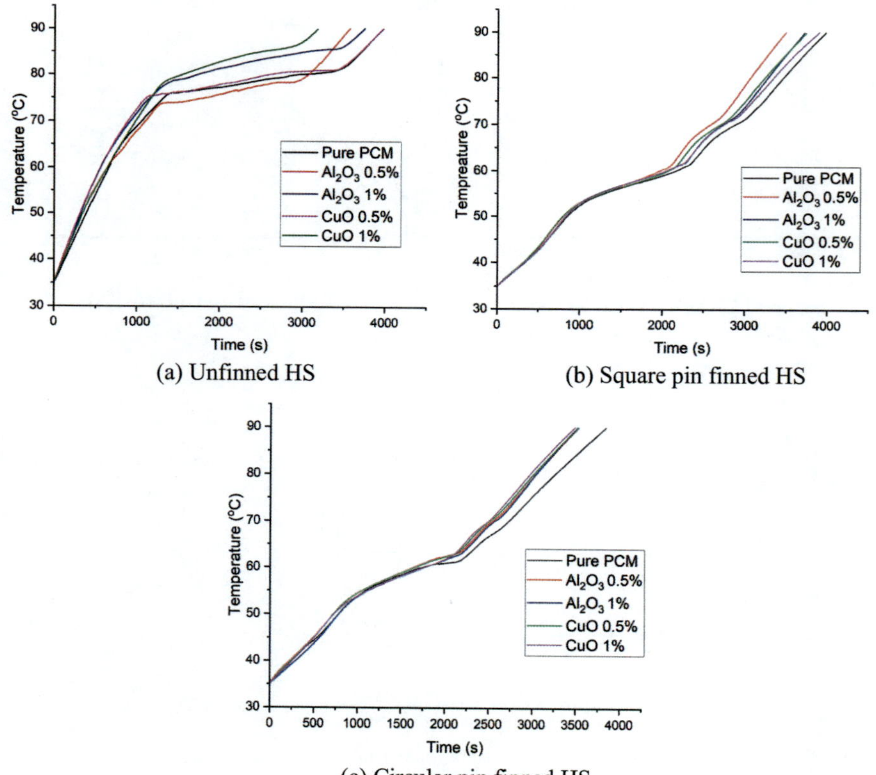

(a) Unfinned HS

(b) Square pin finned HS

(c) Circular pin finned HS

Fig. 7 Time-temperature distribution for different heat sinks configurations

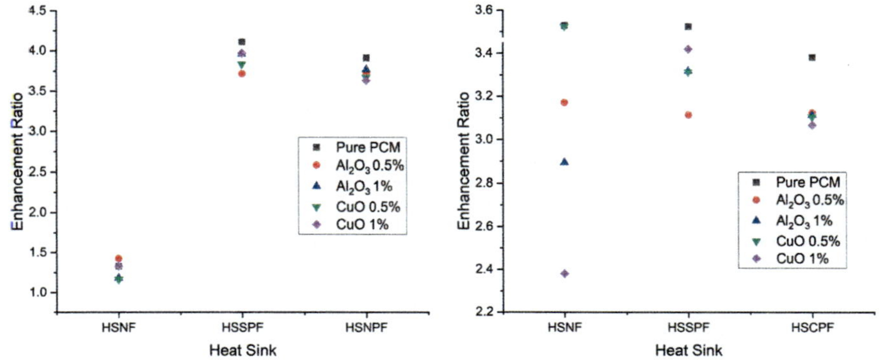

(a) 65°C (b) 85°C

Fig. 8 Enhancement ratio

4.4 Improvement in Operating Time

To clearly distinguish the increase in the working time in respect to the addition of fins and nano-particles in the heat sinks, the enhancement ratio is defined as:

$$Enhancement\ ratio\ (\delta) = \frac{Time\ to\ reach\ critical\ temperature\ using\ fins/nanoperticles\ and\ PCM}{Time\ to\ reach\ critical\ temperature\ without\ fins/nanoperticles\ and\ PCM} \quad (1)$$

For the present analysis, the critical temperature is taken as 65 and 85 °C. In the case of 65 °C, the highest enhancement is observed for the case of pure PCM with square pin finned heat sink. The high value of enhancement represents the increase in the operation time of the electronic component when pure PCM is used with fins i.e., a more better heat absorption rate is obtained. For the fixed temperature of 65 °C the heat sink with square pin fins is beneficial, whereas for a higher temperature range, the enhancement ratio decreases. At 65 °C peak enhancement ratio is 4.1, while for 85 °C, the value is found to be 3.52 (Fig. 8).

5 Conclusion

An experimental investigation of the thermal performance of PCM (paraffin wax) based heat sinks of three different configurations (HSNF, HSSPF, and HSCPF) was performed in this work. The conclusions from this study are as follows.

(i) Thermal conductivity of the PCM is increased by the addition of nano-particles.
(ii) CuO gives better results than Al_2O_3 in case of square pin fins while in case of circular pin fins, Al_2O_3 gives better results than CuO.
(iii) Pure PCM gives the best results in all the three heat sinks considering the time taken to melt completely.
(iv) NePCM give better performance as compared to PCM upto 85 °C and pure PCM gives better results above this temperature.
(v) HSSPF (Heat sink with square pin fins) performs better than the other two sinks, considering pure PCM as well as all NePCM composites. Having the PCM volume as same for all experiments, HSSPF has the highest surface area and thus better results.
(vi) The NePCM composite melts at the center first and later at the side walls in case of both the pin finned heat sinks.

The findings from the present investigation may be useful in designing a PCM enhanced thermal management system by using nano-particles.

Acknowledgements The first author would like to thank and acknowledge the research grant from the Prime Minister's Research Fellowship (PMRF), Govt. of India.

References

1. Viswanath R, Wakharkar V, Watwe A, Lebonheur V (2000) Thermal performance challenges from silicon to systems. Intel Tech J Q3
2. Karus AD, Bar-Cohen A (1983) Thermal analysis and control of electronic equipment, vol 1. Hemisphere Publishing Corporation, Washington DC, p 633
3. Grimes R, Wlash E, Walsh P (2010) Active cooling of a mobile phone handset. Appl Therm Eng 30(16):2363–2369
4. Lu X, Hua T, Liu M, Cheng Y (2009) Thermal analysis of loop heat pipe used for high power LED. Thermochem Acta 493:25–29
5. Saxena V, Kothari R, Sahu SK, Kundalwal SI (2021) An analytical approach for predicting the effective thermal conductivity of coated metal foams infiltrated with phase change materials. In: Proceedings of the 26th national and 4th international ISHMT-ASTFE heat and mass transfer conference. IIT Madras, Chennai, pp 2039–2044. https://doi.org/10.1615/IHMTC-2021.3080
6. Rehman T, Ali HM (2020) Experimental study on thermal behavior of RT-35HC paraffin within copper and iron-nickel open cell foams: energy storage for thermal management of electronics. Int J Heat Mass Transf 146:118852
7. Kothari R, Mahalkar P, Sahu SK, Kundalwal SI (2018) Experimental investigation on thermal performance of PCM based heat sink for passive cooling of electronic components. In: ASME 16th international conference on nanochannels, microchannels, and minichannels, Dubrovnik, Croatia, 10–13 June 2018
8. Baby R, Balaji C (2012) Experimental investigation on phase change material based finned heat sink for electronic equipment cooling. Int J Heat Mass Transf 55:1644–1649
9. Saha SK, Dutta P (2010) Heat transfer correlations for PCM-based heat sinks with plate fins. Appl Therm Eng 30(16):2485–2491
10. Kumar A, KothariR, Sahu SK, Kundalwal SI, Paulraj MP (2021) Numerical investigation of cross plate fin heat sink integrated with phase change material for cooling application of portable electronic devices. Int J Energy Res 1–18. https://doi.org/10.1002/er.6404

11. Hosseinizadeh SF, Tan FL, Moosania SM (2011) Experimental and numerical studies on performance of PCM-based heat sink with different configurations of internal fins. Appl Therm Energy 31:3827–3838
12. Lv Y, Liu G, Zhang G, Yang X (2020) A novel thermal management structure using serpentine phase change material coupled with forced air convection for cylindrical battery modules. J Power Sources 468:228398
13. Ali HM, Arshad A (2017) Experimental investigation of n-eicosane based circular pin-fin heat sinks for passive cooling of electronic devices. Int J Heat Mass Transf 112:649–661
14. Arshad A, Ali HM, Ali M, Manzoor S (2017) Thermal performance of phase change material (PCM) based pin-finned heat sinks for electronics devices: effect of pin thickness and PCM volume fraction. Appl Therm Eng 112:143–155
15. Arshad A, Ali HM, Khushnood S, Jabbal M (2018) Experimental investigation of PCM based round pin-fin heat sinks for thermal management of electronics: effect of pinfin diameter. Int J Heat Mass Transf 117:861–872
16. Ashraf MJ, Ali HM, Usman H, Arshad A (2017) Experimental passive electronic cooling: parametric investigation of pin-fin geometries and efficient phase change material. Int J Heat Mass Transf 115:251–263
17. Ali HM, Ashraf MJ, Giovanneilli A, Irfan M, Irshad TB, Hamid HM, Hassan F, Arshad A (2018) Thermal management of electronics: an experimental analysis of triangular rectangular, and circular pin-fin heat sinks for various PCMs. Int J Heat Mass Transf 123:272–284
18. Ali HM, Arshad W (2015) Thermal performance investigation of staggered and inline pin fin heat sinks using water based rutile and anatase TiO2 nanofluids. Energy Convers Manag 106:793–803
19. Motahar S, Khodabandeh R (2018) An experimental assessment of nanostructured material embedded in a PCM based heat sink for transient thermal management of electronics. Trans Phenom Nano Micro Scales 6(2):96–103
20. Bayat M, Faridzadeh MR, Toghraie D (2018) Investigation of finned heat sink performance with nano enhanced phase change material (NePCM). Therm Sci Eng Prog 5:50–59
21. Baby R, Balaji C (2013) Thermal optimization of PCM based pin fin heat sinks: an experimental study. Appl Therm Eng 54(1):65–77. https://doi.org/10.1016/j.applthermaleng.2012.10.056
22. Pakrouh R, Hosseini MJ, Ranjbar AA (2015) A parametric investigation of a PCM-based pin fin heat sink. Mech Sci 6(1):65–73. https://doi.org/10.5194/ms-6-65-2015
23. Rukh S, Pasha RA, Nasir MA (2019) Heat transfer enhancement of round pin heat sinks using N-eicosane as PCM: an experimental study. Heat Mass Transf/Waerme- Und Stoffuebertragung 55(2):309–325. https://doi.org/10.1007/s00231-018-2411-6
24. Srikanth R, Balaji C (2017) Experimental investigation on the heat transfer performance of a PCM based pin fin heat sink with discrete heating. Int J Therm Sci 111:188–203. https://doi.org/10.1016/j.ijthermalsci.2016.08.018
25. Usman H, Ali HM, Arshad A, Ashraf MJ, Khushnood S, Janjua MM, Kazi SN (2018) An experimental study of PCM based finned and un-finned heat sinks for passive cooling of electronics. Heat Mass Transf/Waerme- Und Stoffuebertragung 54(12):3587–3598. https://doi.org/10.1007/s00231-018-2389-0
26. Yang XH, Tan SC, Ding YJ, Wang L, Liu J, Zhou YX (2017) Experimental and numerical investigation of low melting point metal based PCM heat sink with internal fins. Int Commun Heat Mass Transf 87:118–124. https://doi.org/10.1016/j.icheatmasstransfer.2017.07.001
27. Kumar A, Kothari R, Sahu SK, Kundalwal SI (2021) Thermal performance of heat sink using nano-enhanced phase change material (NePCM) for cooling of electronic components. Microelectron Reliab 121:114144
28. Kothari R, Sahu SK, Kundalwal SI (2021) Investigation on thermal characteristics of nano enhanced phase change material based finned and unfinned heat sinks for thermal management system. Chem Eng Process 162. https://doi.org/10.1016/j.cep.2021.108328

29. Ho CJ, Gao JY (2013) An experimental study on melting heat transfer of paraffin dispersed with Al2O3 nano-particle in a vertical enclosure. Int J Heat Mass Transf 62:2–8
30. Teng TP, Cheng CM, Cheng CP (2013) Performance assessment of heat storage by phase change materials containing MWCNTS and graphite. Appl Therm Eng 50(1):637–644

Experimental Investigation of Bubble Rising in Newtonian and Non-Newtonian Fluids: A Comparative Assessment

Kapil Dev Kumawat, Sachin Balasaheb Shinde, and Lalit Kumar

1 Introduction

Since the early years, the air-water two-phase phenomenon has taken attention in various different fields like metallurgical, chemical, and biochemical due to its application in multiphase reactors, metal refining, metal making, etc. in form of injected bubbles in liquids. Figure 1a shows the Macondo/Deepwater horizon 2010 incident that occurred due to hydrocarbon gas blasting through a concrete core that was recently installed to seal an oil well for later use. This catastrophe in Mexico resulted in significant human, economic, and environmental losses. Following this heinous tragedy, investigators began to capture the dynamics of the gas bubbles to avoid such incidents in the future.

The yield stress fluids have drawn the attention of researchers not only due to the fundamental nature of the problem but also their applications in form of entrapped bubbles and/or rising bubbles in various fields. From the application point of view, bubbles are desirable in the form of trapped state in food processing and cosmetic products [3], column reactors used in chemicals, bubbling mud in Rotorua (NZ) and other geothermal tourist spots (as shown in Fig. 1b) [3]. In contrast, bubbles are undesirable in the case of construction of concrete using vibration to remove the bubble [3], bubble removal to stabilization of cement slurries and oil wells, and other geo-materials [4]. Our study is inspired by carbon dioxide (CO_2) emission in the form of bubbles rising in oil-sand tailing ponds, cement failures, permeable slurries, and poor interfacial bonding problems due to invasion and flow of gas through the slurries in the cementing operations.

K. D. Kumawat · S. B. Shinde (✉) · L. Kumar
Department of Energy Science and Engineering, IIT Bombay, IIT Area, Market Gate Road, Powai, Mumbai, Maharashtra 400076, India
e-mail: sachin1639@gmail.com

L. Kumar
e-mail: lalit.kumar@iitb.ac.in

S. Doolla et al. (eds.), *Advances in Clean Energy and Sustainability*, Green Energy and Technology, https://doi.org/10.1007/978-981-99-2279-6_68

(a) (b)

Fig. 1 Importance of bubbles, **a** deepwater horizon oil spill in Mocando (2010) [1], and **b** application of desirable bubbles in bubbling mud in Rotorua (NZ) [2]

There is a wide range of literature in which researchers have theoretically and experimentally studied the feasible dynamics of tiny, intermediate, and large size bubbles rising in Newtonian and Non-Newtonian fluids [5–10]. Mainly, they are trying to study practicable bubbles rising in fluids in different situations, such as trapezoidal and rectangular columns, etc. [8].

Several different techniques, such as viscosity regularization, augmented Lagrangian technique (ALM), and ALM with a Lagrangian prefactor and quadratic term in the corresponding variational inequality, have been used by Dimakopoulos et al. [3] for the theoretical and computational study of bubbles dynamics in yield stress fluids. While in the experimental study, researchers reported the influence of yield stress, inertia, buoyancy, and elasticity on bubble shape and dynamics. With the difficulty of the independent effect on these parameters, they trendified the different dimensionless numbers of their interest and found the bubble tail and cusped shaped due to the viscoelastic effects of yield stress fluids [9–13].

In a series of experimental and theoretical studies, Dubash and Frigaard have theoretically tried to find out the conditions for static bubbles in viscoelastic fluids using Prager's two variational principles [4, 5]. In recent experiments, Pourzahedi et al. [10] have tried to eliminate the injection and memory effects using the multi-layered experimental design in their bubble rise experiments within the yield stress fluids. However, they found the pointed tail shape bubble due to the rheology and steady fluid deformation around the rising bubble. A current detailed review on a single bubble motion dynamic rising in viscoelastic fluids has also concluded the important results of forming a cusp, negative wake, and velocity jump discontinuity [14].

Researchers have studied bubble rising in Newtonian and Non-Newtonian fluids such as water and different concentration of polymeric solutions to date. However, the previous work has not investigated comparative studies on bubble rising in a different solution. In the present study, the bubble rise in Newtonian (water and glycerol) and non-Newtonian fluids (Carbopol solutions) are experimentally studied. In addition, the effect of shape and size on the bubble dynamic is examined for water, glycerol, and Carbopol solution. Finally, a comparative assessment (both qualitative and quantitative) of bubble dynamics in Newtonian, non-Newtonian fluids (yield stress) fluid was carried out.

2 Experimentation

2.1 Experimental Apparatus

As shown in Fig. 2, the designed experimental setup is a tall acrylic transparent Plexiglas box of square cross-sections with dimensions of 65 × 9.5 × 9.5 cm. The base has a needle-changing feature at the center to inject desirable sizes of bubbles. The needle base is connected to the glass valve, whose other end is connected to the syringe (60 ml) via a flexible tube of 0.5 cm inside diameter. The horizontal side hole is created for drainage intent. A transparent scale is also used for calibration purposes. We have used sufficiently long needles to avoid the base and side walls effect on injection. A camera is utilized to capture the bubble motion. Besides these things, the setup is kept on the transparent acrylic robust stand. All predictable leakages are sealed using silicone sealant.

2.2 Materials and Solution Preparation

The Carbopol 940 powder was used to prepare viscoelastic solutions at different concentrations. The Carbopol solution had transparency and negligible thixotropy properties, so these were ideal solutions for our experiments. Four beakers with a five-liter pouring capacity were used to prepare Carbopol solution at different concentrations. Each set of 5-L Carbopol solutions was prepared by first dispersing weighed Carbopol-940 powder in five liters of water and mixing at 1100–1300 rpm using a lab stirrer with a Remi motor until the powder was dissolved homogeneously. After that, solutions were neutralized by mixing NaOH and were left after covering the beakers to get a proper steady state.

2.3 Experiment Procedure

In each experiment, the column was filled gently with different solutions such as water, pure glycerol, and neutralized Carbopol 0.10 and 0.15 wt/wt%. The bubbles might form inside the Carbopol solution, so we tried to keep away these bubbles as

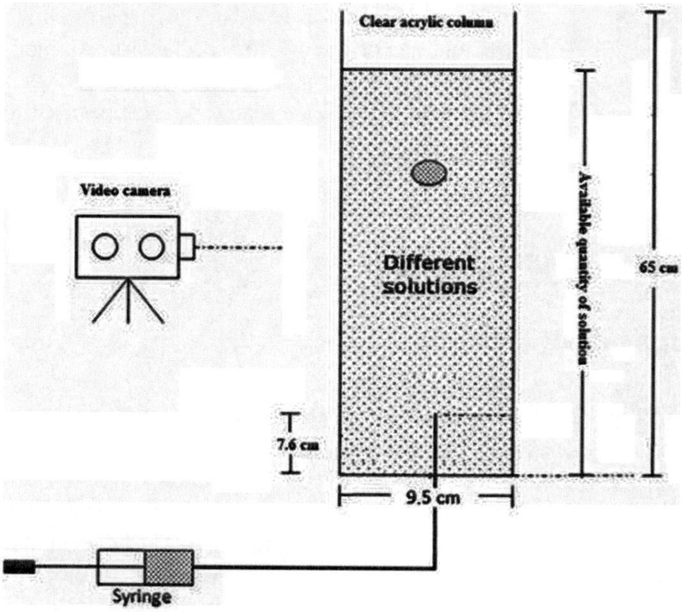

Fig. 2 Schematic of experiment setup for bubble rise in various fluids

much as possible. The syringe plunger was pressed to inject the bubbles into each solution. The glass valve is also employed to inject different sizes of bubbles by controlling the pressure. The bubble in Carbopol smaller than a specific volume is not supposed to rise in solution, while bubbles more giant than a certain maximum volume broke up during the injection process to form two and more than two bubbles. Therefore, we have tried to keep our volume range of bubbles in between stoppable bubbles to breakable bubbles.

The bubble volume V (cm^3) is evaluated by considering an axis-symmetric bubble which has also been also used in [7, 9, 11]. The bubble Radius ($R = (3\ V/4\pi)^{1/3}$) is characterized as the radius of the spherical bubble of the same volume for all cases of Newtonian and Non-Newtonian fluids.

2.4 *Rheological Measurements*

The rheological characterization of all solutions used in our experiments was done by rotational rheometer (Anton-Paar's Modular Compact Rheometer 302). This rheometer utilizes some specific techniques such as an EC motor, low friction bearing, and an optimized normal force sensor. It is able to do rotational tests, oscillatory tests, and a combination of both these tests. A parallel cone plate geometry was used for tests. The RheoCompass software is used to control this rheometer's test and recording of data. The rheological properties are measured for two Carbopol concentrations; 0.10 and 0.15 wt/wt% solutions, whose flow curves are shown in Fig. 3. It can be analyzed that both the solutions show viscoelastic behavior with the fixed yield stress (τ_y). Additionally, a higher viscosity value is noticed for higher Carbopol concentrations. The data for both concentrations of Carbopol is fitted very well with the Herschel-Bulkley equation. Similar rheological trends for Carbopol solutions have been found in [3, 7, 9, 11]. The Herschel-Bulkley constitutive model is well described by the following Eqs. (1) and (2).

$$\tau = \tau_y + k \cdot \dot{\gamma}^n \qquad \text{if } \tau \geq \tau_y \tag{1}$$

$$\dot{\gamma} = 0 \qquad \text{if } \tau < \tau_y \tag{2}$$

where τ and $\dot{\gamma}$ are the second in-variants of the deviatoric stress tensors and rate of strain; τ_y, k and n the yield stress, consistency and power-law index, respectively.

The visco-elastic behavior of Carbopol solutions is evaluated by performing oscillatory tests. The observed storage modulus (G') is higher than the loss moduli (G''), as shown in plotted curves in Fig. 3. This higher value of storage moduli (G') indicates that before yielding, the solution behaves like a solid. Additionally, it is seen that both moduli increase as the concentration of the solution increases and that for all concentrations, the loss moduli lie below the storage moduli, indicating that the solutions are viscoelastic regimes below the yield stress. In these observations, we have also found that the viscoelastic behavior of the Carbopol solution increases with concentration.

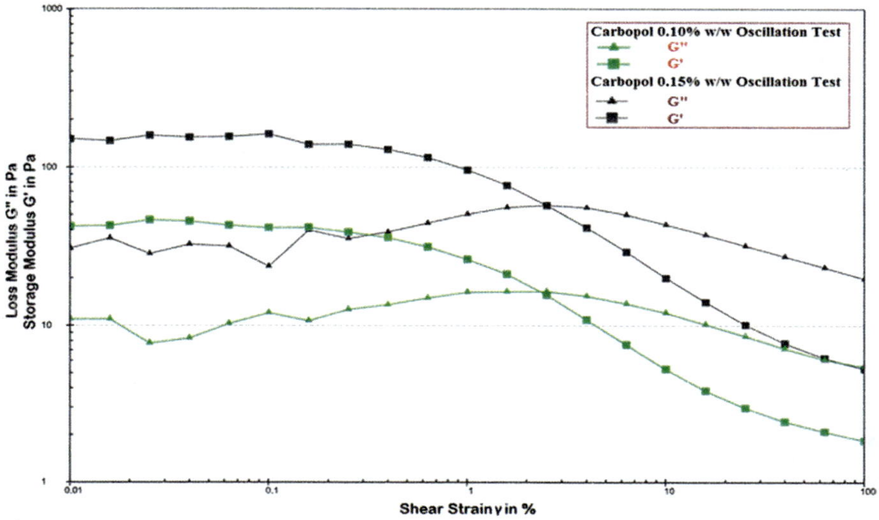

Fig. 3 Storage modulus and loss modulus versus shear rate for Carbopol 0.10 and 0.15%

2.5 Image Post-Processing Work

The bubble motion during each experiment was recorded using the NIKON D3400 DSLR Camera at 60 fps. A fine background and light source was also used for clear visualization of bubbles. The adhesive transparent measurement scale is employed for calibration purposes. The various bubble parameters such as time, displacements, and various velocities were measured by tracking the upper tip of the bubble using TRACKER open-source physics software.

3 Results and Discussion

In this section, we have compared the qualitative and quantitative results between Newtonian and Non-Newtonian fluids obtained with bubble rise experimentation.

3.1 Qualitative Comparison

Figure 4 shows the various shapes and sizes of bubbles rising in Newtonian and Non-Newtonian fluids on a 2 cm scale. It is observed that bubbles are more elongated in the case of water compared to bubbles rising in other solutions due to the dominant nature of surface forces over the other forces as the viscosity of water is less. The bubble reaches a steady state earlier for higher viscosity, at steady state an upper tip oblate shape and transient from lower cusp spherical shape to oblate in glycerol and Carbopol 0.10 wt/wt%, respectively. The shape of bubbles is more inverted teardrop in Carbopol 0.15 wt/wt% due to its high viscoelastic nature compared to Carbopol 0.10 wt/wt%. An inverted teardrop shape is a result of the negative wake phenomenon discussed in [14]. These inverted teardrop shapes are only observed in the case of Carbopol solutions which is similar to what has been reported by Pourzahedi et al. [10].

The entrapment of a bubble significantly affects the quality and life of the final product; thus, bubble entrapment phenomena are a tremendous interest of study in

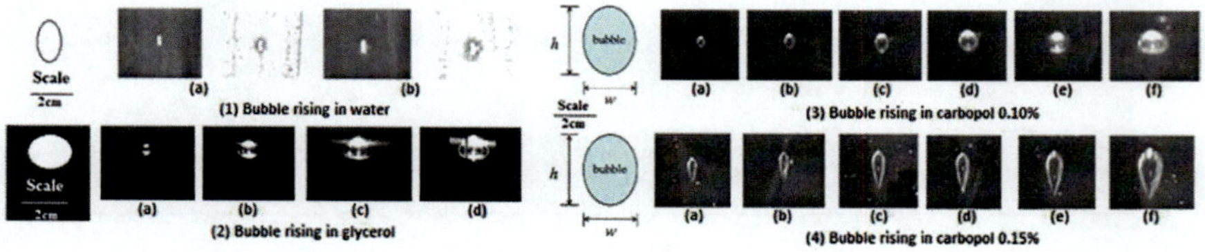

Fig. 4 Post-processed images of different shapes and sizes of bubbles rising in (1) water (2) glycerol, (3) Carbopol 0.10 wt/wt%, and (4) Carbopol 0.15 wt/wt%

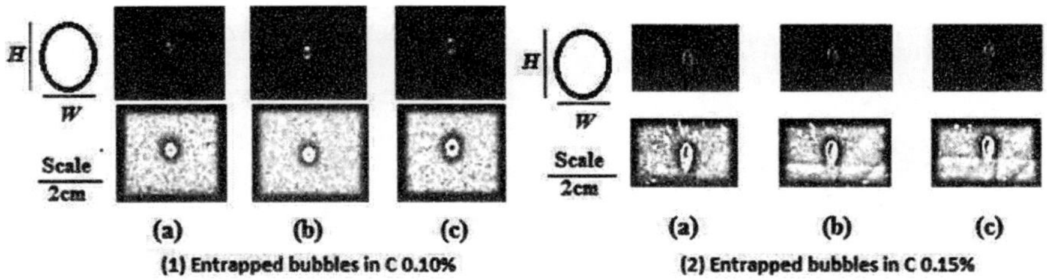

Fig. 5 Different entrapped bubbles shapes with 2 cm scaled measurement; (1) Carbopol 0.10 wt/wt% and (2) Carbopol 0.15 wt/wt%

food processing and cosmetic products, etc. The bubble does not move in the fluid when the yielding dominates over the buoyancy force. The critical ratio of yield stress to buoyant stress defines this situation precisely [4]. Figure 5 shows the shapes and sizes of different entrapped bubbles in Carbopol 0.10 wt/wt% and Carbopol 0.15 wt/wt%. There is no bubble entrapment phenomenon in Newtonian fluids as they do not have any yield stress. It is observed that bubbles are prolate in Carbopol 0.10 wt/wt% while improper inverted teardrop shapes in Carbopol 0.15 wt/wt% because they immediately entrapped before getting a steady-state of inverted teardrop shape.

3.2 *Quantitative Comparison*

Measured velocities of bubbles rising in Newtonian and Non-Newtonian fluids are presented in Fig. 6. It is observed that the net velocity (v) of bubbles rising in water is higher compared to other cases due to less flow resistance. The path of the bubble in water is zigzag due to the significant magnitude of the horizontal component of velocity (v_x). Although net bubble velocity (v) only depends on the vertical component of velocity (v_y) due to the dominant nature of buoyancy force over the viscous drag and gravitational force., etc. There is no discrepancy in both cases of Carbopol solutions, as bubbles rising in their solution can be characterized by the vertical component of velocity (v_y).

Bubble rising in water attained net velocity having a magnitude of 16.69 $\pm$ 0.78 cm/s and 28.60 $\pm$ 0.85 cm/s at a radius of 0.28 cm and at 0.87 cm, respectively. The minimum and maximum velocities for bubbles in glycerol are 5.31 $\pm$ 0.03 cm/s and 16.01 $\pm$ 0.06 cm/s at 0.35 $\pm$ 0.02 cm and 0.78 $\pm$ 0.06 cm radius, respectively. Whereas in the case of Carbopol solutions, the maximum measured velocity is 25.30 $\pm$ 0.23 cm/s for 0.89 $\pm$ 0.08 cm radius and the minimum measured velocity is 1.51 $\pm$ 0.01 cm/s for 0.32 $\pm$ 0.01 cm radius. From these measured velocities, it can be concluded that bubbles have higher velocity in less viscous fluid due to the dominant buoyancy force over the gravitation and viscous drag.

Figure 7a shows an interesting trend in which bubbles rising in water have a higher velocity compared to other solutions, although they have smaller sizes, concluding

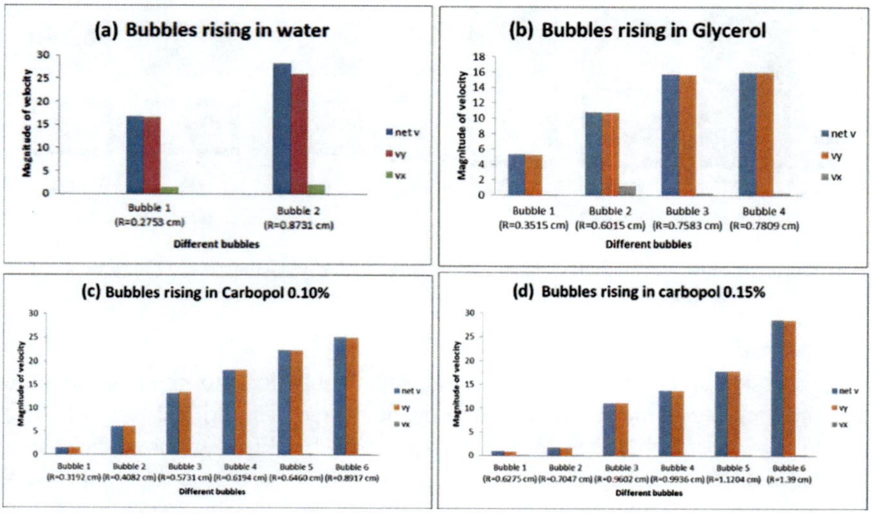

Fig. 6 Comparative study measured velocities of various bubble rising in, **a** water, **b** glycerol, **c** Carbopol 0.10 wt/wt%, and **d** Carbopol 0.15 wt/wt%

that viscosity is an important parameter for comparison of terminal velocity in Newtonian and Non-Newtonian solutions. For each bubble rising in a particular solution, we see a consistent increase in terminal velocity with the volume of bubbles due to the dominant buoyancy force. Similar to the literature [14], the velocity jump discontinuity phenomenon was observed for Carbopol solutions.

Finally, we plotted the aspect ratio (*H/W*) of rising bubbles and entrapped bubbles against the volume of bubbles for Carbopol solutions. As the aspect ratio decreases, the volume of the bubble increases, which shows that bubbles are initially elongated teardrop shape, and after that, their shape becomes wider teardrop shape in 0.15

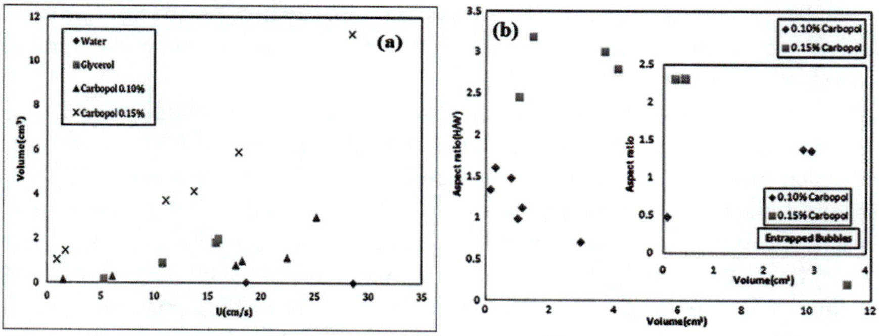

Fig. 7 **a** Terminal velocity versus volume of bubbles rising in water, glycerol, Carbopol 0.10 wt/wt% and Carbopol 0.15 wt/wt%, **b** the aspect ratio (*H*/*W*) of rising bubbles (as shown in the main graph) and entrapped bubbles (as shown in the subgraph) plotted against volume for each experiment

wt/wt% Carbopol solution. The result obtained from Fig. 7b depicts that the aspect ratio has a greater value in the case of higher concentration Carbopol solution than lower concentration solution for both trapped and rising bubbles. Similar trends were also reported by Sikorski et al. [11].

4 Conclusions

In the present research study, we have performed a series of experiments in which gas bubbles rise through Newtonian and Non-Newtonian solutions. The formation of bubbles depends on the injection method and surface tension due to the viscous nature of the fluid. We designed and developed an advanced experimental setup to study the bubble dynamics in various fluids. From the result and discussion section, we have come up with the following remarks:

1. The water bubble has an elongated prolate shape. However, the bubble shape in glycerol is an upper tip spherical shape owing to higher viscosity than water.
2. In yield stress fluids, i.e., Carbopol 0.10 wt/wt%, the bubble shape changes from prolate to upper tip spherical shape with the disappearance of the lower cusp. In contrast, the bubble in Carbopol 0.15 wt/wt% has an inverted teardrop shape for all cases because of its high viscoelastic nature.
3. The observed bubble path in water is zigzag or fish-like due to the significant horizontal component of velocity (v_x). It can be characterized by deformation during injection and a dominant surface force regime due to lower viscosity. There is a rectilinear path when the bubble rises in pure glycerol, 0.10 and 0.15 wt/wt% Carbopol solutions because of the dominant nature of vertical velocity (v_y) over the horizontal velocity (v_x).
4. The terminal velocity of the bubble increases with the size or volume because of a higher buoyant force than other forces. However, bubbles in water have a higher terminal velocity (16.69 ± 0.78 cm/s and 28.60 ± 0.85 cm/s) than the other case due to lower viscosity.
5. The bubble entrapment phenomena were observed only in the case of yield stress fluids because of higher flow resistance due to yielding characteristics. Bubbles are entrapped when yield stress dominates over the buoyant stress, and the aspect ratio below which this phenomenon takes place is called the critical aspect ratio.

Overall, the finding of the present study will help to understand the bubble kinematics in different Newtonian and non-Newtonian fluids used in various industrial process applications.

References

1. Mackowsky RM, Kende CB, Benett RC, Ziemianski JA (2010) The deepwater horizon catastrophe: a factual overview and preliminary first-party analysis. Cozen O'Connor, U.S.A.
2. Lewy Z (2012) Banded iron formations (BIFs) and associated sediments do not reflect the physical and chemical properties of early Precambrian seas. Int J Geosci 3:226
3. Dimakopoulos Y, Pavlidis M, Tsamopoulos J (2013) Steady bubble rise in herschel–bulkley fluids and comparison of predictions via the augmented lagrangian method with those via the papanastasiou model. J Nonnewton Fluid Mech 200:34–51
4. Dubash N, Frigaard I (2007) Propagation and stopping of air bubbles in carbopol solutions. J Nonnewton Fluid Mech 142:123–134
5. Dubash N, Frigaard I (2004) Conditions for static bubbles in viscoplastic fluids. Phys Fluids 16:4319–4330
6. Gnyloskurenko S, Byakova A, Raychenko O, Nakamura T (2003) Influence of wetting conditions on bubble formation at orifice in an inviscid liquid. Transformation of bubble shape and size. Colloids Surf A: Physicochem Eng Aspects 218:73–87
7. Hussein EQ, Rashid FL, Hussein AK, Younis O (2021) Hydro-dynamics of single bubble rising through water column using volume of fluid (vof) method. J Therm Eng 7:2107–2114
8. Islam M, Ganesan P, Sahu J, Uddin M, Mannan A (2013) A single air bubble rise in water: a CFD study. Mech Eng Res J 9:1–6
9. Lopez WF, Naccache MF, de Souza Mendes PR (2018) Rising bubbles in yield stress materials. J Rheol 62:209–219
10. Pourzahedi A, Zare M, Frigaard I (2021) Eliminating injection and memory effects in bubble rise experiments within yield stress fluids. J Nonnewton Fluid Mech 292:104531
11. Sikorski D, Tabuteau H, de Bruyn JR (2009) Motion and shape of bubbles rising through a yield-stress fluid. J Nonnewton Fluid Mech 159:10–16
12. Zare M, Daneshi M, Frigaard I (2021) Effects of non-uniform rheology on the motion of bubbles in a yield-stress fluid. J Fluid Mech 919
13. Wang Z, Lou W, Sun B, Pan S, Zhao X, Liu H (2019) A model for predicting bubble velocity in yield stress fluid at low reynolds number. Chem Eng Sci 201:325–338
14. Wang W (2021) Review of single bubble motion characteristics rising in viscoelastic liquids. Int J Chem Eng 2021

Thermal Analysis of Corrugated-Single Pass Solar Air Heater Integrated with PCM

Vaibhav Dilip Nagale, Satyender Singh, and Sanjay Kumar

Nomenclature

dy_{pcm}	Distance between two adjacent cell centres in y-direction (m)
H	Enthalpy (kJ/kg)
h_c	Convective heat transfer coefficient (W/m^{2}K)
k	Thermal conductivity (W/m.K)
L	Latent heat of fusion (J/Kg)
Q_{cond}	Rate of heat conduction (W)
T_a	Ambient temperature (K)
$T_{liquidus}$	Temperature indicating complete melting of PCM (K)
$T_{solidus}$	Temperature indicating beginning of the melting of PCM (K)
T_{up}	Temperature at the upper boundary of PCM (K)
V	Airflow speed (m/sec)

Greek letters

α	Absorptivity
ε	Emissivity
δs	Distance between cell centres (m)
Δt	Time step (sec)
δt	Distance between cell corners (m)
ΔV	Volume of control volume (m^3)
ρ	Density (kg/m^3)

V. D. Nagale (✉) · S. Singh · S. Kumar
Department of Mechanical Engineering, Dr. B. R. Ambedkar National Institute of Technology, Jalandhar, Punjab 144011, India
e-mail: vaibhavdn.te.20@nitj.ac.in

S. Doolla et al. (eds.), *Advances in Clean Energy and Sustainability*, Green Energy and Technology, https://doi.org/10.1007/978-981-99-2279-6_69

Subscripts

a	Ambient
c	Convection
r	Radiation
o	Outlet
i	Inlet
ap	Absorber plate
w	Wind

1 Introduction

The goal of this study is to use renewable energy sources to meet the expanding population's need. The cost of depleting traditional energy sources is so great that the recent focus of research in field of renewable energy is shifting on finding efficient ways to utilize energy from renewable energy sources like solar energy to meet energy demands. Solar air heaters make use of solar energy that is abundantly available to heat the air at a moderate temperature that is used to support a variety of thermal applications. Major disadvantage of solar application devices is the unavailability of energy during off-sunshine hours, aforementioned disadvantage can be tackled by employing thermal energy storage devices like PCM units.

In comparison to traditional designs, there is a significant improvisation in the thermal performance of recent solar air heater designs. Following which many researchers like Dhiman and Singh [1] are working on changing absorber plate design, whereas Singh and Dhiman [2, 3] used attachment of turbulators or extended surfaces. Furthermore, many studies such as of Arfaoui et al. [4] use microencapsulation for PCM storage for single pass solar air heater in form of spherical balls, metallic tubes were studied by Dheep and Sreekumar [5] and cylindrical and rectangular shells by Raj et al. [6] where 10-h backup was achieved with 6 kg of PCM. Moreover, Moradi et al. [7] performed numerical and experimental studies for flat plate collector under single air pass flow configuration presenting 24-h transient performance with 4.5 °C average temperature throughout the night. In addition, an improvised absorber plate with V-corrugation under single air pass is experimentally investigated by Kabeel et al. [8], reporting 15% improved thermal efficiency compared to flat collector. A numerical investigation by Verma et al. [9] presenting modelling of phase change material by energy enthalpy method, for study of counterflow solar air heater is reported.

Interesting facts that could be highlighted from reported literature are that providing extended surfaces to absorber plate result in lower hydraulic efficiency. Also, very less study in area of corrugated PCM units as thermal energy storage systems for single pass solar air heaters was reported. There is scope of improvisation of thermal output of solar collectors by increasing collector area within same space without affecting hydraulic efficiency by providing corrugated collector.

2 Single Pass Solar Air Heater

In present study a solar air heater with wavy corrugated type absorber plate under single pass flow configuration is considered with integrated PCM unit. The corrugation increases the incident area availing more flux to the absorber plate. Air at ambient temperature enters from one side and exits from another gaining heat from the heat source. Figure 1 shows the schematic of the design (Tables 1 and 2).

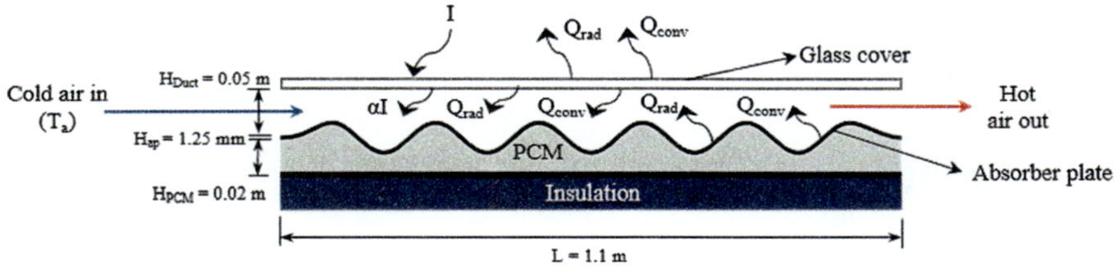

Fig. 1 Schematic 2-D design of single-pass solar air heater

Table 1 Geometrical configuration of solar air heater components [7]

Geometric configuration of collector	Value (mm)
Length (L)	1100
Width (W)	600
Depth of air duct (H_{Duct})	50
Height of PCM unit (H_{PCM})	20
Thickness of absorber plate (H_{ap})	1.25

Table 2 Thermal properties of paraffin wax [7]

Thermal conductivity (W/mK)	Latent heat of fusion (KJ/Kg)	Density in liquid state (Kg/m^3)	Density in solid state (kg/m^3)	Specific heat (KJ/KgK)	Solidus-liquidus temperatures (K)
0.2	250	890	850	2.5	325–329

3 Mathematical Modelling

Figure 2 elaborates the network of the thermal resistances by means of convective and radiative heat transfer and their interaction with each other.

Energy balancing: Steady state energy balancing is considered and the boundary conditions for collector components are formed as follows [9]:

- Glass:

$$I\alpha_g = h_w(T_g - T_a) + h_{rgam}(T_g - T_a) + h_{cgf1}(T_g - T_{f1}) + h_{rapg}(T_{up} - T_g) \tag{1}$$

- Air Duct:

$$\dot{m}_1 C_{p\_air}\frac{dT}{Wdx} = h_{cgf}(T_g - T_f) + h_{capf}(T_{ap} - T_f) \tag{2}$$

- Absorber Plate:
  - Top wall:

$$I\alpha_p\tau_g = h_{cupf}(T_{ap} - T_f) + h_{rapg}(T_{ap} - T_g) + k_{ap}\left(\frac{dT}{dy_{ap}}\right) \tag{3}$$

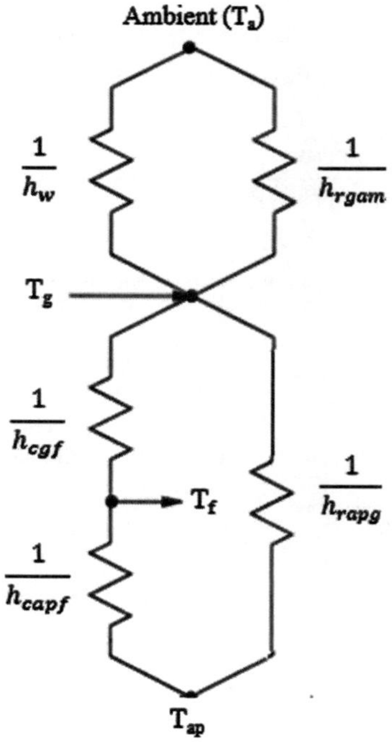

Fig. 2 Thermal resistance network for the proposed design

– Bottom wall:

$$k_{ap}\left(\frac{dT}{dy_{ap}}\right)_{y=0} = k_{pcm}\left(\frac{dT}{dy_{pcm}}\right)_{y=h} \quad (4)$$

- PCM unit:
  – Top wall:

$$k_{ap}\left(\frac{dT}{dy_{ap}}\right)_{y=0} = k_{pcm}\left(\frac{dT}{dy_{pcm}}\right)_{y=h} \quad (5)$$

  – Bottom wall:

$$\left(\frac{dT}{dy_{pcm}}\right)_{y=0} = 0 \quad (6)$$

Boundary Conditions

The above equations are solved simultaneously by matrix inversion method and are then applied to the computational domain on north and south boundary respectively as Dirichlet boundary conditions whereas left and right walls are considered to be insulated for all SAH components to get the temperature distribution inside solar air heater components.

The correlations used in the present study are the radiation and convection heat transfer coefficients which are calculated on basis of correlations presented in the literature.

- Radiative heat transfer coefficient between glass and surrounding:

$$\mathrm{h_{rgam}} = \varepsilon_{\mathrm{g}}\sigma\left(\mathrm{T_g} + \mathrm{T_{am}}\right)\left(\mathrm{T_g^2} + \mathrm{T_{am}^2}\right) \quad (7)$$

- Radiative heat transfer coefficient between absorber plate and glass cover:

$$\mathrm{h_{rapg}} = \frac{\sigma\left(\mathrm{T_{up}} + \mathrm{T_g}\right)\left(\mathrm{T_{up}^2} + \mathrm{T_g^2}\right)}{\left(\frac{1}{\varepsilon_p}\right) + \left(\frac{1}{\varepsilon_g}\right) - 1} \quad (8)$$

- Convective heat transfer between wind and glass cover:

$$\mathrm{h_w} = (2.8 + 3.3\mathrm{V_w}) \quad (9)$$

- Convective heat transfer coefficient inside duct:

$$h_{cgf} = \frac{Nu_m k_a}{D_h} \tag{10}$$

where, Nu for flat and corrugated surface is given by Gao et al. [10]:

$$Wavy\,surface.\ \ Nu_{pf} = 0.0743Re^{0.76} \tag{11}$$

$$Flat\,surface.\ \ Nu_m = 0.018Re^{0.8}Pr^{0.4} \tag{12}$$

Further, FVM solution technique with explicit time marching scheme is employed to find the numerical solution for the governing equation for solar air heater components, viz., absorber plate and PCM unit. Therefore, the governing equations are as follows given by Sharma [11],

$$\rho c_p \frac{T_p^{n+1} + T_p^n}{\partial t} = Q_{cond_p} \tag{13}$$

$$\begin{aligned} Q_{cond_p} = {} & \left(k\frac{T_E^N - T_P^N}{\delta s_e}\Delta S_{s,e} - k\frac{T_P^N - T_W^N}{\delta s_w}\Delta S_{s,w}\right) \\ & + \left(k\frac{T_N^N - T_P^N}{\delta s_n}\Delta S_{s,n} - k\frac{T_P^N - T_S^N}{\delta s_s}\Delta S_{s,s}\right) \\ & + \left(k\frac{T_{ne}^N - T_{sw}^N}{\delta t_e}\Delta S_{t,e} - k\frac{T_{nw}^N - T_{sw}^N}{\delta t_w}\Delta S_{t,w}\right) \\ & + \left(k\frac{T_{ne}^N - T_{nw}^N}{\delta t_n}\Delta S_{t,n} - k\frac{T_{se}^N - T_{sw}^N}{\delta t_s}\Delta S_{t,s}\right) \end{aligned} \tag{14}$$

PCM unit [12]

$$.H(T)_p^{n+1} = H(T)_p^n + \frac{\Delta t}{\rho \Delta V} Q_{cond_p} \tag{15}$$

$$H = \begin{cases} C_p T & if T < T_s(solid) \\ C_p T + \frac{T - T_s}{T_l - T_s} L & if T_s < T < T_l(mushy) \\ C_p T + L & if T > T_l(liquid) \end{cases} \tag{16}$$

4 Results and Discussion

This section focuses on depicting the results of variation of geometrical parameters on the thermal performance of the solar air heater. Figure 3 shows the effect of variation of amplitude on the thermal performance of the solar air heater when $\dot{m} =$

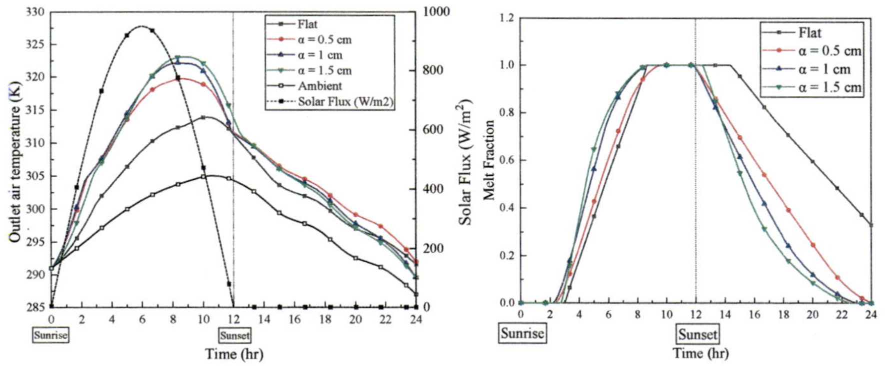

Fig. 3 Results of variation of α when $\lambda = 7.5$ cm is fixed when $\dot{m} = 0.0128\text{kg/s}$ on thermal performance of solar air heater

0.0128kg/s and $\lambda = 7.5$ cm which is the wavelength of the commercially available corrugated sheet. It was observed that with increase in amplitude the maximum air outlet temperature increased from 313.91 K for flat absorber plate to 319.7 K, 322.1 K and 323 K when $\alpha = 0.5$ cm, $\alpha = 1$ cm and $\alpha = 1.5$ cm, respectively. Higher amplitude gives higher temperature during day reason being, higher availability of heat with higher area and increased convective heat transfer coefficient. Although, at the end of 24 h higher amplitude gave lowest air outlet temperature of 289.58 K and vice-versa. It can be seen from the melt fraction plot in Fig. 3 that higher amplitude has high charging rate but at the same time it gives high discharging rate resulting in poor thermal backup. It can be seen that geometry with amplitude 0.5 cm gave optimum performance with hot air at the outlet as compared to flat collector throughout the night with highest air outlet temperature of 292 K after 24 h.

Figure 4 delineates the effects of variation of λ when $\dot{m} = 0.0128\text{kg/s}$ and amplitude is 0.5 cm on the thermal performance of solar air heater. It was observed that just like results of variation of amplitude, shorter wavelength gave higher air outlet temperature of during day time. Reason being shorter wavelength again gives higher area and vice-versa. It was observed geometry with larger wavelength of 10 cm gave highest temperature of 292.3 K at the end of 24 h maintain low charging-discharging rates as compared to other corrugated designs under consideration and higher temperature i.e., by 1 °C that the preexisting flat plate collector with hotter air throughout the night.

Figure 5 presents the contours of PCM temperature distribution and melt fraction when $\dot{m} = 0.0128\text{kg/s}$, $\alpha = 0.5$ cm and $\lambda = 10$ cm for different duration of time of day. It was observed that PCM completely melted after 3:00 PM and the PCM unit started giving thermal backup after 6:00 PM as solar flux becomes zero after that. PCM melt front as well as solidification front can be clearly seen to be propagating from the absorber plate as heat is available at absorber plate during day time whereas, heat is being extracted by air from the absorber plate during off-sunshine hours.

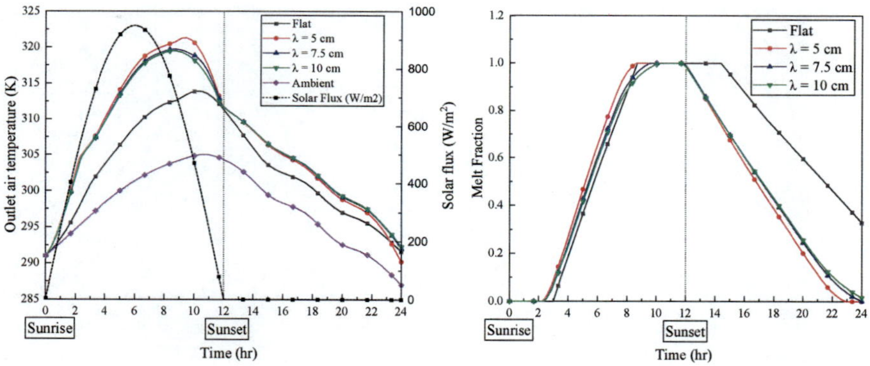

Fig. 4 Results of variation of λ when $\alpha = 0.5$ cm is fixed when $\dot{m} = 0.0128\text{kg/s}$ on thermal performance of solar air heater

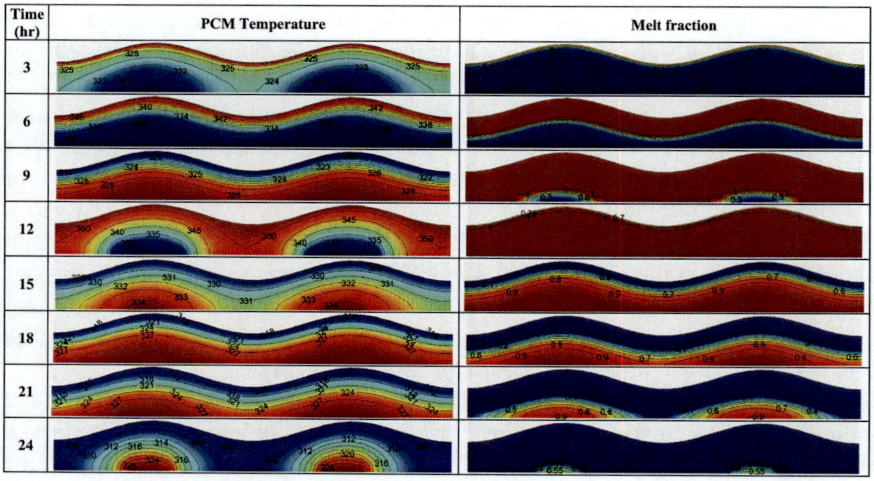

Fig. 5 Contours of temperature distribution and melt fraction for different interval of time $\alpha = 0.5$ cm and $\lambda = 10$ cm

5 Conclusion

In this present study solver is developed on MATLAB employing FVM technique to predict the transient performance of the solar air heater. The conclusions drawn from the discussed results are as follows:

- Charging-discharging rate of PCM is highly dependent on amplitude and wavelength of corrugation and should not be higher in order to achieve long thermal backup

- Wavy corrugated unit gives higher air outlet temperature compared to flat collector in same space.
- Geometry having low amplitude of 0.5 cm and wavelength of 10 cm gave most optimum charging-discharging rate with highest air outlet temperature of 292.32 K at the end of 24-h cycle which is 5.3 °C higher than the ambient.

References

1. Dhiman P, Singh S (2015) Recyclic double pass packed bed solar air heaters. Int J Therm Sci 87:215–227. https://doi.org/10.1016/j.ijthermalsci.2014.08.017
2. Singh S, Dhiman P (2014) Thermal and thermohydraulic performance evaluation of a novel type double pass packed bed solar air heater under external recycle using an analytical and RSM (response surface methodology) combined approach. Energy 72:344–359
3. Singh S, Dhiman P (2015) Using an analytical approach to investigate thermal performance of double-flow packed-bed solar air heaters with external recycle. J Energy Eng 141(3):04014031. https://doi.org/10.1061/(asce)ey.1943-7897.0000211
4. Arfaoui N, Bouadila S, Guizani A (2017) A highly efficient solution of off-sunshine solar air heating using two packed beds of latent storage energy. Sol Energy 155:1243–1253
5. Raam Dheep G, Sreekumar A (2019) Thermal reliability and corrosion characteristics of an organic phase change materials for solar space heating applications. J Energy Storage 23:98–105
6. Raj AK, Srinivas M, Jayaraj S (2019) A cost-effective method to improve the performance of solar air heaters using discrete macro-encapsulated PCM capsules for drying applications. Appl Therm Eng 146:910–920
7. Moradi R, Kianifar A, Wongwises S (2017) Optimization of a solar air heater with phase change materials: Experimental and numerical study. Exp Therm Fluid Sci 89:41–49
8. Kabeel AE, Khalil A, Shalaby SM, Zayed ME (2017) Improvement of thermal performance of the finned plate solar air heater by using latent heat thermal storage. Appl Therm Eng 123:546–553
9. Verma G, Singh S, Chander S, Dhiman P (2022) Numerical investigation on transient thermal performance predictions of phase change material embedded solar air heater. J Energy Storage 47:103619
10. Gao W, Lin W, Liu T, Xia C (2007) Analytical and experimental studies on the thermal performance of cross-corrugated and flat-plate solar air heaters. Appl Energy 84(4):425–441. https://doi.org/10.1016/j.apenergy.2006.02.005
11. Sharma A (2016) Introduction to computational fluid dynamics: development, application and analysis. Wiley
12. Ozisik MN (1994) Finite difference methods in heat transfer. CRC, Boca Raton
13. Salih SM, Jalil JM, Najim SE (2019) Experimental and numerical analysis of double-pass solar air heater utilizing multiple capsules PCM. Renew Energy 143:1053–1066

Experimental Study of PCM Based Latent Heat Thermal Energy Storage System Using Fins

Badal Kudachi, Bipin Mashilkar, Nilesh Varkute, Omkar Mawalankar, Ashish Shanbhag, Shalom Gaikwad, and Antony Maria Camillus

1 Introduction

A sustainable and fulfilling future has long been needed by the world. As a result of the world's population's constant growth, there is a sharp increase in the need for energy. However, the supply of energy is constantly under an imposition to provide the needs of every living being on this planet. The search for new and renewable energy sources still goes on today as well. This makes development of energy storage devices for those new energy sources as crucial as developing new sources of energy. By storing the energy in a different form, there would be no need to maintain schedules to generate energy as per the energy demand at a given time, such as varying demand in energy from nuclear power plants during day time and night time, due to the lack of energy storage method. The intermittent availability of thermal energy resources, especially solar energy can be precisely utilized by developing a simplified and effective model of a storage system for a sufficient accumulation and supply of energy as per demand. One of the most efficient approaches for recovering heat from industrial waste and utilising solar energy is latent heat storage in phase transition materials. These systems' principal benefit appears to be their capacity to store a significant quantity of energy in very tiny volumes at a constant transition temperature. Research into heat transfers/exchanges in PCMs during their phase transition in the requisite operating temperature range is thus necessary for the creation of a latent heat thermal energy storage system.

For an implicit achievement of the knowledge of this trending field and to study the amendment of the properties of PCM, several papers from different journals were referred and studied precisely. Amrouche et al. [1] have made a concise review

B. Kudachi · B. Mashilkar (✉) · N. Varkute · O. Mawalankar · A. Shanbhag · S. Gaikwad · A. M. Camillus
Department of Mechanical Engineering, Fr. C. Rodrigues Institute of Technology, Vashi, Navi Mumbai, Maharashtra 400703, India
e-mail: bipin.m@fcrit.ac.in

S. Doolla et al. (eds.), *Advances in Clean Energy and Sustainability*,
Green Energy and Technology, https://doi.org/10.1007/978-981-99-2279-6_70

of various storage methods for renewable energy systems. They have included a precise study of chemical, electro-chemical, mechanical, thermal and electrical storage systems. Eslami and Bahrami [2] studied the thermal energy storage systems in particular. They asserted that thermal energy can be stored in three different ways: as perceptible heat, latent heat, or thermochemical energy. Thermal energy storage (TES), whether latent or sensible, has been the primary area of study in this work. Zalba et al. [3] has summarized by reviewing many papers based on thermal energy storage using PCMs. It's been determined that PCM makes it possible to store a lot of energy in relatively little spaces. Consequently, it has emerged as one of the least expensive storage media among many storage systems. A. Sharma et al. [4] has reviewed the classification of PCM. While inorganic PCMs can include metallics and hydrated salts, organic PCMs can include a variety of materials like various types of paraffin, eutectic mixtures, fatty acids, esters and numerous other organic compounds. They have compared the different properties of organic and inorganic PCMs and noted that the organic PCMs prove better than the inorganic ones due to their chemical stability, non- corrosive nature, harmonious melting, etc. Boda M. A. et al. [5] discussed various advantages and disadvantages of organic as well as inorganic PCMs and conducted a precise comparison between them. They noted that organic PCMs are more stable and non-corrosive than inorganic ones. However, their low thermal conductivity needs to be taken under concern for improvisation. Various applications of PCMs in the field of solar, medical, textiles, etc. are also included in this paper. Nair et al. [6] focused on the A PCM storage unit's experimental analysis for thermal energy storage. A thermal energy storage system is an essential tool for balancing out the energy supply and demand. It permits the short-term thermal energy storage at either high or low temperatures. These are also essential for energy conservation. Comparisons are made between the charging and discharging (storage) times of two PCMs, paraffin wax and myristic acid. This is done in a straightforward and cheaply constructed experimental setup. Hosseini et al. [7] carried out an experiment to determine the change in heat transfer rate that happened as a result of the phase change that occurs in PCMs when heated over their melting point. They looked into how varied intake temperatures affected how long it took for the PCM to completely melt overall. Their results showed that the convective mode heat transfer becomes more pronounced when phase change occurred, and convection had a higher influence in terms of heat transfer rate as compared to conduction mode of heat transfer. By utilizing both internal and external fins, Al-Abidi et al. [8] investigated how to improve heat transfer for a triplex tube heat exchanger. Various design and operating characteristics including number of fins, fin length, thickness of the fins, PCM unit geometry and Stefan number were examined. The simulation findings show that these parameters significantly affect the amount of time needed for complete melting; the influence of fin thickness on melting rate time is less significant than that of fin length and density. Tiari et al. [9] used a transient 2-Dimensional system to simulate charging a finned heat pipe, latent heat thermal energy storage system with a square-shaped high melting temperature PCM. Numerical studies show that natural convection strongly influences PCM melting.

According to a number of researchers, the main disadvantage is that slow charging and discharging rates are caused by the low thermal conductivities of many PCMs. The PCM's heat transfer rate can be maximized by selecting the fins with the most appropriate geometry and dimensions. The goal of the current work is to examine the charging, discharging, and efficiency of PCM-based thermal energy storage devices with and without fins (annular).

2 Theoretical Analysis

2.1 PCM Characteristics

The comparative study between different PCM materials was done. It was found that Paraffin wax OM46 has the most suited temperature range for the present experiment as the operating temperature range for the heat transfer fluid taken in the current study is around 45 °C -55°C. Because of its easy availability, cheap price as well as the chemical stability, OM46, proves to be the most appropriate and hence selected for this project study. Table 1 shows the thermo physical characteristics of PCM. Figure 1 shows the actual image of PCM OM-46 that was procured for use in the current experimentation.

Table 1 PCM properties

S. No.	Material properties	Value of quantity	Unit
1	Melting temperature	321	K
2	Density	917	kg/m^3
3	Specific heat	2020	J/kg-K
4	Thermal conductivity	0.22	W/m–K
5	Viscosity	0.00534	kg/m-s
6	Thermal expansion coefficient	0.00012	K^{-1}
7	Pure solvent melting heat	165,000	J/kg
8	Solidous temperature	321	K
9	Liquidous temperature	324	K

Fig. 1 Phase change material (OM 46)

2.2 Fin Selection

Various types of fins are available for application in heat exchanger such as annular fins, longitudinal fins and pin fins. Literature review shows that among annular, longitudinal and pin fin, the best suited fin was found to be annular fin due to more surface area, radial symmetry [8–10]. The material of tube and fin has taken as copper because of high thermal conductivity.

2.3 Fin Effectiveness and Efficiency

Effectiveness justifies the applicability of fin, and is defined as the ratio of heat transfer with fins to the heat transfer without fins. The calculation of fin effectiveness discussed below.

Outer Diameter of pipe (d_1) = 0.0127 m.
Length of fin (L) = 0.01 m.
Diameter of fins (d_2) = 0.0127 + 2*0.01 = 0.0327 m.
Taking thickness of fins (t) = 0.003 m.
Number of fins (n) = 31.
Spacing between the fins (s) = 0.029 m.
Base Temperature (T_0) = 60 °C.
Surrounding Temperature = T_∞ = 30 °C.
Thermal Conductivity of Copper material = 324 W/m °C (Fig. 2).

The efficiency has been calculated from above graph using two parameters, which are mentioned below.

$$\frac{r_2 + 0.5t}{r_1} = 2.8897 \tag{1}$$

$$(L + 0.5t)\sqrt{\frac{h}{kt}} \tag{2}$$

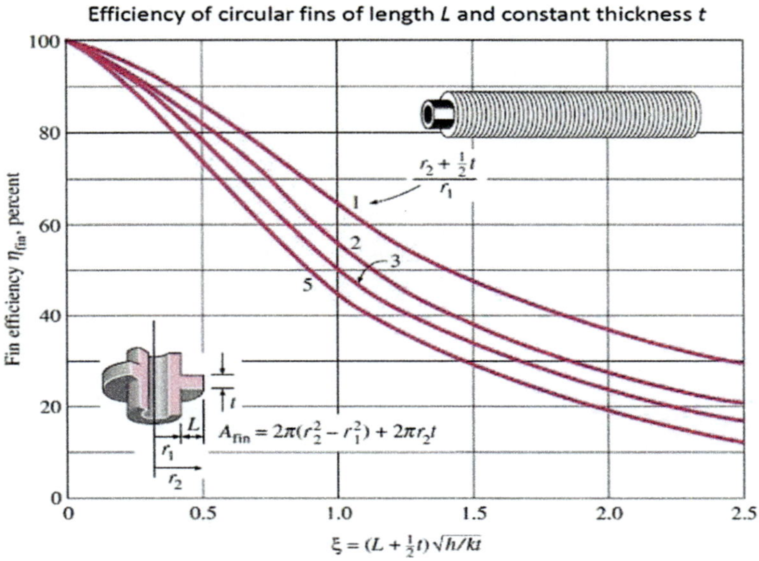

Fig. 2 Variation in efficiency of annular fins [10]

$$\text{Effectiveness of fins} = \frac{Q_{fin}}{Q_{withoutfin}} = \frac{A_{fin}}{A_{withoutfin}} \tag{3}$$

The effectiveness of the fin for present work is found to be 2.004, application of fin justified as it will enhance the heat transfer rate through surface.

2.4 *Fabrication of Fins*

As previously stated, copper was chosen as the fin material, and fins are fabricated as per the requirement from copper rod with an exterior diameter of 36 mm and a 16.5 mm center hole cut on a conventional lathe machine. Circular rings of varying thickness were then cut and brazed to a conventional copper pipe with a diameter of 0.5 in. Figures 3 and 4 show actual image of a fin manufactured, and finned pipe respectively.

Fig. 3 Fin

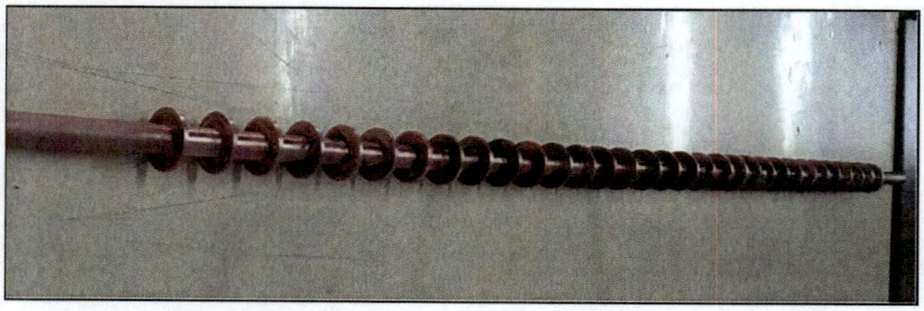

Fig. 4 Finned pipe

3 Experimentation

3.1 Physical Model

The inner tube diameter of the horizontal shell and tube heat exchanger is 0.5 in., and its length is 35 in. Due to its advantageous thermal characteristics, copper has been considered as an inner tube material. The exterior of the shell is covered in asbestos insulating tape and is made of galvanised iron (GI), which helps to keep heat inside. Conventional PCM was used to fill the annular space between the two pipes (GI and copper tube). One configuration includes a fin, whereas the other does not. The inner tube was penetrated by HTF water.

Experiments in the present work were done using two distinct heat cycles: charging and discharging. Figure 5 displays the experimental setup for this study. Although discharging takes place when the PCM returns heat to the working fluid and solidifies, charging takes place when the PCM absorbs heat from the HTF and melts. Initial charging cycle experiments were conducted with a constant 55 °C intake water temperature. The heated water was forced through the copper tube at a rate of 360 L/h until all thermocouples indicated a temperature just above PCM's melting point (Fig. 6).

With a steady intake flow rate of 360 L/h and temperature of 30 °C, the process was repeated for the discharging cycle until the readings of all the thermocouple reached

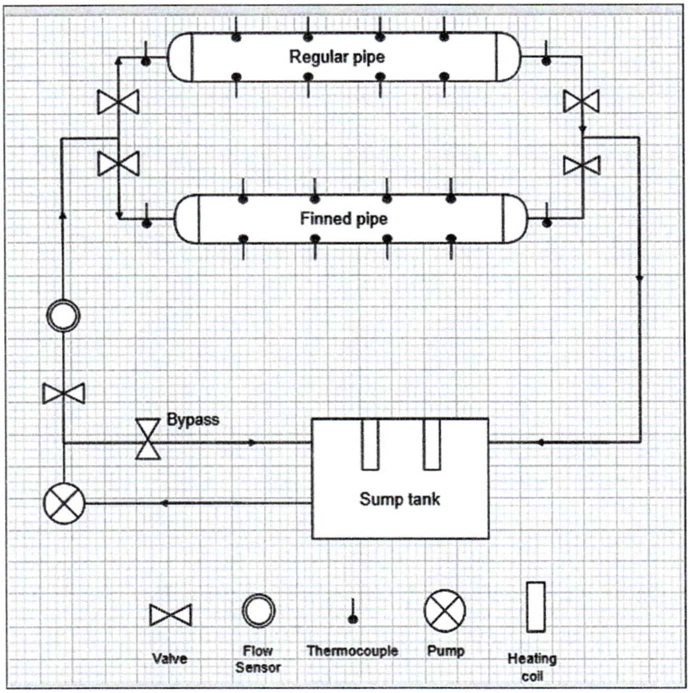

Fig. 5 Line diagram

Fig. 6 Experimental setup-top view

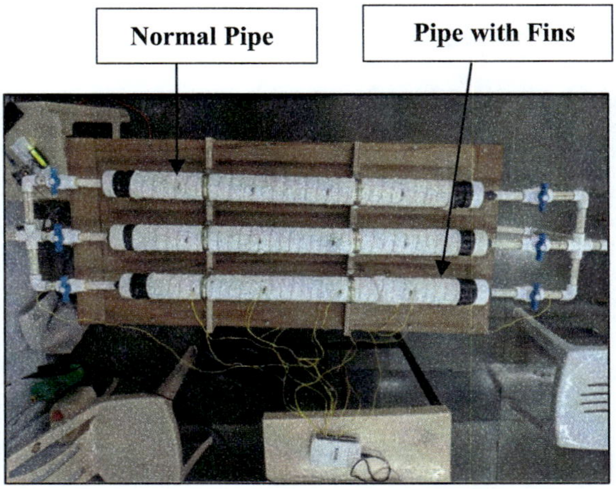

room temperature, signifying that the PCM had solidified and the discharging cycle was over. Using the system's Fin Enhanced PCM (FEPCM), the procedure was carried out, and results for the charging and discharging cycles were obtained.

K-type thermocouples were inserted throughout the system at specific points to measure temperature. The thermocouples were connected to the NI DAQ system (National Instruments modelled as NI-DAQ USB 6211) and readings were obtained using the correct pin diagrams. The system was operated for the allotted period of time, with readings shown continuously every second. The positions and arrangement of thermocouples are depicted schematically in Fig. 7, and the DAQ channel notations are listed in Table 2.

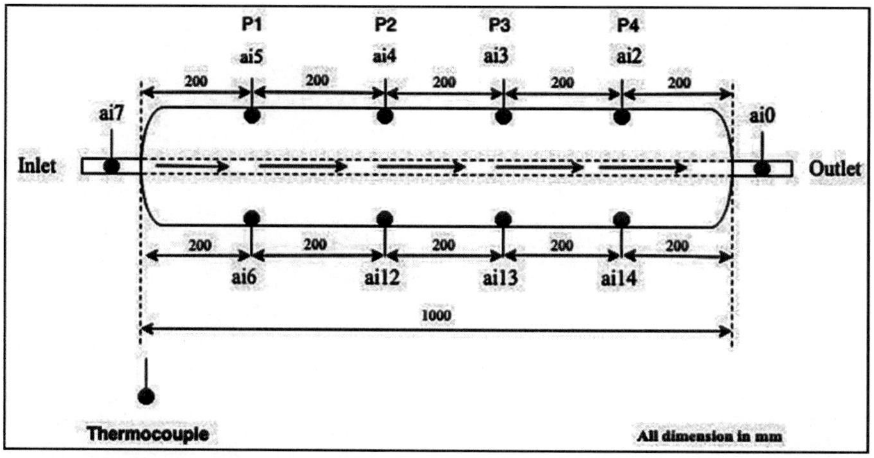

Fig. 7 Location of thermocouples across tube

Table 2 DAQ channels notation with respect to thermocouples position

Point	Position	Thermocouple channel
Water	Intake water position	ai7
	Outlet water position	ai0
P1	Top thermocouple	ai5
	Bottom thermocouple	ai6
P2	Top thermocouple	ai4
	Bottom thermocouple	ai12
P3	Top thermocouple	ai3
	Bottom thermocouple	ai13
P4	Top thermocouple	ai2
	Bottom thermocouple	ai14

4 Results and Discussions

4.1 *Charging and Discharging of PCM System (Without Fins)*

Figure 8 displays thermocouple temperature data during the charging cycle of a typical PCM system (without fins). Close examination of the results reveals that the charging time for the system with regular PCM is around 160 min at each thermocouple station (P1, P2, P3, P4). It also reveals that the PCM melting temperature takes the long time to reach at the temperature point closest to the outlet. Additionally, because the PCM moves upward and the colder portion remains below it, the upper temperatures are always higher than the lower temperatures in any site.

The temperature data for the dis-charging cycle were taken after the charging system. Figure 9 shows the temperature discharge cycle graphs as a function of time. It illustrates that the system takes around 105 min to fully discharge and reach room

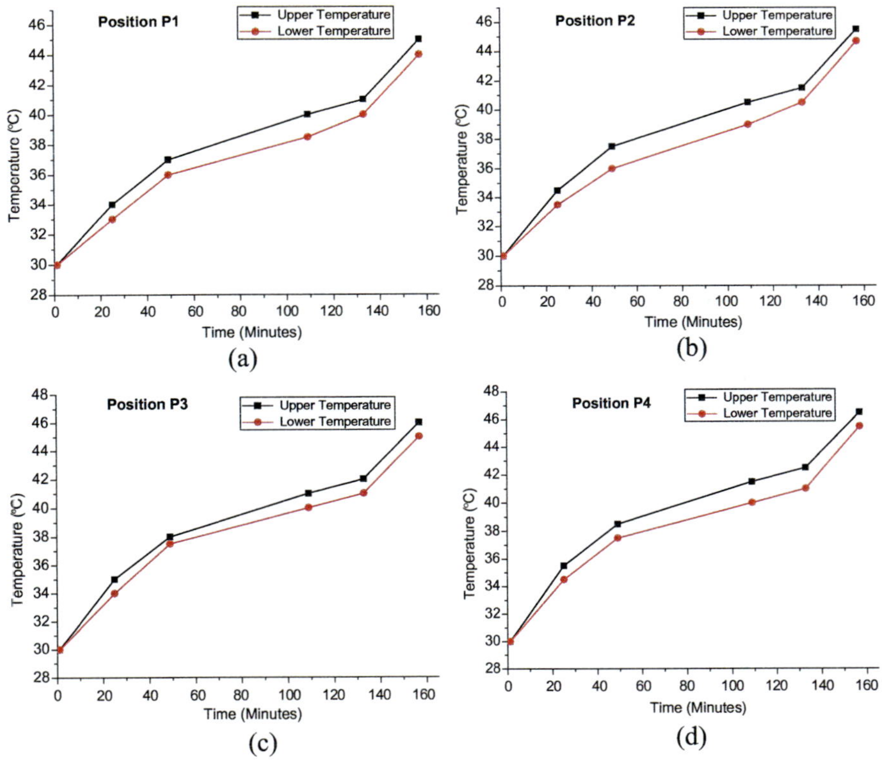

Fig. 8 Charging cycle characteristics of a standard PCM system (without fins) at specific locations **a** P1, **b** P2, **c** P3, **d** P4

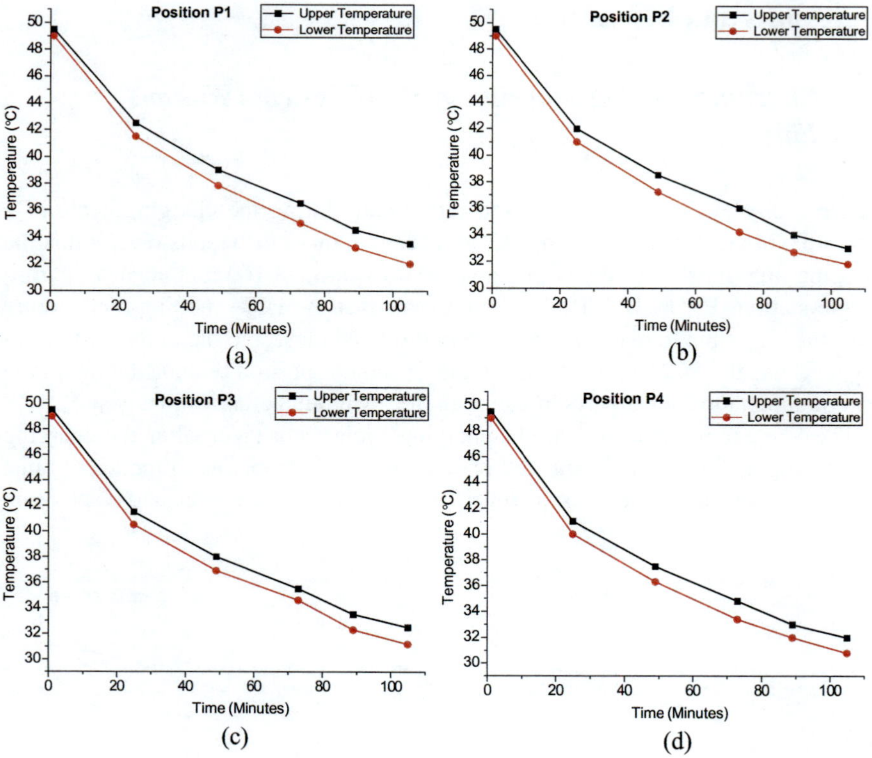

Fig. 9 Discharging cycle characteristics of a standard PCM system (without fins) at specific locations **a** P1, **b** P2, **c** P3, **d** P4

temperature at all thermocouple positions. It can be concluded that the PCM has solidified at this stage.

4.2 *Charging and Discharging of Fin-Enhanced PCM (FEPCM) System*

Figure 10 shows the temperature readings obtained for each thermocouple position (P1, P2, P3, P4) throughout the charging cycle of the FEPCM system. The charging duration is found to be only about 116 min, which is much less than that of a standard PCM system without fins. Furthermore, the discharging duration of this system is only about 62 min, which is significantly less than the first system. Between this example, too, the difference in top and bottom temperatures at each position may be seen in Fig. 11.

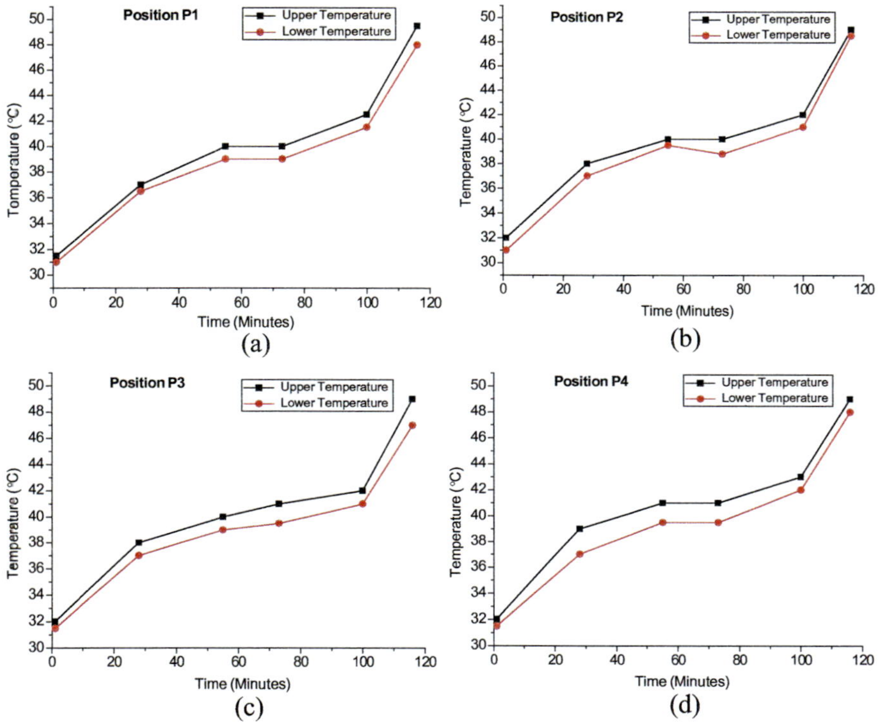

Fig. 10 Charging cycle characteristics of a FEPCM system (with fins) at specific locations **a** P1, **b** P2, **c** P3, **d** P4

Besides that, outlet temperature for both systems is consistently lower than the water inlet temperature during the charging cycle, indicating that the PCM is absorbing heat from the water.

5 Conclusion

A precise interpretation of the results showed that charging time for the heat exchanger with fins was 27.09% lesser than the regular heat exchanger (without fins). Also, regardless of the presence of fins, The heat exchanger's PCM temperature is always higher at the upper point than it is at the bottom point. Solid PCM with lower temperatures settles down while liquid PCM with higher temperatures rises to the top. This is because solid PCM has a higher density than liquid PCM.

The discharging cycle revealed that the time required for complete PCM discharge is 26.36 percent less for heat exchangers with fins, implying that heat retrieval occurs at a faster rate. Additionally, the efficiencies for both the charging and discharging cycles were found to be higher for the finned heat exchanger, demonstrating that the

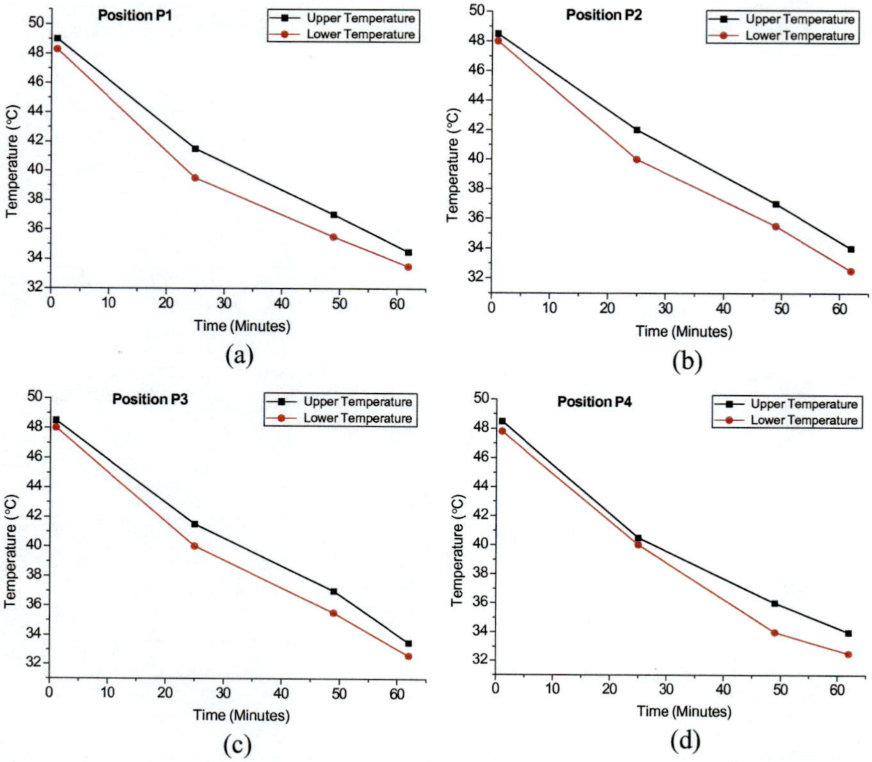

Fig. 11 Discharging cycle characteristics of a FEPCM system (with fins) at specific locations **a** P1, **b** P2, **c** P3, **d** P4

use of fins to increase heat transfer rate is justified because it allows the system to work with higher efficiencies in both cycles.

References

1. Amrouche SO, Rekioua D, Rekioua T, Bacha S (2016) Overview of energy storage in renewable energy systems. Int J Hydrogen Energy 41(45):20914–20927
2. Eslami M, Bahrami MA (2017) Sensible and latent thermal energy storage with constructal fins. Int J Hydrogen Energy 42(28):17681–17691
3. Zalba B, Marın JM, Cabeza LF, Mehling H (2003) Review on thermal energy storage with phase change: materials, heat transfer analysis and applications. Appl Therm Eng 23(3):251–283
4. Sharma A, Veer Tyagi V, Chen CR, Buddhi D (2009) Review on thermal energy storage with phase change materials and applications. Renew Sustain Energy Rev 13(2):318–345
5. Boda MA, Phand RV, Kotali AC (2017) Various applications of phase change materials: thermal energy storing materials. Int J Emerg Res Manag Technol 6(4):167–171

6. Nair AM, Vinod Kumar Naidu P (2018) Comparison of charging and discharging period analysis of phase change materials-paraffin wax and myristic acid. Int J Curr Eng Technol 8(1):44–4ior
7. Hosseini MJ, Ranjbar AA, Sedighi K, Rahimi M (2012) A combined experimental and computational study on the melting behavior of a medium temperature phase change storage material inside shell and tube heat exchanger. Int Commun Heat Mass Transfer 39(9):1416–1424
8. Al-Abidi AA, Mat S, Sopian K, Sulaiman MY, Mohammad AT (2013) Internal and external fin heat transfer enhancement technique for latent heat thermal energy storage in triplex tube heat exchangers. Appl Therm Eng 53(1):147–156
9. Tiari S, Qiu S, Mahdavi M (2015) Numerical study of finned heat pipe-assisted thermal energy storage system with high temperature phase change material. Energy Convers Manage 89:833–842
10. Hosseinizadeh SF, Tan FL, Moosania SM (2011) Experimental and numerical studies on performance of PCM-based heat sink with different configurations of internal fins. Appl Therm Eng 31(17–18):3827–3838
11. Kudachi B, Varkute N, Mashilkar B, Guthulla S, Jayaprakash P, Aaron A, Joy S (2021) Experimental and computational study of phase change material based shell and tube heat exchanger for energy storage. Mater Today: Proc 46:10015–10021
12. Varkute N, Kudachi B, Mashilkar B, Kamble P, Rane A, Harad A, Rebello G (2021) Proceedings of 2nd international congress on advances in mechanical and systems engineering (CAMSE-2021), Dr. B. R. Ambedkar National Institute of Technology, Jalandhar, India. 17–19 July, 2021

Thermal Performance Evaluation of Indian Standard Solar Box Cooker (SBC) with Retrofitted Radiative Control

Md. Rahbar Jamal, S. K. Samdarshi, P. S. Panja, Sandip Kumar Maurya, and Santosh Tigga

Nomenclature

SBC	Solar box cooker
TPPs	Thermal performance parameters
COR	Cooker opto-thermal ratio
F_1	First figure of merit
F_2	Second figure of merit
PCM	Phase change material
η_0	Optical efficiency
U_l	Over all heat loss factor (W/ (m^2 °C))
T_{ps}	Absorber plate saturation temperature (°C)
T_a	Ambient temperature (°C)
M	Mass of load (kg)
C_p	Specific heat capacity of load (J K^{-1} kg^{-1})
$(t_2\text{-}t_1)$	Time interval during which water temperature rises from T_{w1} to T_{w2} (seconds)
T_{w1}	Initial water temperature (°C)
T_{w2}	Final water temperature (between 90–95 °C)
$\overline{G_T}$	Average solar insolation during period of $(t_2\text{-}t_1)$ (W/m^2)

Md. Rahbar Jamal (✉) · S. K. Samdarshi · P. S. Panja · S. K. Maurya · S. Tigga
Department of Energy Engineering, Central University of Jharkhand, Ranchi 835205, India
e-mail: rahbarjamal10@gmail.com

Md. Rahbar Jamal · S. K. Samdarshi
Centre of Excellence in Green and Efficient Energy Technology, Central University of Jharkhand, Ranchi 835205, India

S. Doolla et al. (eds.), *Advances in Clean Energy and Sustainability*,
Green Energy and Technology, https://doi.org/10.1007/978-981-99-2279-6_71

1 Introduction

Affordable and clean energy for all, is one among the 17 sustainable development goals of the 2030 agenda for sustainable development adopted by all united nations member countries in 2015 [1, 2]. Solar cooking, due to the equitable nature and free availability of solar energy, could be pivotal to achieve this goal, mainly in developing nations where solar insolation is in abundance [3, 4]. Researcher and solar cooking enthusiasts around the globe are continuously working to design and develop new systems; and also making the existing system better to meet the need of the user. Through a detailed literature survey, it could be concluded that solar box cookers are being made better each passing day by optimizing the cooker design [5–8], employing different pot designs [9–11], integrating thermal storage systems [8, 12] and better radiation augmentation arrangements [7, 8, 13, 14]. Prior to all these, selection of a stable as well as user friendly opto-thermal performance parameter and test procedure are also necessary for satisfactory and acceptable evaluation; and mass propagation of solar cooking for its acceptance. Mullick et al. presented thermal performance parameter in term of figures of merit which are used.

For thermal performance evaluation of SBC [15]. Other thermal performance parameters associated with its performance evaluation are standard cooking power [16], characteristic and specific boiling time [17], utilizable efficiency [17], thermal efficiency [18], effective concentration ratio [19] and cooker opto-thermal ratio (COR) [20]. Buddhi et al. [8] designed and tested the performance of a solar box cooker with three reflectors and a cooking vessel with latent heat storage arrangement. They reported successful late evening (20.00 h) cooking with 4 kg of PCM charged during sunshine period. Coccia et al. [13] designed and tested a prototype SBC with multiple reflectors arrangement (of high concentration ratio) and reported high value of figures of merit and a steady value of cooker opto-thermal ratio for different quantities of load. Mirdha et al. [7] theoretically investigated various designs of box cooker with different arrangements of booster mirror to get an optimum design. Based on these, an improved design was finalised and fabricated with north and south facing booster mirror and a better performance was reported in comparison to conventional box cooker fabricated out of the same material. Sagade et al. [11] developed and tested hybrid cooking pot with glass lid and reported an improvement in the value of COR and an increase in the typical value of maximum achievable temperature in comparison to conventional cooking pot. Vengadesan et al. [10] investigated the effect of adding fins of varying lengths to the lid of cooking vessel. They reported a better performance of finned vessel compared to conventional vessel and best performance was reported for the longest fin and ascribed it to the high contact area between load and fin. Ali [5] fabricated and tested an SBC for Sudanese condition using internal as well as external reflector and reported a better thermal performance (figures-of-merit) in comparison to same cooker performance without reflectors' arrangement. Amer [6] introduced a modified solar cooker in which bottom insulation of conventional solar cooker's absorber plate was removed and was exposed to solar radiation through an arrangement of the reflectors. The achievable temperature was predicted

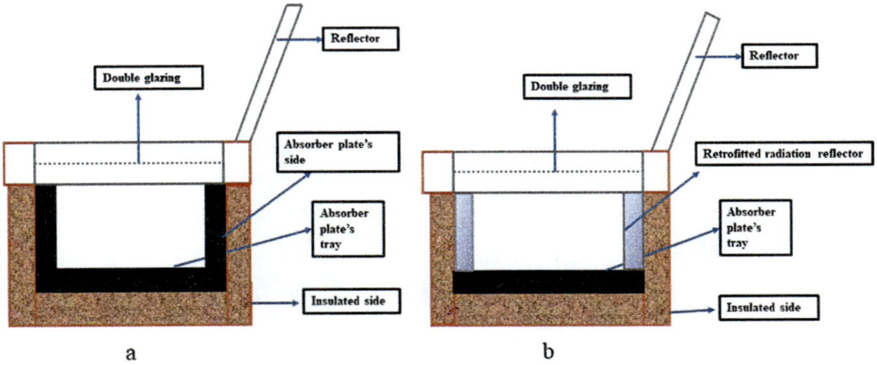

Fig. 1 Schematic of **a** Indian conventional solar box cooker (SBC) **b** retrofitted solar box cooker (SBC)

with the help of energy balance equation and it agreed well with the measured value with a very low variance. Maximum achievable temperature of absorber plate and oven air was 165 and 155 °C, respectively. Tawfik et al. [14] constructed and evaluated the thermal performance of cooker with tracking type bottom reflector and reported a higher value of the mean cooker opto-thermal ratio, effective concentration ratio, first figure of merit and an average reduced cooking time in comparison to the performance of the same cooker without assembling of tracking type bottom reflector.

In present work retrofitting has been done to Indian standard solar box cooker as shown in Fig. 1a. Herein the sides of absorbing plate have been replaced by an anodized aluminium reflecting sheet of dimension equal to the specifications of the absorber plate's sides. Figure 1 schematically shows the specifications of testing cooker with two different conditions taken into account. Part (a) shows the conventional SBC's construction and its parts, and part (b) shows the retrofitting arrangement within conventional design.

Figure 2a shows the one side of retrofitting done as per the reflector's shape and size, Fig. 2b shows SBC after retrofitting has been done, and Fig. 2c schematically shows absorber plate structure (top view). Tests are performed on the aforementioned cooker with different arrangements and is discussed in the following sections.

2 Experimental Procedure

As stated, many TPPs are available to test thermal performance parameter of solar box cooker, namely, figures of merit, standard cooking power, thermal efficiency, utilizable efficiency, specific and characteristic time and COR. Among all these available TPPs, Bureau of Indian Standards has accepted figures of merit to rate the thermal performance of a solar box cooker. F_1 tests the performance through

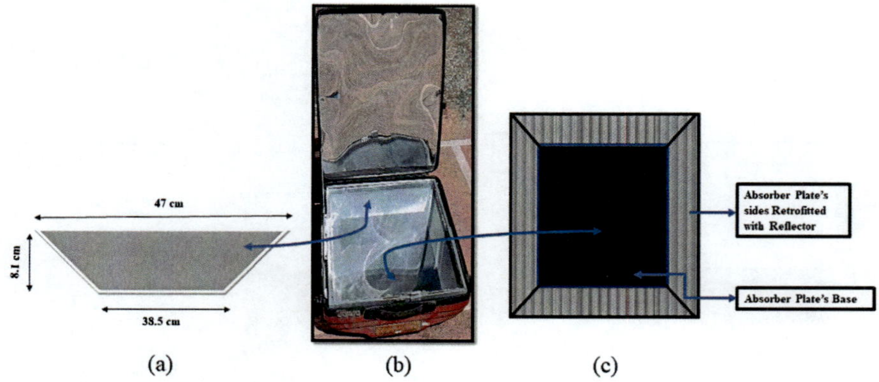

Fig. 2 **a** Reflector sheet with specification (side view), **b** retrofitted solar box cooker, and **c** schematic of retrofitted absorber plate (top view)

stagnation temperature test of absorber plate and the second figure of merit, obtained through sensible heating test of a test load, predicts the performance based on the rate of heat transfer to the load in the cooking pot. In this work test procedure of Bureau of Indian Standards IS13429 part 3 [21] has been followed. All the tests have been performed between 1 h 30 min before and after solar noon (around11:50 am) at Central university of Jharkhand, Ranchi (23.34°N,85.30°E). For the F_1 calculation ambient temperature, solar insolation at the cooker aperture level and temperature of the absorber plate measured at the mid-point between its centre and edge at an interval of 5 min each. F_1 is calculated for steady state condition of SBC. Steady state is defined as a duration of 15 min around which variation in plate temperature is ± 1 °C, solar radiation is ± 20 w/m^2, and ambient temperature is ± 0.2 °C. F_1 is defined as the ratio of optical efficiency to heat loss coefficient.

$$F_1 = \frac{\eta_0}{U_l} = \frac{T_{PS} - T_a}{G_t} \tag{1}$$

For the calculation of F_2 [15, 21] cooker is loaded with cooking pot with a mass equivalent to 8-L of water per square metre of aperture area. Solar insolation, ambient temperature, and water temperature are measured at an interval of 5 min each. Data is recorded till water temperature reaches 95 °C. F_2 is calculated from equation given as

$$F_2 = \frac{F_1(MC)_w}{A(t2 - t1)} \ln \frac{1 - \frac{1}{F_1}\left(\frac{T_{w1} - \overline{T_a}}{G_T}\right)}{1 - \frac{1}{F_1}\left(\frac{T_{w2} - \overline{T_a}}{G_T}\right)} \tag{2}$$

3 Result and Discussion

Experimental data of the performance tests are given in Tables 1 and 2. The tests for first figure of merit were performed for two different conditions:

1. Condition A: test on conventional solar box cooker, and
2. Condition B: same cooker with retrofitted absorber plate's sides with reflector (radiative control).

Hereafter, these conditions have been mentioned as condition A and condition B. Average value of F_1 without retrofitting i.e., condition A, the recorded value is 0.09, whereas with retrofitting i.e., condition B, the recorded value is 0.0942 and at the same time for condition B temperature of plate rose by about 8 °C in comparison to the condition A. A clear gain of 4.67% in the value of F_1 is found for condition B w.r.t condition A.

Average ambient temperature and solar insolation for condition A was 34.40 °C, 1046.54 W/m^2 and for condition B was 35.93 °C, 1057 W/m^2, respectively. The geographical and environmental parameters involved in the calculation of F_1 are average solar insolation ($\overline{G_t}$) and average ambient temperature ($\overline{T_a}$). And F_1 is directly proportional to difference of average plate stagnation temperature and ambient temperature ($\overline{T_{ps}} - \overline{T_a}$) and inversely proportional to average solar insolation ($\overline{G_t}$). So, logically a test condition with a low value of $\overline{T_a}$ and $\overline{G_t}$ (within acceptable

Table 1 Data and test result of test condition A

Date	$\overline{T_p}$	$\overline{T_a}$	$\overline{G_t}$	F_1
15/3/2022	130.67	31.07	1113	0.089
16/3/2022	126.6	31.7	1067	0.089
30/03/2022	131.67	34.97	1053.33	0.092
01/4/2022	124.67	35.87	996	0.089
02/4/2022	128	36.7	1014.33	0.09
04/4/2022	130.2	36.1	1035.6	0.091
Average				0.09

Table 2 Data and test result of test condition B

Date	$\overline{T_p}$	$\overline{T_a}$	$\overline{G_t}$	F_1
14/3/2022	135.25	28.58	1130.84	0.093
05/4/2022	136.5	36.35	1059	0.095
12/4/2022	136.33	37.83	997.33	0.094
14/4/2022	134.67	39.17	1019.67	0.094
25/4/2022	140	37.7	1078.25	0.095
Average				0.0942

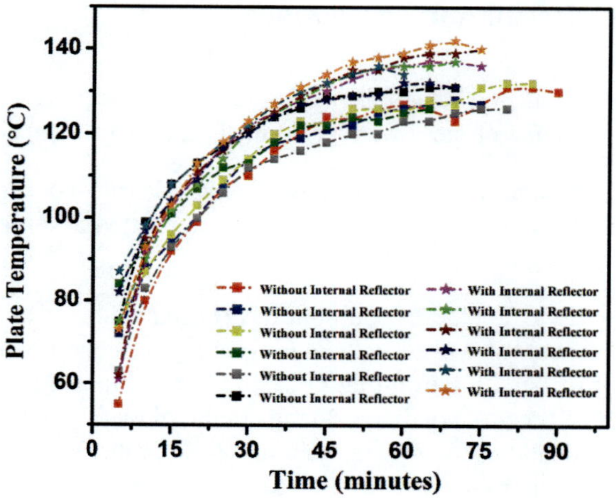

Fig. 3 Plate temperature versus time graph

range, ≥ 600 W/m^2) has a more favourable conditions to expect better performance. But contrary to the expectations the result is otherwise. It clearly indicates a positive additionality in performance due to the retrofitting.

Further, from Fig. 3, it is more evident that time taken to reach stagnation temperature is lesser and at the same time average stagnation plate temperature is higher in condition B compared to condition A. It indicates a faster heating rate too. All these favourable gains could be attributed to retrofitted internal absorber reflector or radiative control, as it produces internal concentrating effect within the enclosure. In a very first appearance a higher stagnation temperature and lesser heating time collectively indicate towards lesser need of insulation to the side wall after retrofitting. Furthermore, the cleaning requirement in this case is limited to the glass cover only. A more detailed study is being carried to support this claim. So, as of now it could be summarised that overall impact of retrofitting is encouraging. A detailed study of result with two more parameters F_2 and COR will give more insight to the results. The results of COR will be crucial one as it will impart a holistic opto-thermal performance of the cooker and retrofitting arrangement.

4 Conclusion

An attempt to achieve a better performance and operational advantage, a hybrid of box cooker and concentrating solar cooker was designed through minimal changes and addition to the cost of the existing box cooker. Retrofitting of radiative control arrangement was made out of anodized aluminium reflector sheet by fixing it to the sides of an SBC. Outcome of retrofitting showed an enhancement in the performance of SBC. F_1 increased by 4.67% and the mean stagnation temperature of plate rose by about 8 °C and at a faster rate, in comparison to the same SBC without any radiative control. More tests are being performed for a better and clear opto-thermal analysis of retrofitting. It promises fairly good carbon reduction potential.

References

1. THE 17 GOALS. Sustainable development. https://sdgs.un.org/goal. Last accessed 2022/05/16
2. TRACKING SDG7 A Joint Report of the Custodian Agencies (2021) World bank
3. Cuce E, Cuce PM (2013) A comprehensive review on solar cookers. Appl Energy 102:1399–1421
4. Lahkar PJ, Samdarshi SK (2010) A review of the thermal performance parameters of box type solar cookers and identification of their correlations. Renew Sustain Energy Rev 14(6):1615–1621
5. Ali BSM (2000) Design and testing of Sudanese solar box cooker. Renew Energy 21(3):573–581
6. Amer EH (2003) Theoretical and experimental assessment of a double exposure solar cooker. Energy Convers Manage 44(16):2651–2663
7. Mirdha US, Dhariwal SR (2008) Design optimization of solar cooker. Renew Energy 33(3):530–544
8. Buddhi D, Sharma SD, Sharma A (2003) Thermal performance evaluation of a latent heat storage unit for late evening cooking in a solar cooker having three reflectors. Energy Convers Manage 44(6):809–817
9. Mawire A et al (2020) Performance comparison of two solar cooking storage pots combined with Wonderbag slow cookers for off-sunshine cooking. Sol Energy 208:1166–1180
10. Vengadesan E, Senthil R (2021) Experimental investigation of the thermal performance of a box type solar cooker using a finned cooking vessel. Renew Energy 171:431–446
11. Sagade AA, Samdarshi SK, Lahkar PJ, Sagade NA (2020) Experimental determination of the thermal performance of a solar box cooker with a modified cooking pot. Renew Energy 150:1001–1009
12. Omara AAM, Abuelnuor AAA, Mohammed HA, Habibi D, Younis O (2020) Improving solar cooker performance using phase change materials: a comprehensive review. Solar Energy 207. Elsevier Ltd., pp 539–563
13. Coccia G, di Nicola G, Pierantozzi M, Tomassetti S, Aquilanti A (2017) Design, manufacturing, and test of a high concentration ratio solar box cooker with multiple reflectors. Sol Energy 155:781–792
14. Tawfik MA, Sagade AA, Palma-Behnke R, El-Shal HM, Abd Allah WE (2021) Solar cooker with tracking-type bottom reflector: an experimental thermal performance evaluation of a new design. Solar Energy 220:295–315
15. Mullick SC, Kandpal TC, Saxena AK (1987) Thermal test procedure for box-type solar cookers. Sol Energy 39(4):353–360
16. Funk PA (2000) Evaluating the international standard procedure for testing solar cookers and reporting performance. Sol Energy 68(1):1–7
17. Khalifa AMA, Taha MMA, Akyurt M (1985) Solar cookers for outdoors and indoors. Energy 10(7):819–829
18. Nahar NM (2003) Performance and testing of a hot box storage solar cooker. Energy Convers Manage 44(8):1323–1331
19. Sagade AA, Samdarshi SK, Panja PS (2018) Experimental determination of effective concentration ratio for solar box cookers using thermal tests. Sol Energy 159:984–991
20. Lahkar PJ, Bhamu RK, Samdarshi SK (2012) Enabling inter-cooker thermal performance comparison based on cooker opto-thermal ratio (COR). Appl Energy 99:491–495
21. Bureau of Indian Standards, IS 13429, part 3 test method, solar cooker-box type.1st revision (2000)

Performance Investigation of Superheat Recovery Water Heater Integrated in Cold Storage

Arunendra K. Tiwari, Harischander, Milind V. Rane, and Adittya M. Rane

Nomenclature

CAGR	Compound annual growth rate
CMP	Compressor
CND	Condenser
COP	Coefficient of performance
CS	Cold storage
EVP	Evaporator
HPL_IITB	Heat pump laboratory IIT Bombay
HE	Heat exchanger
LMTD	Logarithmic mean temperature difference
NE	Northeast
NW	Northwest
PCM	Phase change material
RAC	Refrigeration, air conditioning
RTD	Resistance temperature detector
SHR_WH	Superheat recovery water heater
S_MUHP	Solar multi-utility heat pump
TEV	Thermostatic expansion valve
VCR	Vapour compression refrigeration

A. K. Tiwari · Harischander · M. V. Rane (✉) · A. M. Rane
Heat Pump Laboratory, Mechanical Engineering Department, IIT Bombay, Mumbai, Maharashtra 400076, India
e-mail: ranemv@iitb.ac.in

Harischander
e-mail: 204100027@iitb.ac.in

S. Doolla et al. (eds.), *Advances in Clean Energy and Sustainability*,
Green Energy and Technology, https://doi.org/10.1007/978-981-99-2279-6_72

Suffix

amb	ambient
cmp	compressor
cnd	condensor
dsh	desuperheater
evp	evaporator
I	inlet
O	outlet
R	refrigerant
rad	radiation
w	water

1 Introduction

Heating, cooling, and air-conditioning account for 40–60% of electricity consumption in India. The heating and cooling demand in India is rising fast at a growth rate of 15–20% and the projected cooling requirement in terms of primary energy is anticipated to be more than 2.2 times more in 2027 than the 2017 baseline [1]. Significance efficiency improvement, cogeneration, and waste heat recovery will be required to meet the target without a significant increase in greenhouse gas emissions. A basic estimate of waste heat available from a refrigeration and air-conditioning (RAC) system reveals that for every kilowatt of energy used by the compressor, about 3–5 kW of waste energy is rejected by the environment [2]. Full or partial recovery of this energy in a usable form for heating applications will help to reduce the overall energy footprint. The expansion of online grocery, pharmaceutical sales, and the ongoing Covid-19 vaccination program in India are predicted to propel the Indian cold store sector to a 14% compound annual growth rate (CAGR) during the next three years [3]. The growth in the cold storage sector can be made sustainable and environment friendly if green refrigerants are used and renewable power for operation. Generating multiple utilities using a single system and further enhancing the technology will make it more sustainable.

One way to improve the overall efficiency of cold storage is to integrate a Superheat Recovery Water Heater (SHR_WH) and recover heat for hot water generation which can be used for sanitation, kitchens, and drying agro-produce. SHR_WH units are located between the compressor and condensers to utilize the high-temperature energy of the superheated refrigerant gas. The SHR_WH is normally designed not to condense any refrigerant. However, some liquid refrigerants may form, depending on operating parameters [4]. The SHR_WH can be used to utilize the superheated refrigerant to heat water to a temperature in the range of 50–90 °C depending on the refrigerant, compressor type, and operating conditions. The SHR_WH integration with cold storage not only traps the waste heat, which otherwise would have been wasted in ambient, but it also results in significant improvement in cold storage COP as the compressor has to operate at lower high-side pressure.

Douglas and Todd [5] presented the sustainable use of primary energy consumed in industrial ammonia refrigeration systems by using heat usually discarded in the ambient. They concluded that the refrigerant discharged from a compressor contains thermal energy in the form of sensible (superheat) and latent (condensing) heat. A typical ammonia-based screw compressor operating at 1.7 bar (25 psi_g), − 11.7 °C suction and 12.5 bar (181 psi_g), 35.0 °C condensing pressure and temperature accounts for, 11.5% of the total system heat rejection is available in the superheat range (higher quality) while the remaining 88.5% of the heat is available in the phase change from vapour to liquid (lower quality) in best case scenario. Kausik et al. [6] theoretical modeling work for the study of the feasibility of low-grade heat recovery from industrial vapour compression refrigeration (VCR) system using a Canopus heat exchanger (HE) before condenser concludes that the use of HE results in increased overall COP of the system. As a result, heat recovery using Canopus HE is possible and may be enhanced by selecting the best water flow rate, inlet temperature, operating conditions, and working fluid. Jia and Lee [7] experimented on room AC with a heat recovery unit for water heating. They used a paraffin wax-based storage tank for recovery. The overall coefficient of performance was improved by 6.9–9.8% using heat recovery.

Xia et al. [8] used condensation heat from cold storage refrigeration systems to provide heat for domestic hot water using a novel phase change material (PCM), a mixture of carnauba wax as the PCM, and expanded graphite as the additive. The paper discusses the procedure to integrate and operate PCM-based thermal storage cum HE in a conventional refrigeration system by using a switch valve in the first part and PCM characterization in the second part. The literature reveals that very few studies [8, 9] have investigated the cold storage heat recovery system. The performance study on heat recovery with inbuild storage from micro cold storage has not been investigated so far.

This paper presents a theoretical design for possible heat recovery from cold storage. An SHR_WH with built-in thermal storage has been integrated into the existing 0.75 TR cold store and the performance of the same is presented in the paper.

2 Materials and Methods

To tap the refrigerant superheat, a tube-tube SHR_WH with thermal storage has been developed and integrated with 0.75 TR nominal capacity cold storage (CS) having two independent R 290 refrigerant circuits. The CS has two compressors of 0.375 TR, identifies as Northeast (NE), and Northwest (NW) compressors. The outlet from two compressors is passed to SHR_WH for heat exchange. The SHR_WH is a counter flow, three-pass heat exchanger with sand-like sensible storage media. The outer dimensions of the SHR_WH are a length of 1.3 m, a width of 0.37 m, and a height

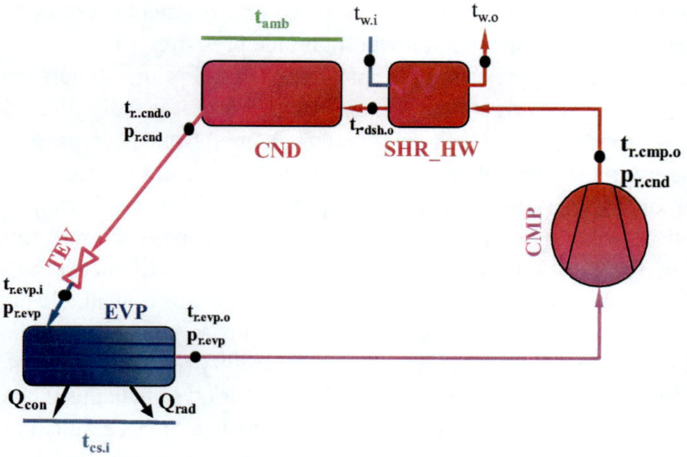

Fig. 1 Experimental setup schematic

of 0.33 m as shown in Fig. 1. The total weight of the SHR_WH is 100.1 kg and its heat capacity is 7.1 kWh per day for 7 h of operation for a temperature change of 30–75 °C. The experiments have been performed at the HPL_IITB during February and March. The refrigerant inlet and outlet temperature, water inlet, outlet temperature, and ambient temperature were recorded using RTD and stored in a multichannel data logger every minute.

A simple water flow measurement scheme was employed which could take care of pressure fluctuation in the water main line from which water was supplied into the SHR_WH. The flow rate of water was set by using a valve at the inlet of the glass tube rotameter. The hot water outlet from SHR_WH was collected in a 30 l plastic jerry can and weighed intermittently to accurately determine the average water flow through the SHR_WH. The experiments were conducted at two flow rates at 0.1 lpm and 0.15 lpm to maintain the hot water outlet at more than 60 °C. The testing of the 0.75 TR cold room has also been done and the performance of refrigerating unit using the SHR_WH is checked but the study is not part of this manuscript because of the constraints of the number of pages.

3 Refrigerating Processes on p–h and t-s Chart

It is useful to evaluate the recoverable heat available in cold storage refrigeration systems for deciding whether heat recovery makes sense for an application. This section presents and discusses the design calculation for the cold store. The refrigerating processes of one of the refrigeration circuits are plotted and presented in Figs. 2 and 3. The results show that out of all heat needed to be condensed 18–21% is in the superheated region when the ambient temperature was in the range of 25–33 °C and can be utilized for water heating. The corresponding heat duty of SHR_WH is ranging from 50 to 80 kJ/kg from a single refrigerating circuit.

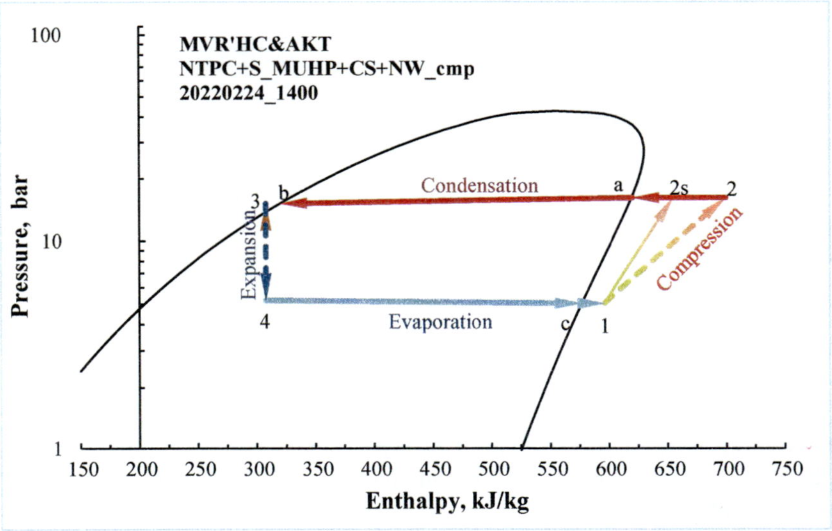

Fig. 2 p–h diagram of NW refrigeration unit at 14:00 h

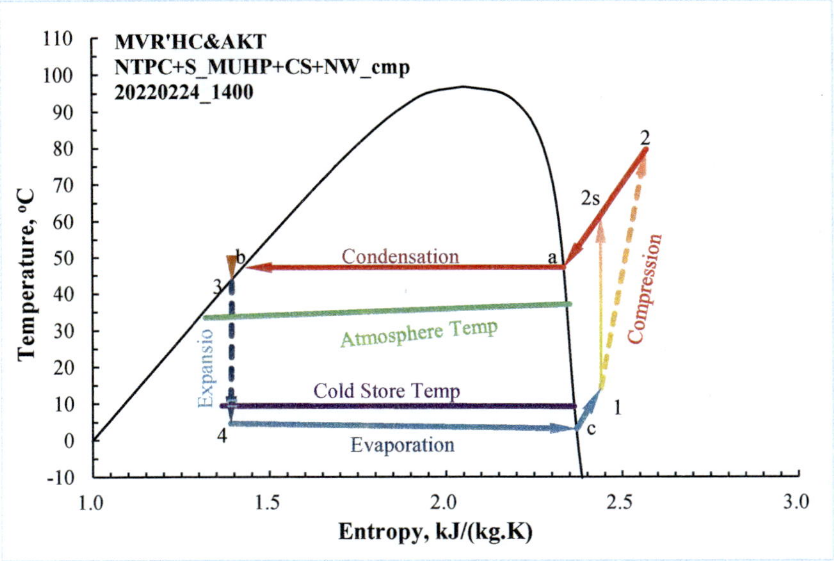

Fig. 3 t-s diagram of NW refrigeration circuit at 14:00 h

Table 1 Uncertainty parameters

Parameter	Uncertainty	Unit
Thermocouple	± 1.5	°C
Temperature recorder	± 0.1	%
Weight	± 5	g

4 Uncertainty Analysis

Commonly, the accuracy of experimental data depends on the measuring tools and measuring techniques. The uncertainties involved in the experiments are presented in Table 1.

5 Experimental Results

In this section, the results obtained from the experiments are presented and discussed. All the temperature recorded at the inlet and outlet of SHR_WH along with ambient temperature obtained on a day is plotted in Fig. 4. SHR_WH testing started at 1130 and the experiment was carried out till 1830. The average flow rate of water throughout was maintained at 0.11 lpm. The average hot water outlet temperature was 68.3 °C with the average water inlet temperature being 30.1 °C during the experiment duration. A total of 47.5 l of hot water was collected in 7 h of operation. Similar performance was achieved during other days of operation. The average performance and total hot water collected during the experimental duration are shown in Fig. 5. In February the average ambient temperature of the HPL_IIB site is in the range of 25–33 °C during hours of operation and the tap water temperature in experimental duration was around 24–30 °C. The water temperature was raised to above 68 °C. The total hot water generated was in the range of 45–50 lpd in 7 h of compressor operation. The output from the SHR_WH would be much more when the compressor operating hours are increased. The hot water can be generated continuously or intermittently as required by the thermal storage unit. Hot water can be generated at higher temperatures, even above 75 °C, when ambient temperatures range from 40 °C to 42 °C. The LMTD of the SHR_WH was calculated to be 9.3 °C to 16.6 °C.

6 Conclusions

The objective was to tap the waste heat of the condenser for heating water for various applications. The heat capacity of SHR_WH is 7.1 kWh per day for 7 h of operation for a temperature change of 30 to 75 °C. Hot water was successfully generated at the temperature range of 60–70 °C, in the integrated SHR_WH and 40–50 lpd in 7 h of operation of the compressor of hot water tapped based on the test runs in HPL_IITB

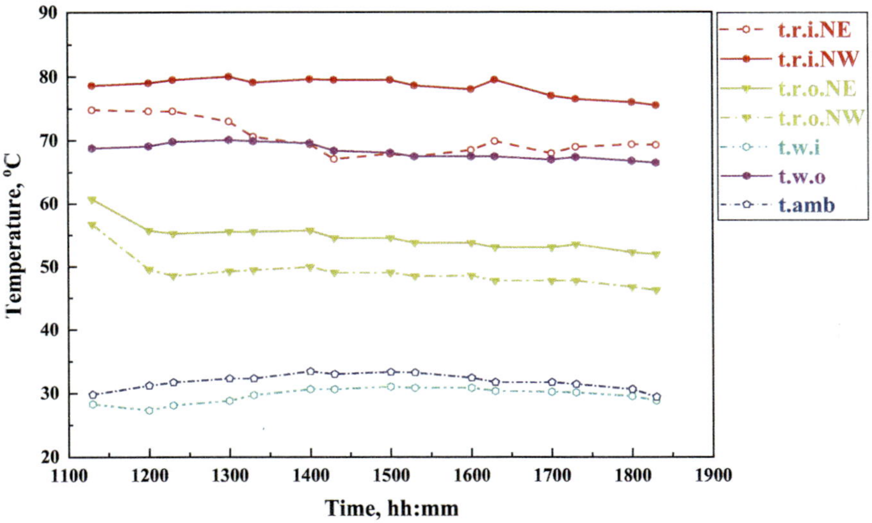

Fig. 4 Temperature versus Time for SHR_WH tested on 20,220,224

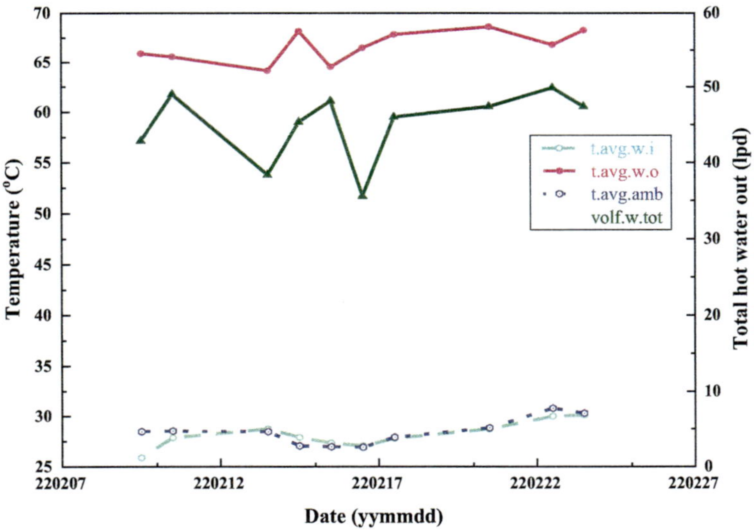

Fig. 5 Total hot water collected during 7 to 7.5 h of operation on various days

with average ambient temperatures limited to 30 °C. Even more hot water can be generated continuously at a low temperature or intermittently at increased flow rates. Water temperatures realized during higher ambient temperatures will be higher than 75 °C.

Acknowledgements The financial support from NTPC through consultancy project: RD/0220-MENTPCL-022 is gratefully acknowledged.

References

1. Kumar S (2018) Demand analysis for cooling by sector in India in 2027, New Delhi
2. Kaushik SC, Singh M (1995) Feasibility and design studies for heat recovery from a refrigeration system with a canopus heat exchanger. Heat Recovery Syst CHP 15:665–673. https://doi.org/10.1016/0890-4332(95)90046-2
3. Khan S (2021) The Indian cold chain sector is expected to grow at 14% CAGR during 2021–2023. https://economictimes.indiatimes.com/industry/services/property-/-cstruction/the-indian-cold-chain-sector-is-expected-to-grow-at-14-cagr-during-2021-2023/articleshow/83426568.cms
4. SWEP: Refrigeration handbook—SWEP. https://www.swep.net/refrigerant-handbook/refrigerant-handbook/. Last accessed 2022/05/17
5. Douglas TR, Todd BJ (2007) Heat recovery in industrial refrigeration. ASHRAE J 49
6. Kaushik SC, Panwar NL, Reddy VS (2011) Thermodynamic evaluation of heat recovery through a Canopus heat exchanger for vapour compression refrigeration (VCR) system. J Therm Anal Calorim. 110:1493–1499. https://doi.org/10.1007/s10973-011-2111-7
7. Jia J, Lee WL (2015) Experimental investigations on using phase change material for performance improvement of storage-enhanced heat recovery room air-conditioner. Energy 93:1394–1403. https://doi.org/10.1016/J.ENERGY.2015.10.053
8. Xia M, Yuan Y, Zhao X, Cao X, Tang Z (2016) Cold storage condensation heat recovery system with a novel composite phase change material. Appl Energy 175:259–268. https://doi.org/10.1016/J.APENERGY.2016.05.001
9. Li M, Li Z, Jiang X, Ye B (2011) Design and performance analysis of the heat pump-based condensing heat of cold storage recovery drying equipment. In: 2011 International conference on computer distributed control and intelligent environmental monitoring. IEEE Xplore, pp 158–161

Effect of Phase Change Material on Thermal Management of Photovoltaic System

Ravita Lamba, Francisco J. Montero, Ramesh Kumar, Arun Kumar Choudhary, Manish Vashishtha, and Sushant Upadhyaya

1 Introduction

Energy demand is continuously increasing at an alarming rate. The globally changing climatic conditions and depletion of fossil fuels motivate the researchers to shift towards renewable energy sources. Solar energy is a sustainable alternative energy source. Photovoltaic technology is the most widely used for harvesting solar energy. Photovoltaic (PV) cells are semiconductor based direct energy conversion devices working on photovoltaic effect. The solar photovoltaic system (PV) converts the short-wavelength radiation into electrical energy. Nevertheless, the long-wavelength fraction of the solar radiation is converted into heat during PV operation that rises its temperature. The PV efficiency is a function of the operating temperature of PV module, solar radiation, and dust accumulation. These parameters also degrade the life of PV module under harsh environmental conditions [1]. The commercial PV modules efficiency is still low (~20%) even after constant efforts in this direction. Almost 45–50% of the input solar energy is converted into heat in the PV system. Therefore, the intense solar radiation and high ambient temperature induce elevated PV operating temperature, and affect its life span and power output. Also, the ambient temperature reaches more than 450 °C in desert areas during harsh summer conditions and thus increasing the PV operation temperature drastically to 800 °C or even higher than 100 °C. Since there is an inverse relationship between the PV efficiency

R. Lamba (✉) · R. Kumar · M. Vashishtha · S. Upadhyaya
Malaviya National Institute of Technology Jaipur, Jaipur, Rajasthan 302017, India
e-mail: ravitalamba.lamba247@gmail.com

F. J. Montero
Pontificia Universidad Católica de Chile, Mackenna, 48607820436 Santiago de Chile, Chile

A. K. Choudhary
Indian Institute of Technology, Delhi, New Delhi 110016, India

Ministry of New and Renewable Energy, C.G.O. Complex, New Delhi 110003, India

S. Doolla et al. (eds.), *Advances in Clean Energy and Sustainability*,
Green Energy and Technology, https://doi.org/10.1007/978-981-99-2279-6_73

and its temperature. The increased PV temperature increases the lattice vibrations which results in decreased charge carrier mobility. Therefore, the PV performance degrades drastically at elevated ambient temperatures in peak summer. It is reported that the crystalline silicon module efficiency drops at a rate of around 0.45%/°C [2]. To stabilize the PV performance, the temperature of the PV system should be reduced. Consequently, effective thermal management of the PV systems is necessary to maintain its conversion efficiency. Researchers have attempted to effectively manage the PV temperature. Various cooling techniques are analyzed to effectively manage the PV temperature, principally passive and active systems like air cooling, water cooling, heat pipe, phase change materials (PCM), thermoelectric cooling, and radiative cooling.

Several cooling methods were proposed in the last years to control the PV systems' heat management and conversion efficiency. Teo et al. [3] studied a hybrid photovoltaic/thermal (PV/T) system for cooling of PV module using a parallel array of ducts for uniform distribution of airflow. A highly concentrated PV/T system with water spray cooling system has been investigated by Chen et al. [4]. The spray cooling solved the problem of temperature uniformity and high heat flux, and thus improved the concentrated PV cell's electrical and thermal efficiency. The analysis presented by Nizetic et al. [5] found that most of the examined passive cooling options are PCM followed by air-based, liquid-based, and finally radiative-based. Nizetic et al. [6] also analyzed the active cooling methods of photovoltaic modules and found that most active cooling option is to utilize water as the coolant for PV/T configurations. Najafi and Woodbury [7] proposed a cooling system in which a thermoelectric cooling system is attached to the back of a photovoltaic module. The power needed for operation of the thermoelectric cooler is provided by the photovoltaic cell itself. Du et al. [2] analyzed various cooling methods for achieving low PV module temperature favoring high module efficiency. They found that passive cooling using a heat spreader or heat sink results in a low PV module temperature even for concentrated PV systems. Bahaidarah et al. [8] studied various technologies to reach uniform cooling in PV panels. The impact of non-uniformity was found to affect all PV systems, but the effect is more pronounced in CPV systems. The use of Phase change material (PCM) as a passive cooling method of PV system was reported long back. In the PCM based PV cooling, a container filled with an appropriately selected PCM is coupled to the PV back side. The heat generated inside the PV module is absorbed by the PCM that stores this heat as latent heat of fusion during the process of melting. This stored heat can be utilized for different heating and power generation applications depending on its temperature. Thus, the phase change material can maintain the PV operating temperature at its melting temperature. It has been reported in the literature that the PV temperature can be controlled within 28–65 °C with PCM and heat exchanger design optimization. Manikandan et al. [9] developed the numerical model of CPV-PCM-heat sink system in COMSOL Multiphysics. Qasim et al. [10] developed the experimental setup of PV-PCM system and studied the effect of number of fins inserted into PCM and the effect of using 2 PCMs having melting points and separated by finned-aluminum plate.

It has been witnessed from the literature that the passive cooling techniques are more attractive as compared to active cooling techniques due to their main advantage of absence of requirement of external electricity. As mentioned earlier, PCM based cooling technique owing to several advantages of PCM, are used for cooling the PV module. Further, studies related to cooling of PV modules used in higher solar radiation and ambient temperature locations such as desert areas, are not available in literature. The objective of this work is to regulate the PV module temperature and increase its efficiency. Thus, the cooling of PV module using PCM has been explored for extreme climatic conditions having ambient temperature exceeding 50 °C and the PV module temperature can be higher than that of ambient up to 40 °C. The effect of PCM thickness on the PV performance has been studied.

2 System Configuration and Geometry

A comprehensive two dimensional model of the PV + PCM + HS system including all the PV module layers, PCM and heat sink as shown in Figs. 1 and 2 is developed and numerically analyzed using COMSOL Multiphysics software. The model includes a photovoltaic module, a PCM and a copper heat sink (HS). The PV module is exposed to a local solar radiation, the PCM module is located between the PV and HS. The dimensions of all the components of PV module are listed in Table 1.

The following assumptions have been taken to simplify the 3-D time dependent analysis:

- All components are contacted directly.
- Constant and isotropic thermal properties are considered for all materials.
- Convective and radiative heat losses are considered.
- The molten PCM is Newtonian and incompressible.

Table 1 Components dimensions in mm (axis)

Component	Length (x)	Width (y)	Height (z)
Photovoltaic cell	156	156	4.5
Glass cover	156	156	3.2
EVA layer	156	156	0.5 * 2
Silicon film	156	156	0.3
Tedlar film	156	156	0.1
Copper plate	156	156	0.4
PCM module	156	156	1.5
Copper Heat Sink	156	156	6
Square grooves (26)	3	156	3

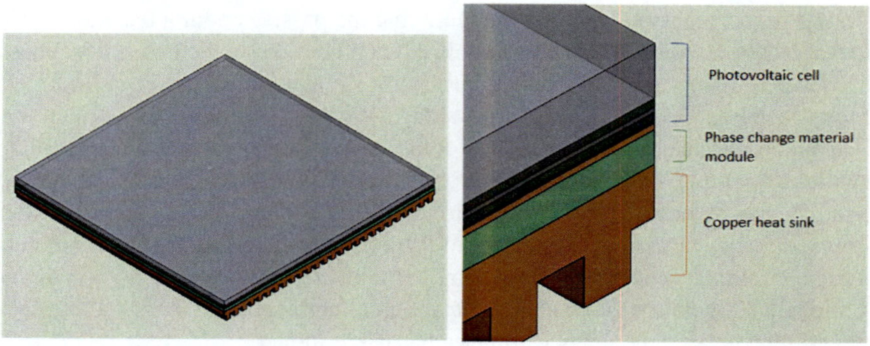

Fig. 1 Model components and configuration

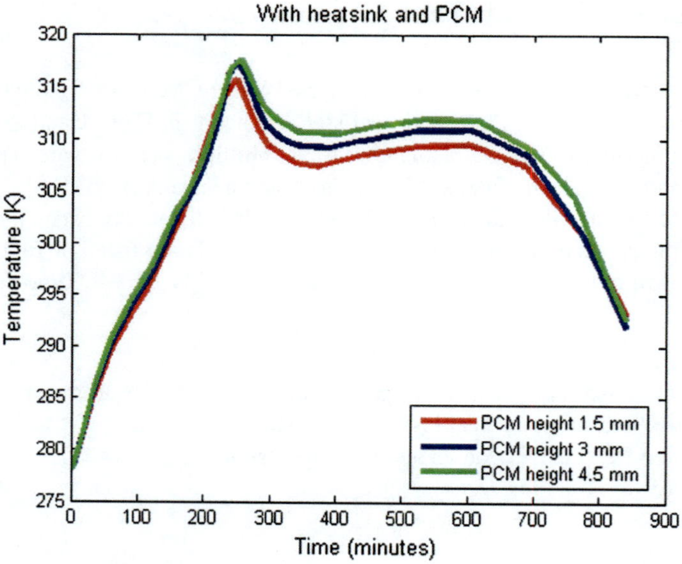

Fig. 2 Transient variation of PV operating temperature for different PCM heights

PV Module

The conductive heat transfer and the convective & conductive heat transfers in the solids and fluid/PCM respectively are the main heat transfer phenomena in the proposed model. The generalized energy conservation equation and Fourier´s law governing the heat transfer are given as:

$$\rho c_p \frac{\partial T}{\partial t} + \rho c_p \mu \nabla T + \nabla . q = Q \tag{1}$$

$$q = -k\nabla T \tag{2}$$

where q, T, c_p and ρ are the heat flux, fluid temperature, constant pressure specific heat capacity and fluid density respectively.

The heat transfer through the different solid layers of PV module is assumed to be conductive and unsteady which can be expressed by the diffusion equation of heat transfer as follows:

$$\rho c_{p,i}\frac{\partial T(x,y)}{\partial t} = k_i\left(\frac{\partial^2 T_i(x,y)}{\partial x^2} + \frac{\partial^2 T_i(x,y)}{\partial x^2}\right) + q_i;\ where,\ i = 1, 2..n \tag{3}$$

where ρ_i, $c_{p,i}$, k_i, q_i and $T_i(x, y)$ represent the density, specific heat capacity, thermal conductivity, the internal local volumetric heat generation and temperature of the i-th layer of the PV module, respectively.

The volumetric internal heat generation inside the i-th layer of the PV module, q_i can be calculated from as:

$$q_i = \frac{(1-\eta_e).G(t).\alpha_i.\tau_j.A_i}{V_i} \tag{4}$$

where η_e is the electrical efficiency of the silicon layer of PV module and it is taken zero for other layers of the PV module, $G(t)$ is the solar radiation, α_i, V_i, A_i, are the absorptivity, volume and area of the layer i-th layer, respectively and τ_j is the transmissivity of the layer above the i-th layer.

The electricity generated by the PV module during the simulation period is calculated using the PV temperature, according to the Eq. (6) as:

$$E_{PV} = E_{sun} \times \eta_{ref} \times [1 - 0.0045(T_{PV} - T_a)] \tag{5}$$

The input solar energy, Esun incident of the PV surface is given as:

$$E_{sun} = G \times A_{PV} \tag{6}$$

where η_{ref}, T_{PV}, T_a, are the PV module efficiency at standard test conditions i.e. reference efficiency, the PV module operating temperature and the ambient temperature, respectively. In this study, the reference efficiency, η_{ref} of 15% and a PV surface area, A_{PV} of 0.02436 m^2 have been taken.

Phase change material

The heat and mass transfer during melting process of the phase change material can be modeled by different available methods. In this analysis, enthalpy-porosity method is used to model the PCM. The governing equation for 2-D transient analysis of phase change process can be expressed as follows:

Continuity equation:

$$\frac{\partial \rho}{\partial t} + \nabla.\left(\rho \overrightarrow{V}\right) = 0 \tag{7}$$

where, V represents the melted PCM velocity vector.

Momentum equation:

$$\rho \frac{\partial \overrightarrow{V}}{\partial t} + \rho\left(\overrightarrow{V}.\nabla\right)\overrightarrow{V} = -\nabla P + \mu \nabla^2 \overrightarrow{V} + \overrightarrow{F_B} + \overrightarrow{S} \tag{8}$$

where, P is the pressure of the PCM, F_B is the buoyancy term vector and S is the source term vector (S_x and S_y are its components).

The liquid fraction, λ is defined as:

$$\lambda\left(= \frac{\Delta H}{L}\right) = \begin{cases} 0 & T \le T_m \\ \frac{T - T_{solidus}}{T_{liquidus} - T_{solidus}} & T_m < T < T_m + \Delta T_m \\ 1 & T \ge T_m + \Delta T_m \end{cases} \tag{9}$$

where, ΔH is the latent heat of the PCM and it varies from zero (solid) to L (liquid); L is the latent heat of the PCM; Tm is the melting temperature of the PCM, and ΔT_m is the phase transition range.

Energy equations

Liquid Phase: The energy equation for molten PCM is given as:

$$\frac{\partial}{\partial t}(\rho H) + \nabla.\left(\rho \overrightarrow{V} H\right) = \nabla.(k \nabla T) \tag{10}$$

Solid Phase: The energy equation for solid PCM is given as:

$$\frac{\partial}{\partial t}(\rho H) = \nabla.(k \nabla T) \tag{11}$$

The enthalpy of the PCM, H is given as:

$$H = h_{ref} + \int_{T_{ref}}^{T} c_p dT + \lambda L \tag{12}$$

The thermal conductivity of PCM, k_{pcm}, depending on the phase change process, can be defined as:

$$k_{pcm} = \begin{cases} k_s & T \le T_m \\ \frac{k_s + k_l}{2} & T_m < T < T_m + \Delta T_m \\ k_l & T \ge T_m + \Delta T_m \end{cases} \tag{13}$$

3 Results and Discussion

The proposed PV + PCM + Heat sink model is analyzed for different heights of PCM. The transient model of the system is developed in COMSOL Multiphysics. The melting behavior of PCMs is simulated using heat transfer model in COMSOL. The effect of PCM height on the PV temperature, power output and efficiency has been discussed. The transient analysis of the proposed model has been carried out with environmental parameters of Atacama Desert.

Figure 2 shows the transient variation of PV operating temperature when the PCM height is varying from 1.5 mm to 4.5 mm with step size of 1.5 mm. The PV operating temperature differences with the consecutive PCM heights increase by 1 K. The maximum PV temperatures corresponding to PCM heights of 1.5, 3.0 and 4.5 mm are 315.66 K, 317.36 K and 317.56 K, respectively. The average PV temperatures corresponding to PCM heights of 1.5, 3.0 and 4.5 mm for are 304.15 K, 305.03 K and 306.05 K, respectively.

Figure 3 shows the PV energy generated during the analyzed period for the different PCM heights. The power peak is found to be around 3.4 W. The best PCM height for maximum PV power is 1.5 mm. It is observed that the period of maximum energy generation corresponds to the maximum solar irradiation and ambient temperature. The maximum PV power output corresponding to PCM heights of 1.5, 3.0 and 4.5 mm are, 3.46 W, 3.43 W and 3.41 W, respectively. The average PV power output corresponding to PCM heights of 1.5, 3.0 and 4.5 mm are 2.66 W, 2.65 W and 2.64 W, respectively.

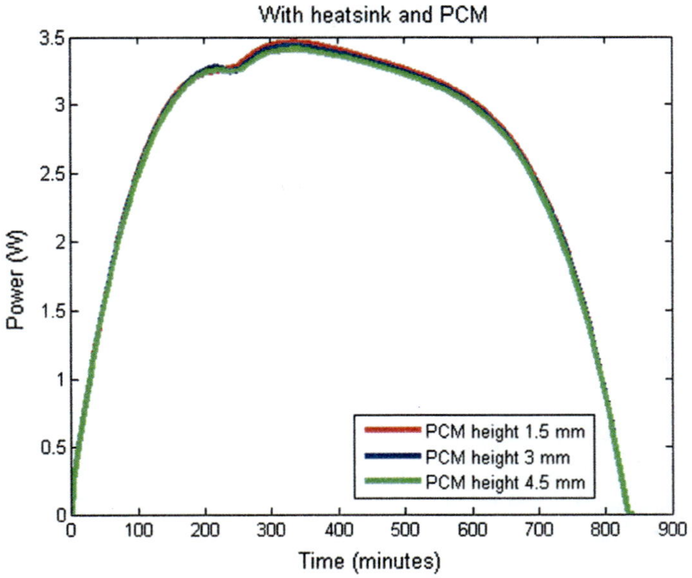

Fig. 3 Transient variation of PV power output for different PCM heights

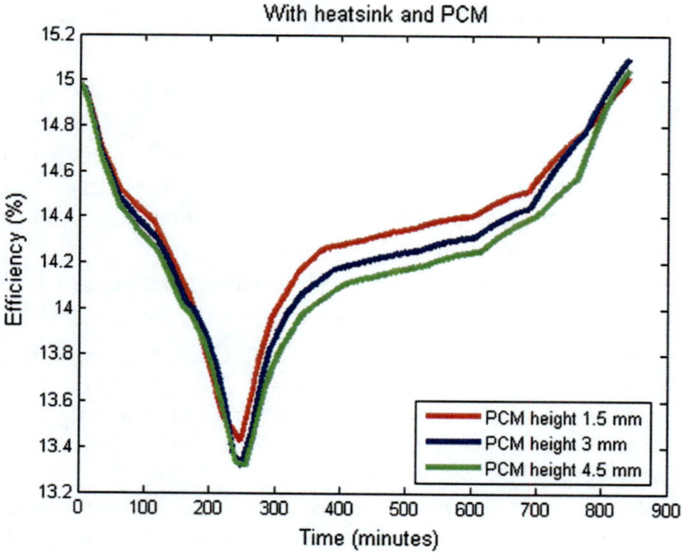

Fig. 4 Transient variation of PV efficiency for different PCM heights

Figure 4 shows the transient variation of PV efficiency during the simulation time. The minimum observed PV efficiency corresponds to 3 mm PCM height, reaching approximately 13.32%. The maximum height simulated also shows a low PV efficiency, 13.33%. PCM height of 1.5 mm causes the same PV efficiency, reaching 13.42% for the proposed system.

4 Conclusions

A 2-D model of the PV + PCM + HS system including a transient analysis and heat transfer simulation is developed and analyzed in COMSOL Multiphysics software using FEM with total simulation time of 840 min. The system is simulated for extremely hot climatic conditions of the Atacama desert. The effect of PCM thickness on the PV performance has been studied. The following conclusion have been drawn:

- The PV energy generation follows the simulated environment conditions.
- The use of a combination of PCM module and heat sink can control the PV operational temperature, maintaining an average of around 312 K.
- The best PCM height is 1.5 mm (minimum height), that is consequence of a less thermal resistance of the PCM module and that allows all the PCM melts during the simulated period.
- A PV power peak of around 3.5 W is obtained and the importance of coupling a heat sink with the PCM module to evacuate the excess heat to the ambient is observed.

References

1. Chandel SS, Nagaraju Naik M, Sharma V, Chandel R (2015) Degradation analysis of 28 year field exposed mono-c-Si photovoltaic modules of a direct coupled solar water pumping system in western Himalayan region of India. Renew Energy 78:193–202. https://doi.org/10.1016/j.renene.2015.01.015
2. Du D, Darkwa J, Kokogiannakis G (2013) Thermal management systems for photovoltaics (PV) installations: a critical review. Sol Energy 97:238–254. https://doi.org/10.1016/j.solener.2013.08.018
3. Teo HG, Lee PS, Hawlader MNA (2012) An active cooling system for photovoltaic modules. Appl Energy 90(1):309–315. https://doi.org/10.1016/j.apenergy.2011.01.017
4. Chen H, Yang J, Zhou N, Chen J, Zhang Y (2019) Performance analysis of a high concentrating photovoltaic/thermal system with a water spray cooling device. IOP Conf Ser Mater Sci Eng 556(1). https://doi.org/10.1088/1757-899X/556/1/012034
5. Nižetić S, Papadopoulos AM, Giama E (2017) Comprehensive analysis and general economic-environmental evaluation of cooling techniques for photovoltaic panels, part I: passive cooling techniques. Energy Convers Manag 149:334–354. https://doi.org/10.1016/j.enconman.2017.07.022
6. Nižetić S, Giama E, Papadopoulos AM (2018) Comprehensive analysis and general economic-environmental evaluation of cooling techniques for photovoltaic panels, part ii: active cooling techniques. Energy Convers Manag 155:301–323. https://doi.org/10.1016/j.enconman.2017.10.071
7. Najafi H, Woodbury KA (2013) Optimization of a cooling system based on Peltier effect for photovoltaic cells. Sol Energy 91:152–160. https://doi.org/10.1016/j.solener.2013.01.026
8. Bahaidarah HMS, Baloch AAB, Gandhidasan P (2016) Uniform cooling of photovoltaic panels: a review. Renew Sustain Energy Rev 57:1520–1544. https://doi.org/10.1016/j.rser.2015.12.064
9. Manikandan S, Selvam C, Poddar N, Pranjyal K, Lamba R, Kaushik SC (2019) Thermal management of low concentrated photovoltaic module with phase change material. J Clean Prod 10(219):359–367
10. Qasim MA, Ali HM, Khan MN, Arshad N, Khaliq D, Ali Z, Janjua MM (2020) The effect of using hybrid phase change materials on thermal management of photovoltaic panels–an experimental study. Sol Energy 1(209):415–423

Aspects of Energy Consumption for Electrochemical Treatment of Tannery Wastewater

Harshika Suman and Vikas K. Sangal

Abbreviations

COD	Chemical Oxygen Demand
TOC	Total Organic carbon
TKN	Total Kjeldahl Nitrogen
TSS	Total Suspended Solid
BDD	Boron Doped Diamond
TN	Total Nitrogen

1 Introduction

Today, the environmental importance of water is considered as essential everywhere in the world.Water is the most basic essential for human survival. These days, the severe environmental burden brought on by water contamination and scarcity is expanding, and there is a corresponding loss of natural water supports due to the scarcity of water. As a result, there is a devaluation in the advancement of economic conditions, human sustenance, and the environment. Water is extremely important for humans to survive. The great majority of the water on the earth, over 97% of all water, is saltwater from oceans or seas, and just approximately 1% of it is readily available to humans [1, 2]. The access to clean water is one of the most basic humanitarian aims, and it is a big global protest for this century. With the expeditious

H. Suman · V. K. Sangal (✉)
Chemical Engineering Department, Malaviya National Institute of Technology Jaipur, Jaipur, Rajasthan 302017, India
e-mail: vksangal.chem@mnit.ac.in

S. Doolla et al. (eds.), *Advances in Clean Energy and Sustainability*,
Green Energy and Technology, https://doi.org/10.1007/978-981-99-2279-6_74

development and expanded extension of factories, industries, and mining operations, wastewater are discharged into the surface water, ground water and environment. Water consumption has risen dramatically as a result of increased urbanisation, climate change, and population growth, as well as agricultural development and a huge increase in oil and gas production. Furthermore, significant quantities of salts, hazardous heavy metals, and organics pollute enormous volumes of industrial effluent. These activities have problems in terms of both water quality and quantity, and they have a significant influence on the energy-efficient management of our country's water resources. For example, centralised water treatment is inefficient in terms of energy use, and resources for efficient treatment are scarce in rural locations. The discharge of untreated wastewater by process industries has become an important environmental concern now-a-days. Many countries throughout the world have strict environmental laws in place to manage industrial discharge [3]. As industry is a vital prerequisite for any country's development, it cannot be wrangled. In many nations, regulatory regulations for the release of toxic wastewater are being strengthened, and discharge fees are being raised time to time. Leather industry considered one of the most polluted industry in the world. The effluents from the leather or tannery industry are very harmful for the aquatic life and human beings if it discharged directly without treatment, before release this wastewater to the environment it need to be treat [4]. Furthermore, treating multi-contaminant waste streams necessitates the use of complicated and energy-intensive treatment trains. To overcome the issues faced by this 'water-energy nexus,' both technical and scientific innovation is required. One of the most widely used process is advanced oxidation process which applied for removal of harmful effluents from contaminated wastewater and according to a recent literature review, has emerged as a promising alternative to traditional methods for the treatment of contaminants. Recent requirements in treatment of effluent have lead to the development of several electrochemical techniques. These techniques provide attractive alternatives that are more environmentally friendly and cost effective processes for treating wastewater containing organic compounds. So, the electrochemical processes are the most adequate tools for the aqueous waste treatment that are ideally suited to the present time. Water treatment technologies based on electrochemistry have emerged as potentially disruptive answers to these problems. Electrochemical reactions can create oxidants, reductants, acids, bases, and coagulants on-site, obviating the need to ship and store vast quantities of corrosive or potentially harmful chemicals, allowing for decentralised and distant water treatment. By decreasing heavy pre-treatment expenditures and chemical inputs, electrochemical desalination and electrocoagulation technologies can achieve greater performance with less energy than traditional methods, lowering the cost and boosting their feasibility for oil and gas water treatment. Electrochemical technology may also be easily integrated with renewable energy sources and include energy recovery mechanisms, reducing their reliance on fossil fuels even more [4–6]. The energy consumption for the treatment of wastewater is a topic of concern. This paper consists of energy consumption aspect of previous work that has been already carried out for treatment of toxic effluents from tannery operations.

2 Electrochemical Treatment

Electrochemical wastewater treatment has been a hot topic in recent years due to its impressive ability to remove a wide spectrum of organic and inorganic pollutants. Although electrochemical treatment of wastewater has been employed as a pre- or post-treatment method, its scale-up use has been hampered by high energy consumption and facility costs. Further development of electrochemical treatment is anticipated since tannery wastewater often contains high salinity and many metal ions that might increase the conductivity of the wastewater that has to be treated. In the electrochemical process, energy consumption is highly important considering that electricity is the energy source.

3 EO/EC/EF Summary

Electro-oxidation involves simultaneous evolution of oxygen at an anode surface or in the vicinity of anode using oxidants generated at anode. The electrochemical cell of Electro-oxidation (EO) and Electro Fenton (EF) consist of anodes and cathodes in a parallel arrangement. The oxidants are generated in the bulk and on the surface of the anode. These oxidants help in the degradation of organic pollutants of the wastewater. The generation of oxidants is fully depended upon electric current, as the electric current pass through electrodes in the electrochemical cell, oxidants are generated [5, 7]. The degradation efficiency of the EO and EF is different from each other. The efficiency of the EO and EF process depends upon the conditions applied during the treatment process. In EO only anode material is required for catalytic activity but in the case of EF addition of catalyst is required. So, the theory behind the degradation of EO and EF is a little bit different. But the oxidants generated in EO and EF processes are same. The concentration of the oxidants generated is variable in EO and EF processes. There are various reactions that are involved in the degradation process of EO and EF [8]. Extensive research has been going on for the electro oxidation of wastewater treatment. Tables 1 and 2 shows the summary of the electro-fenton and electro-oxidation treatment of tannery industry effluents in the terms of energy consumption and removal efficiency. The EF process is a potent remediation technology that uses extremely reactive hydroxyl radicals to oxidise recalcitrant chemicals that are difficult to remove using traditional wastewater treatment methods. The adsorption of organic and inorganic pollutants into generated iron hydroxide particles, primarily $Fe(OH)_3$, that is continually synthesised in the medium solution by electrically dissolving sacrificial anodes, is the basis of this approach [9, 10].

Electrocoagulation (EC) uses electrodes, which are metal sheets that are organised in pairs of two—anodes and cathodes. The cathode is oxidised (loses electrons) while the water is reduced (gains electrons) using electrochemistry principles, resulting in better wastewater treatment. The metal is discharged into the device when the cathode electrode comes contact with the wastewater. The particles are neutralised

Table 1 Summary for treatment of tannery wastewater by Electro Fenton method

Target pollutant	Process	Mode of reaction	Parameters	Electrodes	Energy consumption	Removal efficiency	References
Black NT2 (BNT2)	Electro-Fenton	Pilot reactor	0.3 mM Fe^{2+}, pH 3.0, J = 30 mA/cm^2, Q = 12 L min^{-1}	BDD electrodes	0.01088 kWh/g TOC	100% TOC in 120 min	[9]
Tannery wastewater	Heterogeneous Electro-Fenton (HEF)	Batch system	pH = 3.5, Catalyst dose 0.06 g/L, J = 7.37 mA/cm^2	$Ti/IrO_2/RuO_2$	10.87 kWh/kg of COD	97.08% COD removal Reaction time: 79.43 min	[12]
Tannery wastewater	Electro-Fenton (EF) and Electro-Persulfate (EP) processes	Batch system	pH = 3–9, distance between electrodes = 0.5–1.5	Graphite	EF 8.357 kwh/gCOD and EP 6.867 kwh/gCOD	EF 67.25% COD removal, EP 95.62% COD removal in time 6 h	[13]
Acid Black 210 dye	Electro-Fenton (EF) process with a $CoFe_2O_4$/NOM magnetic hybrid catalyst (Hb200)	Batch system	pH = 3, Time = 7 h, J = 14.1, 28.2, and 42.2 mA cm^{-2}	BDD plate coated on both sides with a niobium substrate	0.34 kWh/g TOC	Mineralization rates 80.3%	[14]
Equalization basin in a common treatment plant of tannery	Electro-Fenton	Batch reactor	pH = 3, 5, 7.2, H_2O_2 = 840, 1670, 3340, and 5010 mg/l	Iron electrode	3.1 kWh/kg COD removed	72% COD removal	[15]

Table 2 Summary for treatment of tannery wastewater by electro-oxidation method

Target pollutant	Process	Mode of Reaction	Parameters	Electrodes	Energy consumption	Removal efficiency	References
Acid Black 210 dye	Electro-oxidation	Batch and continuous	pH = 6.5 Current = 1.6A	MMO electrode	6.3 Kwh/m^3 in 35 min	Decolorization 97%, Degradation 65.9%	[16]
Tannery wastewater	Electro-oxidation	Pilot-scale	J = 0.010 A/cm^2	$TiO_2/RuO_2/IrO_2$	13.5 KWh/m^3 and 1.45 KWh per Kg of COD removed	BOD, COD, TKN and TSS removal 92%, 87.5%, 96.2% and 94.6% respectively	[17]
Tannery wastewater after activated sludge pre-treatment	Electrochemical oxidation	Batch	J = 66.7 mA/cm^2	SnO_2/Ti PbO_2/Ti	117.6 kWh/kg COD	80% of COD and TN after 90 min	[18]
Lime wastewater	Photoassisted electro-oxidation	Batch	J = 15, 20 and 25 mA/cm^2	Ti/IrO_2-RuO_2-TiO_2	0.30 $kWhL^{-1}$, 18.6 kWh/gCOD and 0.33 $kWhL^{-1}$, 24.33 kWh/gCOD for PEO and EO respectively	92% of COD in Photoelectrochemical oxidation 62% of COD in Electrooxidation	[19]
Leather dyeing wastewater	Electrochemical oxidation	Batch	J = 18.70 mA/cm^2 T = 286.18 K	Ti/Pt anode	93.85 kWh/kg of COD	81.2% of COD removal	[20]
Tannery wastewater	Electrochemical oxidation	Batch	J = 17–83 mA/cm^2 pH = 3–11	Ti/RuO_2 Ti/IrO_2 Ti/BDD	7.85 kWh/kg COD (Ti/IrO_2) < 8.54 kWh/kg COD(Ti/BDD) < 8.61 kWh/kg COD(Ti/RuO_2)	Colour, COD, TOC, and total nitrogen removal efficiencies Ti/RuO_2 (88.8%, 88.40%, 64.0%, 96.4%), Ti/IrO_2 (85.40%, 85.9%, 52.3%, 51.4%), Ti/BDD (90.60%, 94.7%, 90.5%, 82.7%)	[21]

(continued)

Table 2 (continued)

Target pollutant	Process	Mode of Reaction	Parameters	Electrodes	Energy consumption	Removal efficiency	References
Tannery wastewater	Electrooxidation	Batch	J = 1.5 A/dm^2	Ti/SnO_2 /PdO_2/RuO_2	78.2–171 kWh/kg of N-NH^{4+}	Full ammonia removal and COD 58.9%	[22]
Evaporated residue of soak liquor (ERSL) generated in Tannery	Electrochemical oxidation	Batch	J = 25–75 mA/cm^2 Concentration of ERSL 30–60% [w/v]	Graphite	0.41 kWh m^{-3} for the removal of COD and 2.57 kWh m^{-3} for the removal of TKN	COD and TKN removal efficiency 93% and 87% respectively	[23]
Saline wastewater	Electro-oxidation	Batch	pH = 9 J = 0.024 A/cm^2 NaCl = 10–40 g	Graphite electrodes	22.45kWh/1 kg of TKN and 0.80kWh/1 kg of COD	–	[24]
Synthetic tannery wastewater	Electrochemical treatment	Batch	J = 25–100 mA cm^{-2}	Si/BDD Ti/SnO_2–Sb Ti/SnO_2–Sb–Ir	Ti/SnO_2–Sb–Ir(2.547 kWh/gTOC) Ti/SnO_2–Sb(0.104 kWh/gTOC) Si/BDD(0.084 kWh/gTOC)	TOC Removal Ti/SnO_2–Sb–Ir(2.1%) Ti/SnO_2–Sb(56.1%) Si/BDD(79.1%)	[25]
Tannery wastewater	Electrochemical oxidation	Batch	Current = 0.1–0.6 A	Ti/Pt Ti/PbO_2 Ti/MnO_2	5.77 kWh/kgCOD and 16.63 kWh/kg NH^{4+}	Ti/Pt(0.802 $kgCOD/hAm^2$ 0.27 $KgNH^4/hAm^2$)	[26]

by the creation of hydroxide complexes, resulting in the development of agglomerates. These agglomerates form at the bottom of the tank and may be filtered away with filtration [11]. The EC process generates coagulants in situ, eliminating the need to add a chemical coagulant. Table 3 shows the summary of the electrocoagulation treatment of tannery industry effluents in the terms of energy consumption and removal efficiency. And Table 4 shows the summary of the combined electrochemical treatment of tannery industry effluents in the terms of energy consumption and removal efficiency.

From the literature review it can be conclude that the energy consumption of a specific process is the factor depended on the process parameters. Time plays biggest role for the treatment energy consumption and can say energy consumption is the directly proportional to the energy consumption. It can be conclude that combined processes has less energy consumption then individual one because in combined process or integrated system removal efficiency is higher in lesser time. Cost of a reactors system and a process is an important parameter. It can be say that energy consumption of a system is play an important role for cost of the system. The major cost of a system is because of energy used by the system and electrodes. Energy consumption also function of treatment time [11]. Most of the literatures not calculate the cost of system but it is an main function for the feasibility of a process for industrial application.

4 Future Aspects

In the era when non - renewable sources are getting exhausted by human greed and need, overuse and misuse, it's time to switch energy sources to renewable ones. Wind, light, water are some of the common renewable sources of energy but in recent time researchers have found out microorganisms as renewable sources of energy. Microorganisms to be specific which are exo-electrogens like geobacter, shewanella putrefaciens. These microbes are being used in techniques like microbial fuel cells (MFC), microbial electrolytic cells (MEC). MFC and MEC is a process where microbes degrade organic matter present in waste water sources and release electrons in external environment. These electrons then move from anode to cathode via circuit and generate electricity. This electricity is then used in treating waste water. Specifically in MEC the electric energy produced by anode oxidation is transferred to cathode for Hydrogen production. The cell voltage required in this process is only 0.123 V as compared to 1.23 V in conventional electrolysis [35, 36]. This energy is derived from fossil fuels which are non—renewable sources of energy. It not only increases the cost but also affects the environment. Studies show that using these techniques can reduce the dependency on non renewable sources, also these techniques are much more efficient and use less energy as compared to conventional technology like electrolysis. Yet more research should be carried out before implementing techniques like Bio-electrolysis on a larger scale. Switching to techniques like MFC and MEC will also help us achieve SDGs 7 established by UNGA in 2015.

Table 3 Summary for treatment of tannery wastewater by electro-coagulation method

Target pollutant	Process	Mode of reaction	Parameters	Electrodes	Cost	Energy consumption	Removal efficiency	References
Bright green dye (BG)	Electrocoagulation (EC)	Continuous	$J = 6$ mA cm^{-2} $Q = 0.69$ cm s^{-1}	Steel plates	0.193 USD m^{-3}	0.077 kWh m^{-3} (0.134 kWh/kg COD)	Decolorization 100% and COD removal 96%	[27]
Tannery wastewaters	Electrocoagulation	Continuous	$J = 0.6$ and 1.2 mA cm^{-2}	Aluminum electrodes	0.239 $ m^{-3} at 0.6A	2.37 kWh m^{-3} in 60 min	84% COD, 98% turbidity, 97% oil-grease, 98% chromium, 68% total nitrogen, 100% phosphate and 79% sulfate removal	[28]
Industrial tannery effluent	Electrocoagulation	Rotating reactor	$J =$ 6 mA cm^{-2} and 100 rpm	Aluminum anodes	$0.7 USD m^{-3}	1.98 kWh m^{-3}	COD 70%, while total suspended solids, chromium (III) and turbidity were almost eliminated (>90%)	[11]
Tannery effluent	Electrocoagulation	Batch	pH = 4–5 $J =$ 1.2 mA cm^{-2}	Iron Plate electrodes	–	2.71 kwh m^{-3}	80.12% of COD 62.91% of TN, 54.78% sulphate and 87.8% of oil-grease	[29]
Leather plant	Electrocoagulation (EC)	Batch	pH = 7 $J = 28$ mA/cm^2	Iron and aluminium electrodes	–	COD—0.37 kWh/m^3 and TOC—0.69 kWh/m^3	COD 72% and TOC 57% with aluminium electrodes and COD 69% and TOC 60% with iron electrodes	[30]

Table 4 Summary for treatment of tannery wastewater by combined electrochemical method

Target pollutant	Process	Mode of reaction	Parameters	Electrodes	Energy consumption	Removal efficiency	References
Tannery wastewater	Electro-Coagulation (EC) and Electro-Fenton (EF)	Batch reactor	J = 33.3 mA m^{-2} pH = 7	Iron plates	EC—1.8 kWh/kg COD removed and 27.7 kWh/kg sulfide removed EF—1.5 kWh/kg COD removed and 8.3 kWh/kg sulfide removed	EC—COD 46% and sulfide 90% EF—COD 54% and sulfide 85%	[31]
Tannery wastewater	Electrocoagulation (EC) coupled with an Electro-Fenton (EF) process	Batch reactor	J = 9.7 mA cm^{-2}	BDD	22.3 kWh m^{-3}	64% TOC removal	[10]
Tannery wastewater	Electro-Fenton and electrocoagulation	Batch	J = 20 mA/cm^2 pH = 7	Aluminum and iron electrode	EC—8.33 kWh/m^3 and EF—6.9 kWh/m^3	EC 82% COD	[32]
Tannery Wastewater	Electrooxidation (EO), electrocoagulation (EC)	Batch	EO J = 42 mA/cm^2 Time = 180 min EC J = 17 mA/cm^2 EC = 10 min pH = 7	EO-Pb/PbO_2 EC-aluminium plate	EC/EO process has low energy consumption of 18.2 kWh/m^3 as compared with EO of 60 kWh/m^3	EO—COD 98.9% and 99.0% color EC—COD 82.2% and 68.3% color	[33]

(continued)

Table 4 (continued)

Target pollutant	Process	Mode of reaction	Parameters	Electrodes	Energy consumption	Removal efficiency	References
Bismark Brown G, Bismark Brown R and Brown DGI	Advanced oxidation processes electrochemical oxidation (EO), Electro-Fenton (EF) and Photoelectro-Fenton (PEF)	Batch	pH = 3 J = 20 mA cm^{-2}	BDD	0.045 kWh/g COD	97% discoloration and 95% COD removal in 60 min	[8]
Tannery wastewater	Electro-oxidation and advanced oxidation as tertiary treatment technique	Batch	Voltage = 3–8 V, Time = 6 h pH = 7	Titanium Mesh coated with IrO_2 (25%), TaO_2 (25%), and TiO_2 (50%)	EO 738 kW h/kg AO 7600 kW h/kg	85% TOC removal by $UV/O_3/H_2O_2$ process, and EO 50% TOC removal	[34]

5 Conclusion

The energy consumption aspect of electrochemical technologies was successfully studied and found the integrated systems have less energy requirement. Electrochemical technologies in combination with renewable energy sources provide new approaches to energy-efficient tannery wastewater treatment. They have the benefit of being more efficient in terms of energy usage. The coupling of these electrochemical advance technologies with renewable sources of energy provides long-term solutions for tackling the water-energy nexus. Fouling management for electrochemical systems, better material systems for electrodes used in electrochemical wastewater treatment, integration schemes between the conventional and advance processes with better energy management methods, sustainable enhanced manufacturing processes and so on are some of the probable future research problems to investigate.

References

1. Zhao C, Chen W (2019) A review for tannery wastewater treatment: some thoughts under stricter discharge requirements. Environ Sci Pollut Res 26:26102–26111. https://doi.org/10.1007/s11356-019-05699-6
2. Weng CH (2021) (2021) Water environment and recent advances in pollution control technologies. Environ Sci Pollut Res 299(29):12462–12464. https://doi.org/10.1007/S11356-021-17392-8
3. Suman H, Sangal VK (2022) An inside for the treatment of tannery industry effluent. Adv Chem Bio Environ Eng 909–925. https://doi.org/10.1007/978-3-030-96554-9_60
4. Suman H, Sangal VK, Vashishtha M (2021) Treatment of tannery industry effluent by electrochemical methods: a review. Mater Today Proc 47:1438–1444. https://doi.org/10.1016/j.matpr.2021.03.300
5. Khatu PM, Suman H, Sangal VK, Vashishtha M, Chaturvedi T (2021) Electro-oxidative decolouration and degradation of amaranth dye wastewater in batch setup using novel Ti/TiO_2-Ru2O-IrO2 anode. Asian J Water, Environ Pollut 18:69–77. https://doi.org/10.3233/AJW210030
6. Panizza M, Cerisola G (2009) Direct and mediated anodic oxidation of organic pollutants. Chem Rev 109:6541–6569. https://doi.org/10.1021/cr9001319
7. Liu X, Song CN, Zhang Y, Sha L, Li Y, Zhang S (2021) Electrochemical pretreatment of coking wastewater by Ti/BTN/RuO2-IrO2-TiO_2: selectivity of chloridion oxidation and multi-response optimization. Sep Purif Technol 276:1–9. https://doi.org/10.1016/j.seppur.2021.119229
8. Medrano-Rodríguez F, Picos-Benítez A, Brillas E, Bandala ER, Pérez T, Peralta-Hernández JM (2020) Electrochemical advanced oxidation discoloration and removal of three brown diazo dyes used in the tannery industry. J Electroanal Chem 873:114360. https://doi.org/10.1016/J.JELECHEM.2020.114360
9. Villaseñor-Basulto D, Picos-Benítez A, Bravo-Yumi N, Perez-Segura T, Bandala ER, Peralta-Hernández JM (2021) Electro-Fenton mineralization of diazo dye Black NT2 using a pre-pilot flow plant. J Electroanal Chem 895:115492. https://doi.org/10.1016/J.JELECHEM.2021.115492

10. Villaseñor-Basulto DL, Picos-Benítez A, Pacheco-Alvarez M, Pérez T, Bandala ER, Peralta-Hernández JM (2022) Tannery wastewater treatment using combined electrocoagulation and electro-Fenton processes. J Environ Chem Eng 10:107290. https://doi.org/10.1016/J.JECE.2022.107290
11. Villalobos-Lara AD, Álvarez F, Gamiño-Arroyo Z, Navarro R, Peralta-Hernández JM, Fuentes R, Pérez T (2021) Electrocoagulation treatment of industrial tannery wastewater employing a modified rotating cylinder electrode reactor. Chemosphere 264:128491. https://doi.org/10.1016/J.CHEMOSPHERE.2020.128491
12. Ozturk D (2022) Fe3O4/Mn3O4/ZnO-rGO hybrid quaternary nano-catalyst for effective treatment of tannery wastewater with the heterogeneous electro-Fenton process: process optimization. Sci Total Environ 828:154473. https://doi.org/10.1016/J.SCITOTENV.2022.154473
13. Doumbi RT, Bertrand Noumi G, Ngobtchok B, Domga (2022) Tannery wastewater treatment by electro-Fenton and electro-persulfate processes using graphite from used batteries as free-cost electrode materials. Case Stud Chem Environ Eng 5:100190. https://doi.org/10.1016/J.CSCEE.2022.100190
14. Cruz DRS, de Jesus GK, Santos CA, Silva WR, Wisniewski A, Cunha GC, Romão LPC (2021) Magnetic nanostructured material as heterogeneous catalyst for degradation of AB210 dye in tannery wastewater by electro-Fenton process. Chemosphere 280:130675. https://doi.org/10.1016/J.CHEMOSPHERE.2021.130675
15. Kurt U, Apaydin O, Gonullu MT (2007) Reduction of COD in wastewater from an organized tannery industrial region by electro-Fenton process. J Hazard Mater 143:33–40. https://doi.org/10.1016/J.JHAZMAT.2006.08.065
16. Suman H, Sangal VK (2022) Tannery dye wastewater treatment in batch and once through continuous mode by electro-oxidation using MMO electrode. J Electrochem Soc 169:043512. https://doi.org/10.1149/1945-7111/AC65B9
17. Sundaramoorthy S, Singh N, Taube CR, Katiyar R, Muralidharan V, Palanivel S (2022) Electro-oxidation of tannery wastewater to achieve zero discharge–a step towards sustainability. Environ Technol (United Kingdom). https://doi.org/10.1080/09593330.2022.2049887
18. Le LuuT (2020) Tannery wastewater treatment after activated sludge pre-treatment using electro-oxidation on inactive anodes. Clean Technol Environ Policy 22:1701–1713. https://doi.org/10.1007/S10098-020-01907-X/FIGURES/8
19. Selvaraj H, Aravind P, George HS, Sundaram M (2020) Removal of sulfide and recycling of recovered product from tannery lime wastewater using photoassisted-electrochemical oxidation process. J Ind Eng Chem 83:164–172. https://doi.org/10.1016/J.JIEC.2019.11.024
20. Oukili K, Loukili M (2019) Electrochemical oxidation treatment of leather dyeing wastewater using response surface methodology. Desalin Water Treat 167:302–312. https://doi.org/10.5004/DWT.2019.24561
21. Le Luu T, Stephane DDF, Minh NH, Canh ND, Thanh BX (2019) Electrochemical oxidation as a post treatment for biologically tannery wastewater in batch reactor. Water Sci Technol 80:1326–1337. https://doi.org/10.2166/WST.2019.380
22. Naumczyk JH, Kucharska MA, Ładyńska JA, Wojewódka D (2019) Electrochemical oxidation process in application to raw and biologically pre-treated tannery wastewater. Desalin Water Treat 162:166–175. https://doi.org/10.5004/dwt.2019.24349
23. Boopathy R, Sekaran G (2013) Electrochemical treatment of evaporated residue of soak liquor generated from leather industry. J Hazard Mater 260:286–295. https://doi.org/10.1016/J.JHAZMAT.2013.05.027
24. Sundarapandiyan S, Chandrasekar R, Ramanaiah B, Krishnan S, Saravanan P (2010) Electrochemical oxidation and reuse of tannery saline wastewater. J Hazard Mater 180:197–203. https://doi.org/10.1016/J.JHAZMAT.2010.04.013
25. Costa CR, Montilla F, Morallón E, Olivi P (2010) Electrochemical oxidation of synthetic tannery wastewater in chloride-free aqueous media. J Hazard Mater 180:429–435. https://doi.org/10.1016/J.JHAZMAT.2010.04.048

26. Rao NN, Somasekhar KM, Kaul SN, Szpyrkowicz L (2001) Electrochemical oxidation of tannery wastewater. J Chem Technol Biotechnol 76:1124–1131. https://doi.org/10.1002/JCTB.493
27. Márquez AA, Coreño O, Nava JL (2022) Removal of brilliant green tannery dye by electrocoagulation. J Electroanal Chem 911:116223. https://doi.org/10.1016/J.JELECHEM.2022.116223
28. Bingul Z, Irdemez S, Yildiz Y, Demircioglu N (2022) Organic and inorganic matter removal from tannery wastewater using the electrocoagulation process. Int J Environ Sci Technol 1–10. https://doi.org/10.1007/s13762-022-04144-5
29. Bingül Z, Irdemez Ş, Kul S, Ekmekyapar Torun F, Demircioğlu N (2021) Investigation of organic and inorganic matters removal from tannery wastewater using iron plate electrode by electrocoagulation process. Int J Environ Anal Chem. https://doi.org/10.1080/03067319.2021.1953002/SUPPL_FILE/GEAC_A_1953002_SM8485.DOCX
30. De La Luz-Pedro A, Martínez Prior EF, López-Araiza MH, Jaime-Ferrer S, Estrada-Monje A, Bañuelos JA (2019) Pollutant removal from wastewater at different stages of the tanning process by electrocoagulation. J Chem 2019. https://doi.org/10.1155/2019/8162931
31. Apaydin O, Kurt U, Gönüllü MT (2009) An investigation on the treatment of tannery wastewater by electrocoagulation. Glob Nest J 11:546–555. https://doi.org/10.30955/gnj.000547
32. Gerek EE, Yılmaz S, Koparal AS, Gerek N (2019) Combined energy and removal efficiency of electrochemical wastewater treatment for leather industry. J Water Process Eng 30:100382. https://doi.org/10.1016/J.JWPE.2017.03.007
33. Azarian G, Miri M, Nematollahi D (2018) Combined electrocoagulation/electrooxidation process for the COD removal and recovery of tannery industry wastewater. Environ Prog Sustain Energy 37:637–644. https://doi.org/10.1002/EP.12711
34. Vijayalakshmi P, Raju GB, Gnanamani A (2011) Advanced oxidation and electrooxidation as tertiary treatment techniques to improve the purity of tannery wastewater. Ind Eng Chem Res 50:10194–10200. https://doi.org/10.1021/IE201039Z
35. Elabed A, El Abed S, Ibnsouda S, Erable B (2019) Sustainable approach for tannery wastewater treatment: bioelectricity generation in bioelectrochemical systems. Arab J Sci Eng 44:10057–10066. https://doi.org/10.1007/S13369-019-04050-Y/FIGURES/6
36. Miran F, Mumtaz MW (2022) Complex waste acclimatized mixed culture enhances bioelectricity generation and tannery wastewater treatment in microbial fuel cells. Biomass Convers Biorefinery 1–7. https://doi.org/10.1007/S13399-021-02093-6/FIGURES/4

Numerical Investigation of Discharging Performance of Paraffin Based Dual Shell Configuration for Latent Heat Storage

Devendra Raut, Arunendra K. Tiwari, and V. R. Kalamkar

1 Introduction

The research looks at how well a dual-shell design for paraffin-based latent heat storage discharges (LHS). During the discharging of LHS, the flowing fluid, also called heat transfer fluid (HTF), gets heated by melted PCM. The heat interaction within the PCM and with other mediums is restricted due to the low thermal conductivity of the PCM. To achieve better heat transfer, the heat transfer area should be larger, which results in better charging and discharging of the system. Thus, the design and configuration of the LHS play an important role in its effectiveness; a design with a larger effective area for heat transfer results in better performance. However, increasing the heat transfer area increases the size of the LHS and thus the cost of the system, so improving storage design would be a more pragmatic approach to improving LHS performance.

In the available literatures, various designs of LHS like helical coiled, flat spiral, multitube, shell and tube are assessed to observe the performance parameters like melting and solidification rate and amount of energy stored and withdrawn. For instance, Mahdi et al. [1] assessed the charging and discharging of paraffin wax in helical coiled LHS and assessed the performance of the respective designs in terms of PCM temperature and melt fraction. In the above-stated works, there was common objective to achieve rapid charging and withdraw maximum heat during the discharging.

D. Raut · V. R. Kalamkar
Visvesvaraya National Institute of Technology, Nagpur 440010, India

A. K. Tiwari (✉)
IIT Bombay, Mumbai, India
e-mail: tiwariarunedra1989@gmail.com

S. Doolla et al. (eds.), *Advances in Clean Energy and Sustainability*,
Green Energy and Technology, https://doi.org/10.1007/978-981-99-2279-6_75

Apart from the configuration-based study, design improvement with authorized alterations leads to improved performance. Internal fins can be used to improve the architecture of the PCM chamber. Numerous researchers have widely accepted the use of fins for performance enhancement. For instance, Mahdi et al. [2] employed the fins for the improvement of PCM solidification. They concluded that using fins alone leads to better results than using a combination of fins and nanoparticles or fins alone. Statistically, using fins alone results in a 55% reduction in solidification time, while combining fins and nanoparticles results in a 30% reduction in solidification time. In independent investigations, Dhaidan et al. [3], Mahdi et al. [4], and Khan et al. [5] give more fascinating data regarding fins in a complete overview. Moreover, research on innovative fins to increase the PCM's thermal conductance for latent heat storage is under progress [6, 7]. References [8, 9] list several works on improving performance through thermal conductivity and heat transfer area enhancement.

According to a literature survey, there are very few works which emphasize design-based performance assessments for specific applications. This motivates the authors to work to assess the applicability of the dual-shell model for solar thermal energy storage by investigating solidification of PCM. The suggested model is applicable where heat source temperatures range between 60 and 95 °C, such as waste heat recovery systems, solar heat storage, and as a thermal regulation unit in space heating and cooling applications.

2 System Description

The proposed configuration is designed for applications in which extreme temperature ranges between 70 and 95 °C like solar heat storage, waste heat recovery systems, and as a thermal regulation unit in space heating and cooling applications confront the same temperature range. For such an application, a PCM, with melting point at the middle of periodic extreme temperature is selected. Considering, the temperature at the beginning of discharging as 80 °C (after charging) and that of cold water at 30 °C (for discharging), a PCM with melting point 56–58 °C is used for this purpose [10]. The other component of the LHS is the storage tank, it is used to enclose the PCM in both molten and solid states. It prevents the mixing of PCM and HTF, thus any kind of chemical reaction. The physical model of the dual shell configuration is presented in Fig. 1. The model consists of 2 shells, the inner contains PCM and the outer is for HTF (water). The wall and fin are made up of a copper sheet of thickness 2 mm. The inlet and outlet are circular openings of a diameter of 10 mm. Three different designs of the LHS model are considered for the performance evaluation (no fin, 2 fins, and 4 fins arrangement) as shown in Fig. 1. For water as HTF, the inlet temperature is 30 °C and the flow rates are 1, 3 and 5 LPM. Mahdi et al. [1] also used a similar values of flow rate and temperature of HTF for the experimental study.

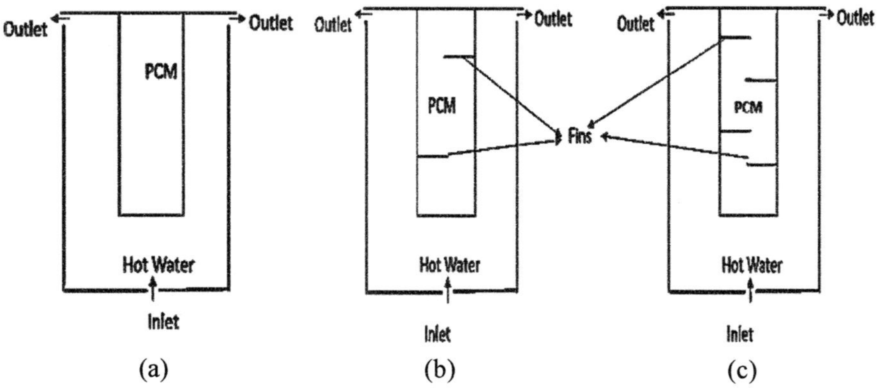

Fig. 1 Schematic of dual shell model, **a** without fin, **b** with two fins, **c** with four fins

3 Results and Discussion

The solidification of paraffin wax is simulated using ANSYS Fluent package. The essential governing equations for PCM solidification are solved using the control volume on a staggered grid. A model of conduction-convection is considered under the assumptions as mentioned in Raut et al. [10].

The enthalpy porosity technique is used to solve the melting and solidification physics. The second-order central difference scheme and the second-order upwind technique are used to discretize the diffusion and convective terms, respectively. Pressure and velocity are linked using the SIMPLEC approach. The governing equations of the used model are given in Raut et al. [10].

3.1 Numerical Model Validation

The numerical model presented in Sect. 3.1 is validated with the experimental work on solidification PCM by Mahdi et al. [1]. The comparison of numerical (present work) and experimental work is shown in Fig. 2. The nature of the computational data is similar to experimental values with some mean deviation of 5–10% in temperature values. This much discrepancy between numerical and experimental work on PCM melting and solidification is accepted in published works [11]. Thus, the proposed model is used for further study.

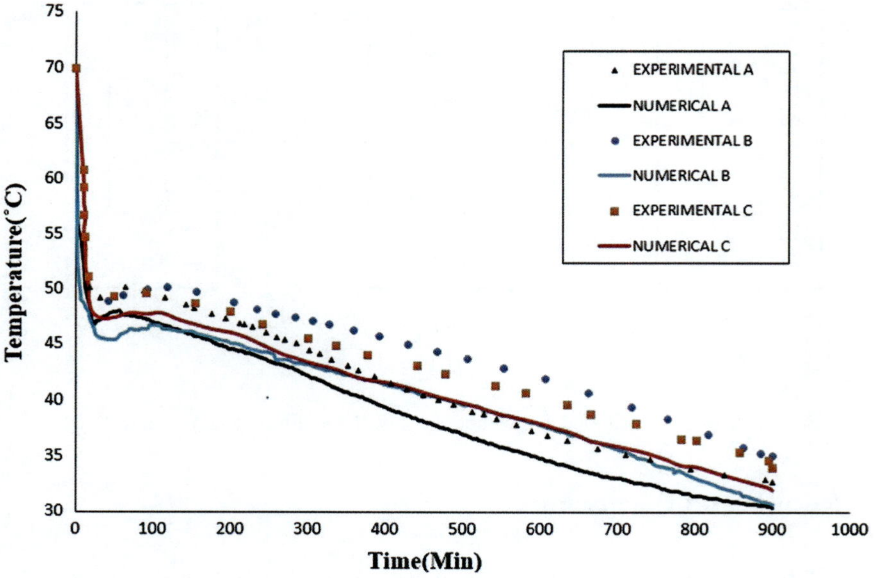

Fig. 2 Comparison of experimental [1] and numerical results for numerical model validation

3.2 *Effect of Fin*

The effect of fins on the solidification behavior of PCM is investigated for 4 fins, 2 fins, and 0 fin arrangement. The case with optimum results (initial temp 80 and HTF flow rate 3 LPM) during the discharging is used to study the effect of fins.

3.2.1 PCM Temperature

Figure 3 shows the solidification of paraffin wax for HTF at a flow rate of 3 LPM and an intake temperature of 25 °C. It is observable from the curves that 4 fins arrangement solidifies faster than others (500 min) and takes lesser time than that in the case of 2 fins (1500 min) and 0 fins (3000 min) conditions.

Table 1 represents the comparison of temperature contours of PCM during the solidification process for 0 fin, 2 fins and 4 fins arrangement at flow rate of 3 LPM. When no heat is lost during the heat transfer, the drop in PCM temperature equals the rise in water temperature. The observations made from the temperature contours are as followed:

After 10 min: At the very first stage of solidification, the temperature distribution in the PCM chamber for 0 fin and 2 fins are almost same. But, in the 4-fin case the temperature contour is more greenish. This behavior is continuing for next 100–150 min.

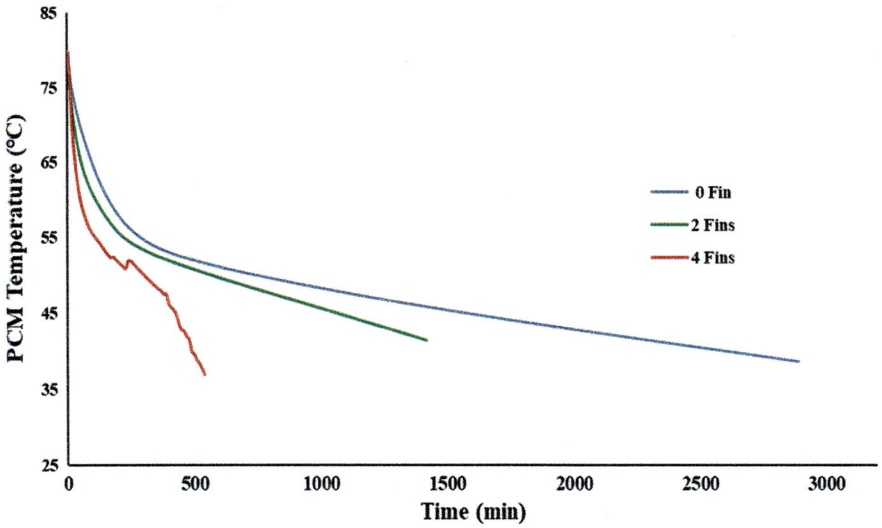

Fig. 3 Average temperature of PCM during the solidification for 0 fin, 2 fins and 4 fins at flow rate of 3 LPM

From 200 min and onwards: In all 3 arrangement, the lower portion cools first as compared to upper portion. 0 fin and 2 fins show similar rate of temperature change, while the temperature drop is faster in 4 fins arrangement.

After 200 min, noticeable difference in temperature distribution can be seen for 0 fin and 2 fins arrangement.

3.2.2 Liquid Fraction

Figure 4 shows the liquid fraction curve of PCM for the discharging when initial temperature of PCM is 80 °C and water flow at rate of 3 LPM for 0 fin, 2 fins and 4 fins arrangement. It shows that 4 fins arrangement undergoes the quickest solidification in time duration of 500 min. Thereafter, 2 fins arrangement solidifies in 1500 min and while 0 fin arrangement solidifies in 3000 min. The liquid fraction map for the solidification is given in Table 2.

From the Table 2, it can be concluded that the lower part of the PCM chamber solidifies earlier than the upper part. 4 fin arrangement solidifies faster than the other 2 arrangements at significant faster rate of solidification. When 90% of the PCM solidifies, there is difference in mushy zone distribution in PCM chamber.

Table 1 Comparison of temperature contours of PCM during the solidification process for 0 fin, 2 fins and 4 fin arrangement at flow rate of 3 LPM

Time (Min)	0 fin	2 fins	4 fins
Temperature	32 37 42 47 52	57 62 67	72 77
10			
200			
400			
600			

3.3 *Effect of Flowrate*

The proposed model's discharging performance is examined to determine its applicability for real-world applications. For the investigation of solidification, the 4 fins configuration is studied and simulated for an initial temperature of the LHS of 80 °C while three distinct flow rates (1, 3 and 5 LPM) are considered. The subsequent subsections present the PCM's average temperature, liquid fraction, and effectiveness during the solidification for the proposed model of LHS.

Figure 5 shows the volume average temperature of PCM, liquid fraction and effectiveness of the heat exchanger during the solidification when the HTF temperature is set to 30 °C and the flow rates are 5, 3, and 1 LPM.

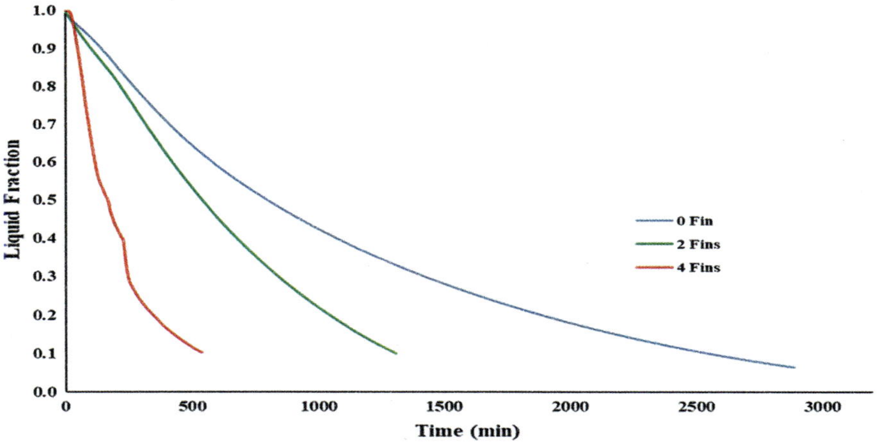

Fig. 4 Liquid fraction of PCM chamber during the solidification process for 0 fin, 2 fins and 4 fins at HTF flow rate of 3 LPM

Figure 5a show that the average temperature of the PCM chamber drops significantly faster for 1 LPM as compared to other flowrates. However, the average temperature of the PCM does not differ much at any of the three flow rates.

Figure 5b represents the liquid fraction curve for the solidification process at flow rates of 5, 3, and 1 LPM when the PCM starts at 80 °C. It can be seen that for the first 200 min, solidification occurs faster with a 1 LPM flow rate. Following that, all three flow rates have a more similar curve tendency. More than 90% of the PCM had solidified after 600 min of flow time.

Figure 5c depicts the change in storage unit efficacy with three different flow rates at an initial temperature of 80 °C. It is clear that the efficacy was at its peak at the start of the solidification (ignoring the first few minutes viewed as a stabilizing period) and then rapidly decreased. The effectiveness is calculated using the relation [1]:

$$\in = \frac{Q_{actual}}{Q_{max}} = \sum \frac{\dot{m}Cp(T_{in} - T_{out})}{\dot{m}Cp(T_{in} - T_m)} = \frac{T_{in}}{T_{out}}$$

Figure 5c shows that the highest effectiveness is achieved at 3 LPM flow rate and the lowest effectiveness is witnessed at 1 LPM flow rate. In all cases, the effectiveness curves decline slowly and reach a constant value when the outlet temperature becomes constant after 600 min of flow time. 3 and 5 LPM curves shows the small at time interval of 250 min, it is due to the time averaging of PCM temperature. The contours show that at 250 min, PCM at height is in the liquid phase for 3 and 5 LPM, but in the semi-solid phase for 1 LPM.

Table 2 Comparison of melting fraction contours in the PCM chamber during the solidification for no fin, 2 fins and 4 fins arrangement at flowrate of 3 LPM

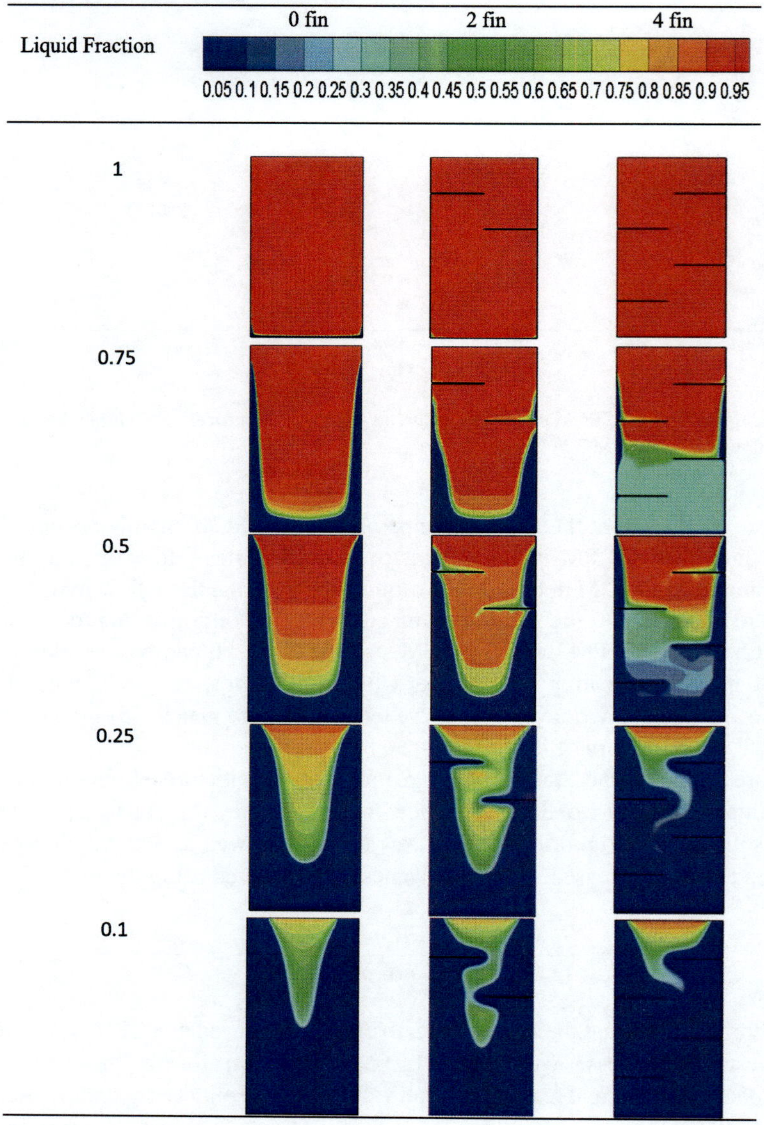

4 Conclusion

The adopted model for simulation is capable enough to capture the solidification of paraffin wax. Secondly, the presented configuration for LHS performance, significantly in different manner when fins are added. The other effect like buoyancy and

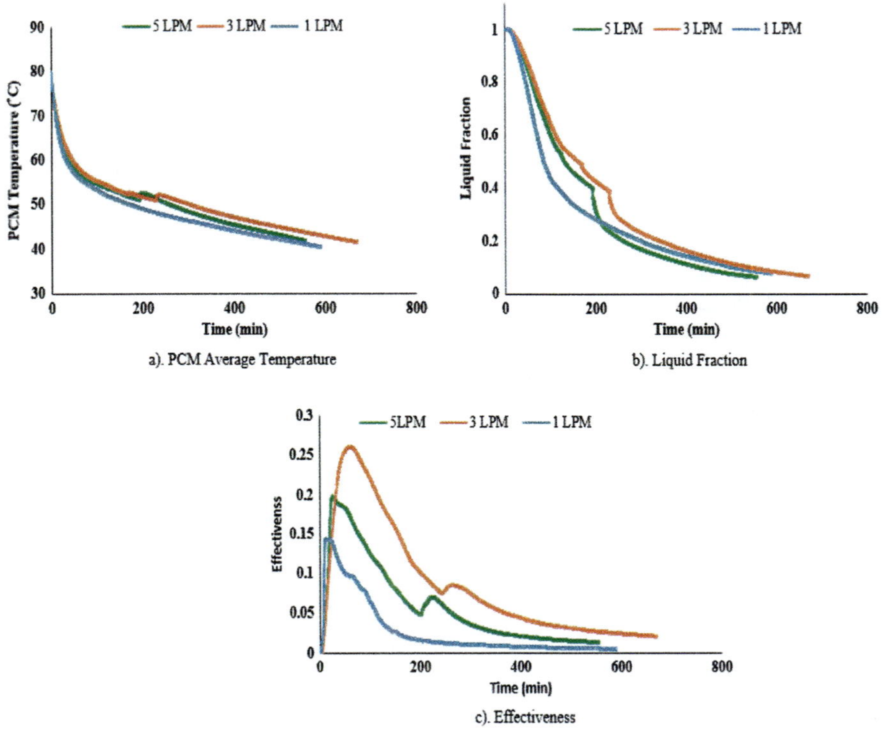

Fig. 5 PCM average temperature, liquid fraction and effectiveness of PCM chamber during the solidification at the initial temperature of PCM 80 °C and HTF flowrates are 1, 3 and 5 LPM

thermal convective currents plays important role during the solidification. The overall results indicate that the presented dual shell configuration is capable enough to work integration with engineering process like solar heat storage, process heat recovery and others. The performance of non-finned and finned arrangement is compared and the summarized results of the study are listed as follows:

- The average temperature of the PCM does not differ significantly at any of the three flow speeds.
- The buoyancy effect due to density difference affects the solidification behavior and rate. Consequently, PCM at higher positions solidifies faster than PCM at lower height.
- Dual shell configuration shows higher effectiveness at the optimized flow rate of HTF. In the current study, 3 LPM shows the highest effectiveness among HTF flowrate of 1, 3 and 5 LPM.

References

1. Mahdi MS, Mahood HB, Khadom AA, Campbell AN, Hasan M, Sharif AO (2019) Experimental investigation of the thermal performance of a helical coil latent heat thermal energy storage for solar energy applications. Therm Sci Eng Prog 10:287–298. https://doi.org/10.1016/j.tsep.2019.02.010
2. Mahdi JM, Nsofor EC (2018) Solidification enhancement of PCM in a triplex-tube thermal energy storage system with nanoparticles and fins. Appl Energy 211:975–986. https://doi.org/10.1016/j.apenergy.2017.11.082
3. Dhaidan NS, Khodadadi JM (2017) Improved performance of latent heat energy storage systems utilizing high thermal conductivity fins: a review. J Renew Sustain Energy 9. https://doi.org/10.1063/1.4989738
4. Mahdi JM, Lohrasbi S, Nsofor EC (2019) Hybrid heat transfer enhancement for latent-heat thermal energy storage systems: a review. Int J Heat Mass Transf 137:630–649. https://doi.org/10.1016/j.ijheatmasstransfer.2019.03.111
5. Khan Z, Khan Z, Ghafoor A (2016) A review of performance enhancement of PCM based latent heat storage system within the context of materials, thermal stability and compatibility. Energy Convers Manag 115:132–158. https://doi.org/10.1016/j.enconman.2016.02.045
6. Tay NHS, Bruno F, Belusko M (2013) Comparison of pinned and finned tubes in a phase change thermal energy storage system using CFD. Appl Energy 104:79–86. https://doi.org/10.1016/j.apenergy.2012.10.040
7. Sheikholeslami M, Lohrasbi S, Ganji DD (2016) Numerical analysis of discharging process acceleration in LHTESS by immersing innovative fin configuration using finite element method. Appl Therm Eng 107:154–166. https://doi.org/10.1016/j.applthermaleng.2016.06.158
8. Koizumi H, Jin Y (2012) Performance enhancement of a latent heat thermal energy storage system using curved-slab containers. Appl Therm Eng 37:145–153. https://doi.org/10.1016/j.applthermaleng.2011.11.009
9. Korth T, Loistl F, Storch A, Schex R, Krönauer A, Schweigler C (2020) Capacity enhancement of air conditioning systems by direct integration of a latent heat storage unit. Appl Therm Eng 167:114727
10. Raut D, Lanjewar S, Kalamkar VR (2022) Effect of geometrical and operational parameters on paraffin's melting performance in helical coiled latent heat storage for solar application: a numerical study. Int J Therm Sci 176:107509. https://doi.org/10.1016/j.ijthermalsci.2022.107509
11. Soni V, Kumar A, Jain VK (2018) Modeling of PCM melting: analysis of discrepancy between numerical and experimental results and energy storage performance. Energy 150:190–204

Effect of Vehicle Parameters on Air:Fuel Ratio and Lambda of the Petrol-Driven Cars

Abhinav Pandey, Govind Pandey, and Rajeev Kumar Mishra

1 Introduction

In the preceding couple of decades, the majority of countries, including developing nations in South-East Asia, have witnessed an improved socio-economic scenario, especially the city areas transforming into urban conglomerations. This situation over the years has led to a substantial increase in vehicular population and the perennially poor urban air quality particularly due to the tailpipe emissions, construction activities, re-suspension of road dust etc. [1, 2]. The corresponding data show that the global car sales growth has increasingly been over 20% between 2010 and 2019 and close to 18% between 2005 and 2019. In figure terms, there were approximately 45 million light-duty private cars sold in 2005 worldwide whereas this number almost doubled to about 90 million units bought in 2019 in less that 15 years' scale [3].

The automobile emission is known to contribute to over 60–70% of the ambient air pollution in the megacities of the developing countries. This unfavorable scenario is further aggravated by the operation of a substantial sum of older cars paired with inadequate vehicle inspection and maintenance, deficient road network infrastructure and relatively inconsistent quality of fuel [4–7]. Recent studies have expressed critical need of exploratory analysis of emission dataset and vehicular traits those helping upgrade of concurrent environmental policy and regulations, especially in the developing countries [8]. It is evident now that the automobile sector (irrespective of vehicle types), contributes to over 90% of the total CO (Carbon Monoxide) emissions in the ambient air. Similarly, the Hydrocarbons (HCs) also get emitted

A. Pandey (✉) · R. K. Mishra
Delhi Technological University, Delhi, India
e-mail: navabhigkp@gmail.com

A. Pandey
Engineering Design and Research Centre, Larsen & Toubro Limited, Chennai, India

G. Pandey
Madan Mohan Malaviya University of Technology, Gorakhpur, India

S. Doolla et al. (eds.), *Advances in Clean Energy and Sustainability*,
Green Energy and Technology, https://doi.org/10.1007/978-981-99-2279-6_76

majorly from the vehicular pollution, alongside the aerosols and solid particles, such as, black carbon (BC), oxides of nitrogen (NOx), and other peculiar masses, for instance, particulate matter (PM_1, $PM_{2.5}$ and PM_{10}) with traces of heavy metals [9].

Air:Fuel Ratio (AFR) is a dimensionless physical entity representing the ratio of mass of air to that of the fuel while the latter is subject to combustion process in an internal engine. This spark-ignition, internal combustion (SI, IC) engine converts the thermal energy to the mechanical energy thereby powering any modern car. Any slight change in AFR may alter the amount of air present in the combustion chamber of the engine, thereby directly influencing the exhaust emission quality. This is the reason why AFR is considered, though as proxy, but a very significant engine tuning parameter. We find the AFR to be also checked alongside the typical tailpipe exhaust emission constituents as far as emission compliance recertification is concerned and also, that the AFR finds place in emission standards. Although few studies have been carried out on how AFR, lambda (λ) and other exhaust emission entities vary in case of petrol-driven vehicle [10], the need of exploring such relationship as regards to the vehicular traits and engine speed is yet to be addressed [8].

A field study was conducted on the gasoline-based end of life (ELV) vehicles in Nigeria collecting emission date from 979 vehicles with an age profile ranging between 7 and 29 years. The field measurement included vehicular variables such as, age, usage, weight class and also the average AFR measurement. The data showed that for the older vehicles ($\geq$20 years) the AFR values were found to be higher than the stoichiometric value, pointing at leaner mixtures in the combustion chamber. This finding was in line with the higher average values observed for O_2 and HC and lower values found for CO_2 in the tailpipe. On the other hand, the AFR readings mostly congregated to the stoichiometric level for younger model vehicles, hinting at optimal or near optimal air and fuel mixture, ensuring efficient combustion characteristics and, therefore, engine performance. [11].

The lambda (λ) calculation is a proxy indicator of the ratio of the amount of oxygen (O_2) actually present in engine's combustion chamber to the amount required to attain an ideal combustion condition. Although it is not measured as part of the volumetric concentration of a typical exhaust from vehicles, it offers a powerful diagnostic tool for fuel mixing conditions and oxygen sensor faults (for example, it can reliably point at an influenced sensor). Also, the lambda analysis and conventional exhaust gas readings can conclusively find out if catalytic converters malfunction almost in a minute. In consideration of the seasonal and slight physio-chemical variations encountered in the petrol (gasoline), it is λ which holds more practical significance compared to AFR. More so, because the modern car engines, when tuned in appropriately, can deal with two or more fuel types (Sinhal and Kumar 2016).

In view of the above, we aimed our study to understand and determine the relationship between various vehicle/engine-related variables on AFR and λ with a larger and manufacturer-wise dataset under a comprehensive in-situ survey.

2 Materials and Methods

The car tailpipe emission testing program attended to 544 light-duty, petrol-powered, privately-owned passenger cars. In the present study we targeted the cars which presented themselves for re-issuance of their Pollution-Under-Control certificate at government-authorized test centres located in the megacity of Delhi. Such cars were investigated for the two tailpipe (emission) parameters, AFR and λ which are considered as a proxy of emission control system, and hence, emission performance of the cars. The basis of selection of high idling speed was sourced from the emission certification test protocol prescribed under the Central Motor Vehicle Rules (CVMR) by the Ministry of Road Transport and Highways (MoRTH), Government of India, which requires AFR and λ both to be tested at high engine idling speeds. The range of high idling engine speed varied from 2240 to 2690 revolutions per minute (RPM). Such values in the range are as per manufacturer's recommendations, for example, car models of Maruti-make have a high-idle engine speed RPMs varying between 2200 and 2800 [12]. Also recorded were actual vehicle age and mileage data along with car body type, status of car's registration life (that is time elapsed sine date of first registration) and ownership details. The vehicular and engine-related characteristics of the cars tested in the present study are underlined in Table 1.

An auto exhaust gas analyzer (brand—Ozone; country of origin—India; constituents type measured—gaseous) was utilized to measure and record the tailpipe values of both AFR and λ (dimension-less entities) in no-load, stationary condition (Fig. 1). The testing protocol followed was in line with the requirements stipulated in the MoRTH guidelines and following the existing emission recertification system in the country. Further, the Statistical Package for the Social Sciences (SPSS, ver. 26) was for quantitative and statistical analysis and processing of the field-collected emission dataset. At few instances, the study also used the MS-Excel tool for data screening.

The key steps of the present research work are outlined in Fig. 2.

3 Results and Discussion

The effect of vehicle age (months) on AFR for the entire dataset considering various fuel mixing conditions under high engine idling speed is presented in Fig. 3. It is seen that the AFR varies significantly with car's age under the high idling engine speed. This is particularly true for the cars having rich fuel mixing conditions compared to the ones running on lean fuel. It is further found that the variation in AFR for the cars having lean fuel mixing conditions lies between 14.75 and 15.02, whereas the same for the other set of cars varies between 14.65 and 13.35. As the car ages and more so after it attains almost half of registration life, the AFR shifts towards the rich fuel operation of the engine. The newer cars operate near stoichiometric or lean conditions, indicating optimal air and fuel mixture and efficient engine performance.

Table 1 Passenger car sample characteristics-key parameters

Item	Particulars
Total numbers	544
Brand-wise numbers	352 (M1 = Hyundai); 192 (M2 = Honda)
Car body type distribution	356 (Hatchback); 188 (Sedan)
Car age scale (months)	0.2–200.2
Average age (months)	54.4
Oldest/newest brand tested	M1
Car mileage range (km)	165–217,017
Average mileage (km)	53,147
Unladen car weight scale (tonne)	0.785–1.21
Cars' transmission type distribution	Manual (523)^; Automatic (21)^
Cars' drivetrain type distribution	Front (524)^; All-Wheel (20)^
Emission norm distribution	BS-II (2)^; BS-III (12)^; BS-IV (530)^
Capacity range (litre)	0.814–1.799
Power scale-maxima (bhp)	55–130
Torque scale-maxima (Nm)	76.5–172
Comp. ratio scale (:1)	8.9–10.8
Bore * stroke scale-maxima/minima (mm)	67–81/71.6–89.4
Cylinder number scale/valve numbers for each cylinder (scale)	3–4/3–4
Engine valve config	SOHC (250)^; DOHC (294)^

BS Bharat Stage; *Comp.* Compression; *Config.* Configuration; *SOHC* Single Overhead Cam; *DOHC* Double Overhead Cam; ^ (Number of cars)

Figure 4 depicts the effect of car's mileage (km) on AFR for the entire dataset considering various fuel mixing conditions under high engine idling speed. The variation in AFR in respect of vehicle mileage and the given fuel mixing conditions closely agrees with the finding reported in the case of car's age. It implies that until approximately between 100,000 and 125,000 km of mileage, the cars are seen to run on both lean and rich fuel conditions; however, post this range, the AFR shifts to rich conditions. The cars with a lower mileage range (not necessarily newer cars) tend to operate on lean or near stoichiometric values (AFR = 14.72–14.93), whereas the cars having accumulated a higher degree of mileage exhibit the rich fuel operating

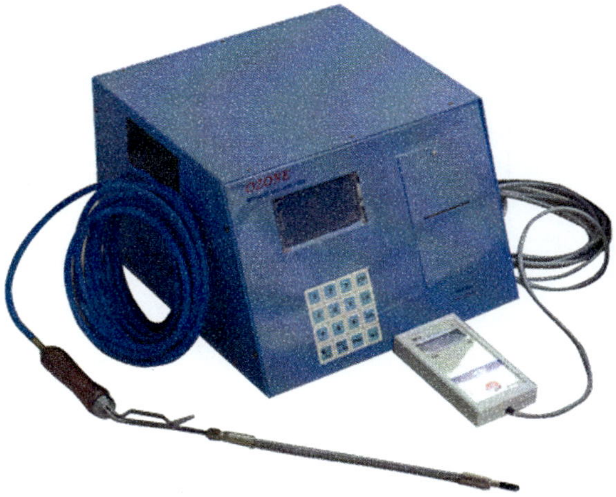

Fig. 1 A view of the auto exhaust gas analyzer used in the study

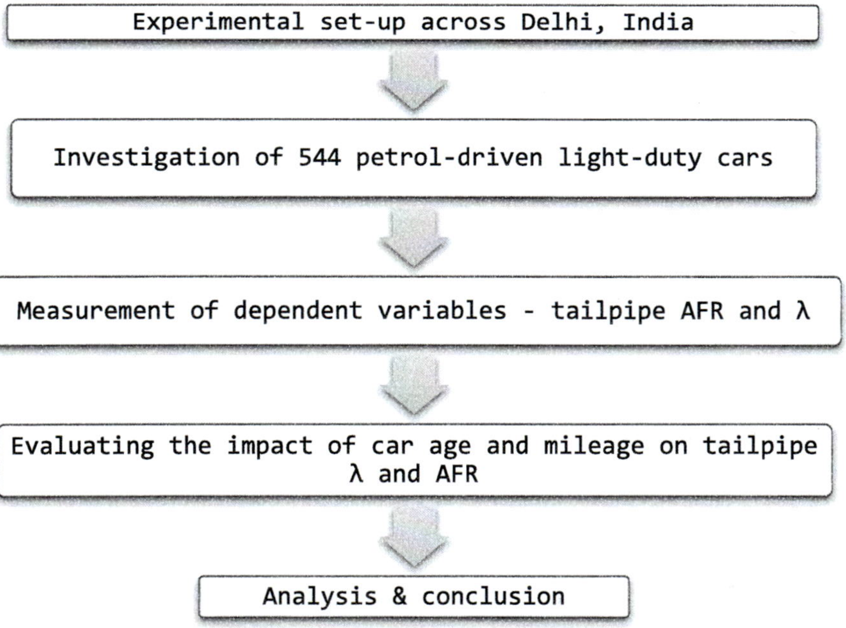

Fig. 2 Flow chart highlighting key research steps

conditions (AFR = 14.68–13.32) thereby indicating sub-par engine performance due to wear and tear.

Figure 5 presents the comparison of the AFR and λ during the high idling engine condition, given the age of the car. For the entire dataset (n = 544), a strong degree of correlation between car's age and AFR and λ was reported ($R^2_{AFR/age} = 0.764$,

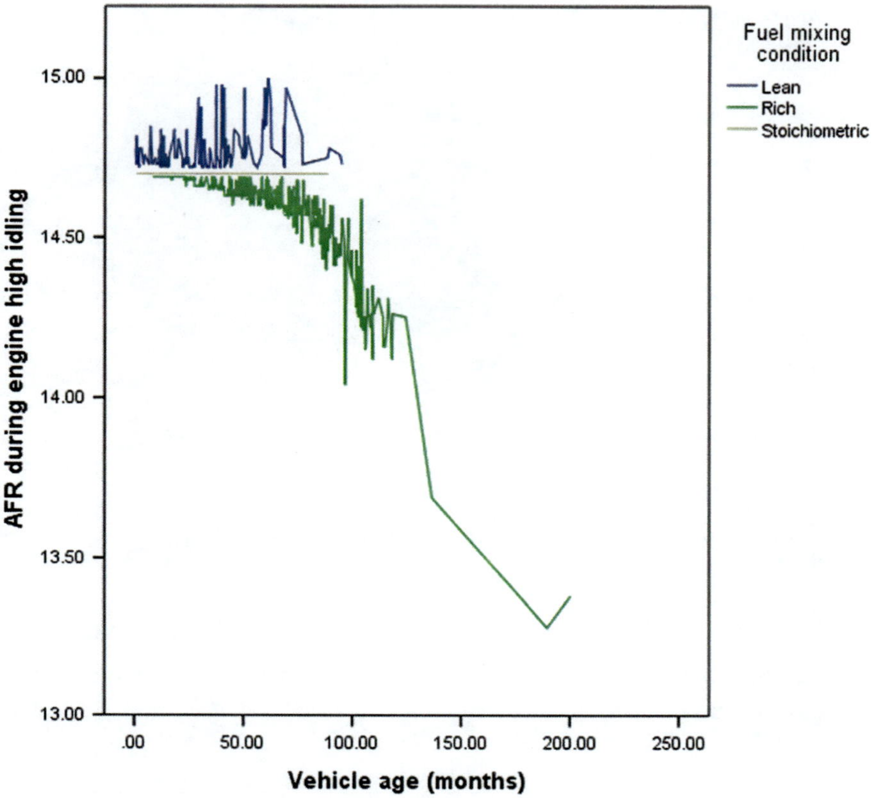

Fig. 3 AFR variation in respect of vehicle age and fuel mixing condition

and $R^2{}_{\lambda/\mathrm{age}} = 0.766$). It indicated that the ageing cars negatively affect both these parameters, which tend to increase with age. Although the average age of the cars which underwent the study was found to be only 54.4 months, the performance of much older cars (age > 15–20 years) has been studied with the finding that AFR increases with the age of vehicles-(particularly for the end-of-life vehicles [11]).

The effect of vehicle mileage (km) on AFR and λ for the entire dataset considering the high engine idling speed scenario is presented in Fig. 6. It is seen that both the AFR and λ vary significantly with car's mileage under the high idling engine speed ($R^2{}_{\mathrm{AFR/mileage}} = 0.659$, and $R^2{}_{\lambda/\mathrm{mileage}} = 0.658$). It showed that the cars accumulating more mileage (again not necessarily older cars) have a negative effect on both these parameters, which were found to increase with the mileage. Unlike the case of car age as independent variable, the scenario of mileage as one such variable has not been extensively studied as regards to its effect on AFR and λ. It is found that car's mileage has a strong negative correlation with both AFR and λ and the cars with higher mileage (particularly above 125,000 km) regardless of their age, should be inspected for proper tuning of engine.

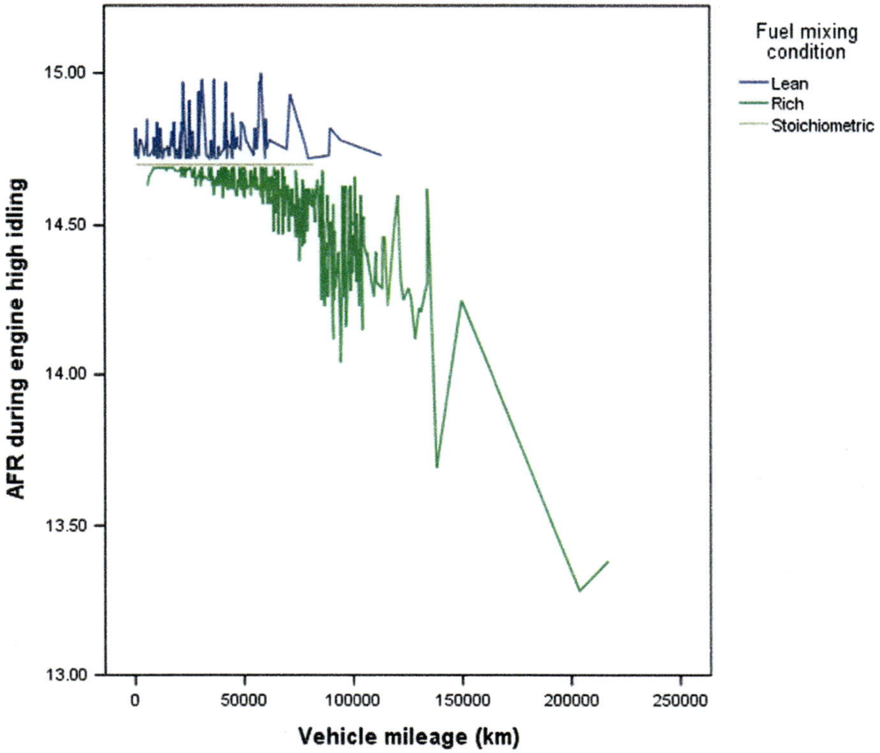

Fig. 4 AFR variation in respect of vehicle mileage and fuel mixing condition

4 Concluding Remarks

In the present research work, an attempt was made to examine the variation of the two tailpipe parameters, namely, AFR and λ, with the age and mileage of 544 petrol-driven light-duty passenger cars plying on the roads of Delhi, India. The study reported, significant correlation between the car's age and mileage as prominent vehicle independent variables influencing the tailpipe parameters ($R^2{}_{AFR/age} = 0.764$, and $R^2{}_{\lambda/age} = 0.766$; and $R^2{}_{AFR/mileage} = 0.659$, and $R^2{}_{\lambda/mileage} = 0.658$ at high idling respectively). It was also found that the newer cars operate near stoichiometric or lean conditions (AFR = 14.75–15.02), indicating towards optimal air and fuel mixture, better combustion, and, therefore, an overall efficient engine performance. As the car ages, the AFR shifts towards the rich fuel operation of the engine, pointing at the requirement of inspection and, if required, overhaul of the engine. In terms of mileage, the cars with a lower mileage range (not necessarily newer cars) tend to operate on lean or near stoichiometric values (AFR = 14.72–14.93), whereas the cars having accumulated a higher degree of mileage exhibit the rich fuel operating conditions (AFR = 14.68–13.32) thereby indicating sub-par engine performance due

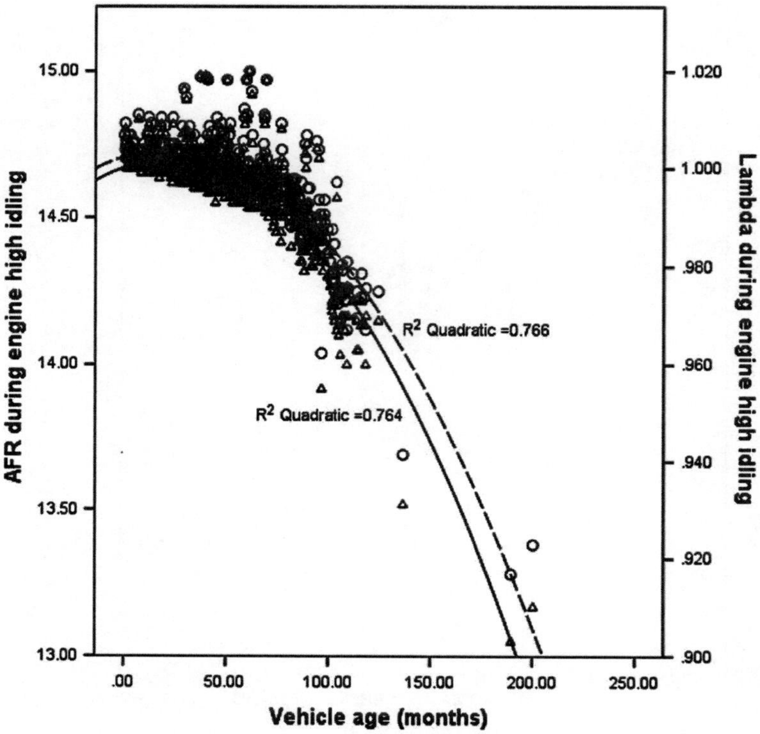

Fig. 5 AFR and Lambda variations in respect of vehicle age (at engine high idling)

to wear and tear. It is recommended that the thorough inspection/maintenance and, if required, the overhaul of the engine be carried-out for aged car (> 7.5 or 8 years) and also for those having accumulated significant mileage (> 100,000 km) exhibiting unusual AFR and λ values in the tailpipe.

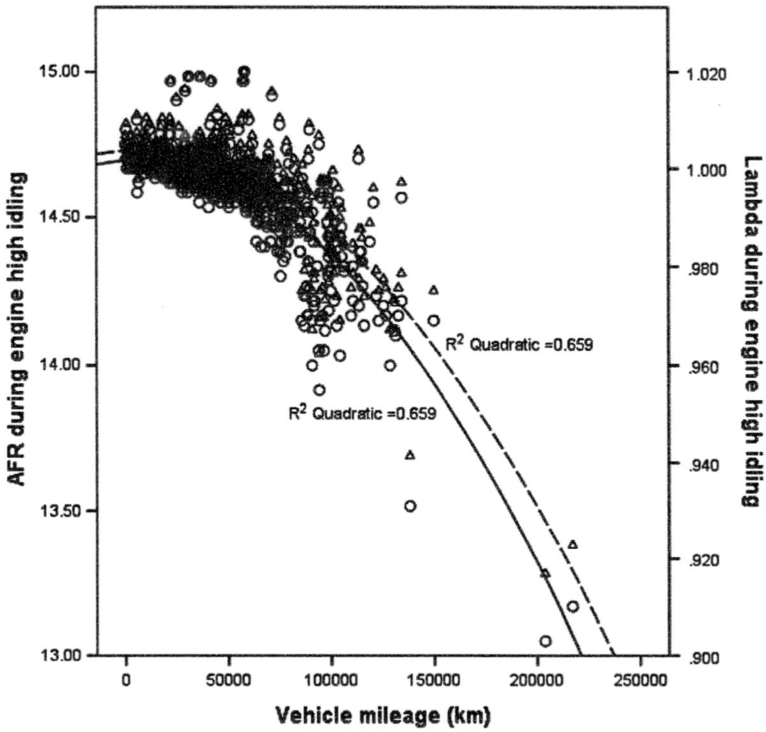

Fig. 6 AFR and Lambda variations in respect of vehicle mileage (at engine high idling)

References

1. Kumar P, Gulia S, Harrison RM, Khare M (2017) The influence of odd–even car trial on fine and coarse particles in Delhi. Environ Pollut 225:20–30
2. Gulia S, Nagendra SS, Barnes J, Khare M (2018) Urban local air quality management framework for non-attainment areas in Indian cities. Sci Total Environ 619:1308–1318
3. OICA (2019) Global sales statistics of passenger cars. Retrieved from https://www.oica.net/category/sales-statistics/. Accessed on 30 July 2020
4. Badami MG (2005) Transport and urban air pollution in India. Environ Manag 36:195–204
5. Singh AK, Gupta HK, Gupta K, Singh P, Gupta VB, Sharma RC (2007) A comparative study of air pollution in Indian cities. Bull Environ Contam Toxicol 78:411–416
6. Wang H, Fu L, Zhou Y, Du X, Ge W (2010) Trends in vehicular emissions in China's mega cities from 1995 to 2005. Environ Pollut 158:394–400
7. Pandey A, Mishra RK, Pandey G (2022) Investigating exhaust emission from in-use passenger cars: an exploratory analysis and policy outlook. J Environ Eng (New York) 148(7):04022035
8. Pandey A, Pandey G, Mishra RK (2022) An in situ exploratory analysis of diesel cars' emission: way forward on policy evaluation. Environ Sci Pollut Res 29(56):84434–84450
9. Rao X, Zhong J, Brook RD, Rajagopalan S (2018) Effect of particulate matter air pollution on cardiovascular oxidative stress pathways. Antioxid Redox Signal 28(9):797–818
10. Al-Arkawazi SAF (2019) Analyzing and predicting the relation between air–fuel ratio (AFR), lambda (λ) and the exhaust emissions percentages and values of gasoline-fueled vehicles using versatile and portable emissions measurement system tool. SN Appl Sci 1:1370

11. Moonsammy S, Oyedotum TDT, Oderinde O, Durojaiye M, Durojaye A (2021) Exhaust determination and air-to-fuel ratio performance of end-of-life vehicles in a developing African country: a case study of Nigeria. Transp Res D Transp Environ 91:102705
12. Pandey A, Pandey G, Mishra RK (2016) Tailpipe emission from petrol driven passenger cars. Transp Res D Transp Environ 44:14–29